Dietary Reference Intakes (DRIs): Recommended Dietary Allowances and Adequate Intakes, Elements
Food and Nutrition Board, Institute of Medicine, National Academies

Life stage group	Calcium (mg/d)	Chromium (µg/d)	Copper (µg/d)	Fluoride (mg/d)	Iodine (µg/d)	Iron (mg/d)	Magnesium (mg/d)	Manganese (mg/d)	Molybdenum (µg/d)	Phosphorus (mg/d)	Selenium (µg/d)	Zinc (mg/d)	Potassium (g/d)	Sodium (g/d)	Chloride (g/d)
Infants															
0 to 6 mo	200*	0.2*	200*	0.01*	110*	0.27*	30*	0.003*	2*	100*	15*	2*	0.4*	0.12*	0.18*
6 to 12 mo	260*	5.5*	220*	0.5*	130*	11	75*	0.6*	3*	275*	20*	3	0.7*	0.37*	0.57*
Children															
1–3 y	700	11*	340	0.7*	90	7	80	1.2*	17	460	20	3	3.0*	1.0*	1.5*
4–8 y	1,000	15*	440	1*	90	10	130	1.5*	22	500	30	5	3.8*	1.2*	1.9*
Males															
9–13 y	1,300	25*	700	2*	120	8	240	1.9*	34	1,250	40	8	4.5*	1.5*	2.3*
14–18 y	1,300	35*	890	3*	150	11	410	2.2*	43	1,250	55	11	4.7*	1.5*	2.3*
19–30 y	1,000	35*	900	4*	150	8	400	2.3*	45	700	55	11	4.7*	1.5*	2.3*
31–50 y	1,000	35*	900	4*	150	8	420	2.3*	45	700	55	11	4.7*	1.5*	2.3*
51–70 y	1,000	30*	900	4*	150	8	420	2.3*	45	700	55	11	4.7*	1.3*	2.0*
> 70 y	1,200	30*	900	4*	150	8	420	2.3*	45	700	55	11	4.7*	1.2*	1.8*
Females															
9–13 y	1,300	21*	700	2*	120	8	240	1.6*	34	1,250	40	8	4.5*	1.5*	2.3*
14–18 y	1,300	24*	890	3*	150	15	360	1.6*	43	1,250	55	9	4.7*	1.5*	2.3*
19–30 y	1,000	25*	900	3*	150	18	310	1.8*	45	700	55	8	4.7*	1.5*	2.3*
31–50 y	1,000	25*	900	3*	150	18	320	1.8*	45	700	55	8	4.7*	1.5*	2.3*
51–70 y	1,200	20*	900	3*	150	8	320	1.8*	45	700	55	8	4.7*	1.3*	2.0*
> 70 y	1,200	20*	900	3*	150	8	320	1.8*	45	700	55	8	4.7*	1.2*	1.8*
Pregnancy															
14–18 y	1,300	29*	1,000	3*	220	27	400	2.0*	50	1,250	60	12	4.7*	1.5*	2.3*
19–30 y	1,000	30*	1,000	3*	220	27	350	2.0*	50	700	60	11	4.7*	1.5*	2.3*
31–50 y	1,000	30*	1,000	3*	220	27	360	2.0*	50	700	60	11	4.7*	1.5*	2.3*
Lactation															
14–18 y	1,300	44*	1,300	3*	290	10	360	2.6*	50	1,250	70	13	5.1*	1.5*	2.3*
19–30 y	1,000	45*	1,300	3*	290	9	310	2.6*	50	700	70	12	5.1*	1.5*	2.3*
31–50 y	1,000	45*	1,300	3*	290	9	320	2.6*	50	700	70	12	5.1*	1.5*	2.3*

NOTE: This table (taken from the DRI reports, see www.nap.edu) presents Recommended Dietary Allowances (RDAs) in **bold type** and Adequate Intakes (AIs) in ordinary type followed by an asterisk (*). An RDA is the average dietary intake level; sufficient to meet the nutrient requirements of nearly all (97–98 percent) healthy individuals in a group. It is calculated from an Estimated Average Requirement (EAR). If sufficient scientific evidence is not available to establish an EAR, and thus calculate an RDA, an AI is usually developed. For healthy breastfed infants, an AI is the mean intake. The AI for other life stage and gender groups is believed to cover the needs of all healthy individuals in the groups, but lack of data or uncertainty in the data prevent being able to specify with confidence the percentage of individuals covered by this intake.

SOURCES: *Dietary Reference Intakes for Calcium, Phosphorous, Magnesium, Vitamin D, and Fluoride* (1997); *Dietary Reference Intakes for Thiamin, Riboflavin, Niacin, Vitamin B₆, Folate, Vitamin B₁₂, Pantothenic Acid, Biotin, and Choline* (1998); *Dietary Reference Intakes for Vitamin C, Vitamin E, Selenium, and Carotenoids* (2000); *Dietary Reference Intakes for Vitamin A, Vitamin K, Arsenic, Boron, Chromium, Copper, Iodine, Iron, Manganese, Molybdenum, Nickel, Silicon, Vanadium, and Zinc* (2001); *Dietary Reference Intakes for Water, Potassium, Sodium, Chloride, and Sulfate* (2C05); and *Dietary Reference Intakes for Calcium and Vitamin D* (2011). These reports may be accessed via www.nap.edu.

Dietary Reference Intakes (DRIs): Tolerable Upper Intake Levels, Elements
Food and Nutrition Board, Institute of Medicine, National Academies

Life stage group	Arsenic[a]	Boron (mg/d)	Calcium (mg/d)	Chromium	Copper (µg/d)	Fluoride (mg/d)	Iodine (µg/d)	Iron (mg/d)	Magnesium (mg/d)[b]	Manganese (mg/d)	Molybdenum (µg/d)	Nickel (mg/d)	Phosphorus (g/d)	Selenium (µg/d)	Silicon[c]	Vanadium (mg/d)[d]	Zinc (mg/d)	Sodium (g/d)	Chloride (g/d)
Infants																			
0 to 6 mo	ND[e]	ND	1,000	ND	ND	0.7	ND	40	ND	ND	ND	ND	ND	45	ND	ND	4	ND	ND
6 to 12 mo	ND	ND	1,500	ND	ND	0.9	ND	40	ND	ND	ND	ND	ND	60	ND	ND	5	ND	ND
Children																			
1–3 y	ND	3	2,500	ND	1,000	1.3	200	40	65	2	300	0.2	3	90	ND	ND	7	1.5	2.3
4–8 y	ND	6	2,500	ND	3,000	2.2	300	40	110	3	600	0.3	3	150	ND	ND	12	1.9	2.9
Males																			
9–13 y	ND	11	3,000	ND	5,000	10	600	40	350	6	1,100	0.6	4	280	ND	ND	23	2.2	3.4
14–18 y	ND	17	3,000	ND	8,000	10	900	45	350	9	1,700	1.0	4	400	ND	ND	34	2.3	3.6
19–30 y	ND	20	2,500	ND	10,000	10	1,100	45	350	11	2,000	1.0	4	400	ND	1.8	40	2.3	3.6
31–50 y	ND	20	2,500	ND	10,000	10	1,100	45	350	11	2,000	1.0	4	400	ND	1.8	40	2.3	3.6
51–70 y	ND	20	2,000	ND	10,000	10	1,100	45	350	11	2,000	1.0	4	400	ND	1.8	40	2.3	3.6
>70 y	ND	20	2,000	ND	10,000	10	1,100	45	350	11	2,000	1.0	3	400	ND	1.8	40	2.3	3.6
Females																			
9–13 y	ND	11	3,000	ND	5,000	10	600	40	350	6	1,100	0.6	4	280	ND	ND	23	2.2	3.4
14–18 y	ND	17	3,000	ND	8,000	10	900	45	350	9	1,700	1.0	4	400	ND	ND	34	2.3	3.6
19–30 y	ND	20	2,500	ND	10,000	10	1,100	45	350	11	2,000	1.0	4	400	ND	1.8	40	2.3	3.6
31–50 y	ND	20	2,500	ND	10,000	10	1,100	45	350	11	2,000	1.0	4	400	ND	1.8	40	2.3	3.6
51–70 y	ND	20	2,000	ND	10,000	10	1,100	45	350	11	2,000	1.0	4	400	ND	1.8	40	2.3	3.6
>70 y	ND	20	2,000	ND	10,000	10	1,100	45	350	11	2,000	1.0	3	400	ND	1.8	40	2.3	3.6
Pregnancy																			
14–18 y	ND	17	3,000	ND	8,000	10	900	45	350	9	1,700	1.0	3.5	400	ND	ND	34	2.3	3.6
19–30 y	ND	20	2,500	ND	10,000	10	1,100	45	350	11	2,000	1.0	3.5	400	ND	ND	40	2.3	3.6
31–50 y	ND	20	2,500	ND	10,000	10	1,100	45	350	11	2,000	1.0	3.5	400	ND	ND	40	2.3	3.6
Lactation																			
14–18 y	ND	17	3,000	ND	8,000	10	900	45	350	9	1,700	1.0	4	400	ND	ND	34	2.3	3.6
19–30 y	ND	20	2,500	ND	10,000	10	1,100	45	350	11	2,000	1.0	4	400	ND	ND	40	2.3	3.6
31–50 y	ND	20	2,500	ND	10,000	10	1,100	45	350	11	2,000	1.0	4	400	ND	ND	40	2.3	3.6

NOTE: A Tolerable Upper Intake Level (UL) is the highest level of daily nutrient intake that is likely to pose no risk of adverse health effects to almost all individuals in the general population. Unless otherwise specified, the UL represents total intake from food, water, and supplements. Due to a lack of suitable data, ULs could not be established for vitamin K, thiamin, riboflavin, vitamin B$_{12}$, pantothenic acid, biotin, and carotenoids. In the absence of a UL, extra caution may be warranted in consuming levels above recommended intakes. Members of the general population should be advised not to routinely exceed the UL. The UL is not meant to apply to individuals who are treated with the nutrient under medical supervision or to individuals with predisposing conditions that modify their sensitivity to the nutrient.

[a] Although the UL was not determined for arsenic, there is no justification for adding arsenic to food or supplements.

[b] The ULs for magnesium represent intake from a pharmacological agent only and do not include intake from food and water.

[c] Although silicon has not been shown to cause adverse effects in humans, there is no justification for adding silicon to supplements.

[d] Although vanadium in food has not been shown to cause adverse effects in humans, there is no justification for adding vanadium to food and vanadium supplements should be used with caution. The UL is based on adverse effects in laboratory animals and this data could be used to set a UL for adults but not children and adolescents.

[e] ND = Not determinable due to lack of data of adverse effects in this age group and concern with regard to lack of ability to handle excess amounts. Source of intake should be from food only to prevent high levels of intake.

SOURCES: *Dietary Reference Intakes for Calcium, Phosphorous, Magnesium, Vitamin D, and Fluoride* (1997); *Dietary Reference Intakes for Thiamin, Riboflavin, Niacin, Vitamin B$_6$, Folate, Vitamin B$_{12}$, Pantothenic Acid, Biotin, and Choline* (1998); *Dietary Reference Intakes for Vitamin C, Vitamin E, Selenium, and Carotenoids* (2000); *Dietary Reference Intakes for Vitamin A, Vitamin K, Arsenic, Boron, Chromium, Copper, Iodine, Iron, Manganese, Molybdenum, Nickel, Silicon, Vanadium, and Zinc* (2001); *Dietary Reference Intakes for Water, Potassium, Sodium, Chloride, and Sulfate* (2005); and *Dietary Reference Intakes for Calcium and Vitamin D* (2011). These reports may be accessed via www.nap.edu.

Aim for success in your course and beyond with these helpful resources

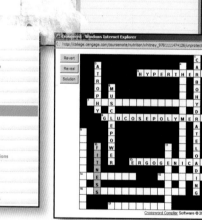
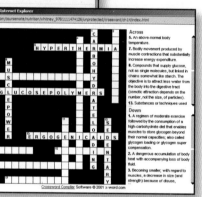

Nutrition
for Sport and Exercise
Second Edition

Marie Dunford
Former Professor and Chair
Department of Food Science and Nutrition
California State University, Fresno

J. Andrew Doyle
Associate Professor and Former Chair
Department of Kinesiology and Health
Georgia State University

WADSWORTH
CENGAGE Learning

Australia • Brazil • Japan • Korea • Mexico • Singapore • Spain • United Kingdom • United States

WADSWORTH
CENGAGE Learning™

Nutrition for Sport and Exercise, Second Edition

Marie Dunford

J. Andrew Doyle

Publisher: Yolanda Cossio

Senior Acquisitions Editor: Peggy Williams

Developmental Editor: Suzannah Alexander

Editorial Assistant: Shana Baldassari

Media Editor: Miriam Myers

Marketing Manager: Laura McGinn

Marketing Communications Manager: Linda Yip

Senior Content Project Manager: Carol Samet

Creative Director: Rob Hugel

Art Director: John Walker

Print Buyer: Judy Inouye

Rights Acquisitions Specialist: Tom McDonough

Production Service: Kelly Keeler/Greg Johnson, PreMediaGlobal

Text Designer: Ellen Pettengill

Photo Researcher: Sara Golden, PreMediaGlobal

Copy Editor: Deborah Bader

Cover Designer: Riezebos Holzbaur/Brie Hattey

Cover Image: © Johannes Kroemer/Corbis

Compositor: PreMediaGlobal

For product information and technology assistance, contact us at
Cengage Learning Customer & Sales Support, 1-800-354-9706.

For permission to use material from this text or product,
submit all requests online at **www.cengage.com/permissions.**
Further permissions questions can be e-mailed to
permissionrequest@cengage.com.

Library of Congress Control Number: 2010942462

ISBN-13: 978-0-8400-6829-3

ISBN-10: 0-8400-6829-8

Wadsworth
20 Davis Drive
Belmont, CA 94002-3098
USA

Cengage Learning is a leading provider of customized learning solutions with office locations around the globe, including Singapore, the United Kingdom, Australia, Mexico, Brazil, and Japan. Locate your local office at **www.cengage.com/global.**

Cengage Learning products are represented in Canada by Nelson Education, Ltd.

To learn more about **Wadsworth** visit **www.cengage.com/Wadsworth**

Purchase any of our products at your local college store or at our preferred online store **www.cengagebrain.com.**

Printed in the United States of America
1 2 3 4 5 6 7 14 13 12 11

Contents

4 Carbohydrates 100

5 Proteins 151

6 Fats   198

7 Water and Electrolytes 240

8 Vitamins 278

9 Minerals 320

10 Diet Planning: Food First, Supplements Second   363

11 Weight and Body Composition   399

12 Disordered Eating and Exercise Patterns in Athletes   442

13 Diet and Exercise for Lifelong Fitness and Health 474

Preface

Sports nutrition is a natural marriage of two fields: nutrition and exercise physiology. These complementary academic disciplines enable us to understand the energy expenditure that is required by exercise and sport, and the energy intake that is vital to support these activities. Exercise challenges the human body to respond and adapt, and proper nutrition supports these processes. Although all people can benefit from proper nutrition and exercise, athletes must pay careful attention to both. Training and nutrition are key elements of excellent athletic performance.

Nutrition for Sport and Exercise is designed primarily as a college-level text for upper-division courses in sports nutrition. It carefully illustrates the links between exercise, nutrition, and, the ultimate goals, optimal performance and health. In addition to explaining the rationale behind the recommendations made to athletes, the text helps instructors and students translate these recommendations to specific plans for the appropriate amount and type of foods, beverages, and/or supplements to support training, performance, and recovery. First and foremost, this book is scientifically sound and evidence based, but it is also filled with practical nutrition information and designed so faculty can easily teach from the text.

To understand sports nutrition, students must understand both nutrition and exercise physiology. For example, carbohydrates are found in food and are used by the body to fuel exercise. The type and amount of carbohydrates in foods are "nutrition" issues. The influences of exercise intensity and duration on carbohydrate usage are "exercise physiology" issues. Sports nutrition requires an understanding and integration of these issues because the timing of carbohydrate intake or the amount needed to delay the onset of fatigue involves both nutrition and exercise physiology. The goal of this book is to integrate the principles of nutrition and exercise physiology in a well-organized, scientifically sound, and practical sports nutrition text.

The Plan of the Text

Chapter 1, *Introduction to Sports Nutrition*, sets the stage. Broad terms such as *athlete* and *exercise* are defined, and basic training and sports nutrition principles are outlined. The intensity and duration of exercise training and the unique demands of competition affect nutrition requirements and food intake. Many recreational athletes require only a good basic diet.

Nearly all athletes have questions about supplements, and the first chapter discusses basic information about dietary supplements.

The first chapter also emphasizes the science behind sports nutrition recommendations. From the beginning students should recognize that the recommendations made throughout the text are evidence based. As part of the critical thinking process, future chapters will reinforce some of the basic concepts introduced in the initial chapter, such as the strength of the scientific evidence, research design, and consensus opinion. Each chapter includes a new feature, *Focus on research*, which examines a specific research study in detail. The feature provides a more in-depth look at a topic relevant to the content of the chapter and uses different types of research studies to explain scientific methods used by the researchers, what was discovered, and the significance of the research.

A unique feature of this chapter is the information on the scope of practice of dietitians, exercise physiologists, athletic trainers, strength and conditioning coaches, and other sports-related professionals. As with any integrated discipline, no one profession "owns" sports nutrition. However, the extent of professional training and licensure can help students understand practice boundaries and when to refer to someone with the appropriate expertise, professional training, and/or credentials.

Chapters 2 and 3 cover energy concepts. Extensive teaching experience has convinced the authors that students more easily understand the difficult area of energy if it is broken into two parts. The first part (*Defining and Measuring Energy*) introduces general energy concepts—what energy is and how it is measured by direct and indirect calorimetry. This leads to a discussion of energy balance and an explanation of factors that affect it, such as resting metabolic rate, physical activity, and food intake.

Once that foundation is established, then students can more easily understand the specific energy systems needed to fuel exercise of varying intensities as presented in Chapter 3, *Energy Systems and Exercise*. The focus of the chapter is an explanation of the three major energy systems used to replenish ATP—creatine phosphate, anaerobic glycolysis, and oxidative phosphorylation. Oxygen consumption, fuel utilization, and the respiratory exchange ratio are described, and the safety and effectiveness of creatine supplements are reviewed.

Chapters 4, 5, and 6 cover three energy-containing nutrients—*Carbohydrates*, *Proteins*, and *Fats*. These topics are at the heart of sports nutrition. Each chapter includes a description of digestion, absorption, and metabolism of these nutrients and explains each as a source of energy based on the intensity and duration of exercise. Current recommendations for athletes are outlined, and the effects of inadequate intake on training and performance are discussed. Type, amount, and timing are important nutrition concepts, and these chapters end with a focus on the translation of current recommendations to appropriate food and beverage choices.

Similar to Chapters 4 through 6, Chapters 7 through 9 are nutrient focused. *Water and Electrolytes* are covered first, followed by *Vitamins* and *Minerals*. These chapters feature a global approach so that students can relate to body systems that are influenced by many different factors. For example, Chapter 7 begins with an overview of water and electrolytes but emphasizes the effect that exercise has on fluid and electrolyte balance by examining water and electrolyte loss and intake during training and competition. The recommendations for replenishment of water and electrolytes are a logical extension of understanding fluid homeostasis.

To avoid the encyclopedic approach that can overwhelm students with detailed information about vitamins and minerals, Chapters 8 and 9 are organized according to function. In the case of vitamins, their major roles in energy metabolism, antioxidant protection, red blood cell function, and growth and development are explained. The minerals chapter is organized according to bone, blood, and immune system function and emphasizes calcium, iron, and zinc, respectively. Each chapter also discusses adequate intake and the potential for clinical and subclinical deficiencies and toxicities. Vitamin- and mineral-rich foods, fortified foods, and supplement sources are covered with special attention paid to the perceived need for supplementation by athletes.

After a solid foundation in principles of sports nutrition has been laid, the text moves into comprehensive diet planning. Chapter 10 is entitled *Diet Planning: Food First, Supplements Second* and helps students take the science-based nutrient recommendations made in the previous chapters and translate them into daily food choices, including food and fluid intake prior to, during, and after exercise. The chapter emphasizes developing a plan for matching dietary intake to the demands imposed by training, with consideration for the athlete's specific sport. This chapter also contains information about caffeine, alcohol, and dietary supplements. Supplements are a complicated issue requiring an understanding of legality, ethics, purity, safety, and effectiveness, and practitioners will have little credibility with athletes if they simply dismiss their use. Exploring the issues surrounding dietary supplements helps students become better critical thinkers.

No sports nutrition book would be complete without a chapter on body composition. Chapter 11, *Weight and Body Composition*, is realistic—it considers measurement techniques, error of measurement, interpretation of body composition results, and the relationship of body composition and weight to performance. The chapter begins with a review of methods for determining body composition and the advantages and disadvantages of each. The role of training and nutrition in increasing muscle mass and decreasing body fat is explained. Minimum and target body weights, based on a body composition that promotes health, are discussed for sports in which making weight or achieving a certain appearance is important. Muscle-building and weight loss supplements are also covered.

Chapter 12 covers disordered eating and exercise patterns in athletes. The philosophy expressed throughout the book is that normal eating is flexible and that food is eaten for fuel and for fun. However, disordered eating and life-threatening eating disorders can touch the lives of anyone who works with athletes, and these problems cannot be ignored. This chapter follows the progression of eating and activity patterns from "normal" to disordered to severely dysfunctional, and explains the interrelated elements of the Female Athlete Triad.

Whereas the focus in most of the chapters is on the trained athlete, the final chapter gives ample coverage to diet and exercise for lifelong fitness and health and their roles in preventing or delaying chronic disease. Many students dream of working with elite athletes, but in reality most will work with many people who are recreational athletes or are untrained, have relatively low fitness levels, eat poorly, and want to lose weight. This chapter addresses the issue of declining physical activity associated with aging and uses scenarios of former athletes to highlight chronic diseases such as obesity, type 2 diabetes, heart disease, metabolic syndrome, osteoporosis, and lifestyle-related cancers. The chapter has been organized to reflect the primary role that overweight and obesity play in the development and progression of many chronic diseases. It also explains the many mechanisms, some of which are not precise, that the body uses to regulate body weight.

Nutrition for Sport and Exercise is a blend of nutrition and exercise physiology and both scientific and practical information. It fully integrates both fields of study. It is not an exercise physiology book with nutrition as an afterthought or a nutrition book with superficial explanations of core exercise physiology principles. The authors, a registered dietitian and an exercise physiologist, have more than 35 years of classroom experience in sports nutrition. They have used that experience to create a text that meets the needs of both nutrition and exercise science majors and faculty.

Features of the Text

Each chapter is designed to guide students through the learning process, beginning with a **Learning Plan** that lists objectives for students to master as they study the material. A **Pre-Test** helps to assess students' current knowledge of the topic to be discussed. At the end of each chapter, a **Post-Test** is given to test what students have learned. The answers to the *Post-Test* can be found in Appendix O, and used to illuminate misconceptions about the topic as well as to pinpoint material that warrants further study.

Glossary terms are highlighted throughout the chapter, giving students immediate access to their definitions as well as helping them identify important terms to study as they prepare for exams. The definitions have also been gathered into an alphabetical glossary at the back of the book.

Numerous sidebars appear throughout the text, exposing students to high-interest information on diverse topics. The sidebars highlight applications of concepts, present the latest findings, and point out controversial ideas without interrupting the flow of the text. **The Internet café** highlights important websites that students can trust to find information on each topic.

Each chapter ends with a **Summary** that restates the major ideas, and a **Self-Test** is provided, which includes multiple-choice, short-answer, and critical thinking questions, so students can test their knowledge of the facts and concepts presented. The answers to the multiple-choice questions can be found in Appendix O. **References** for the major articles discussed throughout the chapter as well as suggested readings are included, so students can further investigate topics on their own. All of these features are designed with the student in mind, to help him or her identify and grasp the important concepts presented in each chapter.

New to the Second Edition

The second edition of *Nutrition for Sport and Exercise* includes a thorough review of the most recent published literature so that the material included in the textbook represents the most current, cutting-edge scientific information, up-to-date guidelines, and evidence-based recommendations.

Three new features were added to each chapter. **Focus on research** is designed to help students understand research methods and results and the significance and application of those results. The studies chosen reflect a topic covered in the chapter and help students see how research and practice are related. The **Application exercise** gives students a brief scenario, along with questions, and encourages them to apply the information that they have read. **Key points** and review questions at the end of each major section assist students in identifying the important information

from that section and test their mastery of that information. Other new or updated content includes:

Chapter 1: Introduction to Sports Nutrition

- Dietary Guidelines for Americans, 2010
- Physical Activity Guidelines for Americans, 2008
- Food Pyramid for Athletes
- More information about the Nutrition Facts label
- Expanded section on dietary supplements, including why athletes choose supplements, issues related to purity, and a summary of supplements that have been shown to be safe and effective

Chapter 2: Defining and Measuring Energy

- Updated graphics and artwork
- Expanded discussion of measurement of energy expenditure with wearable, portable devices
- Updated section on estimating energy intake
- Revised section on estimating energy expenditure of individual physical activities
- Enhanced table for estimating daily energy need for male and female athletes

Chapter 3: Energy Systems and Exercise

- Updated, reorganized, and streamlined graphics and artwork
- Expanded review of high-energy phosphate use by exercising skeletal muscle
- Updated discussion of ATP yield from oxidative phosphorylation
- Updated discussion of creatine loading and supplementation
- Reorganized and expanded discussion of metabolism and fuel utilization
- New chart summarizing effects of feeding and fasting on metabolic pathways
- Expanded section on oxygen consumption and skeletal muscle fiber types

Chapter 4: Carbohydrates

- Updated and improved graphics
- Reorganized and expanded section on digestion and absorption of carbohydrates, with particular emphasis on glucose and fructose transport
- New graphic on carbohydrate absorption
- Updated information and discussion on glycemic index
- Expanded section on carbohydrate metabolism and exercise training
- Expanded and updated section on carbohydrate use before, during, and after exercise
- New table outlining optimal carbohydrate intake during various types of exercise and sports
- New tables detailing the characteristics of the most current sports carbohydrate products
- Discussion of carbohydrate products developed and marketed for use by athletes in various sports and exercise situations

Chapter 5: Proteins

- Extensive revision of pre-, during, and postexercise protein recommendations
- More information about whey and casein
- Expanded section on the role of protein in the immune system and the impact of endurance exercise
- Explanation of how amino acids act as regulators of critical metabolic pathways
- Sidebar that explores the strategies used by some athletes to maximize skeletal muscle mass
- Expanded section on protein and amino acids supplements, including summary table of safety and effectiveness. New supplements include beta-alanine, growth hormone releasers, and nitric oxide

Chapter 6: Fats

- More information on *trans* fats and omega-3 fatty acids
- Discussion of the role that omega-3 fatty acids in foods and as supplements may play in offsetting acute inflammation and chronic immune dysfunction
- Updated caffeine recommendations for athletes

Chapter 7: Water and Electrolytes

- Added information on water content of various beverages
- New information and table on electrolytes
- Enhanced information on hyponatremia
- Updated discussion on exercise-associated muscle cramping
- Updated information and table on sodium-containing products
- New tables detailing the composition of various pre-exercise beverages, and beverages used during and after exercise
- Updated, reorganized, and expanded section on fluid intake strategies before, during, and after exercise
- Updated and reorganized section on individualized planning to meet fluid and electrolyte needs

Chapter 8: Vitamins

- New section added on the function of vitamins in growth and development
- Substantial amount of material added about vitamin D, including the 2010 revisions to the DRI
- Added information about quercetin

Chapter 9: Minerals

- Chapter reorganization to match the organization of the vitamin chapter, to the extent possible
- Expansion and reorganization of the section on the role of minerals in bone formation
- New information about average daily calcium intake and inclusion of the 2010 revisions to the DRI for calcium and vitamin D
- User-friendly chart explaining iron-related blood tests

Chapter 10: Diet Planning: Food First, Supplements Second

- Updated estimate of daily energy need for male and female athletes
- Explanation of the role that solid fats and added sugars (SoFAS) play
- Completely revised sections on food and fluid intake before, during, and after exercise
- Updated and expanded section on caffeine
- Table summarizing the safety and effectiveness of more than 20 dietary supplements

Chapter 11: Weight and Body Composition

- Graphics and artwork updated and streamlined for improved readability
- Updated information on body fat ranges for athletes in selected sports
- Updated references for norms for body composition
- Revised and expanded section on the relationship of body composition and weight to athletic performance
- Updated section on changing body composition to enhance performance with new research
- Reorganized discussion of weight certification in sports
- Revised and updated section on supplements used to change body composition
- New section on the supplement bitter orange

Chapter 12: Disordered Eating and Exercise Patterns in Athletes (*New Chapter Order*)

- Updated incidence and prevalence statistics
- Updated information on the Female Athlete Triad, including the American College of Sports Medicine position paper issued in 2007

Chapter 13: Diet and Exercise for Lifelong Fitness and Health (*New Chapter Order*)

- Updated guidelines including Dietary Guidelines for Americans, 2010, and Physical Activity Guidelines for Americans, 2008
- Repositioning and expanding the section on overweight and obesity to better illustrate the fundamental role of weight in chronic-disease prevention and treatment
- More information about childhood and adolescent obesity
- New section about how the body regulates weight
- Updates on all chronic diseases, including hypertension, diabetes, heart disease, metabolic syndrome, osteoporosis, and lifestyle-related cancers
- Information about the role that solid fats and added sugars (SoFAS) play in chronic disease, particularly heart disease
- Added section on the role of population-level changes that may help to initiate or sustain individual behavior changes

Appendixes

- The DASH (Dietary Approaches to Stop Hypertension) Eating Plan updated with 2010 Dietary Guideline for Americans information
- New USDA Food Patterns including lacto-ovo vegetarian and vegan adaptations
- U.S. Food Exchange System updated to 2008 edition
- New step-by-step directions, enhanced equations, and clear examples for determining energy expenditure rate and total energy expenditure for specific physical activities from the Compendium of Physical Activities
- Updated normative percentile values for maximal oxygen consumption for men and women
- Updated detailed graphics for glycolysis, the Krebs cycle, the electron transport chain, and β-oxidation
- NCAA bylaw on use of banned drugs updated to 2010–11 version
- Updated normative percentile values for percent body fat for males and females

Instructor and Student Resources

PowerLecture

This convenient tool makes it easy for instructors to create customized lectures. Each chapter includes the following features, all organized by chapter:

- Lecture slides
- All chapter art and photos
- Animations and videos
- *Instructor's Manual*, featuring summaries of chapter concepts, chapter outlines, and suggested activities and assignments
- *Test Bank*, including more than 1,400 questions in multiple-choice, true/false, matching, fill-in-the-blank, and essay formats
- *ExamView®* testing software preloaded with the Test Bank items

This single disc places all the media resources at your fingertips.

Nutrition CourseMate

Cengage Learning's Nutrition CourseMate brings course concepts to life with interactive learning, study, and exam preparation tools that support the printed textbook, or the included eBook. With CourseMate, professors can use the included Engagement Tracker to assess student preparation and engagement. Use the tracking tools to see progress for the class as a whole or for individual students. Students can access an interactive eBook, chapter-specific interactive learning tools, including flashcards, quizzes, videos, and more in their Nutrition CourseMate, accessed through **CengageBrain.com**.

Diet Analysis+

We have updated *Diet Analysis+*, the market-leading diet assessment program for Nutrition, to make it more useful for Exercise and Health Science courses. The user can easily create a personalized profile based on height, weight, age, sex, and activity level, including additional features to measure body frame, BMI, girth in centimeters, skinfold in millimeters, and exercise and resting heart rates. Its dynamic interface makes it easy to track calories, carbohydrates, fiber, proteins, fats, vitamins, and minerals in foods, as well as determine whether nutrient needs are being met. The program's Enhanced Search functionality allows users to filter food by category—improving search precision and making it easier to find certain foods.

Global Nutrition Watch

Updated several times a day, Global Nutrition Watch is an ideal resource for classroom discussion and research projects. You and your students get access to information from trusted academic journals, news outlets, and magazines, as well as videos, primary sources, podcasts, and much more.

Walk4life Elite Model Pedometer

This pedometer tracks steps, elapsed time, distance, and calories expended. The pedometer includes an extra-large digital display with a hinged protective cover, and comes with instructions outlining how to use the tool most effectively. It can be used as part of an in-class activity or as a tool to increase awareness and encourage students to simply track their steps and walk toward better fitness. This is a valuable resource for everyone, *and* at $15 when bundled with the text, this pedometer is a deal!

Acknowledgments

From initial conceptualization to final product, this book, and now the second edition, has required several years and the efforts and inspiration of many people. The authors would like to thank those people, both together and individually, who have either directly or indirectly helped make this book a reality.

Many thanks to all of the people at Cengage Learning and associated companies who were able to take all our words and ideas and turn them into the professional work you see here. It takes an astonishing number of talented and creative people to produce a book like this and we want to personally thank them all.

A very special thanks goes to our developmental editor, Suzannah Alexander, for seeing the second edition through to its final form. We also thank Carol Samet, Senior Content Project Manager at Cengage Learning, and Kelly Keeler, Senior Project Manager at PreMediaGlobal, who both shepherded the manuscript through the many production stages to final product. Thanks to Elesha Feldman, Assistant Editor, who managed the development of the print supplements; Miriam Myers, Media Editor, for her development of the PowerLecture DVD and CourseMate website; and Shana Baldassari, Editorial Assistant, for managing a thousand details with grace and good humor. We also extend our gratitude to John Walker for his guidance on the book design and cover, and to photo researcher Sara Golden at PreMediaGlobal for her hard work in securing all the photographs in the book.

We are particularly appreciative of those who reviewed the text. Their time, effort, and suggestions have helped make this a much better book. We appreciate your insights and your suggestions.

Reviewers

Second Edition

Dawn E. Anderson, *Winona State University*

Kathleen M. Laquale, *Bridgewater State University*

Rebecca Mohning, *George Washington University*

Dave Pavlat, *Central College*

Kimberli Pike, *Ball State University*

Jack L. Smith, *University of Delaware*

Stacie L. Wing-Gaia, *University of Utah*

First Edition

Charles Ash, *Kennesaw State University*

John Bergen, *University of West Florida*

Laura Burger, *Grossmont College*

Joseph Chromiak, *Mississippi State University*

Kristine Clark, *Penn State University*

Edward Coyle, *University of Texas, Austin*

Kim Crawford, *University of Pittsburgh*

Robert Cullen, *Illinois State University*

Susan Fullmer, *Brigham Young University*

Kathe A. Gabel, *University of Idaho*

Charlene Harkins, *University of Minnesota, Duluth*

Ronnie Harris, *Jacksonville State University*

Joshua Hingst, *Florida State University*

Michael E. Houston, *Virginia Tech*

Thomas Kelly, *Western Oregon University*

Laura Kruskall, *University of Nevada, Las Vegas*

Lonni Lowery, *University of Akron*

Karen Mason, *Western Kentucky University*

Michael C. Meyers, *West Texas A&M University*

Mary P. Miles, *Montana State University*

Cherie Moore, *Cuesta College*

Joseph A. O'Kroy, *Florida Atlantic University*

Kimberli Pike, *Ball State University*

Robert Skinner, *Georgia Institute of Technology*

Joanne Slavin, *University of Minnesota*

Teresa Snow, *Georgia Institute of Technology*

Tom R. Thomas, *University of Missouri*

Helen Ziraldo, *San Jose State University*

In addition to our appreciation of the work done by our editorial and production teams, each of us wishes to express special thanks as follows:

MD: This book actually began in the 1980s, although I didn't know it at the time, when some insightful faculty at California State University, Fresno, supported the development of a new course—Nutrition and the Athlete. The course evolved over the many years that I taught it, in large part due to feedback from students, and I would like to thank them for challenging me to be a better teacher. I also met Andy Doyle during this time, a fellow member of the faculty, who is a wonderful co-author. I thank him for adding his considerable expertise to this book, bringing the best out in me, and always maintaining his sense of humor.

It takes many years to write the first edition of a textbook, and it is such an arduous task that it would not be possible without support from family, friends, and colleagues. It is a thrill to revise and write the second edition, but it is no less of an arduous task. Heartfelt thanks goes to all the reviewers and colleagues who made suggestions. There are too many to mention by name but I am most appreciative to all who have encouraged me over the course of my career.

JAD: I would like to thank my co-author, Marie, for her patience, persistence, discipline, and good humor. My wife, Colleen, my sons, Patrick and Jackson, and my sister, Liz Doyle, have always been supportive of my education and my career, and I would like to thank them for their love and support. They have been very patient and supportive when this project has demanded a lot of my time and attention. Many thanks are due also to the students who have been an integral part of my courses and research over the years. In particular, I'd like to thank Ryan Luke for his research assistance on this book, and Rob Skinner for the many conversations we've had in which he has shared his ideas and experience in sports nutrition. Finally, I would like to thank the faculty and staff of the Department of Kinesiology and Health at Georgia State University for their support.

Marie Dunford, Ph.D., R.D.
Former Professor and Chair
Department of Food Science and Nutrition
California State University, Fresno

J. Andrew Doyle, Ph.D.
Associate Professor and Former Chair
Department of Kinesiology and Health
Georgia State University

About the Authors

MARIE DUNFORD, Ph.D., R.D., has been involved in sports nutrition since the mid-1980s. In 1985, while a faculty member at California State University, Fresno, she created the curriculum for an upper division course entitled, Nutrition and the Athlete. She taught the course for a total of 16 years during which time she interacted with thousands of student-athletes. This direct exposure to nutrition and exercise science majors and NCAA Division I athletes helped her to develop an understanding of how students learn and the sports nutrition topics that are the most difficult for students to master. In addition to this textbook, Dr. Dunford has written three other books—*Fundamentals of Sport and Exercise Nutrition, The Athlete's Guide to Making Weight: Optimal Weight for Optimal Performance*, and *Nutrition Logic: Food First, Supplements Second*—and numerous online sports nutrition courses for nutrition and exercise professionals. She is an active member of SCAN, the Sports, Cardiovascular, and Wellness Nutritionists, a dietetic practice group of the American Dietetic Association, and a member of the American College of Sports Medicine. She is an avid recreational tennis player and a struggling student of French.

J. ANDREW DOYLE, Ph.D., FACSM, is an Associate Professor of Exercise Physiology and the Director of the Applied Physiology Laboratory in the Department of Kinesiology and Health at Georgia State University where he formerly served as the Department Chair. He received a B.S. in Zoology from Clemson University, an M.S. in Exercise Science from Georgia State University, and his doctorate in Exercise Physiology from the Ohio State University. He has taught exercise physiology, exercise testing and fitness assessment, and exercise programming at the undergraduate and graduate level for over 20 years. His research interests include carbohydrate metabolism and exercise and the role of physical activity, exercise, and fitness in health. He has conducted, published, and presented numerous research studies with cyclists, runners, and triathletes, and has extensive experience testing elite athletes from cycling, running, gymnastics, rowing, canoe and kayak, and basketball. Dr. Doyle is a Fellow of the American College of Sports Medicine.

To my husband, Greg. *C'est le ton qui fait la chanson*. It's the melody that makes the song.
MD

In memory of my mother, Ann Shiver Lundquist.
JAD

Introduction to Sports Nutrition

Learning Plan

- **Define** key terms such as *exercise physiology, nutrition, athlete, physical activity, exercise,* and *sport.*
- **List** and explain basic training and sports nutrition goals.
- **Explain** basic nutrition principles and how they might be modified to meet the needs of athletes.
- **Outline** the basic issues related to dietary supplements, such as legality, ethics, purity, safety, and effectiveness.
- **Distinguish** between types of research studies, weak and strong research designs, and correlation and causation.
- **Explain** the importance of using recommendations based upon current scientific evidence and ways that research results may be misinterpreted.
- **Discuss** the accuracy of sports nutrition information obtained on the Internet.
- **Compare** and contrast the academic training and experience necessary to obtain various exercise and nutrition certifications.

Proper nutrition supports training and performance.

Pre-Test   Assessing Current Knowledge of Sports Nutrition

Read the following statements and decide if each is true or false.

1. An athlete's diet is a modification of the general nutrition guidelines made for healthy adults.

2. Once a healthy diet plan is developed, an athlete can use it every day with little need for modification.

3. In the United States, dietary supplements are regulated in the same way as over-the-counter medications.

4. The scientific aspect of sports nutrition is developing very quickly, and quantum leaps are being made in knowledge of sports nutrition.

5. To legally use the title of sports nutritionist in the United States, a person must have a bachelor's degree in nutrition.

Welcome to the exciting world of sports nutrition. This relatively new field is a blend of nutrition and exercise physiology. These fields are complementary academic disciplines that help us to understand the energy expenditure that is required by exercise and sport, and the energy and nutrient intake that is vital to support excellent **training** and performance. Exercise and sport challenge the human body to respond and adapt, and proper nutrition supports these processes. Training and nutrition are keys to athletic performance at any level.

The Olympics motto is *Citius, Altius, Fortius*, Latin for "swifter, higher, stronger." To achieve the highest level of success, athletes must be genetically endowed, and they must train optimally to meet their genetic potential. Proper nutrition supports the demands of training, and the field of sports nutrition emerged to help athletes train, perform, and recover to the best of their abilities. To run faster, jump higher, and be stronger, athletes must use genetics, training, and nutrition to their advantage.

physiology principles that support and enhance training, performance, and recovery. These principles also help athletes attain and maintain good health.

First and foremost, these disciplines are based on sound scientific evidence. But there is also an art to applying scientific principles to humans. For example, scientists identify nutrients found in food that are needed by the body, but food is sometimes eaten just because it tastes delicious or smells good. Exercise physiologists know from well-controlled research studies that the size and strength of athletes' muscles can be increased with overload training, but choosing the appropriate exercises, the number of sets and repetitions, the amount of resistance, the rest intervals, and the exercise frequency for optimal response by each individual athlete is as much of an art as a science. Because sports nutrition is a relatively young field, the knowledge base is continually expanding and our understanding of the field is constantly evolving. There is more research to be done and much more to be learned, presenting an exciting opportunity for exercise science- and nutrition-oriented students.

1.1 Training, Nutrition, and the Athlete

Sports nutrition is a blend of exercise physiology and nutrition.

Exercise physiology is the science of the response and adaptation of bodily systems to the challenge imposed by movement—physical activity, exercise, and sport. Nutrition is the science of the ingestion, digestion, absorption, metabolism, and biochemical functions of nutrients. **Sports nutrition** is the integration and application of scientifically based nutrition and exercise

The term *athlete* is very broad and inclusive.

The word *athlete* describes a person who participates in a sport. Using that definition, professional, collegiate, and weekend golfers are all athletes. Clearly there are differences among them. One difference is skill and another is training. Elite athletes are exceptionally skilled and dedicated to their training regimens. Their lives are planned around their training and competition schedules because athletic competition is their profession. Collegiate athletes are also trained athletes, although the level of their training is probably less than that of their professional counterparts. Dedication to training is important because proper training is necessary to improve and maintain performance. Many people are recreational

Anyone who participates in a sport can be called an athlete. As a means of distinction, the terms *elite athlete, well-trained athlete,* and *recreational athlete* are often used.

athletes. Some of them are former competitive athletes who continue to train, albeit at a lower level, to remain competitive within their age group or in masters events. They are sometimes referred to as performance-focused recreational athletes. However, many recreational athletes train little, if at all, and their primary focus is not improving performance. They participate in sports to be physically active, to maintain a healthy lifestyle, and for enjoyment.

Physical activity, exercise, and sport differ from each other.

Physical activity is bodily movement that results in an increase in **energy** expenditure above resting levels. Examples can include activities of daily living such as bathing, walking the dog, raking leaves, or carrying bags of groceries. Exercise and sport are very specific types of physical activity. Exercise has been defined as "physical activity that is planned, structured, repetitive, and purposive in the sense that improvement or maintenance of one or more components of physical fitness is the key" (Caspersen, Powell, and Christensen, 1985). For example, running is a specific type of physical activity that is often done regularly by people who hope to improve their **cardiovascular fitness**. Sports can be thought of as competitive physical activities. Track, cross country, or road running (for example, marathon) are examples of running as a sport.

Exercise may be described as **aerobic** or **anaerobic**. Aerobic means "with oxygen" and is used in reference to exercise or activity that primarily uses the oxygen-dependent energy system—oxidative phosphorylation (see Chapter 3). These types of activities can be sustained for a prolonged period of time and are referred to as endurance activities. Those who engage in them are referred to as endurance athletes. Some endurance athletes are better described as ultraendurance athletes because they engage in sports that require hours and hours of continuous activity, such as triathlons. Endurance and ultraendurance athletes are concerned about the same issues, such as adequate carbohydrate and fluid intake, but there are enough differences between them that their concerns are often addressed separately.

Anaerobic means "without oxygen" and is used in reference to exercise that primarily uses one or both of the energy systems that are not dependent on oxygen—creatine phosphate or anaerobic glycolysis (see Chapter 3). These types of activities are short in duration and high in exercise **intensity**. Athletes in high-intensity, short-duration sports are often called strength athletes or strength/power athletes. Although few sports are truly anaerobic, and weight lifting to strengthen muscles is usually a part of an endurance athlete's training, *strength athlete* and *endurance athlete* are terms that are commonly used.

Training: A planned program of exercise with the goal of improving or maintaining athletic performance.

Sports nutrition: The application of nutrition and exercise physiology principles to support and enhance training and performance.

Energy: The capacity to do work. In the context of dietary intake, defined as the caloric content of a food or beverage.

Cardiovascular fitness: Ability to perform endurance-type activities, determined by the heart's ability to provide a sufficient amount of oxygen-laden blood to exercising muscles and the ability of those muscles to take up and use the oxygen.

Aerobic: "With oxygen." Used in reference to exercise that primarily uses the oxygen-dependent energy system, oxidative phosphorylation.

Anaerobic: "Without oxygen." Used in reference to exercise that primarily uses one or both of the energy systems that are not dependent on oxygen, creatine phosphate or anaerobic glycoloysis.

Intensity: The absolute or relative difficulty of physical activity or exercise.

Although each participates in the same sport, the training and nutritional needs of recreational and elite athletes are very different.

Training and nutrition go hand in hand.

The longtime columnist, book author, and running philosopher George Sheehan (1980) once wrote that everyone is an athlete; only some of us are not in training. Athletes improve their sports performance through skill development and training. Skill development is enhanced through practice and instruction or coaching. Success in many sports is directly related to fitness levels achieved by sport-specific training. For example, to be successful, competitive distance runners must have a high level of cardiovascular fitness, which is developed through following a rigorous running training program.

As advances in exercise and sports science have become more widely recognized and adopted, athletes from a wide variety of sports have begun to use improved physical conditioning as a way to further improve their performance. Even athletes in sports such as golf and car racing have begun physical training as a strategy to improve personal performance. Physical training to improve specific components of fitness must be taken into account when considering nutritional needs, such as total energy and carbohydrate intakes. Nutrition supports training and good health—two factors that are essential to excellent performance.

Although nutrition by itself is important, it may have the greatest performance impact by allowing athletes to train consistently. Proper nutrition during the recovery period is essential for replenishing nutrient stores depleted during training, for example, muscle **glycogen**. Inadequate replenishment of energy, fluid, carbohydrates, proteins, and/or vitamins and minerals limits the potential for full recovery after training. Limited recovery can result in **fatigue** during the next training session, and consistent lack of nutritional replenishment can lead to **chronic** fatigue (Maughan, 2002).

Athletes perceive that nutrition is important, but they sometimes fail to realize or acknowledge that it is a factor that needs daily attention. This often leads to **crash diets** and other quick fixes, which may interfere with training and undermine performance. Nutrition and training are similar in that each is a process that needs a well-developed plan (Macedonio and Dunford, 2009).

Athletes can also get so focused on one small aspect of their diet that they neglect their comprehensive daily nutrition requirements. For example, athletes may concentrate on the best precompetition meal, but if they fail to address their day-to-day nutrition needs, then their training will suffer. Inadequate training that is a result of inadequate nutrient replenishment is much more detrimental to performance than the precompetition meal is beneficial to performance (Maughan, 2002).

Nutrition supports training and performance.

The main goal for any competitive athlete is to improve performance. Improvements in sport performance can come as a result of many factors: skill enhancement, psychological changes, specialized equipment and clothing, or physiological improvements due to training. All aspects of training should support this primary goal of improving performance. However, in the quest for excellent performance, the importance of good health should not be disregarded or overlooked. General training goals are listed below:

- Improving performance
- Improving specific components of fitness
- Avoiding injury and overtraining
- Achieving top performance for selected events (that is, peaking)

To support training and improve performance, athletes need to establish both long- and short-term nutrition goals. Some of these goals are listed below (Maughan, 2002).

Long-term sports nutrition goals:

- Adequate energy intake to meet the energy demands of training
- Adequate replenishment of muscle and liver glycogen with dietary carbohydrates
- Adequate protein intake for growth and repair of tissue, particularly skeletal muscle
- Adequate hydration
- Adequate overall diet to maintain good health and support a healthy immune system
- Appropriate weight and body composition

Short-term sports nutrition goals:

- Consumption of food and beverages to delay fatigue during training and competition
- Minimization of dehydration and **hypohydration** during exercise
- Utilization of dietary strategies known to be beneficial for performance, such as precompetition meal, appropriately-timed caffeine intake, or carbohydrate loading
- Intake of nutrients that support recovery

It is important to understand basic training principles.

As the athlete trains, the body responds to the individual exercise sessions and gradually adapts over time. The nature and degree of the adaptation(s) depends upon the type of training the athlete does, and follows general principles derived from the results of many research studies.

The principle of progressive overload. Adaptation occurs as a result of a stimulus that stresses the body. The stimulus must be of sufficient magnitude to cause enough stress to warrant longer-term changes by the body. Stimulus of this magnitude is called **overload**. If exposed to an overload stimulus repeatedly, the body will adapt over time to that level of stimulus. For further adaptation to occur, the overload stimulus must be progressively increased.

For example, in order for the biceps muscles to get stronger, an athlete must perform a weight-lifting exercise like an arm curl. The muscles will not get stronger curling the weight of a pencil; rather, the weight must be heavy enough to achieve overload. Once the muscles have adapted to that weight, they will not get any stronger until the overload stimulus is progressively increased (that is, the weight is increased further).

© Felicia Martinez/PhotoEdit

An overload stimulus, such as an arm curl, is required for the biceps muscles to get stronger.

The principle of individuality. Although general training principles apply to all people, individuals may respond and adapt slightly differently, even when exposed to the same training stimulus. Two similar athletes that follow the same strength-training program will both improve their strength, but it is likely that the amount and rate of change in strength will be slightly different. People do not respond in precisely the same way or time frame, so individual differences must be taken into account when considering an athlete's training program.

The principle of specificity. The type of physiological responses and eventual adaptations will be specific to the type of stimulus and stress imposed on the body. In the most general sense, aerobic exercise will result primarily in cardiovascular adaptations and resistance training will result in neuromuscular adaptations. Adaptations can be more subtle and specific, such as the effect intensity and duration of aerobic exercise may have on changes in energy system pathways such as carbohydrate and fat metabolism (see Chapters 4 and 6).

Glycogen: Storage form of glucose in the liver and muscle.

Fatigue: Decreased capacity to do mental or physical work.

Chronic: Lasts for a long period of time. Opposite of acute.

Crash diet: Severe restriction of food intake in an attempt to lose large amounts of body fat rapidly.

Hypohydration: An insufficient amount of water; below the normal state of hydration.

Overload: An exercise stimulus that is of sufficient magnitude to cause enough stress to warrant long-term changes by the body.

The principle of hard/easy. The stimulus part of training receives the most attention, but often neglected are the rest and recovery that are required for the adaptation to occur. Training programs are usually designed so that hard physical efforts are followed by training sessions with less physical stress to allow for the rest necessary for optimal adaptation.

The principle of periodization. Adhering to the principle of **specificity**, training programs are also often arranged in time periods according to the specific adaptation that is sought. For example, competitive long distance runners may spend a portion of their yearly training time concentrating on running longer distances to improve their maximal aerobic capacity and endurance, and another portion of their training time running shorter distances at higher intensity to improve their speed. Within this principle of **periodization**, training programs are generally arranged according to different time periods:

Macrocycle: A macrocycle is an overall time period that begins at the onset of training and includes the time leading up to a specific athletic goal, such as an important competition. For an athlete seeking to peak at the annual national championship, the macrocycle may be a calendar year. A macrocycle may be longer (for example, 4 years for an athlete concentrating on the Olympics) or shorter (for example, 6 months for a distance runner training for a springtime marathon), depending upon the specific competitive goals of the athlete.

Mesocycle: A macrocycle is subdivided into time frames called mesocycles, each having a specific training purpose. As with the macrocycle, the mesocycles may be of varying lengths of time, depending upon the athlete's goals, but typically are weeks or months in duration. The competitive distance runner may have a mesocycle focused on improving aerobic capacity and endurance and another mesocycle focused on improving speed.

Microcycle: Each mesocycle is made up of repeated time intervals called microcycles. Microcycles are often designed to coincide with the weekly calendar, but can vary from the standard 7-day week, depending upon the athlete's specific needs. Weekly training mileage for the competitive distance runner is an example of a microcycle.

The principle of disuse. Just as the body adapts positively in response to training stress, it can adapt negatively, or **atrophy**, if stress is insufficient or absent. Gradual erosion of physiological capacity over time is often observed in individuals as a result of sedentary lifestyles. Athletes who have improved function through training can experience the loss of function, either intentionally for short periods (for example, resting during the "off-season") or unintentionally due

A registered dietitian can help an athlete develop a diet plan that is well matched to the demands of training.

to forced inactivity from injury. This is the physiological equivalent of the aphorism "Use it, or lose it."

In addition to a training plan, an athlete needs a nutrition plan.

Training periodization involves changing the intensity, **volume**, and specificity of training to achieve specific goals. It is imperative that a parallel nutrition plan be developed to support the various training cycles. This plan may be referred to as nutrition periodization. The nutrition plan should match the training plan. If the training macrocycle is one year, then the athlete should also have an annual nutrition plan. Each mesocycle will have specific nutrition goals as well. For example, weight loss by an endurance athlete is usually planned to take place during a recovery period ("off-season") and early in the preparation period so a restricted-calorie diet can be avoided during high-volume training periods or during the competitive season. During each microcycle, refinements are made to dietary intake (Seebohar, 2011).

Figure 1.1 simply illustrates the concept of having a nutrition plan that matches the demands imposed by various training periods. In this example of a male collegiate 800 meter (m) runner, the plan covers a year (that is, the macrocycle), starting in September, when school begins, through the following August. The training and nutrition goals of each mesocycle vary. During the early months of the preparation period (September through October) the primary focus is on aerobic training. This

	Prior to season					Pre-season			Racing season		Off-season	
	Sept	Oct	Nov	Dec	Jan	Feb	Mar	Apr	May	June	July	Aug
Training goals:	Training volume increasing; emphasis on aerobic base training with some speed/anaerobic training		Training volume high; maintain aerobic base training and increase high-intensity/speed/anaerobic training			Training volume decreased to emphasize speed/anaerobic training			Training volume decreased to emphasize speed training and tapering for competitive races		No formal training; physical activity and exercise for recreation	
Body composition goals:	Reduce 5 lb body fat		Increase skeletal muscle mass by 3–5 lb			Maintain the increased skeletal muscle mass			Maintain body composition		3–5 lb loss of skeletal muscle mass and 5 lb increase in body fat are acceptable to this athlete	
Energy (caloric) intake:	Decrease energy intake from food and increase energy expenditure from training for a slow loss of body fat over 2 months		Increase caloric intake to support muscle growth and an increase in training volume			Caloric intake should equal caloric expenditure so body composition can be maintained					If caloric intake not reduced, body fat will increase	
Nutrient intake:	Adequate carbohydrate and fluid to support a return to training. Compared to the off-season diet, current diet has fewer high-fat, high-sugar foods and more water, fruits, vegetables, and whole grains.		Compared to the past 2 months, slight increase in carbohydrate and protein intakes			For sufficient glycogen stores, a high-carbohydrate diet is recommended. Diet is generally high carbohydrate, moderate protein, and moderate/low fat. In the pre-season, diet plan is fine-tuned to make sure it is realistic (especially on travel days/away meets) and well tolerated.					A nutritious diet that meets the Dietary Guidelines is recommended	

Figure 1.1　A training and nutrition periodization plan for a male 800 m runner

athlete also wants to decrease 5 pounds of body fat that has been gained during the summer. Energy (calorie) and carbohydrate intakes must be sufficient to support training, but energy intake must be reduced from baseline so that some of the energy needed is provided from stored fat. The second part of the preparation period (November through January) focuses on maintaining aerobic fitness, increasing strength and power, and technique. This athlete also wants to increase muscle mass by 3 to 5 pounds. The volume of training is increased and is equally divided between aerobic (for example, running) and anaerobic (for example, high-repetition lifting and **plyometric** exercise) activities. Proper energy, carbohydrate, protein, and fat intakes are needed to support both his training and body composition goals.

During the precompetition period (February through April), most of the training takes place on the track. Training is approximately 40 percent anaerobic and 60 percent aerobic. Weight lifting is decreased

because the goal is maintenance of gained muscle rather than a continued increase in muscle mass. There is an emphasis on plyometric training and an alternating schedule—Monday, Wednesday, and Friday feature

Specificity: A training principle that stresses muscles in a manner similar to which they are to perform.

Periodization: Dividing a block of time into distinct periods. When applied to athletics, the creation of time periods with distinct training goals and a nutrition plan to support the training necessary to meet those goals.

Atrophy: A wasting or decrease in organ or tissue size.

Volume: An amount; when applied to exercise training, a term referring to the amount of exercise usually determined by the frequency and duration of activity.

Plyometric: A specialized type of athletic training that involves powerful, explosive movements. These movements are preceded by rapid stretching of the muscles or muscle groups that are used in the subsequent movement.

hard workouts whereas Tuesday and Thursday involve easy recovery runs as the athlete prepares for competition on Saturday. During the competitive season (May through mid-June), more emphasis is placed on anaerobic training (~75 percent) and less on aerobic training (~25 percent). Almost all of the training is on the track and the athlete does no weight lifting. Friday is a rest and travel day in preparation for racing on Saturday. A new period begins once the competitive season ends and the school year is complete. For about three weeks (mid-June to early July), the athlete does no training in an effort to recuperate mentally and physically from the rigorous months of training and competition. Through most of July and August the focus is on moderate-duration, low-intensity running. Energy expenditure over the summer is the lowest of the entire year and this runner will need to reduce food intake to match reduced expenditure to prevent excessive weight gain as body fat. If he does not, he will likely gain unwanted weight and body fat.

Some athletes create elaborate nutrition plans. The plan can be as simple or detailed as the athlete feels is necessary but the fundamental issues are the same: For optimal training, performance, and recovery, proper nutrition intake is important, changes in weight or body composition need to be appropriately timed, and good health should not be overlooked.

Key points

■ Sports nutrition requires an understanding of the physiological challenges of training and competition and the scientific and applied principles of nutrition.

■ The physical demands of activity, exercise, and sport can vary dramatically between athletes and for individual athletes over a given time period.

■ Training and nutrition go hand in hand.

■ An organized training plan that takes into account specific goals and incorporates basic principles of training is critical for excellent performance.

■ Athletes need a nutrition plan that complements the physical demands of training and performance and supports good health.

What would be some specific training goals of a collegiate-level soccer player?

Fiber: A component of food that resists digestion (for example, pectin, cellulose).

Electrolyte: A substance in solution that conducts an electrical current (for example, sodium, potassium).

Dietary Reference Intakes: Standard for essential nutrients and other components of food needed by a healthy individual.

1.2 Basic Nutrition Standards and Guidelines

Sports nutrition principles are based on sound general nutrition principles that have been modified to reflect the demands of training and competition. General guidelines help all people, including athletes, to achieve optimal nutritional health over a lifetime. An optimal diet is one in which there are neither deficiencies nor excesses.

The early focus of nutrition research was on the amount and type of nutrients needed to prevent deficiencies. Once nutrient deficiency diseases were well understood the research focus changed to the amount and type of nutrients that help prevent chronic diseases. A chronic disease is one that progresses slowly, such as heart disease or osteoporosis (that is, loss of bone mineral density). These diseases are a reflection of long-term, not short-term, nutrient intake. Keeping in mind the need to prevent nutrient deficiencies as well as nutrient excesses, guidelines have been established for energy (calories), carbohydrates, proteins, and fats, **fiber**, vitamins, minerals, **electrolytes** (for example, sodium or potassium), and water. These guidelines are known as the **Dietary Reference Intakes** (Institute of Medicine, 1997, 1998, 2000, 2001, 2002, 2003, 2004, 2010).

The Dietary Reference Intakes (DRI) is a standard used to assess nutrient intake.

The Dietary Reference Intakes (DRI) is a standard used to assess and plan diets for individuals and groups (Institute of Medicine, 2001; Otten, Hellwig, and Meyers, 2006). The DRI expands on and replaces the 1989 Recommended Dietary Allowances (RDA) and the Recommended Nutrient Intakes (RNI) of Canada. The DRI is a general term that includes four types of reference values—Recommended Dietary Allowances, Adequate Intake, Estimated Average Requirement, and Tolerable Upper Intake Level. These terms are defined in Figure 1.2.

The DRI are based on the Recommended Dietary Allowance whenever possible. When an RDA cannot be determined, the Adequate Intake (AI) becomes the reference value for the DRI. The AI is not as scientifically strong since it is based on estimates or approximations derived from scientific research. The Dietary Reference Intakes and the reference value used for each vitamin and mineral are found on the inside gatefold of this textbook. Values for other nutrients are found in Appendix A. The use of the term *RDA* has caused some confusion. For many years, the RDA was the standard, but now is one of the reference values used to compile the DRI, the current standard.

Dietary Reference Intakes (DRI) Definitions

The Dietary Reference Intake (DRI) is a standard used to assess and plan diets. This standard is made up of the four reference values shown below.

Recommended Dietary Allowance (RDA): the average daily dietary intake that is sufficient to meet the nutrient requirement of nearly all (97 to 98%) healthy individuals in a particular group according to stage of life and gender.

Adequate Intake (AI): a recommended intake value based on observed or experimentally determined approximations or estimates of nutrient intake by a group (or groups) of healthy people, that are assumed to be adequate; AI is used when an RDA cannot be determined.

Estimated Average Requirement (EAR): a daily nutrient intake value that is estimated to meet the requirements of half of the healthy individuals in a group according to life stage and gender—used to assess dietary adequacy and as the basis for the RDA.

Tolerable Upper Intake Level (UL): the highest daily nutrient intake that is likely to pose no risk of adverse health effects for almost all individuals in the general population. As the intake increases above the UL, the potential risk of adverse effects increases.

Regarding vitamin and mineral intake, the EAR is used only when planning diets for groups. For individual diet planning, the RDA or the AI is used to guard against inadequate vitamin and mineral intakes and the UL is used to guard against excess intakes.

Institute of Medicine (2003). *Dietary Reference Intakes: Applications in Dietary Planning* (Food and Nutrition Board). Washington, DC: National Academies Press.

Figure 1.2 The Dietary Reference Intakes (DRI) reference values defined

Spotlight on...

The Physical Activity Guidelines for Americans

In 2008, the U.S. Department of Health and Human Services published the first-ever Physical Activity Guidelines for Americans, a series of recommendations for individual physical activity that complements the Dietary Guidelines for Americans. Being physically active and consuming a healthy diet promote good health and reduces the risk of various chronic diseases, such as cardiovascular disease and certain types of cancer (Laukkanen et al., 2001). These two documents provide science-based nutrition and physical activity guidance that can help people obtain long-term health benefits.

The following are the key Guidelines included in the Physical Activity Guidelines for Americans (http://www.health.gov/paguidelines):

Key Guidelines for Children and Adolescents

- Children and adolescents should do 60 minutes (1 hour) or more of physical activity daily.

- Aerobic: Most of the 60 or more minutes a day should be either moderate- or vigorous-intensity aerobic physical activity, and should include vigorous-intensity physical activity at least 3 days a week.

- Muscle-strengthening: As part of their 60 or more minutes of daily physical activity, children and adolescents should include muscle-strengthening physical activity on at least 3 days of the week.

- Bone-strengthening: As part of their 60 or more minutes of daily physical activity, children and adolescents should include bone-strengthening physical activity on at least 3 days of the week.

- It is important to encourage young people to participate in physical activities that are appropriate for their age, that are enjoyable, and that offer variety.

Key Guidelines for Adults

- All adults should avoid inactivity. Some physical activity is better than none, and adults who participate in any amount of physical activity gain some health benefits.

- For substantial health benefits, adults should do at least 150 minutes (2 hours and 30 minutes) a week of moderate-intensity, or 75 minutes (1 hour and 15 minutes) a week of vigorous-intensity aerobic physical activity, or an equivalent combination of moderate- and vigorous-intensity aerobic activity. Aerobic activity should be performed in episodes of at least 10 minutes, and preferably, it should be spread throughout the week.

- For additional and more extensive health benefits, adults should increase their aerobic physical activity to 300 minutes (5 hours) a week of moderate intensity, or 150 minutes a week of vigorous-intensity aerobic physical activity, or an equivalent combination of moderate- and vigorous-intensity activity. Additional health benefits are gained by engaging in physical activity beyond this amount.

- Adults should also do muscle-strengthening activities that are moderate or high intensity and involve all major muscle groups on 2 or more days a week, as these activities provide additional health benefits.

Additional guidelines are provided for older adults, women during pregnancy or postpartum, adults with disabilities, and children and adolescents with disabilities. See http://www.health.gov/paguidelines/.

The Dietary Guidelines Recommendations Encompass Two Over-Arching Goals:

- Maintain calorie balance over time to achieve and sustain a healthy weight
- Focus on consuming nutrient-dense foods and beverages

Key Recommendations (See full document at: http://www.cnpp.usda.gov/DGAs2010-PolicyDocument.htm)

Balancing Calories to Manage Weight

- Prevent and/or reduce overweight and obesity through improved eating and physical activity behaviors
- Control caloric intake to manage body weight
- Increase physical activity and reduce time spent in sedentary behaviors

Foods and Food Components to Reduce

- Reduce daily sodium intake to less than 2,300 milligrams (mg) and further reduce intake to 1,500 mg among African Americans, those who have hypertension, diabetes, or chronic kidney disease, and persons who are 51 and older
- Consume 7–10 percent of calories from saturated fatty acids by replacing them with monounsaturated and polyunsaturated fatty acids
- Consume less than 300 mg per day of dietary cholesterol
- Keep *trans* fatty acid consumption as low as possible
- Reduce the intake of calories from solid fats and added sugars
- If alcohol is consumed, it should be consumed in moderation

Foods and Nutrients to Increase

- Increase vegetable and fruit intake
- Consume at least half of all grains as whole grains
- Increase intake of fat-free or low-fat milk and milk products
- Choose a variety of protein foods
- Increase the amount and variety of seafood consumed by choosing seafood in place of some meat and poultry
- Use oils to replace solid fats where possible

Building Healthy Eating Patterns

- Select an eating pattern that meets nutrient needs at an appropriate calorie level
- Account for all foods and beverages consumed and assess how they fit within a total healthy eating pattern
- Follow food safety recommendations when preparing and eating foods

Figure 1.3 Dietary Guidelines for Americans, 2010
U.S. Department of Agriculture and U.S. Department of Health and Human Services, Dietary Guidelines for Americans, 2010. 7th ed., Washington, DC: U.S. Government Printing Office, December 2010.

Athletes in training may wonder how the DRI apply to them because they were developed for the general population. Since the goal of the DRI is to guard against both nutrient inadequacies and excesses, athletes use the DRI to assess the adequacy of their current diets and to plan nutritious diets. For example, the DRI is used to evaluate an athlete's vitamin intake even though there is some evidence that moderate to strenuous exercise may increase the need for some vitamins. The reason is that the DRI has a built-in margin of safety that likely exceeds the increased need associated with exercise (see Chapter 8).

On the other hand, some of the DRI, such as the estimated energy requirement or the need for water and sodium intake, may not be appropriate to use with athletes in training because athletes' energy, fluid, and electrolyte needs may be greater than those of the general population. In such cases other standards and guidelines are used.

The Dietary Guidelines for Americans provide basic dietary and exercise advice.

The Dietary Guidelines for Americans (U.S. Department of Health and Human Services and U.S. Department of Agriculture, 2010) are published every five years by the U.S. Department of Health and Human Services and the U.S. Department of Agriculture. The purpose of the Dietary Guidelines is to provide dietary and exercise advice to Americans over the age of 2 that will promote health and reduce the risk for chronic diseases. The Dietary Guidelines' key recommendations are listed in Figure 1.3 and the DASH-style dietary patterns that reflect these recommendations are in Appendix B.

Even though they were developed for the general population, most of the recommendations in the Dietary Guidelines apply to athletes, such as consuming nutrient-dense foods, eating fiber-rich fruits, vegetables, and whole grains to meet carbohydrate needs, and focusing on a "total diet" approach, which is a pattern of eating healthy foods and beverages. But some of the recommendations may not apply. For example, for those athletes who lose large amounts of sodium in sweat, limiting sodium intake to 2,300 milligrams (mg) daily may be detrimental. Athletes engaged in regular training will usually easily meet and exceed the physical activity recommendations contained in the Dietary Guidelines. However, some athletes concentrating on sports involving very specific components of fitness (for example, muscular strength for weight lifting or bodybuilding) may need to be conscious of including other components of fitness (for example, cardiovascular exercise) necessary for long-term health. The Dietary Guidelines are a good starting point for people who want to improve their health and fitness. The general nutrition principles can then be modified to fit the demands of training.

MyPyramid is a tool that can be used to create a nutritious diet.

MyPyramid, which is sometimes called the Food Guide Pyramid, is a graphic that reflects the principles outlined in the 2005 Dietary Guidelines (Figure 1.4). It is a food guidance system that can be used to teach consumers about basic nutrition and help them to plan a nutritious diet. Suggested amounts of food to consume can be found in Appendix C.

MyPyramid is designed to convey several general messages: physical activity, variety, proportionality, moderation, gradual improvement, and personalization. Physical activity is represented by a figure climbing steps. This is symbolic of the need for daily physical activity. The colored bands represent variety, with each band depicting a different food group. The size of the band suggests how much food should be chosen from that group in proportion to the other groups. For example, the largest band is orange, which represents grains. The message is that grains should be the largest proportion of food in the total diet. The yellow band, which represents oils, is the smallest band. Moderation is depicted by the narrowing of the bands from the bottom to the top of the pyramid. The foods at the bottom of each group (except oils) represent those foods with little solid fat or sugar. As the band narrows, the foods in that group contain more fat and sugar. The slogan is *steps to a healthier you*, a phrase that suggests that improvement will be gradual. Finally, the pyramid may be personalized by going to the website, http://MyPyramid.gov.

A food pyramid has been developed for athletes.

The Swiss Forum for Sport Nutrition has developed a Food Pyramid for Athletes (Figure 1.5). The purpose of the pyramid is to help athletes translate scientific recommendations, which are frequently made in grams per kilogram (g/kg) of body weight, into the amounts and kinds of foods that meet those recommendations. This pyramid has been scientifically validated for athletes 20–35 years old, weighing 50–85 kg (110–187 pounds [lb]), who train 5–28 hours per week.

The guidelines for the Basic Food Pyramid are listed on the left-hand side of the graphic. These foods make up a lifelong healthy diet. Additional servings are recommended for those in training based on the number of hours of moderate-intensity exercise per day. Each column under the Sports heading represents 1 hour. In general, each hour of exercise requires additional fluid, whole grains and legumes, and oils and nuts. When added to an already healthy diet, these foods provide the additional energy and nutrients that an athlete in training needs. Serving sizes are given as a range, with the smaller portions being appropriate for lower weight athletes (Mettler, Mannhart, and Colombani, 2009).

There are several other meal-planning tools available.

Another diet-planning tool is the Food Exchange System, often referred to as the exchange lists. This system categorizes foods based on their carbohydrate, protein, and fat contents (American Diabetes Association and American Dietetic Association, 2008). There are three groups—carbohydrate, meat and meat substitutes, and fat—with the carbohydrate and meat groups containing several subgroups. The foods on each list can be "exchanged" for another food on the same list because each has approximately the same **macronutrient** content for the portion size listed. For example, one small banana (~4 ounces [oz]) has approximately 15 g of carbohydrates and 60 kilocalories (kcal), about the same as one small orange.

However, foods are broadly categorized according to macronutrient content and there can be substantial **micronutrient** (that is, vitamin and mineral) differences between foods on the same list. For example, an orange is an excellent source of vitamin C (~70 mg) whereas a banana has little (~10 mg). The starch list contains whole wheat and white bread, foods with equivalent amounts of carbohydrates, proteins, and fats. However, whole wheat bread is a nutritionally superior food to white bread because of the fiber and trace mineral contents. Additionally, each food listed does not have the same portion size. On the fat exchange list, the portion size for avocado is 2 tablespoons whereas the portion size for oil

---

Macronutrient: Nutrient needed in relatively large amounts. The term includes energy, carbohydrates, proteins, fats, cholesterol, and fiber but frequently refers to carbohydrates, proteins, and fats.

Micronutrient: Nutrient needed in relatively small amounts. The term is frequently applied to all vitamins and minerals.

MyPyramid
STEPS TO A HEALTHIER YOU
MyPyramid.gov

GRAINS Make half your grains whole	VEGETABLES Vary your veggies	FRUITS Focus on fruits	MILK Get your calcium-rich foods	MEAT & BEANS Go lean with protein
Eat at least 3 oz. of whole-grain cereals, breads, crackers, rice, or pasta every day 1 oz. is about 1 slice of bread, about 1 cup of breakfast cereal, or ½ cup of cooked rice, cereal, or pasta	Eat more dark-green veggies like broccoli, spinach, and other dark leafy greens Eat more orange vegetables like carrots and sweetpotatoes Eat more dry beans and peas like pinto beans, kidney beans, and lentils	Eat a variety of fruit Choose fresh, frozen, canned, or dried fruit Go easy on fruit juices	Go low-fat or fat-free when you choose milk, yogurt, and other milk products If you don't or can't consume milk, choose lactose-free products or other calcium sources such as fortified foods and beverages	Choose low-fat or lean meats and poultry Bake it, broil it, or grill it Vary your protein routine — choose more fish, beans, peas, nuts, and seeds

For a 2,000-calorie diet, you need the amounts below from each food group. To find the amounts that are right for you, go to MyPyramid.gov.

Eat 6 oz. every day	Eat 2½ cups every day	Eat 2 cups every day	Get 3 cups every day; for kids aged 2 to 8, it's 2	Eat 5½ oz. every day

Find your balance between food and physical activity
- Be sure to stay within your daily calorie needs.
- Be physically active for at least 30 minutes most days of the week.
- About 60 minutes a day of physical activity may be needed to prevent weight gain.
- For sustaining weight loss, at least 60 to 90 minutes a day of physical activity may be required.
- Children and teenagers should be physically active for 60 minutes every day, or most days.

Know the limits on fats, sugars, and salt (sodium)
- Make most of your fat sources from fish, nuts, and vegetable oils.
- Limit solid fats like butter, stick margarine, shortening, and lard, as well as foods that contain these.
- Check the Nutrition Facts label to keep saturated fats, *trans* fats, and sodium low.
- Choose food and beverages low in added sugars. Added sugars contribute calories with few, if any, nutrients.

MyPyramid.gov
STEPS TO A HEALTHIER YOU

U.S. Department of Agriculture
Center for Nutrition Policy and Promotion
April 2005
CNPP-15

Courtesy USDA

Figure 1.4 MyPyramid is a tool that can be used to create a healthy diet

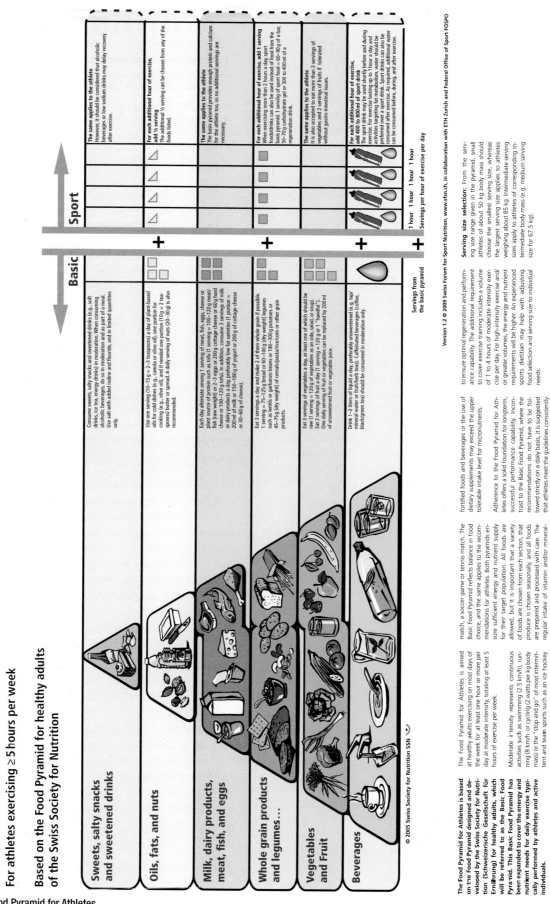

Figure 1.5 Food Pyramid for Athletes

is considerably smaller, 1 teaspoon. The Food Exchange System is found in Appendix D.

Another method that some athletes use is carbohydrate counting. The amount of carbohydrates needed daily is determined and then distributed throughout the day in meals and snacks. Although carbohydrate intake is emphasized to ensure adequate muscle glycogen for training, it is part of a larger plan that considers daily energy (calorie), protein, fat, and alcohol intakes.

A meal-planning system is useful, especially when athletes are learning about the nutrient content of foods and beginning to plan a diet that supports training. Over time, athletes typically want more precise information about the nutrient content of food and this leads to use of nutrient analysis software, such as the dietary analysis program that accompanies this textbook.

The Nutrition Facts label provides specific nutrition information.

One of the best nutrition education tools available to consumers is the Nutrition Facts label. This label is required for most prepared foods. Figure 1.6 shows an actual label except that it has been colored to highlight the information explained below. More information about food labels can be found in Chapter 10.

- Serving Size (shown in blue). Take particular note of the serving size, which is smaller than most Americans normally consume.
- Calories and Calories from fat (shown in red). Based on serving size. In this example, there are two servings per container so total Calories and Calories from fat would need to be doubled if the entire package was eaten.
- Nutrients (shown in gold, white, and green). Americans tend to overconsume total fat, saturated fat, *trans* fat, cholesterol, and sodium, which are shown in gold in this example. Potassium, dietary fiber, vitamins A and C, calcium, and iron tend to be underconsumed.
- % Daily Value* (shown in purple). Daily Value (DV) is an estimate of the amount needed each day based on a 2,000 Calorie diet. The DV is not specific for age or gender but it is a good ballpark figure.

Key points

- The Food Pyramid for Athletes is an excellent meal-planning tool for athletes.
- The Dietary Guidelines, MyPyramid, and the Nutrition Facts label are tools that can be used to develop a nutritionally sound diet plan.
- The Dietary Reference Intakes (DRI) is a standard used to assess and plan diets.

Are the various meal-planning tools substantially different from each other?

Nutrition Facts

Serving Size 1 cup (228g)
Servings Per Container 2

Amount Per Serving	
Calories 250	Calories from Fat 110

	% Daily Value*
Total Fat 12g	18%
Saturated Fat 3g	15%
Trans Fat 3g	
Cholesterol 30mg	10%
Sodium 470mg	20%
Potassium 700mg	20%
Total Carbohydrate 31g	10%
Dietary Fiber 0g	0%
Sugars 5g	
Protein 5g	

Vitamin A	4%
Vitamin C	2%
Calcium	20%
Iron	4%

* Percent Daily Values are based on a 2,000 calorie diet. Your Daily Values may be higher or lower depending on your calorie needs.

	Calories	2,000	2,500
Total Fat	Less than	65g	80g
Sat Fat	Less than	20g	25g
Cholesterol	Less than	300mg	300mg
Sodium	Less than	2,400mg	2,400mg
Total Carbohydrate		300g	375g
Dietary Fiber		25g	30g

U.S. Department of Health and Human Services

Figure 1.6 Nutrition Facts label

Application exercise

Scenario: *Five years after graduation, a former collegiate tennis player decided to return to competitive play in a local league. Based on his skills, he is ranked 5.0, the highest recreational ranking. To his great disappointment, he has not been very competitive in this league. Although his tennis skills are well matched to his opponents, he realizes that his fitness has declined and that he consistently fades in the third set due to fatigue. One of his opponents commented that the same thing used to happen to him until he "bit the bullet and started to eat better."*

1. Analyze this athlete's 1-day dietary intake using the Food Pyramid for Athletes or MyPyramid. Assume 2 hours of exercise and a 24-hour food intake as follows:
   Breakfast: 16 oz strong black coffee
   Lunch: Double cheeseburger, large fries, large soft drink (typical fast-food meal)
   Prematch snack: two bananas, 16 oz water
   Postmatch meal: four slices of a large pepperoni pizza, two 12 oz bottles of beer

2. What conclusions can be drawn about his overall intake as well as intake in each category?

3. If you were a registered dietitian, in what ways might you use the Food Pyramid for Athletes or MyPyramid to help this athlete "bite the bullet" and start eating better?

1.3 Basic Sports Nutrition Guidelines

Sports nutrition recommendations build upon and refine basic nutrition guidelines. Athletes need to understand and apply general nutrition principles before making modifications to reflect their training and sport-specific nutrient demands. Ultimately, sports nutrition recommendations are fine-tuned and are as precise as possible to closely meet the unique demands of training and competition in the athlete's sport and reflect the needs of the individual athlete. Here is a brief overview of some key sports nutrition recommendations (Rodriguez et al., 2009).

Energy: An adequate amount of energy is needed to support training and performance and to maintain good health. Appropriate amounts of food should be consumed daily to avoid long-term energy deficits or excesses. Adjustments to energy intake for the purpose of weight or fat loss should be made slowly and started early enough in the training mesocycle (for example, off-season or well prior to the competitive season) so as not to interfere with training or performance.

Carbohydrates: An intake of 6 to 10 grams (g) of carbohydrates per kilogram (kg) of body weight per day is recommended. The daily amount needed depends on the sport, type of training, gender, and need for carbohydrate loading. Timing is also important, and recommendations for carbohydrate intake before, during, and after exercise are made.

Proteins: An intake of 1.2 to 1.7 g of protein per kg of body weight per day is generally recommended. This recommendation assumes that energy intake is adequate. The daily amount of proteins needed depends on the sport, type of training, and the desire to increase or maintain skeletal muscle mass. Timing of protein intake is also important. For example, postexercise protein ingestion aids in muscle protein resynthesis.

Fats: After determining carbohydrate and protein needs, the remainder of the energy intake is typically from fats, although adult athletes may include a small amount of alcohol. Fat intake should range from 20 to 35 percent of total calories. Diets containing less than 20 percent of total calories from fat do not benefit performance and can be detrimental to health.

Vitamins and minerals: Athletes should meet the DRI for all vitamins and minerals. The DRI can be met if energy intake is adequate and foods consumed are **nutrient dense** (that is, abundant nutrients in relation to caloric content). Any recommendation for vitamin or mineral supplementation should be based on an analysis of the athlete's usual diet.

Fluid: Athletes should balance fluid intake with fluid loss. A number of factors must be considered, including the sweat rate of the athlete and environmental conditions such as temperature, humidity, and altitude. A body water loss in excess of 2–3 percent of body mass can decrease performance and negatively affect health. Similarly, an intake of water that is far in excess of fluid lost (primarily through sweat) puts the athlete at risk for a potentially fatal condition, known as **hyponatremia**, due to low blood sodium.

Food and fluid intake prior to exercise: Many athletes consume a beverage, snack, or meal prior to exercise to relieve hunger and to help with hydration. The volume of food and fluid depends to a large extent on the amount of time prior to exercise and the athlete's gastrointestinal tolerance. In general, a meal or snack should be relatively high in carbohydrate, moderate in protein, and low in fat and fiber.

Food and fluid intake during exercise: Food or beverage intake during exercise can be beneficial for athletes engaged in prolonged exercise because it can help to replace fluid lost in sweat and to provide carbohydrate.

Food and fluid intake after exercise: After exercise the goal is to consume foods and beverages that will help replenish the nutrients lost during exercise and to speed recovery. Athletes focus on carbohydrate to replenish muscle glycogen, fluid to restore hydration status, electrolytes such as sodium if large amounts of sodium have been lost in sweat, and protein to build and repair skeletal muscle tissue.

In addition to the above recommendations, there are a number of other critical areas that involve diet. Some athletes focus on scale weight since weight may be a sport participation criterion, but attaining a particular weight should be done in a healthy manner. **Disordered eating** (that is, abnormal eating patterns) and **eating disorders**, such as anorexia or bulimia, are concerns for individual athletes as well as teammates, coaches, parents, and anyone else who works with athletes. Athletes need a tremendous amount of information about dietary supplements, since the decision to use them should be based on safety, effectiveness, purity, legality, and ethics. All of these issues are covered in depth in the chapters of this text.

With so many details to consider, some athletes find that they begin to follow a rigid daily diet. The key is to meet nutrient needs and support training and performance while maintaining dietary flexibility. Athletes need to keep their diet in perspective: food is needed to fuel the body and the soul (see Keeping It in Perspective).

Adhering to a very rigid eating plan can lead to social isolation and can be a sign of compulsive behavior,

Nutrient dense: A food containing a relatively high amount of nutrients compared to its caloric content.

Hyponatremia: Low blood sodium level.

Disordered eating: A deviation from normal eating but not as severe as an eating disorder.

Eating disorders: A substantial deviation from normal eating, which meets established diagnostic criteria (for example, anorexia nervosa, bulimia nervosa, anorexia athletica).

both of which can create problems for athletes. Some find themselves eating the same foods every day and the joy of eating is diminished. The key is to have a flexible eating plan that is nutritious and includes a variety of foods. Flexibility usually results in short-term over- and undereating, but long-term weight stability, proper nutrition, and enjoyment of eating.

Flexible eating is not the same as unplanned eating. Sports nutrition is complicated and the failure to plan a nutritious diet often results in poor nutrient intake, which may hamper performance and undermine long-term health. But eating according to a rigid schedule is a problem, too. Food is for fuel and fun, and athletes must find the right balance.

The demands of an athlete's sport must be carefully considered.

The basic sports nutrition guidelines are a good starting point for both strength and endurance athletes, but they must be modified according to the athlete's sport. The sport of running provides an excellent example. Sprinters, such as 100 m and 200 m runners, spend hours in training but only seconds in competition and have no need for food or fluid during exercise. They need a moderate amount of carbohydrate daily to support training and performance. In contrast, long distance runners, such as marathon runners, spend hours in training and competition and must determine the right combination of food and beverages during exercise to provide them the fuel and fluid they need. The amount of carbohydrate used and the volume of fluid lost during a training session can be substantial and they need to replenish both before the next training session. A high carbohydrate intake daily is important and in the days prior to competition carbohydrate intake is monitored to make sure that muscle glycogen stores are maximized. The precompetition meal is very important but must be tested during training because gastrointestinal distress could interfere with the race. Cycling and swimming are other sports in which the distance covered influences food and fluid requirements.

In track and field, the nutrition-related issues of "throwers" and "jumpers" are different, in part because one group is throwing an object and the other is hurling the body. Shot putters and discus, hammer, and javelin throwers benefit from being large-bodied and strong. Body fat adds to total body weight, which can be a performance advantage because it adds mass. However, some of these athletes readily gain abdominal fat, which may not affect performance but may be detrimental to health. In contrast, jumpers generally do not want excess body fat because it negatively affects their ability to perform sports such as high jump, long jump, triple jump, or pole vault. Both throwers and jumpers are strength/power athletes but they focus on different nutrition-related issues.

Many sports are team sports and it is important to consider the demands of each position. For example, weight and body composition vary in baseball outfielders, in part, because of their offensive skills. Some hit for power whereas others hit to get on base, steal bases, and score on base hits. Power hitters benefit from greater body weight whereas fast base runners are likely to have a lower body weight when compared to players who hit for power. Sports in which position plays a major role include baseball, basketball, football, ice and field hockey, lacrosse, rugby, soccer, and softball (Macedonio and Dunford, 2009).

Key points

- Sports nutrition recommendations include guidelines for energy, nutrient and fluid intake, and nutrient timing.

- It is important to understand the specifics of the athlete's sport and position played when making nutrition recommendations.

Why might a lineman, linebacker, and wide receiver on the same football team have very different nutrition needs?

1.4 Dietary Supplements

Athletes typically have as many questions about dietary supplements as they have about diet. Supplementation is a complicated topic, and athletes as well as the professionals who work with them, need correct, unbiased information. This section provides an introduction to the topic of dietary supplements. Specific supplements will be discussed in later chapters.

In the United States, the law that governs dietary supplements is the Dietary Supplement Health and Education Act.

The Dietary Supplement Health and Education Act (DSHEA), passed in 1994, provides a legal definition for the term *dietary supplement* in the United States. A dietary supplement is defined as a "vitamin, mineral, herb, botanical, amino acid, metabolite, constituent, extract, or a combination of any of these ingredients" (Food and Drug Administration, 1994). This broad definition results in supplements that have very different functions and safety profiles being grouped together. This legislation also provides labeling guidelines, such as the requirements for the Supplement Facts label (see Chapter 10).

It is important to know what the law does not cover. DSHEA does not ensure safety or effectiveness. The Food and Drug Administration (FDA) does not have the authority to require that a dietary supplement be approved for safety before it is marketed. In other words, any dietary supplement that appears on the market is *presumed* to be safe. The FDA must prove that a supplement is unsafe or **adulterated** before it can be removed from the market. The law also does not require that a dietary supplement be proven to be effective.

In 2007, quality standards for supplements were mandated by the FDA. These standards were developed to ensure that dietary supplements contain the intended ingredients, are free from contamination, and are accurately labeled. **Good Manufacturing Practices (GMP)** are intended to bring dietary supplement manufacturing standards more in line with pharmaceutical standards. Some dietary supplements have been reported as tainted and have caused athletes to test positive for banned substances. In some cases, these supplements contained tainted ingredients due to poor manufacturing practices, which the quality standards were developed to alleviate. However, there is also evidence that some dietary supplements contain ingredients that are intentionally added but not labeled (Maughan, 2005).

Many products fall under the umbrella known as dietary supplements.

Because the legal definition for a dietary supplement is so broad, it may be helpful to further divide this vast category into three subcategories: (1) vitamins, minerals, and amino acids; (2) botanicals and (3) herbals. These categories are different in several ways. Vitamins, minerals, and amino acids are all nutrients that are found in food. Most of these compounds have an established standard for how much is needed by humans, such as a Dietary Reference Intake. Some also have a Tolerable Upper Intake Level established, the highest level taken daily that is not likely to cause a health problem. Therefore, it is not difficult to determine if one's diet is lacking in these nutrients and if the amount in a supplement exceeds the amount needed or is safe. In surveys of athletes, vitamins are among the most frequently used supplements.

Botanicals are typically compounds that have been extracted from foods and then concentrated into liquid or solid supplements. These supplements have a link to both food (that is, original source) and medications (that is, concentrated dose). However, botanicals may not provide the same benefit as the food from which they were extracted even though the dose is more concentrated. Botanical supplements often become popular because of scientific studies of diet. For example, heart disease rates tend to be lower in Mediterranean countries where lots of garlic is used in cooking. Garlic contains allicin, a biologically active ingredient that may influence blood cholesterol concentration, and garlic supplements are sold as a concentrated source of allicin. However, after years of study there is a lack of scientific evidence that garlic supplementation is

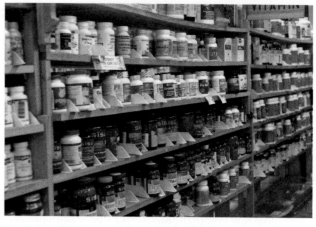
How to choose? Dietary supplements line the shelves in grocery and drug stores in the United States.

beneficial for lowering blood cholesterol (Khoo and Aziz, 2009). Usage of botanicals among athletes is not known, as most surveys of athletes do not distinguish between botanical and herbal supplements.

The majority of the most widely used herbal supplements in the United States (for example, ginkgo biloba, St. John's wort, echinacea, saw palmetto) do not contain nutrients that are found in food. In fact, these herbal products are typically being used as alternative medications. Although the DSHEA prohibits manufacturers from making claims that herbal products can treat, prevent, diagnose, or cure a specific disease, such claims are made frequently, especially when these supplements are marketed via bulk mailings and the Internet (Temple, 2010; Morris and Avorn, 2003).

Preliminary studies of athletes suggest that herbal supplement use is higher in athletes than the general public. Typically, athletes use herbal supplements to improve health or prevent illness. Ginseng and echinacea are the most widely used herbal supplements by athletes (Senchina et al., 2009).

Dietary supplement use among athletes is high.

Athletes regularly use dietary supplements. Surveys suggest that 60 percent of all elite athletes and 85 percent of all elite track and field athletes use one or more dietary supplement (Tscholl et al., 2010; Schroder et al., 2002). Collegiate athletes also frequently use supplements, particularly energy drinks, calorie replacement beverages, multivitamins, and vitamin C. Creatine, protein powders,

Adulterate: To taint or make impure.

Good Manufacturing Practices (GMP): Quality control procedures for the manufacture of products ingested by humans to ensure quality and purity.

and protein drinks are popular with male collegiate athletes (Froiland et al., 2004). Adolescent athletes often emulate the practices of collegiate and elite athletes so it is no surprise that approximately 70 percent consume at least one supplement. Energy drinks, protein supplements, and creatine are popular with high school football and baseball players (Hoffman et al., 2008).

Athletes consume supplements for many reasons.

Limited research suggests that athletes consume dietary supplements for many reasons. Frequently mentioned are the consumption of a poor diet, the physical demands of training and competition, the fact that teammates and competitors are taking supplements, and the recommendation of a physician, coach, or parent (Petróczi et al., 2007, 2008). In a study of British athletes, nearly 73 percent took multivitamin supplements, primarily to avoid sickness. Other reasons mentioned less frequently were advice from a physician, lack of time to prepare meals, and to overcome injury (Petróczi et al., 2007). Athletes consume herbal supplements to prevent or heal illness or injury, provide support for the immune system, and to enhance overall health (Senchina et al., 2009).

Unfortunately, there appears to be a large disconnect between the supplements being consumed and the reasons that athletes give for consuming them. Many athletes consume supplements to meet broad goals such as increasing strength or improving endurance, but without knowing whether the supplements they take are effective in reaching such goals. Athletes lack knowledge about the mechanisms of action and the effectiveness of many supplements and often take supplements based on a personal recommendation, advertising, or information found on the Internet. In many cases, an athlete's approach to choosing a supplement is poorly thought out (Petróczi et al., 2007, 2008).

Knowledge of a supplement's legality, safety, purity, and effectiveness is crucial.

Determining if, and how much of, a dietary supplement should be consumed requires athletes to gather unbiased information. This information then needs to be carefully considered before a judgment is made. Particularly important to consider are the legality, safety, purity, and effectiveness of the dietary supplement.

Legality of dietary supplements. Many athletes are governed by the rules of the International Olympic Committee (IOC) or the National Collegiate Athletic Association (NCAA). Other sports governing bodies may adopt these rules or rules of their own, so each athlete is responsible for knowing the current rules as they pertain to dietary supplements. Banned substances may be intentionally or unintentionally added to some dietary supplements. It is considered unethical to circumvent the testing of banned substances with the use of masking agents or other methods that prohibit detection of banned substances.

Safety of dietary supplements. Before consuming any dietary supplement, the athlete should ask, Is it safe? Safety refers not only to the ingredients in the dietary supplement but also to the dose. For example, vitamin supplements are considered safe at recommended doses, but may be unsafe at high doses.

At recommended doses, many dietary supplements are safe for healthy adults including vitamins, minerals, protein powders, and amino acid supplements. Creatine and caffeine at recommended doses have good safety profiles, although the athlete may experience some adverse effects with their use. Androstenedione, dehydroepiandrosterone (DHEA), and ephedrine are banned substances in many sports, in part because of safety concerns. The FDA does not review or approve any dietary supplement before it is available for sale so it is imperative that the athlete be aware of any potential safety issues (Dunford and Coleman, in press).

There is growing concern about the safety of herbal supplements. These concerns include lack of standardization of the active ingredients, the risk for contamination, and potential interactions with medications (Gershwin et al., 2010). Prior to 1994, herbal preparations were considered neither a food nor a drug, but DSHEA reclassified them as dietary supplements. This is in contrast to most European countries, which regulate herbals and botanicals as medications (Van Breemen, Fong, and Farnsworth, 2008).

Athletes need to be cautious of both the recommended dosage and the ingredients found in herbal weight loss supplements. Some may contain ephedrine (for example, Ma Huang), a nervous system stimulant, which has a narrow safe dose range. Others may contain herbal sources of caffeine. Caffeine is also a nervous system stimulant and is typically safe at moderate doses. However, the use of caffeine-containing weight loss supplements along with energy drinks containing a concentrated amount of caffeine could put an individual at risk for caffeine intoxication (Reissig, Strain, and Griffiths, 2009). Some herbal weight loss supplements, such as Citrus aurantium (bitter orange), contain the drug synephrine, which may be a banned substance.

Purity of dietary supplements. Purity is related to a lack of contamination and accurate labeling. Consumers assume that the ingredients and the amounts listed on the Supplement Facts label are accurate. They should be, but it is not true in all cases. Gurley, Gardner, and Hubbard (2000) tested the ephedrine (ephedra) content of 20 dietary supplements and compared it to the amount listed on the label. Half of the supplements varied by more than 20 percent, including one that contained none and one that contained 150 percent of the amount stated. The authors also detected five dietary supplements that contained nor-pseudoephedrine, a controlled substance (that is, drug). Unfortunately, some dietary supplements do contain substances banned by sports governing bodies, and athletes are subject to disqualification even if the banned substances are not listed on the label and were consumed unintentionally.

It is not known how many dietary supplements are mislabeled or impure. A 2009 report by the FDA found that more than 70 weight loss supplements contained prescription drugs (Food and Drug Administration: FDA uncovers additional tainted weight loss products). The IOC tested 634 supplements and found that ~15 percent contained substances not listed on the label that would have led to a positive test for banned substances (Geyer et al., 2004). Judkins, Hall, and Hoffman (2007) analyzed randomly chosen supplements commonly used by athletes. Results showed that 25 percent of the 52 supplements analyzed contained a small amount of steroids. The presence of banned substances in dietary supplements prompted the National Football League (NFL) and its players association (NFLPA) to begin a supplement certification program in 2004. Under this program, dietary supplement manufacturers can be certified by an independent testing organization and players can be confident that the labels are accurate and the dietary supplements do not contain any banned substances. For athletes who choose to use supplements, one of the best tools available is to check that the supplement manufacturer is a participant in one of the dietary supplement certification programs. These programs include:

- ConsumerLab.com (Athletic Banned Substances Screened Products): http://www.consumerlab.com
- NSF (Certified for Sport™): http://www.nsf.org
- United States Pharmacopeia: http://www.usp.org/ USPVerified
- Informed-Choice: http://www.informed-choice.org

One of the problems, particularly with herbal dietary supplements, is the lack of standardization. Standardization means that the amount found in the supplement is the same as the amount found in the laboratory standard. For example, assume that a supplement is supposed to contain 1 percent active ingredient. That means that every supplement produced in every batch should contain 1 percent active ingredient. Diligent manufacturers use good manufacturing processes to ensure standardization. However, not all

Clockwise: Courtesy of NSF International; Courtesy of Consumer Lab; Used with permission of The United States Pharmacopeial Convention; Courtesy of NSF International

When these certifications appear on dietary supplements, it means that they are produced using good manufacturing practices.

manufacturers do so. Standardization is particularly important when using herbs. The amount of the active ingredient depends on the plant's species, the part of the plant used (the leaf has a different amount than the stem), the age at harvest, the way the plant is prepared (cutting gives different results than mashing), the processing (dried or not dried), and the methods used for extraction. It is easy to see why the amount of the active ingredient could vary tremendously (Dunford, 2003).

There is evidence that some herbal dietary supplements are contaminated with minute amounts of lead, mercury, cadmium, and arsenic, heavy metals that can be toxic (Government Accountability Office, 2010). Consumer Reports ("Alert: Protein Drinks," 2010) found some protein supplements contained low levels of these metals. In each case, the amounts found were low and not acutely toxic. However, these low amounts are part of a person's total exposure.

Effectiveness of dietary supplements. Most dietary supplements sold are not effective for improving performance, increasing muscle mass, or decreasing body fat. Scientific research suggests that the following supplements are safe and effective at recommended doses:

- **Caffeine.** Effective as a central nervous system stimulant and for improving endurance and high-intensity activities lasting up to 20 minutes (see Chapter 10).
- **Creatine.** Effective in conjunction with vigorous training for increasing lean body mass in athletes performing repeated high-intensity, short-duration (<30 seconds) exercise bouts. Effective for increasing performance in weight lifters (see Chapter 3).
- **Vitamins and minerals.** Effective as a way to increase nutrient intake and to reverse nutrient deficiencies, if present (see Chapters 8 and 9).
- **Protein.** Effective as a source of protein (neither superior nor inferior to food proteins). Whey protein may be more effective than casein for stimulating skeletal muscle protein synthesis (see Chapter 5).

Some supplements under scientific investigation have shown promise of safety and effectiveness including (1) beta-alanine for buffering of muscle pH, (2) branched chain amino acids (BCAA) for immune system support and reduction of postexercise fatigue, and (3) quercetin for anti-inflammatory effects. Given the number of dietary supplements available for purchase, which is estimated to be more than 30,000 in the United States, few have been shown to be effective. Individual supplements are discussed in detail in the appropriate chapters of this textbook (for example, creatine in Chapter 3, Energy Systems and Exercise).

In a few cases, supplements may be effective but have questionable safety profiles. Green tea extract (which is not the same as green tea) and ephedrine in combination with caffeine have been shown to be effective for a small weight loss in obese individuals. However, there have been reports of liver toxicity with green tea extract, and the safety of ephedrine has been hotly debated. For the vast majority of dietary supplements, there is little or no scientific evidence of effectiveness (Dunford and Coleman, in press; Temple, 2010; Jenkinson, and Harbert, 2008). Interestingly, Temple (2010) found that supplements with good evidence of effectiveness cost ~\$3–4/month whereas those with little evidence of effectiveness cost ~\$20–60/month.

Quackery. Quackery is the practice of making false claims about health-related products, and some dietary supplements fall under this category. It is very difficult to combat quackery, but some resources are listed in the sidebar (see The Internet Café). Many dietary supplements are highly advertised or are sold using multilevel marketing (MLM), and unscrupulous companies or distributors may exaggerate their value because they will be financially rewarded if sales increase. Consumers can reduce their risk for being a victim of quackery by critically evaluating products before purchasing them. One method for evaluating dietary supplements is shown in the Spotlight on supplements: Evaluating Dietary Supplements feature.

Keeping it in perspective

Food Is for Fuel and Fun

Many athletes follow rigorous training programs. To support such training, diet planning becomes very important because food provides the fuel and nutrients that are needed to train hard. Endurance and ultraendurance athletes must carefully plan their food and beverage intake before, during, and after training and competition or they risk running out of fuel and

becoming hypohydrated. This need for constant, nutritious food and drink sometimes means that athletes get very rigid about their dietary intake. Rigid meal planning might meet the scientific requirements of nutrition, but it falls short when it comes to the art of eating, which also involves pleasure and enjoyment. In other words, sometimes athletes need to eat food just for fun.

Key points

- In the United States, dietary supplements are not required to be proven safe or effective before being sold.

- The majority of athletes at all levels use at least one dietary supplement.

- In many cases, athletes use a random approach to choosing dietary supplements.

- Some supplements may contain banned substances.

- Many dietary supplements are safe, but some are not.

- Some supplements are contaminated, particularly weight loss and muscle-building supplements.

- Only a handful of supplements have been proven to be effective.

What specific things should an athlete consider before choosing to take a dietary supplement?

1.5 Understanding and Evaluating Scientific Evidence

Although sports nutrition is a fairly new academic discipline, there have always been recommendations made to athletes about foods that could enhance athletic performance. Ancient Roman athletes were encouraged to eat meat before competing. One ancient Greek athlete is reported to have eaten dried figs to enhance training. There are reports that marathon runners in the 1908 Olympics drank cognac (brandy) to improve performance (Grandjean, 1997). The teenage running phenomenon, Mary Decker (Slaney), surprised the sports world in the 1970s when she reported that she ate a plate of spaghetti noodles the night before a race. Such practices may be suggested to athletes because of their real or perceived benefits by individuals who excelled in their sports. Obviously, some of these practices, such as drinking alcohol during a marathon,

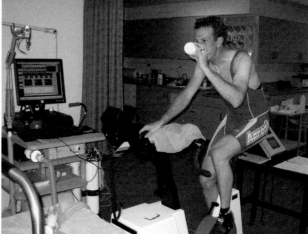

An experimental study of carbohydrate consumption and endurance cycling performance.

© Andy Doyle

are no longer recommended, but others, such as a high-carbohydrate meal the night before a competition, have stood the test of time.

Today, sports nutrition recommendations are **evidence based**. Evidence-based practice is the review and use of scientific research to determine the most effective outcome. The scientific evidence plays a central role, although clinical judgment and the athlete's personal preferences and values must also be considered. Because research findings are fundamental in forming recommendations, the quality of the research is very important (Amonette, English, and Ottenbacher, 2010; Gray and Gray, 2002).

There are three basic types of research studies.

Most research studies fall into one of three categories: case studies, epidemiological studies, or experimental studies. **Case studies** are observational records. They provide information about an individual in a particular situation. Gathering information is an important first step because it helps researchers to form hypotheses, but case studies are the weakest of all scientific findings.

Epidemiological studies help to determine the distribution of health-related events in specific populations. Such studies highlight nutrition and exercise patterns and help to show associations and **correlations**. These studies are stronger than case studies because

Evidence-based recommendations: Recommendations based on scientific studies that document effectiveness.

Case study: An analysis of a person or a particular situation.

Epidemiological study: The study of health-related events in a population.

Correlation: A relationship between variables. Does not imply that one causes the other.

large groups of people are studied and data are statistically analyzed. However, these are observational studies and they lack control of all the variables.

The strongest studies are experimental studies. These studies follow strict protocols and control most variables except the ones being studied. This is how cause-and-effect relationships are established. One example of each type of study is reviewed here.

Morton and colleagues (2010) published a case study of the training and nutritional intake of a 25-year-old professional boxer who was trying to make weight for the superfeatherweight division. They reported the nutritional and conditioning interventions and outcomes over a 12-week period including changes in energy intake, energy expenditure, total body weight, lean mass, and body fat. The boxer studied followed a low-calorie, high-protein diet over the three-month period and reduced body weight by 9.4 kg (~21 lb) and body fat from 12.1 percent to 7.0 percent. The subject was able to meet his goals with a gradual weight loss that did not include large or rapid losses in body water. These are important observations but they do not provide a scientific basis for making recommendations to other boxers or to athletes in other sports who want to make weight.

The relationship between physical activity or physical fitness levels and improved health is well known as a result of a number of epidemiological studies. A notable example is the large well-designed study of Blair et al. (1989) from the Institute for Aerobic Research. In a study that included over 13,000 subjects, the authors showed that there was a strong relationship between aerobic fitness and decreased all-cause **mortality**, primarily from decreased premature mortality from cardiovascular disease and cancer.

An **experimental study** that has become a "classic" in sports nutrition was conducted by Coyle et al. (1983) to determine if drinking a carbohydrate beverage helped endurance cycling performance. Well-trained cyclists rode to exhaustion on two occasions, once while drinking a beverage containing carbohydrates and once while drinking a **placebo**. Although muscle glycogen utilization was no different when consuming the carbohydrate drink, blood glucose was maintained, a high rate of carbohydrate oxidation was also maintained, and the cyclists were able to ride at the prescribed intensity for an additional hour before becoming exhausted. This study showed that there was a cause-and-effect relationship between consuming a carbohydrate-containing beverage and the ability to cycle for a longer period of time.

Spotlight on supplements

Evaluating Dietary Supplements

Everyone agrees that a proper diet is necessary for good health. However, opinions vary among professionals in sport-related fields and athletes themselves about dietary supplement use. Professionals emphasize safety because they honor the principle: First, do no harm. Ethics are also a consideration because a harmful supplement directly harms the athlete, not the person recommending it. Athletes are concerned about safety too, but the decision to use supplements will affect them directly—the athlete will reap the benefits and/or suffer the consequences.

Some athletes are risk takers. When safety data are not available, which is the case for many dietary supplements, they feel comfortable adopting a "probably will not be harmful" approach. Such an approach does not pose an ethical issue for the athlete because of a willingness to accept the consequences. However, it is beyond the comfort and ethical limits of many sports-related professionals, because of the potential for harm to the athlete.

Evaluating dietary supplements requires a systematic, multistep process of gathering, weighing, and judging information that leads to decision making. In other words, it requires critical thinking. An athlete and a sports-related professional

Judging Information

- Is it legal?
- Is it ethical?
- Is it safe (both the ingredients and the recommended dosage)?
- Is it effective?
- Is it pure?
- What is not known?

There are fundamental questions that should be asked about any dietary supplement.

may examine the same information but come to different conclusions, in part because of each person's perspective. In the end, nearly everyone agrees on the general approach—consume a healthy diet—but philosophies about whether, and which, supplements should be taken vary widely among individuals. "Should I recommend this supplement to an athlete/client?" is a different question than "Would I use this supplement myself?"

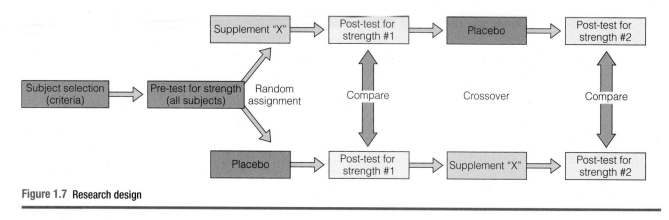

Figure 1.7 Research design

The basis of good research is strong research design and methodology.

The hallmark of good scientific research is the use of strong research design and methodology. Well-designed studies reduce bias and help to ensure accurate results. The strongest research protocol is a randomized, double-blind, placebo-controlled, crossover study performed on humans. It may include a familiarization trial. The number of subjects in the study should be as large as possible and their characteristics (for example, age, fitness, training, health status) should be similar.

Randomization is part of the subject selection process. It is usually difficult to study 100 percent of the population of interest, so a sample is chosen. Randomization tries to ensure that all people in the study population will have the same chance of being selected for the sample. The study subjects are also randomly assigned to either the treatment or the placebo group. The placebo group receives an inactive substance that resembles the treatment in every way possible. A double-blind study is one in which neither the researchers nor the study participants know which group they are in or which treatment they are receiving. In a crossover study, subjects will be in both the treatment and placebo groups. For example, if four trials were scheduled, subjects would receive the treatment in two of the trials. For the other two trials they would "cross over" and be in the placebo group.

Randomized double-blind, placebo-controlled study designs help to reduce bias, which can lead to inaccurate results and erroneous conclusions. If subjects aren't randomized, then selection bias will be present. Think about what might happen if researchers knew that the subjects were in the treatment group. They might subtly influence the participants to ensure that the treatment works. If the subjects knew that they were receiving a treatment, they may try to perform better.

Before data are gathered, subjects should complete a familiarization trial. This is critical in studies where the subjects' performance is being measured. Consider a study held in an exercise physiology lab on a stationary bicycle. If subjects are not familiar with the bike or other laboratory equipment, their performance might not be as fast during the first trial. A familiarization trial gives the subjects a chance to practice on equipment and also understand what is expected of them (for example, length of time in the lab, intensity of exercise). When data collection begins, unfamiliarity with equipment or the study protocol should not be a factor.

The strongest results come from studies in which data are obtained from both an experimental (treatment) and a control (nontreatment/placebo) group. Most of the time, the treatment and control groups comprise different people. In a crossover design, subjects are in both groups so they serve as their own controls. For example, a study may be designed to have subjects perform four trials. In the first and fourth trials, the subject is in the experimental group and receives a treatment (for example, a carbohydrate-containing beverage). In the second and third trials, the subject is in the control group and receives a placebo (for example, an artificially sweetened beverage that contains no carbohydrates). The results for each person can be directly compared because each subject received both the treatment and the placebo. Crossover studies are advantageous, but researchers must be careful to account for any carryover effects. For example, the effect of creatine loading may last for a month or more whereas bicarbonate loading may show effects for only a day or two. When designing a crossover study, ample time must be allowed before the next phase of testing begins. An example of a research design for an experimental study is shown in Figure 1.7.

Strong research design and methodology is fundamental to unbiased and accurate scientific information.

Experimental study: A research experiment that tests a specific question or hypothesis.

Mortality: Death; the number of deaths in a population.

Placebo: An inactive substance.

The recommendations made to athletes are only as strong as the research studies on which those recommendations are based. But how does a person who does not conduct research know if a research study is well designed and accurate? One safeguard is the peer review process.

Peer review is an important safeguard in the publication of scientific research.

Every scientific study should be peer reviewed, which means that it is scrutinized by a group of similarly trained professionals (peers) before publication. The peer review process begins when the editor of the publication receives a written manuscript from the researchers. The researchers' names are removed and the manuscript is sent to two or more reviewers. They carefully read the study protocol, review the data, and evaluate the authors' conclusions. If the study design and methodology are not scientifically sound, the peer reviewers will recommend that it not be published. If the method is sound, they will make suggestions to ensure clarity and accuracy in reporting the data and drawing conclusions, which the authors will incorporate into a

Focus on research

Designing a Research Study to Test the Effect of a Sports Drink on Performance

Scientific research is a systematic process of gathering information (data) for the purpose of answering a specific question or solving a specific problem. Researchers have the best opportunity to obtain accurate and relevant information, and thus to most accurately answer the research question, if the question/problem is clearly defined, the research study is well-designed and carefully-conducted, and the results are appropriately analyzed and interpreted. The first Focus on Research in this text will examine the process of designing a research study to answer a specific research question in sports nutrition.

Consumption of sports drinks—carbohydrate-containing beverages—has been shown repeatedly in the research literature to help athletes in certain sports to improve their performance. Use of sports drinks in training and competition has become routine practice for many athletes, but athletes in endurance sports of relatively long duration, for example, marathon running, benefit the most from ingesting carbohydrate during exercise. Muscles use carbohydrate in the form of glucose, and most research studies have shown that sports drinks that contain mostly glucose are the most effective. There appears to be a limit as to how much glucose an athlete can consume during exercise—more is not necessarily better. Too much glucose intake does not help performance and may even lead to gastrointestinal (GI) problems.

Recent research, however, has shown that adding fructose to a glucose-containing sports drink likely results in more total carbohydrate being absorbed from the GI tract and used during exercise. This is thought to occur because fructose is absorbed by a different transport mechanism than glucose (see Chapter 4) and may provide a route for more carbohydrate to enter the body once the glucose transporters are saturated. Although it seems promising that more carbohydrate can be taken up and used during exercise when fructose is added to a sports drink, it is not known if this in turn would result in improved athletic performance.

Scientists and sports nutritionists are confident the consumption of glucose (for example, a sports drink) during prolonged endurance exercise will help improve performance in events such as marathon running, bicycling races, and triathlons. A valid research question is: will consuming a sports drink containing fructose and glucose improve performance more than when the athletes consume a sports drink containing only glucose?

We will examine how a research study was designed specifically to answer that research question:

Triplett, D., Doyle, J. A., Rupp, J. C., & Benardot, D. (2010). An isocaloric glucose-fructose beverage's effect on simulated 100-km cycling performance compared with a glucose-only beverage. *International Journal of Sport Nutrition and Exercise Metabolism*, 20: 122–131.

What was the measure of performance? If the purpose of the study is to determine the effect of a sports drink on performance, one of the first tasks in designing the study is to determine the specific method for testing performance. Because sports drinks are most effective in long-duration endurance events where carbohydrate stores may be depleted, the performance task needed to be one that would take the subjects in excess of 2 hours to complete. For the results to be most applicable to competitive athletes, the performance task should also mimic the physical demands of a competitive event as closely as possible while still maintaining the ability to control the exercise conditions in the laboratory. Rather than invent one on their own, Triplett et al. (2010) chose an endurance performance test performed on a cycle ergometer in the laboratory (100 km/62 mile time trial) that had been used in previous studies and had been shown to be a reliable and accurate measure of cycling performance that simulated the demands of a competitive bicycling road race.

What subjects were selected? In order to be able to physically complete the prolonged endurance performance test, and for the results of the study to be most applicable to competitive athletes, subjects recruited for this study needed to have current experience in competitive cycling races. Because of the likely difference in performance between male and female

revised manuscript. The revised manuscript is reviewed by the editor and then scheduled for publication. Like the scientific studies, the peer review process should be blind—researchers should not know who reviewed the article and reviewers should not know who conducted the research and wrote the article.

Readers can have confidence in the quality of an article in a peer-reviewed journal. The peer review process should be a rigorous one. Journals that have the strongest peer review processes have the best reputations. Examples of peer-reviewed journals that publish sports nutrition-related articles are listed in Figure 1.8.

Levels of evidence and grades of recommendations put the scientific body of literature in perspective.

The most conclusive evidence comes from studies that are randomized, double-blind, placebo-controlled, and published in peer-reviewed journals. The results of such studies should be given greater weight than results from other study designs or those published in non-peer-reviewed journals. But even the results of the best-designed study cannot stand alone. Reproducible results are an important part of the scientific process. Recommendations must be based on the cumulative

cyclists, the researchers chose to restrict the subject population to a single gender to reduce the potential variability in subjects' performance results. In other words, the more similar all of the subjects are to one another in performance, the better chance the researchers have to see if an experimental intervention such as a new sports drink has an effect on the results.

What was the experimental sports drink? A typical research study such as this might test the effect of an experimental sports drink against a placebo beverage. In this case, the researchers were interested in the effect a new sports drink (one with fructose and glucose) might have compared to a traditional sports drink (one with glucose only). The composition of the two drinks was designed to be similar to those used in the studies that showed improved carbohydrate uptake when fructose was added. In order to make sure that the total amount of carbohydrate in the drinks was not an influencing factor, the two drinks had the same amount of carbohydrate and differed in only one respect—one contained only glucose and the other contained glucose and fructose in equal amounts. The drinks looked the same and had very similar taste. The researchers chose not to include a placebo beverage because it is already very clear from the research literature that this type of performance will improve when consuming a sports drink over consuming a placebo, and the inclusion of a placebo trial would require the subjects to perform yet another rigorous performance trial.

What was the experimental design? Nine male cyclists with current bicycling road race experience were subjects in this study. They performed a trial run of the cycling performance test to familiarize them with the procedures and because previous research has shown this type of familiarization trial to improve reliability of endurance performance tests. During the familiarization trial the subjects drank only water. After the familiarization trial, the subjects completed two experimental trials of cycling endurance performance, once

while drinking a sports drink with only glucose and once while drinking a beverage with both fructose and glucose. The type of sports drink was given in random order (some got glucose-only first; others got glucose-fructose first) and in a double-blind fashion (neither the researcher nor the subjects knew which beverage they were consuming). The subjects acted as their own control subjects by completing a performance trial with both experimental sports drinks (crossover design). At the conclusion of the study, the average (mean) time it took the subjects to complete the performance test when they drank the glucose-only sports drink was statistically compared to the average time it took the subjects to complete the same performance test when they drank the glucose and fructose sports drink.

What did the researchers find? Triplett et al. (2010) reported that when subjects consumed the sports drink with fructose and glucose in it, they improved their endurance cycling performance, completing the performance test nearly 3.5 minutes faster on average than when they consumed the sports drink that contained only glucose. They also found that the mean power output throughout the performance test on the cycle ergometer (a measure of cycling intensity) was higher when they consumed the glucose and fructose sports drink.

Answering the question: Will consuming a sports drink containing fructose and glucose improve performance more than when the athletes consume a sports drink containing only glucose? Based upon the results of their research study, Triplett et al. (2010) concluded that the answer to this research question is yes, endurance exercise performance is improved with a sports drink that contains fructose along with glucose. Researchers and athletes can have a high degree of confidence in this conclusion because the research study carefully considered previous research, adhered to good experimental design principles, and was published in a peer-reviewed journal.

American Journal of Clinical Nutrition

European Journal of Applied Physiology

International Journal of Sports Medicine

International Journal of Sports Physiology and Performance

International Journal of Sport Nutrition and Exercise Metabolism

Journal of the American Dietetic Association

Journal of the American Medical Association

Journal of Applied Physiology

Medicine and Science in Sports and Exercise

Sports Medicine (reviews)

Figure 1.8 Examples of peer-reviewed journals

body of scientific literature and not on the results of one study. Just as the strength of each study must be established, the quality of the body of literature must also be determined. This process involves levels of evidence and grades of recommendations.

Level of evidence refers to the relative strength or weakness of the current collective body of scientific research. The strongest evidence comes from a review of all the randomized controlled trials. Such reviews compare the results of high-quality research studies. Many of these reviews involve meta-analysis, a statistical method of comparison. Articles reviewing the collective body of scientific research give rise to the strongest recommendations.

As noted previously, sports nutrition is a relatively young scientific field, so abundant, high-quality research is lacking in many areas. Practitioners must make recommendations based on the current body of literature, knowing full well the limitations of the current scientific knowledge. Grading the scientific evidence is important because it indicates the relative strength and quality of the body of scientific research. Four grades, designated either by Roman numerals or by letters, are generally accepted as described below.

Grade I (Level A): The conclusions are supported by good evidence, known as a rich body of data. The evidence is based on consistent results of well-designed, large randomized research studies. Confidence in the accuracy of these studies is high.

Grade II (Level B): The conclusions are supported by fair evidence, known as a limited body of data. The evidence is less convincing because either the results of well-designed studies are inconsistent or the results are consistent but obtained from a limited number of randomly controlled trials or studies with weaker designs.

Grade III (Level C): The conclusions are supported by limited evidence. Confidence in the results of the research studies is limited by their size and design (for example, nonrandomized trials or observational studies) or by the size of the cumulative body of literature that consists of a small number of studies.

Grade IV (Level D): The conclusions are supported by expert opinion, known as panel consensus judgment, as a result of the review of the body of experimental research. This category includes recommendations made by sports nutrition experts based on their clinical experience (Myers, Pritchett, and Johnson, 2001; see also National Heart, Lung and Blood Institute evidence categories at http://www.nhlbi.nih.gov).

In a perfect world, Grade I (Level A) evidence would be available to answer all questions regarding the nutrition and training needs of athletes. But in many cases, recommendations are supported by only fair or limited evidence. In some cases, expert opinion is relied upon until more research can be conducted. Although dietary supplements are widely used by athletes, the scientific evidence available may be limited or nonexistent in many cases. Lack of scientific research makes it difficult to evaluate claims regarding safety and effectiveness. When making sports nutrition, training, or dietary supplement recommendations to athletes, it is important to indicate the relative strength of the research or the absence of scientific studies.

Anecdotal evidence. Anecdotes are personal accounts of an incident or event and are frequently used as a basis for testimonials. Anecdotal evidence is based on the experiences of one person and then stated as if it had been scientifically proven. Often anecdotal evidence is cited to show that the current recommendations are not correct. Anecdotal evidence is not necessarily false (it may be proven in the future), but it should not be used as proof.

Anecdotal evidence and testimonials are often used to market dietary supplements. For example, a well-known athlete may appear in a supplement advertisement and endorse the product. It is not illegal to include endorsements in advertisements, but it is deceptive if the consumer is led to believe that the endorsement is made voluntarily when the person is being paid to promote the product. The Federal Trade Commission (FTC) is responsible for regulating the advertisement of dietary supplements, and more information can be found at http://www.ftc.gov/.

Conclusions from scientific studies can be misinterpreted.

One scientific study does little by itself to answer critical questions about sports nutrition issues. As shown above, it is both the quality and quantity of scientific research that allows practitioners to make sound nutrition and training recommendations to athletes. Critical thinking skills are needed to correctly interpret scientific research and properly communicate their

Many ads for dietary supplements include testimonials by elite athletes.

results to athletes. Here are some issues that need special attention when drawing conclusions from scientific studies.

Distinguish between correlation and causation. One of the fundamental differences between epidemiological and experimental studies is the establishment of **causation**. Epidemiological studies can establish only a correlation (that is, that a relationship exists between two variables, and the strength of that relationship). It takes experimental research to establish causation—that the variable studied produces a particular effect. It is very important in both written and oral communications that professionals do not use the word *causes* if in fact the study or the body of research shows only an association or correlation. For example, epidemiological studies clearly show that people with lower aerobic fitness levels have an increased risk of premature death, particularly from cardiovascular disease and cancer (Blair et al., 1989; Fogelholm, 2010). This knowledge is based upon the strong inverse relationship between fitness levels and premature all-cause mortality. So it is correct to say that there is an association between aerobic fitness levels and premature death. However, based upon this type of research, it is not valid to suggest that low fitness levels cause premature death.

Understand the importance of replicating results. Recommendations should not be made based on the results of one research study. Preliminary studies, many of which are performed in very small study populations, often produce surprising results, but many of these studies are never replicated. Unfortunately, the results of single studies are often widely reported as "news" by the media and then become falsely established as fact.

Extrapolate results of scientific research with caution, if at all. Extrapolation takes known facts and observations about one study population and applies them to other populations. This can lead to erroneous conclusions because only the original study population was tested directly. In the area of sports nutrition there are many ways that data may be extrapolated, including animals to humans, males to females, adults to adolescents, or children and younger adults to older adults. Sports nutritionists work with athletes of all levels, from recreational to elite, and must carefully consider the validity of extrapolating research results to these populations. For example, a dietary intervention or manipulation that shows positive results in a sedentary population with diabetes may not be applicable to well-trained runners. Other factors that are powerful influences and should be extrapolated with caution, if at all, are the presence or absence of training, the type of sport, laboratory or field conditions, and competition or practice conditions.

Interpret results correctly. The results of research studies are often misinterpreted or applied to an inappropriate population. Professionals must be able to evaluate the results of research studies, and be able to recognize when recommendations are being made in an inappropriate way. One important consideration is to evaluate the characteristics of the subject population studied and to determine if the results observed in this group can reasonably be expected in other groups. Many studies of specific sport or exercise performance have used physically active college students as subjects. Results of these studies should be considered with great caution when making recommendations to other groups, such as highly trained athletes. Many sports nutritionists and physiologists who work with elite athletes will use only the results of studies that have used highly trained athletes who are similar to their client population.

Another important consideration in evaluating research results is how closely the laboratory experimental design mimics the "real" demands of the athletic event. Scientists study athletes in the laboratory so that they can carefully control as many experimental conditions as possible. In studying endurance performance, it is common to use "time to exhaustion" protocols in which athletes run or ride at a fixed exercise intensity until they are no longer able to maintain the required pace. Although these types of protocols are useful for studying metabolic responses during these types of activities, they do not reflect the demands or strategy of a real race, and may therefore be less useful for predicting performance. Field-based research is not as common because it is more difficult to conduct and does not offer the same control of research conditions.

Causation: One variable causes an effect. Also known as causality.

Does the laboratory research study accurately reflect the demands or strategy of a competitive event?

Focus on cumulative results and consensus. There is much excitement when a new study is published, especially when the results contradict current sports nutrition recommendations or long-held theories. But startling breakthroughs are the exception, not the rule. Cumulative data, not single studies, are the basis for sound recommendations. It is imperative that any new study be considered within the context of the current body of research.

Whereas some topics may be subject to healthy debate among experts, many topics have good scientific agreement, known as **consensus**. One of the best ways to know the consensus opinion is to read review articles or position statements. Review articles help students to understand the body of literature on a particular topic. These articles also help practitioners put the results of new research studies in the proper context and remain up-to-date with the current body of research. For example, Burke, Kiens, and Ivy (2004) published an article on carbohydrates and fats for training and recovery in which the authors reviewed the scientific literature on postexercise glycogen storage published between 1991 and 2003. Such articles consider the cumulative body of scientific evidence over a long period of time and help students and practitioners become familiar with the consensus opinion.

Recognize the slow evolution of the body of scientific knowledge. Occasionally, landmark research studies are published that increase knowledge in a particular field in a quantum leap, such as Watson and Crick elucidating the structure of DNA. However, for the most part, knowledge in scientific areas such as sports nutrition increases gradually as additional research studies are completed and evaluated in the context of the existing research. The process can move slowly, as it takes time for research studies to be proposed, funded, completed, published, and evaluated by the scientific community. At times it may seem to be a slow and cumbersome process. However, a deliberate, evaluative approach is an important safeguard for the integrity of the information. The scientific process is similar to building a brick wall; it takes time and must be done one brick at a time, but if done correctly, the end product has considerable strength.

Not only scientists read scientific studies.

In the past, only other scientists read articles published in scientific journals. Today, results of scientific studies are widely reported in the print and visual media. Athletes may hear results of a research study before the journal is received and read by professionals. Consumers like to hear about new studies, but few consumers can interpret the research or put the newest results in the proper context. The role of the professional is threefold: (1) provide sound science-based information, (2) recognize and correct misinformation, and (3) address the effects of misinformation (Wansink and American Dietetic Association, 2006).

Surveys suggest that the primary source of nutrition information for consumers is the media. The top two media sources are television and magazines, with newspapers a distant third. All of these media routinely report the results of research studies. Athletes are also consumers, so these sources are likely popular with athletes too. In addition to media sources, studies of collegiate athletes suggest that nutrition information is also obtained from fellow athletes, friends, family members, coaches, strength and conditioning coordinators, and certified athletic trainers (Froiland et al., 2004; Jacobson, Sobonya, and Ransone, 2001).

Concensus: General agreement among members of a group.

One of the problems with the coverage of scientific studies in the media is that preliminary research data are reported. Preliminary data often raise more questions than they answer, and such data are not considered sound until replicated. Additionally, research results may not be put in the larger context of the known body of literature on the subject. Other concerns are that cause and effect are reported or inferred when an association or correlation was actually found. Distinctions between probability and certainty are rarely made (Wansink and American Dietetic Association, 2006).

The results of research studies are frequently used as a marketing tool, some of them prior to being reviewed and published in peer-reviewed journals, and some from unpublished studies. Clever advertising copy can make it appear that there is a direct link between scientific research and the product being sold, as shown in Spotlight on supplements: Use of Scientific Studies as a Marketing Tool. Testimonials from athletes—a form of anecdotal, not scientific, evidence—are widely used to sell sports nutrition products.

Much of the nutrition-, exercise-, and health-related information on the Internet is inaccurate.

The Internet is a major source of information about health, exercise, and nutrition. It is estimated that 60 percent of all people who access the Internet look

The Internet is a popular source for nutrition, exercise, and health information, much of which can be inaccurate and/or misleading, and therefore requires consumers to be discriminating.

© Rob Wilkinson/Alamy

for health-related information. This information is more often sought by women, those with more education, those with chronic disease, and those without health care insurance (Bundorf et al., 2006).

Unfortunately, much of the information is inaccurate or of poor quality. A small study of online physical activity websites found that less than one-quarter of the sites were judged as high quality and none were accurate (Bonnar-Kidd et al., 2009). Similarly,

Spotlight on supplements

Use of Scientific Studies as a Marketing Tool

The picture on the supplement bottle was eye-catching. A good-looking couple was running down the beach. He was shirtless, which highlighted his impressive upper body muscle mass. She had a tight-fitting white sundress, a perfect figure, and long blond hair. The supplement was advertised as a nutritional means to control weight. The accompanying materials were also impressive. Two research studies were called into evidence. The first indicated that those who took the supplement had a much greater weight loss than those who were given a placebo. The second study was cited as evidence that taking the supplement could boost energy. The results of these studies were reduced to a single sentence prominently displayed in the advertising: "University studies show that this supplement helps people lose weight and feel less fatigued." A closer look at the studies (something most consumers cannot do) tells a more complete story. In the first study, the 14 obese subjects were all consuming a liquid diet of 1,000 Calories daily. They were living in an experimental research ward where they had no access to food other than the liquid diet. Although the women receiving the

supplement did lose more weight than the ones not receiving it, they were still obese at the end of the 21-day study period.

The second study involved a different population—20- to 30-year-old males. The eight subjects did not receive a calorie-restricted diet. The exercise that the study subjects performed was on a stationary bike in the laboratory. Those who received the supplement experienced less fatigue than those who did not.

These studies were published in peer-reviewed journals and their results are an important contribution to the body of literature about this particular compound. But small study populations and tightly controlled food and exercise conditions limit their applicability outside the laboratory setting. Using these studies to sell supplements to the general population is stretching the scientific literature beyond its application limits. The study citations give the supplement more scientific credibility than it deserves. It is safe to say that consumers will not look like the man or woman in the picture by just taking this supplement.

nutrition-related websites were often unreliable sources of information (Sutherland et al., 2005).

Consumers should look for the HONcode certification. HONcode is a code of conduct for medical and health websites. A HONcode certification means that the website abides by HONcode principles including backing up claims with scientific research and clearly distinguishing advertising from editorial content. Some reliable sources of diet-, exercise-, and health-related information are listed in The Internet Café sidebar.

The Internet café

Where do I find reliable information about diet, exercise, and health?

Although there are many reliable websites, those listed here are government agencies or professional organizations.

Healthfinder: A guide to reliable health information sponsored by the U.S. Department of Health and Human Services. **http://www.healthfinder.gov**

MedlinePlus®: Trusted health information sponsored by the National Library of Medicine and the National Institutes of Health. **http://www.medlineplus.gov**

Nutrition.gov: "Smart Nutrition Starts Here" sponsored by the National Agricultural Library and the Food and Nutrition Information Center. **http://www.nutrition.gov**

American College of Sports Medicine: Professional organization whose mission is to advance health through science, education, and medicine has resources for the general public at **http://www.acsm.org**

American Dietetic Association: Professional organization committed to helping people enjoy healthy lives through good nutrition. Food and nutrition information for consumers can be found at **http://www.eatright.org**

Key points

- Sports nutrition recommendations should be evidence based.

- The strength of any scientific recommendation depends on the quality of the research conducted.

- A strong research design is fundamental to obtaining accurate results.

- Epidemiological and experimental studies provide different types of data.

- Individual research studies are important, but knowledge of the body of literature is necessary to understand a topic.

- Much of the information on the Internet about exercise, nutrition, and dietary supplements is inaccurate.

How would you design a research study to determine the effect of Supplement XYZ on vertical jump height?

1.6 Exercise and Nutrition Credentials and Certifications

Many people work with athletes in the areas of exercise and nutrition; however, their backgrounds and training may be very different. Some are health care professionals, such as a medical doctor (MD), registered dietitian (RD), physical therapist (PT), and Registered Clinical Exercise Physiologist® (RCEP). These licenses and certifications require at least a bachelor's degree and successful completion of a written exam. In some cases higher degrees and more training are needed—an RD must complete an internship, an RCEP must complete a master's degree in exercise science, exercise physiology, or kinesiology, and an MD must complete medical school and postgraduate training such as internships and residencies. Health care professions generally have rigorous standards because patients can be harmed if practitioners are not properly trained. However, athletes at all skill levels may work with those who are not health care professionals. Such practitioners may or may not possess certifications that are professionally rigorous.

There are many types of practitioners in the area of exercise science.

There are a wide variety of practitioners in the exercise and fitness field: personal trainers working at local health clubs, individual sports coaches, strength and conditioning coaches, fitness or life coaches, athletic trainers, exercise physiologists, and clinical exercise specialists in a clinic or hospital working with patients with diagnosed cardiac, pulmonary, or metabolic disease. The background, experience, academic training, and preparation of practitioners in the exercise/fitness field may also vary dramatically. In today's popular culture, a "fitness expert" might be anyone from a reality TV celebrity with his or her own exercise DVD to a research scientist with a PhD in exercise physiology presenting results of a study at a national conference. Consumers may be further confused by the various terminology and titles used to describe practitioners (for example, personal trainer, exercise physiologist, physical therapist, exercise specialist, and so on) and the lack of a single national or international organization that oversees the credentials of practitioners.

At present, there are very few regulations governing who can give exercise advice or work directly with individuals or groups, but the public is best served by practitioners who have an appropriate background in

A personal trainer shows an athlete how to check her pulse.

anatomy, physiology, and principles of exercise and fitness. They should also have experience in the assessment of fitness and the design and implementation of safe and effective fitness programs for individuals and groups. One way to determine if exercise/fitness practitioners have the necessary and appropriate knowledge is to look at their academic training. College and university degrees in exercise science, kinesiology, physical education, and other related degrees are available from many institutions at the undergraduate and graduate level. There is a growing movement for accreditation of these academic programs through the Commission on Accreditation of Allied Health Education Programs (CAAHEP), which provides evidence that an academic program meets rigorous educational standards.

Certification programs can also attest to the skills and experience of the fitness practitioner. There are many certification programs available, many with minimal requirements, but there are some that require specific educational degrees and thorough preparation in the field. For example, the American College of Sports Medicine (ACSM) certifies Clinical Exercise Specialists®, Health Fitness Specialists, and Personal Trainers. The academic preparation of each is different. Clinical Exercise Specialists® must have a bachelor's degree in an allied health field, such as exercise physiology, physical therapy, or nursing. Health Fitness Specialists must have an associate's or bachelor's degree in any number of allied health fields, whereas a Personal Trainer must have a high school degree or the equivalent to be certified. Because academic training is different, scope of practice is also different. Brief explanations of these exercise certifications can be found in Figure 1.9.

In addition to those already mentioned, other practitioners such as a certified strength and conditioning specialist (CSCS) and a certified athletic trainer (ATC) may work closely with athletes. The CSCS is a certification by the National Strength and Conditioning Association, which tests and certifies individuals who can design and implement safe and effective strength

The ACSM Certified Personal Trainer[SM] **(CPT)** is a fitness professional involved in developing and implementing an individualized approach to exercise leadership in healthy populations and/or those individuals with medical clearance to exercise.

The ACSM Health Fitness Specialist (HFS) is a degreed health and fitness professional qualified to assess, design, and implement individual and group exercise and fitness programs for apparently healthy individuals and individuals with medically controlled diseases.

The ACSM Clinical Exercise Specialist® (CES) is a health care professional certified by the ACSM to deliver a variety of exercise assessment, training, rehabilitation, risk factor identification, and lifestyle management services to individuals with or at risk for cardiovascular, pulmonary, and metabolic disease(s).

The ACSM Registered Clinical Exercise Physiologist® (RCEP) is an allied health professional who works in the application of exercise and physical activity for those clinical and pathological situations in which exercise has been shown to provide therapeutic or functional benefit. Patients for whom services are appropriate may include, but not be limited to, those with cardiovascular, pulmonary, metabolic, immunological, inflammatory, orthopedic, and neuromuscular diseases and conditions.

Figure 1.9 Examples of exercise specialist certifications

and conditioning programs. To be eligible, a candidate must have a bachelor's or chiropractic degree or be enrolled as a senior at an accredited college or university and have a current CPR (cardiopulmonary resuscitation) certification. ATCs are allied health care professionals who provide for risk management and injury prevention, acute care of injury and illness, and assessment and evaluation of injury, and who conduct therapeutic modalities and exercise. To be eligible for certification, candidates must have a degree from an accredited college or university athletic training program that includes both academic and clinical requirements, and pass the certification exam.

There are many types of practitioners in the area of nutrition.

Similar to the field of exercise science, there are many types of practitioners in the field of nutrition. A registered dietitian (RD) is a food and nutrition expert. An RD must earn a bachelor's degree with specialized dietetics courses, complete 6 to 12 months of supervised practice (internship), pass a national exam, and meet continuing professional educational requirements. In addition to the RD, which is a national certification, dietitians may also be subject to state licensure laws. More information about registered dietitians and the American Dietetic Association can be found at http://www.eatright.org.

Registered dietitians have clinical training, counseling experience, and a solid background in nutrition assessment. They translate the science of nutrition into practical solutions for healthy living. In many states, anyone can legally give general nutrition advice, but only certain practitioners can give medical nutrition therapy (MNT). MNT is nutrition advice that is intended to prevent, treat, or cure a disease or disorder. MNT requires extensive training and usually falls under the scope of practice of physicians, registered dietitians, and those with master's degrees in clinical nutrition.

Nutritionist refers to someone who has studied nutrition. It is a very general term and academic training can range from marginal to rigorous. For example, one online university certifies nutritionists after they complete just six online courses and pass an exam. Contrast that academic training with a person with a master's degree in nutrition science. A science-based bachelor's degree is a prerequisite, and the master's degree requires at least two years of course work and research study in the area of nutrition. When the term *nutritionist* is used, consumers should inquire about the practitioner's academic background.

Certification in the field of sports nutrition is available. A registered dietitian (RD) who has specialized knowledge and experiences in sports nutrition is eligible to be Board Certified as a Specialist in Sports Dietetics (CSSD). To be eligible to take the board certification exam the individual must be an RD for a minimum of 2 years with at least 1,500 hours of sports nutrition experience in the past 5 years. CSSDs are referred to as sports dietitians. More information can be found at http://www.cdrnet.org.

The International Society of Sports Nutrition offers the CISSN (Certified Sports Nutritionist from the International Society of Sports Nutrition). This certification requires the individual to possess a 4-year undergraduate degree, preferably in exercise science, kinesiology, physical education, nutrition, biology, or related biological science. There is an alternative pathway that does not require a 4-year degree and is based on 5 years of experience and possession of two other certifications. The CISSN requires successful completion of the CISSN certification exam. More information can be found at http://www.sportsnutritionsociety.org/site/cert_cissn.php.

Scope of practice helps establish professional boundaries.

Who is qualified to make nutrition and training recommendations to athletes? Because health, nutrition, and exercise are interrelated, the lines between

Scope of practice: Legal scope of work based on academic training, knowledge, and experience.

This logo indicates the individual is a registered dietitian who is Board Certified as a Specialist in Sports Dietetics (CSSD).

these disciplines are sometimes blurry and there is some overlap among various practitioners. **Scope-of-practice** definitions help establish professional boundaries by outlining the skills, responsibilities, and accepted activities of practitioners. Such definitions take into account academic training and professional knowledge and experiences. Certifications and licenses are one way to formally define scope of practice, but not all health, exercise, and nutrition practitioners are formally licensed and some certifications are voluntary. In some states, scope of practice may be legally defined. Many professional organizations have clear statements regarding scope of practice.

Scope of practice protects both consumers and practitioners. Consumers can be assured that practitioners have been properly trained and are qualified to practice. Likewise, practitioners are aware of their professional boundaries and can avoid areas in which they do not have appropriate training and skills. When a client's needs fall outside a practitioner's scope of practice, the appropriate response is to acknowledge these limits and make a referral to a qualified practitioner. Practitioners may be professionally and personally liable if they work outside their scope of practice and their clients are harmed.

Practitioners must recognize the limitations of their training, skills, and knowledge. Many practitioners have basic but not advanced knowledge of nutrition and physical fitness. In such cases professional organizations often suggest using and promoting only general nutrition and exercise materials that are in the public domain. Public domain documents can be freely copied and distributed. There are excellent public domain general nutrition and fitness materials produced by federal government agencies and large health-oriented organizations such as the American Heart Association. The Dietary Guidelines for Americans (U.S. Department of Health and Human Services and U.S. Department of Agriculture, 2010), MyPyramid, and ACSM's position statements are examples. The use of public domain general nutrition and exercise materials is unlikely to cause harm, and consumers receive consistent health-related messages from professionals.

Specific nutrition and training recommendations for athletes should be made by practitioners who are qualified to do so. To most effectively help athletes and to avoid potentially harming them, it is important to know and respect professional boundaries and to make referrals to other qualified professionals.

Key points

- Certifications vary widely in their requirements.
- Many exercise- and nutrition-related certifications do not require a bachelor's degree.
- Practitioners must recognize the limits of their knowledge, training, and expertise or athletes can be harmed.

What's the difference between a sports dietitian, a sports nutritionist, and a nutritionist who works with athletes?

Post-Test   Reassessing Knowledge of Sports Nutrition

Now that you have more knowledge about sports nutrition, read the following statements and decide if each is true or false. The answers can be found in Appendix O.

1. An athlete's diet is a modification of the general nutrition guidelines made for healthy adults.
2. Once a healthy diet plan is developed, an athlete can use it every day with little need for modification.
3. In the United States, dietary supplements are regulated in the same way as over-the-counter medications.
4. The scientific aspect of sports nutrition is developing very quickly and quantum leaps are being made in knowledge of sports nutrition.
5. To legally use the title of sports nutritionist in the United States, a person must have a bachelor's degree in nutrition.

Summary

Sports nutrition combines the scientific disciplines of exercise physiology and nutrition. The ultimate goal is improved performance, which involves both skill development and training. Proper nutrition helps to support training and recovery as well as good health. Sports nutrition principles are based on sound general nutrition principles that have been modified to reflect the demands of training and competition for the athlete's sport and position. Dietary supplements are widely used by athletes but are not well regulated in the United States. Athletes should take into account the legality, ethics, purity, safety, and effectiveness of any dietary supplement, with the understanding that many are not effective and some may contain banned substances.

Sports nutrition recommendations should be evidence based. The most conclusive evidence comes from studies that are randomized, double-blind, placebo-controlled, and published in peer-reviewed journals. The results of research studies are often misinterpreted or misapplied. The role of the professional is to provide sound science-based information and to correct erroneous information. Athletes are subject to much misinformation via the Internet, especially concerning dietary supplements.

Practitioners must understand and respect the limitations of their training, skills, and knowledge. Scope-of-practice definitions help establish professional boundaries and protect athletes and practitioners. A referral should be made when an athlete's needs fall outside a practitioner's scope of practice. Many people who work with athletes are certified or licensed. Some certifications are rigorous but others are not.

Self-Test

Multiple-Choice

The answers can be found in Appendix O.

1. The term *endurance athlete* is generally interpreted to mean that the athlete:
   a. predominantly uses the oxygen-dependent energy system.
   b. trains for many hours.
   c. engages in only moderate-intensity activity.
   d. does not lift weights as part of training.

2. The nutrition goals of each mesocycle are:
   a. always the same to ensure consistency.
   b. a reflection of the intensity and volume of activity.
   c. unplanned and very flexible.
   d. based on each athlete's aerobic capacity.

3. Regulation of dietary supplements in the United States is:
   a. among the strictest of all countries in the world.
   b. similar to regulation of prescription medications.
   c. minimal.
   d. nonexistent.

4. What does it mean when an active ingredient in a dietary supplement is *standardized*?
   a. Strict quality control measures have been followed.
   b. The amount consumed needs to be adjusted based on body weight.
   c. The same amount is found in each pill or tablet.
   d. Only natural extracts are used.

5. Which of the following certifications does NOT require a bachelor's degree?
   a. Registered Dietitian
   b. Registered Clinical Exercise Physiologist
   c. Certified Athletic Trainer
   d. Certified Personal Trainer

Short Answer

1. What is sports nutrition?
2. What are the general goals of training and the short- and long-term nutrition goals that may improve performance?
3. Explain the ways in which the Dietary Supplement Health and Education Act regulates dietary supplements in the United States. In what ways are dietary supplements not well regulated?
4. What are the advantages of experimental research? What are the limitations of epidemiological research and case studies?
5. What are the elements of research design that give strength to a scientific study?
6. Who is qualified to make nutrition and training recommendations to athletes?
7. What is scope of practice?

Critical Thinking

1. "Garbage in, garbage out" is a saying that indicates that if the input is bad, a good result cannot be expected. Explain how this saying applies to research design and results.
2. What is meant by the statement, "Food is for fuel and fun?"
3. When do the roles of exercise physiologists, sports dietitians, athletic trainers, and strength and conditioning specialists overlap and when are they distinct?

References

Alert: Protein drinks. You don't need the extra protein or the heavy metals our tests found. (2010). *Consumer Reports, 75*(7), 24–27.

American Diabetes Association and American Dietetic Association. (2008). *Choose your foods: Weight management.* Chicago: Author.

Amonette, W. E., English, K. L., & Ottenbacher, K. J. (2010). Nullius in verba: A call for the incorporation of evidence-based practice into the discipline of exercise science. *Sports Medicine, 40*(6): 449–457.

Blair, S. N., Kohl, H. W., III, Paffenbarger, R. S., Jr., Clark, D. G., Cooper, K. H., & Gibbons, L. W. (1989). Physical fitness and all-cause mortality: A prospective study of healthy men and women. *Journal of the American Medical Association, 262*(17), 2395–2401.

Bonnar-Kidd, K. K., Black, D. R., Mattson, M., & Coster, D. (2009). Online physical activity information: Will typical users find quality information? *Health Communication, 24*(2), 165–175.

Bundorf, M. K., Wagner, T. H., Singer, S. J., & Baker, L. C. (2006). Who searches the Internet for health information? *Health Services Research, 41*(3, Pt. 1), 819–836.

Burke, L. M., Kiens, B., & Ivy, J. L. (2004). Carbohydrate and fat for training and recovery. *Journal of Sports Sciences, 22*(1), 15–30.

Caspersen, C. J., Powell, K. E., & Christensen, G. M. (1985). Physical activity, exercise, and physical fitness: Definitions and distinctions for health-related research. *Public Health Reports, 100*(2), 126–131.

Coyle, E. F., Hagberg, J. M., Hurley, B. F., Martin, W. H., Ehsani, A. A., & Holloszy, J. O. (1983). Carbohydrate feeding during prolonged strenuous exercise can delay fatigue. *Journal of Applied Physiology, 55*(1, Pt. 1), 230–235.

Dunford, M. (2003). *Nutrition logic: Food first, supplements second.* Kingsburg, CA: Pink Robin.

Dunford, M., & Coleman, E. (in press). Ergogenic aids, dietary supplements and exercise. In Rosenbloom, C. *Sports nutrition: A practice manual for professionals.* (5th ed). Chicago: American Dietetic Association.

Fogelholm, M. (2010). Physical activity, fitness and fatness: Relations to mortality, morbidity and disease risk factors. A systematic review. *Obesity Reviews,* 11(3), 202–221.

Food and Drug Administration. (1994). Dietary Supplement Health and Education Act (DSHEA). Retrieved August 22, 2010 from http://www.cfsan.fda.gov/~dms/dietsupp.html

Food and Drug Administration. (2009). FDA uncovers additional tainted weight loss products. Retrieved August 22, 2010, from http://www.fda.gov/NewsEvents/Newsroom/PressAnnouncements/2009/ucm149547.htm

Froiland, K., Koszewski, W., Hingst, J., & Kopecky, L. (2004). Nutritional supplement use among college athletes and their sources of information. *International Journal of Sport Nutrition and Exercise Metabolism, 14*(1), 104–120. (Erratum in 14[5] following 606).

Gershwin, M. E., Borchers, A. T., Keen, C. L., Hendler, S., Hagie, F., & Greenwood, M. R. (2010). Public safety and dietary supplementation. *Annals of the New York Academy of Science, 1190*(1), 104–117.

Geyer, H., Parr, M. K., Mareck, U., Reinhart, U., Schrader, Y., & Schänzer, W. (2004). Analysis of non-hormonal nutritional supplements for anabolic-androgenic steroids—results of an international study. *International Journal of Sports Medicine, 25,* 124–129.

Government Accountability Office (GAO). (2010). Herbal dietary supplements. Examples of deceptive or questionable marketing practices and potentially dangerous advice. Testimony before the Special Committee on Aging, U.S. Senate. Retrieved August 2, 2010 from http://www.gao.gov/new.items/d10662t.pdf

Grandjean, A. C. (1997). Diets of elite athletes: Has the discipline of sports nutrition made an impact? *Journal of Nutrition, 127*(5), 874S–877S.

Gray, G. E., & Gray, L. K. (2002). Evidence-based medicine: Applications to dietetic practice. *Journal of the American Dietetic Association, 102*(9), 1263–1272.

Gurley, B. J., Gardner, S. F., & Hubbard, M. A. (2000). Content versus label claims in ephedra-containing dietary supplements. *American Journal of Health-System Pharmacy, 57*(10), 963–969.

Hoffman, J. R., Faigenbaum, A. D., Ratamess, N. A., Ross, R., Kang, J., & Tenenbaum, G. (2008). Nutritional supplementation and anabolic steroid use in adolescents. *Medicine & Science in Sports & Exercise, 40,* 15–24.

Institute of Medicine. (1997). *Dietary Reference Intakes for calcium, phosphorus, magnesium, vitamin D and fluoride* (Food and Nutrition Board). Washington, DC: National Academies Press.

Institute of Medicine. (1998). *Dietary Reference Intakes for thiamin, riboflavin, niacin, vitamin B_6, folate, vitamin B_{12}, pantothenic acid, biotin and choline* (Food and Nutrition Board). Washington, DC: National Academies Press.

Institute of Medicine. (2000). *Dietary Reference Intakes for vitamin C, vitamin E, selenium and carotenoids* (Food and Nutrition Board). Washington, DC: National Academies Press.

Institute of Medicine. (2001). *Dietary Reference Intakes for vitamin A, vitamin K, arsenic, boron, chromium, copper, iodine, iron, manganese, molybdenum, nickel, silicon, vanadium, and zinc* (Food and Nutrition Board). Washington, DC: National Academies Press.

Institute of Medicine. (2002). *Dietary Reference Intakes for energy, carbohydrate, fiber, fat, fatty acids, cholesterol, proteins and amino acids* (Food and Nutrition Board). Washington, DC: National Academies Press.

Institute of Medicine (2003). *Dietary Reference Intakes: Applications in dietary planning* (Food and Nutrition Board). Washington, DC: National Academies Press.

Institute of Medicine. (2004). *Dietary Reference Intakes for water, potassium, sodium, chloride, and sulfate* (Food and Nutrition Board). Washington, DC: National Academies Press.

Institute of Medicine (2010). *Dietary Reference Intakes for calcium and vitamin D* (Food and Nutrition Board). Washington, DC: National Academies Press.

Jacobson, B. H., Sobonya, C., & Ransone, J. (2001). Nutrition practices and knowledge of college varsity athletes: A follow-up. *Journal of Strength and Conditioning Research, 15*(1), 63–68.

Jenkinson, D. M., & Harbert, A. J. (2008). Supplements and sports. *American Family Physician, 78*(9), 1039–1046.

Judkins, C., Hall, D., & Hoffman, K. (2007). Investigation into supplement contamination level in the US market Retrieved August 2, 2010, from http://www.usatoday.com/sports/hfl-supplement-research-report.pdf

Khoo, Y. S., & Aziz, Z. (2009). Garlic supplementation and serum cholesterol: a meta-analysis. *Journal of Clinical Pharmacy and Therapeutics, 34*(2), 133–145.

Laukkanen, J. A., Lakka, T. A., Rauramaa, R., Kuhanen, R., Venalainen, J. M., Salonen R., & Salonen, J. T. (2001). Cardiovascular fitness as a predictor of mortality in men. *Archives of Internal Medicine, 161*(6), 825–831.

Macedonio, M. A., & Dunford, M. (2009). *The athlete's guide to making weight*: Optimal weight for optimal performance. Champaign, IL: Human Kinetics.

Maughan, R. (2002). The athlete's diet: Nutritional goals and dietary strategies. *Proceedings of the Nutrition Society, 61*(1), 87–96.

Maughan, R. (2005). Contamination of dietary supplements and positive drug tests in sport. *Journal of Sports Science, 23,* 883–889.

Mettler, S., Mannhart, C., & Colombani, P. C. (2009). Development and validation of a food pyramid for Swiss athletes. *International Journal of Sport Nutrition and Exercise Metabolism, 19*(5), 504–518.

Morris, C. A., & Avorn, J. (2003). Internet marketing of herbal products. *Journal of the American Medical Association, 17*(11), 1505–1509.

Morton, J. P., Robertson, C., Sutton, L., & MacLaren, D. P. (2010). Making the weight: A case study from professional boxing. *International Journal of Sport Nutrition and Exercise Metabolism, 20*(1), 80–85.

Myers, E. F., Pritchett, E., & Johnson, E. Q. (2001). Evidence-based practice guides vs. protocols: What's the difference? *Journal of the American Dietetic Association, 101*(9), 1085–1090.

Otten, J. J., Hellwig, J. P., & Meyers, L. D. (Eds.). (2006). *Dietary Reference Intakes: The essential guide to nutrient requirements.* Washington, DC: National Academies Press.

Petróczi, A., Naughton, D. P., Mazanov, J., Holloway, A., & Bingham, J. (2007). Limited agreement exists between rationale and practice in athletes' supplement use for maintenance of health: A retrospective study. *Nutrition Journal, 30*(6), 34.

Petróczi, A., Naughton, D. P., Pearce, G., Bailey, R., Bloodworth, A., & McNamee, M. (2008). Nutritional supplement use by elite young UK athletes: Fallacies of advice regarding efficacy. *Journal of the International Society of Sports Nutrition, 5*, 22.

Reissig, C. J., Strain, E. C., & Griffiths, R. R. (2009). Caffeinated energy drinks—a growing problem. *Drug & Alcohol Dependence, 99*, 1–10.

Rodriguez, N. R., DiMarco, N. M., Langley, S., American Dietetic Association, Dietitians of Canada, & American College of Sports Medicine. (2009). Position of the American Dietetic Association, Dietitians of Canada, and the American College of Sports Medicine: Nutrition and athletic performance. *Journal of the American Dietetic Association 109*(3), 509–527

Schroder, H., Navarro, E., Mora, J., Seco, J., Torregrosa, J. M., & Tramullas, A. (2002). The type, amount, frequency and timing of dietary supplement use by elite players in the First Spanish Basketball League. *Journal of Sports Science, 20*, 353–358.

Seebohar, B. (2011). *Nutrition periodization for endurance athletes: Taking traditional sports nutrition to the next level.* Boulder, CO: Bull.

Senchina, D. S., Shah, N. B., Doty, D. M., Sanderson, C. R., & Hallam, J. E. (2009). Herbal supplements and athlete immune function—what's proven, disproven, and unproven? *Exercise Immunology Review, 15*, 66–106.

Sheehan, G. (1980). *This running life.* New York: Simon and Schuster.

Sutherland, L. A., Wildemuth, B., Campbell, M. K., & Haines, P. S. (2005). Unraveling the web: an evaluation of the content quality, usability, and readability of nutrition websites. *Journal of Nutrition Education and Behavior, 37*(6), 300–305.

Temple, N. J. (2010). The marketing of dietary supplements in North America: The emperor is (almost) naked. *Journal of Alternative and Complementary Medicine, 16*(7), 803–806.

Triplett, D., Doyle, J. A., Rupp, J. C., & Benardot, D. (2010). An isocaloric glucose-fructose beverage's effect on simulated 100-km cycling performance compared with a glucose-only beverage. *International Journal of Sport Nutrition and Exercise Metabolism, 20*, 122–131.

Tscholl, P., Alonso, J. M., Dollé, G., Junge, A., & Dvorak, J. (2010). The use of drugs and nutritional supplements in top-level track and field athletes. *American Journal of Sports Medicine, 38*, 133–140.

U.S. Department of Agriculture and U.S. Department of Health and Human Services. *Dietary Guidelines for Americans, 2010.* (7th ed.). Washington, DC: U.S. Government Printing Office, December, 2010.

van Breemen, R. B., Fong, H. H., & Farnsworth, N. R. (2008). Ensuring the safety of botanical dietary supplements. *American Journal of Clinical Nutrition, 87*, 509S–513S.

Wansink, B., & American Dietetic Association (2006). Position of the American Dietetic Association: Food and nutrition misinformation. *Journal of the American Dietetic Association, 106*(4), 601–607.

Defining and Measuring Energy

2

- **Define** and explain bioenergetics, ATP, calorie, kilocalorie, and other energy-related terms.
- **Explain** the concept of conservation of energy and how this concept applies to energy utilization in the body.
- **Identify** the primary source of energy in the body and explain how it is used by skeletal muscle during exercise.
- **Explain** the resynthesis of ATP, including the role of enzymes, and name the major energy systems involved.
- **Explain** how the energy content of food and energy expenditure are measured directly and indirectly, and how estimates can be made more accurately.
- **List** and explain the components of the energy balance equation.
- **Explain** resting metabolic rate, the factors that influence it, and how it is measured or predicted in athletes and nonathletes.
- **Explain** the impact of physical activity on energy expenditure.
- **Calculate** an estimated energy requirement for a 24-hour period using a simple formula.

Sport and exercise require the use of energy.

Pre-Test   Assessing Current Knowledge of Energy

Read the following statements and decide if each is true or false.

1. The body creates energy from the food that is consumed.
2. The scientific unit of measure of energy is the calorie.
3. A person's resting metabolic rate can change in response to a variety of factors such as age, food intake, or environmental temperature.
4. Physical activity is responsible for the largest amount of energy expended during the day for the average adult in the United States.
5. The energy source used by all cells in the body is adenosine triphosphate (ATP).

For the body to remain viable and to function properly, thousands of physiological and biochemical processes must be carried out on a daily basis. Each requires energy. A relatively large amount of energy is needed just to sustain life throughout the day. Physical activity, exercise, and sport require energy beyond that needed for daily subsistence. Where do humans get the energy for these activities? The simple answer is food. **Bioenergetics,** the process of converting food into biologically useful forms of energy, is the focus of the next two chapters.

Energy is such a large topic that it can quickly become overwhelming, so the discussion of energy is covered in two chapters in this textbook. This chapter will focus on the introductory concepts of energy primarily, defining and measuring it, and energy balance, "Energy in = Energy out." On one side of the energy balance equation is energy intake in the form of food (that is, "energy in"). On the other side of the energy balance equation is energy expenditure (that is, "energy out"), primarily from resting metabolism and physical activity, exercise, and sport. Chapter 3 will cover the specific energy systems that are used to fuel the body at rest and during exercise.

2.1 Energy and Energy Concepts

Energy is the ability to perform work.

Energy is difficult to define, but simply stated, it is the ability to perform work. Energy exists in different forms: atomic, chemical, electrical, mechanical, radiant, and thermal. It is easy to think of many different tasks that need to be performed in the body that require work and, therefore, energy. Examples include:

- Chemical work (for example, storage of carbohydrates by forming glycogen for later use)
- Electrical work (for example, maintenance of the distribution of ions across cell membranes)
- Mechanical work (for example, **force production** by skeletal muscle)
- Transportation work (for example, circulation of blood throughout the body to deliver oxygen, nutrients, and other compounds to tissues)

To understand the various energy systems in the body, it is necessary to understand some basic concepts of energy. An important concept, commonly known as the law of "Conservation of Energy," is the First Law of Thermodynamics, which states, "Within a closed system, energy is neither created nor destroyed." It can, however, be transformed from one form of energy to another. An old-fashioned steam locomotive is an excellent example of how thermal, or heat energy, is transformed into mechanical energy to drive the locomotive wheels (Figure 2.1).

How does the First Law of Thermodynamics apply to the human body and the study of sports nutrition? Humans do not create energy or lose energy; rather, they have a variety of processes that are used to transfer energy from one form to another for use by the body. Energy is consumed in the form of food and transformed into different chemical forms that can be used immediately or stored for later use.

Humans are relatively inefficient in the process of energy conversion, at least compared to machines. Electric motors and steam turbine engines can convert

Figure 2.1 Conversion of thermal (heat) energy to mechanical energy

approximately 85 percent of electrical or thermal energy to mechanical work. However, biological systems are much less efficient, directing a much lower percentage of available energy to the performance of useful work. For example, when one is using muscle for exercise, all of the energy expended is not used for force production, but a large amount of the energy is transferred to heat. This is why muscle and body temperature rise during exercise.

Bicycling is an excellent physical activity to study the energy efficiency of human exercise. First, with the use of a precision cycle ergometer, the amount of work performed can be accurately measured. Second, because of the mechanical design of the gearing, cycling on a bicycle or cycle ergometer is at the upper end of the efficiency range for a human to perform work. As seen in the example in Figure 2.2, exercising on a cycle ergometer results in only approximately 25 percent of the energy expended being converted to useful work. Bicycling is one of the more energy-efficient activities for humans; other activities such as walking, running, and swimming are even less energy efficient.

Storing and releasing energy. Energy can exist in a state of **potential energy** when it is being stored for future use. As stored (potential) energy is released to perform some type of work, it is referred to as **kinetic energy**. The water captured in a reservoir behind a dam has a tremendous amount of potential energy as shown in Figure 2.3. If water is allowed to flow through pipelines in the dam to turbines, the kinetic energy of the moving water turns the blades of the turbines, which then drives a generator to produce electricity. A hydroelectric dam (Figure 2.4) is an excellent example of (1) storing energy in a potential state to be used later (water in the reservoir), and (2) the conversion of energy from one form to another useful form of energy

(the mechanical energy of the moving water and the spinning turbine to electrical energy).

These same concepts can be applied to energy processes in the body. Carbohydrates consumed by an athlete can be stored for later use. The energy contained in carbohydrate foods can be stored in skeletal muscle as glycogen (that is, a carbohydrate reservoir). Glycogen can then be used as a fuel source during exercise; its stored potential energy can be converted to chemical energy (that is, ATP), which can be used by skeletal muscles for force production.

There are processes and reactions that store energy and ones that release energy. Those that store energy are referred to as **endergonic** reactions; those that release energy are referred to as **exergonic** reactions. The setting and use of a mousetrap illustrates the complementary processes of endergonic and exergonic reactions in a mechanical fashion as illustrated in Figure 2.5. In order for the mousetrap to be useful, the spring trap must be set. To set the trap, the wire bail must be forced back into position against the spring and locked into place, which requires the input of energy. This is the endergonic portion of the process; energy is put into the process and can be stored for

Bioenergetics: The process of converting food into biologically useful forms of energy.

Force production: The generation of tension by contracting muscle.

Potential energy: Stored energy.

Kinetic energy: Energy of motion.

Endergonic: Chemical reactions that store energy.

Exergonic: Chemical reactions that release energy.

Male cyclist weighing 150 pounds (68.2 kg) riding on a cycle ergometer at 180 Watts (for perspective, riding at a power output of 180 Watts would be a moderate-to-hard training pace for a recreational cyclist, and a relatively easy pace for a well-trained competitive cyclist).

Exercise oxygen consumption = 36.0 ml/kg/min or 2.455 L/min

Resting oxygen consumption = 3.5 ml/kg/min

Conversion factors (see Conversion Tables in the front of the book)

1 Watt = 0.0143 kcal/min

5 kcal = 1 L oxygen

1 L = 1,000 ml

External work performed:

180 Watts × 0.0143 kcal/min = 2.574 kcal/min

Total energy expended during exercise:

36.0 ml/kg/min × 1000 ml/L = 0.036 L/kg/min

0.036 L/kg/min × 68.2 kg = 2.455 L/min

2.455 L/min × 5 kcal/L = 12.28 kcal/min

Energy expended (above resting):

36.0 ml/kg/min – 3.5 ml/kg/min = 32.5 ml/kg/min

32.5 ml/kg/min × 1,000 ml/L = 0.0325 L/kg/min

0.0325 L/kg/min × 68.2 kg = 2.217 L/min

2.217 L/min × 5 kcal/L = 11.08 kcal/min

Efficiency = External Work ÷ Energy Expended

Efficiency = (2.574 kcal/min ÷ 11.08 kcal/min) × 100 = 23.2%

Every minute this person cycles at this intensity, he expends 11.08 kilocalories of energy, but accomplishes only 2.574 kilocalories of external work.

Figure 2.2 Example of energy efficiency during exercise
Legend: kg = kilogram; ml = milliliter; min = minute; L = liter; kcal = kilocalorie

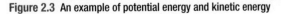

Figure 2.3 An example of potential energy and kinetic energy

Water stored in a reservoir behind a dam is an example of potential energy. The force of the moving water released from the dam is an example of kinetic energy.

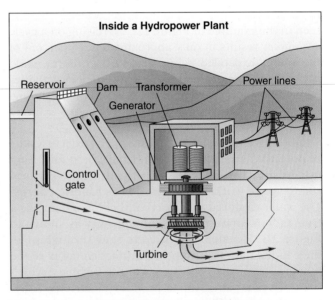

Figure 2.4 Conversion of kinetic energy to electrical energy

The kinetic energy of the flowing water is transformed to electrical energy.

(a) Input of energy required   (b) Storing energy as potential energy   (c) Release of energy—kinetic energy

Figure 2.5 A mechanical example of endergonic and exergonic processes

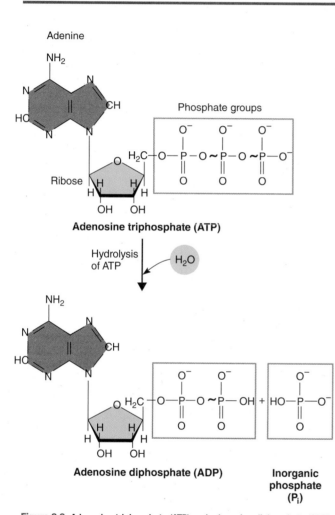

Figure 2.6 Adenosine triphosphate (ATP) and adenosine diphosphate (ADP)

The energy currency of all living things, ATP consists of adenine, ribose, and three phosphate groups. The hydrolysis of ATP, an exergonic reaction, yields ADP and inorganic phosphate. The black wavy lines indicate unstable bonds. These bonds allow the phosphates to be transferred to other molecules, making them more reactive.

device is triggered, the energy that has been stored in the spring is released, causing the wire bail to snap. This is the energy-releasing, or exergonic, part of the process.

The mousetrap is a mechanical example of input, storage, and release of energy; this same process can occur in chemical reactions. Some chemical reactions require the input of energy, but that energy can be stored for later use. Other chemical reactions can release the energy, which can then be used to power activities such as skeletal muscle contraction. Many of the chemical reactions that store and release energy in the body utilize chemical compounds called **high-energy phosphates**.

High-energy phosphate compounds store and release energy.

The primary example of a chemical compound used in reactions in the body that can store and release energy is **adenosine triphosphate (ATP)**. Adenosine triphosphate is a high-energy phosphate compound, a chemical that can store energy in its phosphate bonds. A protein molecule (adenine) combines with a sugar molecule (ribose) to form adenosine, which then has three phosphate groups attached, thus the name adenosine triphosphate (Figure 2.6). Energy is released from ATP in a very rapid one-step chemical reaction when a phosphate group is removed. The catalyst for the reaction is the **enzyme** ATPase. As shown

High-energy phosphate: A chemical compound that stores energy in its phosphate bonds.

Adenosine triphosphate (ATP): A chemical compound that provides most of the energy to cells.

Enzyme: A protein-containing compound that catalyzes biochemical reactions.

later use. Once the mousetrap is set, it can be used immediately, or it might be placed somewhere in the house where it can perform its useful task at any time over the next few hours, days, or weeks. When the

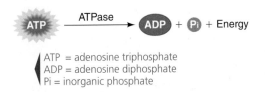

ATP = adenosine triphosphate
ADP = adenosine diphosphate
Pi = inorganic phosphate

Figure 2.7 Breakdown of ATP and release of energy

in Figure 2.7, ATP is broken down to yield **adenosine diphosphate (ADP)**, inorganic phosphate (Pi), and energy. The role of enzymes in catalyzing reactions is explained in Spotlight on…The Role of Enzymes.

Use of ATP by muscle. Adenosine triphosphate is the source of energy that is common to all cells in the body. It is often referred to as the common energy currency, much like the euro has become the common currency of most countries in the European Union. Although ATP is used by all cells of the body, the use of ATP by skeletal muscle is of major interest for those engaged in activity, exercise, and sport.

Any physical activity requires energy expenditure beyond what is needed at rest. As the intensity of the activity increases, the energy that is necessary to support that activity increases as well. Exercise or sports activities may require a large total amount of energy (for example, marathon running) or may require a very high **rate** of energy expenditure (for example, sprinting or weight lifting).

For skeletal muscles to produce force, the globular heads of the thick contractile protein, myosin, must form attachments (crossbridges) with the thin contractile protein, actin (Figure 2.8). Once this crossbridge is formed, the myosin heads swivel in a power stroke similar to a rower pulling on an oar. This power stroke causes the actin strands to be pulled so that they slide over the myosin filaments, creating tension or force. This is known as the Sliding Filament Theory of muscle contraction. After the contraction, known as a force-producing phase, the myosin heads must detach and reset, allowing relaxation of the muscle and preparation for the next contraction.

Where does the energy come from for skeletal muscles to produce force for exercise? The direct source of energy for force production and relaxation of skeletal muscle comes from ATP. Molecules of ATP are stored directly on the myosin head at the site where energy is needed for force production. When ATP is split (hydrolyzed), the energy released from the phosphate bonds puts the myosin heads in an energized state in which they are capable of forming a crossbridge with actin and performing a power stroke. In order to prevent rigor or sustained contraction of the skeletal muscle, and to allow relaxation, ATP must be reloaded on the myosin head, which allows it to detach from

Adenosine diphosphate (ADP): A chemical compound formed by the breakdown of ATP to release energy.

Rate: Speed.

Static: Not moving or changing.

Spotlight on…

The Role of Enzymes

Many chemical reactions in the body are catalyzed by enzymes. Enzymes are protein structures that speed up chemical reactions, primarily by lowering the energy required to activate the chemical reaction. Enzymes do not change the chemical compounds or the outcome of the reaction; they just allow the reaction to proceed more rapidly. Enzymes themselves are not **static**, but can change their activity or the degree to which they influence the speed of the reactions that they catalyze. A number of factors can influence enzymatic activity; two of the most common are temperature and pH.

If enzymes are warmed slightly, their activity increases. Exercise results in an increase in skeletal muscle temperature, which in turn results in an increase in enzymatic activity for those chemical reactions that support the exercise. Excessive changes in temperature, however, can reduce enzymatic

activity. Excessive heating causes an enzyme to be broken down or denatured; it is literally cooked. Cooling an enzyme reduces the activity to the point that the enzyme cannot affect the rate of the chemical reaction. Enzymes also have an optimal pH range, and too much acidity or alkalinity can reduce their activity and slow the reaction.

Enzymes may also have different isoforms, slightly different versions of the enzyme for specific areas or functions in the body. For example, the enzyme ATPase catalyzes the breakdown of ATP to release energy for use by cells. When this enzyme is found in skeletal muscle, the enzyme is a specific isoform, called myosin ATPase. In this way, enzymes can be "optimized" to work in specific situations such as when large amounts of energy are needed very quickly by skeletal muscle for exercise.

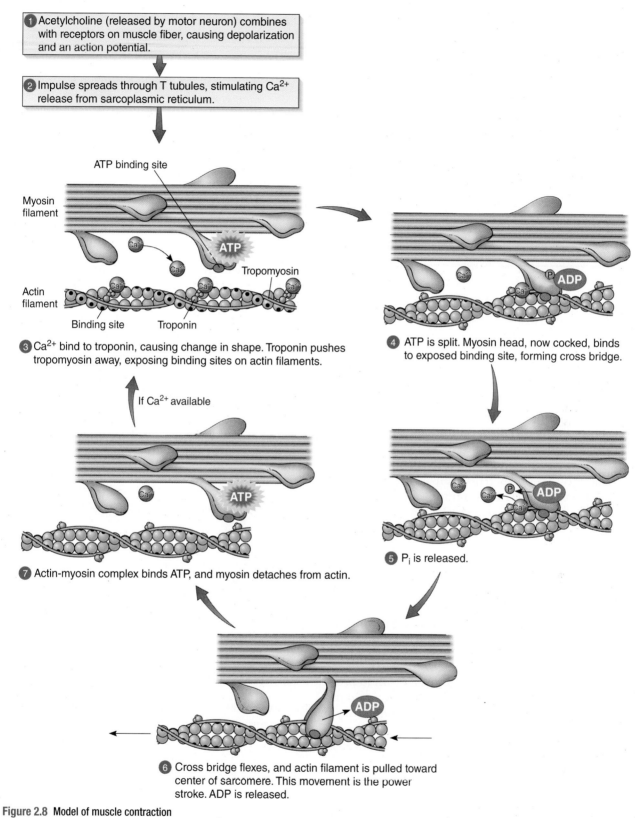

1 Acetylcholine (released by motor neuron) combines with receptors on muscle fiber, causing depolarization and an action potential.

2 Impulse spreads through T tubules, stimulating Ca^{2+} release from sarcoplasmic reticulum.

ATP binding site

Myosin filament

Ca^{2+}

ATP

Tropomyosin

Actin filament

Binding site   Troponin

3 Ca^{2+} bind to troponin, causing change in shape. Troponin pushes tropomyosin away, exposing binding sites on actin filaments.

If Ca^{2+} available

7 Actin-myosin complex binds ATP, and myosin detaches from actin.

ATP

4 ATP is split. Myosin head, now cocked, binds to exposed binding site, forming cross bridge.

Ca^{2+} P_i ADP

Ca^{2+} P   ADP

5 P_i is released.

ADP

6 Cross bridge flexes, and actin filament is pulled toward center of sarcomere. This movement is the power stroke. ADP is released.

Figure 2.8 Model of muscle contraction

Contraction results when actin filaments slide toward the center of individual sarcomeres of a myofibril. After Step 7, the cycle repeats from Step 4.

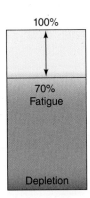

Figure 2.9 ATP concentrations in muscle during very high-intensity exercise

ATP concentration may fall to 70–75 percent of original levels, but fatigue will soon occur. ATP does NOT approach depletion in muscle cells during voluntary exercise.

the actin. Therefore, ATP is crucial for both the force production and relaxation of skeletal muscle.

As an athlete exercises, skeletal muscles are using ATP as the direct source of energy. One can picture how a high rate of ATP utilization might result in a rapidly declining concentration of ATP in the skeletal muscle, potentially leading to complete depletion if the exercise is intense enough or lasts long enough. Research studies show, however, that ATP concentrations in exercising muscle rarely drop more than 20-30 percent, even during the highest exercise intensity that an athlete can voluntarily perform (Hirvonen et al., 1987; see Figure 2.9). Once the ATP concentration in an exercising muscle is reduced to this degree, the force production ability of the skeletal muscle is reduced and the muscle starts to fatigue.

The response of ATP concentration in the muscle to exercise reveals two things: (1) a relatively large proportion of the ATP stored in the muscle is not able to be used for force production, even during very high-intensity exercise and (2) the ATP that is used to provide energy for muscle force is replaced very rapidly.

Resynthesis of ATP. After ATP is broken down to provide energy, it must be resynthesized for use again in the future. This process is known as **rephosphorylation**, an endergonic reaction requiring an input of energy. In this reaction, phosphate (Pi) is chemically joined to ADP to produce ATP (Figure 2.10).

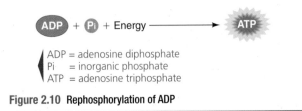

ADP = adenosine diphosphate
Pi = inorganic phosphate
ATP = adenosine triphosphate

Figure 2.10 Rephosphorylation of ADP

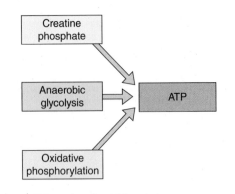

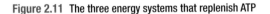

ATP = adenosine triphosphate

Figure 2.11 The three energy systems that replenish ATP

The body utilizes several different energy systems to rephosphorylate ADP (that is, to resynthesize ATP). The major energy systems used during exercise are the creatine phosphate, anaerobic glycolysis, and oxidative phosphorylation energy systems (Figure 2.11). These energy systems will be discussed in detail in Chapter 3.

Key points

■ Energy contained in food is transferred to other forms that can be used in the body.

■ High-energy phosphates such as adenosine triphosphate (ATP) are chemicals that can store and release energy.

■ ATP is the common energy source used by all cells in the body.

■ Energy for muscle contraction is provided directly by ATP.

■ Three major energy systems are used to replace ATP that is used for muscle contraction.

What happens when ATP levels in muscle decrease too much?

2.2 Measuring Energy

Energy is not only hard to define; it is also difficult to measure. There are a variety of techniques used to measure energy, and different units are used to express energy, adding to the confusion. The scientific community has adopted a standardized system of weights and measures that is based on metric measurements and is known as the International System of Units (SI units). In the United States, most people outside the scientific community are familiar with and use the term *calorie*, a unit of energy measurement that is *not* the designated SI unit for energy.

The unit of measure for energy in SI units is the **joule (J)**, named for British physicist James Prescott

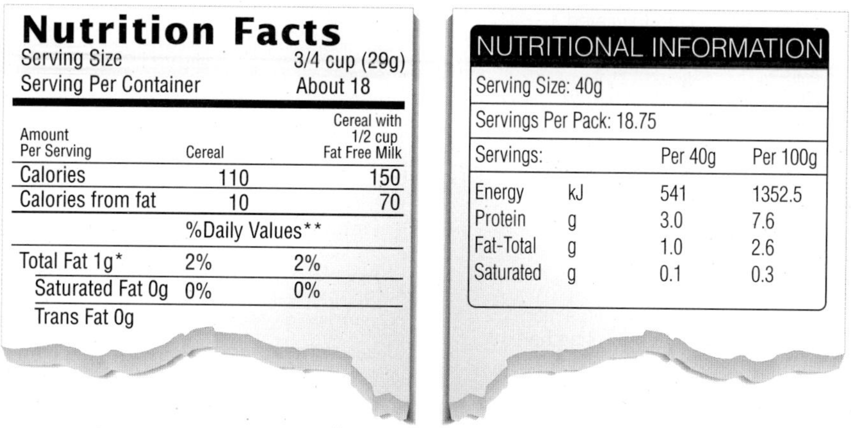

Figure 2.12 Energy measurements used on food labels in the United States and Australia

Labels in the United States use the term *Calories*, but most countries in the world use *kilojoules* (kJ).

Joule (1818–89). The technical definition of a joule is the work done by a force of 1 Newton acting to move an object 1 meter, or 1 Newton-meter. This is a definition of mechanical energy, which is better suited to physicists and scientists working with physical rather than biological systems. However, Joule was renowned for a series of innovative and meticulous experiments that demonstrated the equivalence between mechanical energy and thermal (heat) energy. Therefore, one of the acceptable and practical ways to study and discuss energy is in terms of heat, which is particularly appropriate for bioenergetics.

In nutrition and exercise physiology, energy is often expressed in calories, an expression of thermal energy. One **calorie** (lowercase *c*) is the heat required to raise the temperature of 1 gram (g) of water by 1°C, and is equal to 4.184 joules. This is a small amount of energy, so when discussing the energy content of food or the energy expenditure of physical activities, it is more practical to express energy in larger units. Usually the equivalent terms **Calorie** (uppercase *C*) and **kilocalorie (kcal)** are used. One kilocalorie (1 Calorie) is equal to 1,000 calories, the energy required to raise the temperature of 1 kilogram or 1 liter of water by 1°C.

Throughout this book the term *kilocalorie* will be used. Food labels in the United States use Calorie as the unit of measure. Keep in mind that kilocalorie and Calorie are equivalent units. However, kilocalorie and Calorie are not terms that most Americans are familiar with. In everyday language and in nonscientific writing the word *calorie* (lowercase *c*) is used. Although not technically correct, *calorie* is often used interchangeably with Calorie and kilocalorie.

Most scientific journals require that energy be expressed as an SI unit, which is the joule. Once again, because a joule is a relatively small amount of energy, the unit **kilojoule (kJ)**, which is equal to 1,000 joules, is used because it is a more practical measure

when discussing large amounts of energy intake or expenditure. To convert kilojoules to kilocalories, divide kJ by 4.2 kcal/kJ. For example, a food intake of 8,400 kJ is equivalent to 2,000 kcal. Table 2.1 lists energy units and their equivalents. Figure 2.12 uses food labels from the United States and Australia to illustrate the different units of measure used to express the energy value of food.

Table 2.1 Energy Units and Equivalents

1 calorie (cal) = 4.184 joules (J)
1 kilocalorie (kcal) = 1,000 calories
1 kilocalorie = 4,184 joules
1 Calorie (C) = 1,000 calories
1 Calorie = 4,184 joules
1 kilojoule (kJ) = 1,000 joules
1 kilocalorie = 4.184 kilojoules

Rephosphorylation: Re-establishing a chemical phosphate bond, as in adenosine diphosphate (ADP) re-establishing a third phosphate bond to become adenosine triphosphate (ATP).

Joule (J): The International System of Units (SI) way to express energy; specifically, the work done by a force of 1 Newton acting to move an object 1 meter, or 1 Newton-meter. 1 calorie is equal to 4.184 joules.

calorie: The amount of heat energy required to raise the temperature of 1 gram of water by 1°C.

Calorie: The amount of heat energy required to raise the temperature of 1 kilogram or 1 liter of water by 1°C. Equal to 1,000 calories.

Kilocalorie (kcal): A unit of expression of energy, equal to 1,000 calories (see calorie).

Kilojoule (kJ): A unit of expression of energy equal to 1,000 joules (see Joule).

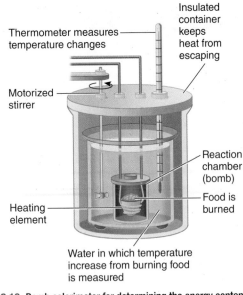

Thermometer measures temperature changes

Insulated container keeps heat from escaping

Motorized stirrer

Reaction chamber (bomb)

Food is burned

Heating element

Water in which temperature increase from burning food is measured

Figure 2.13 Bomb calorimeter for determining the energy content of food

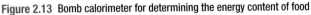

The energy content of food is measured by calorimetry.

Only one factor accounts for the "energy in" side of the energy balance equation: food. Most people are familiar with the concept that the energy content of food is its caloric content. How does one measure the energy content of food? The caloric content of food is determined through a process of calorimetry, during which food samples are burned and the resulting liberation of heat energy is precisely measured as a change in temperature. Because the thermal energy of the food is directly measured, this process is referred to as **direct calorimetry**.

Direct calorimetry analysis of food involves a bomb **calorimeter** (Figure 2.13), a device that determines energy by the amount of heat produced. The "bomb" is a metal container in which the food sample is burned in a pressurized, pure oxygen atmosphere. It is constructed to withstand high temperature and pressure, and is surrounded by a water bath that is insulated from outside temperature changes. As the food sample burns, the temperature change in the surrounding water bath is recorded with a sensitive thermometer and determines the thermal energy of the food.

Caloric content of carbohydrates, fats, proteins, and alcohol. Using direct calorimetry, the average caloric content of a wide variety of foods has been determined. When the amount of carbohydrates, fats, proteins, and alcohol in a food is known, the *approximate* caloric (energy) value can be calculated. The average energy values for each are listed in Table 2.2. Notice there are two columns of values, one for a bomb calorimeter and one for a human calorimeter (that is, the body). The human calorimeter refers to the amount of

Table 2.2 Energy Content of Carbohydrates, Fats, Proteins, and Alcohol

	Bomb calorimeter (kcal/g)	Human calorimeter (kcal/g)
Carbohydrates	4.2	4.2
Fats	9.4	9.4
Proteins	5.7	4.2
Alcohol	7.0	7.0*

Legend: kcal/g = kilocalorie per gram

*Under normal circumstances. However, when alcohol is a high percentage of total caloric intake, some of the energy is not available to the body.

energy a person can utilize from the food. The values are the same for carbohydrates and fats, but differ for proteins and may differ for alcohol.

Carbohydrate foods contain approximately 4.2 kilocalories of energy per gram (kcal/g) of food, whereas fats contain ~9.4 kcal/g. When burned completely in a bomb calorimeter, protein foods yield, on average, 5.7 kcal/g. However, when proteins are metabolized in the body (the human calorimeter), the full potential energy of these foods is not available because nitrogen, an important constituent of proteins, is not metabolized but is excreted. Therefore, when proteins are metabolized in the body the average caloric value is estimated to be 4.2 kcal/g (World Health Organization, 1991).

Alcohol (ethanol) yields approximately 7.0 kcal/g when burned in a bomb calorimeter. Under normal circumstances, the human calorimeter value is the same as the bomb calorimeter. For all practical purposes this value is used to estimate the caloric value of alcohol. However, when alcohol intake represents a large percentage of an individual's total caloric intake, some of the energy in the alcohol is not available to the body, resulting in alcohol yielding less than 7 kcal/g. This is most likely a result of damaged mitochondria in liver cells, an undesirable health circumstance (Lieber, 2003).

From a practical perspective, the values for carbohydrates and proteins (4.2 kcal/g) and fats (9.4 kcal/g) are too cumbersome, so they are rounded to the nearest whole number. That is why the energy content of carbohydrates and proteins is estimated to be 4 kcal/g and the energy content of fats is estimated at 9 kcal/g. For example, 1 tablespoon of oil has 14 grams of fat, or approximately 126 kcal (14 g × 9 kcal/g). That estimate is slightly low compared to the 131.6 kcal that is calculated if the more precise figure of 9.4 kcal/g is used. However, all the values obtained are estimates, and the error introduced by using the rounded off numbers is considered small and acceptable. At best, the amount of energy in food can only be estimated; the true energy value—when the potential energy from food is converted to useful forms of energy in the body—cannot be directly measured.

This difficulty in determining the "exact" caloric content of any food illustrates the imprecision and potential problems with strict "calorie counting" as a weight loss strategy. The caloric content of an individual food, a combination of foods in a meal, or the total amount of food consumed in a day can only be estimated. Tracking energy intake over weeks and months is especially difficult. Relatively small differences in energy intake exert influence and bring about change in body composition (for example, body fat loss or gain) but over relatively long periods of time. The person who diligently counts calories may calculate the amount of body fat that should be lost in a given time period, but this may be counterproductive. If caloric intake is underestimated and body fat is not lost according to "schedule," then the individual may conclude that the weight loss diet is not effective. In fact, counting calories gives the dieter a false sense of precision and an erroneous time frame for the loss of body fat. Although estimates of the amount of energy consumed daily can be useful, the precise caloric content of the foods consumed cannot be determined.

The amount of energy expended can be measured directly or indirectly.

There are a variety of ways to measure the amount of energy expended—"energy out." However, because people are active, free-living beings, it can be difficult to obtain measures of energy expenditure. The basic methods of measuring the energy expended by an individual are covered here. Some of the most accurate measurement techniques have limited use with athletes because of their impracticality.

Direct calorimetry. As the body utilizes the potential energy contained in food, a large portion of that energy is converted to heat energy. The amount of heat that is produced is proportional to the amount of energy expended; this heat energy can be measured as a change in temperature. Therefore, energy expenditure in humans and small animals can be determined by the principle of direct calorimetry in a process roughly similar to the determination of the energy content of food.

In the late 1700s, the French chemist Antoine Lavoisier developed a direct calorimeter in which he measured the amount of water produced by ice melting from an animal's body heat. The energy expenditure of the animal was determined from the amount of heat required to melt a measured quantity of ice. Over time, more sophisticated calorimeters have been developed, similar in principle to the bomb calorimeters used to measure the energy in food, except that they have been built large enough to completely enclose a human. Heat emitted from the human subject is recorded as a temperature change in an insulated water layer around the room-like calorimeter. In the late 1800s and early 1900s Atwater and colleagues (1899, 1905) conducted a series of elegant and precise metabolic experiments in which they used direct (and indirect) calorimetry to elucidate the metabolic effects of diet and exercise and the balance of energy intake and energy expenditure. Direct calorimetry is still used today but for research purposes only.

Indirect calorimetry. Another of Lavoisier's important discoveries was the relationship of **oxygen consumption** (O_2) and **carbon dioxide production** (CO_2) to energy expenditure and heat production. As the body's energy expenditure increases, the use of oxygen and the production of carbon dioxide by the aerobic energy system (see Chapter 3) increase proportionately. If the oxygen consumption and carbon dioxide production is measured, then the amount of energy expended can be calculated. This determination of energy expenditure by gas exchange is termed **indirect calorimetry**.

Indirect calorimetry can be used in conjunction with the direct method in room-size calorimeters. Air is circulated through the chamber of the calorimeter, and the difference in the oxygen and carbon dioxide in the air entering and leaving the chamber is calculated, which indicates the amount of oxygen used and the amount of carbon dioxide produced. Sophisticated, room-size calorimeters can measure both direct and indirect calorimetry (Figure 2.14). The size and complexity of these devices make them unsuitable for measuring energy expenditure for short time intervals, particularly less than 30 minutes. They are more appropriate for the determination of energy expenditure over long periods of time of at least 24 hours and up to several days (Seale, Rumpler, and Moe, 1991).

As with the energy content of food, energy expenditure is expressed as either kilocalories or kilojoules. It can be expressed as an absolute amount

Direct calorimetry: A scientific method of determining energy content of food or energy expenditure by measuring changes in thermal or heat energy.

Calorimeter: A device that measures energy content of food or energy expenditure.

Oxygen consumption ($\dot{V}O_2$): The amount of oxygen used by the body in aerobic metabolism.

Carbon dioxide production ($\dot{V}CO_2$): The amount of carbon dioxide that is produced and eliminated by the body through the lungs.

Indirect calorimetry: A scientific method of determining energy expenditure by measuring changes in oxygen consumption and/or carbon dioxide production.

Figure 2.14 Whole-room calorimeter with direct and indirect calorimetry capability

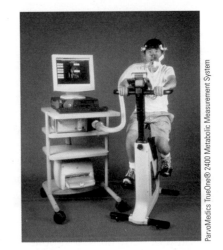

Figure 2.15 An open-circuit metabolic measurement system

expended (for example, 200 kcal or 837 kJ) or as an amount over a given time period, generally a day (for example, 1,500 kcal/day or ~6,300 kJ/day).

Although room-size calorimeters can precisely measure temperature and changes in oxygen consumption and carbon dioxide production over long time intervals, they require a confined space for observation. Studies of exercise activities by direct calorimetry may be further complicated by relatively slow response time of the measurements and the heat produced by exercise equipment, such as a motorized treadmill or friction-braked cycle ergometer (Schoffelen et al., 1997). Room-size calorimeters are another example of a measurement method best suited for research purposes. Fortunately, there are other methods that can be used to determine energy expenditure during exercise and other free-living activities.

Metabolic measurement systems measure energy expenditure through indirect calorimetry without the need to enclose the subject inside the measurement apparatus (Figure 2.15). The most commonly used systems are open circuit, in which the subject breathes in **ambient** room air and is not confined to breathing the air within a sealed environment. These computerized systems are compact and can be contained within a mobile cart, often referred to as a metabolic cart. The cart can easily be positioned beside a hospital bed for medical studies or next to a subject on a treadmill, cycle ergometer, or other type of ergometer for exercise studies. The metabolic cart has a flow meter to measure the amount of air breathed, analyzers for measuring the percentage of oxygen and carbon dioxide in the air, and computer hardware and software to perform the oxygen consumption and carbon dioxide production calculations.

A metabolic measurement system determines energy expenditure indirectly by measuring the amount of oxygen consumed and the amount of carbon dioxide produced during different activities. Oxygen consumption is converted to energy expenditure in kilocalories using conversion factors based upon Respiratory Exchange Ratio (see Chapter 3). Oxygen consumption can be expressed in relative terms as the milliliters of oxygen consumed per kilogram of body weight per minute (ml/kg/min). It can also be expressed as an absolute number of liters of oxygen consumed each minute (L/min). Although it varies slightly depending upon exercise intensity, on average, 1 liter of oxygen consumed is equivalent to 5 kcal of energy expended. Therefore, as shown in the previous example (Figure 2.2), a 150-pound (68.2 kg) cyclist riding at a power output of 180 Watts, with an exercise oxygen consumption of 36.0 ml/kg/min or 2.455 L/min expends 12.28 kcal of energy each minute (2.455 L/min × 5 kcal/L = 12.28 kcal/min). Note this is an expression of *total* oxygen consumption and *total* caloric expenditure, not just the amount of energy expended for the exercise task above **resting oxygen consumption**, which was also shown in Figure 2.2. Measurements of energy expenditure during exercise are commonly expressed as total oxygen consumption, without subtracting the individual's resting oxygen consumption.

Metabolic carts are also commonly used in exercise physiology laboratories to determine the oxygen consumption and energy expenditure response to exercise, in some cases to determine maximal oxygen consumption ($\dot{V}O_{2max}$). Because of their rapid response times and high degree of accuracy, these devices have advantages for these purposes over a room-size calorimeter. However, full metabolic measurement systems are expensive, require trained personnel to operate, and require time for setup, cleanup, and maintenance.

A metabolic measurement system can also be used for measurements of resting oxygen consumption to

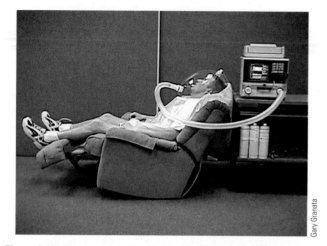

Figure 2.16 Measurement of resting metabolic rate with indirect calorimetry

Figure 2.17 A simplified portable system for measuring resting metabolic rate

To obtain the most accurate measure of resting metabolic rate in healthy adults, before beginning an indirect calorimetry procedure, the subject should:

- Fast at least 5 hours after a meal or snack
- Abstain from alcohol and nicotine for at least 2 hours
- Abstain from caffeine for at least 4 hours
- Abstain from vigorous resistance exercise for at least 14 hours
- Abstain from moderate aerobic or anaerobic exercise for at least 2 hours

C. Compher, D. Frankfield, N. Keim, and L. Roth-Yousey. Evidence Analysis Working Group (2006). Best practice methods to apply to measurement of resting metabolic rate in adults: A systematic review. *Journal of the American Dietetic Association,* 106(6), 881–903.

Figure 2.18 Obtaining the most accurate estimate of resting metabolic rate by indirect calorimetry

restricted to measurements at rest; they are not valid for activity studies. Studies have shown that these devices provide a more accurate determination of RMR than commonly used **prediction equations** (for example, Harris-Benedict) yet do not provide the same degree of accuracy as the full metabolic measurement systems (Melanson et al., 2004; Nieman, Trone, and Austin, 2003).

Although the metabolic measurement systems are much smaller and more portable than the room-size calorimeters, there is a limit to their portability and their use requires the subject to be tethered to the equipment by a breathing tube. There is, therefore, a limitation in using these systems for many sports, exercise activities, and other free-living activities. Advancements in technology have led to the development of small, fully portable versions of metabolic measurement systems that can be carried by an individual in a backpack or vest and can be used to measure energy expenditure during a wide range of activities (Figure 2.19). These devices can either store data for later analysis or transmit the information instantly by telemetry (radio waves) to a receiver attached to a computer for calculation. As

determine **Resting Metabolic Rate (RMR)**. These measurements are performed with a subject resting quietly in a reclining, mostly **supine** position while connected to the metabolic cart through a breathing tube with a mouthpiece, face mask, or ventilated hood (Figure 2.16). Data are collected and averaged for time periods of 30 minutes to several hours to obtain the most accurate results.

Smaller, easier to use, and less expensive systems have been developed specifically to determine resting metabolic rate in a nonresearch setting (Figure 2.17). These devices are much more portable, their use requires less training, maintenance, and troubleshooting, and results can be obtained in less time. To obtain the most accurate results, subjects should follow the guidelines summarized in Figure 2.18 (Compher et al., 2006). Because of their limitations, these devices are

Ambient: In the immediate surrounding area.

Resting oxygen consumption: Measurement of energy expenditure while a person is awake, reclining, and inactive.

Resting Metabolic Rate (RMR): The amount of energy per unit time required by the body to maintain a nonactive but alert state.

Supine: Lying on the back with the face upward and the palms of the hands facing upward or away from the body.

Prediction equation: A statistical method that uses data from a sample population to predict the outcome for individuals not in the sample.

Courtesy of CareFusion

Figure 2.19 A portable metabolic measurement system

the technology and resulting accuracy and reliability of portable metabolic systems improve, the energy expended by athletes in a wide variety of activities can be more accurately determined (Macfarlane, 2001).

Direct and indirect calorimeters are generally used to determine energy expenditure in time periods of minutes, hours, or days. The **Doubly Labeled Water (DLW)** technique allows for the indirect determination of energy expenditure over much longer periods of time (usually 1 to 3 weeks) and allows subjects to participate in their normal, free-living activities. This method makes use of the known pathways for the elimination of hydrogen and oxygen, the two constituents of water (H_2O), from the body. Hydrogen and oxygen are eliminated as water in urine, sweat, respiratory water vapor, and other avenues. Oxygen is also eliminated as carbon dioxide (CO_2). A person who is more active and expends more energy over the observation period will consume more oxygen and will produce and eliminate more carbon dioxide than an individual with a lower energy expenditure.

The amount of hydrogen and oxygen eliminated is measured with radioactively labeled water using two safe and stable isotopes, one for oxygen and one for hydrogen (thus the term *doubly labeled water*). A known amount of each of the stable isotopes, oxygen-18 (^{18}O) and deuterium (^2H) is mixed with water and consumed by the subject. Within a few hours the radioactively

labeled water is distributed throughout the various water compartments in the body. The amount of radioactivity being eliminated as water is measured in urine samples at specified time intervals. The difference in the rates of excretion of ^{18}O and ^2H allows the calculation of long-term carbon dioxide production and energy expenditure. Subjects who are more physically active expend more energy, consume more oxygen, and produce and expel more carbon dioxide, thus increasing the rate of excretion of the radioactively labeled oxygen. Although this method has shown good **validity** and **reliability,** the high cost of the measurement equipment and the high cost of the individual tests have prevented the widespread use of this technique of energy expenditure assessment outside research studies.

Key points

- The SI unit of measure for energy is the Joule (J).

- Because the amount of energy expressed by a Joule is small, the kilojoule (kJ) is commonly used when expressing the energy contained in food. In countries such as the United States, energy in food is expressed as kilocalories (kcal), with 1 kcal being equal to 4.2 kJ.

- The energy content of foods is determined by direct calorimetry in which the heat energy released is measured when they are burned in a bomb calorimeter.

- The energy content of carbohydrates, fats, proteins, and alcohol is expressed as the number of kilocalories per gram of that food, and is approximately 4, 9, 4, and 7 kcal/g, respectively.

- Energy expenditure by individuals can be measured by direct calorimetry in room-size calorimeters or by indirect calorimetry by measuring the amount of oxygen consumed and carbon dioxide produced.

What type of energy expenditure measurement device would be best suited to obtaining accurate measurements while an athlete is performing in his or her sport (for example, playing soccer)?

What is meant by the statement, "Food = fuel = exercise?"

2.3 Concepts of Energy Balance

So far this chapter has examined the ways in which the energy content of food and the energy expended via activity and metabolism are measured and expressed. Many of these measurement techniques are largely research related and focus on the discrete components of energy intake or energy output. Outside the research setting, practical methods that are reasonably accurate are needed to estimate energy intake and energy expenditure. In addition, these two components

Doubly Labeled Water (DLW): A measurement technique for determining energy expenditure over a long time period using radioactively labeled hydrogen and oxygen.

Validity: Ability to measure accurately what was intended to be measured.

Reliability: Ability to reproduce a measurement and/or the consistency of repeated measurements.

Determining the Accuracy of a Device to Measure Daily Energy Expenditure

Accurate assessment of daily energy expenditure is important whether the goal is weight maintenance, weight/fat loss, or lean weight gain. This assessment is further complicated for athletes whose daily energy expenditure can vary dramatically due to their training and competition. Quite often, highly accurate methods used in the laboratory—the "gold standard"—are not suitable for use outside of research studies because of their cost and difficulty, and because they cannot be used by people engaged in their normal daily activities outside of the lab. When equations or devices are developed to predict or estimate energy expenditure their accuracy must be determined by comparing to the gold standard (criterion) measurement before they can be used with confidence.

This research study describes how accurately energy expenditure is measured by a portable device worn on an armband in comparison to the criterion method: St-Onge, M., Mignault, D., Allison, D. B., & Rabasa-Lhoret, R. (2007). Evaluation of a portable device to measure daily energy expenditure in free-living adults. *American Journal of Clinical Nutrition*, 85, 742–749.

What did the researchers do? A large number of subjects were recruited in order to simultaneously measure energy expenditure over a 10-day period using the Doubly Labeled Water (DLW) method and a portable measurement device worn on an armband (HealthWear Bodymedia). The DLW method is considered to be the criterion method for long-term (for example, days) measurement of energy expenditure, and the results determined by the portable armband device were compared to this standard. The device uses a combination of data gathered from a two-axis accelerometer, a heat flux sensor, a galvanic skin response sensor, a skin temperature sensor, and a near body ambient temperature sensor to estimate energy expenditure.

A total of 50 subjects were recruited and tested for this study, representing a broad range of healthy adults: both genders, young and older (age range from 20 to 78), lean to obese, but not morbidly obese (Body Mass Index [BMI] range from 18 to 34), and low to high physical activity level and daily energy expenditure. The researchers excluded the data from 5 of the subjects because they did not comply well with the directions for participation, specifically the amount of time wearing the portable device.

The DLW was administered to the subjects, and the next day they came into the laboratory to give urine samples for the determination of the baseline DLW levels on the first day of the measurement period. They also put on the armband and had their resting metabolic rate (RMR) measured by indirect calorimetry. Subjects were trained in the use of the armband and were given a daily diary to record any problems with the device and document all times the armband was removed. Subjects then wore the armband for the next 10 days as they pursued their normal daily activities. They were instructed to take the armband off only for bathing or any other water activity, for example, swimming. The subjects came back to the laboratory on the 4th day to download data, replace batteries, and to have their daily diaries checked. After 10 days the armbands were turned in and urine samples were collected for the final determination of DLW levels.

What did the researchers find? When determined over the 10-day period by the Doubly Labeled Water method, energy expenditure averaged 2,492 kcal per day. For that same time period, the armband underestimated total daily energy expenditure by 117 kcal/day, approximately a 5 percent difference. Despite this difference, the armband device was more accurate than two commonly used energy expenditure prediction equations.

The energy expenditure associated with physical activity each day was determined by subtracting the resting metabolic rate measured by indirect calorimetry from the total energy expenditure measured by DLW. The researchers found that the armband tended to overestimate RMR and underestimate the energy expenditure associated with physical activity, although the differences were not severe. For example, the RMR measured by indirect calorimetry was 0.96 kcal/min, whereas the armband estimated it as 1.05 kcal/min, which could lead to a difference in estimating resting energy expenditure of approximately 130 kcal per day.

What was the significance of this research study? Researchers, practitioners, and individuals need to have a method to determine energy expenditure with reasonable accuracy that is inexpensive, easy to use, portable, unobtrusive, and can be used during normal daily activities. After testing a portable device worn on an armband, St-Onge et al. (2007) found that the Bodymedia device provided a "reasonable level" of accuracy when compared to DLW, the criterion measurement, particularly for total daily energy expenditure.

Answering the question: Can a portable, wearable device accurately measure energy expenditure? This study provided evidence that a portable device can estimate energy expenditure on a daily basis with "reasonable" accuracy. However, it also illustrated the difficulty in measuring energy expenditure with great precision. Devices can sometimes give people a false sense of a high degree of accuracy and precision. As this study showed, devices such as these, or the commonly used prediction equations, may give results that vary from actual energy expenditure by 100 kcal or more. Practitioners and individuals need to recognize that estimates of energy expenditure are just that—estimates.

Figure 2.20 Energy balance

When "energy in" balances with "energy out," a person's body weight is stable.

must be considered from the perspective of how they relate to each other—the balance of energy intake and output—rather than as distinct entities.

One of the simplest ways to illustrate the concept of energy balance as it relates to the human body is to use a balance scale, shown in Figure 2.20. On one side of the scale is energy intake in the form of food. On the other side of the scale is energy expenditure. The primary influences on energy expenditure are resting metabolism and physical activity in the form of activity, exercise, and sport. A small influence on energy expenditure is the process of digesting food.

Estimating the amount of energy consumed (that is, food) is relatively simple, and the application of that information is easy to understand. In contrast, measuring the amount of energy expended by the body is difficult and often involves complicated prediction equations and formulas. Adding to the difficulty is the use of various terms, each slightly different from the other (for example, basal metabolic rate and resting metabolic rate). Knowledge of terminology and the concepts on which the equations are based helps dispel confusion about estimating energy expenditure. Precise terminology and detailed prediction equations are necessary for research purposes, but practitioners also need to apply energy expenditure information in a way that is easy for consumers to understand. To accomplish this goal, complicated terminology and equations have been simplified for use with individuals outside research settings.

Energy intake is estimated by analyzing daily food and beverage consumption.

Daily energy intake is usually estimated by having individuals self-report their food intake for 1, 3, or 7 days by using a food diary in which they list all the foods and beverages consumed in each 24-hour period. A sample form is found in Appendix E. These foods and beverages are then entered into a computer

program that estimates energy (and nutrient) intake. The greatest source of error with this method is the accurate recording of both the types and amounts of foods and beverages consumed. Recording *all* foods and beverages consumed is a tedious task. It is often difficult to estimate fluid intake. For example, athletes sip sports drinks throughout training or consume water from drinking fountains. Many people also snack frequently and may forget to record snacks in their food diaries. It is estimated that about one-third of adults who record food intake underreport it (Poslusna et al., 2009). Studies that compare reported energy intake with actual energy expenditure (using doubly labeled water) suggest that individuals, including athletes, *underestimate* their energy intake by approximately 15 percent and some as high as 20 percent (Poslusna et al., 2009; Trabulsi and Schoeller, 2001).

A major source of error is underestimating portion size (Subar et al., 2010; Magkos and Yannakoulia, 2003). Many consumers are unfamiliar with standard portion sizes, such as those used on a food label, and others cannot conceptualize or remember the amount consumed when they record intake at a later time. Visual cues, such as photographs or computer imaging, can be helpful. Attention to detail in the recording stages, especially accurate portion size, is needed to reduce the error when estimating energy intake.

Not only do athletes underreport their food intake, but they may consciously or unconsciously undereat during the period when they are keeping a food diary (Burke et al., 2001; Jonnalagadda, Benardot, and Dill, 2000). The act of writing down what is eaten often changes eating behavior. There may be a number of subtle influences and reasons why athletes underreport food intake. For example, in a study of elite female gymnasts, those who underreported their energy intake more often had a higher percent body fat than the gymnasts who more accurately reported their food intake (Jonnalagadda, Benardot, and Dill, 2000). Underreporting is not unique to females; males appear to underreport their intakes as well (Trabulsi and Schoeller, 2001).

Error can also be introduced if athletes, sports dietitians, or others erroneously enter the data into the computer (Braakhuis et al., 2003). Nutrient analysis databases are large, and each food or beverage must be matched as closely as possible to those found in the database. The amount must be entered accurately, but many athletes are unfamiliar with portion sizes, especially those expressed as tablespoons or ounces. Careful attention to recording and computer-coding food and beverage intake will help to reduce the error associated with estimating energy intake. Although the methods used to assess energy intake are known to underestimate actual intake, they are helpful. Food diaries will continue to be used until better methods can be developed.

Components of energy expenditure can be estimated by different methods.

The other side of the energy balance equation is the amount of energy expended, or "energy out." The total amount of energy required by the body over the course of a day is termed **Total Energy Expenditure (TEE)**. Sometimes the term used is Total Daily Energy Expenditure (TDEE). These terms are used interchangeably and are estimates of the amount of energy expended over a 24-hour period.

Total energy expenditure is broken down into three discrete components for study and analysis—metabolism, thermic effect of food, and physical activity. Figure 2.21 illustrates the contribution of each of the three factors for a sedentary individual. On average, resting metabolism makes up approximately 70 percent of TEE, whereas thermic effect of food (~10 percent) and physical activity (~20 percent) are much smaller contributors. There are times when it is beneficial to calculate the individual components of total energy expenditure, particularly resting metabolic rate and energy expended from physical activities. However, much of the time it may be best to measure total energy expenditure or the amount of energy expended in a 24-hour period so that it can be compared to 24-hour food intake.

Basal and resting metabolism. The major component of TEE is basal metabolism. Basal metabolism refers to the energy necessary to keep the body alive at complete rest. Many life-sustaining body processes require energy (that is, ATP). Breathing is an obvious one, but energy is also needed to circulate blood throughout the body, move food through the digestive system, absorb nutrients, conduct nerve signals, maintain body temperature, and so on. In other words, basal metabolism is the minimal energy expenditure compatible with life. It is typically measured in the morning soon after waking after an overnight fast, with the person lying supine at complete rest in a temperature-controlled room. When determined under these conditions in the laboratory the measurement is referred to as **Basal Metabolic Rate (BMR)**.

Most people are studied in a state of wakefulness at different times throughout the day, which requires slightly more energy than the basal level (for example, food must be digested, and body temperature must be maintained in a room where the temperature is not precisely controlled). This is referred to as resting metabolism and its measurement is known as resting metabolic rate (RMR). Although BMR and RMR are often used interchangeably, there is a slight difference between them in measurement methodology and the energy required. Resting metabolism is about 10

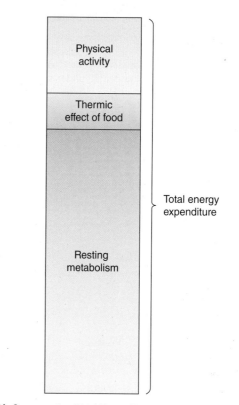

Figure 2.21 Components of Total Energy Expenditure (TEE) for a sedentary individual

For a sedentary individual, resting metabolism typically makes up approximately 70 percent of the day's total energy expenditure. The thermic effect of food accounts for approximately 10 percent, whereas the physical activity component accounts for approximately 20 percent.

percent greater than basal metabolism. Nutrition and exercise professionals should take care to use the terms BMR and RMR correctly. Because the majority of a person's time is spent in an active, awake state, it is more common and accurate outside a research setting to discuss resting metabolism and resting metabolic rate. **Resting Energy Expenditure (REE)** is an equivalent term to resting metabolism and these terms are used interchangeably.

Resting metabolism comprises approximately 70 percent of a sedentary person's total daily energy expenditure, although the total amount and the percentage can vary substantially among individuals. Resting metabolism can be affected by a wide variety of factors, including age, gender, genetics, hormonal

Total Energy Expenditure (TEE): The amount of energy that is required by the body, typically determined over the course of a 24-hour day.

Basal Metabolic Rate (BMR): A measure of the amount of energy per unit of time necessary to keep the body alive at complete rest.

Resting Energy Expenditure (REE): The amount of energy required by the body to maintain a nonactive but alert state.

Table 2.3 Influences on Resting Metabolic Rate

Factor	Influence not under voluntary control	Substantial influence, under some voluntary control	Subtle or temporary influence, under voluntary control
Gender	X		
Genetics	X		
Age	X		
Body size (height)	X		
Thyroid hormones	X		
Starvation (self-restricted food intake)		X	
Amount of fat-free tissue		X	
Exercise			X
Environmental temperature			X
Ascending to high altitude			X
Caffeine			X

changes, body size, body composition (especially the amount of skeletal muscle), exercise, environmental temperature, altitude, food and caffeine intake, and cigarette smoking (Table 2.3). Some of these factors are under voluntary control, so resting metabolism can be altered to some degree.

Gender and genetics are two factors that influence a person's baseline RMR and cannot be changed. When males and females are compared and the differences in skeletal muscle mass are accounted for, it appears that the metabolic rate of females is less than that of males by about 100 kcal/d (Ferraro et al., 1992). Metabolic rate tends to be similar among family members, which suggests that it is genetically influenced (Bouchard, 1989). As with any other physiological process, individuals show large variations within the normal range. Thus some people tend to have higher metabolic rates whereas others have lower metabolic rates, and some of this difference is probably the influence of genetics.

In the healthy individual, two factors are known to decrease RMR: (1) age and (2) starvation (for example, famine or severe dieting). The 1–2 percent decline in RMR seen per decade that is a result of aging is not under voluntary control. Stopping the aging process is not an option! However, severe self-restriction of food, which unfortunately is a method some people use to lose weight, is voluntary.

A starvation state forces the body to adapt. One of the problems with the very low calorie "starvation" diets that people, including some athletes, often employ to lose weight rapidly is the reduction in resting metabolic rate that occurs as a result of a dramatically reduced energy intake. Studies have shown that a starvation state can reduce resting metabolic rate by 20 percent or more (Bray, 1969). This "famine response" can result in an individual expending many fewer kilocalories over the course of the day. Ironically, the reduction in RMR may actually impede weight loss.

Predictably, the body adapts relatively quickly to starvation (that is, within a couple of days); RMR declines and energy expenditure is reduced. The impact of this response is to protect both fat and lean tissue from being substantially reduced. Although a substantial reduction in body fat may be the goal of an individual who adopts a "starvation diet," the physiological response is to protect against fat loss because the body has no mechanism for determining if starvation will be short-term or prolonged. Prolonged starvation can lead to death of the organism so a decline in resting metabolic rate is a survival mechanism.

Perhaps not as obvious is the body's metabolic response upon refeeding (that is, consuming a normal amount of food after food restriction). Naïve dieters may think that RMR increases as soon as they stop the food restriction. In studies of normal-weight men who were starved and then refed, RMR was still lower after 12 weeks of refeeding (Dulloo and Jacquet, 1998; Keys et al., 1950). In other words, the effects of starvation on RMR persisted even after the starvation was no longer present. Of note, the men in this very famous starvation study received 50 percent of their usual food intake.

One of the greatest influences on RMR is the amount of body mass, specifically, fat-free mass (Hulbert and Else, 2004; Leonard et al., 2002). Fat-free mass refers to all the tissues in the body that are not fat (for example, muscle, bones, organs). Fat is a tissue with low metabolic activity. In other words, it does not take much energy to maintain fat stores (**adipose tissue**). In contrast, fat-free tissues are more metabolically active, even when the body is at rest. Remember that resting metabolism refers to the energy necessary to keep the body alive at complete rest. Thus it is logical that the amount of fat-free mass a person has would greatly influence RMR. Studies have shown that people with more fat-free mass (measured in kg) have higher resting metabolic rates than those with less fat-free mass (Leonard et al., 2002). This holds true for both males and females and for all races. Because people can change the amount of fat-free mass by increasing the size of their skeletal muscles through strength training, body composition is a factor that can increase RMR and is under some voluntary control.

Body size can influence RMR (Hulbert and Else, 2004). One of the body's basic functions is the maintenance of body temperature and this is reflected in RMR. Those with a larger body (for example, taller and broader) have more surface area than those with a smaller body. More body heat is lost when the surface area is larger, thus larger body size is associated with a higher RMR when compared to smaller body size.

Most of the factors discussed thus far have an enduring effect. Gender and genetics are not temporary factors, so their influence on RMR is constant. The influence of aging is progressive and permanent. Most people do not dramatically increase or decrease their fat-free mass, although it is possible to gradually increase it over time. Other factors can cause more temporary alterations in RMR. They include hormonal changes, exercise, environmental temperature, altitude, food and caffeine intake, and cigarette smoking.

Thyroid hormones influence many metabolic processes throughout the body, including fat and carbohydrate metabolism and growth. These hormones affect nearly every cell in the body and have a constant effect on resting energy expenditure. Temporary alterations occur if thyroid concentrations fall outside the normal range. When thyroid hormones are elevated above normal concentrations, RMR is abnormally high. Conversely, when thyroid hormones are below normal concentrations, RMR is abnormally low. The inadequate or excessive production of thyroid hormones is a medical condition that needs treatment.

Other hormones cause temporary increases in RMR. For example, at certain times during a woman's menstrual cycle and during pregnancy, hormone concentrations (for example, estradiol and progesterone) fluctuate and RMR may be increased. Exercise also temporarily increases some hormones (for example, epinephrine and norepinephrine) that increase RMR. These are all temporary conditions, and such hormonal changes do not permanently increase resting metabolic rate.

Variations in environment, such as temperature and altitude, can have a measurable effect on RMR. Studies of populations residing in tropic and polar environments show differences in metabolic rate, with RMR increasing when the environmental temperature gets colder. This effect also occurs with seasonal variations in temperature and when individuals travel to warmer or colder climates (Leonard et al., 2002). The range of increase is approximately 3–7 percent. Ascending to high altitudes (for example, above 10,000 feet) also increases resting metabolic rate by an estimated 15–25 percent. This increase is transient, however, and RMR returns to the rate seen at sea level within 1 to 3 weeks (Butterfield et al., 1992; Mawson et al., 2000).

It is clear that exercise results in an increase in energy expenditure. Does exercise also result in a change in the subsequent resting metabolic rate?

Unfortunately, the answer is not entirely clear. Energy expenditure is elevated in the immediate aftermath of exercise as an individual rests and recovers from the exercise bout. The level of this "postexercise energy expenditure" is dependent upon the intensity and duration of the exercise session—harder and/or longer exercise results in metabolism being elevated for a longer period of time during recovery. Although some studies have shown metabolism to be elevated for hours after exercise, most studies show a return to pre-exercise resting levels within 10 to 90 minutes (Molé, 1990). Regular chronic exercise training may also affect resting metabolic rate. The addition of exercise training by obese subjects restricting food intake results in RMR increasing, and the cessation of exercise by trained runners results in a decrease in RMR, suggesting that regular exercise training may slightly increase daily RMR. The interpretation of this research is complicated, however, by the variability in research study designs, methods, and subject populations (Molé, 1990).

Caffeine also increases RMR but for very short periods of time (that is, up to a couple of hours) (LeBlanc et al., 1985). Cigarette smoking has a more dramatic effect, and the change in RMR may be part of the reason that people who quit smoking experience a small weight gain (Perkins et al., 1989). Of course, the health risk of gaining weight is minimal compared to the risk of continuing to smoke.

Although there are many factors that influence resting metabolic rate, there are only a few that are under voluntary control. Many of the factors have subtle rather than dramatic influences. Athletes would be wise to focus on the two factors they can influence that have the strongest effects. The first is to avoid declines in resting metabolic rate by avoiding severe starvation states. The second is to build and maintain skeletal muscle mass at a level that is compatible with their sport. Maintaining skeletal muscle mass is especially important as one ages because some of the decline attributed to aging is due to the loss of skeletal muscle. These two factors have a substantial and long-lasting influence on RMR.

Estimating resting metabolic rate. Resting metabolic rate can be measured directly, but in many cases direct measurement is impractical. Thus RMR is often calculated using a formula (that is, prediction equation). The formulas were originally developed as a way to estimate energy expenditure in hospital patients (Harris and Benedict, 1919). Studies of nonhospitalized individuals showed that these same formulas could reasonably predict RMR in healthy people (Lee and

---

Adipose tissue: Fat tissue. Made up of adipocytes (fat cells).

Mifflin-St. Jeor Equation

Men: RMR (kcal/day) = (9.99 × wt) + (6.25 × ht) − (4.92 × age) + 5

Women: RMR (kcal/day) = (9.99 × wt) + (6.25 × ht) − (4.92 × age) − 161

Where: wt = weight (kg)

ht = height (cm)

age = age (years)

To convert weight in pounds (lb) to weight in kilograms (kg): Divide weight in lb by 2.2 lb/kg.

To convert height in feet (ft) and inches (in) to height in centimeters (cm): 1) determine total height in inches by multiplying height in feet by 12 in/ft and adding remaining inches, and 2) multiplying height in inches by 2.5 cm/in.

Example:

Estimating resting metabolic rate in a 23-year-old Caucasian nonobese female who is 5'6" and weighs 135 lb.

Step 1: Convert weight in pounds to weight in kilograms

135 lb ÷ 2.2 lb/kg = 61.36 kg

Step 2: Convert height in feet and inches to height in centimeters

5 ft × 12 in/ft = 60 in; 60 in + 6 in = 66 in

66 in × 2.5 cm/in = 165 cm

Step 3: Calculate formula

RMR (kcal/day) = (9.99 × wt) + (6.25 × ht) − (4.92 × age) − 161

RMR (kcal/day) = (9.99 × 61.36) + (6.25 × 165) − (4.92 × 23) − 161

RMR (kcal/day) = 612.99 + (6.25 × 165) − (4.92 × 23) − 161

RMR (kcal/day) = 612.99 + 1,031.25 − (4.92 × 23) − 161

RMR (kcal/day) = 612.99 + 1,031.25 − 113.16 − 161

RMR (kcal/day) = 1,644.24 − 113.16 − 161

RMR (kcal/day) = 1,531.08 − 161

RMR (kcal/day) = approximately 1,370

Figure 2.22 Using the Mifflin-St. Jeor equation to estimate resting metabolic rate

Nieman, 1993). Frankenfield, Roth-Yousey, and Compher (2005) reviewed four prediction equations (Harris-Benedict, Mifflin-St. Jeor, Owen, WHO/FAO/ UNU) and found the Mifflin-St. Jeor equation to be the most appropriate equation to use with healthy Caucasian adults, both nonobese and obese. Unfortunately, not enough research has been conducted to validate these prediction equations in nonwhite populations.

Of the four equations examined, the Mifflin-St. Jeor equation most accurately predicted resting metabolic rate, typically within 10 percent of the RMR that had been determined under laboratory conditions. When used with nonobese individuals, 82 percent of the estimates are considered "accurate" with the remaining equally divided between overestimation (as much as 15 percent) and underestimation (as much as 18 percent). When used with obese individuals, the Mifflin-St. Jeor equation is considered accurate about 70 percent of the time. When inaccurate, the estimate tends to be an *underestimate* of RMR (by up to 20 percent), but in some cases RMR may be *overestimated* by 15 percent. It is important to understand that the calculations are just estimates of the amount of kilocalories required to meet resting metabolic needs (Frankenfield, Roth-Yousey, and Compher, 2005).

The example in Figure 2.22 uses the Mifflin-St. Jeor equation (Mifflin et al., 1990) to estimate the RMR of a 23-year-old Caucasian nonobese female who is 5'6" and weighs 135 pounds. All of the mathematical calculations are shown, but in most workplace settings the formula is part of a computer program that performs the calculations once the person's age, weight, and height are entered. The Mifflin-St. Jeor equation estimates RMR for this individual at 1,370 kcal/d. For the purposes of this discussion, assume that this estimate is accurate but recognize that it may be underestimated by up to 18 percent (that is, actual RMR is 1,617 kcal) or overestimated by up to 15 percent (that is, actual RMR is 1,165 kcal).

The Mifflin-St. Jeor equation can be used with confidence with healthy Caucasian adults (the majority of whom are sedentary), but which prediction equation is best for use with athletes? Athletes represent a specialized subpopulation because they typically have a larger amount of fat-free mass (for example, skeletal muscle) than nonathletes. The amount of skeletal muscle influences RMR because muscle tissue has a higher metabolic activity than adipose tissue. Thompson and Manore (1996) compared four equations

Cunningham Equation

RMR = 500 + 22 (FFM)

Where: FFM = fat-free mass (kg)

Example:

Estimating resting metabolic rate in a 23-year-old Caucasian nonobese female endurance athlete who is 5'6" and weighs 135 lb. At 20 percent body fat, she has approximately 49 kg of fat-free mass (135 lb × 0.20 = 27 lb; 135 lb − 27 lb = 108 lb; 108 lb/2.2 lb/kg = 49.1 kg)

RMR (kcal/d) = 500 + 22(FFM)

RMR (kcal/d) = 500 + 22(49.1)

RMR (kcal/d) = 500 + 1,080

RMR (kcal/d) = 1,580

Figure 2.23 Using the Cunningham equation to estimate resting metabolic rate

Simplified Resting Metabolic Rate Formula

Men: 1 kcal per kilogram body weight per hour

Women: 0.9 kcal per kilogram body weight per hour

To convert weight in pounds (lb) to weight in kilograms (kg): Divide weight in lb by 2.2 lb/kg

Example:

Estimating resting metabolic rate in a 23-year-old Caucasian nonobese female who is 5'6" and weighs 135 lb.

Step 1: Convert weight in pounds to weight in kilograms

135 lb divided by 2.2 lb/kg = 61.36 kg

Step 2: Determine kcal used per hour

61.36 kg × 0.9 kcal/kg/hr = 55.22 kcal/h (rounded off to 55 kcal/h)

Step 3: Determine kcal used per day (24 hours)

55 kcal/h × 24 hr = 1,320 kcal per day

Figure 2.24 Using a simplified formula to estimate resting metabolic rate

(Harris-Benedict, Mifflin-St. Jeor, Owen, Cunningham) and found that the Cunningham equation most accurately predicted RMR in their study population, 24 male and 13 female endurance athletes. Although more research is needed, those who work with athletes, particularly endurance athletes, might use the Cunningham equation because it may better account for the higher amount of lean body mass in trained athletes.

The Cunningham equation is shown in Figure 2.23. Note that to use the Cunningham equation the athlete must have an estimate of body composition (see Chapter 11). In this example, the estimate of RMR by the Cunningham equation (1,580 kcal/d) was ~14 percent higher than the Mifflin-St. Jeor equation (1,370 kcal/d).

It may be impractical to calculate the equations discussed above, for example, when body composition, and therefore, fat-free mass (FFM) is unknown. A simple and commonly used estimate of resting energy expenditure for men is 1 kcal per kilogram body weight per hour. For women, RMR may be estimated as 0.9 kcal per kilogram body weight per hour. Although gender and body weight are considered, this simple calculation does not account for age. The calculations, which can be done using a hand calculator, are shown in Figure 2.24. Using this simple formula, the RMR of the 23-year-old, 5'6", 135 lb female is estimated to be 1,320 kcal per day. Comparing the simple formula to the Mifflin-St. Jeor equation, the estimates differed by 50 kcal or approximately 4 percent. The simple formula produced a "ballpark" figure but it likely *underestimated* resting metabolic rate. Compared to the Cunningham equation, the simple formula *underestimated* RMR by 260 kcal or approximately 20 percent.

Until better prediction equations are developed and tested, especially in athletes, practitioners will continue to use some of the methods described above

to estimate resting metabolic rate. Caution must be used in the application of the estimates obtained since RMR may be over- or underestimated by up to 20 percent. Furthermore, RMR is only one of three factors that accounts for total daily energy expenditure so it should not be overemphasized to the exclusion of the other factors.

Thermic Effect of Food (TEF). When food is consumed, it must be mechanically digested and moved through the gastrointestinal tract. Nutrients must also be absorbed and transported across cell membranes from the gut into the blood for distribution throughout the body. All of these processes require energy, and an increase in energy expenditure can be measured in a time period after a meal is consumed. This increase in energy expenditure due to food consumption is called the **Thermic Effect of Food (TEF)**. The TEF can vary slightly depending upon the frequency and energy content of the meals, but generally makes up a fairly small proportion of the day's energy expenditure. Because TEF is a small part of the energy expenditure equation, the emphasis is generally on the two predominant factors, resting metabolism and physical activity.

Direct measurement of TEF can be done in a research laboratory, but there is no practical way to measure it in a nonresearch setting. In many cases, TEF is not calculated or included in estimates of total daily energy expenditure because the increase in energy expenditure occurs for only an hour or two after eating. If TEF is estimated, it is typically calculated by multiplying daily caloric intake by 10 percent. For example, if a person consumed 2,000 kcal in a 24-hour period, the amount that is attributed to TEF is 200 kcal (2,000 × 0.10). Proteins have a greater effect on TEF than carbohydrates, whereas the effect of fats on TEF is very small. The composition of the diet may, therefore, have a small effect on TEF, but the contribution of TEF to total daily energy expenditure is small, especially when compared to resting metabolism.

Physical activity. Any kind of physical activity requires energy expenditure above resting metabolism. **Activities of Daily Living (ADL)**, such as walking or moving about (ambulation), bathing, grooming, dressing, and other personal care activities, result in modest expenditures of energy and may be the majority of energy expended as activity by sedentary people. Daily energy expenditure can be increased by voluntary physical activities such as housework, yard work, climbing stairs, or walking for transportation. Sports and exercise activities can dramatically increase energy expenditure. Whereas physical activity may comprise only approximately 20 percent of the sedentary person's daily energy expenditure, an athlete engaged in intense training dramatically increases daily energy expenditure through exercise.

Of the three constituents of total energy expenditure—resting metabolic rate, thermic effect of food, and physical activity—the one that can be influenced most readily and to the largest extent by the individual is physical activity. Figure 2.25 shows the differences in energy expenditure due to physical activity for a person on different days when sedentary, moderately physically active, or exercising at a vigorous level. Again, this example is based on a 23-year-old nonobese female who is 5'6" and weighs 135 pounds.

On her sedentary day, only a small amount of her daily energy expenditure is a result of activity. If she increases daily physical activity to 30 minutes, the minimum recommended by many health organizations, the physical activity portion of TEE increases by approximately 63 kcal per day. This is based upon walking at a moderate pace for 30 minutes, and represents an increase in daily energy expenditure of approximately 5 percent. If one is calorie counting, this may not appear to be much of a difference, but over time a slight energy deficit can have a noticeable effect. A difference of ~60 kcal each day could account for a change in body weight of about 1 pound every two months (assuming that food intake is the same and no other changes are made in physical activity). Incorporating 30 minutes of moderate exercise daily can lead to a small but steady weight loss of about 6 pounds per year.

The total energy expenditure of the example subject is increased more dramatically if she pursues an exercise activity at a more vigorous intensity and/or for a longer duration. For example, if she runs or jogs for 45 minutes, this activity adds approximately 300 kcal of energy expenditure to the physical activity component of her total energy expenditure. This increases her TEE by nearly 25 percent.

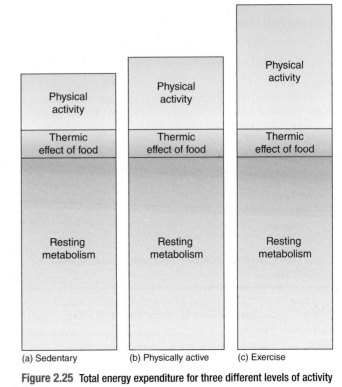

(a) Sedentary   (b) Physically active   (c) Exercise

Figure 2.25 **Total energy expenditure for three different levels of activity**

(a) Sedentary; (b) Moderately physically active—active to the minimum recommended by health organizations, that is, accumulates 30 minutes of moderate-intensity physical activity daily (for example, walking at a moderate pace); (c) Exercise—regular moderate to vigorous activity (for example, running/jogging for 45 minutes).

Daily total energy expenditure is increased if a person engages in physical activity or exercise.

What about the energy requirements of extreme physical labor or very intense exercise training or competition? For relatively short periods of time, humans can participate in activities that require extreme levels of energy expenditure. An example is the Tour de France bicycle race, in which competitors race nearly 2,500 miles in stages over a 3-week period. The energy demand of this race has been estimated to be in excess of 5,300 kilocalories per day (Saris et al., 1989), representing a 3,000 kcal difference in TEE from that of a sedentary person! It is thought that these extreme levels of energy expenditure can be sustained for relatively short periods (for example, several weeks) before there are performance and health consequences.

Estimating daily energy expenditure through physical activity. Although it is more accurate to have daily energy expenditure assessed by one of the direct or indirect calorimetry methods described previously in the chapter (for example, whole-room calorimeter or doubly labeled water), in most cases these assessments cannot be performed because of a lack of time, money, or access to the equipment. A common and practical method is to use self-reported physical activity logs or questionnaires to keep track of activities, including household, occupational, transportation, sport, exercise, and leisure (see Appendix F). Once the type and amount of activity has been determined, it can be entered into a computer program that can calculate total daily energy expenditure.

Most computer programs, including the one used with this text, allow an individual to input all physical activities for a 24-hour period. The energy values given for each activity and for the 24-hour period include the energy needed for metabolism. In other words, resting metabolic rate is not a separate calculation. From a practical point of view, it is not necessary to separate the amount of energy used to support metabolism from the amount used to support physical activity, especially when total energy expenditure is being compared to total energy intake.

The most variable aspect of daily energy expenditure is the amount of energy expended through physical activity. The same individual may have markedly different activity patterns from day to day, as well. Over a weekend, a person may have a day with a large amount of physical activity by participating in recreational sports, working in the yard, or other leisure-time activities. The next day may be largely sedentary, reading the paper, lying on the couch watching TV, and taking a nap. An athlete pursuing a "hard-easy" training program may have a day of heavy, intense training followed by a day of rest or low-intensity training. Therefore, keeping activity logs for more than one day will add to the accuracy of the assessment.

Estimating energy expended by a single physical activity. People often want to know the amount of energy expended when they perform a specific physical activity. For example, those who work out in health clubs may ask, "How many calories do I burn when I lift weights for an hour?" Or, "How many calories do I burn if I walk for 30 minutes?" In these cases, the person wants to know only the energy expended by one activity, not the energy expended by all the physical activities done in a day. How is the energy expended by a single activity determined? As discussed earlier in this chapter, energy expenditure can be measured by direct or indirect calorimetry, but these measurement methods are cumbersome, expensive, and time-consuming. There are a number of methods to estimate energy expenditure of various physical activities with acceptable degrees of accuracy.

One way is to enter a single activity into a computer program or online physical activity calculator, however not everyone has access to such tools. Another way is to use a reference such as the Compendium of Physical Activities (Ainsworth et al., 1993, 2000; see Appendix G). This compendium provides a coding scheme to describe each activity, both generally (for example, walking) and specifically (for example, walking at 4 miles per hour). The Compendium was developed more for providing

an activity classification system than as a means of determining specific energy expenditure levels, but it is sometimes used in this way. Whereas the energy expenditure levels of many of the listed activities are based upon studies using indirect calorimetry, the amount of energy expended for a number of the activities has been estimated from activities with similar movement patterns. Therefore, caution should be used when using the energy expenditure values in the Compendium.

The Compendium expresses the energy expended in **metabolic equivalents (MET)**, not kilocalories. One MET is equal to the energy expenditure of an *average* resting metabolic rate. Activity intensity expressed in MET, therefore, is a multiplication of energy expenditure at rest. For example, a task that requires 3 MET requires an energy expenditure level three times that of resting. How is this converted to the more familiar kilocalories of energy expenditure?

Returning to the example of the 23-year-old female, assume she is a sedentary person and wants to increase her daily physical activity to 30 minutes of walking at a moderate pace. If she walks at a moderate pace, the Compendium suggests a MET level of 3.3. Her resting metabolic rate has been previously determined to be ~55 kcal per hour (Figure 2.24). At 3.3 MET she would expend approximately 181.5 kcal per hour (55 kcal/hr × 3.3). But she walks for only 30 minutes, or 0.5 hours, so she expends about 91 kcal via this activity (181.5 kcal/h × 0.5 h).

This ~91 kcal estimate represents the total energy expenditure for that 30 minutes of exercise and includes her resting energy expenditure. If she wants to know how much energy was expended just for the activity of walking, she must subtract the estimate of her resting energy rate for those 30 minutes. As mentioned above, the estimate of RMR for this woman is 55 kcal/h or 27.5 kcal for 30 minutes. By subtracting 27.5 kcal from 90.75 kcal, it can be estimated that she expended approximately 63 kcal by walking at a moderate pace for 30 minutes (90.75 kcal − 27.5 kcal = 63.25 kcal). Many charts that list the energy expended through activity or exercise include RMR, so unless RMR is accounted for, the figure *overestimates* the amount of energy expended via the movement alone.

Thermic Effect of Food (TEF): The amount of energy required by the body to digest and absorb food.

Activities of Daily Living (ADL): Personal care activities (for example, bathing, grooming, dressing) and the walking that is necessary for day-to-day living.

Metabolic equivalents (MET): Level of energy expenditure equal to that measured at rest. 1 MET = 3.5 ml/kg/min of oxygen consumption.

Estimated Energy Requirement (EER): The estimated amount of energy that needs to be consumed to maintain the body's energy balance.

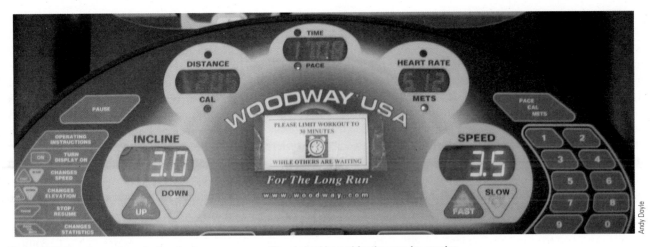

Figure 2.26 Exercise equipment often displays energy expenditure in kcal burned for the exercise session.

Energy expenditure information is commonly provided, usually in the form of kilocalories burned, by various types of exercise equipment such as treadmills and by fitness-tracking devices used by individuals when they exercise. For example, it is common to find a portion of the display on a treadmill in a fitness center that displays "calories burned" during an exercise session (Figure 2.26). These devices use the exercise intensity setting (for example, current speed and grade of the treadmill) and estimate the rate of caloric expenditure based upon general equations for that mode of exercise. Total caloric expenditure is calculated using the total time of the exercise session. Accuracy of this estimation for energy expenditure of weight-bearing exercises can be enhanced when the exercise device allows for the input of the exercising person's body weight.

As technological advances have both improved and become less expensive, a wide array of devices has been developed for individuals to track their physical activity and fitness activities. Through the use of Global Positioning Systems (GPS) or accelerometers, these devices can determine distance traveled, time of the exercise session, and speed of movement, and through these measures energy expenditure can be estimated. Some of the more sophisticated devices may also include elevation change, which can add to the accuracy of energy expenditure estimation. Examples include bike computers for cyclists and motion sensors attached to shoes for runners.

A number of devices have been developed to be worn throughout the day to estimate total daily energy expenditure in addition to individual exercise sessions. These devices may couple the information gained about body movement from an accelerometer with other information such as sweating response and body temperature changes to estimate and track daily

Figure 2.27 Energy expenditure can be estimated using portable fitness monitors.

energy expenditure. Research studies indicate that these devices have reasonable accuracy in estimating energy expenditure (St-Onge et al., 2007).

Athletes, coaches, and trainers often use heart rate as an indication of exercise intensity, and therefore as an indirect prediction of energy expenditure. Heart rate is easy to measure, either with an electronic heart rate monitor or using the fingertips and a watch to count the pulse rate. It provides a reflection of energy expenditure because heart rate has a mostly linear relationship with oxygen consumption. The specific relationship of heart rate to oxygen consumption is highly individualized, however, and standardized heart rates cannot be used accurately to predict energy expenditure. For example, one person exercising at a heart rate of 120 beats per minute (bpm) might be expending 180 kcal per hour, whereas another person at that same rate of energy expenditure might have a heart rate of 130 bpm.

Determination of resting metabolic rate and energy expenditure associated with an exercise session:
Subject: 20 year-old male runner, weight = 139 lb (63 kg), height = 5'7" (170 cm)

1. How many kilocalories does this person burn if he lies on the couch and watches TV for 1 hour (use the simplified formula in Figure 2.24)?

2. How many kilocalories does this person burn if he runs for 1 hour at a 9.0 minute per mile pace (use the Compendium of Physical Activities in Appendix G)?

3. How many additional kilocalories does he burn by running for an hour instead of watching TV?

The Internet café

Where do I find information about estimating energy intake and expenditure?

Shape Up America! has three online calculators available to estimate energy intake and expenditure: Physical activity calculator (**http://www.shapeup.org/interactive/phys1.php**), resting metabolic rate calculator (**http://www.shapeup.org/interactive/rmr1.php**), and meal and snack calculator (**http://www.shapeup.org/interactive/msc1.php**).

Shape Up America! is a nonprofit organization committed to raising awareness of obesity as a health issue. The calculators are part of the effort to educate people about attaining and maintaining a healthy body weight throughout life.

Estimated Energy Requirement is a daily balance of energy intake and expenditure.

Energy balance is simply defined as "Energy in = Energy out." It is important to be able to estimate how much energy is being consumed, how much is being expended, and how these two estimates compare, but as this chapter has shown, it is difficult to measure energy. Most nutritional analysis computer programs will calculate an **Estimated Energy Requirement (EER)**, which is the average dietary energy intake that will maintain energy balance. EER is based on age, gender, weight, height, and physical activity, all of which are self-reported. Determining EER helps to answer the often-asked question: "How many calories do I need to consume each day?"

The answer to that question depends, to a large degree, on activity level. Computerized nutritional analysis programs ask the user to designate a category that best reflects usual physical activity (for example, sedentary, light, moderate, heavy, exceptional activity) and uses that and other information to calculate EER. However, not everyone has access to or time to enter information into the computer. Thus a very simple method has been developed to estimate daily energy intake in adult males and nonpregnant females based on gender, weight in kilograms, and level of physical activity (Table 2.4). This very simple estimate is the

Keeping it in perspective

Food = Fuel = Exercise

The word *energy* typically has a positive connotation. It denotes both power and vigor and is the opposite of fatigue. The body uses energy, in the form of ATP, to fuel all of its activities from sleep to vigorous exercise. Without the consumption of food and the transformation of the energy contained in that food into biologically useful energy, life would not be possible. Curiously, the measure of energy, or calories, often has a negative connotation. Athletes should think of energy and calories in positive terms because energy is needed to fuel activity. Many sports nutritionists use the words *energy intake* instead of *calorie intake* to help athletes understand the connection between food, fuel, and exercise.

The "energy in" side of the energy balance equation receives much attention (that is, the amount of food eaten), but it is the "energy out" side of the equation that needs first consideration. In other words, the amount of physical activity should determine the amount of food consumed. Athletes who wish to maintain their body composition must consume enough food to match their energy expenditure. Those who wish to lose body fat must still consider daily energy output and reduce food intake somewhat, forcing the body to use some of its stored energy to fuel activity. However, too severe of an energy restriction can be counterproductive, particularly if resting metabolic rate is reduced or a high volume of training cannot be sustained. When viewed from the perspective of activity and exercise, the energy (calories) contained in food has a positive connotation. Food = fuel = exercise.

Table 2.4 Estimating Daily Energy Need for the General Population

Level of activity	Energy expenditure—females (kcal/kg/d)	Energy expenditure—males (kcal/kg/d)
Sedentary (Activities of daily living [ADL] only)	30	31
Light activity (ADL + walking 2 miles per day or the equivalent)	35	38
Moderate activity (ADL + moderate exercise 3 to 5 days per week)	37	41
Heavy activity (ADL + moderate to heavy exercise on most days)	44	50
Exceptional activity (ADL + intense training)	51	58

Legend: kcal/kg/d = kilocalorie per kilogram body weight per day

These "ballpark" figures can be used to quickly calculate an estimate of daily energy need, taking into account gender and activity.

Note: These estimates have generally been derived from surveys and clinical observations.

least accurate way to estimate an individual's daily energy requirement; however, it is often used because of its simplicity. It is truly a "ballpark" figure.

It is estimated that sedentary people use about 30 kilocalories of energy per kilogram body weight daily. For example, a 110 lb (50 kg) sedentary female may not need more than 1,500 kcal daily to maintain energy balance (30 kcal/kg × 50 kg = 1,500 kcal). A sedentary 165 lb (75 kg) male is estimated to need about 2,325 kcal per day to maintain energy balance (31 kcal/kg × 75 kg = 2,325 kcal). As Table 2.4 and Figure 2.25 clearly show, as activity increases, the amount of energy expended increases. Those who are lightly active, generally defined as activities of daily living plus walking 2 miles a day or the equivalent, need between 35 (women) and 38 (men) kcal/kg/d. Moderate, heavy, and exceptional activity take into account an increasing intensity and/or duration of activity, and the estimate of the amount of energy needed to maintain energy balance increases proportionately. Although these estimates are just ballpark figures, they can help an individual quickly determine the amount of energy (that is, caloric intake) needed each day to maintain energy balance.

There are also "ballpark" figures that can be used to quickly calculate an estimate of daily energy need for male and female athletes (Table 2.5). Like the estimates used for the general population, these figures have been derived from surveys and clinical observations. For each activity level, examples are given to help athletes accurately choose an activity category that reflects their level of training.

Of course not all people, including athletes, are or want to be in energy balance. The information obtained from food and activity diaries and the estimates of energy intake and expenditure that are based on these records or other methods of measuring energy expenditure will be useful information for athletes and practitioners as weight, body composition, training, and performance goals are set. The application of the concepts of "energy in" and "energy out" to maintaining or changing weight and body composition is covered later in this text, specifically in Chapters 10–13.

Key points

- Energy intake is determined by recording and analyzing food intake, which must be done accurately for the results to be meaningful.

- The major influences on energy expenditure are resting metabolism and physical activity.

- Resting metabolism accounts for approximately 70 percent of a sedentary person's daily energy expenditure and can be influenced by a variety of factors. The two major factors under voluntary control are avoiding a starvation diet and building and maintaining skeletal muscle mass.

- Resting metabolic rate and total energy expenditure can be measured by direct or indirect calorimetry, and/or estimated by using prediction equations.

- The factor contributing to daily energy expenditure that can be influenced the most by voluntary behavior is physical activity.

- Energy balance refers to the relationship between the amount of energy consumed and the amount of energy expended, usually on a daily (24-hour) basis.

- Athletes should first consider their energy expenditure and then plan their energy intake to meet the demands and goals of their training and competition.

What are the advantages and disadvantages of the various methods for estimating energy expenditure for physical activities such as walking or jogging?

Table 2.5 Estimating Daily Energy Need for Male and Female Athletes

Level of activity	Example of activity level	Energy expenditure—females (kcal/kg/d)	Energy expenditure—males (kcal/kg/d)
Sedentary (little physical activity)	During an acute recovery from injury phase	30	31
Moderate-intensity exercise 3–5 days/week or low-intensity and short-duration training daily	Playing recreation tennis (singles) 1–1½ hr every other day; practicing baseball, softball, or golf 2½ hr daily, 5 days/week	35	38
Training several hours daily, 5 days/week	Swimming 6,000–10,000 m/day plus some resistance training; Conditioning and skills training for 2–3 hr/day such as soccer practice	37	41
Rigorous training on a near daily basis	Performing resistance exercise 10–15 hr/week to maintain well-developed skeletal muscle mass, such as a bodybuilder during a maintenance phase; Swimming 7,000–17,000 m/day and resistance training 3 days/week, such as an elite swimmer; typical training for college, professional and elite football, basketball, and rugby players	38–40	45
Extremely rigorous training	Training for a triathlon (nonelite triathlete)	41	51.5
	Running 15 mi (24 km)/day or the equivalent (elite triathlete or equivalent)	50 or more	60 or more

Legend: kcal/kg/d = kilocalorie per kilogram body weight per day; hr = hour; m = meter; mi = mile; km = kilometer

Adapted from Macedonio, M. A., & Dunford, M. (2009). *The Athlete's Guide to Making Weight: Optimal Weight for Optimal Performance*. Champaign, IL: Human Kinetics.

Post-Test   Reassessing Knowledge of Energy

Now that you have more knowledge about energy read the following statements and decide if each is true or false. The answers can be found in Appendix O.

1. The body creates energy from the food that is consumed.

2. The scientific unit of measure of energy is the calorie.

3. A person's resting metabolic rate can change in response to a variety of factors such as age, food intake, or environmental temperature.

4. Physical activity is responsible for the largest amount of energy expended during the day for the average adult in the United States.

5. The energy source used by all cells in the body is adenosine triphosphate (ATP).

Summary

Energy exists in a number of forms and can be transferred from one type to another. The energy contained in food can be converted in the body to other forms of energy for immediate use or storage. The primary form of energy used by the body is in chemical form, in the high-energy phosphate compound adenosine triphosphate (ATP). Skeletal muscles directly use the energy from ATP to produce force for exercise and physical activity, and the purpose of the major energy systems of the body is to replenish ATP.

The energy contained in food can be measured through direct calorimetry by measuring the heat given off when the food is burned in a bomb calorimeter. Similarly, the energy expended by the body can be determined through direct calorimetry by measuring the heat emitted. Energy expenditure can also be determined indirectly by measuring oxygen consumed and carbon dioxide produced using a variety of techniques. Units of energy include kilojoules, kilocalories, and Calories.

The energy balance equation, "Energy in = Energy out," is one of the simplest ways to illustrate energy balance. Food is the only factor on the "energy in" side of the equation. The "energy out" side includes resting metabolism, thermic effect of food, and physical activity. Although many factors influence resting metabolism, the two factors under voluntary control that have the strongest influence are self-starvation (for example, severe dieting) and building and maintaining skeletal muscle mass through strength training. The thermic effect of food accounts for only a small amount of the total energy expended in a day. Physical activity is the factor on the "energy out" side that is the most variable and is under the most voluntary control.

Resting metabolic rate can be estimated by using appropriate prediction equations. Computer-based nutritional analysis programs provide reasonable estimates of the amount of energy consumed, the amount expended through physical activity, and the total amount of energy needed daily to maintain energy balance. Very simple calculations are sometimes used in practice settings. In such cases, some accuracy is sacrificed for ease of use.

Self-Test

Multiple-Choice

The answers can be found in Appendix O.

1. The best definition of energy is the:
   a. absence of physical or mental fatigue.
   b. ability to do work.
   c. generation of tension by contracting muscle.
   d. thermic effect of movement.

2. If a food contains 350 Calories, how many kilocalories does it contain?
   a. 148
   b. 350
   c. 1,480
   d. 3,250

3. Indirect calorimetry works on the principle that:
   a. a rise in body temperature reflects the amount of energy expended.
   b. the body heat that is produced is proportional to the energy expended.
   c. oxygen consumption and carbon dioxide production are related to energy expenditure.
   d. energy expenditure can be predicted if respiration and perspiration are carefully measured.

4. The "energy out" factor that can be increased to the greatest degree is:
   a. resting metabolism.
   b. thermic effect of food.
   c. physical activity.

5. Which of the following statements is NOT true regarding intake when athletes are keeping a food record?
   a. Food intake is consciously reduced.
   b. Food intake is unconsciously reduced.
   c. Food intake is difficult to record due to frequent snacking.
   d. Food intake is underreported by females but not by males.

Short Answer

1. Name at least three forms of energy and give examples of how energy might be transferred between these different forms.

2. Is ATP ever used up completely by exercising muscle? How does the muscle respond to reduced ATP concentrations?

3. What are the commonly used units of measure for energy in the field of nutrition, and how do they relate to the SI unit of measure for energy?

4. How is the energy content of food determined?

5. How is energy expenditure measured by direct calorimetry?

6. What is meant by indirect calorimetry and how is energy expenditure measured by this method?

7. What is the energy balance equation? What are the three components of "energy out" and what is the magnitude of their contribution to daily energy expenditure?

8. What is resting metabolic rate? Which factors influence RMR? Which have the greatest influence?

9. What effect does severe food restriction have on resting metabolic rate? Why? What happens to resting metabolism after food is reintroduced? Why?

10. How accurate are prediction equations for estimating resting metabolic rate?

11. What is the most accurate way of measuring total daily energy expenditure? What are the most practical ways in a university setting? In a health and fitness club?

Critical Thinking

1. You have a client who meets his afternoon hunger and craving for sweets by getting a candy bar from the vending machine in the break room at his office. He justifies this dietary choice by adjusting his exercise sessions to "burn" the 280 kilocalories contained in the candy bar, yet he finds that he has been slowly gaining weight. How would you explain his weight gain despite his attempts to balance his energy intake and expenditure?

References

Ainsworth, B. E., Haskell, W. L., Leon, A. S., Jacobs, D. R., Jr., Montoye, H. J., Sallis, J. F., et al. (1993). Compendium of physical activities: Classification of energy costs of human physical activities. *Medicine and Science in Sports and Exercise, 25*(1), 71–80.

Ainsworth, B. E., Haskell, W. L., Whitt, M. C., Irwin, M. L., Swartz, A. M., Strath, S. J., et al. (2000). Compendium of physical activities: An update of activity codes and MET intensities. *Medicine and Science in Sports and Exercise, 32*(9), S498–S516.

Atwater, W. O., & Benedict, F. G. (1905). *A respiration calorimeter with appliances for the direct determination of oxygen.* Washington, DC: Carnegie Institute of Washington.

Atwater, W. O., & Rosa, E. B. (1899). *Description of a new respiration calorimeter and experiments on the conservation of energy in the human body* (Bulletin 63). Washington, DC: U.S. Government Printing Office, Office of Experiment Stations.

Bouchard, C. (1989). Genetic factors in obesity. *Medical Clinics of North America, 73*(1), 67–81.

Braakhuis, A. J., Meredith, K., Cox, G. R., Hopkins, W. G., & Burke, L. M. (2003). Variability in estimation of self-reported dietary intake data from elite athletes resulting from coding by different sports dietitians. *International Journal of Sport Nutrition and Exercise Metabolism, 13*(2), 152–165.

Bray, G. A. (1969). Effect of caloric restriction on energy expenditure in obese patients. *Lancet, 2*(7617), 397–398.

Burke, L. M., Cox, G. R., Cummings, N. K., & Desbrow, B. (2001). Guidelines for daily carbohydrate intake: Do athletes achieve them? *Sports Medicine, 31*(4), 267–299.

Butterfield, G. E., Gates, J., Fleming, S., Brooks, G. A., Sutton, J. R., & Reeves, J. T. (1992). Increased energy intake minimizes weight loss in men at high altitude. *Journal of Applied Physiology, 72*(5), 1741–1748.

Compher, C., Frankenfield, D., Keim, N., Roth-Yousey, L., & Evidence Analysis Working Group. (2006). Best practice methods to apply to measurement of resting metabolic rate in adults: A systematic review. *Journal of the American Dietetic Association, 106*(6), 881–903.

Dulloo, A. G., & Jacquet, J. (1998). Adaptive reduction in basal metabolic rate in response to food deprivation in humans: A role for feedback signals from fat stores. *American Journal of Clinical Nutrition, 68*(3), 599–606.

Ferraro, R., Lillioja, S., Fontvieille, A. M., Rising, R., Bogardus, C., & Ravussin, E. (1992). Lower sedentary metabolic rate in women compared with men. *Journal of Clinical Investigation, 90*(3), 780–784.

Frankenfield, D., Roth-Yousey, L., & Compher, C. (2005). Comparison of predictive equations for resting metabolic rate in healthy nonobese and obese adults: A systematic review. *Journal of the American Dietetic Association, 105*(5), 775–789.

Harris, J. A., & Benedict, F. G. (1919). A biometric study of basal metabolism in man (Rep. No. 279). Washington, DC: Carnegie Institute of Washington.

Hirvonen, J., Rehunen, S., Rusko, H., & Harkonen, M. (1987). Breakdown of high-energy phosphate compounds and lactate accumulation during short supramaximal exercise. *European Journal of Applied Physiology and Occupational Physiology, 56*(3), 253–259.

Hulbert, A. J., & Else, P. L. (2004). Basal metabolic rate: History, composition, regulation, and usefulness. *Physiological and Biochemical Zoology, 77*(6), 869–876.

Jonnalagadda, S. S., Benardot, D., & Dill, M. N. (2000). Assessment of under-reporting of energy intake by elite female gymnast. *International Journal of Sport Nutrition and Exercise Metabolism, 10*(3), 315–325.

Keys, A., Brozek, J., Henschel, A., Mickelsen, O., & Taylor, H. L. (1950). *The biology of human starvation.* Minneapolis: University of Minnesota Press.

LeBlanc, J., Jobin, M., Cote, J., Samson, P., & Labrie, A. (1985). Enhanced metabolic response to caffeine in exercise-trained human subjects. *Journal of Applied Physiology, 59*(3), 832–837.

Lee, R. D., & Nieman, D. C. (1993). *Nutritional assessment.* Dubuque, IA: Brown & Benchmark.

Leonard, W. R., Sorensen, M. V., Galloway, V. A., Spencer, G. J., Mosher, M. J., Osipova, L., & Spitsyn, V. A. (2002). Climatic influences on basal metabolic rates among circumpolar populations. *American Journal of Human Biology, 14*(5), 609–620.

Lieber, C. S. (2003). Relationships between nutrition, alcohol use, and liver disease. *Alcohol Research & Health, 27*(3), 220–231.

Macedonio, M. A., & Dunford, M. (2009). *The Athlete's Guide to Making Weight: Optimal Weight for Optimal Performance.* Champaign, IL: Human Kinetics.

Macfarlane, D. J. (2001). Automated metabolic gas analysis systems: A review. *Sports Medicine, 31*(12), 841–861.

Magkos, F., & Yannakoulia, M. (2003). Methodology of dietary assessment in athletes: Concepts and pitfalls. *Current Opinion in Clinical Nutrition and Metabolic Care, 6*(5), 539–549.

Mawson, J. T., Braun, B., Rock, P. B., Moore, L. G., Mazzeo, R., & Butterfield, G. E. (2000). Women at altitude: Energy requirement at 4,300 m. *Journal of Applied Physiology, 88*(1), 272–281.

Melanson, E. L., Coelho, L. B., Tran, Z. V., Haugen, H. A., Kearney, J. T., & Hill, J. O. (2004). Validation of the BodyGem hand-held calorimeter. *International Journal of Obesity, 28*(11), 1479–1484.

Mifflin, M. D., St. Jeor, S. T., Hill, L. A., Scott, B., Daugherty, S. A., & Koh, Y. O. (1990). A new predictive equation for resting energy expenditure in healthy individuals. *American Journal of Clinical Nutrition, 51*(2), 241–247.

Molé, P. A. (1990). Impact of energy intake and exercise on resting metabolic rate. *Sports Medicine, 10*(2), 72–87.

Nieman, D. C., Trone, G. A., & Austin, M. D. (2003). A new handheld device for measuring resting metabolic rate and oxygen consumption. *Journal of the American Dietetic Association, 103*(5), 588–592.

Perkins, K. A., Epstein, L. H., Stiller, R. L., Marks, B. L., & Jacob, R. G. (1989). Acute effects of nicotine on resting metabolic rate in cigarette smokers. *American Journal of Clinical Nutrition, 50*(3), 545–550.

Poslusna, K., Ruprich, J., de Vries, J. H., Jakubikova, M., & van't Veer, P. (2009). Misreporting of energy and micronutrient intake estimated by food records and 24 hour recalls, control and adjustment methods in practice. *British Journal of Nutrition, 101*(Suppl. 2), S73–S85.

Saris, W. H. M., van Erp-Baart, M. A., Brouns, F., Westerterp, K. R., & ten Hoor, F. (1989). Study on food intake and energy expenditure during extreme sustained exercise: The Tour de France. *International Journal of Sports Medicine, 10*(1), S26–S31.

Schoffelen, P. F., Westerterp, K. R., Saris, W. H., & ten Hoor, F. (1997). A dual-respiration chamber system with automated calibration. *Journal of Applied Physiology, 83*(6), 2064–2072.

Seale, J. L., Rumpler, W. V., & Moe, P. W. (1991). Description of a direct-indirect room-sized calorimeter. *The American Journal of Physiology, 260*(2, Pt. 1), E306–E320.

St-Onge, M., Mignault, D., Allison, D. B., & Rabasa-Lhoret, R. (2007). Evaluation of a portable device to measure daily energy expenditure in free-living adults. *American Journal of Clinical Nutrition, 85,* 742–749.

Subar, A. F., Crafts, J., Zimmerman, T. P., Wilson, M., Mittl, B., Islam, N. G., et al. (2010). Assessment of the accuracy of portion size reports using computer-based food photographs aids in the development of an automated self-administered 24-hour recall. *Journal of the American Dietetic Association, 110*(1):55–64.

Thompson, J., & Manore, M. M. (1996). Predicted and measured resting metabolic rate of male and female endurance athletes. *Journal of the American Dietetic Association, 96*(1), 30–34.

Trabulsi, J., & Schoeller, D. A. (2001). Evaluation of dietary assessment instruments against doubly labeled water, a biomarker of habitual energy intake. *American Journal of Physiology, Endocrinology and Metabolism, 281*(5), E891–E899.

World Health Organization. (1991). *Energy and protein requirements.* World Health Organization Technical Report Series 724. Geneva, Switzerland: Author. Retrieved [December 16, 2010] from http://www.fao.org/docrep/003/AA040E/AA040E00.HTM

Energy Systems and Exercise

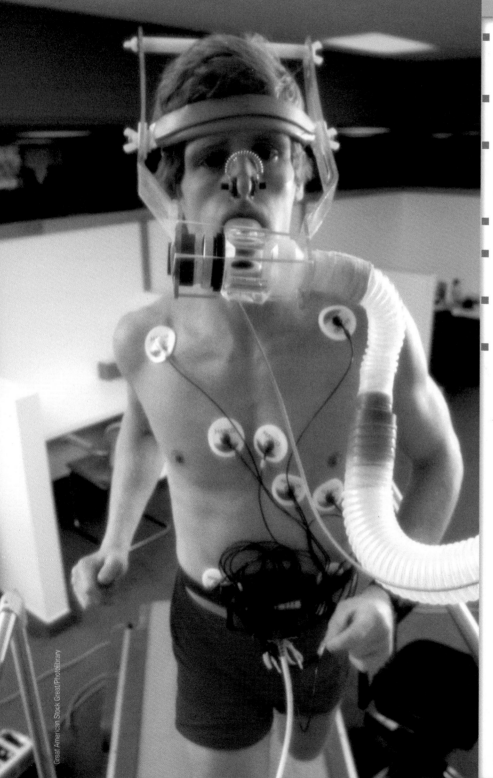

Learning Plan

- **Describe** the characteristics of the creatine phosphate, anaerobic glycolysis, and oxidative phosphorylation energy systems.
- **Evaluate** creatine supplements based on legality, ethics, safety, and effectiveness.
- **Compare** and contrast the three energy systems and give examples of physical activities, exercise, or sports in which each is the predominant energy system.
- **Outline** the process of carbohydrate, fat, and protein (amino acid) oxidation.
- **Explain** the response of oxygen consumption to steady-state, submaximal exercise.
- **Explain** the concept of maximal oxygen consumption and become familiar with relative values of $\dot{V}O_{2max}$.
- **Explain** the concept of respiratory exchange ratio and describe the process of determining the percentage of fat and carbohydrate fuel utilization.

Three major energy systems provide energy for exercise.

Pre-Test | **Assessing Current Knowledge of Energy Systems and Exercise**

Read the following statements and decide if each is true or false.

1. The direct source of energy for force production by muscle is ATP.

2. Creatine supplements result in immediate increases in strength, speed, and power.

3. Lactate is a metabolic waste product that causes fatigue.

4. The aerobic energy system is not active during high-intensity anaerobic exercise.

5. At rest and during low levels of physical activity, fat is the preferred source of fuel for the aerobic energy system.

In the previous chapter, the high-energy phosphate compound adenosine triphosphate (ATP) was shown to be the immediate source of energy in the body for activity and exercise. This chemical compound can store potential energy in its phosphate bonds, and when broken down to ADP (adenosine diphosphate), can release energy to be used for a wide variety of tasks such as force production by muscle. ATP is found in all cells in the body and can be thought of as the body's primary energy currency. This important energy compound is found in surprisingly small amounts in the body and can be used at very high rates in some cases, such as during high-intensity exercise. Because of the importance of ATP, the body must have ways to restore ATP after it has been used. The process of resynthesizing ATP from ADP is called rephosphorylation. This chapter reviews the three major energy systems used for rephosphorylation of ATP, the interactions of these systems, and how exercise intensity and duration influences the utilization of each energy system.

3.1 Overview of Energy Systems

ADP is rephosphorylized to form ATP.

As discussed in Chapter 2, the chemical breakdown of ATP to ADP is an exergonic reaction; that is, it releases energy (Figure 3.1) for the body to perform work such as force production by skeletal muscle. The process of rephosphorylation of ADP back to ATP therefore is an endergonic reaction; it requires the input of energy, which is then stored in the phosphate bond that is re-established to form ATP (Figure 3.2). One objective of this chapter is to examine the various bioenergetic

Figure 3.1 Hydrolysis of ATP

ATP (adenosine triphosphate) is split chemically, leaving adenosine diphosphate (ADP) and inorganic phosphate (Pi) and releasing energy in an exergonic reaction.

processes (for example, converting food to energy in the body) used to resynthesize ATP. In other words, the focus is on the connections between food, the energy that is found in food, and the transformation of that energy into ATP, which is the body's source of energy during exercise and at rest.

Sports and exercise provide an excellent model to study the energy systems that lead to the restoration of ATP concentrations in the body. There are few situations encountered by the human body that utilize ATP at a faster rate than exercise, which in turn creates a substantial demand for rapid replacement of ATP stores. A useful schematic for picturing this process can be seen in Figure 3.3. If an individual muscle fiber is stimulated to produce force, energy is required. The direct source of energy for force production by muscle comes from splitting ATP molecules that are stored in the muscle cells. This process can occur very rapidly, particularly during high-intensity exercise when the amount of force produced by muscles is very high.

Figure 3.2 Rephosphorylation of ADP to form ATP

An inorganic phosphate (Pi) is joined to adenosine diphosphate (ADP) to re-form ATP. This endergonic process requires the input of energy, which is then stored as potential energy in the phosphate bonds of ATP.

Muscle fiber

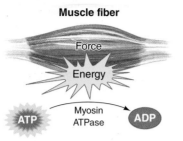

Figure 3.3 Schematic of ATP and energy use by exercising muscle

The production of force by a muscle fiber requires energy. The direct source of energy for muscle force production comes from ATP (adenosine triphosphate), which is stored in the muscle cell. When ATP is broken down to ADP (adenosine diphosphate), energy is released and can be used for muscle contraction.

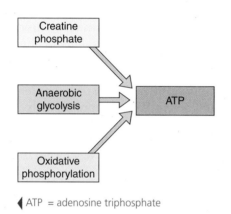

ATP = adenosine triphosphate

Figure 3.4 The three major energy systems that replenish ATP

The purpose of the three energy systems—creatine phosphate, anaerobic glycolysis, and oxidative phosphorylation—is to use chemical energy to re-form ATP, the direct source of energy used by cells in the body.

Just as important as the force production phase, the muscle cells must be able to relax and prepare for the next contraction. This process requires the replenishment of ATP within the muscle. As discussed in Chapter 2, failure to replenish ATP in the muscle may result in fatigue and rigor, a persistent contracted state. How is it that ATP is replenished?

The focus of this chapter is an examination of the three major energy systems that are used to replenish ATP: creatine phosphate, anaerobic glycolysis, and oxidative phosphorylation energy systems (Figure 3.4). Each of the three energy systems has its advantages (for example, speed of action, amount of energy released) and disadvantages (for example, duration of action) as shown in Table 3.1.

A preliminary word must be said about the use and interaction of these three energy systems. Because of their characteristics (for example, speed, duration), each energy system may be used by the body

Table 3.1 Characteristics of the Three Energy Systems

	Speed of action	Amount of ATP replenished	Duration of action
Creatine phosphate	Very fast	Very small	Very short
Anaerobic glycolysis	Fast	Small	Short
Oxidative phosphorylation	Very slow	Large	Very long

Legend: ATP = adenosine triphosphate

under different exercise or activity conditions. In other words, certain activities may require the use of the creatine phosphate system, whereas the energy demands of other activities may be met using the aerobic, or oxidative phosphorylation energy system. The tendency, therefore, is to think of each of these energy systems operating to the exclusion of the other systems. It is tempting to think of these systems as light switches: when during a certain activity one of the energy systems is "switched on" to support the activity, the others must therefore be switched off. This image incorrectly depicts the interaction of the energy systems.

A better analogy for the operation of the three energy systems might be a series of dimmer switches. All three are "on" all the time, but in certain situations, one of the systems may be "turned up" more than the other two. Certain exercise or sports activities may have an energy requirement that results in one of the energy systems being the predominant energy system used for that activity. The other energy systems are active, but to a lesser degree. As each energy system is described, examples of activities that *predominantly* use that source of energy will be given, but one must recognize that the other energy systems may be in use to some degree at the same time.

Key points

- ATP is hydrolyzed, or split, to release energy for exercise, leaving adenosine diphosphate (ADP).

- ATP is split at a fast rate during exercise.

- ADP is rephosphorylated to re-form ATP in an endergonic, or energy-requiring reaction.

- The three major energy systems—creatine phosphate, anaerobic glycolysis, and oxidative phosphorylation—are used to replenish ATP.

- All three energy systems are active at all times; however, one may be the predominant system depending upon the intensity and duration of the exercise activity.

What energy system is likely to be used for exercise that is of short duration but very high intensity?

3.2 The Creatine Phosphate Energy System

The most prevalent high-energy phosphate in the body is ATP, but there is another high-energy phosphate compound that is stored in muscle and other tissues—**creatine phosphate (CrP)**. Just as with ATP, potential energy is stored within the phosphate bond of creatine phosphate, and this energy can be released when the phosphate bond is broken and the inorganic phosphate (Pi) released. Unlike ATP, however, this energy is not used directly to power muscle contraction. Instead, the energy released from the breakdown of creatine phosphate is used to rephosphorylate ADP into ATP (Figure 3.5). Thus the creatine phosphate stored in muscle acts as a readily accessible reservoir of energy for the re-formation of ATP.

It should be noted that a number of slightly different terms and abbreviations are used interchangeably for creatine phosphate including **phosphocreatine**, PC, PCr, and CP. In this textbook, creatine phosphate, abbreviated as CrP, will be used. Because creatine phosphate is a high-energy phosphate compound like ATP, this energy system is referred to by many sports nutrition and exercise physiology texts as a combined system—the ATP-PC energy system. As has been pointed out, ATP directly provides the energy for force production by muscle and is present in such limited amounts that it may provide energy for only a few seconds of very high-intensity exercise. The role of creatine phosphate, like the two other major energy systems (anaerobic glycolysis and oxidative phosphorylation), is to replenish ATP so that it is available for muscle relaxation and subsequent force production. In this text, the term *creatine phosphate energy system* is used rather than the term *ATP-PC energy system*.

Creatine is consumed in the diet or synthesized in the body from amino acids.

Creatine is an **amine**, a nitrogen-containing compound similar to a protein, constructed from the amino acids arginine, glycine, and methionine. It can be either consumed in the diet via food or supplements or produced by the body. The major food sources of creatine are beef and fish. Those who eat beef and fish consume approximately 1 to 2 g of creatine daily, whereas nonmeat and nonfish eaters consume negligible amounts. Creatine can also be consumed as a dietary supplement, usually 3 to 5 g/day. Creatine as a supplement generally is found as creatine monohydrate in a white powdered form, which is mixed with water for consumption.

Even if creatine is not consumed directly in the diet, the liver and kidneys can synthesize it in adequate amounts if the amino acids arginine, glycine, and

Figure 3.5 The creatine phosphate energy system

Creatine phosphate is split chemically, releasing energy that is used to rephosphorylate ADP and re-form ATP, leaving unphosphorylated creatine. The reaction is catalyzed by the enzyme creatine kinase (CK).

methionine are present in sufficient quantities (that is, protein intake is adequate). A creatine deficiency in humans would be extremely rare.

Whether consumed or produced by the body, creatine is transported in the blood to tissues throughout the body. Tissues such as skeletal muscle will take up creatine from the blood and store it, approximately one-third as creatine (Cr) and two-thirds as creatine phosphate (CrP). Excess creatine is filtered by the kidneys and is excreted as the chemical compound **creatinine**. The body's "turnover" of creatine is approximately 2 g per day. That is, an average nonvegetarian who does not use creatine supplements will consume and/or synthesize approximately 2 g of creatine daily and the body will excrete approximately the same amount as creatinine (Figure 3.6). Since the dietary intake of creatine by a vegetarian is near or at

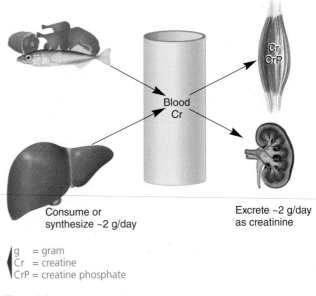

g = gram
Cr = creatine
CrP = creatine phosphate

Figure 3.6 Creatine metabolism—consumption, synthesis, uptake, excretion

For the majority of people, creatine either is consumed in the diet from meat or fish sources and/or is synthesized by the liver and other tissues. Approximately 2 g are consumed or synthesized daily (this figure assumes that creatine supplements are not taken). Creatine is distributed throughout the body in the blood and is taken up by tissues such as skeletal muscle. In muscle, approximately two-thirds is phosphorylated as creatine phosphate, whereas one-third remains as creatine. Approximately 2 grams per day is excreted in the urine as creatinine.

zero, the amount of creatine in a vegetarian's body is dependent on how much the body can make. Although research is limited in this area, there is some consistent evidence that the total amount of creatine in the muscles of vegetarians is lower than for nonvegetarians (Watt, Garnham, and Snow, 2004; Burke et al., 2003). Vegetarians are not creatine deficient (they may be on the lower end of the normal range), but they do have relatively less creatine than nonvegetarians who complement the production of creatine in the body with creatine found in meat and fish.

Because creatine is an important energy source, particularly for higher-intensity exercise, strength athletes and vegetarians have attempted to manipulate the intake of creatine through supplementation. Creatine loading and supplementation will be discussed in a later section of this chapter (see Spotlight on supplements).

The creatine phosphate energy system rephosphorylates ADP to ATP rapidly.

As seen in Figure 3.5, creatine phosphate is broken down, releasing its energy and Pi, which is then used to rephosphorylate ADP into ATP. This is a very rapid, one-step chemical reaction, catalyzed by the enzyme **creatine kinase (CK)**. If ATP concentrations in a muscle cell start to decline, the drop in ATP and the concomitant rise in ADP in the cell result in an increase in the activity of CK, allowing the reaction to proceed even faster. The reaction does not depend upon the presence of oxygen, so this energy system is considered to be one of the anaerobic ("without oxygen") energy systems. Each molecule of CrP can rephosphorylate 1 molecule of ADP to ATP, so the ratio of ATP energy produced is 1:1.

Because of the speed with which the CrP can act to replenish ATP, it is the preferred energy system during very high-intensity exercise when ATP is utilized very rapidly. As discussed in Chapter 2, it is potentially disastrous to a muscle cell to have ATP concentrations drop to very low levels so the muscle uses fatigue as a protective mechanism to prevent ATP depletion. When a muscle begins to fatigue, it fails to produce force at the same level or the same rate. Because it is not producing force at the same rate, the requirement for ATP declines and it is not used up as rapidly.

As with ATP, CrP is stored in finite amounts in the muscle, but unlike ATP, CrP can be used to the extent that it will decrease to very low concentrations. A normal resting level of total creatine (creatine + creatine phosphate) in muscle is approximately 120 mmol/kg (millimoles per kilogram of muscle), which can be reduced below 20 mmol/kg with very intense exercise. The term *depletion* is used, even though the muscle creatine levels do not actually drop to zero.

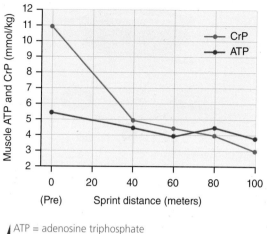

ATP = adenosine triphosphate
CrP = creatine phosphate
mmol/kg = millimoles per kilogram

Figure 3.7 Utilization of ATP and creatine phosphate in muscle during short-term, very high-intensity exercise

Well-trained sprinters had muscle ATP and creatine concentrations measured immediately before and after sprinting various distances to examine the use of these energy sources during very high-intensity exercise.

Hirvonen, J., Rehunen, S., Rusko, H., & Harkonen, M. (1987). Breakdown of high-energy phosphate compounds and lactate accumulation during short supramaximal exercise. *European Journal of Applied Physiology and Occupational Physiology, 56*(3), 253–259.

A number of studies have demonstrated this near depletion of CrP in muscle during high-intensity exercise, and one research study (Hirvonen et al., 1987) investigating the levels of ATP and CrP in the skeletal muscles of sprinters is illustrated here. As seen in Figure 3.7, as the sprinters ran as fast as they could for 40, 60, 80, and 100 meters, the concentration of ATP dropped approximately 20 percent in the first 40 meters and then essentially reached a plateau. Although there was some modest variation in the ATP concentration over the last 60 meters, the overall concentration did not go down further than the initial 20–25 percent decline. The level of CrP,

Creatine phosphate (CrP): Organic compound that stores potential energy in its phosphate bonds.

Phosphocreatine: See creatine phosphate.

Creatine: An amine, a nitrogen-containing chemical compound.

Amine: An organic compound containing nitrogen, similar to a protein.

Creatinine: Waste product excreted in the urine.

Creatine kinase (CK): Enzyme that catalyzes the creatine phosphate energy system.

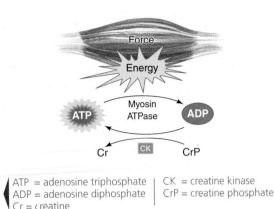

Figure 3.8 Creatine phosphate and ADP rephosphorylation

Creatine phosphate is used to rephosphorylate ADP to re-form ATP and provide energy for exercising muscle.

The predominant energy system used for very high-intensity, power activities is creatine phosphate.

however, declined substantially throughout the high-intensity exercise, particularly within the first 5 seconds (that is, ~40 meters). Very low levels of CrP in the muscle are associated with muscle fatigue. At very high intensities of exercise, it takes approximately 5 to 10 seconds for CrP in muscle to be depleted and fatigue to occur, so it is said that the duration of the creatine phosphate energy system is approximately 5 to 10 seconds.

Creatine phosphate is an important high-energy phosphate that is stored in muscle where it is readily available to replenish ATP very rapidly (Figure 3.8). Because this energy system can be used for rapid replenishment of ATP, it is the energy system that is predominantly used during very high-intensity, short-duration exercise when ATP is being used very rapidly. What types of exercise or sports activities meet this description?

- Short, fast sprints, such as the 100-meter sprint in track or the 40-yard sprint in football
- Short, powerful bursts of activity such as the shot put or jumping to dunk a basketball
- Activities requiring large amounts of force, such as very heavy weight lifting, for example, maximal effort bench press or a maximal Olympics lift such as a clean and jerk

In summary, the characteristics of the creatine phosphate energy system are:

- One chemical step
- Catalyzed by creatine kinase (CK)
- Very fast reaction
- One ATP per CrP molecule
- 5- to 10-second duration

- Anaerobic
- Fatigue associated with CrP depletion
- Predominant energy system in very high-intensity exercise, for example, "power" events

Rephosphorylation of creatine to creatine phosphate depends upon aerobic metabolism.

Using the same schematic depicting energy and force production by exercising muscle, Figure 3.8 shows the role of creatine phosphate in the rephosphorylation of ADP to ATP. When this reaction proceeds, one of the end products is creatine without the phosphate group. If creatine phosphate is an important energy source, particularly for high-intensity exercise, how does the body recover its stores of creatine phosphate?

The answer lies in the realization that the creatine phosphate energy system is not completely anaerobic. The presence of oxygen is not required when CrP is used to replenish ATP. However, the recovery of CrP from creatine depends upon aerobic metabolism in the cell, in a process referred to as the Creatine Shuttle (Bessman and Carpenter, 1985). As discussed in greater detail later in this chapter, aerobic metabolism takes place within mitochondria, small organelles in cells. A series of chemical reactions requiring oxygen are continuously carried out to produce ATP within the mitochondria. Creatine molecules can use the energy from ATP produced aerobically in the mitochondria to restore creatine to creatine phosphate (Figure 3.9). Once restored, creatine phosphate is ready as a reservoir of energy for high-intensity exercise.

Focus on research

Determining the Use of ATP and Creatine Phosphate in Skeletal Muscle during Exercise

High-energy phosphates contained in skeletal muscle, adenosine triphosphate (ATP) and creatine phosphate (CrP), provide energy to power very high-intensity exercise. These energy reactions take place inside muscle cells and are therefore difficult to study, particularly during exercise. A variety of research techniques are available to scientists to study skeletal muscle using either animal models or human subjects. Certain research techniques offer good laboratory control and precise measurements, but these studies may involve surgically exposing (*in situ*) or removing whole muscles or muscle cells (*in vitro*) from an animal. Studies performed on whole, living organisms such as humans are called *in vivo* experiments, but when performed in the laboratory may not accurately represent the response of these energy systems to a real exercise or sport challenge. How can the use of ATP and creatine phosphate be studied realistically during very high-intensity exercise?

The following publication describes ATP and creatine phosphate use during maximal effort sprinting by human subjects: Hirvonen, J., Rehunen, S., Rusko, H., & Härkönen, M. (1987). Breakdown of high-energy phosphate compounds and lactate accumulation during short supramaximal exercise. *European Journal of Applied Physiology and Occupational Physiology, 56*(3), 253–259.

What did the researchers do? In order to examine the use of high-energy phosphates during a common sporting activity, these researchers designed their study to measure ATP and CrP in the muscles of competitive sprinters before and after a series of short sprints. To determine the concentration of ATP and CrP in muscle, a very small piece of muscle has to be obtained by a procedure known as a muscle biopsy. After a local anesthetic is used, a small incision is made in the skin and connective tissue layer over the muscle. A biopsy needle is passed through this incision into the muscle, and a small piece of muscle (approximately 10 mg or 0.0004 ounces) is clipped off and removed. Subjects are able to exercise shortly after this procedure with no adverse consequences. Once the muscle sample is removed, it can be treated and analyzed in many different ways, including chemical assays to determine ATP and CrP concentrations.

Muscle biopsies were obtained in this study while the subjects were at rest and then after warming up to determine their pre-exercise or baseline ATP and CrP concentrations. It is not possible to take muscle biopsies during exercise, so the study was designed to have the well-trained sprinters run all-out sprints of 40, 60, 80, and 100 meters in random order and to obtain the muscle samples at the end of exercise. Because these high-energy phosphates are metabolites that may change concentration quickly, the muscle samples needed to be collected as soon after exercise as possible. According to the researchers, "a mattress was placed at the end of the track for the sprinters to fall onto at full speed so that blood and muscle samples could be taken immediately after each run." Once researchers had obtained muscle samples before and immediately after the sprints, the amount and rate of ATP and CrP used during this high-intensity exercise could be calculated.

What did the researchers find? The duration of the high-intensity exercise was from approximately 4 to 11 seconds for the 40- to 100-meter sprints. ATP concentration declined slightly during the sprints, but never fell below 60 percent of the pre-exercise values. In contrast, the subjects used approximately 88 percent of their creatine phosphate stores. The response of ATP and CrP in skeletal muscle during very high-intensity exercise is shown in Figure 3.7.

What was the significance of this research study? This research study provided evidence that the use of high-energy phosphates during "real" high-intensity exercise was consistent with what had been demonstrated by other laboratory studies. These results support the fact that ATP concentrations in skeletal muscle cells do not decline very much, even during very high-intensity exercise, whereas creatine phosphate levels can be dramatically depleted.

Answering the question: How can the use of ATP and creatine phosphate be studied realistically during very high-intensity exercise? There are many technical challenges and potential limitations to performing research studies that closely mimic "real world" exercise or sports activities. This study demonstrated that researchers with ingenuity and good technical expertise, with the help of subjects willing to endure arduous study procedures, are capable of designing and completing research studies that add substantially to our understanding of the physiology of exercise.

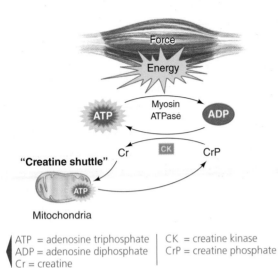

ATP = adenosine triphosphate   CK = creatine kinase
ADP = adenosine diphosphate   CrP = creatine phosphate
Cr = creatine

Figure 3.9 The creatine shuttle

Rephosphorylation of creatine to creatine phosphate via aerobic metabolism. Creatine that remains from chemically splitting creatine phosphate is rephosphorylated using energy from ATP produced aerobically in the mitochondria of the muscle. This process requires an increase in aerobic metabolism (oxidative phosphorylation) and returns creatine phosphate concentrations to normal within a minute or two.

How long does it take for creatine to be rephosphorylated to creatine phosphate? It depends upon how much creatine phosphate was used, but it can take up to 1 to 2 minutes if the exercise has been intense and long enough to substantially deplete creatine phosphate in the muscle. Ever wonder why athletes breathe hard for a time after short duration high-intensity exercise? Because the body uses the aerobic energy system to restore creatine phosphate, it must temporarily increase oxygen intake and aerobic metabolism to meet the increased aerobic task of restoring creatine to creatine phosphate. This is another example of the integration of the body's energy systems.

The major advantage of the creatine phosphate energy system is that it is a very rapid way to replace ATP. However, it has significant disadvantages in that creatine phosphate is stored in very limited amounts in the muscle, can be depleted rapidly, and has a very short duration as an energy system. Fortunately, the body has additional energy systems to replace ATP under different circumstances.

Spotlight on supplements

Creatine Loading and Supplementation

Creatine phosphate is an important high-energy phosphate that can be used to replenish ATP very rapidly. Its major disadvantage as an energy system, however, is its limited supply in muscle: it can be depleted relatively quickly during very high-intensity exercise. It is not unexpected that athletes competing in events that rely heavily upon this energy source would try to improve their performance by increasing the amount of creatine phosphate available to the muscle (American College of Sports Medicine, 2000).

How can the muscle's supply of creatine phosphate be increased above normal levels? Muscle creatine and creatine phosphate concentrations can be increased by approximately 20 percent by consuming creatine supplements (Hultman et al., 1996). Creatine loading is a strategy used to accomplish this increase in a short period of time by consuming a large amount of creatine (20 to 25 g per day) in supplement form over a period of 5 to 6 days. The Hultman et al. study also showed that a similar 20 percent increase in muscle creatine phosphate concentrations could be achieved by supplementing the diet with smaller amounts of a creatine supplement (3 g per day) over a longer period of time (1 month). It should be pointed out that some athletes respond to creatine supplements with substantial increases in muscle creatine concentrations whereas others have small increases and some are nonresponders. This is likely explained by the amount of creatine in the muscle prior to supplementation. In one study, both vegetarians and nonvegetarians increased the amount of creatine in muscles after 5 days of creatine supplementation, but the vegetarians saw relatively larger increases (76 percent increase) than nonvegetarians (36 percent increase) because they had lower levels prior to the supplementation (Watt, Garnham, and Snow, 2004).

Increasing the amount of creatine phosphate in the muscle is similar in concept to increasing the size of the gas tank in a race car. Will increasing the size of the fuel tank necessarily allow the race car to go faster? Will increasing the size of the creatine "gas tank" in the muscle immediately make the athlete faster or more powerful? Merely increasing the size of the gas tank won't make the race car go faster. It will, however, allow the car to maintain its top speed for a longer period of time. This appears to be the long-term benefit of creatine loading or supplementation for the strength and power athlete.

There is insufficient evidence to suggest that creatine loading or supplementation results in an immediate increase in an athlete's strength, speed, or power or that it has a direct effect on performance for most athletes. Though many creatine studies show an **ergogenic** effect for a variety of athletes, a performance effect has been shown only for weight lifters (Volek and Rawson, 2004). The studies do suggest that creatine loading or supplementation allows an athlete to train harder, for example, by completing more weight-lifting repetitions (Volek et al., 1999). The increase in the training stimulus over time allows the athlete

Key points

- Creatine phosphate is a high-energy phosphate that is stored in muscle and can be used to replenish ATP very quickly.

- Creatine is synthesized in the body and/or consumed in the diet (meat or fish) or as a supplement.

- Creatine phosphate is the preferred energy system during short duration, very high-intensity exercise, and its depletion is associated with muscle fatigue.

- Creatine is rephosphorylated to creatine phosphate using the aerobic energy system.

What are the major advantages and disadvantages of the creatine phosphate energy system?

3.3 The Anaerobic Glycolysis Energy System

Anaerobic glycolysis is the process of taking carbohydrate in the body and putting it through a series of chemical reactions that release enough energy to

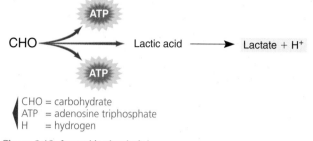

CHO = carbohydrate
ATP = adenosine triphosphate
H = hydrogen

Figure 3.10 Anaerobic glycolysis

Carbohydrates in the form of glucose are broken down through a series of chemical reactions that result in a net formation of ATP. The final product, lactate, is an important metabolic compound.

rephosphorylate ADP and re-form ATP (Figure 3.10). The final product of this series of chemical reactions is often referred to as lactic acid, so this energy system is often called the lactic acid system. In reality, lactic

Ergogenic: Ability to generate or improve work. Ergo = work; genic = formation or generation.

Anaerobic glycolysis: A series of chemical steps that break down glucose without the use of oxygen to rephosphorylate ADP to ATP.

to potentially become stronger, faster, or more powerful but only in weight lifters is it directly related to performance. In the case of runners, the weight gain that accompanies creatine supplementation could have a detrimental effect on performance (Volek and Rawson, 2004; Volek et al., 1999).

Numerous studies have documented that creatine supplementation results in a small but significant increase in lean body mass with repeated high-intensity, short-duration (<30 seconds) exercise bouts (Tipton and Ferrando, 2008; Volek and Rawson, 2004; Branch, 2003). However, the exact mechanisms are not clear. Creatine supplementation increases intracellular water in the muscle, which may stimulate muscle glycogen storage. Increases in intracellular water also influence protein metabolism. These effects are intriguing, but more studies are needed to determine the effect creatine supplementation has on either glycogen storage or muscle protein synthesis or breakdown.

The safety of creatine supplementation has always been subject to much debate, but currently available research suggests that creatine supplementation is safe. When creatine supplements first became popular, it was widely reported in the press that they caused dehydration, muscle cramps, and, possibly, kidney damage. There is no scientific evidence that creatine supplementation causes dehydration or muscle cramps (Lopez et al., 2009). Studies have also shown that creatine supplements do not adversely affect kidney or liver function, hormone levels, lipids, or sperm count and

mobility (Crowe, O'Connor, and Lukins, 2003; Mayhew, Mayhew, and Ware, 2002; Schilling et al., 2001; Poortmans and Francaux, 2000). One clinical sign of kidney damage is excess creatinine in the urine. Because excess creatine is eliminated in the form of creatinine, urinary creatinine concentrations may be elevated after creatine loading, but studies have shown these levels to rise only to the high end of the normal range (Poortmans and Francaux, 2000). Some athletes may experience some minor side effects with supplements, such as gastrointestinal upset or cramps.

Creatine Supplementation Recommendations:

- Application is very specific to strength and power activities.

- Creatine is not a "magic pill"—only effective in conjunction with vigorous training.

- Response may be related to initial creatine levels.

- Do not use if kidney disease or dysfunction is present.

- Supplement dose is typically 3 to 5 grams per day.

- Loading dose is typically 20 to 25 grams per day in four to five doses for five to seven days.

- Loading is not necessary unless time urgent.

- Consume with carbohydrates to enhance uptake.

- Be well-hydrated.

- Document any adverse effects and re-evaluate use.

acid is a weak acid and under normal conditions in the body is rapidly **dissociated**, separating into the **lactate** molecule and a hydrogen ion (H⁺). It would therefore be more accurate to use the term lactate energy system; however, in exercise physiology the term *anaerobic glycolysis* is also widely used. In this text, the term *anaerobic glycolysis energy system* is used to identify the biochemical pathways involving the conversion of glucose to lactate in contracting muscle under anaerobic conditions. The term *glycolysis*, not anaerobic glycolysis, is used in most biochemistry textbooks.

Carbohydrates will be covered in much greater detail in Chapter 4, but a summary is needed at this point as background for understanding anaerobic glycolysis. Virtually all carbohydrates consumed are converted to and used as **glucose**, or are stored as glycogen in the muscle and liver for later use. Glycogen is a large molecule composed of many glucose molecules linked together. **Glycolysis** is the breakdown of glucose through a specific series of chemical steps to rephosphorylate ADP and form ATP. When the starting point is glycogen, the process is called **glycogenolysis** (lysis means to separate).

Glycolysis uses the energy contained in glucose to rephosphorylate ADP to ATP.

The complete process of glycolysis involves 18 chemical reactions but only 12 chemical compounds, because 6 reactions are repeated or duplicated. The glycolytic pathway begins with glucose and ends with lactate, and the chemical compounds in this pathway are called glycolytic intermediates. Each chemical step is catalyzed by an enzyme. When a series of chemical reactions is catalyzed by enzymes, one of the enzymes is considered to be the **rate-limiting enzyme**. In other words, there is an enzyme that controls the speed of all the reactions in the same way that the overall speed of an assembly line is governed by the speed of the slowest worker. The rate-limiting enzyme for glycolysis is **phosphofructokinase (PFK)**, which catalyzes the third step. If the activity of PFK increases, the entire reaction speeds up, and if PFK activity decreases, the entire reaction slows down. As mentioned in Chapter 2, the activity of enzymes can be affected by temperature and pH, among other factors. For example, if the environment becomes too acidic, the activity of PFK will decline, slowing the entire glycolytic reaction, and ultimately slowing the rate of ATP replacement. The activity of PFK is dramatically increased, however, by a drop in the concentration of ATP and the concomitant rise in ADP concentration—this is an important signal that ATP is being used up and must be replenished.

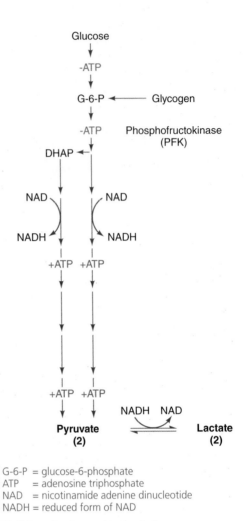

G-6-P = glucose-6-phosphate
ATP = adenosine triphosphate
NAD = nicotinamide adenine dinucleotide
NADH = reduced form of NAD

Figure 3.11 Schematic of anaerobic glycolysis

Glucose proceeds through a series of 18 chemical steps, 6 of which are repeated, ending as lactic acid or lactate. In the initial steps, energy in the form of two ATP is needed to allow the reaction to proceed, and sufficient energy is released in subsequent chemical reactions to re-form four ATP. When the process begins with stored glycogen, it is called glycogenolysis and it proceeds from the point of G-6-P.

Oxygen is not required for these chemical reactions to proceed, so glycolysis is one of the anaerobic energy systems. Knowledge of the details of glycolysis is important for a thorough understanding of sports nutrition, but at this point in the discussion the important details and concepts have been condensed as represented in Figure 3.11. For now, the simplified schematic will be used to develop an understanding of the fundamental concepts of glycolysis. A detailed description of anaerobic glycolysis can be seen in Appendix J, which illustrates the specific chemical reactions that occur, the names of the intermediate compounds, and the cofactors involved.

Beginning with glucose, two out of the first three reactions require ATP; that is, they require the input of energy for the reaction to proceed. In the first step, energy and a phosphate group from an ATP is added to glucose to form glucose-6-phosphate (G-6-P), one of the notable glycolytic intermediate compounds. Glucose is a compound that contains six carbon molecules, and in the fourth step it is split into 2 three-carbon molecules. After a side step reaction to DHAP (dihydroxyacetone phosphate) the three-carbon molecules are identical, so the remaining reactions of glycolysis are duplicated. In the sixth chemical reaction, sufficient energy is released to rephosphorylate ADP into ATP. This is one of the reactions that is repeated, so two ATP are produced at this point. In the ninth reaction the same thing occurs, so two additional ATP are produced. The result of this ninth reaction is the production of **pyruvate**, an important intermediate compound. In anaerobic glycolysis, pyruvate then proceeds through the final chemical reaction to lactate. Glycolysis produces four ATP but two ATP are used in the process, so the final (net) ATP production from anaerobic glycolysis is two ATP.

In Figure 3.11, notice that if the beginning point of the reaction is glycogen, the pathway skips the first reaction that requires the use of an ATP, and the reaction beginning with glycogen goes directly to glucose-6-phosphate. Therefore, when glycogenolysis (that is the breakdown of glycogen) is used instead of glycolysis (that is, the breakdown of glucose), four ATP are produced and only one is used in the process, so the final or net ATP production is three ATP. Glycogenolysis is obviously more energetically efficient, and in fact, exercising muscle will generally use stored glycogen in preference to glucose because the energy yield is higher. This helps during exercise, but the ATP that is not used when glycogen is broken down during exercise is used eventually when glucose is stored as glycogen during rest or recovery.

Given the number of chemical steps, it is readily apparent that anaerobic glycolysis will produce ATP more slowly than the one-step creatine phosphate system. Anaerobic glycolysis is a relatively fast-acting energy system, however, and becomes the preferred or predominant system to supply energy during high-intensity or repeated exercise lasting approximately one to two minutes. The overall speed of glycolysis and the subsequent rate of ATP replenishment are primarily governed by the activity of PFK, the rate-limiting enzyme. The body has a considerable amount of carbohydrate energy in the form of blood glucose and muscle glycogen, so unlike the creatine phosphate system, the anaerobic glycolytic energy system is rarely limited by depleted energy stores.

The major disadvantage of this energy system is the increasing acidity (that is, decline in pH) within the muscle cell that occurs when anaerobic glycolysis is used at a high rate, as occurs during moderately high to high-intensity exercise. Because the acidity is a result of a very high rate of metabolism, it is referred to as **metabolic acidosis**. If this acidosis occurs in exercising muscle, the drop in pH can result in the decrease in activity of key metabolic enzymes and can interfere directly with the process of force production, resulting in muscle fatigue. When exercising at high-intensity, this metabolic acidosis results in muscle fatigue in approximately one to two minutes, so the duration of this energy system is said to be one to two minutes.

Anaerobic glycolysis utilizes only carbohydrate as a fuel source, and can replace ATP rapidly during moderately high to high-intensity exercise. It begins to function at the onset of high-intensity exercise, but becomes the predominant energy system after 5 to 10 seconds when the creatine phosphate energy system begins to reach its limit. To return to the light controller analogy, after 5 to 10 seconds the dimmer switch is turned down on the CrP energy system and turned up on the anaerobic glycolysis energy system. If used continuously at a high rate, anaerobic glycolysis remains the predominant energy system for one to two minutes, or longer if the activity is intermittent. What types of exercise or activities meet this description?

- Long sprints such as the 400-meter sprint in track
- Repeated high-intensity sprints such as the intermittent sprints by a soccer or basketball player
- Repeated high-force activities such as 10 to 15 repetitions of weight lifting
- Regular, repeated intervals such as 50- to 100-meter swimming intervals

Dissociated: The breakdown of a compound into simpler components, such as molecules, atoms, or ions.

Lactate: The metabolic end product of anaerobic glycolysis.

Glucose: A sugar found in food and in the blood.

Glycolysis: Metabolic breakdown of glucose.

Glycogenolysis: Metabolic breakdown of glycogen.

Rate-limiting enzyme: In a series of chemical reactions, the enzyme that influences the rate of the entire series of reactions by changes in its activity.

Phosphofructokinase (PFK): The rate-limiting enzyme for glycolysis.

Pyruvate: Chemical compound that is an important intermediate of glycolysis.

Metabolic acidosis: Decrease in pH associated with high-intensity exercise and the use of the anaerobic glycolysis energy system.

Long or repeated sprints use anaerobic glycolysis as the predominant energy system.

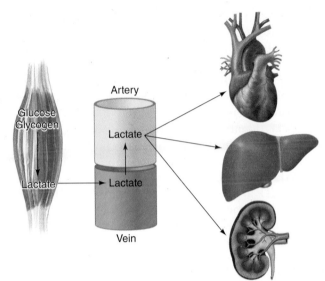

Figure 3.12 The fate of lactate

Lactate is transported out of exercising muscle into the venous circulation and is then distributed throughout the body via the arterial circulation. Highly aerobic tissues such as heart, liver, and kidney can remove lactate from the blood and use it as an energy source in aerobic metabolism.

To summarize the characteristics of the anaerobic glycolytic energy system:

- 18 chemical steps/reactions, 6 are repeated
- 12 chemical compounds, 11 enzymes
- Rate-limiting enzyme phosphofructokinase (PFK)
- Fast, but not as fast as the creatine phosphate (CrP) system
- Two ATP produced via glucose, three ATP produced via glycogen
- Anaerobic
- 1- to 2-minute duration
- Fatigue associated with decreased pH (metabolic acidosis)
- Predominant energy system in high-intensity exercise, for example, sustained, repeated sprints

Lactate is metabolized aerobically.

When anaerobic glycolysis is utilized as an energy system, the concentration of glucose or glycogen declines and the concentration of lactate increases within the cell. What happens to this metabolic by-product? Previously, lactate had been thought of as a metabolic "waste product," something that "poisons" the muscle. However, an important distinction must be made between the lactate molecule and the acidity that is associated with high-intensity exercise when glycolysis is used as the predominant energy system. The body has several processes to buffer acidity, but the body's buffering capacity may be overwhelmed by acidosis that occurs during high-intensity exercise. This acidity may result in an impairment of cell processes. Therefore, the acidity associated with high-intensity exercise may be detrimental to the cells, but far from being a waste product, the lactate molecule

is used by cells in the body as an important fuel source (Gladden, 2004).

Notice in Figure 3.11 that the reaction between pyruvate and lactate is a two-way reaction: that is, pyruvate can be converted to lactate, but when concentrations of lactate are high, lactate can also be converted to pyruvate. This is especially important in highly aerobic cells (for example, muscle, liver), as pyruvate can be taken into mitochondria and metabolized aerobically to produce ATP. In Figure 3.11, pyruvate is identified as a key glycolytic intermediate compound because it represents an important metabolic crossroad. Pyruvate can be metabolized either anaerobically (converted to lactate) or aerobically (taken into a mitochondria and sent through oxidative phosphorylation). Therefore, highly aerobic tissues in the body (for example, liver, heart muscle, slow-twitch muscle fibers) can take up lactate molecules and use them as a fuel source to produce ATP via aerobic metabolism.

When lactate is produced in cells, such as those in exercising muscle, the increase in lactate concentration results in the movement of lactate molecules out of the cell and into the blood. Once in the blood, lactate molecules can be distributed throughout the body where they can be taken up by other highly aerobic tissues (for example, heart, liver, kidney) and metabolized aerobically (Figure 3.12). The more lactate that needs to be oxidized, the more aerobic metabolism must be increased to accomplish this task. Just as the creatine phosphate system is not completely anaerobic, neither is the anaerobic glycolysis energy system. Anaerobic glycolysis relies upon the aerobic energy system to metabolize lactate, the final product of glycolysis. This is another example of the integration of the body's energy systems.

In addition to aerobic metabolism of lactate for energy, the body can use lactate in other ways. The necessary enzymes are not present in muscle to take lactate through the chemical reactions to make glucose (that is, gluconeogenesis). This process is referred to as the Cori cycle and must take place in the liver, which does have the necessary enzymes. During exercise, lactate that is produced by exercising muscle can be transported to the liver, then used to create glucose, which is then released from the liver into the blood, helping to maintain blood glucose concentration and providing carbohydrates to other cells in the body.

The major advantages of the anaerobic glycolysis energy system are the relatively fast ATP production, the reliance upon a fuel source that is present in large amounts (glucose and glycogen), and the potential usefulness of lactate, its end product. The major disadvantage is the relatively short duration of this energy system due to the metabolic acidosis associated with its use at a high rate.

Key points

- Glycolysis is a series of chemical steps, beginning with glucose or glycogen, that result in the formation of ATP.
- Glycolysis replenishes ATP quickly, although not as quickly as the creatine phosphate energy system.
- Glycolysis is the predominant energy system used for high-intensity or repeated exercise bouts and results in the formation of lactate.
- Lactate produced in exercising muscle can be metabolized aerobically by a variety of tissues.

If exercise intensity is increased gradually, why does the concentration of lactate in the blood not increase until after approximately 50–60 percent of maximal oxygen consumption?

Spotlight on...

Lactate Threshold

Lactate can be produced by exercising muscle and is transported into the blood where it is often monitored by exercise physiologists as an indication of exercise intensity. As exercise increases in intensity, there is usually little increase in the concentration of lactate in the blood initially. Once the intensity reaches a certain level, however, the amount of lactate in the blood begins to increase dramatically. This point of sudden increase is called the Lactate Threshold (LT) (see figure).

This point of exercise intensity is often erroneously referred to as the Anaerobic Threshold (AT), with the explanation that this is the point of exercise intensity where aerobic metabolism is insufficient and anaerobic glycolysis suddenly becomes active, producing lactate. However, lactate is produced at lower levels of exercise intensity in small amounts, but the lactate concentration in blood does not increase because lactate is being removed from the blood by highly aerobic tissues and oxidized as fast as the exercising muscles are releasing lactate into the blood. The lactate threshold more accurately represents the level of exercise intensity in which lactate production has increased to the point where it has overwhelmed the body's lactate removal mechanisms. Although the term *Anaerobic Threshold* has achieved widespread usage, it should be eliminated in favor of the more accurate term *Lactate Threshold*. Another acceptable term is *Ventilation Threshold (VT)*, if this point of exercise intensity is determined using measures of breathing instead of taking blood samples and measuring for lactate.

What is the practical significance of the LT? It has been associated with an exercise intensity that can be sustained for

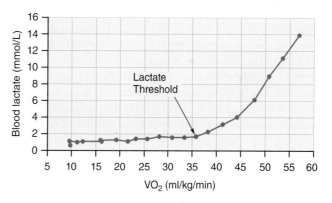

mmol/L　　= millimoles per liter
VO₂　　　 = oxygen consumption
ml/kg/min = milliliters oxygen per kilogram body weight per minute

The concept of the lactate threshold

As exercise intensity and oxygen consumption increase, initially, the concentration of blood lactate does not change much, indicating lactate is being removed from as fast as it is being transported into the blood. At a certain point of exercise intensity, however, the production of lactate exceeds its rate of removal, and the lactate concentration continually climbs. This point is referred to as the Lactate Threshold.

long periods during endurance exercise. For example, LT correlates highly with the race pace for a distance runner: a pace too much faster results in fatigue and poor results, whereas a pace too much slower results in less than optimal performance.

3.4 The Oxidative Phosphorylation Energy System

The major limitation of the two anaerobic energy systems is their relatively short duration of action, on the order of seconds or minutes. The aerobic energy system, known as **oxidative phosphorylation**, is the energy system that can be used to supply ATP on a virtually limitless basis, as long as oxygen and sources of fuel are available. Oxidative phosphorylation is a process by which carbohydrates, fats, or proteins can be metabolized through a series of chemical reactions to release the energy necessary to rephosphorylate ADP to ATP (Figure 3.13).

The oxidative phosphorylation process, which takes place in the mitochondria, is made up of three major phases. In the first phase, carbohydrates, fats, and proteins are prepared to be metabolized aerobically. The second major phase is the **Krebs cycle**, also known as the tricarboxylic acid (TCA) cycle or the citric acid cycle. The Krebs cycle produces a very limited amount of ATP directly, so its primary function is to oxidize, or remove electrons from, the compounds produced from the breakdown of carbohydrates, fats, and proteins. The electrons removed in the Krebs cycle are used in the final phase, the **electron transport chain**. Electrons are passed through a series of chemical reactions, which release energy to rephosphorylate ADP into ATP. As the electrons pass through the electron transport chain, the final electron acceptor is oxygen, making this the aerobic energy system.

Even more so than anaerobic glycolysis, oxidative phosphorylation is a complex series of chemical reactions involving many steps, intermediates, enzymes, and cofactors. In fact, if the starting point is glucose, complete aerobic metabolism involves 124 steps, 30 chemical compounds, and 27 enzymes. Compare that to glycolysis, where the starting point is also glucose, but the process involves only 18 steps, 12 chemical compounds, and 11 enzymes. It is not just the number of steps, compounds, or enzymes that makes aerobic metabolism complicated. Once the reaction progresses to pyruvate, the pyruvate molecules must be transported into mitochondria for oxidative phosphorylation to proceed. In addition, oxygen must be transported from the lungs via the blood and through the muscle cells into the mitochondria to be available to pick up the electrons. Once again, a detailed knowledge of this metabolic pathway is important for a thorough understanding of sports nutrition and is explained in the next column, but the basics of oxidative phosphorylation have been condensed and represented in Figure 3.14.

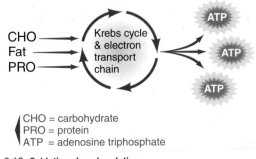

CHO = carbohydrate
PRO = protein
ATP = adenosine triphosphate

Figure 3.13 Oxidative phosphorylation

Carbohydrates, fats, and proteins are broken down through a series of chemical reactions that result in a net formation of ATP. The final step requires oxygen, making this the aerobic energy system.

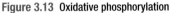

Refer to Figure 3.14 and follow the steps of oxidative phosphorylation as glucose (carbohydrates) is completely metabolized aerobically. Glycolysis has already been discussed from the perspective of anaerobic glycolysis in which glucose proceeds to pyruvate and on to lactate. In aerobic metabolism, glycolysis prepares glucose and glycogenolysis prepares glycogen for aerobic metabolism by producing pyruvate, which is shuttled into the mitochondria for the second phase of oxidative phosphorylation. Later in this chapter the preparation of fats and proteins for aerobic metabolism will be discussed. At this point the discussion of glucose will continue with an explanation of the second phase of aerobic metabolism, the Krebs cycle.

Carbohydrates are oxidized in the Krebs cycle.

Once glucose (or glycogen) has been broken down into pyruvate, the pyruvate molecules are shuttled into one of many mitochondria in a typical muscle cell. Remember that for each glucose molecule metabolized, two pyruvate molecules are formed by glycolysis, and therefore two "turns" of the Krebs cycle can be completed. Although a high-energy phosphate (guanosine triphosphate [GTP]) is produced by the Krebs cycle, the major function of the Krebs cycle is to oxidize, or remove, electrons from the compounds going through the cycle for later use in the electron transport chain.

Once in the mitochondria, pyruvate is converted to another important intermediate compound, **acetyl CoA**, as shown in Figure 3.14. In this process a molecule of carbon dioxide is created, the first of three for each complete turn of the Krebs cycle. One can begin to see the production of carbon dioxide that goes along with the utilization of oxygen in aerobic metabolism.

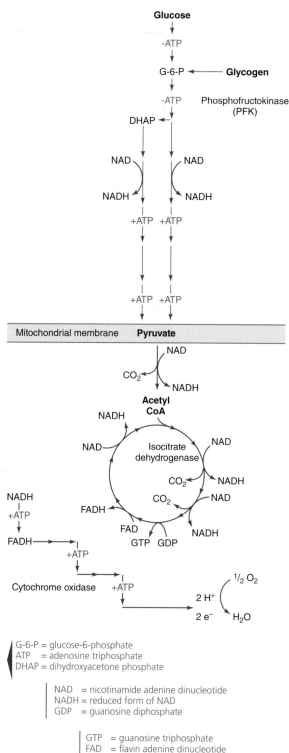

Figure 3.14 Schematic of glycolysis, Krebs cycle, and electron transport chain of oxidative phosphorylation

Glucose follows the steps of glycolysis, except that rather than being converted to lactate, pyruvate is transported into a mitochondrion for aerobic metabolism. Pyruvate goes through the Krebs cycle, a series of chemical reactions that oxidize, or remove, the electrons from the intermediate compounds in the process. The electrons are transported to the electron transport chain where they participate in a series of reactions that release sufficient energy to rephosphorylate ADP to ATP. Oxygen is the final electron acceptor and forms water.

The Krebs cycle is a series of 10 chemical reactions that begin with acetyl CoA (a two-carbon compound) joining with **oxaloacetate** (four carbons) to form a six-carbon compound, **citric acid** (see Appendix J for greater detail of the Krebs cycle including names and structures of intermediate compounds, enzymes, and cofactors). For this reason, the Krebs cycle is often referred to as the citric acid cycle. The rate-limiting enzyme for this series of reactions is **isocitrate dehydrogenase (IDH)**. During the course of the series of reactions, two carbons are lost due to the formation of two carbon dioxide molecules. The reactions eventually return to the four-carbon oxaloacetate, which is then available to combine with another acetyl CoA—a repeating cycle of chemical reactions.

Beginning with pyruvate, six **oxidation-reduction** reactions take place for each turn of the Krebs cycle. An aspect of chemical reactions that was not previously discussed even though it occurs in glycolysis is oxidation-reduction reactions. Some chemical compounds need to have electrons removed in order to be transformed chemically to another compound. The process of giving up electrons is called **oxidation**, and the compound that has electrons removed has been **oxidized**. The process of accepting electrons is called **reduction**, and the compounds that receive the electrons have been **reduced**. When electrons are removed they can be destructive if allowed to remain free within the cell. Therefore, oxidation-reduction reactions are coupled. That is, when one compound needs to give up electrons and is oxidized, another compound simultaneously accepts electrons and is reduced. The two most

Oxidative phosphorylation: The aerobic energy system.

Krebs cycle: A series of oxidation-reduction reactions used to metabolize carbohydrates, fats, and proteins.

Electron transport chain: A series of electron-passing reactions that provides energy for ATP formation.

Acetyl CoA: A chemical compound that is an important entry point into the Krebs cycle.

Oxaloacetate: Chemical compound that is one of the intermediate compounds in the Krebs cycle.

Citric acid: Chemical compound that is one of the intermediate compounds in the Krebs cycle; the first compound formed in the Krebs cycle by the combination of oxaloacetate and acetyl CoA.

Isocitrate dehydrogenase (IDH): The rate-limiting enzyme for the series of chemical reactions in the Krebs cycle.

Oxidation-reduction: The giving up of (oxidation) and acceptance of (reduction) electrons in chemical reactions; these reactions typically occur in pairs.

Oxidize/oxidation: Chemical process of giving up electrons.

Reduce/reduction: Chemical process of accepting electrons.

common electron-accepting compounds are **nicotin-amide adenine dinucleotide (NAD)** (Figure 3.15) and **flavin adenine dinucleotide (FAD)**. Although most electrons are coupled in oxidation-reduction reactions, some can remain free within a cell and they may form damaging chemicals called **free radicals** (see Spotlight on...Free Radicals).

One of the Krebs cycle reactions is exergonic and releases enough energy to produce a high-energy phosphate compound, **guanosine triphosphate (GTP)**, which is then used to form an ATP. In four of the oxidation-reduction reactions NAD is the electron acceptor, and in one, FAD is the electron acceptor. The vitamins niacin (that is, nicotinic acid and nicotinamide) and riboflavin are major components of NAD and FAD respectively. NAD and FAD shuttle the electrons to the final major component of oxidative phosphorylation, the electron transport chain.

The electron transport chain uses the potential energy of electron transfer to rephosphorylate ADP to ATP.

The electron transport chain is yet another series of chemical reactions that take place within the inner membrane of the mitochondria. Electrons are passed

Nicotinamide adenine dinucleotide (NAD): Molecule involved in energy metabolism that contains a derivative of the vitamin niacin.

Flavin adenine dinucleotide (FAD): A molecule involved in energy metabolism that contains a derivative of the vitamin riboflavin (vitamin B_2).

Free radicals: A highly reactive molecule with an unpaired electron. Also known as reactive oxygen species (ROS).

Guanosine triphosphate (GTP): A high-energy phosphate compound produced in the Krebs cycle used to replenish ATP.

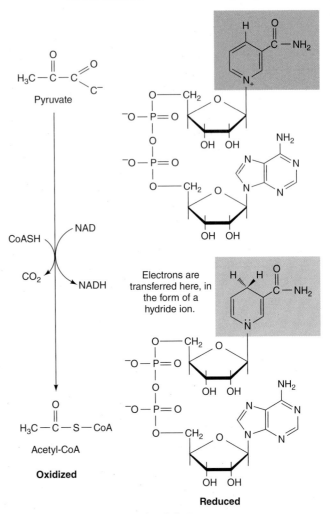

Figure 3.15 NAD and the transfer of electrons

In the first step of the Krebs cycle, electrons are transferred from pyruvate to NAD so that pyruvate can be converted to Acetyl CoA. In this oxidation-reduction reaction, pyruvate has lost electrons, so it has been oxidized, and NAD has gained electrons and has therefore been reduced.

Spotlight on...

Free Radicals

Electrons are traded between molecules in the course of oxidation-reduction reactions, particularly in the Krebs cycle and the electron transport chain. In some instances, individual electrons may "leak" out of the process, leaving a molecule with an unpaired electron. These molecules (or fragments of molecules) with unpaired electrons are called free radicals or reactive oxygen species. The formation of a small amount of free radicals is normal wherever oxidative phosphorylation or aerobic metabolism takes place.

The concern about free radical formation is related to the damage that an excessive amount of these molecules can do, referred to as oxidative stress. These reactive oxygen species

can damage cells in the body by inactivating enzymes, breaking DNA strands, binding NAD (thus preventing it from assisting in other oxidation-reduction reactions), and damaging the fatty acids in cell membranes. Excessive oxidative stress has been associated with aging and the development of cancer, atherosclerosis, and other chronic diseases.

Fortunately, the body has defense mechanisms against free radicals, including enzymes for removal and chemicals that can react with the free radicals (antioxidants) to produce a more stable molecule. Free radicals and antioxidants will be discussed in more detail in Chapter 8.

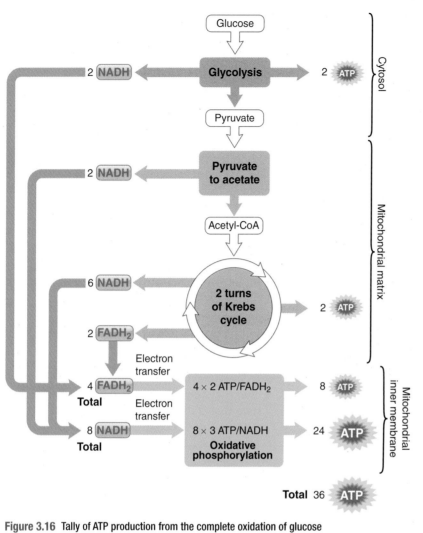

Figure 3.16 Tally of ATP production from the complete oxidation of glucose

Recent evidence suggests that there may be an overestimation of the total ATP yield from the oxidation of glucose and the total may be as low as 32 ATP.

tissues use NAD to shuttle these electrons into the mitochondria, which affects the final tally of ATP. The electrons picked up by NAD in the glycolysis phase can be shuttled into the mitochondria as FADH and entered into the reactions of the electron transport chain. Detailed steps of the electron transport chain can be found in Appendix J.

It is a relatively simple matter to account for the ATP produced from the aerobic metabolism of glucose (Figure 3.16). In the glycolysis phase, 2 ATP are used and 4 are produced, giving a net production of 2 ATP. In addition, 2 NADH are formed in glycolysis. In the Krebs cycle, 1 ATP is produced via GTP, and with two turns of the Krebs cycle for each glucose molecule, 2 ATP are the result. With the two turns of the Krebs cycle, 6 NADH are formed, giving a total of 8 NADH when combined with the 2 formed in the conversion of pyruvate to acetate. Using the traditional energy yield for NADH and FADH in the electron transport chain, 24 ATP are formed from the 8 NADH. With the two turns of the Krebs cycle 2 FADH are formed. The 2 NADH from glycolysis are shuttled into skeletal muscle mitochondria as FADH, resulting in 4 total $FADH_2$ and 8 additional ATP. Therefore, the net ATP for aerobic metabolism of glucose in skeletal muscle is 36 ATP, although the ATP yield may be as low as 32. In other tissues the net ATP may be 38. If the starting point is glycogen rather than glucose, 1 ATP is saved, bringing the total ATP produced to 37 or 39, depending upon the type of tissue.

In comparison to the two other energy systems, oxidative phosphorylation has significant advantages in that it can produce many more ATP and has a virtually limitless duration. However, the series of reactions are very slow compared to creatine phosphate or anaerobic glycolysis, and are dependent upon the provision of oxygen. Due to these limitations, oxidative phosphorylation is the predominant energy system

along a series of complexes containing compounds that shuttle the electrons down the chain. Oxygen is the final electron acceptor, picking up electrons in the form of hydrogen to form water. The rate-limiting enzyme for the electron transport chain is **cytochrome oxidase (CO)**. When the electrons are deposited into the electron transport chain, the resulting series of reactions release enough energy to power the ATP resynthesis reaction. When NAD is the electron carrier, enough energy is released to re-form 3 ATP, and when FAD is the electron carrier, the resulting ATP formation is 2. There is a growing consensus based upon recent research that these figures may slightly overestimate the ATP produced and that the estimates may be closer to 2.5 and 1.5 ATP for NAD and FAD, respectively, although further studies are needed to confirm this view. In skeletal muscle, these electrons are transferred into the mitochondria via FAD; other

Cytochrome oxidase (CO): The rate-limiting enzyme of the Krebs cycle.

Oxidative phosphorylation is the predominant energy system for endurance exercise, such as marathon running.

To summarize the characteristics of the oxidative phosphorylation energy system:

- 124 chemical steps/reactions
- 30 compounds, 27 enzymes
- Rate-limiting enzymes: PFK, IDH, CO
- Slow
- 36 ATP via glucose, 37 ATP via glycogen (in skeletal muscle; may be as low as 32 and 33, respectively)
- Potentially limitless duration
- Aerobic
- Fatigue associated with fuel depletion (for example, muscle glycogen)
- Predominant energy system in endurance exercise, for example, long-distance running

Application exercise

Examine each of the exercise or sports scenarios below and determine the utilization of each of the three major energy systems during that activity. Is there one system that is the predominate source of energy (ATP) replacement? Keep in mind the intensity and the length of time it takes to complete the activity.

1. A baseball player hits a double and sprints from home plate to second base.
2. A soccer midfielder runs back and forth alternately playing offense and defense as the teams change possession of the ball.
3. A triathlete completes the Hawaii Ironman Triathlon®.
4. A track athlete runs a 1,500-meter race.
5. An Olympic swimmer swims in a 200-meter freestyle race.

used at rest and to support low- to moderate-intensity exercise. Examples of activities where oxidative phosphorylation is the predominant energy system include "aerobic" activities, such as:

- Walking
- Jogging and distance running
- Cycling (longer distances, not sprint cycling)
- Swimming (longer distances, not sprints)
- Dance aerobics

Returning to the light controller analogy, the oxidative phosphorylation bulb burns brightly for many activities since they last longer than two minutes and the intensity of the exercise is lower when compared to anaerobic activities. However, the other systems are dimmed, not off. For example, when a distance runner sprints to the finish line the anaerobic energy systems are used.

The examination of the aerobic metabolism, or oxidation, of glucose is a good introduction to oxidative phosphorylation, and gives a consistent basis of comparison to anaerobic glycolysis, as they have the same starting point. What about the metabolism of other nutrients? In addition to carbohydrates, proteins and fats can be metabolized for energy. Figure 3.17 illustrates the overall scheme of the metabolism of carbohydrates, fats, and proteins (the example shows two amino acids) through oxidative phosphorylation. Remember that the first step in the process is the preparation of the compounds for aerobic metabolism. Once acetyl CoA is formed, the steps of the Krebs cycle and the electron transport chain are the same regardless of whether the original source was carbohydrates, proteins, or fats. Carbohydrates,

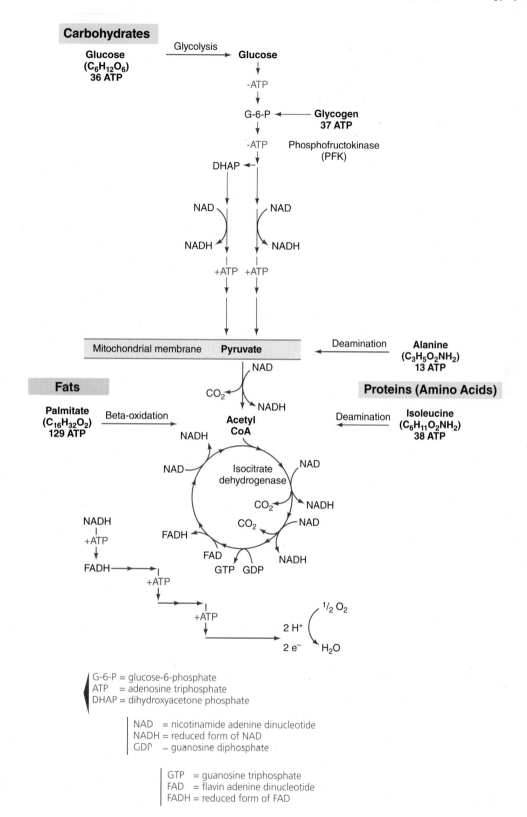

Figure 3.17 Oxidation of carbohydrates, proteins, and fats

Carbohydrates are metabolized as glucose via glycolysis to pyruvate, and produce 36 ATP through oxidative phosphorylation in skeletal muscle. Fats are metabolized in a variety of ways; the fatty acid palmitate is shown. The process of beta-oxidation converts two-carbon portions of palmitate to acetyl CoA where it enters oxidative phosphorylation, eventually producing 129 ATP. Metabolism of alanine and isoleucine are two examples of protein metabolism. After removing the nitrogen group, alanine can enter the metabolic pathway as pyruvate, producing 13 ATP, whereas isoleucine enters as acetyl CoA, resulting in the production of 38 ATP.

proteins, and fats will be discussed in greater detail in Chapters 4, 5, and 6, respectively.

Fats are metabolized aerobically by the oxidation of fatty acids.

Metabolism of fat is complex and varied. The principal fats metabolized for energy in the body are fatty acids containing chains of either 16 or 18 carbons (for example, **palmitate** and stearate, respectively). These fatty acid chains are stored as **triglycerides** in fat cells (**adipocytes**) and in other tissues such as muscle. A triglyceride is composed of a **glycerol** molecule with three fatty acid chains attached. When fat is needed for energy, triglycerides are broken down and the fatty acid chains are transported to mitochondria for aerobic metabolism. **Lipolysis** is the term used for the breakdown of triglycerides.

When fats are to be metabolized, they must first be "mobilized" from the adipocytes. Lipolysis occurs, breaking the triglycerides down into the glycerol and the three fatty acids. The fatty acids are then released from the adipocytes into the blood for transportation throughout the body. At this point, they are usually referred to as free fatty acids (FFA). This term is a bit of a misnomer, as fatty acids must be bound to a protein in order to be carried in the blood. Albumin is the most common plasma protein that binds FFA in the circulation.

As FFA are distributed throughout the body, they are taken up by various tissues such as muscle. After an activation step that requires the use of ATP, the fatty acids are transported into the mitochondria for aerobic metabolism. Before entering the Krebs cycle, however, the fatty acids must pass through another series of reactions called β- (beta-) oxidation.

Beta-oxidation is a process comprising four chemical steps that occur in the mitochondrial matrix. The overall function of β-oxidation is to remove 2-carbon segments from the fatty acid chains and convert them to acetyl CoA. Once the acetyl CoA molecules have been formed, they can enter the Krebs cycle and be oxidized. The process of β-oxidation also results in FAD and NAD picking up electrons, which can be used in the electron transport chain. Eight 2-carbon segments can be cleaved from a 16-carbon fatty acid such as palmitate, one of which is acetyl CoA, seven of which go through the series of chemical steps to form acetyl CoA. Compared to the 36 ATP produced from the aerobic metabolism of glucose, complete metabolism of the fatty acid palmitate results in the production of 129 ATP (Figure 3.17). Details of fat oxidation can be found in Chapter 6.

The clear advantage of fat metabolism is the large ATP yield. In addition, even a person with a relatively low percentage of body fat has a large number of kilocalories stored as fatty acids in adipocytes, which are accessible for metabolism. The disadvantages of fat metabolism include the increased number of steps required and the amount of oxygen needed for aerobic metabolism. This increased oxygen requirement will be discussed in an upcoming section.

Proteins are metabolized aerobically by oxidation of amino acids.

Protein is not "stored" as an energy source in the same sense as are carbohydrates (as glycogen) and fats (as triglycerides, predominantly in adipose tissue). Proteins are typically incorporated into functional (for example, enzymes) or structural (for example, skeletal muscle) elements in the body. Many proteins can, however, be broken down into their amino acid building blocks under certain metabolic conditions. Amino acids can be metabolized aerobically once the amino- or nitrogen-containing group has been removed. Nitrogen groups are typically removed by either being transferred to another compound (**transamination**) or being removed completely (**deamination**). The nitrogen is eliminated from the body as urea in the urine.

Figure 3.17 illustrates the path of oxidation of two amino acids, **alanine** and **isoleucine**. Protein metabolism is complex because there are 20 amino acids and a variety of points of entry into the oxidative process. Alanine and isoleucine are amino acids that are readily oxidized, and are given as examples because each has a slightly different way of being metabolized. Alanine is a 3-carbon amino acid, which when deaminated can be converted to a single pyruvate. The pyruvate can then be oxidized via the Krebs cycle as previously described. Isoleucine, one of the branched chain amino acids, contains six carbons. Each 2-carbon fragment can be converted to acetyl CoA, which can then be oxidized in the Krebs cycle. Three acetyl CoA can be formed from each isoleucine, giving three turns of the Krebs cycle, resulting in a higher ATP yield than for alanine. The branched chain amino acids, leucine, isoleucine, and valine, can be relatively easily metabolized by skeletal muscle.

Of the three energy-containing nutrients, protein is the least preferred as a fuel source. Although amino acids can certainly be metabolized, they are not stored for ready access as are carbohydrates and fats. Along with the additional steps required to remove and dispose of the nitrogen groups, these limitations prevent protein from making up a large portion of the fuel sources for metabolism, particularly during exercise. Under metabolically stressful conditions, such as starvation or exercise in a glycogen-depleted state (for example, latter stages of a marathon), protein may compose a larger percentage of the fuel utilization, but rarely exceeds 10 percent of total energy production.

The Internet café

Where do I find reliable information about energy systems?

It may come as a surprise that when the term *oxidative phosphorylation* is entered into one of the popular Internet search engines, nearly half a million matches are found. Many of the websites containing information about oxidative phosphorylation are university based (look for websites that end in .edu); professors have posted information, and in some cases, animations, to help students understand this complicated aspect of metabolism. The chemical reactions involved in oxidative phosphorylation are well defined, so the scientific information is the same, but it sometimes helps students in their understanding of complicated processes to read different explanations. Similarly, there are university-affiliated sites with information about anaerobic glycolysis, although in much smaller numbers.

In contrast, a search term such as *creatine phosphate energy system* yields many commercial sites (.com). The majority of these sites are selling creatine phosphate supplements, and finding .edu sites that are devoted to explaining the metabolic processes involved in the creatine phosphate energy system is harder.

The respiratory exchange ratio (RER) indicates utilization of carbohydrate and fat as fuels.

Carbohydrates, proteins, and fats can be used as fuel sources, both at rest and during exercise. Which fuel source is used and under what circumstances? How is the source of the fuel used for aerobic metabolism determined?

Although three fuel sources are available, protein is used only to a small degree under normal circumstances. The preferred fuel sources are carbohydrates and fats; which fuel source predominates is largely dependent upon activity level and exercise intensity, and can be modified by diet, activity duration, fitness level or exercise training status, and other factors.

At rest, the preferred fuel source is fat. As described earlier, even a person with a relatively low percentage of body fat has ample fat stores, and fat oxidation has a very high ATP yield. Because the rate of energy expenditure at rest is low and there is no need to replace ATP quickly, there is sufficient time for lipolysis, fat mobilization, and beta-oxidation. Even though fat oxidation requires additional oxygen consumption, there is sufficient time at rest for the pulmonary and cardiovascular systems to supply all of the oxygen that is needed. The body does not rely one hundred percent on fat oxidation at rest, however. Some tissues, such as brain cells, rely on carbohydrates (glucose) as their primary source of

fuel. In addition, excessive reliance on fat metabolism results in the production of keto acids. When produced in excess, keto acids can disturb the acid-base balance in the body. An example of this can be seen in those who have difficulty metabolizing carbohydrates. People with diabetes can be susceptible to diabetic ketoacidosis if their diabetes is not managed well and they rely too extensively on fat metabolism. Under normal circumstances, at rest, fat is generally the predominant fuel source, making up approximately 85 percent of energy expenditure, whereas carbohydrates make up the balance (15 percent).

As activity increases, the body begins to rely more on carbohydrates and less on fat, as a percentage of total energy expenditure. As exercise intensity increases, the percentage of kilocalories derived from fat metabolism decreases and the percentage from carbohydrate oxidation increases. At fairly modest levels of physical activity or exercise, such as walking at a moderate pace, the mixture of fuels is about 50 percent from fats and 50 percent from carbohydrates. As intensity increases above these modest levels, carbohydrate becomes the predominant fuel source. At moderate to hard exercise intensities (for example, a distance runner's race pace during a 10-kilometer/6.2-mile road race), nearly all of the energy expenditure is derived from carbohydrates. There are fewer chemical steps involved in carbohydrate oxidation and less oxygen consumption is required for the amount of ATP produced, making carbohydrates the preferred fuel source at higher intensities of exercise.

The distribution of fat and carbohydrate metabolism can be modified to a certain degree by other factors such as diet. The chronic consumption (that is,

Palmitate: One of the most widely distributed fatty acids found in food and in stored body fat.

Triglyceride: Major storage form of fat in the body; consists of a glycerol molecule and three fatty-acid chains.

Adipocytes: A fat cell.

Glycerol: A structural component of triglycerides, the major storage form of fat in the body.

Lipolysis: Breakdown of fat.

Beta-oxidation: Chemical process of breaking down fatty acid chains for aerobic metabolism.

Transamination: Removal and transfer of a nitrogen group to another compound.

Deamination: Process of removing and eliminating a nitrogen group.

Alanine: An amino acid.

Isoleucine: A branched chain amino acid.

over a period of days or weeks) of a high-carbohydrate diet influences the body to rely more heavily on carbohydrate metabolism to a certain degree, even at rest. Consuming a high-carbohydrate meal before exercise can also influence the body to rely slightly more on carbohydrates at the same exercise intensity compared to not eating or fasting before exercise. Similarly, chronic consumption of a high-fat diet influences the body to rely more on fat oxidation at rest and during low- and moderate-intensity exercise. Higher intensities of exercise, however, still rely predominantly on carbohydrates.

Fuel utilization may change somewhat over the course of a single exercise bout, even if the exercise is performed at the same intensity. At the onset of a moderate-intensity exercise session (for example, jogging), carbohydrates may provide the majority of energy. Given the steps involved in fat oxidation (that is, lipolysis, mobilization, beta-oxidation) there may be a lag time between the onset of exercise and when fat metabolism reaches its peak. There is often a 10- to 20-minute period of time at the beginning of exercise when carbohydrate metabolism takes precedence until fat oxidation can catch up. Duration can also affect the fuel utilized in moderate to higher exercise intensity if the exercise continues for long periods of time. If the body is relying predominantly on carbohydrates, eventually the carbohydrate stores are reduced to the point that fat oxidation becomes the predominant fuel source.

An athlete's training status also affects the mixture of fuels utilized during exercise. Individuals that have engaged in extensive endurance training improve their ability to metabolize fat. At the same absolute exercise intensity (for example, the same running pace), they will utilize a higher percentage of fat and a lesser percentage of carbohydrate than lesser-trained athletes. When exercise intensity is increased (for example, faster running pace), even the well-trained athlete will still reduce the percentage of energy from fat and increase the percentage of energy from carbohydrates.

Use of indirect calorimetry allows for the measurement of both oxygen consumption and carbon dioxide production. The proportion of energy coming from carbohydrates and fats can be estimated accurately by use of the **respiratory exchange ratio (RER)**, a ratio of the amount of carbon dioxide produced to the amount of oxygen consumed (Figure 3.18). This ratio can be determined from $\dot{V}O_2$ and $\dot{V}CO_2$ measured from respiratory gases with a metabolic measurement system. This procedure is based upon a tissue-level exchange of gases during oxidative phosphorylation called the respiratory quotient (RQ).

Complete oxidation of glucose, a six-carbon sugar, requires 6 oxygen molecules (Figure 3.19). In the

$$RER = \frac{\dot{V}CO_2}{\dot{V}O_2}$$

$\dot{V}CO_2$ = carbon dioxide production
$\dot{V}O_2$ = oxygen consumption

Figure 3.18 Respiratory exchange ratio (RER)

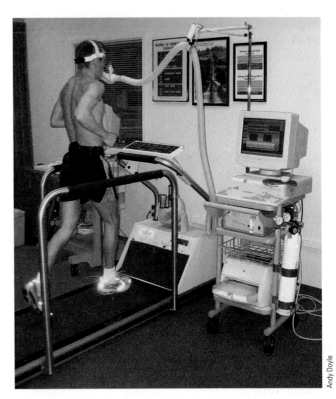

Measurement of the ratio of the amount of carbon dioxide produced to oxygen consumed can be used to estimate carbohydrate and fat fuel utilization during exercise.

$$C_6H_{12}O_6 + 6\,O_2 \rightarrow 6\,CO_2 + 6\,H_2O$$

$$RER = \frac{\dot{V}CO_2}{\dot{V}O_2} = \frac{6\,CO_2}{6\,O_2} = 1.00$$

RER = respiratory exchange ratio
$\dot{V}CO_2$ = carbon dioxide production
$\dot{V}O_2$ = oxygen consumption

Figure 3.19 Respiratory exchange ratio of carbohydrate metabolism—glucose

The metabolism of one molecule of glucose ($C_6H_{12}O_6$) requires 6 oxygen molecules and results in the production of 6 carbon dioxide molecules. The respiratory exchange ratio that results from metabolizing pure carbohydrates is 1.0.

$$C_{16}H_{32}O_2 + 23\ O_2 \rightarrow 16\ CO_2 + 16\ H_2O$$

$$RER = \frac{\dot{V}CO_2}{\dot{V}O_2} = \frac{16\ CO_2}{23\ O_2} = 0.70$$

RER = respiratory exchange ratio
$\dot{V}CO_2$ = carbon dioxide production
$\dot{V}O_2$ = oxygen consumption

Figure 3.20 Respiratory exchange ratio of fat metabolism—palmitate

The metabolism of one molecule of fat (palmitate, $C_{16}H_{32}O_2$) requires 23 oxygen molecules and results in the production of 16 carbon dioxide molecules. The respiratory exchange ratio that results from metabolizing pure fat is 0.7.

Table 3.2 Nonprotein Respiratory Exchange Ratio and Percentages of Energy from Carbohydrates and Fats

RER	Percent CHO	Percent fat
0.70	0	100
0.75	15	85
0.80	32	68
0.85	49	51
0.90	66	34
0.95	83	17
1.00	100	0

RER = respiratory exchange ratio; CHO = carbohydrate

The RER calculated from measured $\dot{V}O_2$ and $\dot{V}CO_2$ can be used to determine the percentage of energy that is being derived from carbohydrate and fat oxidation. The full table can be seen in Appendix I.

Carpenter, T. M. (1964). *Tables, factors, and formulas for computing respiratory exchange and biological transformations of energy* (4th ed., p. 104). Washington, DC: Carnegie Institution of Washington.

process of oxidizing glucose, 6 carbon dioxide molecules are produced. Therefore, if only carbohydrate (glucose) is being metabolized for energy, the RER = 1.0 (6 CO_2 produced divided by 6 O_2 consumed). Fatty acid chains such as palmitate, however, contain many more carbons (16) than oxygen (2) (Figure 3.20). Complete oxidation of fat, therefore, requires significantly more oxygen to be consumed. In the example of palmitate, 23 oxygen molecules are needed for complete oxidation, resulting in the production of 16 carbon dioxide molecules. The resulting RER = 0.70 (16 CO_2 produced divided by 23 O_2 consumed), indicating that fat is the only source of fuel. Because protein is not completely oxidized in the body due to the nitrogen group, and because protein metabolism generally comprises a very small percentage of the fuel utilized for energy, it is generally ignored as a part of this fuel utilization analysis. RER is therefore referred to as the nonprotein RER.

The source of fuel is rarely exclusively from carbohydrates or fats, but is usually a mixture of the two. The percentage that may come from each fuel source can depend on a variety of conditions such as previous diet, exercise intensity, and exercise duration. A metabolic measurement system can be used to determine oxygen consumption, carbon dioxide production, and RER. A table can then be used to determine the relative contribution of each fuel source to the energy expenditure (Table 3.2). For example, if the RER = 0.75, approximately 85 percent of the total energy expenditure is coming from fat oxidation, whereas 15 percent is contributed by the metabolism of carbohydrates. This is the approximate RER and fuel distribution that is often found in humans at rest. At an RER of 0.95, the fuel utilization is nearly reversed, with approximately 83 percent of energy coming from carbohydrates and 17 percent coming from fats. An RER above 0.90 is common for physical activity or exercise that is in the moderate- to hard-intensity range, indicating that carbohydrate is the predominant fuel source for these activities.

Dietary intake influences carbohydrate, fat, and protein metabolism.

This chapter has explained how carbohydrate, fat, protein, and alcohol are metabolized individually to produce energy and the effect that exercise has on the usage of the energy-containing nutrients, especially carbohydrates and fats. However, all metabolic pathways are integrated, and the liver plays a major role in regulating their metabolism. The discussion of energy metabolism is incomplete without considering overall food intake or "energy in." Food provides both short-term and long-term energy for the body. For example, distance athletes consume small amounts of carbohydrates during competition (provides immediate energy) as part of a large amount of dietary carbohydrates daily (provides future energy by increasing muscle glycogen stores). The body has many ways it can adjust its metabolic pathways to meet its short- and long-term energy needs but it has certain pathways that it prefers to use. Some of these pathways cannot be used if the body is in a starvation state.

Respiratory exchange ratio (RER): Ratio of carbon dioxide production to oxygen consumption; used to determine percentage of fats and carbohydrates used for metabolism.

Table 3.3 Metabolic Pathways Favored under Normal and Starvation Conditions

	Liver	Muscle	Adipose tissue	Central nervous system (CNS)
Fed (absorptive) state	Glucose used as energy, stored as glycogen, and converted to fatty acids if energy intake is greater than expenditure; amino acids metabolized; fatty acids transported to adipose tissue for storage as triglycerides	Glucose used for energy or stored as glycogen	Fatty acids are stored as triglycerides (three fatty acids + glycerol)	Glucose from food used to provide energy
Postabsorptive state	Glycogen broken down to provide glucose; manufacture of glucose from lactate and alanine (provided by muscle) and glycerol (provided by the breakdown of fat from adipose tissue) begins	Glucose used for energy, some glycogen storage continues; lactate and alanine released to liver to make glucose; fatty acid uptake (provided by the breakdown of fat from adipose tissue) for use as energy	Triglycerides are broken down to provide fatty acids to muscle and liver; glycerol to liver to be used for glucose	Glucose comes predominantly from liver glycogen
Fasting (18 to 48 hours without food)	Liver glycogen is depleted; glucose made from lactate and amino acids provided by muscle; red blood cells also provide some lactate	Muscle protein degraded to provide amino acids to liver; lactate to liver for glucose synthesis	Same as above	Glucose provided by the liver (from lactate and amino acids)
Starvation (>48 hours without food)	Liver continues to manufacture glucose, predominantly from glycerol (from adipose tissue) to prevent muscle from providing amino acids and lactate; fatty acids broken down to produce ketones (for use by CNS and muscle)	Muscle depends predominantly on fatty acids and ketones for energy	Triglycerides are broken down to provide fatty acids to muscle and liver; glycerol to liver to be used for glucose	CNS depends primarily on ketones produced by the liver for energy

Metabolism is influenced by the fed-fast cycle.

Substrate utilization will depend, in part, on whether the body is in a fed (absorptive) state or a postabsorptive state, the two states humans experience under normal conditions. Other conditions, such as fasting or long-term starvation, force the body to reprioritize its metabolic processes. Table 3.3 shows the metabolic pathways favored under various conditions (Gropper, Smith, and Groff, 2009).

The fed (absorptive) state refers to the time period that surrounds a meal when food is eaten and the nutrients in it are absorbed. It is assumed that a meal is usually a mixture of carbohydrate, fat, and protein and that it will take approximately 3 to 4 hours for full absorption. Insulin is the hormone that primarily affects substrate metabolism at this time. In general (assuming the individual is not obese or diabetic), as a result of insulin, the liver and muscle cells take up glucose, synthesize glycogen, and produce energy (that is, glycolysis). Additionally, fatty acids are synthesized in the liver and stored in adipose tissue, and protein synthesis is increased.

The postabsorptive state is the period of time after the meal is fully absorbed but before the next meal is eaten, which can be minutes but is usually at least a couple of hours when the person is awake. If one goes to sleep at midnight and does not eat breakfast until 10:00 a.m., then the individual has been in a postabsorptive state for 10 hours. Other metabolic pathways are favored in the postabsorptive state, which begins about 3 to 4 hours after eating and may last as long as 12 to 18 hours. The liver provides glucose to the blood by breaking down liver glycogen. Some glucose is also produced from noncarbohydrate sources (that is, gluconeogenesis). Fatty acids are released from adipose cells and are used by both the liver and muscle to produce ATP. Glucagon, a counter-regulatory hormone to insulin, stimulates protein degradation in skeletal muscle, although it also stimulates the synthesis of liver proteins.

Fasting and starvation are extended periods of time when food is not consumed. These different states influence the usage and storage of carbohydrate, fat, and protein, and the hormones that help regulate energy metabolism. For the purposes of discussing energy and metabolism, fasting is defined as 18 to 48 hours without food. Fasting forces the body to adapt its energy systems to an uncommon and threatening circumstance. Glycogen is nearly depleted in the liver, and gluconeogenesis must be increased. This forces the liver to use more amino acids as a source to produce glucose. Eighteen of the 20 amino acids are glucogenic, meaning that they are biochemically capable of being used to manufacture glucose. Some of these amino acids are in muscle cells, and the fasting state increases the rate at which skeletal muscle is broken down. Degrading muscle proteins results in a mixture of amino acids being released; however, one of the most prominent is alanine. Alanine and lactate, also present in muscle cells, and glycerol, obtained from the breakdown of fatty acids stored in adipose tissue, will be transported to the liver for the manufacture of glucose.

Starvation is defined as a total lack of food intake for more than 48 hours but may last for weeks or a few months. Long-term starvation poses a clear threat to the body, and more metabolic adjustments are made. Amino acids must be protected from being used for glucose so that they can be available for the synthesis of vital compounds such as the plasma proteins and enzymes. To meet its glucose needs the body depends on glycerol (released from adipose tissue) to manufacture glucose. The brain uses some of the glucose produced but because the amount is small, it will use ketones predominantly for energy under starvation conditions (it cannot use fatty acids for fuel). Fatty acids become the primary fuel source for skeletal muscle and liver. All these adaptations are made to prevent or delay the breakdown of body proteins.

The fed state favors nutrient storage.

What happens when a person eats a candy bar, which is made up of carbohydrate, fat, and a small amount of protein? Does all the energy contained get oxidized immediately for fuel or does some get stored? If it gets stored, does it get stored as body fat? Those simple questions have surprisingly complicated answers.

In the fed state the most important hormone is **insulin**, and the body's energy priority is delivering glucose from carbohydrate-containing foods. The red blood cells and the cells of the central nervous system cannot store glucose, so some of the glucose from food (in this example, the carbohydrates in the candy bar) gets utilized immediately by those cells. But much of the carbohydrates consumed does not get used immediately for energy; instead, it is stored in muscle and liver cells as glycogen. The body is looking out for both its short-term and long-term energy needs. When nonobese people are in energy balance (that is, energy intake = energy expenditure), carbohydrate is primarily used for immediate energy or glycogen storage. Essentially none is converted to fatty acids and stored as fat.

The reason is the hormone insulin. In some respects insulin is a traffic cop directing the flow of metabolic traffic. Insulin affects cell membranes so glucose can be taken up by the cells and used immediately for energy. It also favors the storage of glycogen by activating the enzymes that help cells store glycogen and inhibiting the enzymes that break glycogen down. Insulin also has an enormous effect on fat storage and creates an environment in which fat from food can be readily stored as body fat. In this example, most of the fat in the candy bar is probably stored in adipose cells.

The fat-storage-friendly environment created in the fed (absorptive) state no longer exists in the postabsorptive state. This is because insulin concentration is low and insulin is no longer acting as a traffic cop. Instead, several other hormones are now directing traffic. Liver glycogen is broken down to provide a slow, steady stream of glucose to the blood. The body is now using the fat that was previously stored in adipose cells for energy. Some of the candy bar fat that was stored in fat cells is now coming out of those cells and being used by other cells for energy. For the nonobese person in energy balance, carbohydrate and fat stores will always be in flux, but that person's overall body composition is relatively stable.

What happens to the amino acids from the small amount of protein in the candy bar? In the fed state the amino acids will go to the liver and this organ will determine how they will be used. Most will be used for anabolic functions, such as building and repairing tissues, not catabolic functions such as using the amino acids for energy. When daily energy intake is balanced with energy expenditure, the carbohydrate and fat in food provide the energy that the body needs. Breaking down amino acids for energy is a low or last priority for the body.

Insulin Hormone produced by the beta cells of the pancreas that helps regulate carbohydrate metabolism among other actions.

Total energy intake is an important factor.

Notice that the example above clearly states that the person is in energy balance and is not obese. If the body is in a fasting, semistarvation, or starvation state, then it must adjust some of the energy pathways to respond to the reduction in food intake. In other words, the body must adapt to a reduced amount of carbohydrate, fat, and protein from food. It still must meet its short- and long-term energy needs but it emphasizes different metabolic pathways when food intake is low or absent.

Metabolic adjustments are also made when energy intake consistently exceeds energy output. In the case of carbohydrate, some is still used for immediate energy, and muscle and liver glycogen stores are still a priority. However, glycogen storage is limited, and sedentary individuals do not use much on a daily basis because they are physically inactive. When total caloric intake consistently exceeds expenditure and carbohydrate intake consistently exceeds the body's need for immediate energy and glycogen storage, carbohydrate undergoes a series of chemical reactions in which it is incorporated into fatty acids and stored in adipose cells (body fat).

Let's go back to the candy bar example and compare what happens in two different circumstances. When nonobese people are in energy balance, carbohydrates are primarily used for immediate energy or glycogen storage. This is what happens to the carbohydrates from the candy bar. Most of the fat from the candy bar get stored in adipose cells in the fed state, but fat gets released in the postabsorptive state and is used for energy. The daily energy intake from food equals daily energy expenditure so body composition remains the same.

Now consider what happens to the person who consistently takes in an excess amount of energy daily. The candy bar is part of the excess energy and carbohydrate intake, so after some carbohydrate is used for immediate energy and some is stored in liver and muscle glycogen, the remainder is converted to fatty acids and stored as body fat. Because the person consumes too much food over the course of the day, more fatty acids are stored in adipose cells than get released in the postabsorptive state. The amount of body fat slowly increases. This outcome is very different from the one in the individual who has an energy balance in which essentially none of the carbohydrates will be converted to fatty acids and fat is released in the postabsorptive state and used for energy by the body. Although this example does not specifically address physical activity, it should be noted that many people who are in energy balance are physically active, and those who

Keeping it in perspective

Understanding the Micro- and Macroaspects of Energy Metabolism

All students of nutrition and exercise physiology need to learn the details of energy metabolism. The study of energy involves understanding ATP and the potential energy that is stored in phosphate bonds. Memorization of chemical pathways, rate-limiting enzymes, cofactors, and all the steps involved in glycolysis, the Krebs cycle, and the electron transport chain helps individuals understand the details of a complicated process that humans cannot see but that sustains life. The microview of energy systems helps athletes understand the ultimate need for carbohydrates, fats, proteins, and vitamins.

In studying the minute details of the individual elements, there is always a risk that students lose sight of the broader perspective on energy. For the athlete, energy systems must be viewed in a larger context—the demands of exercise on the various energy systems and what may be done to manipulate the systems to the competitive athlete's advantage, which in many cases means delaying fatigue. Understanding the details of the creatine phosphate system is important for athletes performing high-intensity exercise but is of minor importance for endurance athletes. Similarly, long-distance athletes are concerned about having an adequate supply of muscle glycogen to provide the carbohydrates for oxidative phosphorylation. Understanding the function of the energy systems is also important to understand the alterations that may need to be made during different phases of an athlete's training (for example, carbohydrate loading prior to an endurance competition). Training can alter the utilization of fats and carbohydrates to some degree. The athlete's perspective on energy systems will be influenced by the physiological demands of the sport.

Another perspective is providing energy substrates through food or supplements. In some cases, the athlete's goal is near-maximum stores of muscle glycogen or creatine phosphate. The broadest perspective is the total amount of energy needed to fuel activity and the consumption of appropriate amounts of carbohydrate, protein, fat, and alcohol. Keeping energy in perspective means focusing sometimes on the smallest details of metabolism and other times focusing on the larger aspects of food intake and exercise output.

are physically active or exercise can store more carbohydrates as glycogen than sedentary people.

The examples used in this section also assume that the individual is nonobese. Obesity complicates the situation because the cells of an obese individual are often insulin resistant and excessive body fat affects the hormonal balance in the body (see Chapter 13). An important point is that athletes should consider not only the individual energy-containing nutrients but also overall energy intake (see Chapter 2).

Key points

- Oxidative phosphorylation (aerobic metabolism) produces ATP relatively slowly but in large amounts.

- Carbohydrates, fats, and proteins can be metabolized via the oxidative phosphorylation energy system, although protein is not a preferred source, particularly for exercise.

- Oxidative phosphorylation is the predominant energy system used at rest and for low- to moderate-intensity activities and is the predominant energy system for endurance exercise.

- Oxidative phosphorylation is referred to as the aerobic energy system because oxygen is used as the final electron acceptor.

- Carbohydrates are metabolized in oxidative phosphorylation as glucose or from the storage form, glycogen.

- Fats, in the form of fatty acid chains, are prepared for oxidative phosphorylation by a series of chemical steps called beta-oxidation.

- The utilization of carbohydrates, fats, and protein depends on the fed-fast cycle.

- The fed state favors nutrient storage due to the powerful influence of insulin.

- When not in a fed state, other hormones influence the body to release nutrient stores.

- Total energy intake influences the use of carbohydrates, fats, and proteins (amino acids) for fuel.

- The utilization of carbohydrate, fat, and protein as fuel is complex and depends upon a variety of factors such as diet composition, activity level, health and/or disease state, hormonal status, and time since the last meal.

Why is protein not a preferred fuel source compared to fat and carbohydrate?

Why is the phrase "you are what you eat" especially true for energy metabolism?

3.5 Oxygen Consumption

The preceding sections described how carbohydrates, fats, and proteins can be metabolized aerobically via the oxidative phosphorylation energy system. All can be broken down chemically to form compounds that can be oxidized by the Krebs cycle, and the removed electrons are subsequently used in the electron transport chain to rephosphorylate ADP to form ATP. As part of this metabolic process, carbon dioxide is produced and oxygen is used as the final electron acceptor. Therefore, aerobic metabolism ultimately depends on an adequate supply of oxygen.

Increased use of aerobic metabolism results in an increase in oxygen consumption.

The body needs a constant supply of ATP, and at rest the vast majority of this energy is provided by oxidative phosphorylation, the aerobic energy system. Any increase in activity requires an increase in energy expenditure, which in turn results in an increase in the activity of oxidative phosphorylation. An increased utilization of the aerobic energy system results in an increase in the amount of oxygen consumed and the amount of carbon dioxide produced. As discussed in Chapter 2, indirect calorimetry can be used to measure energy expenditure by determining oxygen consumption ($\dot{V}O_2$) and carbon dioxide production ($\dot{V}CO_2$) with a metabolic measurement system.

It is common in exercise physiology to study the metabolic response to physical activity and exercise, particularly with incremental exercise tests. In this type of test, exercise intensity is increased every few minutes in stages or increments. A wide variety of physiological responses can be measured in response to the gradually increasing exercise stress. As exercise intensity increases, oxygen consumption increases in a linear fashion until maximum oxygen consumption is reached. That is, with each increment in exercise intensity, oxygen consumption increases by a similar amount (Figure 3.21).

Submaximal exercise. If exercise intensity is increased to a level that is below the person's maximal ability (submaximal) and continued for some time, oxygen consumption rises to a level that provides sufficient energy to support that level of activity, and then plateaus (Figure 3.22). Think of an exercise session lasting 30 minutes, such as a person jogging. At the onset of exercise, oxygen consumption increases, and if the exerciser maintains a steady pace, oxygen consumption will reach a "steady state." Once the person stops the exercise task, the need for increased energy is removed, and oxygen consumption will gradually return to pre-exercise levels.

Note in Figure 3.22 that if exercise begins after 5 minutes of resting, it takes a couple of minutes for

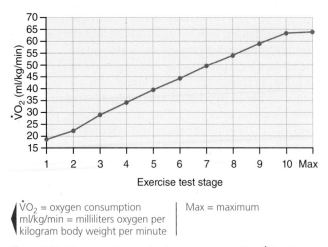

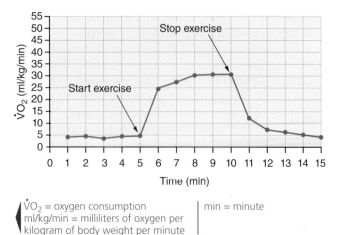

$\dot{V}O_2$ = oxygen consumption
ml/kg/min = milliliters oxygen per
kilogram body weight per minute

Max = maximum

Figure 3.21 The concept of maximal oxygen consumption ($\dot{V}O_{2max}$)

Maximal oxygen consumption is measured with a maximal effort exercise test in which exercise intensity is increased at each stage. As exercise intensity increases, oxygen consumption increases in a linear fashion. Eventually, oxygen consumption does not increase further, although exercise intensity has continued to increase. This plateau in oxygen consumption is called $\dot{V}O_{2max}$

$\dot{V}O_2$ = oxygen consumption
ml/kg/min = milliliters of oxygen per
kilogram of body weight per minute

min = minute

Figure 3.22 Response of oxygen consumption to steady-state exercise

After 5 minutes of resting, exercise begins and oxygen consumption increases and eventually plateaus at the level that is necessary to supply sufficient energy for the given intensity of exercise. Notice that it takes several minutes for oxygen consumption to rise to the required steady-state level; this is the oxygen deficit. Exercise ends at minute 10 and oxygen consumption declines, but it takes several minutes to return to the pre-exercise resting levels; this is the excess postexercise oxygen consumption (EPOC).

oxygen consumption to reach the steady-state level. If one considers the large number of steps that are required in oxidative phosphorylation and the need to transport oxygen to the exercising muscle, it is understandable that some degree of "lag time" exists between the onset of exercise and the ability of the aerobic energy system to fully meet this elevated energy requirement. Where does the energy come from to support the exercise activity during this lag time? Fortunately, the anaerobic energy systems, creatine phosphate and anaerobic glycolysis, exist to provide ATP in these types of circumstances. This is yet another example of the interactions between the various energy systems. Because ATP energy has been "borrowed" from the anaerobic energy systems for what is primarily an aerobic exercise activity, the term **oxygen deficit** is typically used to describe this lag in oxygen consumption at the onset of exercise.

Also, note in Figure 3.22 that exercise is stopped at minute 10 and oxygen consumption begins to decline immediately, but does not reach its original pre-exercise level for several minutes. Even though exercise has ceased and the person is now resting, oxygen consumption remains slightly elevated for some period of time after exercise. This gradual decline in oxygen consumption after exercise has historically been referred to as the **oxygen debt**, but is now more accurately termed the **excess postexercise oxygen consumption (EPOC)** (Gaesser and Brooks, 1984). It is easy to understand what the majority of this extra oxygen consumption is for by

recalling the aerobic recovery portions of the anaerobic energy systems.

When creatine phosphate is used to replenish ATP, creatine remains and must be eventually restored to creatine phosphate by the creatine shuttle. The creatine shuttle uses the energy from ATP that is produced aerobically in the mitochondria to rephosphorylate creatine into creatine phosphate. Therefore, any creatine phosphate used during the aerobic lag time in the exercise example would require some additional oxygen consumption after exercise to restore creatine to pre-exercise creatine phosphate levels. Similarly, lactate that may be produced by use of anaerobic glycolysis during the aerobic lag time also depends upon subsequent oxygen consumption. Lactate molecules are eventually taken up by highly aerobic tissues, converted to pyruvate, and metabolized aerobically in mitochondria. There are a number of additional factors that are responsible for the elevated oxygen consumption after exercise, but the majority of this excess oxygen consumption is devoted to "paying back" the energy borrowed from creatine phosphate and anaerobic glycolysis at the onset of the exercise task.

Each individual has a maximal ability to consume oxygen, or $\dot{V}O_{2max}$.

If an incremental exercise test continues to increase in intensity, oxygen consumption will continue to increase as well, up to a point. When this point is reached, if exercise intensity is increased again, oxygen consumption does not rise any further; rather, it plateaus (Figure 3.21). Each individual has an inherent maximal ability to consume oxygen, a characteristic known as **maximal oxygen consumption**. The exercise physiology term is $\dot{V}O_{2max}$.

A person's maximal ability to consume oxygen is a reflection of one's capacity to take up oxygen from the atmosphere, circulate it throughout the body, and use oxidative phosphorylation in muscles and other

Oxygen deficit: The lag in oxygen consumption at the beginning of an exercise bout.

Oxygen debt: See *excess postexercise oxygen consumption*.

Excess postexercise oxygen consumption (EPOC): The elevated oxygen consumption that occurs for a short time during the recovery period after an exercise bout has ended; replaces the older term *oxygen debt*.

Maximal oxygen consumption ($\dot{V}O_{2max}$): Highest amount of oxygen that can be utilized by the body; the maximal capacity of the aerobic energy system.

Spotlight on...

Alcohol Metabolism

Too often alcohol is left out of discussions of energy metabolism. Certainly carbohydrates and fats deserve the most attention because they are the primarily sources of energy at rest and during activity. However, when alcohol is consumed, one must account for its metabolism.

The chemical term for alcohol found in alcoholic beverages is *ethanol*. Ethanol is absorbed rapidly beginning in the stomach. If the stomach does not contain food, the absorption is especially rapid because the ethanol molecules will diffuse into the blood as soon as they touch the stomach cells. The presence of food in the stomach slows the absorption of alcohol because there is less surface area available for contact. Food also results in the contents of the stomach being emptied more slowly into the intestine. Once in the intestine alcohol is quickly absorbed. Although food slows the relative rate of absorption, it is important to remember that when alcohol is consumed under any conditions, ethanol absorption is rapid.

Ethanol is transported to the liver where it is metabolized. There are two primary pathways. When ethanol is consumed at low to moderate levels (defined below) the primary pathway is the alcohol dehydrogenase (ADH) pathway. In this pathway, ethanol enters the liver cells, and the enzyme ADH breaks down ethanol into acetaldehyde by removing hydrogen and transferring it to a niacin-containing molecule, NAD, to form NADH. Acetaldehyde is highly toxic to cells and must be converted to acetate, which is then converted to acetyl CoA. These reactions take place in the cytoplasm of the cell. For low to moderate alcohol consumption, this pathway can metabolize all the ethanol consumed.

When alcohol (that is, ethanol) consumption is high, the ADH pathway is overwhelmed because the rate at which ADH breaks down ethanol does not increase as ethanol

consumption increases. Thus a second pathway is used to metabolize ethanol when alcohol consumption is high. This pathway is known as the microsomal ethanol-oxidizing system (MEOS) and takes place in the microsome. The microsome is a small structure found in the cellular fluid. Ethanol is converted to acetaldehyde, but the enzymes (for example, cytochrome P450) and the by-products (for example, NADP, reactive oxygen species) are different. Of significance, the reactive oxygen species, which are also known as free radicals, damage the liver cells (Lieber, 2003; Murray et al., 2003).

When alcohol is present, the breakdown of ethanol is the body's highest priority because an intermediary compound, acetaldehyde, is toxic. Alcohol contains energy (that is, 7 kcal/g) because it is converted to acetyl CoA and oxidized via the Krebs cycle. Ethanol must be used immediately for energy; it cannot be stored as glycogen or in adipose tissue. Sometimes people say that alcohol "makes you fat" but an increase in stored fat due to alcohol intake is indirect. When alcohol intake is part of excess energy intake, the energy contained in the fat consumed via food is diverted to storage (that is, adipocytes) rather than being used as energy immediately. Athletes often remove alcohol from their diets when they are trying to restrict caloric intake as part of an effort to lose weight as body fat.

As stated above, the pathway(s) used to metabolize ethanol depend on the amount consumed. Moderate alcohol intake is defined as the consumption of up to one drink per day for women and up to two drinks per day for men. A drink is defined as ½ ounce (oz) of ethanol. In practical terms, each of the following is considered a drink: 3 to 4 oz wine; 10 oz wine cooler; 12 oz beer; 1.5 oz hard liquor (for example, rum, whiskey).

tissues to produce ATP. This ability is influenced by genetics, age, gender, and a host of other factors, the most influential of which is aerobic exercise training. Maximal oxygen consumption is not a static figure—it can be increased markedly if a person engages in regular aerobic exercise or can diminish over time if a person is sedentary.

As discussed in Chapter 2, there are two common ways to express oxygen consumption, in absolute or relative terms. Absolute oxygen consumption (in liters per minute or L/min) refers to the total amount of oxygen consumed every 60 seconds. In addition to aerobic training, maximal oxygen consumption is largely influenced by body size—larger people consume more oxygen. In order to draw accurate comparisons between people of different body sizes, $\dot{V}O_{2max}$ is often expressed in terms relative to body weight (milliliters per kilogram body weight per minute or ml/kg/min). Sports nutrition professionals should be familiar with approximate values for maximal oxygen consumption such as those shown in Table 3.4. Another approach to understanding $\dot{V}O_{2max}$ is to look at normative values and percentile ranks obtained by testing large numbers of people. One such set of normative values was developed from the Aerobic Center Longitudinal Study at the Cooper Institute and are used in the *ACSM's Guidelines for Exercise Testing and Prescription* (Appendix H). These "norms" are listed by percentile rank for males and females and by age ranges by decade. For example, the 50th percentile $\dot{V}O_{2max}$ for a male between 20 and 29 years old is 43.9 ml/kg/min and 37.4 ml/kg/min for a female in the same age range.

The descriptions and ranges used in Table 3.4 are general categories, some of which do not have precise definitions. For both males and females, the average resting oxygen consumption is estimated to be approximately 3.5 ml/kg/min (similar to resting heart rate is 72 beats per minute and resting blood pressure is 120/80 mm Hg), but there are large individual variations. The Health/Fitness category represents a *minimum* level of aerobic fitness that has been associated with improved health and a decrease in risk of premature death from chronic disease (Blair et al., 1989). The results of this and other studies imply that for good health, all people should have a $\dot{V}O_{2max}$ that at the very minimum meets this level (and preferably exceeds it). Average $\dot{V}O_{2max}$ for young adults (less than 40 years of age) is approximately 40–45 ml/kg/min for males and 35–40 ml/kg/min for females. These figures were derived from the 50th percentile rank from normative values in *ACSM's Guidelines for Exercise Testing and Prescription* (8th edition, 2010), as were the values for the 90th percentile and above, designated as "Superior." Designations of "well-trained" and "elite" are often associated with endurance athletes or those with high levels of aerobic training and fitness. Although no universally accepted definition exists, it is reasonable to suggest that males with a $\dot{V}O_{2max}$ in excess of 60 ml/kg/min and females exceeding 50 ml/kg/min can be considered well-trained from an aerobic or endurance perspective. The highest measured values for $\dot{V}O_{2max}$ in humans, those athletes who combine the genetic predisposition with many years of intense training, can be slightly in excess of 90 ml/kg/min for males and just over 80 ml/kg/min for females.

Oxygen consumption is influenced by different skeletal muscle fiber types.

Skeletal muscle fibers have a range of anatomical, biochemical, and performance characteristics. Some muscle fiber types produce high amounts of force, produce that force very quickly, and are known as fast-twitch fibers. Because fast-twitch fibers produce a high amount of force quickly, they utilize large amounts of ATP quickly, and therefore must rely on creatine phosphate and anaerobic glycolysis, our fastest energy systems to replace ATP. Fast-twitch fibers, sometimes referred to as type IIb fibers, are also called glycolytic for their reliance on anaerobic energy systems. It is not hard to discern that these fibers have a lesser ability to use oxidative phosphorylation and therefore to consume oxygen. Individuals who have a higher percentage of fast-twitch muscle fibers will have a limitation to the amount of oxygen they can consume, and will have a limited ability

Table 3.4 Relative Values for $\dot{V}O_{2max}$ Expressed in Relative Terms as Milliliters of Oxygen per Kilogram Body Weight per Minute (ml/kg/min)

	Male	Female
Health/fitness[a]	35	32.5
"Average"[b]	40–45	35–40
"Superior"[c]	54+	47+
"Well-trained"[d]	60+	50+
"Highest" elite[e]	90+	80+

[a]Minimum value associated with decreased risk of premature death from chronic disease.

[b,c]50th and 90th percentile rank from population norms (Appendix H; Thompson, W. R., ed. [2010]. *ACSM's Guidelines for Exercise Testing and Prescription* [8th ed.]. Baltimore: Lippincott Williams & Wilkins.)

[d,e]Observed values of individual athletes.

to increase their maximal oxygen consumption with training.

On the other hand, slow-twitch muscle fibers produce less force and produce that force more slowly, making them ideal for endurance-type activities. Because force is produced more slowly, these fibers, also known as type I fibers, can take advantage of the slower process of oxidative phosphorylation as the energy system to replenish ATP. These fibers have a greater capillary network, more myoglobin, a larger number and larger size mitochondria, and greater oxidative enzyme activity than fast-twitch fibers, greatly enhancing their capacity to use our aerobic energy system and consume oxygen. An individual with a higher percentage of slow-twitch muscle fibers will display an enhanced ability to consume oxygen and increase their $\dot{V}O_{2max}$ with training.

Key points

- As exercise intensity increases and the use of oxidative phosphorylation increases to replenish ATP, the amount of oxygen consumed by the body increases.

- At the onset of exercise there is often a lag between energy demand and energy provided by oxidative phosphorylation, creating an "oxygen deficit."

- When exercise is stopped, oxygen consumption remains elevated during the recovery period for some time in order to rephosphorylate creatine phosphate used and oxidize lactate produced at the onset of exercise, known as excess postexercise oxygen consumption (EPOC).

- Individuals have a maximal ability to consume oxygen, known as maximal oxygen consumption or $\dot{V}O_{2max}$.

- The respiratory exchange ratio (RER), the ratio of carbon dioxide produced to oxygen consumed, can be used to determine fuel utilization, or the proportion of energy derived from fat and carbohydrate.

Why is there a lag time in the rise in oxygen consumption at the onset of exercise?

Post-Test   Reassessing Knowledge of Energy Systems and Exercise

Now that you have more knowledge about energy systems and exercise, read the following statements and decide if each is true or false. The answers can be found in Appendix O.

1. The direct source of energy for force production by muscle is ATP.

2. Creatine supplements result in immediate increases in strength, speed, and power.

3. Lactate is a metabolic waste product that causes fatigue.

4. The aerobic energy system is not active during high-intensity anaerobic exercise.

5. At rest and during low levels of physical activity, fat is the preferred source of fuel for the aerobic energy system.

Summary

The direct source of energy for most cellular processes is ATP. Creatine phosphate, anaerobic glycolysis, and oxidative phosphorylation are the three major energy systems that are used to replenish ATP as it is being utilized. Each of these energy systems is active at all times, but one of the systems may be the predominant source of energy for ATP replenishment, depending upon a variety of factors related to energy need, such as duration and intensity of

activity and force requirements. The advantage of the creatine phosphate system is its rapid production of ATP; however, the amount of ATP produced is very small. Anaerobic glycolysis can produce ATP relatively rapidly, but this system is limited by the accumulation of lactate and the associated cellular pH and muscle fatigue. These energy systems are also advantageous because they do not rely on the use of oxygen.

Oxidative phosphorylation can provide a virtually unlimited amount of ATP, albeit slowly, because many chemical steps and the presence of oxygen are required. Carbohydrates, proteins, and fats can be metabolized aerobically through the oxidative phosphorylation energy system. Protein metabolism generally accounts for a very small proportion of energy expenditure, except under relatively extreme conditions such as starvation. At rest, fat metabolism is the source of approximately 85 percent of energy expenditure. The proportion of carbohydrate and fat fuel utilization can change, depending upon a number of factors such as previous diet and intensity and duration of activity. As exercise intensity increases above moderate levels, carbohydrates become the predominant fuel source for energy expenditure.

Self-Test

Multiple-Choice

The answers can be found in Appendix O.

1. Which major factor limits the use of the creatine phosphate energy system?
   a. The pH of the cell increases.
   b. Creatine phosphate can be depleted.
   c. Muscle glycogen can be depleted.
   d. Oxygen consumption increases.

2. Which of the following is a true statement regarding creatine supplementation?
   a. It has been shown to be beneficial for both strength and endurance athletes.
   b. It has a direct effect on improving strength, speed, and power for strength athletes.
   c. It causes dehydration and muscle cramping in strength athletes.
   d. It allows strength and power athletes to sustain high-intensity training.

3. The preferred energy source by exercising muscle for the process of anaerobic glycolysis is:
   a. glucose.
   b. glycogen.
   c. glycerol.
   d. creatine.

4. How does the amount of ATP produced from the breakdown of one molecule of carbohydrate compare to the breakdown of one molecule of a fatty acid?
   a. The amount produced is about the same.
   b. The amount produced via carbohydrate is about two times greater.
   c. The amount produced via fat is about three and a half times greater.
   d. The amount produced via fat is about five times greater.

5. The respiratory exchange ratio (RER) is a way of determining the body's metabolism of:
   a. creatine phosphate and ATP.
   b. amino acids.
   c. carbohydrate and fat.
   d. lactate.

Short Answer

1. How is ATP replenished by ADP?
2. What energy system supplies the predominant amount of energy during very high-intensity, short-term exercise (for example, baseball player sprinting to first base [30 yards])?
3. How is creatine rephosphorylated after its use in skeletal muscle?
4. What is the predominant energy system during high-intensity exercise lasting 1 to 2 minutes?
5. How is lactate that is produced as a result of anaerobic glycolysis disposed of or used in the body?
6. What is the predominant energy system during moderate-intensity, long-term exercise?
7. What is meant by the term $\dot{V}O_{2max}$?
8. When aerobic exercise begins, why does it take a few minutes for oxygen consumption to rise to a level that will supply sufficient energy to support the exercise? Where does the energy for exercise come from during this period of time?
9. Why does oxygen consumption remain elevated for a period of time after exercise?
10. Why are proteins not a preferred fuel source when compared to carbohydrates and fats?
11. Compare the advantages and disadvantages of carbohydrates and fats as fuel sources, particularly for exercise.
12. Briefly explain what happens to the carbohydrate, fat, and protein eaten by a nonobese person in energy balance in the fed and postabsorptive periods.

Critical Thinking

1. A group of competitive runners is at the starting line of a 1,500-meter (metric mile) race. Describe the use of each of the major energy systems from the time the gun goes off to start the race until they cross the finish line approximately four minutes later.
2. Why are anaerobic energy systems not entirely anaerobic?

References

American College of Sports Medicine. (2000). The physiological and health effects of oral creatine supplementation. *Medicine and Science in Sports and Exercise, 32*(3), 706–717.

Bessman, S. P., & Carpenter, C. L. (1985). The creatine-creatine phosphate energy shuttle. *Annual Review of Biochemistry, 54*, 831–862.

Blair, S. N., Kohl, H. W., Paffenbarger, R. S., Clark, D. G., Cooper, K. H., & Gibbons, L. W. (1989). Physical fitness and all-cause mortality: A prospective study of healthy men and women. *Journal of the American Medical Association, 262*(17), 2395–2401.

Branch, J. D. (2003). Effect of creatine supplementation on body composition and performance: A meta-analysis. *International Journal of Sport Nutrition and Exercise Metabolism, 13*(2), 198–226.

Burke, D. G., Chilibeck, P. D., Parise, G., Candow, D. G., Mahoney, D., & Tarnopolsky, M. (2003). Effect of creatine and weight training on muscle creatine and performance in vegetarians. *Medicine and Science in Sports and Exercise, 35*(11), 1946–1955.

Carpenter, T. M. (1964). *Tables, factors, and formulas for computing respiratory exchange and biological transformations of energy* (4th ed., p. 104). Washington, DC: Carnegie Institution of Washington.

Crowe, M. J., O'Connor, D. M., & Lukins, J. E. (2003). The effects of beta-hydroxy-beta-methylbutyrate (HMB) and HMB/creatine supplementation on indices of health in highly trained athletes. *International Journal of Sport Nutrition and Exercise Metabolism, 13*(2), 184–197.

Gaesser, G. A., & Brooks, G. A. (1984). Metabolic bases of excess post-exercise oxygen consumption: A review. *Medicine and Science in Sports and Exercise, 16*(1), 29–43.

Gladden, B. (2004). Lactate metabolism: A new paradigm for the third millennium. *Journal of Physiology, 558*(1), 5–30.

Gropper, S. S., Smith, J. L., & Groff, J. L. (2009). *Advanced nutrition and human metabolism*. Belmont, CA: Thomson/Wadsworth.

Hirvonen, J., Rehunen, S., Rusko, H., & Härkönen, M. (1987). Breakdown of high-energy phosphate compounds and lactate accumulation during short supramaximal exercise. *European Journal of Applied Physiology and Occupational Physiology, 56*(3), 253–259.

Hultman, E. S., Timmons, J. A., Cederblad, G., & Greenhaff, P. L. (1996). Muscle creatine loading in men. *Journal of Applied Physiology, 81*, 232–237.

Lieber, C. S. (2003). Relationships between nutrition, alcohol use, and liver disease. *Alcohol Research & Health, 27*(3), 220–231.

Lopez, R. M., Casa, D. J., McDermott, B. P., Ganio, M. S., Armstrong, L. E., & Maresh, C. M. (2009). Does creatine supplementation hinder exercise heat tolerance or hydration status? A systematic review with meta-analyses. *Journal of Athletic Training, 44*, 215–223.

Mayhew, D. L., Mayhew, J. L., & Ware, J. S. (2002). Effects of long-term creatine supplementation on liver and kidney functions in American college football players. *International Journal of Sport Nutrition and Exercise Metabolism, 12*(4), 453–460.

Murray, R. K., Granner, D. K., Mayes, P. A., & Rodwell, V. W. (2003). *Harper's illustrated biochemistry* (26th ed.). New York: McGraw-Hill.

Poortmans, J. R., & Francaux, M. (2000). Adverse effects of creatine supplementation: Fact or fiction? *Sports Medicine, 30*(3), 155–170.

Schilling, B. K., Stone, M. H., Utter, A., Kearney, J. T., Johnson, M., Coglianese, R., et al. (2001). Creatine supplementation and health variables: A retrospective study. *Medicine and Science in Sports and Exercise, 33*(2), 183–188.

Thompson, W. R. (Ed.). (2010). *ACSM's guidelines for exercise testing and prescription* (8th ed.). Baltimore: Lippincott Williams & Wilkins.

Tipton, K. D., & Ferrando, A. A. (2008). Improving muscle mass: Response of muscle metabolism to exercise, nutrition and anabolic agents. *Essays in Biochemistry, 44*, 85–98.

Volek, J. S., Duncan, N. D., Mazzeitti, S. A., Staron, R. S., Putukian, M., Gomez, A. L., et al. (1999). Performance and muscle fiber adaptations to creatine supplementation and heavy resistance training. *Medicine and Science in Sports and Exercise, 31*(8), 1147–1156.

Volek, J. S., & Rawson, E. S. (2004). Scientific basis and practical aspects of creatine supplementation for athletes. *Nutrition, 20*(7–8), 609–614.

Watt, K. K., Garnham, A. P., & Snow, R. J. (2004). Skeletal muscle total creatine content and creatine transporter gene expression in vegetarians prior to and following creatine supplementation. *International Journal of Sport Nutrition and Exercise Metabolism, 14*(5), 517–531.

4 Carbohydrates

Learning Plan

- **Classify** carbohydrates according to their chemical composition.
- **Describe** the digestion and absorption of carbohydrates.
- **Explain** the metabolism of glucose.
- **Describe** how muscle glycogen and blood glucose are used to fuel exercise.
- **State** carbohydrate recommendations for athletes, including specific guidelines for intake before, during, and after exercise.
- **Discuss** the glycemic response of carbohydrate-containing foods and the use of the glycemic index by athletes.
- **Determine** the daily carbohydrate needs of an athlete and select carbohydrate-containing foods to meet the recommended intake.
- **Compare** and contrast carbohydrate loading protocols.
- **Assess** an athlete's carbohydrate intake and determine if it meets guidelines for performance and health.

Pasta is a good source of carbohydrate, a nutrient critical to sports and exercise performance.

Pre-Test | **Assessing Current Knowledge of Carbohydrates**

Read the following statements and decide if each is true or false.

1. The body uses carbohydrates primarily in the form of fruit sugar, or fructose.

2. Sugars such as sucrose (table sugar) are unhealthy and should rarely be a part of an athlete's diet.

3. Low levels of muscle glycogen and blood glucose are often associated with fatigue, particularly during moderate- to high-intensity endurance exercise.

4. A diet that contains 70 percent of total kilocalories as carbohydrates will provide the necessary amount of carbohydrates for an athlete.

5. Most athletes consume enough carbohydrates daily.

Carbohydrates are compounds that contain carbon, hydrogen, and oxygen. The presence of these three atoms also gives the word **carbohydrate** its common abbreviation—CHO. Carbohydrates are found in food as **sugars**, **starches**, and **cellulose**. Carbohydrates are found in the body predominantly in the form of **glucose** (mostly in the blood) and in the storage form of **glycogen** (in many tissues, predominantly muscle and liver).

Carbohydrates are the primary energy source for moderate to intense exercise and provide approximately 4 kcal/g. The largest amount of carbohydrate in the body is stored in the form of muscle glycogen. Smaller amounts are stored as liver glycogen, which helps to maintain normal concentrations of glucose (a sugar) in the blood. Carbohydrates found in food replenish the carbohydrates used, although the body has a limited ability to make glucose from other substances.

The sugars and starches found in food provide energy because the body can digest and absorb these kinds of carbohydrates. The cellulose found in starchy foods does not provide energy because humans do not possess the enzymes necessary to digest it. However, cellulose and other **fibers** are important forms of carbohydrates since fiber is needed for good health. In addition to energy, starchy foods also contain vitamins, minerals, and other nutrients. Sugars provide only energy and do not contain vitamins, minerals, or fiber.

Athletes use carbohydrates as a source of energy and nutrients. Training and competition reduce carbohydrate stores, which must be replenished on a daily basis. The timing of carbohydrate intake can be important, especially immediately

Christina Micek

Carbohydrates can be found in a wide variety of foods including breads, cereals, pastas, beans, fruits, vegetables, milk, and nuts.

after exercise when muscle glycogen resynthesis begins. Consumption of carbohydrates may be necessary before and during training or competition, and athletes can choose from a variety of liquid, solid, or semisolid products containing the proper amount and type of carbohydrates. Carbohydrate-containing foods also taste sweet, making them a palatable energy source.

4.1 Carbohydrates in Food

Carbohydrates are found in various forms in food.

To understand the differences in the various forms of carbohydrates in food, one must look more closely at their chemical composition. Carbohydrates are

Table 4.1 Characteristics of Monosaccharides

Chemical name	Sweetness (100 = Sweetness of table sugar)	Glycemic index (based on 100)	Miscellaneous
Glucose	75	100	In the body, found circulating in the blood and stored as glycogen. In food, generally found as part of disaccharides and polysaccharides (starches). When added to food, glucose is referred to as dextrose.
Fructose	170	19	In the body, found temporarily in the liver before being converted to glucose. In food, found naturally in fruits and vegetables and added to processed foods typically as high-fructose corn syrup.
Galactose	30	Unknown	Found in food only as part of lactose.

generally classified as **monosaccharides**, **disaccharides**, or **polysaccharides.** It helps to know that *saccharide* means sugar, and *mono* means one, *di* means two, and *poly* means many. Therefore, a monosaccharide consists of one sugar molecule, a disaccharide two sugar molecules, and a polysaccharide many sugar

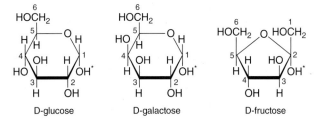

Figure 4.1 Chemical structure of the monosaccharides—glucose, galactose, and fructose

The chemical structures are shown using the Haworth model. The carbons are numbered, and the * refers to the anomeric carbon, which forms the ring structure, is the reducing end of the molecule, and can react with the OH group.

Carbohydrates: Sugars, starches, and cellulose. Chemical compound made from carbon, hydrogen, and oxygen.

Sugar: Simple carbohydrates (mono- or disaccharides); in everyday language, used interchangeably with *sucrose.*

Starch: A polysaccharide.

Cellulose: The main constituent of the cell walls of plants.

Glucose: Sugar found naturally in food, usually as a component of food disaccharides and polysaccharides. Glucose is a monosaccharide.

Glycogen: A highly branched glucose chain. The storage form of carbohydrates in humans and animals.

Fiber: An indigestible carbohydrate. Fiber is a polysaccharide.

Monosaccharide: A one-sugar unit. Mono = one, saccharide = sugar. Glucose, fructose, and galactose are monosaccharides.

Disaccharide: A two-sugar unit. Di = two, saccharide = sugar. Sucrose, lactose, and maltose are disaccharides.

Polysaccharide: Chains of glucose molecules such as starch. Poly = many, saccharide = sugar.

Fructose: Sugar found naturally in fruits and vegetables. May also be processed from corn syrup and added to foods. Fructose is a monosaccharide.

Galactose: Sugar found naturally in food only as part of the disaccharide lactose. Galactose is a monosaccharide.

Lactose: Sugar found naturally in milk. May also be added to processed foods. Lactose is a disaccharide made up of glucose and galactose.

Sucrose: A disaccharide made of glucose and fructose.

Maltose: Sugar produced during the fermentation process that is used to make beer and other alcoholic beverages. Maltose is a disaccharide made up of two glucose molecules.

molecules. Sugar alcohols are derived from mono- and disaccharides and are discussed separately (see Spotlight on...Sugar Alcohols).

The three monosaccharides found in foods are glucose, **fructose**, and **galactose**. Their structures are shown in Figure 4.1, and their characteristics are outlined in Table 4.1.

In which foods are these monosaccharides found? Although glucose can be found by itself in foods, most of the time it is a component of food disaccharides and polysaccharides. Fructose is naturally found in fruits and vegetables, but the largest amount of fructose in American diets is added to foods when they are processed, such as the addition of high-fructose corn syrup as a sweetener. Galactose is a monosaccharide but it is found naturally in food only as part of the disaccharide, **lactose**.

There are three disaccharides found in food—**sucrose**, lactose and **maltose**. Their chemical structures and characteristics are shown in Figure 4.2 and Table 4.2. Sucrose is made of one molecule of glucose and one molecule of fructose. Lactose is a combination of one molecule each of glucose and galactose. Maltose is made up of two glucose molecules.

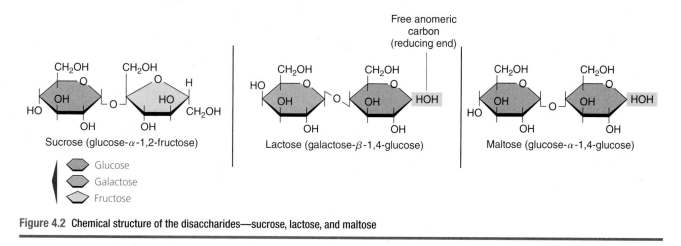

Figure 4.2 Chemical structure of the disaccharides—sucrose, lactose, and maltose

Sucrose is found naturally in fruits, vegetables, honey, and maple syrup. It is also found in sugar beets and sugar cane, which are processed into white and brown sugar. Sucrose is added to many processed foods. Lactose is naturally found in milk. It is often referred to as milk sugar and is sometimes added to processed foods. Maltose is produced during the **fermentation** process that is used to make maltose syrup (commonly used in Asian cooking), beer, and other alcoholic beverages.

Polysaccharides are chains of glucose. These glucose chains are known as starch, fiber, and glycogen. Starches in food may be straight chains (amylose) or branched chains (amylopectin). Enzymes in the digestive tract help to break down these chains into their basic component, glucose. Fiber is a tightly packed polysaccharide that is the structural component of plants. Humans lack the enzymes needed to break down fiber. Glycogen is a highly branched glucose chain and is the form in which humans and animals store carbohydrates in their bodies (Figure 4.3). Although glycogen is found in muscle and liver tissue of live animals, it is not considered a food source of carbohydrates for humans, because it degrades rapidly.

Starch is found in many foods including grains, **legumes** (beans), and tubers. Grains are grasses that

bear seeds and include wheat, corn, rice, rye, oats, barley, and millet, and foods that are made from them such as breads, cereals, and pasta. Legumes are plants that have a double-seamed pod containing a single row of beans. Examples of legumes are lentils, split peas, black-eyed peas, and many kinds of beans such as soy, kidney, lima, and northern beans. Beans are also known by their color—white, pink, red, or black. Tubers, such as white or sweet potatoes and yams, have underground stems and are often referred to as starchy vegetables.

Dietary fiber is found naturally in grains. Whole grains, which contain the **endosperm**, the **germ**, and the **bran**, have more fiber than grains that are highly refined. When whole grains are processed, the germ and the bran are removed. Since these two parts contain most of the fiber, the processing results in a substantial loss of fiber. Other sources of dietary fiber include legumes, seeds, fruits, and vegetables, including the starchy vegetables.

In addition to dietary fiber, some processed foods may contain added fibers that have been extracted from plants. Ingredients such as cellulose, **guar gum**, or pectin are often added to foods in small amounts. These are known as functional fibers and are generally added to affect the food's texture by making it more

Table 4.2 Characteristics of Disaccharides

Chemical name	Monosaccharide composition	Sweetness (100 = sweetness of table sugar)	Glycemic index (based on 100)	Miscellaneous
Sucrose	Glucose + fructose	100	68	Found in fruits, vegetables, honey, and maple syrup; sugar beets and sugar cane are processed into white and brown sugar.
Lactose	Glucose + galactose	15	46	Most adults lose their ability to digest lactose (milk sugar).
Maltose	Glucose + glucose	40	105	Minor disaccharide in most diets.

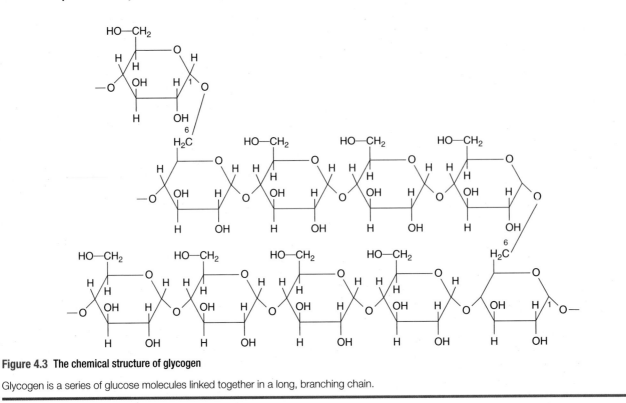

Figure 4.3 The chemical structure of glycogen

Glycogen is a series of glucose molecules linked together in a long, branching chain.

solid. Fiber supplements may also contain these fibers or another functional fiber, **psyllium**.

Carbohydrates are classified in different ways.

There is no single, perfect way to classify the various carbohydrates found in foods, so a number of terms have been used to distinguish them. Often carbohydrates are divided into two categories: sugars and starches. Sugars are also known as simple sugars or simple carbohydrates, and starches are referred to as complex carbohydrates. The widespread processing of carbohydrate-containing foods has given rise to new terms. The term *highly processed* refers to foods that are primarily sugar (for example, sugared beverages) or products made from grains

that have been highly refined and sweetened (for example, sugared cereals). In contrast, whole grains and foods made from them are referred to as minimally processed, fiber-containing, or quality carbohydrates. All of these terms are an attempt to distinguish carbohydrate foods based on their nutrient content.

Unfortunately, carbohydrates have also been referred to as *good* and *bad*. *Bad* has been used to describe highly processed, fiber-deficient, and/or highly sweetened carbohydrate foods and beverages. Although the original intention was to distinguish more nutritious foods from less nutritious ones, this terminology is unfortunate since the words *good* and *bad* carry powerful language and cultural connotations. Rather than simply labeling carbohydrate foods as good or bad, one should consider the context and frequency of their consumption. There are circumstances that make it appropriate to consume carbohydrates typically referred to as bad. For example, the consumption of large amounts of simple sugars is generally discouraged because they are considered "empty calories." *Empty* refers to a lack of nutrients, not a lack of calories, and excess consumption of such foods and beverages may contribute to obesity and malnutrition. But, there are circumstances when sugars (for instance, sucrose or table sugar) may be a "good" choice. For example, the most appropriate, rapid source of energy for athletes, particularly during or immediately after endurance exercise, may be sucrose.

Fermentation: The breaking down of a substance into a simpler one by a microorganism, such as the production of alcohol from sugar by yeast.

Legumes: Plants that have a double-seamed pod containing a single row of beans. Lentils and beans are legumes.

Endosperm: Tissue that surrounds and nourishes the embryo inside a plant seed.

Germ: When referring to grains, the embryo of the plant seed.

Bran: The husk of the cereal grain.

Guar gum: A polysaccharide added to processed foods as a thickener.

Psyllium: The seed of a fleawort that swells when moist and is a functional fiber.

Vegetables are considered quality carbohydrates because they are minimally processed and contain fiber.

In such cases the immediate focus is an energy source that is rapidly digested, absorbed, and metabolized, not its nutrient content. At times, the consumption of high-sugar, low-fiber foods and beverages may be necessary (for example, during a long race or training session) and the immediate energy needs of exercise will outweigh the longer-term "healthy diet" perspective. From the perspective of health, the majority of carbohydrates consumed in the diet should be minimally processed fiber-containing carbohydrates, such as whole grains, beans, legumes, fruits, and vegetables. Highly processed carbohydrate foods lose fiber and nutrients in the processing, and they often have sugar, fat, and/or salt added. They are not the nutritional equivalents of minimally processed carbohydrate-containing foods, and this distinction in nutrient content is important.

Spotlight on...

Sugar Alcohols

Sugar alcohols can be formed from monosaccharides and dissacharides. They are less sweet and typically have fewer kilocalories than sucrose (table sugar). Common sugar alcohols are glycerol, sorbitol, mannitol, and xylitol.

Glycerol is part of the structure of triglycerides, the most common type of fat found in food and in the body. A triglyceride is a glycerol molecule with three fatty acids attached. Because it is an integral part of the structure of a triglyceride, glycerol may be thought of as a fat, but this is erroneous—glycerol is a sugar alcohol. When triglycerides are metabolized, the fatty acids are broken down and oxidized for energy. The majority of glycerol produced from the breakdown of triglycerides will be used to resynthesize other triglycerides, but a small amount may be converted to glucose (see Gluconeogenesis).

Glycerol may be added to food, in which case the terms *glycerin* or *glycerine* are used. Glycerin provides sweetness and moisture and has become popular in foods sold to consumers who are following a low-carbohydrate diet. Glycerol contains approximately 4.3 kcal/g, so its caloric content is approximately the same as sucrose (~4 kcal/g). However, only a small amount of glycerin is typically used in a product, and its absorption from the gastrointestinal tract is slow, so glucose and **insulin** concentrations are not rapidly elevated. Such products are often advertised as containing "low-impact carbs," a marketing term that is used to describe a slow rise in **blood glucose** and insulin concentrations after a carbohydrate-containing food is consumed.

Sorbitol and mannitol are found in some foods naturally, and are used in sugar-free foods because they contain fewer kilocalories per gram (2.6 and 1.6, respectively) than sucrose or fructose. They are also incompletely absorbed. When absorbed, the rise in blood glucose is much slower when compared to sucrose-containing

Examples of foods that contain sugar alcohols such as glycerol, sorbitol, mannitol, or xylitol.

foods. When these sugar alcohols have been added to foods, the label must carry a warning stating that excess consumption may have a laxative effect, because the unabsorbed portion remains in the digestive tract and may cause diarrhea. Excess consumption is defined as 20 g/day of mannitol or 50 g/day of sorbitol, although some people have symptoms at lower doses.

The sugar alcohol xylitol (2.4 kcal/g) is frequently used in chewing gum because it can help prevent dental caries. Although the sugar alcohols are found in foods, they receive less attention than the sugars and starches because sugar alcohol intake is typically a small part of total carbohydrate intake (Nabors, 2002).

However, describing carbohydrate-containing foods as good or bad is confusing and does not help athletes determine the quality or quantity of carbohydrates that they need in their diets for training, performance, and health.

Key points

- Carbohydrates are found as single sugars (monosaccharides), two linked sugar molecules (disaccharides), or many sugar molecules linked together (polysaccharides).

- Carbohydrates vary in their sweetness and in the blood glucose and insulin response to their consumption.

- Carbohydrates are often classified as simple sugars and starches, or complex carbohydrates.

- Carbohydrates of different types should not be considered good or bad, but appropriate or inappropriate considering the athletes' goals and general health.

Under what circumstances might it be appropriate for an athlete to consume simple sugars?

Why is it important for people to consume fiber if humans lack the enzymes to digest it?

4.2 Digestion, Absorption, and Transportation of Carbohydrates

Digestion is the breakdown of foods into smaller parts so the body can absorb them. Absorption involves taking these smaller parts into the cells of the intestine where they will then be transferred into the blood for transportation to other parts of the body. The digestive tract starts at the mouth and includes the stomach, small intestine (that is, duodenum, jejunum, ileum), large intestine (that is, colon [ascending, transverse, and descending], cecum, appendix, rectum), and anus (Figure 4.4). The majority of digestion and absorption takes place in the small intestine, although the mouth, stomach, and large intestine do account for some digestion and for absorption of some nutrients. The walls of the small intestine contain numerous villi and microvilli to increase surface area for absorption of nutrients (Figure 4.5). These small projections into the intestinal lumen contain a dense capillary and lymphatic network to further aid the uptake of nutrients into the body.

As mentioned previously, starches are polysaccharides, which are chains of glucose. When a starch-containing food is consumed, these chains must be

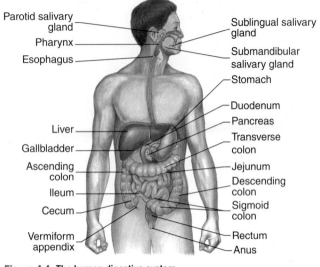

Figure 4.4 The human digestive system

broken down to yield glucose. This process begins in the mouth with the action of salivary amylase and continues in the lumen of the digestive tract with pancreatic amylase, but digestion of starch occurs predominantly in the small intestine. Carbohydrates consumed as polysaccharides will first be broken down into the disaccharides sucrose, maltose, and lactose (Figure 4.6).

The three disaccharides—sucrose, lactose, and maltose—are digested in the small intestine (not the mouth or stomach). Enzymes specific to each disaccharide are found in the brush border of epithelial cells lining the interior border of the small intestine. The enzyme sucrase breaks down sucrose to yield glucose and fructose. Sucrase can also break down maltose to two molecules of glucose, but dietary maltose intake is usually small. The enzyme lactase breaks down lactose into glucose and galactose. People who lack a sufficient amount of lactase are unable to break down the lactose or milk sugar, resulting in a condition known as lactose intolerance. Lactose intolerance is discussed later in this chapter.

After the breakdown of starches and disaccharides, only the monosaccharides—glucose, fructose, and galactose—remain. Glucose is typically present in the largest amount since it is the only monosaccharide found in starch and maltose and it is half of each molecule of sucrose and lactose. Galactose is absorbed by the same mechanism as glucose, whereas fructose absorption occurs by a different process.

Glucose and fructose are absorbed by different mechanisms.

The absorption of glucose is a two-step process, with each step involving a carrier protein embedded in

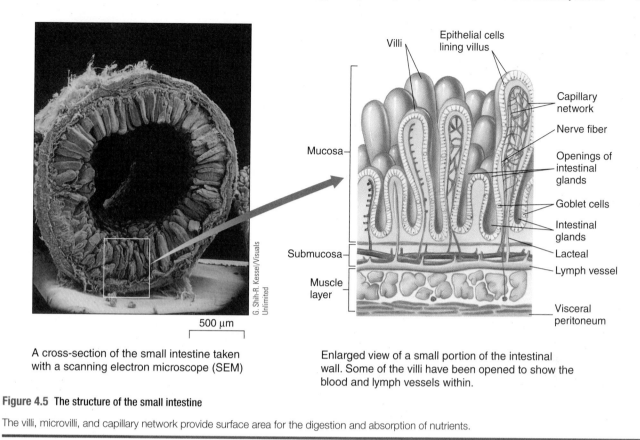

A cross-section of the small intestine taken with a scanning electron microscope (SEM)

Enlarged view of a small portion of the intestinal wall. Some of the villi have been opened to show the blood and lymph vessels within.

Figure 4.5 The structure of the small intestine

The villi, microvilli, and capillary network provide surface area for the digestion and absorption of nutrients.

the membrane of the epithelial cells lining the small intestine. Glucose is transported by secondary active transport from the intestinal lumen into the epithelial cell by a sodium and glucose cotransporter (SGLT). Because sodium is typically found in higher concentration outside the cell, it moves down its concentration gradient into the cell using the SGLT transporter. Sodium binding to the transporter increases the affinity of the SGLT for binding glucose molecules, so glucose is transported into the cell along with sodium. The sodium-potassium pump then uses energy to pump sodium back out of the cell to maintain the concentration gradient. As glucose enters the epithelial cell, its concentration increases relative to the concentration in the capillaries on the basal side of the cell. Glucose then moves from the cell into the blood by facilitated diffusion through another membrane-bound glucose transporter, GLUT-2 (Figure 4.6). Galactose uses the same carrier and process for absorption. However, once transported to the liver, the galactose is immediately trapped by the liver cells and converted to glucose.

Fructose must also attach to a specific carrier to cross the wall of the small intestine. However, the carrier that transports fructose from the intestinal lumen into the epithelial cell, GLUT-5, is different from the glucose carrier. Fructose is also able to be absorbed into the epithelial cell by passive facilitated diffusion. Once in the epithelial cell, fructose is transported into the blood by the same process and membrane-bound transporter, GLUT-2, as glucose (Figure 4.6).

The number of glucose and fructose carriers is limited; therefore, if the amount of these carbohydrates in the small intestine is greater than the number of carriers present, then some of these sugars will not be absorbed. The small intestine can adapt somewhat to increased amounts of glucose and fructose in the diet by upregulating (increasing) the number of transporters in order to increase absorption ability.

Absorption of fructose also depends on a concentration gradient, but when fructose is present in the small intestine, it is greater than the concentration in the blood since there is no fructose circulating in the blood. Sport beverage manufacturers are careful not to include too much fructose in their drinks, since fructose that is not absorbed from the small intestine passes through to the colon, ferments, and causes gastrointestinal distress such as bloating and gas.

Carbohydrate is transported as blood glucose.

Once absorbed, glucose and fructose are carried directly to the liver via the portal vein. The liver is extremely

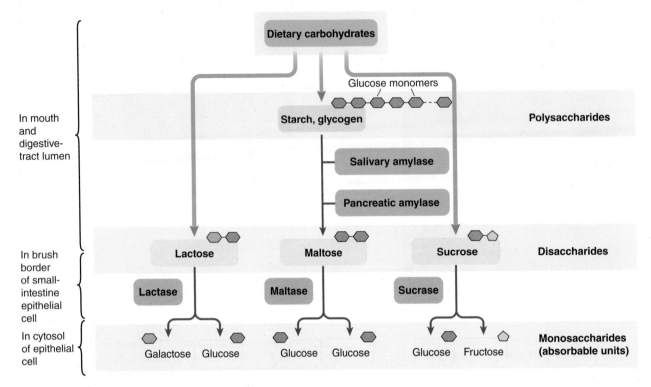

(a) Carbohydrate digestion

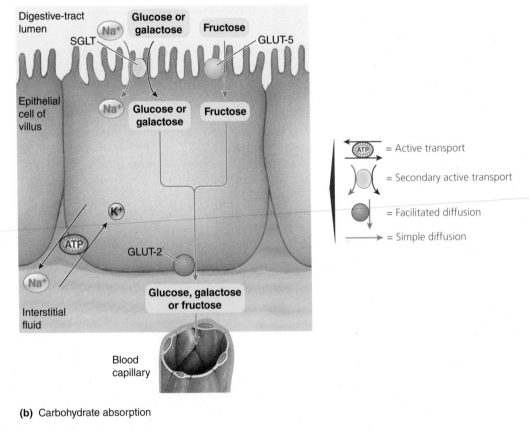

(b) Carbohydrate absorption

Figure 4.6 Carbohydrate digestion and absorption

efficient in capturing fructose, which will eventually be converted to glucose by the liver cells. The liver's ability to capture all of the fructose and galactose that is absorbed prevents either of these monosaccharides from leaving the liver and circulating in the blood. Ultimately, all of the sugars and starches found in food are broken down to monosaccharides and absorbed. Glucose is the end point, the eventual form of sugar that the body uses, regardless of whether the original compound was a polysaccharide, a disaccharide, or a monosaccharide.

Carbohydrate is then circulated throughout the body in the blood as glucose. As glucose circulates through capillary beds in various tissues, it can be taken up and used to produce ATP (energy) or stored for later use. The amount of glucose in the blood is carefully regulated by the body to provide a steady supply of this energy source to tissues that rely heavily on glucose, such as nerve cells and red blood cells.

Key points

- Digestion of carbohydrates begins in the mouth but mostly occurs in the small intestine.

- Specific digestive enzymes break complex carbohydrates down into disaccharides and, finally, monosaccharides.

- The monosaccharides glucose, fructose, and galactose are absorbed from the small intestine into the blood using specific protein transporters found in the intestinal cell membrane.

- Glucose and galactose are absorbed into the intestinal epithelial cells by active transport by a sodium and glucose transporter (SGLT) and into the blood by facilitated diffusion through the GLUT-2 glucose transporter.

- Fructose is absorbed into the intestinal epithelial cell by facilitated diffusion by the GLUT-5 transporter and into the blood by the same process and transporter (GLUT-2) as glucose.

- Regardless of the type of carbohydrate consumed, most carbohydrate is transported through the body in the blood as glucose.

What happens when someone consumes carbohydrate in excess of what can be absorbed from the small intestine?

What happens to fructose and galactose molecules that are absorbed from the small intestine into the blood?

4.3 Metabolism of Glucose in the Body

Metabolism refers to all of the physical and chemical changes that take place within the cells of the body. Glucose is needed for cellular energy, and the metabolism of glucose is regulated by a number of hormones. The most predominant of the glucose-regulating (glucoregulatory) hormones are insulin and **glucagon**. Glucose metabolism is intricate and involves many metabolic pathways. These pathways include: (1) the regulation of blood glucose concentration, (2) the immediate use of glucose for energy, (3) the storage of glucose as glycogen, (4) the use of excess glucose for fatty acid synthesis, and (5) the production of glucose from lactate, amino acids, or glycerol.

Blood glucose is carefully regulated.

The normal concentration of blood glucose is approximately 70 to 110 mg/dl (3.89–6.06 mmol/L), and sensitive hormonal mechanisms are used to maintain equilibrium **(homeostasis)** within this fairly narrow range. For example, when blood glucose concentration is elevated after a carbohydrate-containing meal, the hormone insulin is secreted from the beta (β) cells of the **pancreas** to stimulate the transport of glucose from the blood into the cells of various tissues. When blood glucose concentration is too low, the hormone glucagon is secreted from the alpha (α) cells of the pancreas to stimulate the release of glucose stored as liver glycogen into the blood. Blood glucose concentration is always in flux, but hormonal mechanisms are in place to bring blood glucose concentration back within the normal range (equilibrium).

A very simple example of how blood glucose is regulated hormonally at rest is illustrated in Figure 4.7. When carbohydrates such as sucrose are consumed, glucose is quickly digested, absorbed, and transported into the blood. Blood glucose concentration rises, and a temporary state of **hyperglycemia** (high blood glucose) exists. The rise of blood glucose stimulates beta (β) cells of the pancreas to increase the secretion of insulin. Insulin mediates the transfer of glucose out of the blood

Insulin: A hormone produced by the pancreas that helps to regulate blood glucose.

Blood glucose: The type of sugar found in the blood.

Metabolism: All of the physical and chemical changes that take place within the cells of the body.

Glucagon: A hormone produced by the pancreas that raises blood glucose concentration by stimulating the conversion of glycogen to glucose in the liver. It also stimulates gluconeogenesis, the manufacture of glucose by the liver from other compounds. Glucagon is counter-regulatory to insulin.

Homeostasis: A state of equilibrium.

Pancreas: An organ that produces and secretes the hormones insulin and glucagon into the blood. It also secretes digestive juices into the small intestine.

Hyperglycemia: Elevated blood glucose. Hyper = excessive, glyc = sugar, emia = blood.

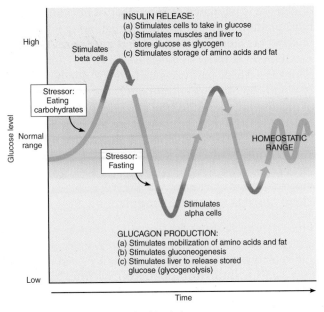

Figure 4.7 Regulation of resting blood glucose

Blood glucose rises in response to eating carbohydrate, which stimulates the beta cells to release insulin. Fasting for more than a few hours results in blood glucose declining, which stimulates the alpha cells to release glucagon.

and into most cells, such as red blood cells, resting muscle cells, and fat cells. As the glucose enters the cells, the concentration of glucose in the blood begins to drop. Depending upon the amount and type of carbohydrate consumed, the rise of blood glucose and its return to its resting level may take from 30 minutes to 1–2 hours.

Not all cells in the body need insulin to take up glucose, however. Exceptions are brain and liver cells, which are not dependent on insulin for the transport of glucose. Exercising muscle cells can also take up glucose from the blood without insulin, a process known as non-insulin-dependent glucose transport. Glucose moves from the blood into cells by the process of facilitated diffusion using plasma membrane carrier glucose transporters, GLUT. There are a number of different GLUT transporters found in different tissues, named numerically in the order in which they were discovered (for example, GLUT-1, GLUT-2, and so on). GLUT-2 was discussed in the previous section. GLUT-4 is the most abundant glucose transporter in the body and is found in tissues such as skeletal muscle and adipose tissue. GLUT-4 transporters respond to insulin by increasing their numbers on the cell membrane to increase glucose uptake into these tissues in what is called insulin-dependent glucose transport.

Blood glucose concentration can drop too low, such as when a person goes without eating for several hours. This condition is known as **hypoglycemia**,

which is usually defined as a blood glucose concentration below 50 mg/dl (2.76 mmol/L). In response to a decreasing blood glucose concentration and hypoglycemia, the hormone glucagon is released from the alpha (α) cells of the pancreas. Glucagon stimulates the breakdown of stored glycogen in the liver and its release into the blood as glucose. Insulin and glucagon counter each other's actions (**counter-regulatory**); they act in opposition to either remove or add glucose to the blood to help keep the concentration of blood glucose within the normal range over time (homeostasis).

Glycemic effect of various carbohydrates and the glycemic index. Since the 1980s, scientists have been studying the effect that different carbohydrate foods have on blood glucose (known as **glycemic response**) and insulin secretion. Hundreds of carbohydrate-containing foods have been tested and their glycemic responses have been quantified (Atkinson, Foster-Powell, and Brand-Miller, 2008). This has led to the classification of carbohydrate foods based on their **glycemic index (GI)**.

Under normal circumstances, the result of carbohydrate consumption, digestion, and absorption is a relatively rapid increase in blood glucose, which reaches a peak and is then followed by a decline due to the secretion of insulin and the subsequent increase in glucose uptake by tissues. The time course and magnitude of this glycemic response are highly variable with different foods and do not fall neatly into categories based on chemical structure (that is, mono-, di-, or polysaccharides) or other descriptions, such as simple or complex carbohydrates. For example, the consumption of the same amount of the two monosaccharides (simple sugars) glucose and fructose results in very different blood glucose responses. Glucose ingestion results in a rapid and large increase in blood glucose, which, in turn, rapidly returns to baseline levels. Fructose consumption, on the other hand, results in a much slower and lower glycemic response—it rises more slowly, does not reach as high a level, and returns more slowly to baseline. The glycemic response of a low-GI food, applesauce, and a moderate-GI food, orange juice, is illustrated in Figure 4.8.

The concept of the glycemic index was created, tested on a variety of foods, and initially published in 1981 by Jenkins et al. The GI is a ranking based on the blood glucose response of a particular food compared to a reference food. Glucose or white bread containing 50 g of carbohydrate is typically used as the reference food. Test foods contain an identical amount of carbohydrate, and the blood glucose response is determined for 2 to 3 hours after consumption of the test food. Extensive testing of foods has resulted in the publication of tables of glycemic indices for a wide

Highly processed starches such as white bread, pasta, or rice have a higher glycemic index compared to beans and legumes.

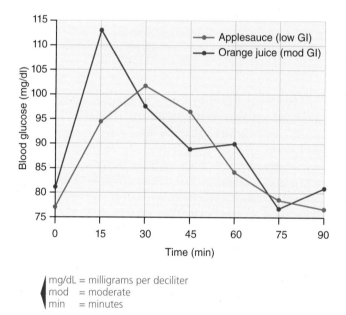

mg/dL = milligrams per deciliter
mod = moderate
min = minutes

Figure 4.8 Glycemic response to a low- and moderate-glycemic index food

Blood glucose and insulin response to 50 g of carbohydrate in unsweetened applesauce (a low-glycemic index [GI] food) and unsweetened orange juice (a moderate glycemic index food). Note that when a food with a higher GI is consumed, blood glucose rises more quickly, reaches a higher level, and declines more quickly than a food with a lower GI. The insulin response is also higher and remains higher for the food with the higher glycemic index, even though the same amount of carbohydrate is consumed.

variety of foods in scientific journals (Atkinson, Foster-Powell, and Brand-Miller, 2008) and online (http://www.mendosa.com).

The variability in glucose and insulin response with different foods has important implications, and the GI may be a useful tool, particularly for those who must carefully control their blood glucose concentration, such as people with **diabetes** (Doyle and Papadopoulos, 2000). Nondiabetic athletes may also benefit from considering the GI of the carbohydrates they consume. There may be specific situations in which an athlete would want to consume foods with a high glycemic index and provoke a large blood glucose and insulin response, such as when attempting to synthesize muscle glycogen quickly after glycogen-depleting exercise.

In general, highly refined starchy foods and starchy vegetables (for example, white bread, corn flakes, and baked potatoes) have a high glycemic index, which means that blood glucose levels rise quickly after their consumption. This rapid rise in blood glucose is followed by a rapid rise in insulin. Legumes, beans,

Hypoglycemia: Low blood glucose. Usually defined as a blood glucose concentration below 50 mg/dl (2.76 mmol/L). Hypo = under, glyc = sugar, emia = blood.

Counter-regulatory: Counter refers to opposing; regulatory is a mechanism that controls a process. Counter-regulatory refers to two or more compounds that oppose each other's actions.

Glycemic response: The effect that carbohydrate foods have on blood glucose concentration and insulin secretion.

Glycemic Index (GI): A method of categorizing carbohydrate-containing foods based on the body's glucose response after their ingestion, digestion, and absorption.

Diabetes: A medical disorder of carbohydrate metabolism. May be due to inadequate insulin production (type 1) or decreased insulin sensitivity (type 2).

fruits, and nonstarchy vegetables have a low glycemic index. Blood glucose rises slowly because the digestion and subsequent absorption of glucose is slower when compared to highly refined starchy foods. A slow rise in blood glucose also results in a slow rise in insulin (Ludwig, 2002). More information about glycemic index can be found in the Spotlight on...Glycemic Index.

Spotlight on...

Glycemic Index (GI)

The glycemic index is a method of categorizing foods based on the body's glucose response after their ingestion, digestion, and absorption.

Table 4.3 Glycemic Index of Selected Foods

Food	Glycemic index (mean value)	Food	Glycemic index (mean value)
Gatorade® (original Gatorade)	111	Sweet potato	63
Glucose	103	Honey	61
Clif bar	101	Soft drinks	59
Potatoes (instant, mashed)	87	Oatmeal (old fashioned)	55
Rice milk	86	Chapati	52
Cornflakes	81	Banana (ripe)	51
Oatmeal (instant)	79	Mango	51
Potatoes (boiled)	78	Ice cream	51
Watermelon	76	Spaghetti noodles (white)	49
Waffle	76	Corn tortilla	46
Met-RX bar	74	Apple juice	41
White bread	73	Chocolate	40
White rice	73	Whole milk	39
Bagel	69	Skim milk	37
Brown rice	68	Apple	36
Pancakes	67	Soy milk	34
Powerade	65	Fructose	15
Sucrose	65		

Atkinson, F. S., Foster-Powell, K., & Brand-Miller, J .C. (2008). International tables of glycemic index and glycemic load values: 2008. Diabetes Care, 31(12), 2281–2283 and Gretebeck, R.J., Gretebeck, K.A., and Titelback, T.J. (2002). Glycemic index of popular sport drinks and energy foods. *Journal of the American Dietetic Association, 102*(3), 415–417.

In Australia, the glycemic index is included on the label.

Foods with a GI <60 are considered low-GI foods, and those >85 are considered high-GI foods, but these numerical values were established arbitrarily. Table 4.3 gives examples of such foods.

In general, high-GI foods are processed grain products with added sugar and low fiber content (for example, sweetened cereals), and starchy vegetables (for example, some kinds of potatoes). It appears that the GI of some foods are increasing due to greater processing, which allows for faster cooking or a texture that is more acceptable to consumers (Atkinson, Foster-Powell, and Brand-Miller, 2008). Beans, legumes, dairy products, and some fruits have a low GI because the carbohydrate contained is more slowly absorbed. Medium-GI foods are less rapidly absorbed than high-GI foods. Some medium-GI foods contain sugar (as sucrose) but also have added fats, which slows the absorption.

Some athletes fine-tune their diets by using the glycemic index, especially the consumption of high-GI foods after exercise (Donaldson et al., 2010). The World Health Organization has endorsed the use of the glycemic index, and countries such as Australia put GI values on food labels. However, GI values are not required on labels in the United States. Glycemic index and another measure, glycemic load, may have health implications, especially for those with diabetes. These issues are discussed in Chapter 13.

Glucose can be metabolized immediately for energy.

As glucose is taken up into a cell, it can be either metabolized or stored for later use, depending upon the current energy state of the cell. If the energy need of the cell is low and the cell has the enzymatic capability, glucose will be stored as glycogen as described below. The metabolism of glucose follows the process of glycolysis as explained in Chapter 3. The energy content of carbohydrate is approximately 4 kcal/g.

In glycolysis, glucose is broken down in a series of chemical steps to form pyruvate, often referred to as a glycolytic intermediate. From pyruvate, the completion of the metabolism of glucose follows one of two pathways: conversion to lactate (anaerobic glycolysis) or oxidation of pyruvate in the mitochondria (oxidative phosphorylation) (Figure 4.9 and Appendix J). How glucose is metabolized is dependent upon a variety of factors: the type of cell, the enzymatic capability of the cell, energy state, hormonal status, training history, and intensity of exercise.

Certain types of cells are more likely to use glucose anaerobically and as a result produce lactate. Some cells lack the organelles or enzymatic capability to oxidize glucose. For example, erythrocytes (red blood cells) have no mitochondria and are therefore incapable of metabolizing glucose aerobically; they derive their energy from anaerobic glycolysis. Fast-twitch muscle fibers do contain mitochondria and the inherent oxidative enzymes, yet they are biased to use glucose anaerobically. They contain a highly active form of lactate dehydrogenase (LDH), the enzyme that catalyzes the conversion of pyruvate to lactate, and are thus likely to use anaerobic glycolysis even when oxygen delivery to the cell is sufficient. Fast-twitch muscle fibers are also typically not recruited for use until exercise intensity is relatively high, indicating that the energy need of the muscle cells is high, further favoring the use of a more rapid energy-producing system such as anaerobic glycolysis.

Cells in the body that are considered to be highly aerobic (for example, heart muscle cells, slow-twitch muscle fibers) will predominantly metabolize glucose aerobically via oxidative phosphorylation. Glucose taken up into these cells will proceed through the steps of glycolysis to pyruvate. Pyruvate is then transported into mitochondria where it is oxidized in the steps of the Krebs cycle, and energy (that is, ATP) is ultimately produced via the electron transport chain.

Glucose can be stored as glycogen for later use.

If the immediate energy needs of a cell are low, glucose is often stored in the form of glycogen for future use. As shown in Figure 4.3, glycogen is a series of glucose

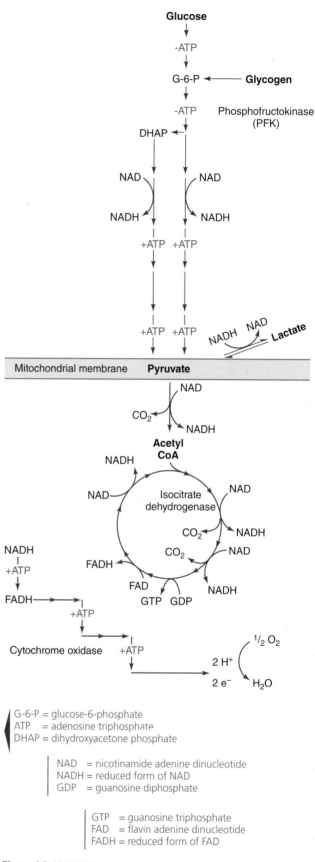

Figure 4.9 Metabolism of carbohydrate

Glucose or glycogen is broken down to pyruvate and is either converted to lactate (anaerobic metabolism) or is oxidized (aerobic metabolism).

molecules that have been linked chemically. Glycogen can be stored in many types of cells, but two of the major storage sites for carbohydrate in the body are skeletal muscle and liver. By far, the largest amount of glycogen is stored in skeletal muscle, approximately 400 g in the average-size person, with approximately 90 g of glycogen stored in the liver of a person in a rested, well-fed state.

The pathway to glycogen formation is favored when conditions exist to activate the primary enzyme that controls this process, **glycogen synthase**. Glycogen storage is favored when the energy need of the cell is low and insulin is elevated, as occurs when a person is resting after a meal, particularly a meal that contains carbohydrate. Glycogen synthesis is further enhanced in muscle when glycogen stores have been reduced through exercise. Thus athletes who exercise regularly and consume sufficient carbohydrate typically have higher muscle glycogen levels at rest than sedentary people.

Products of glucose metabolism can be used to synthesize fatty acids.

For the athlete or physically active person, the majority of carbohydrate that is consumed is either stored as glycogen or metabolized. However, carbohydrate consumed in amounts in excess of what can be stored as glycogen can be converted to other stored forms of energy, namely fat. The term for the synthesis of fatty acids is **lipogenesis**.

As discussed in Chapters 3 and 6, fats that are used by the body for energy are composed of long chains of carbon molecules. The fatty acid chains most commonly used by humans for metabolism contain either 16 or 18 carbons. They are synthesized by attaching carbons two at a time in the form of acetyl CoA until a chain of 16 or 18 carbons is formed. Lipogenesis takes place in the liver or in adipocytes (fat cells).

Excess glucose can be taken up by the liver or adipocytes, and through glycolysis, can be a source of acetyl CoA for fatty acid synthesis. The chemical pathway is somewhat indirect, but acetyl CoA formed from glucose metabolism can then be used to synthesize fatty acid chains. Fatty acid chains can be attached in groups of three to a glycerol molecule to form a triglyceride, which is the major storage form of fat in the body. Excess glucose can also be used as a source to form glycerol that can be used in triglyceride synthesis.

Athletes who consume a high-carbohydrate diet based upon their training demands (within their total caloric needs for the day) typically do not need to worry about the formation of fat from carbohydrate because they are not consuming *excess* carbohydrate. Exercise and training result in acute decreases in muscle glycogen, which then stimulates the storage of dietary carbohydrate as glycogen. Consumption of a high-carbohydrate diet also results in the metabolism of a higher percentage of carbohydrate than fat at rest, so a larger proportion of the dietary carbohydrate consumed is metabolized. Athletes do need to be aware, however, of excess carbohydrate intake during periods when training intensity, frequency, and/or duration is decreased, such as during the off-season or when injured. The primary population that needs to be concerned about lipogenesis from excess carbohydrate intake is the sedentary individual who is consuming too many kilocalories and excess carbohydrate.

Glucose can be produced from lactate, amino acids, and glycerol by a process called gluconeogenesis.

Although most of the body's glucose needs are supplied by dietary carbohydrate, the body does have a limited ability to produce glucose from other sources. The process of producing glucose from other sources is called **gluconeogenesis**. The major sources for the production of glucose are lactate, amino acids, and glycerol. The production of glucose from lactate was outlined in Chapter 3.

During periods of fasting or starvation, proteins in the body can be broken down into amino acids, which can then be metabolized (see Chapters 3 and 6). Amino acids that can be converted to certain intermediates in glycolysis or the Krebs cycle can also be used in gluconeogenesis to form glucose in the liver. Alanine is an example of an amino acid that can be used to form glucose. In fact, 18 of the 20 amino acids are biochemically capable of being converted to glucose.

When stored triglycerides are broken down, the fatty acid chains and glycerol can be metabolized for energy. Glycerol can be converted to an intermediate in glycolysis, and can be used to form glucose via gluconeogenesis. This occurs in the liver. However, the liver has a limited capacity to form glucose from glycerol in this way. Therefore, the major sources for glucose formation by gluconeogenesis are lactate and amino acids.

Glycogen synthase: The primary enzyme that controls the process of glycogen formation.

Lipogenesis: The production of fat.

Gluconeogenesis: The manufacture of glucose by the liver from other compounds such as lactate, protein, and fat. Gluco = glucose, neo = new, genesis = beginning.

Key points

- The body normally maintains blood glucose within a fairly narrow range, approximately 70 to 110 mg/dl.

- The pancreas is a major organ controlling blood glucose through its ability to secrete insulin and glucagon.

- Glucose circulating in the blood is taken up into various tissues through glucose transporters (GLUT) found in their cell membranes.

- Glucose transport into tissues is dependent upon the presence and action of insulin in some instances but is not dependent on insulin in other instances. Glucose uptake into skeletal muscle at rest is insulin-dependent, but during exercise glucose uptake is not dependent upon insulin.

- Consumption of different carbohydrate foods results in varying responses of blood glucose and insulin, and the scale developed to describe these responses is the glycemic index.

- When glucose is taken up into a cell, it can be either used immediately for energy or stored as glycogen for later use.

- If carbohydrates are consumed in excess, products of glucose metabolism may be used to produce fatty acids and store energy as fat.

- The liver can produce glucose from a number of sources such as lactate, amino acids, and glycerol by gluconeogenesis.

How does the body respond when blood glucose rises after consuming a carbohydrate meal?

How does the body maintain blood glucose when a person goes for a long period of time without eating, such as when sleeping?

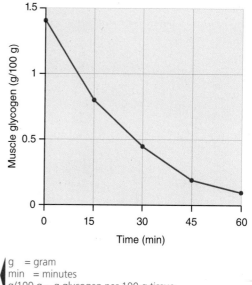

g   = gram
min = minutes
g/100 g = g glycogen per 100 g tissue

Figure 4.10 Muscle glycogen utilization during exercise

Subjects rode at a hard aerobic intensity on a cycle ergometer and had their muscle glycogen measured at 15-minute intervals. Subjects repeated the 15-minute intervals until they were too fatigued to continue at the required intensity, which corresponded to a very low muscle glycogen concentration.

Redrawn from: Bergström, J., and Hultman, E. (1967). A study of the glycogen metabolism during exercise in man. *Scandinavian Journal of Clinical Laboratory Investigation, 19*(3), 218–228.

4.4 Carbohydrates as a Source of Energy for Exercise

As discussed in Chapter 3, muscle can use a variety of fuel sources to provide the energy necessary for exercise (for example, creatine phosphate, carbohydrates, fats, and/or proteins). The fuel source utilized depends on a variety of factors, with exercise intensity playing a major role. Very high-intensity, very short-duration anaerobic exercise typically uses creatine phosphate as the energy source to replenish ATP. Carbohydrate is used as the predominant source of energy via anaerobic glycolysis during high-intensity, short-duration anaerobic exercise or through oxidative phosphorylation during moderate- to high-intensity aerobic exercise.

Carbohydrate used by exercising muscle can come from stored muscle glycogen or from glucose that is brought into the muscle from the blood. Glucose is made available in the blood from the liver as a result of at least three processes: the breakdown of liver glycogen, the production of glucose from other sources (gluconeogenesis), or the ingestion of carbohydrates as food or fluids, which are absorbed and passed through the liver.

Exercising muscle first uses carbohydrate stored as glycogen.

Exercising muscle preferentially uses carbohydrate from stored glycogen. A study conducted by Bergström and Hultman (1967) has become the classic description of muscle glycogen utilization during exercise. Subjects rode on a cycle ergometer at a moderately hard aerobic intensity and their muscle glycogen utilization was determined every 15 minutes by muscle biopsy. Results of this study clearly show the decline in muscle glycogen as exercise time progressed (Figure 4.10). After 60 minutes of exercise, muscle glycogen fell to very low levels that corresponded with fatigue, an inability of the subjects to complete the next 15-minute exercise period. Although not shown in Figure 4.10, the study also demonstrated that the rate of muscle glycogen usage was related to exercise intensity; muscle glycogen was used at a higher rate at higher exercise intensities.

Low muscle glycogen stores can lead to fatigue in prolonged endurance events such as a marathon.

AP Photo/Corrado Giambalvo

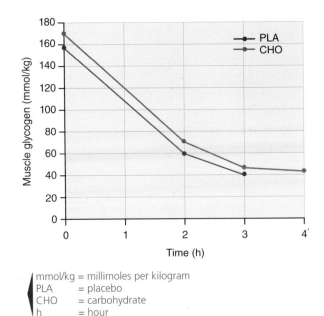

mmol/kg = millimoles per kilogram
PLA = placebo
CHO = carbohydrate
h = hour

Figure 4.11 Muscle glycogen usage when carbohydrate is consumed during exercise

Well-trained, competitive cyclists rode at a hard aerobic intensity while consuming either a carbohydrate drink or a placebo beverage. Muscle glycogen utilization was the same regardless of the carbohydrate intake during the ride. When consuming the placebo, the cyclists fatigued at the 3-hour time point, but when they consumed carbohydrate, they were able to maintain this exercise intensity for an additional hour.

Redrawn from: Coyle, E. F., Coggan, A. R., Hemmert, M. K., and Ivy, J. L. (1986). Muscle glycogen utilization during prolonged strenuous exercise when fed carbohydrate. *Journal of Applied Physiology, 61*(1), 165–172.

The results of the Bergström and Hultman study, as well as many subsequent studies, demonstrate that muscle glycogen is a primary source of energy during moderate to intense aerobic exercise and during higher intensity, repeated intervals. Fatigue is associated with muscle glycogen depletion during prolonged endurance exercise and can often occur in intermittent high-intensity exercise, such as the repeated sprints that occur during soccer matches.

Exercising muscle takes up and metabolizes blood glucose.

Exercise stimulates the uptake of glucose from the blood because it has a very strong insulin-like effect. As glucose is being taken out of the blood by exercising muscle, a fall in blood glucose is prevented by two metabolic adjustments, both stimulated by the release of the hormone glucagon by the pancreas. Liver glycogen is broken down and released into the blood as glucose, a process called **glycogenolysis**. Glucagon also stimulates the process of gluconeogenesis by the liver to make glucose available to maintain blood glucose.

Glycogenolysis: The breakdown of liver glycogen to glucose and the release of that glucose into the blood. -lysis = the process of disintegration.

Initially, this response acts to maintain or even slightly elevate blood glucose, but if exercise continues for a prolonged duration, reductions in blood glucose may occur as liver glycogen is reduced and gluconeogenesis fails to produce glucose at the rate that it is being utilized.

Although the exercising muscles begin to take up and use glucose, they rely most heavily on muscle glycogen if sufficient stores are available. An excellent example of this preference for glycogen over glucose is illustrated by the study of Coyle et al. (1986) in which muscle glycogen and blood glucose concentrations were determined when well-trained cyclists rode a prolonged endurance trial at a relatively hard aerobic intensity (Figure 4.11). During the trial when subjects consumed only a placebo beverage, blood glucose eventually began to fall and fatigue was associated with the point at which muscle glycogen fell to a very low concentration. In a subsequent cycling trial, the subjects consumed a carbohydrate drink that resulted in maintenance of blood glucose throughout the exercise trial. Muscle glycogen utilization was no different from that during the placebo trial, however, indicating a preference to use muscle glycogen even when blood glucose remained high.

Exercise training increases the capacity for carbohydrate metabolism.

Regular exercise training, specifically aerobic training, stimulates an increase in the oxidative (aerobic) capacity of muscle, primarily through an increase in the number and size of mitochondria and an increase in oxidative enzyme activity. This increase in oxidative capacity increases the muscle's maximal capacity to utilize carbohydrates and to oxidize lactate. When a person increases exercise intensity to the same relative level (for example, to the same percentage of maximum aerobic capacity), the utilization of carbohydrate

Focus on research

How Does Carbohydrate Consumption Improve Endurance Exercise Performance?

Athletes and scientists have known for years that consuming carbohydrate during prolonged endurance exercise helps performance. A significant amount of research since the late 1970s and 1980s has helped increase our knowledge about this performance improvement—the amount and type of carbohydrate consumed, the timing on carbohydrate intake, and the factors associated with fatigue during prolonged exercise. A classic study by Coyle et al. (1986) provided significant insight into the utilization of various carbohydrate sources during endurance exercise, the effect on performance, and the mechanisms associated with improved performance and fatigue: Coyle, E. F., Coggan, A. R., Hemmert, M. K., & Ivy, J. L. (1986). Muscle glycogen utilization during prolonged strenuous exercise when fed carbohydrate. *Journal of Applied Physiology, 61*(1), 165–172.

What did the researchers do? This study was designed to examine carbohydrate utilization and fatigue during prolonged endurance exercise when the subjects consumed a carbohydrate beverage or a placebo that contained no carbohydrate. The subjects consumed a relatively large amount of concentrated carbohydrate as a priming dose early in the exercise (approximately 144 grams) and then approximately 85 grams of carbohydrate each hour of exercise in a 10 percent carbohydrate beverage. The researchers recruited well-trained cyclists who were accustomed to riding for hours at a time. Subjects followed a consistent exercise and diet schedule for the 2 days before each experimental trial. Muscle glycogen was determined by a muscle biopsy before exercise and at several time points during the exercise trial, blood samples were taken to determine glucose concentration, and a metabolic measurement system was used to determine the rate of carbohydrate metabolism. The subjects rode a cycle ergometer in the laboratory at an exercise intensity of approximately 70–74 percent of their maximal aerobic capacity for as long as they could. The exercise trial was stopped when the subjects were fatigued and could not continue at the prescribed exercise intensity.

What did the researchers find? The first important finding of this study was the improvement in endurance exercise ability when the cyclists consumed the carbohydrate. When consuming no carbohydrate (placebo) they fatigued after 3 hours of riding, but when they received the carbohydrate they were able to exercise for an additional hour before fatiguing. The use of muscle glycogen was the same for each of the endurance exercise trials. In other words, consuming the carbohydrate did not "spare" muscle glycogen for use later in exercise (Figure 4.11). The final hour of the carbohydrate exercise trial was completed with very low muscle glycogen levels. Analysis of blood glucose showed significant declines during the placebo trial, but the consumption of the carbohydrate resulted in the maintenance of blood glucose levels during 4 hours of exercise. The rate of carbohydrate metabolism also stayed high, particularly in the late stages of exercise when carbohydrate was consumed leading to the improvements in endurance exercise performance.

What was the significance of the research? Results from this study provided significant insight into the process of fatigue during prolonged endurance exercise—muscle glycogen levels become depleted, blood glucose levels decline, and carbohydrate metabolism is unable to be maintained at a high rate leading to fatigue. Consumption of carbohydrate delays fatigue significantly, not by delaying the use of muscle glycogen but by maintaining blood glucose and keeping carbohydrate metabolism at a high level.

As with all good research studies important questions were answered, but others were raised. The results of the Coyle et al. study demonstrated that carbohydrate intake during exercise allowed the athletes to maintain their blood glucose and carbohydrate oxidation levels that significantly aided their endurance exercise performance. Despite blood glucose and carbohydrate oxidation remaining high, the subjects eventually fatigued, prompting questions and further research into causes or mechanisms of fatigue during endurance exercise other than carbohydrate metabolism.

as a percentage of the total energy expenditure is approximately the same.

The increase in ability to metabolize carbohydrate is due mostly to the increased oxidative or aerobic capacity of the muscle. The activity of the enzymes catalyzing the anaerobic pathway, glycolysis, do not change much, probably because they are already at a high level of intrinsic activity in the muscle before training. Another adaptation that increases the muscle's total capacity to utilize carbohydrate is the increase in stored muscle glycogen as a result of regular exercise training. As demonstrated elegantly in a one-legged cycling study by Bergström and Hultman (1966), muscles that exercised and reduced muscle glycogen synthesized and stored significantly more glycogen afterward than the leg muscles that were not exercised. Resting muscle glycogen levels in average, sedentary adults are approximately 20–30 percent lower than those of trained athletes.

The increase in oxidative capacity due to aerobic exercise training also increases the ability of the muscle to metabolize fat for energy. Therefore, after weeks to months of aerobic exercise training, it is common for a person exercising at the same absolute intensity (for example, the same running pace) to metabolize less carbohydrate and more fat (fat metabolism will be explained in greater detail in Chapter 6).

A training strategy that has become popular in the attempt to enhance fat metabolism is to purposely reduce the amount of carbohydrate in the diet for a period of time or withhold it prior to exercise in order to train with low carbohydrate stores. Research studies do indicate that there may be some metabolic training enhancements, such as an increase in mitochondrial function, as a result of these "depletion workouts" or by "training low" (Hawley and Burke, 2010). However, training with low carbohydrate availability may be difficult, and athletes commonly choose lower exercise intensities during these workouts. In addition, immune system function may be compromised, potentially exposing the athlete to increased risk of illness or injury. Additional research is needed to determine if the enhanced metabolic adaptations that occur with this type of training lead to improvements in performance without detrimental effects on the athlete.

Glucose metabolism during exercise is controlled by hormones.

Glucose metabolism during exercise is regulated by several overlapping or redundant hormonal mechanisms. Glucagon is a hormone secreted by the alpha (α) cells of the pancreas that is a counter-regulatory hormone to insulin. It stimulates essentially the opposite effect of insulin (that is, glycogen breakdown instead of glycogen synthesis). Glucagon is secreted during periods of fasting or starvation, and exercise is metabolically similar to these conditions. During exercise, insulin secretion is suppressed and glucagon secretion is stimulated.

Glucagon stimulates glycogen breakdown in the liver and the release of glucose into the blood, thus acting to maintain or increase blood glucose. It also stimulates the process of gluconeogenesis by the liver, in which the liver can take precursors such as lactate or amino acids and make new glucose that can subsequently be released into the blood. This process begins early in exercise and can be thought of as a preemptive response to prevent a fall in blood glucose, rather than as a reactive response to blood glucose declining (one of the rare feed-forward mechanisms in the body, which usually operate in a feedback fashion).

The stress of higher exercise intensity results in a substantial stimulation of the sympathetic nervous system; this leads to a release of two important stress hormones, the catecholamines epinephrine and norepinephrine. Epinephrine (also known as adrenaline) is secreted from the adrenal glands (adrenal medulla), whereas the majority of the norepinephrine originates from the nerve endings of sympathetic nervous system cells. The catecholamines also stimulate glycogen breakdown and gluconeogenesis and thus provide an additional mechanism to regulate blood glucose. Another stress hormone released from the adrenal cortex, cortisol, aids blood glucose regulation indirectly by stimulating the breakdown of proteins into amino acids that may be used in gluconeogenesis (Figure 4.12).

Exercise intensity affects carbohydrate metabolism.

Exercise intensity and duration have a substantial effect on metabolism, both on the metabolic rate and on the source of fuel utilized. Carbohydrates, fats, and proteins can be utilized as sources of energy for metabolism and are used in different proportions under different conditions. As exercise intensity increases, the proportion of energy that is derived from fat decreases and the proportion from carbohydrate increases. Therefore, carbohydrate is the main source of energy for moderate- to high-intensity exercise. The interplay between carbohydrate and fat metabolism is complex and is discussed in greater detail in Chapters 3 and 6. Proteins normally make up a small percentage of energy metabolism and are discussed in more detail in Chapter 5.

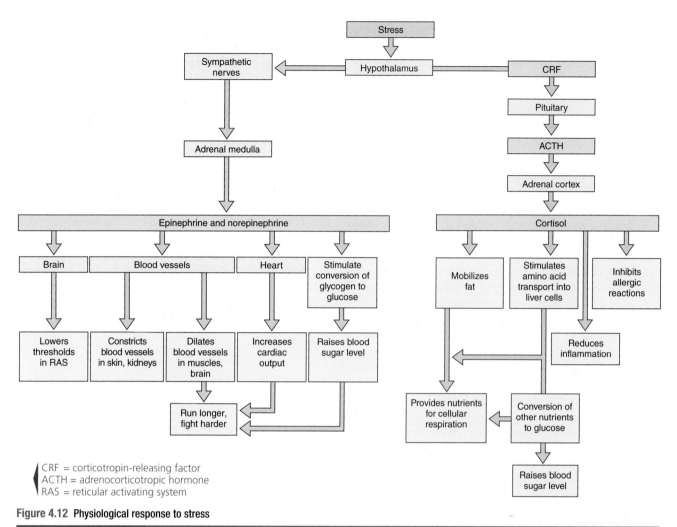

CRF = corticotropin-releasing factor
ACTH = adrenocorticotropic hormone
RAS = reticular activating system

Figure 4.12 Physiological response to stress

Key points

■ Exercising muscle prefers to use muscle glycogen as its source of carbohydrate for metabolism.

■ Muscle glycogen may be depleted and fatigue may occur after 60 or more minutes of exercise at a hard aerobic intensity.

■ Exercise has a strong insulin-like effect, stimulating the uptake of glucose from the blood.

■ Exercise stimulates the breakdown of liver glycogen to release more glucose into the blood and stimulates the production of more glucose by the liver by gluconeogenesis.

■ The breakdown of liver glycogen, glycogenolysis, and the increase in gluconeogenesis during exercise is stimulated by the hormone glucagon, secreted by the pancreas.

■ As exercise intensity increases, the proportion of energy derived from carbohydrate metabolism increases.

What effect does the consumption of a sports drink have on the use of glycogen by skeletal muscle during endurance exercise?

What happens to the level of insulin in the blood during endurance exercise? Why?

4.5 Carbohydrate Recommendations for Athletes

It has been established that carbohydrate exists in a variety of different forms in food and is an extremely important source of energy for exercise, sports training, and performance. How much carbohydrate should an athlete eat? What types of carbohydrate should be consumed? Given the specific recommendations for carbohydrate consumption, what are the practical applications for athletes with different training and competitive requirements? This section focuses on the answers to these questions.

Daily carbohydrate intake is based upon individual needs to meet the long-term demands of training and competition.

The current recommendation for total carbohydrate intake for athletes in training is 5 to 10 g/kg body weight daily (American Dietetic Association, Dietitians of Canada, and the American College of Sports Medicine, 2009; Burke et al., 2001). This recommendation assumes that energy (caloric) intake is adequate. This broad recommendation should be fine-tuned to meet the individual needs of the athlete based on the intensity and duration of training as shown in Table 4.4. Carbohydrate intake may need to be nearer the higher end of the recommended range during periods of rigorous training to ensure adequate muscle glycogen resynthesis. Further adjustments to the general recommendation may be made based on body size, gender, and the amount of dietary proteins and fats needed.

Athletes must consume enough carbohydrates daily to replenish muscle glycogen used during training. Prolonged moderate-intensity exercise, such as distance running and distance cycling, depletes muscle glycogen, and studies have shown that at least 5 g/kg/d is needed to replenish it to a level that allows for training on consecutive days (Sherman et al., 1993). However, an intake of 5 to 7 g/kg/d does not restore muscle glycogen levels of endurance athletes to pre-exercise or near maximum capacity. Endurance athletes likely need to consume 8 to 10 g/kg/d to maintain high levels of muscle glycogen over weeks and months of rigorous training (Sherman et al., 1993).

The majority of studies conducted on muscle glycogen depletion and replenishment have used endurance athletes as subjects (Jacobs and Sherman, 1999). The body of scientific literature clearly shows that prolonged endurance exercise depletes muscle glycogen and daily carbohydrate intake is needed to restore it. What about athletes in other sports? Limited studies have shown that the recommendations made for endurance athletes also apply to athletes in intermittent, high-intensity sports such as soccer and ice hockey (Bangsbo, Norregaard, and Thorsoe, 1992; Akermark et al., 1996; Balsom et al., 1999). The depletion of muscle glycogen when performing intermittent, high-intensity exercise during practice and games is similar to the depletion seen after prolonged moderate-intensity exercise.

Sprinters, weight lifters, and hurdlers are examples of athletes who perform very high intensity exercise for a very short period of time. Muscle glycogen is reduced during training and competition, but it can typically be restored with a moderate carbohydrate intake (that is, 5 to 7 g/kg/d). There is a lack of evidence that athletes who depend on strength and power but not endurance would benefit from high-carbohydrate diets (that is, >8 g/kg) on a daily basis (Lamb et al., 1990; Vandenberghe et al., 1995).

The needs of **ultraendurance** athletes may, at times, exceed the general range for daily carbohydrate

intake for other athletes. Examples of ultraendurance sports include Ironman® triathlons and multiple-day cycling events (stage races) such as the Tour de France. These are grueling events that require huge amounts of energy, much of which comes from muscle glycogen. Ultraendurance athletes may need more than 10 g/kg daily because of the higher carbohydrate needs associated with heavy training and ultraendurance competitions (Burke, Kiens, and Ivy, 2004). Although many elite ultraendurance athletes successfully consume the recommended amount of carbohydrate (Coleman, in press b), it can be difficult for ultraendurance athletes to meet these recommendations. They must eat frequently while awake and may need to consume carbohydrate during the night (for example, after waking to go to the bathroom or setting the alarm during the night to wake and eat). In addition to carbohydrate-containing meals, ultraendurance athletes may include sports bars, beverages, and gels before, during, and after training to try and reach their daily carbohydrate goals.

Expressing carbohydrate recommendations. All of the carbohydrate recommendations for athletes mentioned so far in this text have been expressed on a gram per kilogram body weight basis (g/kg). In other words, recommendations are stated as an absolute amount of carbohydrate. Recommendations may also be stated as a relative amount—a percentage of total energy intake (for example, 70 percent of total kilocalories as carbohydrate).

Recommendations given as a percentage, relative to the amount of energy consumed, can be misinterpreted. For example, a female endurance athlete states that she consumes a high-carbohydrate diet consisting of 70 percent of her total calories as carbohydrate. Is this an adequate amount? It depends on whether she consumes enough energy, as shown in Table 4.5. This example shows how an athlete could meet the *percentage* of carbohydrate recommended, but fall short of the minimum *amount* of carbohydrate recommended. Also notice in this example that both athletes state that they consume a "high" carbohydrate diet. *High* and *low* are relative terms, and unless such terms are defined, using these words to describe dietary carbohydrate intake can be misleading. To avoid misinterpretations, it is best to express recommendations on an absolute basis: grams of carbohydrate per kilogram body weight (g/kg). However, it is not uncommon to see carbohydrate recommendations made to athletes as a percentage of total kilocalories. The usual recommendation for most athletes is 50–65 percent of total caloric intake, increasing to 70 percent for those athletes with higher carbohydrate needs (for example, athletes in endurance sports). These percentages assume that total daily energy intake is adequate.

Ultraendurance: Very prolonged endurance activities such as the Ironman-length triathlons. Ultra = excessive.

Table 4.4 General Carbohydrate Recommendations Based on Exercise Intensity and Duration

Exercise intensity and duration	Examples of sports	Daily carbohydrate recommendation (energy intake must be adequate)
Very high intensity, very short duration (less than 1 minute)	Field events such as shot put, discus, or high jump Track sprints (50–200 m) Swimming sprints (50 m) Sprint cycling (200 m) Weightlifting Power lifting Bobsled (running start)	5–7 g/kg
High intensity, short duration (1 to 30 minutes continuous)	Track (200 to 1,500 m) Swimming (100 to 1,500 m) Cycling (short distance) Rowing (crew) Canoeing/Kayaking (racing) Skiing (downhill racing) Figure skating Mountain biking	5–7 g/kg
High intensity, short duration (1 to 30 minutes with some rest periods)	Gymnastics Wrestling Boxing Fencing Judo Tae kwon do	5–8 g/kg
Moderate intensity, moderate duration (30 to 60 minutes)	10 km running (elite runners finish in <30 minutes)	6–8 g/kg
Intermittent high intensity, moderate to long duration (more than 1 hour)	Soccer (football) Basketball Ice hockey Field hockey Lacrosse Tennis Water polo	6–8 g/kg; 8–10 g/kg during heavy training and competition
Moderate intensity, long duration (1 to 4 hours)	Distance running (marathon) Distance swimming Distance cycling Nordic (cross country) skiing	8–10 g/kg during periods of heavy training and competition
Moderate intensity, ultralong duration (more than 4 hours)	Ultradistance running Ultradistance swimming Ultradistance cycling Triathlon Adventure sports	8–10 g/kg or more depending on the stage of training
Low intensity, long duration (more than 1 hour)	Golf Baseball Softball	5–7 g/kg
Other	Bodybuilding American football	5–10 g/kg depending on the stage of training 5–8 g/kg; Varies according to position

Legend: m = meter; g/kg = gram per kilogram body weight; km = kilometer

Note: Athletes may need to consume near the higher end of the range during periods of rigorous training to ensure adequate muscle glycogen resynthesis.
Dunford, M. (Ed.). (2006). *Sports nutrition: A practical manual for professionals*. Chicago: American Dietetic Association.

Table 4.5 Two 70 Percent Carbohydrate Diets Compared

	Jennifer	Liza
Total carbohydrate intake	350 g	245 g
Total energy intake	2,000 kcal	1,400 kcal
% energy intake as carbohydrate	70 ("high")	70 ("high")
Carbohydrate (g) per kg of body weight	6.4 g/kg	4.5 g/kg

Is a 70 percent carbohydrate diet adequate? In this example, two female athletes, Jennifer and Liza, both weigh 121 pounds (55 kg). Each correctly states that she is consuming a diet consisting of 70 percent carbohydrate. Each also states that she is consuming a "high" carbohydrate diet. When their actual intakes are compared, the differences between these two diets are dramatic. In Jennifer's case, the answer to the question, "Is a 70 percent carbohydrate diet adequate?" is yes. In Liza's case, the answer is no. Liza's carbohydrate intake is high in relation to her caloric intake, but because her caloric intake is so low, her carbohydrate intake is below the minimum recommended for athletes.

Athletes need to plan their carbohydrate intake before, during, and after training and competition.

Once the daily carbohydrate need is established, the focus turns to dividing the total carbohydrate intake appropriately over the course of the day. The athlete's training and conditioning program and the unique demands of the competitive environment before, during, and after exercise will dictate the amount and timing of carbohydrate intake.

Intake prior to training and competition. The dietary goals of the athlete prior to exercise include avoiding hunger, delaying fatigue, minimizing gastrointestinal distress, and preventing hypohydration (below a normal state of hydration). With the exception of preventing hypohydration, all of these goals involve carbohydrate. Athletes should determine both the amount and timing of carbohydrate consumption (as well as proteins, fats, and fluids) needed to support exercise. For most athletes, dietary intake before training becomes the basis for fine-tuning the amount and timing of carbohydrate and other nutrients consumed prior to competition. Precompetition intake can be tricky because start times for some events may not be known, familiar foods may not be available, environmental conditions may be different from usual training conditions, and the stress of competition may result in increased gastrointestinal distress.

As is the case with total daily carbohydrate recommendations, the majority of research on pre-exercise carbohydrate intake has been conducted on endurance athletes while training. Studies show that carbohydrate intake up to 3 or 4 hours prior to endurance exercise is beneficial (Hargreaves, Hawley, and Jeukendrup, 2004). Muscle glycogen stores have been replenished with carbohydrates consumed over the past 24 hours, so the benefit comes primarily from enhanced liver glycogen and the glucose that eventually appears in the blood after the breakdown of the carbohydrates in the pre-exercise meal. The effect is similar to "topping off" a car's gas tank before starting on a long trip. Pre-exercise carbohydrates are also recommended for athletes in intermittent high-intensity sports such as soccer. Food intake prior to very high intensity short-duration sports such as sprinting has not been studied extensively because no performance benefit would be expected. However, most athletes do eat between 1 and 4 hours prior to exercise to avoid hunger and to ensure that they consume sufficient carbohydrates over the course of the day.

The amount of carbohydrate in the pre-exercise meal depends upon how close the meal is consumed to the start of exercise. Gastrointestinal distress can be caused by exercise, especially at the intensities at which athletes train and compete, because blood flow to the gastrointestinal tract is reduced. It is recommended that approximately 1 g of carbohydrate per kilogram body weight (1 g/kg) be consumed 1 hour prior to exercise. As the time before exercise increases, the amount of carbohydrate can be increased; for example, 2 hours prior to exercise, 2 g of carbohydrate per kilogram body weight can typically be tolerated (Coleman, in press a). Larger amounts of carbohydrate (for example, 3–4.5g/kg eaten 3–4 hours prior to exercise) may be appropriate for athletes under certain circumstances and will depend on the athlete's tolerance. The adjustment of carbohydrate amount based on time prior to exercise helps athletes prevent gastrointestinal distress and avoid hunger.

An active area of research regarding endurance athletes is whether low-glycemic index foods are preferred for a precompetition meal. Most of the studies have been conducted on trained distance cyclists who consumed 1 g/kg low-GI carbohydrate approximately 1 hour before exercise. Some studies showed low-GI carbohydrates to be beneficial because blood glucose concentrations were maintained during 1 to 2 hours of exercise due to the slow absorption of glucose from the low-GI carbohydrate. Other studies did not find a benefit, but no studies found that low-GI foods were detrimental (Donaldson et al., 2010; Siu and Wong, 2004). Although the research is not conclusive, some endurance athletes include low-GI carbohydrate-containing foods as part of their pre-exercise meal. However, the benefit of a low-GI carbohydrate prior to exercise may be small if carbohydrate is consumed during exercise.

In the past, athletes were advised against the consumption of high-GI carbohydrates (for example, sugar, white bread, sugary cereals, and many sports drinks) in the 1-hour period prior to exercise (Foster,

Stokely-Van Camp, Inc.

Commercial products are formulated to provide carbohydrate and other nutrients for specific time periods before, during, and after exercise.

Costill, and Fink, 1979). It was speculated that the ingestion of a high-GI food resulted in high blood glucose and insulin concentrations that in turn caused low blood glucose with the onset of exercise. But subsequent research demonstrated that high-GI carbohydrates are not a problem for the majority of athletes (Sherman, Peden, and Wright, 1991). Exercise has a very strong insulin-like effect; that is, it stimulates glucose uptake. If exercise is initiated at the time insulin is high and blood glucose is being lowered, the additional glucose uptake by exercising muscles could result in blood glucose being temporarily lowered too much. For most endurance athletes, however, the effect of high-GI foods on blood glucose would be transient and the body would reestablish a normal blood glucose concentration as it usually does during exercise. Thus performance would not likely be impaired.

However, a small number of athletes may be prone to reactive (rebound) hypoglycemia, a low blood glucose concentration that follows food intake. When these athletes consume a food with a high glycemic index 1 hour or less before prolonged exercise, blood glucose and insulin concentrations rise rapidly, which then shortly results in low blood glucose (hypoglycemia). The reestablishment of a blood glucose concentration within the normal range takes longer and performance may be affected. Athletes who respond in this way will need to experiment with the amount and timing of high-glycemic index foods prior to exercise (Hargreaves, Hawley, and Jeukendrup, 2004).

The timing of the carbohydrate meal before exercise is a major issue when considering recommendations about the glycemic nature of the meal and its potential impact on performance. Virtually every study that has shown a detrimental impact of high-GI meals on endurance performance has been timed so that the meal was consumed 45 to 60 minutes before exercise. From a practical perspective, few athletes would consume a meal this close to a competitive event or hard training session. Two or more hours is a more likely time period for food consumption prior to exercise for most athletes because of concerns of gastrointestinal upset during exercise. Whether a high- or low-glycemic meal is consumed several hours before exercise, insulin and blood glucose have usually returned to normal during this time frame. Carbohydrate consumed within approximately 10–15 minutes of the onset of exercise will also not likely cause blood glucose or insulin problems. This may be an important strategy for carbohydrate intake for events where consumption during exercise is difficult, for example, the swimming leg of a triathlon. The bottom line: for most athletes the glycemic index of the pre-exercise meal should rarely be of concern, and using low-glycemic foods during this time is primarily based on the athlete's personal preference (Donaldson et al., 2010).

Pre-exercise meals may contain protein and fat in addition to carbohydrate. Protein and fat may be components of foods that are favored by the athlete but should not be large in amount in order to allow for adequate gastric emptying and digestion. The inclusion of foods with protein and/or fat may aid in satiety and prevent feelings of hunger that may be felt soon after carbohydrate-only meals. Although suggested by some research studies, the inclusion of protein/amino acids in the pre-exercise meal is not likely to aid performance over the consumption of carbohydrate alone (see further discussion in Chapter 5). Particularly for the endurance athlete, carbohydrate should make up the majority of the pre-exercise meal.

Athletes need fluid prior to exercise, thus liquid pre-exercise meals that contain carbohydrates are popular. Liquid meals may also be better tolerated since they move out of the stomach and into the gastrointestinal tract faster than most solid meals. Tolerability is always important but is especially so prior to competition when the gastrointestinal tract is prone to additional upset due to psychological stress and nervousness.

Specific products continue to be developed and marketed as pre-exercise meals to provide carbohydrate prior to exercise. These pre-packaged products provide convenience and a known amount of carbohydrate, but athletes should realize that pre-exercise carbohydrate needs can be met in a variety of ways including the consumption of appropriate foods and/or other beverages found in a grocery store. Many

Table 4.6 Examples of Prepackaged Pre-Exercise Products for Athletes

Product	Serving size	Energy (kcal)	CHO (g)	Fiber (g)	Protein (g)	Fat (g)	Recommended timing
Gatorade G Series, 01 Prime, Pre-Game Fuel	118 ml pouch	100	25	0	0	0	15 minutes before exercise
G Series Pro, 01 Prime, Carb Energy Drink	335 ml	330	82	0	0	0	60 minutes or more before exercise
G Series Pro, 01 Prime, Nutrition Shake	330 ml	360	54	1	20	8	3–4 hours before exercise

Legend: kcal = kilocalories; CHO= carbohydrate, g = gram; ml = milliliter

companies offer a range of pre-exercise products that contain varying amounts of carbohydrate (see Table 4.6 for an example). Some of these products need to be consumed hours before exercise, whereas others can be consumed within 15 minutes of exercise. In all cases, the athlete will need to determine individual tolerance through trial and error, which may include fine-tuning both the amount and timing.

The pre-exercise meal is important but it cannot completely offset the lack of muscle and liver glycogen that results from repeated days of insufficient carbohydrate intake. For athletes who restrict carbohydrates and energy to compete in a specific weight category (for example, wrestlers, lightweight rowers, kick boxers), the precompetition meal does provide an opportunity to replenish some fluid and glycogen stores and to increase blood glucose concentrations. However, the time between weight certification and the start of the competition is probably too short to adequately replenish depleted glycogen stores. Nonetheless, these athletes try to consume as much carbohydrate prior to competition as they can tolerate.

Intake during exercise training and competition. Athletes who perform endurance exercise or intermittent high-intensity exercise for more than 1–2 hours are at risk for glycogen depletion, low blood glucose, and fatigue during training and competition (Coyle, 2004). Ultradistance racers, triathletes, marathon runners, distance cyclists, and other long-duration athletes must consume both carbohydrates and fluids during heavy training and competition or they may fail to finish. Although the need is not as great, intermittent high-intensity athletes such as soccer and basketball players also benefit from carbohydrate (and fluid) intake during practices and games. Carbohydrate intake during training and competition helps these athletes to spare muscle glycogen, maintain blood glucose concentrations, delay fatigue, and reduce the athlete's perception of fatigue.

It is typically recommended that 30 to 60 g of carbohydrate be consumed each hour during prolonged exercise as either fluid or food (Coleman, in press a).

Lance Armstrong (wearing the yellow jersey) eats while racing to increase carbohydrate intake.

JOEL SAGET/AFP/Getty Images

This recommendation is based on the cumulative results of research studies (reviewed by Coyle, 2004) and the maximum rate of glucose absorption from the gastrointestinal tract, which is estimated to be 1 g/min or 60 g/hour (Guezennec, 1995). However, during ultraendurance exercise, such as distance cycling, it may be advantageous to consume up to 90 g/hour (Jeukendrup, 2007). This increased consumption of carbohydrate is possible by consuming different types of carbohydrates (for example, glucose, sucrose, fructose), which use different transporters in the small intestine. Table 4.7 summarizes carbohydrate recommendations during exercise for a variety of sports.

Each athlete must find the concentration of carbohydrate that is tolerable during exercise. Too great a carbohydrate concentration will slow gastric emptying (that is, how quickly the contents of the stomach pass into the intestine) and may cause gastrointestinal distress. During exercise many athletes consume beverages (sports drinks) that contain 6–8 percent carbohydrate. At those concentrations, 1,000 ml (a little more than four 8-ounce cups) would contain 60 to

Table 4.7 Carbohydrate Intake during Exercise to Enhance Performance

Exercise intensity and duration	Sport or event	Recommended carbohydrate intake to enhance performance*
High-intensity exercise less than 45 minutes	Running (sprints up to 10 km); cycling (track cycling, short criteriums); swimming (sprints up to 1500 m); crew (rowing)	None
High-intensity exercise (continuous or intermittent) approximately 45–60 minutes	Team sports, such as basketball, lacrosse, water polo, or ice hockey; cycling time trials	0–30 g/h
High-intensity exercise (intermittent) approximately 90 minutes	Team sports, such as soccer; skilled recreational tennis players; team or individual handball, racquetball, or squash	30–50 g/h
Moderate to vigorous exercise more than 2 hours	Backpacking, hiking; recreational cycling	30–60 g/h**
High-intensity exercise more than 2 hours	Marathon running; sprint and Olympic distance triathlon; 50 km ski racing; professional tennis match	50–70 g/h**
Ultraendurance competitions lasting many hours or repeated over days	Ironman length triathlons; cycling stage races, adventure racing	60–90 g/h**

g/h = grams per hour

*Assumes the athlete can tolerate this amount; athletes should experiment with amounts and types of carbohydrates before using in competition.

**The maximal rate of carbohydrate absorption from the gastrointestinal tract is estimated to be 60 g/h. If need approaches or exceeds this amount, it may be helpful to consume sports beverages that contain a variety of sugars, such as glucose, fructose, sucrose, galactose, maltodextrin, isomaltulose, amylose, amylopectin, and/or trehalose.

80 g of carbohydrate. Distance runners report greater gastrointestinal problems than distance cyclists, and it is imperative that athletes use trial and error during training to gauge their tolerance. However, strategies that work well during training often need to be adjusted for competition because training can rarely simulate the stress associated with competing.

There has been speculation that moderate- to high-glycemic index foods are beneficial during prolonged exercise because such foods are rapidly digested and absorbed (Burke, Collier, and Hargreaves, 1998). However, there is a lack of scientific studies to confirm or dispute the benefits of moderate- or high-GI foods during exercise greater than 1 hour (Donaldson et al., 2010).

A wide variety of commercial products have been developed and marketed as vehicles to increase carbohydrate intake during exercise (see Spotlight on... Sports Drinks, Bars, and Gels) although athletes can take in carbohydrates through food (for example, bananas) or many types of fluids. Carbohydrate beverages, or sports drinks, intended to be consumed during exercise are typically formulated to contain less than 10 percent carbohydrate. One example is the original Gatorade, rebranded in 2010 as Gatorade G Series 02 Perform, which contains 6 percent carbohydrate. Another example is Powerade, rebranded as Powerade ION4. Previously 8 percent carbohydrate, Powerade was reformulated in 2010 to contain 14 grams of carbohydrate in 8 fluid ounces, or 6 percent carbohydrate (6 g carbohydrate per 100 ml), the same concentration of carbohydrate as the original Gatorade.

Sports drinks often contain nutrients other than carbohydrate, such as sodium, potassium, vitamins, protein, and so on. The inclusion of sodium may aid in carbohydrate absorption from the gastrointestinal tract but it also serves to replace electrolytes lost in sweat. Versions of sports drinks containing additional electrolytes (calcium and magnesium as well as sodium and potassium) have been formulated for participation in long-duration endurance sports where sweat and electrolyte loss is high. These additional components address fluid and electrolyte loss as discussed in Chapter 7, but the amount and delivery of carbohydrate is the same.

Another recent trend in sports drink formulation is the lowering of carbohydrate content for those participants whose exercise is not as demanding in intensity and/or duration as that of other athletes, and therefore whose carbohydrate needs are not as great. For example, Gatorade G Series G2 Low Cal contains approximately one-third of the carbohydrate found in original Gatorade (5 grams in 240 ml or 2 percent carbohydrate). Powerade Zero contains no carbohydrate and no calories. These "lite" versions of sports drinks may be useful for fluid and electrolyte replacement but will obviously not provide sufficient carbohydrate for athletes under more rigorous training and competition situations.

Developing a carbohydrate intake plan during exercise must take into account a number of practical issues:

- Duration and intensity of the exercise
- Type of exercise and opportunity or ease of food/fluid consumption
- Food/fluid preferences of the athlete
- Availability of carbohydrate sources

As contrasting examples, consider a triathlete and a bicycle racer. A triathlete has virtually no opportunity to consume carbohydrate during the swim portion of the race and must therefore develop a plan for consumption during the bike and run portions. Although food and fluid can be carried on the bike, an Ironman-length triathlon is too long for the athlete to carry all the necessary food and fluid, so a triathlete must depend upon products supplied by the race organizers. During a long bicycle race, however, a cyclist can consume carbohydrate in the preferred form throughout the race typically from sports drinks carried in water bottles on the bike and from food supplied by team support vehicles.

Application exercise

Carbohydrate Intake Plan for a Marathon Runner

A 46-year-old male runner is planning to run the Chicago Marathon. He is a well-trained runner, having had some success in 10-kilometer and half-marathon distance races but has fatigued and slowed down in the late stages of the two previous marathons he has run. Develop a specific plan for carbohydrate consumption for this athlete to ensure that he consumes sufficient carbohydrate during the marathon. Assume he weighs 152 pounds and that Gatorade G Series Pro 02 Perform ("endurance formula" Gatorade) is available on the race course at aid stations every 1–2 miles. Bananas are also available at aid stations between miles 20 and 24. The plan should include instructions for timing, amount, and type of carbohydrate intake during the race, including practical suggestions.

Intake after training and competition. Long, intense training sessions or competitive events may leave athletes with substantially reduced or depleted liver and muscle glycogen stores. Athletes need to consider a variety of factors to optimally replenish those stores. As discussed earlier, one of the conditions that stimulate the synthesis of glycogen is its depletion. Some muscle glycogen may be resynthesized after hard exercise even if the athlete does not eat, although the amount is minimal. The glucose used for glycogen synthesis in

this case comes from the liver through gluconeogenesis, particularly from lactate. To optimize glycogen replacement, however, two things are needed: carbohydrate and insulin.

Glucose molecules are needed to re-form the glycogen chains and are typically obtained by consuming carbohydrate-rich foods. Insulin plays an important role by facilitating uptake of glucose into muscle cells and by activating the enzyme principally responsible for glycogen resynthesis, glycogen synthase. Consumption of foods or beverages containing carbohydrate provides the source of glucose and will also stimulate the release of insulin from the pancreas. A substantial amount of research supports the following guidelines for optimal muscle glycogen resynthesis:

Timing—To maximize the rate at which muscle glycogen is replaced, carbohydrate should be consumed as soon after the exercise bout as possible. Studies show that waiting as little as 2 hours after exercise to begin consuming carbohydrate will significantly slow the rate of muscle glycogen resynthesis (Ivy, Katz et al., 1988). Athletes should therefore begin consuming carbohydrate as soon as is practical after the exercise session or competition is over.

Meal size—Consumption of carbohydrate in smaller, more frequent meals appears to further aid the rate at which muscle glycogen is replaced in the hours after exercise (Doyle, Sherman, and Strauss, 1993). With large single meals, blood glucose and insulin rise rapidly and then return to baseline relatively quickly. Elevations in blood glucose and insulin can be sustained for a longer period of time with smaller, more frequent feedings, which maintains the appropriate environment for muscle glycogen synthesis. It is also likely to be more palatable for the athlete to consume smaller amounts of food and/or beverages over several hours than trying to consume a large meal soon after fatiguing exercise.

Type of carbohydrate—Carbohydrate beverages that are consumed after exercise to replace glycogen should contain mostly glucose and/or sucrose as the carbohydrate source. Studies clearly show that beverages containing mostly fructose do not result in glycogen synthesis rates that are as high as those beverages with glucose and sucrose (Blom et al., 1987). Athletes may consume fructose because it is found naturally in fruit juices and because it is often added to beverages to enhance flavor and sweetness. However, fructose-containing beverages should not be the primary recovery beverage because of the reduced effect on muscle glycogen resynthesis and the potential for gastrointestinal upset.

Although carbohydrate beverages may be adequate for glycogen recovery, athletes may be hungry after a hard workout or competition and may prefer to consume solid food. It has been shown that solid foods that contain carbohydrates can be as effective as beverages in replenishing muscle glycogen (Keizer et al., 1987). This is one situation in which athletes may want to pay particular attention to the glycemic index of foods, as the consumption of high-GI foods in the postexercise (recovery) period can enhance the resynthesis of muscle glycogen (Burke, Collier, and Hargreaves, 1993). By definition, consumption of high-glycemic index foods results in higher blood glucose and insulin responses, conditions that favor the rapid synthesis of muscle glycogen.

Along with commercial products formulated and marketed for consumption before and during exercise, recovery products are available as well. Examples include Gatorade G Series 03 Recover, G Series Pro 03, and Endurox R4 Recovery Drink. These recovery drinks typically provide carbohydrate to stimulate the resynthesis of muscle glycogen, as well as protein to aid in reducing protein breakdown and stimulating protein synthesis during recovery from exercise (see discussion in Chapter 5). These products provide convenience and a known amount of carbohydrate and protein, but sufficient carbohydrate and protein can easily be provided through other foods as well.

Amount of carbohydrate—The highest rates of muscle glycogen synthesis have been observed in the hours after fatiguing exercise when approximately 1.5g/kg of carbohydrate were consumed in the first hour immediately after exercise (Doyle, Sherman, and Strauss, 1993; Ivy, Lee et al., 1988). In these studies, subjects consumed approximately 120 g of carbohydrate in the first hour postexercise. This might be considered the "priming" dose of carbohydrate to initiate the glycogen synthesis process, with more carbohydrate being consumed over the next few hours depending upon the need for rapid resynthesis. For athletes needing maximal rates of muscle glycogen synthesis, 0.75 to 1.5g/kg of carbohydrate should be consumed each subsequent hour until approximately 4 hours after exercise. It is important to recognize that the higher end of this range is a large amount of carbohydrate and may cause gastrointestinal upset.

Addition of protein/amino acids—A number of studies suggest that adding proteins to the meal after exercise may increase the rate at which muscle glycogen is replaced (Van Loon et al., 2000; Zawadzki, Yaspelkis, and Ivy, 1992). Some amino acids and proteins may have an additional effect on insulin secretion, acting to increase and prolong the insulin response that stimulates muscle glycogen synthesis.

Whey protein was initially studied as an addition to carbohydrate for postexercise consumption, and resulted in the recommendation of a 4:1 ratio of carbohydrates to proteins for recovery drinks (for example, Endurox and Accelerade). However, more recent studies have called this recommendation into question (Jentjens et al., 2001). Studies that have shown higher rates of muscle glycogen synthesis when combining proteins with carbohydrates may be confounded by the addition of extra kilocalories with the proteins in comparison to the carbohydrates alone. When the amount of kilocalories consumed after exercise was equalized by increasing the amount of carbohydrate, the protein-and-carbohydrate beverage was no different from the carbohydrate-only beverage. The addition of protein to the recovery beverage or food certainly does not impede glycogen recovery over carbohydrate alone, and may result in an increase in postexercise protein synthesis (see Chapter 5). Because athletes need to consume both carbohydrate and protein after exercise, they may choose foods that contain both (for example, chocolate milk, fruit-in-the-bottom yogurt, turkey sandwich), but the ratio of carbohydrates to proteins needed has not been established definitively.

Muscle glycogen stores can be maximized by diet and exercise manipulation.

Carbohydrate loading (also known as carbohydrate supercompensation) is a technique that some athletes use to attain maximum glycogen stores prior to an important competition. This technique is appropriate for endurance and ultraendurance athletes who perform 90 minutes or more of continuous exercise, and it may be used by some bodybuilders as part of their precontest preparations. In the case of endurance and ultraendurance athletes, without maximum levels of glycogen when the race begins these athletes could run out of stored carbohydrate as a fuel source and be forced to reduce the intensity of their exercise or drop out of the race. In the case of bodybuilders, maximum glycogen storage is a strategy used to promote muscle definition, one feature on which contestants are judged. In these circumstances, performance could be enhanced by carbohydrate loading.

A carbohydrate-loading protocol was first reported in 1967 (Bergström, Hermansen, and Saltin, 1967), and this method is still used today by some bodybuilders and some endurance athletes (Figure 4.13). Seven days prior to the competition, exhaustive exercise is performed for 3½ days in combination with an extremely low carbohydrate diet. During this phase, known as the depletion stage, carbohydrate stores are severely depleted by exercise and remain low due to the lack of dietary carbohydrate. Side effects include irritability, hypoglycemia, fatigue, inability to accomplish the required exercise, and risk for injury. The depletion

Carbohydrate loading: A diet and exercise protocol used to attain maximum glycogen stores prior to an important competition.

Sports Drinks, Bars, and Gels

Manipulation of carbohydrate intake during exercise to improve performance has been studied extensively, particularly for endurance or intermittent high-intensity sports or activities. Numerous research studies have sought to determine the optimal amount, type, form, concentration, and timing of carbohydrate intake during exercise. The results of these studies have led to the development and marketing of a variety of commercially available carbohydrate products intended for use during exercise.

The first "sports drink" (Gatorade) was developed with a dual purpose—fluid and electrolyte replacement for sweating athletes along with energy replacement. Gastric emptying, and therefore water absorption, may be hampered by too much carbohydrate in the beverage, so many sports drinks are formulated to provide carbohydrate in a concentration that will not interfere with fluid balance and thermoregulation. This aspect of beverages is discussed in more detail in Chapter 7, but Table 4.8 lists some of the carbohydrate-related characteristics of popular sports beverages. Athletes need to be aware of the carbohydrate content of sports beverages so they can consume the recommended amount and type of carbohydrate during exercise.

Sports drinks containing the majority of their carbohydrate in the form of glucose or glucose polymers (maltodextrins) have been shown to have a more beneficial effect on performance than other types of carbohydrates such as fructose. However, a small amount of fructose is usually added to sports drinks to improve their flavor, which may encourage athletes to consume more fluid. Although consuming carbohydrate after fatigue has occurred can be helpful, it is more beneficial to endurance performance to delay fatigue by consuming carbohydrate at the onset of exercise and continuing throughout exercise to maintain blood glucose and carbohydrate availability.

Although consistent research has shown that when individual sugars are tested, glucose and polymers of glucose have the most

Sandra Mu/Getty Images

beneficial effect on endurance exercise performance. A number of research studies (Jentjens and Jeukendrup, 2005; Triplett et al., 2010) have shown that adding fructose to a glucose beverage may enhance the ability to take up and utilize carbohydrate during exercise and may lead to further improvements in endurance exercise performance. As discussed earlier in this chapter, glucose and fructose are absorbed by different transporter mechanisms in the gut. When too much glucose is consumed, the glucose transporters may become saturated limiting glucose absorption. The addition of fructose takes advantage of the different absorption

Table 4.8 Carbohydrate and Energy Content of Sports Beverages

Beverage	Serving size (oz)	Energy (kcal)	CHO (source)	CHO (g)	CHO (%)
Hydrade®*	8	40	HFCS	9	4
Gatorade® G Series, 02 Perform (formerly Original Thirst Quencher)	8	50	Sucrose, dextrose	14	6
Gatorade G Series Pro, 02 Perform, Endurance Formula	8	50	Sucrose syrup; glucose-fructose syrup	14	6
Accelerade®	8	80	Sucrose, trehalose, fructose, maltodextrin	14	6
All Sport Body Quencher®	8	40	HFCS	16	6.5
Powerade® ION4®	8	50	HFCS, glucose polymers	14	6

Legend: oz = fluid ounces; kcal = kilocalories; CHO = carbohydrate; g = grams; HFCS = high-fructose corn syrup

Nutrient information was obtained from company websites and product labels.

*Also contains 5.1 percent glycerol, a sugar alcohol.

Table 4.9 Carbohydrate and Energy Content of Sports Gels and Bars

Food	Serving Size	Energy (kcal)	CHO (source)	CHO (g)	Fiber (g)
GU Energy Gel®	1 package (32 g)	100	Maltodextrin (glucose polymers), fructose	25	0
GU Chomps	4 pieces (30 g)	90	Tapioca syrup, cane sugar, maltodextrin	23	0
Clif Shot Energy Gel®	1 package (34 g)	110	Organic maltodextrin, organic evaporated cane juice	22	1
Clif Shot Bloks®	3 pieces (30 g)	100	Organic brown rice syrup, organic evaporated cane juice	24	0
Balance Bar (various flavors)	1 bar	200	Fructose, sugar	20–22	>1–2
Clif Bar® (various flavors)	1 bar	230–250	Organic brown rice syrup, rolled oats, cane juice, fig paste, and/or dried fruits	44–48	4
EAS AdvantEdge Carb Control Bar	1 bar	240	Maltitol syrup, maltitol*, dextrin, glycerine*	27	6
PowerBar (Harvest Energy Bar)	1 bar	250	Brown rice syrup, evaporated cane juice syrup	43	5
Gatorade G Series Pro, 01 Prime, Nutrition Bar	1 bar	230	Maltodextrin (glucose polymers), fructose	39	2

Legend: kcal = kilocalories; CHO = carbohydrate; g = grams

*Glycerine and maltitol are sugar alcohols.

Product formulations are subject to change.

pathway and may provide for increased carbohydrate uptake during exercise that may in turn aid performance.

Carbohydrate can be consumed during exercise in forms other than a beverage (Table 4.9). For example, carbohydrate is available in a semiliquid, concentrated gel form. These gels are usually packaged in small pouches or carried in small containers, making the carbohydrate convenient to carry, open, and consume during exercise. A typical packet provides 100 kcal and 25g of carbohydrate. Two to three packages of gel consumed each hour during prolonged exercise provides the recommended 30–60 g of carbohydrate per hour. Gels are consumed with water, which provides the fluid that will also be needed during exercise. Some products also contain caffeine, a central nervous system stimulant that masks fatigue.

The carbohydrate used in a product is sometimes chosen because of its glycemic response. Glucose polymers provide a slow, sustained delivery of glucose into the blood. Other products contain sugars such as sucrose that are rapidly absorbed. Some contain sugar alcohols, which are slowly absorbed and have little impact on glucose and insulin concentrations. Reading the ingredient list will help athletes choose the most appropriate product for a given situation.

Carbohydrate can be consumed easily and comfortably in liquid and gel form during exercise, but some endurance athletes can also tolerate solid carbohydrate foods. For example, distance running is not very compatible with chewing and swallowing solid food, but a marathon runner may be able to tolerate a ripe banana. Other endurance and ultraendurance athletes (for example, distance cyclists) may desire some solid food in their stomach because it lessens the feeling of hunger.

Because athletes generally need more carbohydrate than sedentary individuals, and some athletes need substantial amounts of carbohydrate daily to replenish muscle glycogen, products containing carbohydrate in a convenient form have been developed. Known as sports or energy bars, these bars often contain substantial amounts of carbohydrates, along with varying amounts of fiber, proteins, and fats. Although an energy bar marketed to athletes may have about the same amount of carbohydrates as two packets of a gel, its fiber and fat content is typically too high to be tolerated during higher intensity exercise such as distance running. However, adventure athletes, who may engage in many hours of mountain biking, trail running, and orienteering, often snack on sports bars during or between events, and the higher carbohydrate and caloric content is beneficial. The important point is that athletes need to determine the appropriate amount, type, and form of carbohydrate and then identify products that best meet these needs.

Carbohydrate in liquid, semisolid, and bar form are convenient ways to obtain the carbohydrates needed, and many athletes consume some of these products at times other than during exercise. Athletes should be aware that convenient and good-tasting carbohydrate foods and beverages are easy to overconsume and can result in an excess caloric intake. This is especially true for recreational athletes, whose carbohydrate needs are typically not much greater than that recommended for the general population, athletes in low-energy-expenditure sports (for example, baseball, softball), and athletes in the off-season when intensity and duration of training are low.

Bodybuilders need to focus on carbohydrates, too. These men are contestants in a drug-free bodybuilding championship.

CHO   = carbohydrate
g/kg/d = grams per kilogram body weight per day

Figure 4.13 "Classical" carbohydrate-loading protocol

Drawn from the methods of: Bergström, J., Hermansen, L., and Saltin, B. (1967). Diet, muscle glycogen, and physical performance. *Acta Physiologica Scandinavica, 71*(2), 140–150.

stage is followed by 3½ days of very light or no exercise and a high-carbohydrate diet (~8 g/kg/d). This phase, known as the repletion stage, supplies large amounts of carbohydrate to glycogen-starved muscles. In response the body "supercompensates," and near maximum glycogen storage is achieved. Typical resting muscle glycogen concentration is approximately 100–125 mmol/kg and with this "classic" protocol, subjects stored muscle glycogen at levels approaching 220 mmol/kg. The repletion stage may also have some adverse effects including gastrointestinal distress due to the reintroduction of sugars, starches, and fiber and an increase in muscle water storage. Although the increase in water storage is a normal part of the glycogen resynthesis process, it can leave athletes feeling heavy, waterlogged, and uncomfortable.

The original carbohydrate-loading protocol was tested and used by endurance athletes but better protocols have been developed. Several aspects of the original depletion stage were problematic. First, carbohydrate intake was so low during the depletion stage that normal training and conditioning could not be maintained. Additionally, the risk for injury was too great. These limitations resulted in the testing and development of other protocols.

In 1981, Sherman et al. published a modified protocol that eliminates the severe depletion stage but still results in high levels of stored muscle glycogen (Figure 4.14). This 6-day plan includes a 3-day depletion stage that consists of a dietary carbohydrate intake of 5g/kg/d. Recall that this level of carbohydrate intake is the minimum amount recommended to athletes in training. It is sufficient to allow athletes to complete the required training during this stage— 90 min of hard training (70 percent of $\dot{V}O_{2max}$) on the first day followed by 2 days of hard training for 40 min. Glycogen depletion is achieved, but the serious side effects associated with the original depletion stage are avoided. The repletion stage calls for a

carbohydrate intake of 10 g/kg/d for 3 days and 2 days of 20 min of exercise followed by a rest day. This repletion phase is similar to the original protocol. This modified approach utilizes a strategy that is more similar to what an endurance athlete would likely do in the week prior to a big event, such as gradually tapering exercise. The only manipulation that differs from a usual training regime is to slightly reduce the amount of carbohydrate early in the week when the exercise duration is longer, and increase dietary carbohydrate later in the week when there is more rest time. Muscle glycogen levels of 205mmol/kg can be obtained without the adverse effects often experienced by athletes following the classical regimen.

Researchers continue to look at ways to modify the carbohydrate-loading protocol, such as by manipulating the time period required. Since 2002, some studies have shown that high levels of muscle glycogen can be attained within 1 day if large amounts of carbohydrate are consumed. These levels can be attained in 1 to 3 days if the athlete refrains from exercise (Bussau et al., 2002) or if the athlete performs 3 minutes of high-intensity exercise, then rests, and consumes a large amount of carbohydrate (Fairchild et al., 2002). It is important to note that these studies do not show higher muscle glycogen supercompensation than Sherman's modified method, and may involve strategies incompatible with the athlete's precompetition preparation (for example, complete cessation from exercise or performing very high intensity exercise the day before). Carbohydrate-loading techniques remain an area of active research.

Training and performance may be impaired if insufficient carbohydrate is consumed.

The amount and timing of carbohydrates before, during, and after exercise must be considered in the context of total dietary intake over days, weeks, and months of training. Daily dietary intake considers not only the 24-hour consumption of carbohydrates but also total caloric intake and the relative contribution of carbohydrates, proteins, fats, and alcohol. Daily

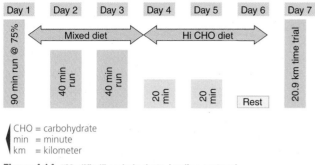

CHO = carbohydrate
min = minute
km = kilometer

Figure 4.14 "Modified" carbohydrate-loading protocol

Drawn from the methods of: Shermfan, W. M., Costill, D. L., Fink, W. J., and Miller, J. M. (1981). The effect of exercise and diet manipulation on muscle glycogen and its subsequent use during performance. *International Journal of Sports Medicine, 2*(2), 114–118.

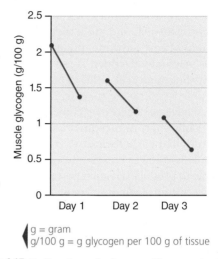

g = gram
g/100 g = g glycogen per 100 g of tissue

Figure 4.15 Decline of muscle glycogen with successive days of training

Well-trained endurance runners had muscle glycogen measured before and after they completed 10-mile runs on a treadmill in the laboratory each day for 3 consecutive days. The runners ate their normal moderate-carbohydrate diet. Note the sequential decline in muscle glycogen.

Costill, D. L., Bowers, R., Branam, G., and Sparks, K. (1971). Muscle glycogen utilization during prolonged exercise on successive days. *Journal of Applied Physiology, 31*(6), 834–838.

intake will vary, but over time both energy and carbohydrate intake must be adequate. If not, training and performance will be negatively affected.

First and foremost, the body must meet its energy needs. Insufficient total energy intake means that some of the carbohydrates consumed will be used for immediate energy and will be unavailable for storage as muscle or liver glycogen (Burke, Kiens, and Ivy, 2004). Surveys of female athletes suggest that energy intake is often low (Burke, 2001). The need to balance energy intake with energy expenditure is covered in Chapter 2.

Surveys also suggest that many athletes do not meet the recommendations for total daily carbohydrate intake. Burke, Kiens, and Ivy (2004) report the following daily mean (average) intake of carbohydrates:

Male nonendurance athletes: 5.8 g/kg

Male endurance athletes: 7.6 g/kg

Female nonendurance athletes: 4.6 g/kg

Female endurance athletes: 5.7 g/kg

Female nonendurance athletes report the lowest daily carbohydrate intake and fail to exceed the minimum amount of carbohydrate recommended (5 g/kg/d). On average, both male and female endurance athletes fall short of the recommended 8–10 g/kg daily. Because female endurance athletes fail to attain even the midrange recommendation for training (6 g/kg/d), it is suspected that many of these athletes do not restore muscle glycogen to pre-exercise levels with repeated days of training. Low energy intake, which is more frequently reported in females than in males, compounds the low-carbohydrate problem.

Insufficient carbohydrate intake can lead to hypoglycemia and both acute and chronic fatigue. As mentioned previously, the symptoms of hypoglycemia are loss of concentration, apathy, light-headedness, shakiness, and hunger. When blood glucose concentration is low during exercise, the uptake of glucose by the brain is reduced and the ability to concentrate is

decreased. Other hypoglycemic symptoms, such as light-headedness, may also affect the athlete's perception of exercise difficulty. Fatigue that results from low carbohydrate intake over a period of days, weeks, or months can have a substantial negative effect on training and performance.

The muscle glycogen levels an athlete has at the beginning of exercise may affect both the intensity and duration of that exercise session. Therefore, athletes should pay attention to the effect that each exercise session has on carbohydrate stores and the dietary steps necessary to maintain adequate glycogen levels on a daily basis. Low muscle glycogen levels may compromise the athlete's ability to complete a training session or work out at the required intensity. Insufficient carbohydrate stores may not be the result of a single, depleting exercise session, but a result of fatiguing exercise on a daily, weekly, or monthly basis.

A classic study published by Costill et al. (1971) demonstrates the daily pattern of declining muscle glycogen when athletes train hard and do not consume enough carbohydrate (Figure 4.15). Distance runners trained 10 miles each day and ate a normal, mixed diet. Muscle biopsies clearly show the decline in resting muscle glycogen levels each day. In fact, the runners began their 10-mile run on the third day with lower muscle glycogen levels than they had when they finished the first 10-mile run. Although they were able to complete their training runs, it can easily be seen how the glycogen losses over a few days of training may adversely affect training or competitive efforts later in the week.

In another study, when runners and cyclists trained hard every day for a week and ate a diet containing a moderate amount of carbohydrate, pre-exercise muscle glycogen levels gradually declined stepwise, similar to the Costill study (Sherman et al., 1993). Once again, these athletes were able to complete their training sessions, but the potential adverse effects may come when the athletes are chronically short of carbohydrate. Rowers that consumed insufficient carbohydrate over the course of a month not only had lower muscle glycogen levels, but were also unable to produce as much power during high-intensity-rowing training (Simonsen et al., 1991).

Months of low-carbohydrate intake, in conjunction with a long-term negative energy balance, are factors that contribute to overtraining syndrome. There is no single cause for this condition, but chronic low-carbohydrate intake is one likely factor. An increase in energy and carbohydrate intakes and rest are all part of the treatment plan (Shephard, 2001).

Carbohydrate and fiber must be consumed in appropriate amounts for good health.

For general health, it is recommended that carbohydrate intake for adults should be 45–65 percent of total energy intake, assuming that energy intake is adequate (Institute of Medicine, 2002). This recommendation is based on scientific studies conducted in the general population. These studies found associations between dietary intake and the prevention of nutrient inadequacies as well as a reduced risk for chronic diseases, such as cardiovascular disease. Recommendations for the general population are made as a percentage of total energy intake, so some calculations are necessary before comparisons can be made to athletes.

An adult woman who engages in daily activity equivalent to walking 2 miles needs an estimated 2,000 kcal per day. Based on this caloric intake and the recommendation that carbohydrate be 45–65 percent of total energy, 225 to 325g of carbohydrate are recommended daily. Assuming this woman is 55kg (121 lb), her recommended carbohydrate intake is ~4 to 6 g/kg/d. General recommendations take into consideration the low physical activity that is characteristic of the majority of U.S. adults. If the woman in this example was more active and needed 2,700 kcal daily, her recommended daily carbohydrate intake would be ~300 to 440 g or ~5.5 to 8 g/kg.

Most athletes consume between 5 and 8 g/kg daily. Those who engage in prolonged endurance exercise need more carbohydrate (and kilocalories). They also have a high level of fitness. There is no evidence that a high-carbohydrate intake that accompanies a high level of physical activity is unhealthy. The recommended carbohydrate intake for athletes—5 to 10 g/kg/d—is consistent with the recommendations made for good health. This amount of carbohydrate is also speculated to reduce the risk for exercise-induced immune system suppression, something that is common in endurance athletes who consume low levels of carbohydrate for several days (Gleeson, Nieman, and Pedersen, 2004).

Another carbohydrate-related recommendation for general health is sufficient fiber intake. It is recommended that women under the age of 50 consume 25 g of fiber daily whereas men 50 and younger should consume 38 g per day. For adults over the age of 50, the recommendation is 21 g daily for females and 30 g daily

There are a wide variety of whole grain breads to choose from, which provide more fiber and nutrients than highly processed white bread.

for males (Institute of Medicine, 2002). The average daily fiber intake of U.S. adults is 15 to 19 g.

Limited surveys of collegiate (Hinton et al., 2004) and elite (Ziegler et al., 2002) athletes suggest that the fiber intake of athletes does not differ from that of the general population. Female athletes report a fiber intake of ~15 to 19 g/d whereas male athletes consume about 18 to 19 g daily. Fiber intake is associated with the consumption of carbohydrate-containing foods such as fruits, vegetables, whole grains, legumes, beans, and nuts. Sugar is devoid of fiber, and grains that are highly refined (for example., whole wheat that is processed to make white bread) lose most of the fiber originally present before processing. Depending on the kinds of foods and beverages consumed, it is possible for athletes to meet carbohydrate recommendations but fall short of meeting fiber recommendations.

Key points

- The recommended daily intake of carbohydrates for athletes is 5 to 10 g/kg, depending upon the intensity and duration of training and competition, although ultraendurance athletes may need more during certain training periods.

- Carbohydrate intake should be adjusted according to sport specific needs as well as training and competition levels.

- The amount of carbohydrate in an athlete's diet should be based upon body weight and not a percentage of the overall diet to ensure adequate intake.

- In order to optimize the body's stores, a carbohydrate-containing meal should be eaten 1–3 hours before training or competition. The timing of the meal and the amount and type of carbohydrate should be individualized.

- Thirty to 60 grams of carbohydrate should be consumed each hour during prolonged or high-intensity intermittent exercise.

- If exercise has been prolonged and/or is intense enough to result in substantial muscle glycogen depletion, carbohydrate should be consumed as soon as possible after exercise to stimulate the resynthesis of muscle glycogen.

- Pre-exercise muscle glycogen stores can be maximized by manipulating exercise and carbohydrate content of the diet in the days before a competitive event in a process called carbohydrate loading.

- Endurance or high-intensity training may be impaired if insufficient carbohydrate is consumed and the body's carbohydrate stores are lowered.

If an athlete needs to replace muscle glycogen quickly, what is the amount, timing, type, and pattern of carbohydrate intake that should be planned?

When is it important for an athlete to consider the glycemic index of foods to consume?

4.6 Translating Daily Carbohydrate Recommendations to Food Choices

Carbohydrate recommendations for athletes are scientifically based and reflect the need to maintain glucose homeostasis, maintain adequate muscle glycogen stores, and to fuel exercise. It is not difficult to understand the physiological need for carbohydrates. The challenge is to translate those recommendations into food choices that will support training, performance, the immune system, and long-term good health.

A carbohydrate-rich diet requires planning.

Many athletes fail to consume an adequate amount of carbohydrate. In some cases this may be due to a lack of knowledge about the amount of carbohydrate needed. Table 4.10 lists the estimated daily total carbohydrate intake needed by athletes based on their weight. For example, the 165 lb (75 kg) cyclist who needs 8 to 10 g/kg will need 600 to 750g of carbohydrate daily. Each gram of carbohydrate contains approximately 4 kcal. Therefore, this athlete will need to consume 2,400 to 3,000 kcal from carbohydrates alone. This is not an easy task!

In addition to the total amount of carbohydrate needed, athletes must also know the amount of carbohydrates found in various foods or food groups. Table 4.11 lists representative foods in each group and the amount of carbohydrate contained in one serving. Because there are so many carbohydrate-containing foods to choose from, it is helpful to have a framework for planning meals. Table 4.12 provides such a framework by listing the number of servings from each food group recommended for various caloric levels. Since the athlete's diet should support good health in addition to supporting training and performance, much of the carbohydrate-containing food eaten should be whole grain, rich in fiber, and nutrient dense. Many foods have all of these characteristics.

Even when athletes have information about the carbohydrate content of food, it is not always easy to consume the proper amount. Athletes may lack the time and skill to prepare carbohydrate-rich meals or might be too tired after training and prefer to sleep. Lack of time, money, and knowledge can be barriers to proper carbohydrate intake. More information on diet planning can be found in Chapter 10.

Table 4.10 Estimated Total Daily Carbohydrate Intake Based on Body Weight

Weight lb (kg*)	5 g/kg	6 g/kg	7 g/kg	8 g/kg	9 g/kg	10 g/kg
85 to 95 (39 to 43)	195–215	234–258	273–301	312–344	351–387	390–430
96 to 105 (44 to 48)	220–240	264–288	308–336	352–384	396–432	440–480
106 to 115 (48 to 52)	240–260	288–312	336–364	384–416	432–468	480–520
116 to 125 (53 to 57)	265–285	318–342	371–399	424–456	477–513	530–570
126 to 135 (57 to 61)	285–305	342–366	399–427	456–488	513–549	570–610
136 to 145 (62 to 66)	310–330	372–396	434–462	496–528	558–594	620–660
146 to 155 (66 to 70)	330–350	396–420	462–490	528–560	594–630	660–700
156 to 165 (71 to 75)	355–375	426–450	497–525	568–600	639–675	710–750
166 to 175 (75 to 79)	375–395	450–474	525–553	600–632	675–711	750–790
176 to 185 (80 to 84)	400–420	480–504	560–588	640–672	720–756	800–840
186 to 195 (84 to 89)	420–445	504–534	588–623	672–712	756–801	840–890
196 to 205 (89 to 93)	445–465	534–558	623–651	712–744	801–837	890–930
206 to 215 (94 to 98)	470–490	564–588	658–686	752–784	846–882	940–980
216 to 225 (98 to 102)	490–510	588–612	686–714	784–816	882–918	980–1,020
226 to 235 (103 to 107)	515–535	618–642	721–749	824–856	927–963	1,030–1,070
236 to 245 (107 to 111)	535–555	642–666	749–777	856–888	963–999	1,070–1,110
246 to 255 (112 to 116)	560–580	672–696	784–812	896–928	1,008–1,044	1,120–1,160
256 to 265 (116 to 120)	580–600	696–720	812–840	928–960	1,044–1,080	1,160–1,200
266 to 275 (121 to 125)	605–625	726–750	847–875	968–1,000	1,089–1,125	1,210–1,250
276 to 285 (125 to 129)	625–645	750–774	875–903	1,000–1,161	1,125–1,161	1,250–1,290

Legend: lb = pound; kg = kilogram; g/kg = gram per kilogram body weight

*Weight in kg is rounded to the nearest whole number.

Table 4.11 Carbohydrate-Containing Foods by Food Group

Food group	Food	One serving	CHO (g)	Fiber (g)	Whole grain
Starches	Bagel, plain	4 in diameter (71 g)	38	1.6	No
	Bagel, whole grain	4 in diameter (85g)	35	6	Yes
	Bread, white	1 slice	14	0.5	No
	Bread, whole grain	1 slice	14	5	Yes
	Cereal, sweet (e.g., Froot Loops, Honey Nut Cheerios)	1 cup (30 g)	24–28	1–2	Not usually
	Cereal, low sugar (e.g., Cornflakes, Cheerios)	1 cup (28 g)	22–24	1–3	Varies
	Cereal, high fiber (e.g., Raisin Bran)	½ cup (30) g	22	4	Usually
	Corn chips	1 oz	16	1	Usually
	Corn bread	1 piece (55g)	18	1	Not usually

Legend: CHO = carbohydrates; g = gram; in = inch; oz = ounce

(Continued)

Table 4.11 Carbohydrate-Containing Foods by Food Group (Continued)

Food group	Food	One serving	CHO (g)	Fiber (g)	Whole grain
	Crackers (e.g., Ritz, Saltines)	5 crackers	10	0.5	Varies
	Energy bars (e.g., CLIF Bars®)	1 bar	35–43	1–5	Varies
	English muffin	Both halves	25	1.5	Not usually
	Grits	1 cup cooked	31	<1	Not Usually
	Granola bar	1 bar (43 g)	29	1	Varies
	Hamburger bun	Both halves	21	1	No
	Oatmeal	1 cup cooked	25	4	Yes
	Pancakes, buttermilk	3 pancakes each 4 in diameter	33	1	Not usually
	Pasta (e.g., spaghetti or macaroni noodles)	½ cup cooked	19	<1	Not usually
	Pita bread	1 (60 g)	33	1	Varies
	Popcorn	1 cup popped	6	1	Yes
	Potato chips	1 oz	15	1	No
	Pretzel sticks	10 (30 g)	20	1	No
	Pretzel, soft	1	43	1	No
	Rice, brown	½ cup cooked	22	2	Yes
	Rice, white	½ cup cooked	22	<0.5	No
	Tortillas (corn)	1 6 in diameter	12	1.5	Yes
	Tortillas (flour)	1 8 in diameter	28	1	Not usually
	Waffle	1 7 in diameter	25	1.5	Not usually
	Wheat germ	2 tablespoons	6	1.5	Yes
Starchy vegetables	Corn	½ cup cooked	15	1.6	
	Peas (green)	½ cup cooked	11	3.5	
	Potatoes (mashed)	⅔ cup cooked	25	2	
	Squash (winter)	½ cup cooked	5	0.9	
	Sweet potatoes	¼ cup cooked	26	2.5	
	Yams	¾ cup cooked	28	4	
Beans/legumes	Dried beans or lentils	½ cup cooked	20	6–8	
	Hummus	¼ cup	8	3.4	
	Miso (soybean) soup	1 cup	8	2	
	Split pea soup	1 cup	19	1.5	

(Continued)

Table 4.11 Carbohydrate-Containing Foods by Food Group (Continued)

Food group	Food	One serving	CHO (g)	Fiber (g)	Whole grain
Fruits	Apple	~2.5 in diameter	19	3.3	
	Applesauce	½ cup	25	1.5	
	Banana	~9 in long	27	3	
	Blueberries or raspberries	1 cup	15–21	3.5–8	
	Cantaloupe	¾ cup or ~¼ of a 5 in diameter melon	10	1	
	Peach	~2.5 in diameter	9	1.5	
	Plum	~3 in diameter	9	1	
	Orange	~2.5 in diameter	15	3	
	Orange juice	½ cup	13	0.5	
	Strawberries	1 cup	11	3	
	Tangerine	2	19	4	
Vegetables	Broccoli	½ cup cooked	6	2.5	
	Cabbage	½ cup cooked	3	1.5	
	Carrot	½ cup cooked or 1 raw carrot ~8 in long	6	2	
	Lettuce (dark green, leafy)	1½ cups (84 g)	2	1	
	Pepper	½ cup raw	3.5	1.3	
	Spinach	½ cup cooked	3.5	2	
	Tomato	~2.5 in diameter	5	1.5	
Milk	Chocolate milk	1 cup	26	1	
	Milk	1 cup	12	0	
	Soy milk	1 cup	18	3	
	Yogurt (plain)	1 cup	17	0	
	Yogurt (sweetened)	1 cup	26	0	
Sugared beverages	6% carbohydrate sports beverage	1 cup	14	0	
	High carbohydrate sports beverage	12 oz	88	0	
	Soft drink	12 oz (1 can)	40	0	
Mixed foods	Cheese pizza, thick crust	2 slices (142 g)	55	2.5	
	Cheese pizza, thin crust	2 slices (166 g)	46	2	
	Cheese lasagna	1 cup	45	3	

Table 4.11 Carbohydrate-Containing Foods by Food Group (Continued)

Food group	Food	One serving	CHO (g)	Fiber (g)	Whole grain
	Chili with beans	1 cup	22	6	
	Chili without beans	1 cup	12	3	
Nuts	Almonds	¼ cup	7	4	
	Peanut butter	2 tablespoons	6	2	
	Pecans	¼ cup	4	2.5	
	Walnuts	¼ cup	4	2	
	Hazelnuts (filberts)	¼ cup	6	3	

Diet planning for carbohydrate intake must consider practical issues.

Laboratory research has helped identify the important roles that carbohydrates play in supporting training and enhancing performance. However, scientific knowledge must be translated into recommendations and these recommendations must be practical for athletes to achieve. In addition to practical issues such as food availability, food choices may be influenced by individual preferences, such as the desire to be a vegetarian, or medical conditions, such as lactose intolerance or diabetes.

Having carbohydrate-rich foods available. Many athletes know the kinds of foods that they should eat but find it difficult to "just do it." They often return home from a hard practice tired and hungry. Planning is key and a well-stocked pantry of easy-to-fix carbohydrate-containing foods increases the likelihood of making good food choices. Table 4.14 gives examples of such foods.

Choosing carbohydrate-rich meals in fast-food and other restaurants. Like most Americans, athletes find restaurant food appealing. Athletes frequently travel to compete, and often the team bus stops at a fast-food restaurant. Eating out may be a favorite way to socialize. For these and other reasons athletes often eat away from home.

The challenge when eating in restaurants is to order foods that fit within the athlete's diet plan. Many

Table 4.12 Guidelines for Number of Servings from Carbohydrate-Containing Food Groups Based on Energy Intake

Food group	Servings/ 2,000 kcal	Servings/ 2,200 kcal	Servings/ 2,400 kcal	Servings/ 2,600 kcal	Servings/ 2,800 kcal	Servings/ 3,000 kcal	Servings/ 3,200 kcal	Servings/ 3,400 kcal
Fruits	4	4	4	4	5	5	6	6
Vegetables*	5	6	6	7	7	8	8	8
Grains**	6	7	8	9	10	10	10	11
Beans	1	1	1	1	1	1	1	1
Milk	2–3	3	3	3	3	3–4	3–4	4
Sugar***	Moderate	Moderate	Moderate	Moderate	Moderate	Moderate	Moderate	Moderate

Legend: kcal = kilocalorie

Based on the DASH Eating Plan, 2005, with slight modifications for use with athletes.

*Vegetables include the starchy vegetables.

**Grains include starches.

***The amount of sugar included in the diet should be moderate. Many athletes consume sports beverages, which contain sugar.

Lucas, a Cross Country Runner

In this section the efforts of a male collegiate cross country runner to meet his recommended carbohydrate intake will be followed. Lucas, who is 20 years old and currently weighs about 138 pounds (~63 kg), wants to consume 8 g/kg of carbohydrate daily. This level of carbohydrate is necessary because he is running 75 to 80 miles per week. How should he plan his diet?

To begin, Lucas needs to determine his total carbohydrate intake for the day. He could calculate this by multiplying his weight in kg by 8 or by looking at the amounts listed for his weight in Table 4.10. Both these methods suggest that his daily carbohydrate intake should be approximately 500g.

Once Lucas establishes a daily goal he would then need to know the kinds of foods that contain carbohydrates: starches, starchy vegetables, beans and legumes, fruits, vegetables, milk, and sugared beverages (Table 4.11). Each person will have preferences for certain foods in each group. For example, Lucas prefers making a smoothie with yogurt rather than drinking milk at breakfast, favors spaghetti noodles over other types of pasta because he knows how to cook them, and is familiar with black beans but not navy beans because he grew up in the Southwest.

Even with the knowledge of the amount of carbohydrates found in various foods, the task of creating a diet plan for the entire day can be daunting. Since Lucas needs approximately 3,400 kcal daily to maintain his present body composition, he can use the framework suggested in Table 4.12 to plan his diet. Using this framework also helps Lucas obtain nutrients other than carbohydrate.

The athlete's schedule and access to food are important considerations. In Lucas's case, he grabs a quick breakfast before going to morning classes. His biggest meal of the day is an early lunch since the team has access to the training table set up for athletes in the dorm dining room. He trains for at least 3 hours in the afternoon. After he gets home from training he takes a nap, and later he has an easy-to-fix dinner. Most evenings he studies and socializes with friends and he usually has a late night snack. Considering all of the issues mentioned above, Lucas can create a skeleton of his daily diet by choosing carbohydrate foods from each group and fitting them into his training and school schedule. The 1-day diet plan that Lucas created is shown in Table 4.13.

The figure at right shows a dietary analysis of Lucas's 1-day diet plan for energy, carbohydrate, and fiber. Lucas's goals included a total energy intake of approximately 3,400 kcal (~54 kcal/kg) and a carbohydrate intake of 500 g (8 g/kg). Lucas's intake was estimated at 3,333 kcal and 532 g of carbohydrate. His intake of dietary fiber, 37 g, nearly met the recommended intake of 38 g for adult men under age 50. On this day Lucas essentially met his established goals. It is not considered a problem that Lucas slightly exceeded or fell short of the recommended amounts. The amounts of energy and fiber were nearly the amounts recommended (remember the amounts listed in nutrient databases are only estimates). The additional carbohydrates are appropriate as long as proteins, fats, and total energy intakes are adequate.

Although the goal for total amount of carbohydrates was exceeded, a closer look at the sources of carbohydrates are necessary. Based on the food groups used by MyPyramid, Lucas exceeded the intake of grains, but fell short of recommended fruit and vegetable consumption (see figure). Vegetable intake was particularly low, a common problem in the United States. An analysis of carbohydrate source (not shown in figure) reveals that nearly 25 percent of Lucas's carbohydrate intake was in the form of sugared drinks (for example, soda, sports

Table 4.13 Example of a 24-Hour Dietary Intake of a Male Collegiate Cross Country Runner

Time/place	Food
7:30 a.m. quick breakfast, at home	Yogurt smoothie: 1 cup low-fat plain yogurt ½ cup frozen, sweetened strawberries 2 T honey ½ cup skim milk
11:15 a.m. at training table for athletes	2 large vegetarian burritos: 2 in low-fat flour tortillas ½ cup low-fat black beans ½ cup cheddar cheese 1½ cup seasoned rice ½ cup grilled peppers 4 T guacamole (avocado) 2 T sour cream ½ cup tomato salsa 16 oz soda 8 oz water
2:00–5:00 p.m. during practice	16 oz 6% carbohydrate drink Orange
After practice	Energy bar
7:00 p.m. dinner, usually at home but sometimes out	Turkey sandwich: 2 slices sourdough bread 3 oz smoked turkey 2 slices of tomatoes 1 T mayonnaise 2 tsp mustard Apple 2 Fig Newtons 1 cup non-fat milk
Late-night snack	2 slices pepperoni pizza 16 oz noncaffeinated soda

Legend: T = tablespoon; in = inch; oz = ounce; tsp = teaspoon

Nutrient	DRI	Intake	0%	50%	100%
Energy					
Kilocalories	3365 kcal	3332.8 kcal			99%
Carbohydrate	379–547 g	532.15 g			
Fat, total	75–131 g	93.86 g			
Protein	84–294 g	124.09 g			
Carbs					
Dietary fiber, total	38 g	37.4 g			98%
Sugar, total	no rec	310.37 g			

MyPyramid.gov
STEPS TO A HEALTHIER YOU

	Goal*	Actual	% Goal
Grains	10 oz. eq.	11.8 oz. eq.	118%
Vegetables	4 cup eq.	1.2 cup eq.	30%
Fruits	2.5 cup eq.	1.7 cup eq.	68%
Milk	3 cup eq.	5.4 cup eq.	180%
Meat & Beans	7 oz. eq.	8.8 oz. eq.	126%
Discretionary	648	1152	178%

*Your results are based on a 3200 calorie pattern, the maximum caloric intake used by MyPyramid.

Dietary analysis of a 24-hour diet of a male collegiate cross country runner

beverages). Thus the quantity of carbohydrate was appropriate, but the quality (for example, nutrient content) could be improved.

Now contrast Lucas's intake over 24 hours with two people that Lucas sometimes trains with—Sophie, who is a female cross country runner, and Jackson, who is an elite distance runner. In doing so, some of the practical problems that athletes face are highlighted.

Sophie is typical of many female collegiate cross country runners. She trains hard, running 50 to 60 miles per week, but does not consume a lot of food. She says she is not hungry, but she also is concerned about gaining body fat, so even when she is hungry she keeps an eye on how much she eats. She lives in the dorm and does not like some of the food offered or how the food is prepared. She finds herself thinking about appearance, weight, and energy (kcal) restriction more frequently than performance, the energy needed to train, and the amount of muscle mass she has. Her typical daily intake is 275 g of carbohydrate and 1,800 kcal. At her current weight of 50 kg, this amount of carbohydrate is 5.5 g/kg, far less than is recommended to adequately replenish glycogen stores in an endurance athlete engaged in hard training.

Jackson is an ultraendurance athlete whose need for carbohydrate represents the extreme end of the carbohydrate spectrum. He runs about 120 miles a week in preparation for the Badwater Ultramarathon, a 135-mile (217 km) race beginning in Death Valley (282 ft [85m] below sea-level) and ending at an altitude of 8,360 ft (2,533 m). His goal is to consume at least 10 g/kg of carbohydrate daily. He has a good appetite but he finds it difficult to eat frequently throughout the day due to work. He eats large carbohydrate-containing meals but he cannot consume sufficient carbohydrate from meals alone. His carbohydrate needs are so high that consuming concentrated carbohydrate supplements is a must. He also struggles with the amount of fiber in his diet, sometimes consuming 60 g a day, which results in frequent bowel movements. To reduce the fiber but obtain the carbohydrate, he finds himself drinking sugared drinks often.

Recall that surveys show that most athletes do not consume the recommended amount of carbohydrate. As Lucas's diet illustrates, it is possible to do so. However, many athletes find themselves in situations similar to Sophie's and must make a concerted effort to eat more carbohydrate-containing foods and fiber, along with more energy. The challenge for those who need greater than 10 g/kg is to be able to consume an adequate volume of carbohydrates day after day and to avoid excessively high fiber intakes. Knowing the carbohydrate recommendation is the easy part; eating a sufficient amount of carbohydrate is harder than it appears at first glance. More information about diet planning can be found in Chapter 10.

Vegetables can be quickly stir-fried in a wok.

Fruit and yogurt both contain carbohydrates and other nutrients.

Table 4.14 Carbohydrate-Containing Foods That Are Easy to Store and Prepare

Carbohydrate-containing food	Storage and preparation tips
Bread or bagels	Keep in freezer. Put in toaster twice.
Waffles	Buy frozen waffles and heat in toaster. Top with syrup or jam.
Pancakes	Buy pourable pancake mix. Add water and cook. Top with syrup or jam.
Cereal	Add shelf-stable (UHT) milk, which does not need refrigeration until it is opened. Cereal can also be used as a topping for yogurt.
Oatmeal or grits	Buy instant oatmeal or grits packages and add hot water.
Pasta	Cook dry or frozen pasta noodles. Heat a jar of tomato-based spaghetti sauce. Combine.
Tortillas (fresh or frozen) and beans (canned)	Spoon canned beans on tortilla (add cheese if desired). Microwave 1 minute. Add salsa and fold.
Fruits	Apples and bananas tend to last longer than fresh berries or stone fruits (for example, peaches or nectarines). If fruits get overripe add to smoothies.
Vegetables	Fresh carrots tend to last a long time when stored in a cool place. Frozen or canned vegetables are easy to store and prepare.
Canned beans	Most kinds of beans can be purchased in cans and only need to be reheated.
Milk	Milk that has been processed using ultrahigh temperature (UHT) pasteurization can remain on the shelf until opened.
Frozen entrees	Several brands specialize in "healthy" frozen entrees that contain high-carbohydrate, moderate-protein, low-fat, and low-sodium meals.
Nuts	Unopened jar or cans can remain on the shelf; after opening nuts can be stored in the freezer.

entrees, especially in fast-food restaurants, are high in fats, sugar, and salt and low in fiber-containing carbohydrates. A double cheeseburger, large order of fries, and a milkshake provide lots of calories and fat but few high-quality carbohydrates. Although a typical fast-food meal may be consumed occasionally, a steady diet of such fare will typically leave the athlete short of recommended carbohydrate and nutrient intakes. Figure 4.16 lists some good choices for athletes when eating in restaurants.

Vegetarian diets. Vegetarians do not eat meat, fish, or poultry, but some consume animal products such as milk or yogurt. **Vegans** avoid any foods of animal origin. Obtaining an adequate amount of carbohydrates is not difficult for vegetarian athletes since so many carbohydrate-containing foods are of plant origin. Those who avoid milk products would still have ample food groups from which to choose—starches, starchy vegetables, beans and legumes, nuts, fruits, vegetables, and sugar. Vegans could also choose foods from these groups but would want to avoid any prepared products that contain an animal-derived ingredient. The American Dietetic Association supports the position that well-planned vegetarian diets are healthy, nutritionally adequate, and able to meet sports nutrition recommendations (Craig, Mangels, and the American Dietetic Association, 2009).

Sugar intake and the use of artificial sweeteners. In 2003, per capita sugar consumption in the United States was 156 lb per year, 36 lb per year greater than in 1970 (Johlin, Panther, and Kraft, 2004; Howard and Wylie-Rosett, 2002). Sugar is commonly consumed as sweetened breakfast cereals, sweetened grains such as cookies, cakes, and other bakery products, soft drinks and other sweetened beverages, candy, jam, and table sugar. Sugars found in fruits and milk are generally not included in sugar consumption figures.

The effect that a high sugar intake may have on chronic disease, such as obesity, has been controversial. The consumption of sugar-sweetened beverages has been scrutinized, in part, because soft drinks and other beverages sweetened with sugar are one of the main sources of sugar in the diets of Americans of all ages (Johnson et al., 2009). Some observational studies suggest an association between a high intake of sugar-sweetened beverages and weight gain. However, there are also studies that do not show an association (van Baak and Astrup, 2009; Gibson, 2008). Not every person who gains weight does so because of excessive sugar intake, but for many people, including some athletes, too much sugar is one major factor in their weight gain. More information about sugar and the role it may play in chronic diseases can be found in Chapter 13.

Vegan: One who does not eat food of animal origin.

Breakfast Menu Items:

Pancakes or waffles with syrup*

Hot cereal, such as oatmeal

Cold cereal with milk

Toast or muffin with jelly

Bagel

Fruit

Juice, such as orange juice*

Milk*

Cocoa*

Lunch or Dinner Menu Items:

Hearty soup, such as minestrone or bean, with crackers or roll

Salad* or other vegetables

Deli sandwiches with lower fat meats and plenty of vegetables

Baked potato*

Chili*

Thick crust pizza with vegetable toppings

Pasta with marinara sauce

Vegetarian burritos (beans, vegetables, and tortillas)

Asian vegetable dishes with noodles or rice, such as vegetable chow mein

Fruit smoothies

Soft-serve low-fat yogurt

Figure 4.16 Carbohydrate-rich choices when eating in restaurants

*Typically available at fast-food restaurants

Sugar intake and weight gain have always been of interest to American consumers. Surveys suggest that approximately 85 percent of all U.S. adults use low-calorie, reduced sugar, or sugar-free foods and beverages at least once every 2 weeks. The most popular sugar-free products are artificially sweetened soft drinks and other beverages. Consumers also add sugar substitutes to coffee and tea and to home-cooked products. The demand for artificially sweetened and sugar-free products remains high (Caloric Control Council, 2010).

Americans report that they use artificial sweeteners to reduce calorie intake; some in an effort to lose weight and others to prevent weight gain. Athletes need carbohydrates, of which sugar is one, but many athletes worry that they consume too much sugar and too many excess calories from sugar. They consider artificially sweetened foods as alternatives to highly sweetened foods and ask the logical questions, "What are artificial sweeteners?" and "Are they safe?"

Artificial sweeteners (technically known as nonnutritive sweeteners) are not found naturally in foods; rather they are laboratory-manufactured compounds that provide a sweet taste but few or no calories. Saccharin (Sweet 'n Low®), aspartame (Nutrasweet®, Equal®), acesulfame potassium (Acesulfame K or Sunett®), sucralose (Splenda®), and neotame are examples of nonnutritive sweeteners. Table 4.15 explains the various artificial sweeteners and their similarities and differences.

Questions about safety have been raised ever since the Food and Drug Administration approved the first artificial sweeteners. The position of the American Dietetic Association (2004) is that "nonnutritive sweeteners are safe for use within the approved regulations." However, concerns about artificial sweeteners are frequently raised,

Table 4.15 Artificial (Nonnutritive) Sweeteners

Artificial sweetener	Description
Acesulfame potassium (Acesulfame K or Sunett®)	• Often mixed with other artificial sweeteners such as aspartame • ADI* is 15mg/kg/d in the United States; 9 mg/kg/d in Europe (~490 to 818 mg/kg/d for a 120 lb person) • Average intake in the United States is below 9 mg/kg/d
Aspartame (Nutrasweet®, Equal®)	• Made of two amino acids, L-aspartic acid and L-phenylalanine • Warning label is required because those with phenylketonuria cannot metabolize the phenylalanine • FDA approved. Opponents question its safety on the basis that it causes seizure, headache, memory loss, and mood change. • ADI* is 50 mg/kg/d in the United States; 40 mg/kg/d internationally (~2,180 to 2,725 mg/d for a 120 lb person) • Diet soft drinks (12 oz) sweetened only with aspartame contain about 200 mg • Average intake in the United States is between 3.0 and 5.2 mg/kg/d
Neotame	• Intensely sweet (7,000 to 13,000 times sweeter than sugar) • ADI* is 18 mg/d in the United States; 2 mg/kg/d internationally (~109 to 981 mg/day for a 120 lb person) • Average intake in the United States is between 0.04 and 0.10 mg/kg/d
Saccharin (Sweet 'n Low®)	• Oldest artificial sweetener (discovered in 1879) • Once thought to be a cause of bladder cancer but studies in humans do not support this association • Often found in restaurants as single-serving packets • ADI* is 5mg/kg/d (~273 mg/d for a 120 lb person) • Average intake in the United States is 50 mg/d
Sucralose (Splenda®)	• Derived from sugar but is not digestible due to the substitution of three chlorine atoms for three OH groups • Also contains maltodextrin, a starch, which gives it bulk so that it will measure like sugar in recipes • ADI* is 5mg/kg/d (~272 mg/day for a 120 lb person) • Average intake in the United States probably does not exceed 2 mg/d

Legend: mg/kg/d = milligrams per kilogram body weight per day; mg/d = milligrams per day; lb = pound; oz = ounce; mg = milligrams

*The Acceptable Daily Intake (ADI) is an estimate of the amount a person could consume daily over a lifetime without appreciable risk. The ADI is based on animal studies and has a large margin of safety. The ADI is expressed as mg/kg/d.

American Dietetic Association (2004). Position paper: Use of nutritive and nonnutritive sweeteners. *Journal of the American Dietetic Association, 104*(2), 255–275.

most often via the media, and individuals who use artificial sweeteners should evaluate any safety concerns raised.

Perhaps the biggest questions are about the effectiveness of artificial sweeteners to reduce total caloric intake, promote weight loss, and prevent or slow weight gain. When artificial sweeteners first came into the market, there was great hope that they would play an important role in weight reduction for overweight and obese people. Theoretically, if a person substituted artificial sweeteners for sugar, then the caloric deficit over time from the substitution would result in weight loss. But the prevalence of obesity and the use of artificial sweeteners have both increased substantially over the past 30 years. It may be that instead of substituting artificially sweetened foods for sugar-sweetened products people are using artificial sweeteners in addition to foods sweetened with sugar. The causes of obesity are multifactorial, but clearly artificial sweeteners are not the weight-related panacea that both industry and consumers hoped they would be.

Theoretically, the use of artificial sweeteners could improve diet quality (American Dietetic Association, 2004). For example, suppose a person normally consumed 12 oz (~360 ml) of a sugared soft drink, which has about 150 kcal. If an artificially sweetened soft drink was consumed and food with the same amount of kilocalories but more nutrients was eaten, an argument could be made that the quality of the diet was improved. To date there have been no studies that have examined if people usually make such substitutions, but athletes may choose to do so as part of their overall diet plan.

Stevia has been touted as an alternative to noncaloric sweeteners. Stevia is an extract from the leaves of the Stevia bush and is 100 to 300 times sweeter than table sugar. It has received GRAS (generally regarded as safe) status in the United States and is used in many other countries. An intake of 0–4 mg/kg body weight per day of steviol glycosides (the compound extracted) is considered safe. There is evidence in humans that Stevia has a beneficial effect on blood glucose and insulin levels and is another choice for consumers as a sweetener (Anton et al., 2010).

Lactose intolerance. The digestion of lactose (milk sugar) requires the enzyme lactase, known scientifically as β-galactosidase. In humans and other mammals, lactase activity is high during infancy. In most humans, lactase activity begins to decline at about age 2, but in some Caucasians the decline does not begin until adolescence. Although the decline is extensive, most adults do have a low level of lactase activity. Lactose maldigestion occurs when there is insufficient lactase relative to the amount of lactose consumed in the diet. In people of Northern European descent, lactase activity remains at infant levels throughout adulthood and their tolerance

of lactose remains high (Vesa, Marteau, and Korpela, 2000; de Vrese et al., 2001).

In the United States, nearly 100 percent of Asian Americans and Native Americans maldigest lactose because their amount of lactase activity is very low. Approximately 75 percent of African Americans and half (53 percent) of the Hispanic population also have low lactase activity. Maldigestion by U.S. Caucasians is estimated to be 6 percent to 22 percent (Jackson and Savaiano, 2001). Altogether, about 80 million people in the United States malabsorb lactose due to low lactase activity.

Lactose malabsorption results in gas, bloating, diarrhea, nausea, and abdominal pain, a condition known as lactose intolerance. Because lactase activity is low but not absent, people who are lactose intolerant can usually digest about 9 to 12 g of lactose. An 8 oz (240 ml) glass of milk has about 12 g of lactose. Chocolate milk appears to be better tolerated than milk not flavored with chocolate (de Vrese et al., 2001).

Those with lactose intolerance typically can digest fermented milk products, such as yogurt or *Acidophilus* milk better than other milks or unfermented milk products, even though the same amount of lactose is consumed. One reason is that fermented products contain bacteria that produce β-galactosidase (lactase). A second reason is that solids, such as yogurt, tend to move more slowly through the gastrointestinal tract than fluids, which allows for more time to digest the lactose (de Vrese et al., 2001). This same effect is found when milk is consumed with a meal (Jackson and Savaiano, 2001).

Those who maldigest lactose typically use trial and error to determine the lactose-containing foods to include in their diets. They experiment with the amount consumed, the form (that is, solid, semisolid, liquid), and the presence of fermentation bacteria. Other tactics include adding lactase tablets to food or drinking milk that has lactase added. These strategies allow people to consume dairy products, a concentrated source of calcium in the diet. In several studies of Caucasians with lactose intolerance, subjects limited their dairy intake, and therefore, their calcium intake. Lactose maldigestion is thought to be one factor that contributes to low calcium intake and low bone mineral density, both of which are associated with osteoporosis (Jackson and Savaiano, 2001).

Gas, bloating, and diarrhea are not unique to lactose intolerance. These same symptoms are associated with irritable bowel syndrome, which affects about 20 percent of the adult population. Athletes who experience such symptoms should seek medical care to determine their probable cause and the most appropriate treatment.

Fructose intolerance. Lactose intolerance has been frequently studied, but the study of fructose intolerance is in its infancy. Some researchers feel that fructose intolerance is underrecognized and therefore not treated (Choi et al., 2003). The symptoms are often the same as lactose intolerance or irritable bowel syndrome—bloating, gas, and diarrhea. Bloating results because unabsorbed fructose draws water into the gastrointestinal tract. Gas is produced when bacteria in the colon ferment the unabsorbed fructose. In the process of fermentation, short-chain fatty acids are produced and this results in increased gastrointestinal motility. Theoretically, all of these symptoms should be resolved if fructose-containing foods in the diet are reduced or eliminated (Choi et al., 2003; Skoog and Bharucha, 2004).

In 1970, high-fructose corn syrup (HFCS) represented only 5 percent of the sugar in the U.S. diet; by 2001 it represented 55 percent (Johlin, Panther, and Kraft, 2004). Fructose is better absorbed when glucose is present in equal amounts. Fructose is not well absorbed when it is accompanied by sorbitol, a sugar alcohol that is found naturally in some foods and added to many sugar-free foods (see Spotlight on…Sugar Alcohols). As sports beverages and other products developed for athletes may contain fructose, fructose intolerance should not be overlooked by athletes as a possible source of gastrointestinal distress.

For those with fructose intolerance, large amounts of soft drinks, fruit juice such as orange juice, honey, or dates may contain too much fructose relative to the amount of glucose. Cherries, apples, pears, and foods with sorbitol added may also be problematic. A fructose-reduced diet is likely to reduce the bloating, gas, and diarrhea. However, such a diet requires a high degree of motivation and careful label reading.

Diabetes. The hormonal response to carbohydrate intake described earlier in the chapter helps the body to keep blood glucose in homeostasis. About 90 percent of Americans have a normal response, but at least 23 million Americans have diabetes. In those with type 1 diabetes, insulin secretion is absent. These individuals will need to match their food intake with the proper amount of insulin, which is injected or released from an insulin pump. In those with type 2 diabetes (about 21 million of the 23 million people with diabetes), insulin secretion is diminished or the cells are resistant to the influence of insulin. These individuals may take medications to stimulate the release of more insulin. They should reduce the intake of high-glycemic index foods that rapidly increase blood glucose concentrations. If insulin resistance is present, reducing excess body fat and engaging in at least 30 minutes of moderately intense exercise 5 days a week will help reduce the body's resistance to the action of insulin. More information about diabetes can be found in Chapter 13.

Being diagnosed with diabetes does not preclude one from excelling in athletics. For example, swimmer Gary Hall Jr., who has type 1 diabetes, has won 10 Olympic medals, 5 of them gold. During training and competitions, he tested his blood glucose 10 to 12 times a day with the possibility that he had to inject insulin each time. The challenge for athletes with

Keeping it in perspective

Carbohydrates Are for Fuel and Fun

Many athletes follow rigorous training programs and must replenish glycogen stores by eating carbohydrate-rich diets daily. Endurance and ultraendurance athletes carefully plan their carbohydrate intake before, during, and after exercise or they risk nearly depleting muscle glycogen during competition or hard training. Many marathon runners and triathletes have stories about "bonking" or "hitting the wall," popular terms that describe glycogen depletion. This need for constant carbohydrate-containing foods sometimes means that athletes get rigid about their food intake. The key to meeting carbohydrate recommendations is to maintain a flexible eating plan.

Because they know the importance of sufficient carbohydrate intake daily, many dedicated athletes find themselves eating the same foods every day. Structure and consistency are important, but food intake can become mechanical and the joy of eating can be diminished. To maintain flexibility, athletes should experiment with new carbohydrate-containing foods during the off-season or when not preparing for an important competition. Learning to cook different kinds of foods or finding new restaurants can be fun and can lead to the inclusion of a greater variety of carbohydrate foods. Eating traditional foods from different ethnic groups helps with variety, since most ethnic diets include a carbohydrate food as a staple and use different spices. For example, Italian food is pasta based, Asian food is rice and soybean based, Mexican food is rice and bean based, and South American food is tuber, bean, and nut based. The key to maintaining a flexible eating plan is the inclusion of a variety of foods daily. Food is for fuel and fun and athletes must find the right balance.

type 1 diabetes is to properly match insulin injections, food intake, and exercise. Registered dietitians who are board certified specialists in sports dietetics (RD, CSSD) work with such athletes, who must adjust carbohydrate and insulin intakes to maintain normal blood glucose concentrations before, during, and after exercise.

Exercise is encouraged for individuals with type 2 diabetes because it improves blood glucose control. Total carbohydrate and energy intakes are also important because many people with type 2 diabetes need to lose body fat. These individuals will also benefit from working with a dietitian to devise a diet plan with adequate but not excessive carbohydrates or calories. Some world-class athletes have type 2 diabetes, including Sir Steve Redgrave, a British rower who was knighted after he won his fifth consecutive Olympic gold medal, an unparalleled Olympic achievement.

A small percentage of the nondiabetic population may have either reactive hypoglycemia or fasting hypoglycemia. Reactive hypoglycemia refers to a low blood glucose concentration that follows food intake. For example, if a person with reactive hypoglycemia were to consume a high-glycemic food, blood glucose would rise rapidly. However, glucagon secretion may be slow or limited and would be insufficient to completely offset the action of insulin. In that case, the blood glucose level would continue to fall and result in hypoglycemia. The symptoms of hypoglycemia are hunger, shakiness, light-headedness, and loss of concentration. Treatment for reactive hypoglycemia involves the avoidance of foods that cause blood glucose to rise rapidly and the consumption of small meals or snacks every 3 hours to help keep blood glucose stable. Fasting hypoglycemia is rare and needs immediate medical attention because it is usually a symptom of an underlying disease such as a tumor.

Some nondiabetic endurance athletes may experience exercise-induced hypoglycemia, which impairs performance. Although the exact number of athletes affected is not known, it is estimated that it is less than 25 percent. Consuming carbohydrate before and, most importantly, during prolonged exercise prevents hypoglycemia. Training helps endurance athletes make adaptations that also reduce the risk. These adaptations include the use of fats as a fuel source, an increase in the production of glucose in the liver, and changes in hormone sensitivity. However, athletes who overtrain actually reverse some of the training adaptations and may experience hypoglycemia that results in fatigue (Brun et al., 2001).

Abnormal glucose or insulin responses require dietary management but do not preclude people from training, competing, or excelling in athletics. These individuals must manage externally what is typically controlled internally in those with normal blood glucose regulation. It may be an inconvenience to have to closely monitor carbohydrate intake but it is not an insurmountable barrier. Sports dietitians can help such athletes devise an appropriate dietary plan.

Key points

- Athletes must plan carefully to ensure adequate intake of carbohydrates on a daily basis.

- Having carbohydrate-rich foods available increases the likelihood of making good food choices and adhering to the dietary plan for carbohydrate intake.

- Athletes must be knowledgeable about making good food and carbohydrate choices when dining out.

- Vegetarian diets typically contain a high percentage of carbohydrate-rich foods but care must be taken to consume an adequate total amount of carbohydrates.

- Carbohydrate intake in the diet may be influenced by individual situations, such as lactose intolerance, fructose intolerance, or altered glucose and/or insulin responses (for example, diabetes or pre-diabetes).

Identify at least five carbohydrate-rich foods that are easy to store and prepare.

Why might a vegetarian athlete who is getting 70 percent of total calories from carbohydrate not be getting a sufficient amount of carbohydrate in his or her diet?

Summary

Carbohydrates are found in food as sugars, starches, and fiber. Sugars and starches are digested and absorbed from the gastrointestinal tract; fiber is indigestible. After absorption, carbohydrates are found as glucose (blood sugar), muscle glycogen, and liver glycogen. The metabolism of glucose is intricate and involves many metabolic pathways, including the immediate use of glucose for energy and the storage of glucose as glycogen for future use as energy. Carbohydrates are the primary energy source for moderate to intense exercise. Exercising muscle prefers to use carbohydrate from stored glycogen rather than from blood glucose so athletes need sufficient glycogen stores.

Exercise reduces glycogen stores, which must be replenished on a daily basis. The general recommendation for carbohydrate intake for athletes in training is 5 to 10 g/kg body weight daily. This recommendation assumes that energy (caloric) intake is adequate. The general recommendation must be individualized based on the intensity and duration of exercise during training and competition. General recommendations are also made for carbohydrate consumption before, during, and after exercise and competition. Athletes fine-tune these recommendations to help meet their training and competition goals. Carbohydrate loading is one technique that endurance athletes and some bodybuilders use before an important competition.

Insufficient carbohydrate intake can lead to hypoglycemia and both acute and chronic fatigue. Many athletes fail to consume enough carbohydrates daily and some consistently fall short of recommended intakes over a period of weeks or months. Sufficient carbohydrate intake is an important element in athletic training and competition.

Even when athletes know and understand the carbohydrate recommendations, meeting those recommendations is not easy. Athletes must purchase or prepare carbohydrate-rich meals daily. Some athletes have lactose intolerance, reactive hypoglycemia, or diabetes, and these conditions influence the choice of carbohydrate-containing foods. Proper carbohydrate intake positively affects training, performance, and health.

Self-Test

Multiple-Choice

The answers can be found in Appendix O.

1. The end product of carbohydrate absorption is:
   a. fructose.
   b. glucose.
   c. lactose.
   d. maltose.

2. Exercising muscle prefers to use carbohydrate from:
   a. blood glucose.
   b. liver glycogen.
   c. muscle glycogen.
   d. any carbohydrate source available (no preference).

3. The recommended minimum amount of daily carbohydrate for athletes in training is:
   a. 4 g/kg.
   b. 5 g/kg.
   c. 6 g/kg.
   d. 7 g/kg.

4. The two factors needed to optimize muscle glycogen resynthesis after exercise are:
   a. carbohydrate and excessive caloric intake.
   b. carbohydrate and lactate.
   c. carbohydrate and protein.
   d. carbohydrate and insulin.

5. Which of the following athletes would benefit from carbohydrate loading?
   a. professional golfer who takes approximately 4 hours to finish a round
   b. professional baseball player who plays ~2.5 hours, 4–5 days per week
   c. marathon runner preparing for an important race
   d. all of the above

Short Answer

1. What types of carbohydrates are found in foods? Where are carbohydrates found in the body?

2. How do monosaccharides, disaccharides, and polysaccharides differ?

3. Compare and contrast the absorption of glucose and fructose.

4. How is blood glucose regulated?

5. What happens to glucose when it is taken into a cell?

6. Who is more likely to convert glucose into fat—active or sedentary individuals? Why?

7. What carbohydrate source does exercising muscle prefer to use?

8. Exercising muscle takes up and uses blood glucose. What keeps blood glucose from dropping too low?

9. What is the current recommendation for total carbohydrate intake each day for endurance athletes in a part of their training cycle that involves a very high volume (long duration and high intensity) of training?

10. What are some practical issues that prevent athletes from meeting the daily recommended intake of carbohydrates?

11. Under which circumstances would the use of the glycemic index be helpful to athletes?

12. What are the advantages of consuming starches rather than sugars? Disadvantages?

13. Compare and contrast the various artificial sweeteners.

14. What is lactose intolerance and who is most likely to have problems digesting lactose?

Critical Thinking

1. A runner consumes a carbohydrate meal before long training runs and races but reports feeling fatigued and sluggish shortly after beginning exercise. What information would you need to determine the cause of these symptoms and what specific recommendations would you give this athlete to avoid this problem?

2. A triathlete complains of bloating and fullness during the triathlon and often experiences diarrhea afterwards. What information would you need to determine the cause of these symptoms and what specific recommendations would you give this athlete to avoid this problem?

References

Akermark, C., Jacobs, I., Rasmusson, M., & Karlsson, J. (1996). Diet and muscle glycogen concentration in relation to physical performance in Swedish elite ice hockey players. *International Journal of Sport Nutrition*, 6(3), 272–284.

American Dietetic Association. (2004). Position paper: Use of nutritive and nonnutritive sweeteners. *Journal of the American Dietetic Association*, 104(2), 255–275. Erratum in 104(6), 1013.

American Dietetic Association, Dietitians of Canada, & the American College of Sports Medicine (2009). Position paper: Nutrition and athletic performance. *Medicine and Science in Sports and Exercise*, 41(3), 709–731.

Anton, S. D., Martin, C. K., Han, H., Coulon, S., Cefalu, W. T, Geiselman, P., et al. (2010). Effects of stevia, aspartame, and sucrose on food intake, satiety, and postprandial glucose and insulin levels. *Appetite*, 55(1), 37–43.

Atkinson, F. S., Foster-Powell, K., & Brand-Miller, J. C. (2008). International tables of glycemic index and glycemic load values: 2008. *Diabetes Care*, 31(12), 2281–2283.

Balsom, P. D., Wood, K., Olsson, P., & Ekblom, B. (1999). Carbohydrate intake and multiple sprint sports: With special reference to football (soccer). *International Journal of Sports Medicine*, 20(1), 48–52.

Bangsbo, J., Norregaard, L., & Thorsoe, F. (1992). The effect of carbohydrate diet on intermittent exercise performance. *International Journal of Sports Medicine*, 13(2), 152–157.

Bergström, J., Hermansen, L., & Saltin, B. (1967). Diet, muscle glycogen, and physical performance. *Acta Physiologica Scandinavica, 71*(2), 140–150.

Bergström, J., & Hultman, E. (1966). Muscle glycogen synthesis after exercise: An enhancing factor localized to the muscle cells in man. *Nature, 210*(33), 309–310.

Bergström, J., & Hultman, E. (1967). A study of the glycogen metabolism during exercise in man. *Scandinavian Journal of Clinical Laboratory Investigation, 19*(3), 218–228.

Blom, P. C. S., Hstmark, A. T., Vaage, O., Kardel, K. R., & Mælum, S. (1987). Effect of different post-exercise sugar diets on the rate of muscle glycogen synthesis. *Medicine and Science in Sports and Exercise, 19*(5), 491–496.

Brun, J. F., Dumortier, M., Fedou, C., & Mercier, J. (2001). Exercise hypoglycemia in nondiabetic subjects. *Diabetes & Metabolism, 27*(2, Pt. 1), 92–106.

Burke, L. M. (2001). Energy needs of athletes. *Canadian Journal of Applied Physiology, 26*(Suppl.), S202–S219.

Burke, L. M., Collier, G. R., & Hargreaves, M. (1993). Muscle glycogen storage after prolonged exercise: Effect of the glycemic index of carbohydrate feedings. *Journal of Applied Physiology, 75*(2), 1019–1023.

Burke, L. M., Collier, G. R., & Hargreaves, M. (1998). Glycemic index—a new tool in sport nutrition? *International Journal of Sport Nutrition, 8*(4), 401–415.

Burke, L. M., Cox, G. R., Culmmings, N. K., & Desbrow, B. (2001). Guidelines for daily carbohydrate intake: Do athletes achieve them? *Sports Medicine, 31*(4), 267–299.

Burke, L. M., Kiens, B., & Ivy, J. L. (2004). Carbohydrates and fat for training and recovery. *Journal of Sports Sciences, 22*(1), 15–30.

Bussau, V. A., Fairchild, T. J., Rao, A., Steele, P., & Fournier, P. A. (2002). Carbohydrate loading in human muscle: An improved 1 day protocol. *European Journal of Applied Physiology, 87*(3), 290–295.

Caloric Control Council (2010) http://www.caloriecontrol.org/

Choi, Y. K., Johlin, F.C., Jr., Summers, R. W., Jackson, M., & Rao, S. S. (2003). Fructose intolerance: An under-recognized problem. *The American Journal of Gastroenterology, 98*(6), 1348–1353.

Coleman, E. (in press a). Carbohydrate and exercise. In C. Rosenbloom (Ed.), *Sports nutrition: A practical manual for professionals* (5th ed.). Chicago: American Dietetic Association.

Coleman, E. (in press b). Nutrition for endurance and ultraendurance sports. In C. Rosenbloom (Ed.), *Sports nutrition: A practical manual for professionals* (5th ed.). Chicago: American Dietetic Association.

Costill, D. L., Bowers, R., Branam, G., & Sparks, K. (1971). Muscle glycogen utilization during prolonged exercise on successive days. *Journal of Applied Physiology, 31*(6), 834–838.

Coyle, E. F. (2004). Fluid and fuel intake during exercise. *Journal of Sports Sciences, 22*(1), 39–55.

Coyle, E. F., Coggan, A. R., Hemmert, M. K., & Ivy, J. L. (1986). Muscle glycogen utilization during prolonged strenuous exercise when fed carbohydrate. *Journal of Applied Physiology, 61*(1), 165–172.

Craig, W. J., Mangels, A .R., & American Dietetic Association. (2009). Position of the American Dietetic Association: Vegetarian diets. *Journal of the American Dietetic Association, 109*(7), 1266–1282.

Donaldson, C.M., Perry, T.L., & Rose, M.C. (2010). Glycemic index and endurance performance. *International Journal of Sports Nutrition and Exercise Metabolism, 20*(2),154-165.

de Vrese, M., Stegelmann, A., Richter, B., Fenselau, S., Laue, C., & Schrezenmeir, J. (2001). Probiotics—compensation for lactase insufficiency. *American Journal of Clinical Nutrition, 73*(2 Suppl.), 421S–429S.

Doyle, J. A., & Papadopoulos, C. (2000). Simple and complex carbohydrates in exercise and sport. In I. Wolinsky & J. A. Driskell (Eds.), *Energy-yielding macronutrients and energy metabolism in sports nutrition* (pp. 57–69). Boca Raton, FL: CRC Press.

Doyle, J. A., Sherman, W. M., & Strauss, R. L. (1993). Effects of eccentric and concentric exercise on muscle glycogen replenishment. *Journal of Applied Physiology, 74*(4), 1848–1855.

Dunford, M. (Ed.). (2006). *Sports nutrition: A practical manual for professionals. 4th ed*. Chicago: American Dietetic Association.

Fairchild, T. J., Fletcher, S., Steele, P., Goodman, C., Dawson, B., & Fournier, P.A. (2002). Rapid carbohydrate loading after a short bout of near maximal-intensity exercise. *Medicine and Science in Sports and Exercise, 34*(6), 980–986.

Foster, C., Costill, D. L., & Fink, W .J. (1979). Effects of pre-exercise feedings on endurance performance. *Medicine and Science in Sports and Exercise, 11*(1), 1–5.

Gibson, S. (2008). Sugar-sweetened soft drinks and obesity: a systematic review of the evidence from observational studies and interventions. *Nutrition Research Reviews, 21*(2), 134-147.

Gleeson, M., Nieman, D. C., & Pedersen, B. K. (2004). Exercise, nutrition and immune function. *Journal of Sports Sciences, 22*(1), 115–125.

Gretebeck, R. J., Gretebeck, K. A., & Titelback, T. J. (2002). Glycemic index of popular sport drinks and energy foods. *Journal of the American Dietetic Association, 102*(3), 415–417.

Guezennec, C. Y. (1995). Oxidation rates, complex carbohydrates and exercise. Practical recommendations. *Sports Medicine, 19*(6), 365–372.

Hargreaves, M., Hawley, J. A., & Jeukendrup, A. (2004). Pre-exercise carbohydrate and fat ingestion: Effects on metabolism and performance. *Journal of Sports Sciences, 22*(1), 31–38.

Hawley, J. A., & Burke, L. M. (2010). Carbohydrate availability and training adaptation: Effects of cell metabolism. *Exercise and Sport Sciences Reviews, 38*(4), 152–160.

Hinton, P. S., Sanford, T. C., Davidson, M. M., Yakushko, O. F., & Beck, N. C. (2004). Nutrient intakes and dietary behaviors of male and female collegiate athletes. *International Journal of Sport Nutrition and Exercise Metabolism, 14*(4), 389–405.

Howard, B. V., & Wylie-Rosett, J. (2002). Sugar and cardiovascular disease: A statement for healthcare professionals from the Committee on Nutrition of the Council on Nutrition, Physical Activity, and Metabolism of the American Heart Association. *Circulation, 106*(4), 523–527.

Institute of Medicine. (2002). *Dietary Reference Intakes for energy, carbohydrate, fiber, fat, fatty acids, cholesterol, protein and amino acids* (Food and Nutrition Board). Washington, DC: National Academies Press.

Ivy, J. L., Katz, A. L., Cutler, C. L., Sherman, W. M., & Coyle, E. F. (1988). Muscle glycogen synthesis after exercise: Effect of time of carbohydrate ingestion. *Journal of Applied Physiology, 64*(4), 1480–1485.

Ivy, J. L., Lee, M. C., Brozinick, J. T., & Reed, M. J. (1988). Muscle glycogen storage after different amounts of carbohydrate ingestion. *Journal of Applied Physiology, 65*(5), 2018–2023.

Jackson, K. A., & Savaiano, D. A. (2001). Lactose maldigestion, calcium intake and osteoporosis in African-, Asian-, and Hispanic-Americans. *Journal of the American College of Nutrition, 20*(2 Suppl.), 198S–207S.

Jacobs, K. A., & Sherman, W. M. (1999). The efficacy of carbohydrate supplementation and chronic high carbohydrate diets for improving endurance performance. *International Journal of Sport Nutrition, 9*(1), 92–115.

Jenkins, D. J., Wolever, T. M., Taylor, R. H., Barker, H., Fielden, H., Baldwin, J. M., et al., (1981). Glycemic index of foods: A physiological basis for carbohydrate exchange. *American Journal of Clinical Nutrition, 34*(3), 362–366.

Jentjens, R. L. P. G., & Jeukendrup, A. E. (2005). High rates of exogenous carbohydrate oxidation from a mixture of glucose and fructose ingested during prolonged cycling exercise. *British Journal of Nutrition, 93*, 485–492.

Jentjens, R. L., van Loon, L. J., Mann, C. H., Wagenmakers, A. J., & Jeukendrup, A. E. (2001). Addition of protein and amino acids to carbohydrates does not enhance postexercise muscle glycogen synthesis. *Journal of Applied Physiology, 91*(2), 839–846.

Jeukendrup, A. (2007). Carbohydrate supplementation during exercise: Does it help? How much is too much? Gatorade Sports Science Institute, SSE #106.

Johlin, F. C., Panther, M., & Kraft, N. (2004). Dietary fructose intolerance: Diet modification can impact self-rated health and symptom control. *Nutrition in Clinical Care, 7*(3), 92–97.

Johnson, R. K., Appel, L. J., Brands, M., Howard, B. V., Lefevre, M., Lustig, R. H., Sacks, F., et al. (2009). Dietary sugars intake and cardiovascular health: A scientific statement from the American Heart Association. *Circulation, 120*(11), 1011–1120.

Keizer, H. A., Kuipers, H., van Kranenburg, G., & Geurten, P. (1987). Influence of liquid and solid meals on muscle glycogen resynthesis, plasma fuel hormone response, and maximal physical working capacity. *International Journal of Sports Medicine, 8*(2), 99–104.

Lamb, D. R., Rinehardt, K., Bartels, R. L., Sherman, W. M., & Snook, J. T. (1990). Dietary carbohydrate and intensity of interval swim training. *American Journal of Clinical Nutrition, 52*(6), 1058–1063.

Ludwig, D. S. (2002). The glycemic index: Physiological mechanisms relating to obesity, diabetes, and cardiovascular disease. *Journal of the American Medical Association, 287*(18), 2414–2423.

Nabors, L. O. (2002). Sweet choices: Sugar replacements for foods and beverages. *Food Technology, 56*(7), 28–32.

Shephard, R. J. (2001). Chronic fatigue syndrome: An update. *Sports Medicine, 31*(3), 167–194.

Sherman, W. M., Costill, D. L., Fink, W. J,. & Miller, J. M. (1981). Effect of exercise and diet manipulation on muscle glycogen and its subsequent use during performance. *International Journal of Sports Medicine, 2*(2), 114–118.

Sherman, W. M., Doyle, J. A., Lamb, D. R., & Strauss, R. H. (1993). Dietary carbohydrate, muscle glycogen, and exercise performance during 7 d of training. *American Journal of Clinical Nutrition, 57*(1), 27–31.

Sherman, W. M., Peden, M. C., & Wright, D. (1991). Carbohydrate feedings 1 h before exercise improves cycling performance. *American Journal of Clinical Nutrition, 54*(5), 866–870.

Simonsen, J. C., Sherman, W. M., Lamb, D. R., Dernbach, A. R., Doyle, J. A., & Strauss, R. (1991). Dietary carbohydrate, muscle glycogen, and power output during rowing training. *Journal of Applied Physiology, 70*(4), 1500–1505.

Siu, P. M., & Wong, S. H. (2004). Use of the glycemic index: Effects on feeding patterns and exercise performance. *Journal of Physiological Anthropology and Applied Human Science, 23*(1), 1–6.

Skoog, S. M., & Bharucha, A. E. (2004). Dietary fructose and gastrointestinal symptoms: A review. *American Journal of Gastroenterology, 99*(10), 2046–2050.

Triplett, D., Doyle, J. A., Rupp, J. C., & Benardot, D. (2010). An isocaloric glucose-fructose beverage's effect on simulated 100-km cycling performance compared with a glucose-only beverage. *International Journal of Sport Nutrition and Exercise Metabolism, 20*, 122–131.

van Baak, M.A., & Astrup, A. (2009). Consumption of sugars and body weight. *Obesity Reviews, 10*(1), 9–23.

van Loon, L. J., Saris, W. H., Kruijshoop, M., & Wagenmakers, A. J. (2000). Maximizing post-exercise muscle glycogen synthesis: Carbohydrate supplementation and the application of amino acid or protein hydrolysate mixtures. *American Journal of Clinical Nutrition, 72*(1), 106–111.

Vandenberghe, K., Hespel, P., Vanden Eynde, B., Lysens, R., & Richter, E. A. (1995). No effect of glycogen level on glycogen metabolism during high intensity exercise. *Medicine and Science in Sports and Exercise, 27*(9), 1278–1283.

Vesa, T. H., Marteau, P., & Korpela, R. (2000). Lactose intolerance. *Journal of the American College of Nutrition, 19*(2 Suppl.), 165S–175S.

Zawadzki, K. M., Yaspelkis, B.B., III, & Ivy, J. L. (1992). Carbohydrate-protein complex increases the rate of muscle glycogen storage after exercise. *Journal of Applied Physiology, 72*(5), 1854–1859.

Ziegler, P. J., Jonnalagadda, S. S., Nelson, J. A., Lawrence, C., & Baciak, B. (2002). Contribution of meals and snacks to nutrient intake of male and female elite figure skaters during peak competitive season. *Journal of the American College of Nutrition, 21*(2), 114–119.

Proteins

Learning Plan

- **Describe** amino acids and how the structure of a protein affects its function.
- **Describe** the digestion, absorption, transportation, and metabolism of amino acids.
- **Describe** when and how the body uses protein to fuel exercise.
- **Explain** the role of protein in the immune system and how endurance exercise can affect its function.
- **State** protein recommendations for athletes and the effects of high and low protein and/or energy intakes on training, performance, and health.
- **Explain** the physiological basis for recommendations related to the amount and timing of protein intake before, during, and after exercise.
- **Identify** sources of dietary protein.
- **Assess** an athlete's dietary protein intake.
- **Evaluate** dietary supplements containing amino acids and proteins, for safety, effectiveness, and purity.

In many sports, well-developed muscle is necessary for successful performance.

Pre-Test   Assessing Current Knowledge of Proteins

Read the following statements and decide if each is true or false.

1. Skeletal muscle is the primary site for protein metabolism and is the tissue that regulates protein breakdown and synthesis throughout the body.

2. In prolonged endurance exercise, approximately 3–5 percent of the total energy used is provided by amino acids.

3. To increase skeletal muscle mass, the body must be in positive nitrogen balance, which requires an adequate amount of protein and energy (calories).

4. Athletes who consume high-protein diets are at risk for developing kidney disease.

5. Strength athletes usually need protein supplements because it is difficult to obtain a sufficient amount of protein from food alone.

Since the days of the ancient Olympiads, protein has held a special place in the athlete's diet. Protein is critical to growth and development, including growth of skeletal muscle tissue, so its place as an important nutrient has been earned. But its role can be overstated if it is elevated above other nutrients. Protein functions optimally only when energy (caloric) intake from carbohydrate and fat is sufficient.

The optimal amount, type, and timing of protein intake by athletes are active areas of research. Scientists are providing research results that help practitioners and athletes better define the amount of protein needed to build and maintain skeletal muscle, to protect skeletal muscle from breakdown during training, and to speed recovery after training. There is an emerging body of literature about the amount of protein needed by athletes in training who are trying to lose body fat while protecting against the loss of skeletal muscle. This chapter covers both the fundamentals of protein metabolism as well as the practical applications for athletes in training.

5.1 Protein Basics

Protein functions in the body in many ways. A primary function of protein is to build and maintain tissues. To increase muscle mass, an athlete must engage in resistance exercise, consume a sufficient amount of energy (kcal), and be in positive nitrogen (protein) balance. Because so much emphasis is put on protein and muscle growth, it can easily be forgotten that protein is also the basis of enzymes and many **hormones**, and of structural, transport, and immune system proteins.

Although it is not the primary function of protein, the **amino acids** that make up protein can be used to provide energy. In prolonged endurance exercise, amino acids are an important energy source even though carbohydrates and fats supply the majority of the energy.

The basic component of all proteins is the amino acid, a nitrogen-containing compound. Proteins found in food and protein supplements are broken down into amino acids through the processes of digestion and absorption. Once absorbed, the amino acids are transported to the liver, which plays a major role in amino acid metabolism. Body proteins are constantly being manufactured and broken down. Dietary protein and degraded body proteins provide a steady stream of amino acids for protein synthesis. Sufficient energy intake is needed to support the growth and maintenance of skeletal muscle; insufficient caloric intake does not support growth and undermines the ability to build and maintain skeletal muscle. In a prolonged starvation state, the body's ability to maintain **nitrogen balance** is compromised and skeletal muscle tissue will be sacrificed to ensure survival.

Proteins are found in both plant and animal foods. Although they differ in quality (that is, amount and type of amino acids), in industrialized countries it is easy to consume an adequate amount of protein with sufficient quality. The amount of protein recommended for athletes in training is higher than for nonathletes, although protein consumption typically exceeds recommendations in both athletes and nonathletes when caloric intake is not restricted. Timing of protein intake is important, especially during the postexercise (recovery) period.

Athletes have numerous protein-containing foods to choose from as well as a range of protein supplements, so it is not difficult for athletes in training to meet the recommended daily protein intake. Athletes typically receive the majority of their protein from foods. Protein supplements

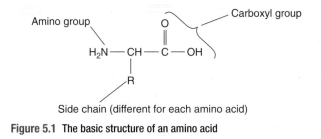

Figure 5.1 The basic structure of an amino acid

Proteins are found in both plant and animal foods.

are popular, convenient to use, and safe for healthy adults. Protein supplements are generally manufactured using food proteins such as milk or soy. These supplements are neither superior nor inferior to the equivalent protein-containing foods. Supplementation with individual amino acids appears to be safe, but has not been shown to be effective for improving performance or health, although research continues in these areas.

Amino acids form the basic structure of proteins.

Proteins are made up of amino acids, which contain carbon, hydrogen, oxygen, and nitrogen. It is the nitrogen that distinguishes them from the composition of carbohydrates, fats, and alcohol, which are made up of only carbon, hydrogen, and oxygen. To understand their functions one must understand the structures of proteins. The basic structural component is an amino acid.

An amino acid is a chemical compound that contains an NH_2 (that is, an amino) group and a COOH (that is, carboxyl) group of atoms. The basic structure of an amino acid is shown in Figure 5.1. The nitrogen content of an amino acid is approximately 16 percent. There are a total of 20 different amino acids that will be used by the body to make various proteins. Amino acids may have side chains (e.g., glycine, leucine), acid groups (e.g., glutamine), basic groups (e.g., lysine), or rings (e.g., tryptophan). Some contain sulfur (e.g., cysteine, methionine). These differences in amino acid structure play critical roles in the functions of the proteins created. Important features of the 20 amino acids are shown in Table 5.1.

Some amino acids cannot be manufactured by the body and must be provided by food.

Of the 20 amino acids needed by healthy adults, 9 are considered **indispensable** because the body cannot manufacture them. The remaining 11 amino acids are termed **dispensable** because they can be manufactured in the liver. Six of these 11 amino acids are referred to as **conditionally indispensable**, because during periods of stress the body cannot manufacture a sufficient amount. Illness, injury, and prolonged endurance exercise are examples of physiologically stressful conditions. In the past, the terms *essential*

and *nonessential* were used to describe indispensable and dispensable amino acids, respectively. *Nonessential* is a misleading term because it implies that such amino acids are not needed. In fact, they are needed but the body has the ability to manufacture them if they are not consumed directly from food. Indispensable and dispensable are now the preferred terms when describing amino acids.

In the United States and other countries where a variety of food, including protein-containing food, is widely available, few healthy adults are at risk for amino acid deficiencies. They consume an adequate amount of protein daily and receive ample amounts of all the indispensable amino acids. Those at risk for low protein intake include those with eating disorders, the frail elderly, and people with liver or kidney disease.

Proteins vary in quality due to the amount and types of amino acids present.

Protein quality is determined based on the amounts and types of amino acids and the extent to which the

Protein: Amino acids linked by peptide bonds.

Hormone: A compound that has a regulatory effect.

Amino acid: The basic component of all proteins.

Nitrogen balance: When total nitrogen (protein) intake is in equilibrium with total nitrogen loss.

Indispensable amino acid: Amino acid that must be provided by the diet because the body cannot manufacture it.

Dispensable amino acid: Amino acid that the body can manufacture.

Conditionally dispensable amino acid: Under normal conditions, an amino acid that can be manufactured by the body in sufficient amounts, but under physiologically stressful conditions an insufficient amount may be produced.

Protein quality: The amounts and types of amino acids contained in a protein and their ability to support growth and development.

amino acids are absorbed. Protein quality is a critical issue in human growth and development. In countries where protein foods are abundant, sufficient protein quality is a near certainty, an assumption that should not be made in countries where protein foods are limited.

Humans must obtain through diet all of the indispensable amino acids, which are found in lower concentrations in plant proteins than in animal proteins. Animal proteins are termed **complete proteins** because they contain all the indispensable amino acids

Table 5.1 Summary of the 20 Indispensable and Dispensable Amino Acids

Amino acid	Figure number	Classification	Glucogenic	Ketogenic	Miscellaneous
Alanine (Ala)	2	Dispensable	Yes	No	Can be produced in the muscle from pyruvate but must be transported to the liver for conversion to pyruvate to produce glucose. An important glucose-generating pathway during starvation
Arginine (Arg)	14	Conditionally indispensable	Yes	No	
Asparagine (Asn)	12	Dispensable	Yes	No	
Aspartic acid (Asp)	10	Dispensable	Yes	No	One of the two amino acids that make up the structure of the artificial sweetener aspartame
Cysteine (Cys)	8	Conditionally indispensable	Yes	No	
Glutamic acid (Glu)	11	Dispensable	Yes	No	
Glutamine (Gln)	13	Conditionally indispensable	Yes	No	Represents about half of all the amino acids in the amino acid pool
Glycine (Gly)	1	Conditionally indispensable	Yes	No	
Histidine (His)	16	Indispensable	Yes	No	
Isoleucine (Ile)	5	Indispensable	Yes	Yes	Branched chain amino acid; muscle can use as an energy source during prolonged endurance exercise when muscle glycogen stores are low
Leucine (Leu)	4	Indispensable	No	Yes	Branched chain amino acid; muscle can use as an energy source during prolonged endurance exercise when muscle glycogen stores are low
Lysine (Lys)	15	Indispensable	No	Yes	
Methionine (Met)	9	Indispensable	Yes	No	
Phenylalanine (Phe)	17	Indispensable	Yes	Yes	One of the two amino acids that make up the structure of the artificial sweetener aspartame
Proline (Pro)	20	Conditionally indispensable	Yes	No	
Serine (Ser)	6	Dispensable	Yes	No	
Threonine (Thr)	7	Indispensable	Yes	Yes	
Tryptophan (Trp)	19	Indispensable	Yes	Yes	
Tyrosine (Tyr)	18	Conditionally indispensable	Yes	Yes	
Valine (Val)	3	Indispensable	Yes	No	Branched chain amino acid; muscle can use as an energy source during prolonged endurance exercise when muscle glycogen stores are low

Table 5.1 Summary of the 20 Indispensable and Dispensable Amino Acids (Continued)

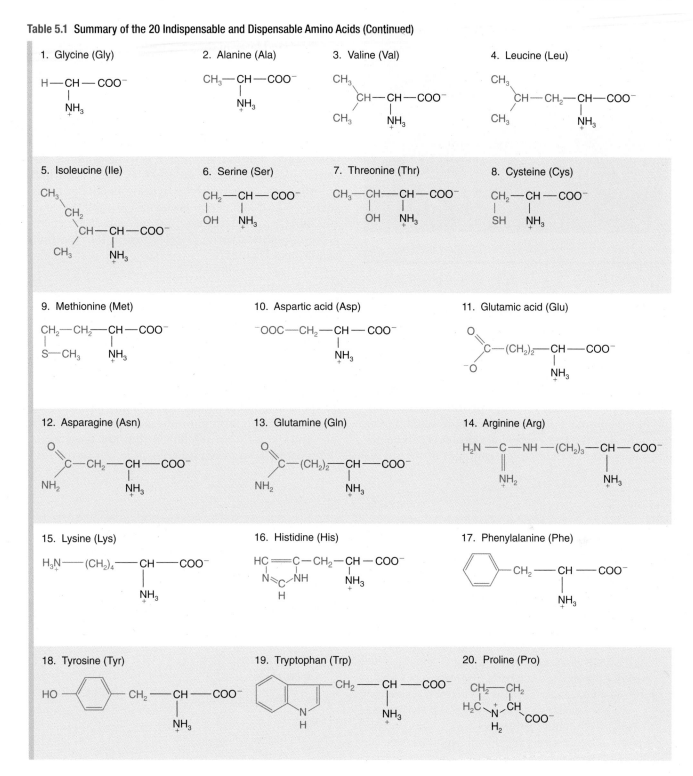

in the proper amounts and proportions to each other to prevent amino acid deficiencies and to support growth. In contrast, plant proteins may lack one or more of the indispensable amino acids or the proper concentrations and are termed **incomplete proteins**. The indispensable amino acids that are of greatest concern are lysine, threonine, and the sulfur-containing

Complete protein: Protein that contains all the indispensable amino acids in the proper concentrations and proportions to each other to prevent amino acid deficiencies and to support growth.

Incomplete protein: Protein that lacks one or more of the indispensable amino acids in the proper amounts and proportions to each other to prevent amino acid deficiencies and to support growth.

Beans and corn tortillas provide complementary proteins in this vegetarian meal.

© Michael Newman/PhotoEdit

amino acids, cysteine and methionine. If the intake of these specific amino acids is limited, then protein deficiencies could occur.

It is possible to pair different plant proteins with each other and bring the total concentration of all the indispensable amino acids to an adequate level. This is the concept of **complementary proteins** or combining two incomplete proteins. When consumed during the same day, the complementary proteins can be nutritionally equal to a complete (animal) protein. This concept is more fully illustrated when vegetarian diets are discussed later in the chapter.

In countries where people are starving, the quality of dietary protein continues to be an important issue to ensure survival and adequate growth, especially in infants and children. In countries where the quantity of protein is sufficient, protein quality is emerging as an issue related to optimal skeletal muscle growth and bone health. For example, there is some evidence that certain high-quality proteins, such as whey, egg white, or milk, are more effective than lower-quality proteins, such as wheat and cereal grains, for stimulating skeletal muscle and bone growth (Millward et al., 2008). Protein quality scoring is explained in the Spotlight on...Protein Quality.

The structure of a protein determines its function.

Peptide refers to two or more amino acids that are combined. Specifically, **dipeptide** refers to two amino acids, **tripeptide** to three amino acids, and **polypeptide** to four or more amino acids. Most proteins are polypeptides and are made up of many amino acids, often numbering in the hundreds or thousands. *Protein* and *polypeptide* are terms that are used interchangeably. *Dipeptide* and *tripeptide* are terms that

Spotlight on...

Protein Quality

The internationally accepted method for determining protein quality is known as protein digestibility-corrected amino acid score (PDCAAS). This method is preferred over other methods such as protein efficiency ratio (PER) or biological value (BV). Using PDCAAS, proteins are scored on a scale from 0 to 1, with 1 being the highest score obtainable. This measure is based on protein absorption and utilization as well as amino acid content.

As shown in the chart below, several proteins have a score of 1 (WHO, 2007; Kreider and Campbell, 2009). These include whey and casein, which are milk proteins, egg, and bovine colostrum. Colostrum is the first secretions from the mammary glands after birth and is sold as a supplement; the others are eaten as foods or are found in protein supplements. Other proteins have a lower PDCAAS because either digestibility is lower or they lack of one or more indispensable amino acids. Although PDCAAS is a measure of protein quality, there is more to consider than just protein score when evaluating protein intake. For example, sufficient energy intake is a substantial influence on protein utilization.

Source of protein	PDCAAS*
Bovine colostrum	1
Casein	1
Egg (Albumin)	1
Milk protein (casein + whey)	1
Whey	1
Soy	.95
Beef	.92
Nuts	.70
Beans and legumes	.60
Wheat	.43

*Protein digestibility-corrected amino acid score

are typically used when discussing digestion and absorption.

Polypeptides are synthesized on ribosomes, organelles found in large numbers in the cytoplasm of cells. The primary structure of the protein is determined at its creation based on information contained in DNA (deoxyribonucleic acid) and RNA (ribonucleic acid). The RNA acts like a blueprint for the type, number, and sequence of amino acids to be included in a particular polypeptide. The differences in amino acids influence the bonding abilities of the polypeptide, which affect the shape of the protein. For example, proteins can be straight, coiled, or folded based on the type, number, and sequence of amino acids in the polypeptide. The primary structure of the polypeptide determines how a protein functions.

The secondary structure of the polypeptide is a result of bonding of amino acids that are located close to each other. These bonds give more rigidity and stability to the protein, an important characteristic for structural proteins such as collagen. The tertiary (third) level of structure is a result of interactions of amino acids that are located far away from each other. These interactions, if present, cause the polypeptide to form a loop. The loop results in a clustering of certain amino acids, which then function in a particular way. For example, a cluster of amino acids may have a positive or negative charge and accept or repel other compounds, such as water. Quaternary (fourth) level structure involves more than one polypeptide, typically two or four. Because of their quaternary structure, these proteins can interact with other molecules. Insulin, which interacts with glucose, is an example of a compound made up of two polypeptides.

Proteins perform many functions in the body.

As explained above, the structure of the protein determines its function. Body proteins are often classified in five major categories: enzymes, hormones, structural proteins, transport proteins, and immune system proteins. A summary of the functions of proteins in the human body is shown in Table 5.2.

Enzymes are polypeptides that are necessary to catalyze (speed up) reactions. It is the structure of the enzyme, particularly the quaternary structure that allows the protein-based enzyme to interact with other compounds. The unique structure of each enzyme interacts with its substrate much like a key fits into and opens a specific lock. The purpose of enzymes is to regulate the speed of chemical reactions (refer to the more detailed description of enzymes in Chapter 2).

Hormones are compounds that act as chemical messengers to regulate metabolic reactions. Many hormones are protein based, although some hormones are made from cholesterol (for example, steroid hormones). Insulin, glucagon, and human growth hormone are just three examples of the hundreds of hormones made from amino acids. Insulin is a relatively small polypeptide, made up of only 51 amino acids, yet it is one of the body's most essential hormones.

Complementary proteins: The pairing of two incomplete proteins to provide sufficient quantity and quality of amino acids.

Peptide: Two or more amino acids linked by peptide bonds.

Dipeptide: Two amino acids linked by peptide bonds.

Tripeptide: Three amino acids linked by peptide bonds.

Polypeptide: Four or more amino acids linked by peptide bonds; often contain hundreds of amino acids.

Table 5.2 Summary of the Functions of Proteins in the Human Body

Protein category	Functions
Component of enzymes	Enzymes are specialized proteins that speed up (catalyze) chemical reactions in cells
Component of hormones and signaling proteins	Hormones, many of which are protein based, regulate metabolic processes; signaling proteins (cytokines) are known as growth factors and can bind to the surface of a cell and influence its cellular processes.
Structural proteins	Component of muscle, connective tissue, skin, hair, and nails.
Transport proteins	Part of molecules that allow compounds to be transported, such as oxygen, carbon dioxide, iron, and fats.
Immune system proteins	Fundamental component of the immune system.
Acid-base regulator	Amino acids have both acid and basic groups, which helps the body to achieve acid-base balance and optimal pH.
Fluid regulator	Proteins, especially those found in the blood, help to maintain fluid balance.
Source of energy	Under normal conditions, a minor energy source; under temporary stressful conditions, a small but important source of energy; under severe or prolonged stress, such as starvation, a major source of energy but to the detriment of health.

Part of the polypeptide chain folds back on itself (due to its secondary structure), and the two protein chains that make up its quaternary structure are linked by disulfide bonds. Because insulin is small, it can move through the blood quickly, and the folded chain and the disulfide bonds give it great stability.

Structural proteins include the proteins of muscle and connective tissue (for example, actin, myosin, and collagen), as well as proteins found in skin, hair, and nails. Structural proteins can be constructed into long polypeptide strands, similar to a long chain. These strands can be twisted and folded into a wide variety of three-dimensional shapes (secondary structure). Elements of the constituent amino acids in the polypeptide chains, such as sulfide groups, may be brought close together by the twisting and folding and may form interconnected bonds (tertiary structure). The secondary and tertiary structures of the polypeptide are responsible for the differences in rigidity and the durability of these polypeptides.

Examples of transport proteins include lipoproteins (lipid carriers) and hemoglobin, which carries oxygen and carbon dioxide in the blood. Without its particular quaternary structure, hemoglobin would not be as efficient. Four polypeptide chains are bonded in such a way that they can work in concert and can change shape slightly when necessary. This structure allows hemoglobin to be flexible and capable of changing its ability to bind and release oxygen. For example, hemoglobin needs a high affinity to bind oxygen in the lungs and carry it throughout the body, but it must reduce its affinity for oxygen so oxygen can be released for use by the tissues.

The immune system is a protein-based system that protects the body from the invasion of foreign particles, including viruses and bacteria. One immune system response is the activation of **lymphocytes**, cells that produce antibodies. All antibodies are compounds that are made of polypeptide chains (usually four) in the shape of a Y. The antibody fits the virus or bacteria like a key in a lock, aiding in their destruction. The shape of the "key" is due to disulfide bonds and the sequence of the amino acids. Some amino acids help to regulate the immune system (see Spotlight on...Amino Acids as Regulators).

All of the compounds described above are proteins, but none of these compounds are provided directly from proteins found in foods. Enzymes, hormones, and the other protein-based compounds are manufactured in the body from indispensable and dispensable amino acids. To understand how food proteins become body proteins one must know how amino acids found in food are digested, absorbed, transported, and metabolized.

Key points

- The basic protein unit is an amino acid.
- In adults, nine amino acids are indispensable.
- Complete and incomplete proteins differ due to their indispensable amino acid content.
- Proteins are a critical element in many body functions including being a part of enzymes and skeletal muscle tissue.

How do structural differences of proteins influence function?

How do plant and animal proteins differ in quality?

Spotlight on...

Amino Acids as Regulators

One of the best-known functions of amino acids is their role as the building blocks of proteins. However, amino acids also regulate critical metabolic pathways. Such amino acids are sometimes termed functional amino acids. Researchers are trying to determine the exact roles these amino acids play as regulators and if biochemical pathways would be positively or negatively affected by an increase in certain amino acids. Although it is too early to use individual amino acid supplements to influence metabolic pathways, it is critical that such studies be conducted (Wu, 2009). Arginine, glutamine, and leucine are among those amino acids that are being studied for their regulatory effects.

Arginine is necessary to regulate the immune system. It is known to stimulate killer cell activity. Arginine also helps to decrease some of the proinflammatory cytokines, signaling proteins that bind to the surface of cells and have hormone-like effects. Glutamine is the most abundant amino acid in the amino acid pool and regulates the immune system. Skeletal muscle is a major site of glutamine synthesis. Glutamine is a potent stimulator of the immune system and is a major source of fuel for certain immune system cells. A lack of glutamine reduces killer cell activity and impairs the immune system response (Roth, 2008).

Leucine is a stimulator of skeletal muscle amino acid transporter expression, which could theoretically improve the delivery of amino acids within skeletal muscle cells. An increase in skeletal muscle protein synthesis is important to athletes, who wish to increase or maintain skeletal muscle mass, and to the elderly, who are at risk for loss of skeletal muscle mass, known as muscle wasting (Little and Phillips, 2009). The regulatory roles of amino acids are a hot topic for researchers.

5.2 Digestion, Absorption, and Transportation of Protein

Digestion begins as soon as proteins found in food arrive in the stomach. Absorption takes place primarily in the middle and lower small intestine by several mechanisms. Once absorbed, the amino acids will be transported to the liver, which acts as a clearinghouse. After a meal the majority of the amino acids absorbed will remain in the liver for metabolism. The remainder will circulate in the blood and be transported to other parts of the body.

Proteins are digested in the mouth, stomach, and small intestine.

Protein digestion begins when a food protein comes in contact with the gastric juice of the stomach. The hydrochloric acid (HCl) in the gastric juice begins to **denature** (change the structure of) the protein. At the same time, the HCl activates pepsin, an enzyme that will break down the polypeptides into smaller units. Pepsin prefers to break the bonds of certain amino acids, such as leucine and tryptophan. This initial stage of digestion generally breaks down very large polypeptides into smaller units, but these smaller units are still very large amino acid chains (Figure 5.2).

As the denatured and partially digested polypeptides move from the stomach to the small intestine, a number of digestive enzymes are activated. Some of these are found in pancreatic juice, which is secreted from the pancreas into the small intestine. Other digestive enzymes are released from the cells of the brush border that line the gastrointestinal tract. Similar to pepsin, these enzymes prefer to break the bonds of specific amino acids. For example, some enzymes break down amino acids with rings whereas other enzymes break down amino acids with side chains.

Due to the action of the various enzymes, the large polypeptides that entered from the stomach are broken down into small polypeptides (usually no more than six amino acids), tripeptides, dipeptides, and free amino acids. The cells of the gastrointestinal tract can normally absorb nothing larger than a tripeptide, so the small polypeptides are broken down further by brush border enzymes.

Proteins are absorbed in the middle and lower part of the small intestine.

Absorption takes place primarily in the **jejunum** and the **ileum**. Two-thirds of the amino acids absorbed are in the form of dipeptides or tripeptides, whereas

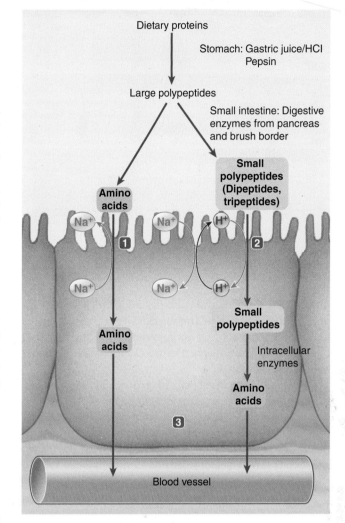

Figure 5.2 Digestion and absorption of dietary protein

Dietary proteins are broken down into large polypeptides in the stomach by the action of gastric juice, hydrochloric acid, and pepsin. In the small intestine, large polypeptides are broken down into smaller polypeptides and amino acids by digestive enzymes secreted from the pancreas and the brush border. Amino acids are absorbed into epithelial cells by Na+ and energy-dependent active transport (1). Small polypeptides are absorbed via H+, Na+, and energy-dependent active transport (2). Once absorbed, most polypeptides are broken down to amino acids by intracellular enzymes. Amino acids then leave the cell by various passive carriers and enter the blood via diffusion (3).

one-third are individual amino acids. Absorption takes place in a variety of ways, but one of the most common is the use of a carrier that moves these compounds across the cell membrane. There are a number

Lymphocytes: Cells that produce antibodies.

Denature: To change the chemical structure of a protein by chemical or mechanical means.

Jejunum: The middle portion of the small intestine.

Ileum: The lowest portion of the small intestine.

A wide array of protein supplements is available. Some are sold as predigested; however, there is no evidence that predigestion is advantageous.

of different carriers and each has an affinity for certain amino acids. Some of the carriers require sodium to load the amino acids. Not surprisingly, indispensable amino acids are absorbed more rapidly than dispensable amino acids.

Since proteins must be broken down into dipeptides, tripeptides, and free amino acids to be absorbed, protein supplements may be sold as "predigested." The predigestion is a result of exposing the food proteins in the supplement to enzymatic action during the manufacturing process. Due to a lack of scientific studies, it is not known if supplements containing predigested dipeptides and tripeptides are absorbed faster or differently than food proteins broken down into dipeptides and tripeptides by digestive enzymes. It is known that free amino acids do compete with each other for absorption because of carrier competition. If some free amino acids are found in higher concentrations than others, it is possible that they would be preferentially absorbed over the amino acids found in lower concentrations. Well-designed research studies are needed in this area.

Protein from food provides about two-thirds of the amino acids absorbed from the small intestine. These amino acids are **exogenous**, meaning they originate from outside of the body. The other one-third of the amino acids is **endogenous**, originating from inside the body). Endogenous proteins include mucosal cells shed into the gastrointestinal tract and gastrointestinal secretions that contain enzymes and other protein-based compounds. These endogenous proteins are broken down and absorbed in a manner similar to proteins originally derived from food, although they are often absorbed lower in the gastrointestinal tract, including the colon. This is the body's way of recycling amino acids, but not all of them can be reclaimed. Those that are not absorbed are excreted in feces, which represents one way that nitrogen is lost from the body and is one reason why

dietary protein must be consumed daily. Once amino acids are absorbed, the body does not distinguish between the amino acids originally obtained from food and those from endogenous sources.

Once inside the mucosal cells, any dipeptides and tripeptides are broken down into free amino acids by cellular enzymes. Some of the amino acids will not leave the intestinal cells because they will be used to make cellular proteins. Those that are not incorporated into cellular proteins will be released into the blood via the **portal** (liver) **vein**.

After absorption, some amino acids are transported to the liver whereas others circulate in the blood.

The liver serves as a clearinghouse for the amino acids by monitoring the supply and dictating which amino acids will be transported to which tissues. Exceptions to this are the **branched chain amino acids** (BCAA)— leucine, isoleucine, and valine. The liver has very low levels of the enzyme BCAA transferase, which is needed to transfer these amino acids to tissues. Therefore, the branched chain amino acids leave the liver, circulate in the **plasma**, and are taken up by skeletal muscle cells, which have high levels of the enzyme that the liver lacks. BCAA transferase is also found in the heart, kidneys, and adipose tissue.

After a protein-containing meal, 50 to 65 percent of the amino acids absorbed will be found in the liver. The remainder of the amino acids absorbed will be immediately released as free amino acids into the blood and **lymph** and become part of the **amino acid pool**, which is shown in Figure 5.3. The concentration of amino acids in the blood is increased for several hours after a protein-containing meal is consumed.

The amino acid pool refers to free amino acids that are circulating in the blood or in the fluid found within or between cells. Half of the amino acid pool is found in or near skeletal muscle tissue whereas the remainder is distributed throughout the body. Some of the amino acids in the amino acid pool have recently been absorbed from the gastrointestinal tract, but most come from a different source—the breakdown of body tissues, including skeletal muscle tissue. The amino acid pool undergoes constant change but on average contains about 150 g of amino acids, of which ~80 g is glutamine. There are more dispensable amino acids in the pool than indispensable amino acids. The amino acid pool is always in flux as a result of food intake, exercise, and the breaking down or building of tissues, especially skeletal muscle. This constant flux is referred to as **protein turnover**. It is thought that a relatively large amount of energy (~20 percent of resting metabolism) is expended each day on synthesizing and degrading proteins.

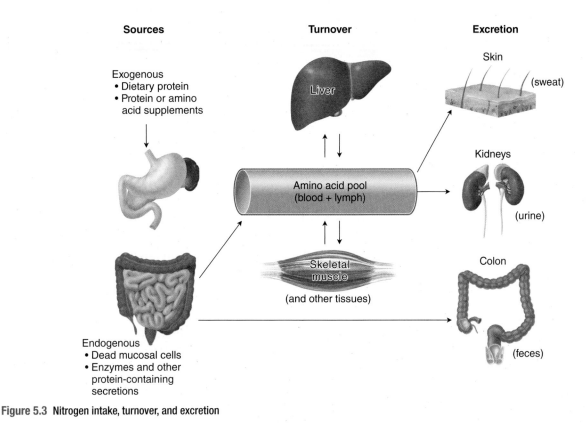

Sources **Turnover** **Excretion**

Figure 5.3 Nitrogen intake, turnover, and excretion

The amino acid pool is a reservoir of amino acids. It is always in flux as dietary protein is consumed, body proteins are broken down and synthesized, and nitrogen is excreted in urine, sweat, and feces.

Key points

- The proteins found in food are eventually broken down into one, two, or three amino acid units for absorption.

- Indispensable amino acids are absorbed more rapidly than dispensable amino acids.

- Once absorbed, the liver monitors the supply and delivery of amino acids.

- The amino acid pool contains amino acids recently absorbed from food as well as amino acids from the breakdown of skeletal muscle tissue.

What happens to a protein from the time that you eat it until the time that the amino acids are absorbed by skeletal muscle cells?

5.3 Metabolism of Proteins and Amino Acids

The liver is a major site for amino acid metabolism. The liver monitors the body's amino acid needs and responds accordingly with anabolic or catabolic processes. **Anabolic** is defined as building complex molecules from simple molecules. An example is the synthesis of a protein. **Catabolic** is the breakdown of complex molecules into simple ones. The use of

protein for energy is a catabolic process, as the protein must be broken down into its amino acid components, some of which are then metabolized for energy. The liver plays the primary role in amino acid metabolism but it functions in concert with other tissues, such as skeletal muscle and kidneys.

What is the fate of the amino acids absorbed from the gastrointestinal tract and transported to the liver? The liver uses approximately 20 percent of these amino acids to make proteins and other nitrogen-containing compounds. The liver

Exogenous: Originating from outside of the body.

Endogenous: Originating from within the body.

Portal vein: A vein that carries blood to the liver; usually refers to the vein from the intestines to the liver.

Branched chain amino acid (BCAA): One of three amino acids (leucine, isoleucine, and valine) that has a side chain that is branched.

Plasma: Fluid component of blood; does not include cells.

Lymph: A fluid containing mostly white blood cells.

Amino acid pool: The amino acids circulating in the plasma or in the fluid found within or between cells.

Protein turnover: The constant change in the body proteins as a result of protein synthesis and breakdown.

Anabolic: Building complex molecules from simple molecules.

Catabolic: The breakdown of complex molecules into simple ones.

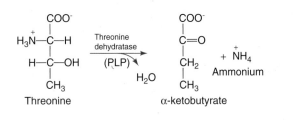

◀ PLP = pyroxidoxal phosphate (Vitamin B$_6$ containing enzyme)

Figure 5.4 An example of deamination

The amino acid threonine is deaminated to form α-ketobutyrate (an α-keto acid).

catabolizes (breaks down) the majority of the amino acids delivered from the gastrointestinal tract. Two important metabolic processes are **deamination** and **transamination**. Deamination refers to the removal of the amino group from the amino acid (Figure 5.4). When the amino group is removed, the remaining compound is an **alpha-keto acid (α-keto acid)**, frequently referred to as the **carbon skeleton**. Transamination involves the transfer of an amino group to another carbon skeleton, whereby an amino acid is formed. Transamination allows the liver to manufacture dispensable amino acids from indispensable amino acids. Deamination and transamination are regulated by enzymes and are part of an intricate system that the liver uses to monitor and respond to the body's amino acid and protein needs.

Amino acids absorbed from food that do not remain in the liver become part of the amino acid pool. The amino acids in the pool are also involved in both anabolic and catabolic processes. For example, the synthesis of skeletal muscle protein is an anabolic process that uses amino acids from the pool to synthesize new proteins. When skeletal muscle proteins are degraded, a catabolic process, the amino acids are returned to the amino acid pool.

The body uses amino acids to build proteins, a process known as anabolism.

One of the major functions of the liver is protein **anabolism**. Some amino acids will be incorporated into liver enzymes. Others will be used to make **plasma proteins**. For example, the liver manufactures **albumin**, a protein that circulates in the blood and helps to transport nutrients to tissues. Many of the proteins made in the liver are synthesized and released in response to infection or injury. As mentioned earlier, the liver continually monitors the body's amino acid and protein needs and responds to changing conditions.

In addition to protein synthesis, the liver uses amino acids to manufacture compounds such as creatine. Recall from Chapter 3 that creatine can be obtained directly from the diet (for example, beef, fish), but it can also synthesized from the amino acids arginine, glycine, and methionine. The creatine synthesis process begins in the kidneys but is completed in the liver.

Of particular interest to athletes is the synthesis and breakdown of skeletal muscle proteins. Figure 5.5 illustrates protein turnover. An anabolic state occurs when the synthesis of proteins is greater than their breakdown. In the anabolic state, amino acids from the amino acid pool are incorporated into the synthesis of proteins.

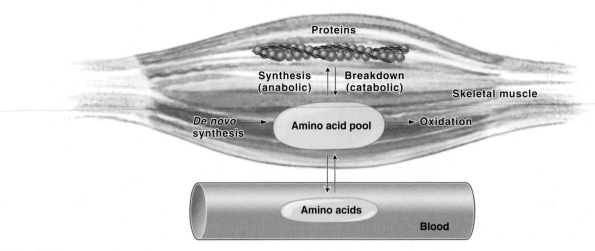

Figure 5.5 Skeletal muscle protein turnover

A general depiction of muscle protein turnover (synthesis and breakdown). The amino acid pool in muscle tissue is derived from amino acids taken up from the blood, those synthesized in the muscle (de novo synthesis), or those from the breakdown of muscle protein. Amino acids from this pool can be used to synthesize muscle proteins, metabolized for energy via oxidative phosphorylation, or released into the blood for distribution to other tissues in the body.

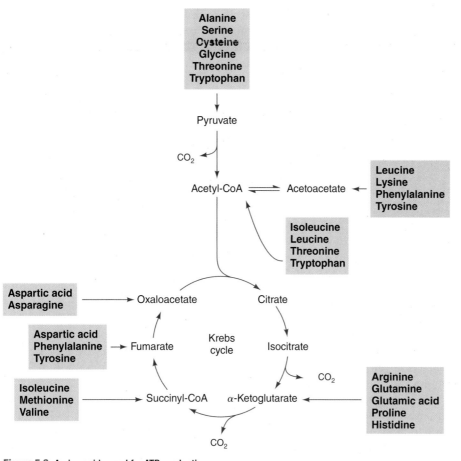

Figure 5.6 Amino acids used for ATP production

The structure of the amino acid determines the entry point into the Krebs cycle.

The body breaks down proteins into amino acids, a process known as catabolism.

Amino acids that are not used for building proteins are catabolized. In other words, excess amino acids are not "stored" for future use in the same way that carbohydrates and fats are stored. Carbohydrate can be stored in liver or muscle as glycogen, and fat can be stored in adipocytes (fat cells) and at a later time removed easily from storage and used as energy. In contrast, the so-called "storage" site for protein is skeletal muscle. Under relatively extreme circumstances, protein can be removed from the skeletal muscle. However, the removal of a large amount of amino acids has a very negative effect on the muscle's ability to function, so the body tries to protect the skeletal muscle from being used in this way.

Amino acids can provide energy. In fact, the source of approximately half of the ATP used by the liver comes from amino acids. When the amino group is transferred or removed, the carbon skeleton (alpha-keto acid) can be oxidized to produce energy (that is, ATP). Although it is commonly written that amino acids are oxidized for energy, this is technically incorrect because the nitrogen is not oxidized. The term *oxidation of amino acids*

is understood to mean that after the nitrogen is removed, the carbon skeleton of the amino acid is oxidized for energy and the nitrogen goes through the urea cycle in the liver.

Similar to carbohydrate, protein yields approximately 4 kcal/g. Energy is best supplied by carbohydrate and fat rather than protein. Sufficient caloric intake in the form of carbohydrate and fat is referred to as having a **protein-sparing effect**. In other words, the carbohydrate and fat provide the energy that the body needs and the protein is "spared" from this function. The protein is available for other important functions that can be provided only by protein. When people consume sufficient energy, all the important protein-related functions can be met and, in fact, some of the protein consumed *will* be metabolized for energy. Problems can result if caloric intake is too low and some protein *must* be used to meet energy needs.

The metabolic pathways for producing energy from protein are not reviewed in detail here; however, the **catabolism** of amino acids to provide ATP is summarized. Under aerobic conditions the carbon skeleton of an amino acid can be used in the Krebs cycle. As shown in Figure 5.6, amino

Deamination: The removal of an amino group.

Transamination: The transfer of an amino group.

α-keto acid: The chemical compund that is a result of the deamination (i.e., nitrogen removal) of amino acids.

Carbon skeleton: The carbon-containing structure that remains after an amino acid has been deaminated (i.e., nitrogen removed).

Anabolism: Metabolic processes involving the synthesis of simple molecules into complex molecules.

Plasma protein: Any polypeptide that circulates in the fluid portion of the blood or lymph (for example, albumin).

Albumin: A protein that circulates in the blood and helps to transport nutrients to tissues.

Protein-sparing effect: The consumption of sufficient kilocalories in the form of carbohydrate and fat, which protects protein from being used as energy before other protein-related functions are met.

Catabolism: Metabolic processes involving the breakdown of complex molecules into simpler molecules.

Prolonged endurance exercise results in increased protein metabolism, particularly in the later stages of a triathlon.

acids have different entry points into the Krebs cycle, which is based on the structure of each amino acid. Some amino acids, such as alanine and glycine, can be converted to pyruvate. Others, such as leucine, are converted to acetyl Co-A. Additional pathways include the conversion of various amino acids to intermediate compounds of the Krebs cycle.

Six amino acids are most commonly broken down in muscle cells to yield energy—the branched chain amino acids (leucine, isoleucine, and valine), aspartate, asparagine, and glutamate. The breakdown of muscle, or **proteolysis**, is stimulated by the stress hormone **cortisol**, which is secreted by the adrenal glands. When the body is stressed, one response is the oxidation of amino acids. Endurance exercise represents an **acute** (short-term) stress so the use of amino acids for energy is not unexpected.

Endurance exercise results in an increased oxidation of leucine. At the beginning of an endurance exercise task, there is usually sufficient carbohydrate stored as muscle glycogen, so little of the energy needed comes from amino acids initially. But as muscle glycogen stores decline substantially, the skeletal muscle uses some amino acids, particularly leucine, for energy. This metabolic response is influenced by the carbohydrate content of the athlete's diet. When the athlete has been consuming a low-carbohydrate diet and is carbohydrate-depleted, the oxidation of leucine is increased, whereas a high-carbohydrate diet and near-maximal muscle glycogen stores decrease the need for leucine oxidation (Howarth et al., 2010).

The percentage of the total energy that comes from amino acids is generally small, about 3–5 percent, although it may be higher during prolonged, intense exercise that results in severe glycogen depletion.

Under the stress of long-duration endurance exercise, amino acids can represent an important energy source for skeletal muscle cells. For this reason, endurance athletes have explored the use of BCAA supplements, which will be discussed later in this chapter.

A discussion of protein catabolism would not be complete without mention of the ammonia (NH_3) or ammonium ions (NH^+_4) that are produced as a result of the catabolism of amino acids. Ammonia, which is toxic to the body, must be converted to urea, a related compound that can be safely transported in the blood to the kidneys where it is excreted in urine. The conversion of ammonia to urea takes place in the liver, so ammonia that is a result of amino acid catabolism in the liver can readily enter the urea cycle. Ammonia is also produced in the muscle as a result of the breakdown of adenosine monophosphate (AMP), which produces a compound that can be oxidized in the Krebs cycle for energy. This ammonia must be transported in the blood to the liver for conversion to urea. Urea contains nitrogen, so a small amount of nitrogen is lost every day via the urine. This loss of nitrogen is one reason that dietary protein must be consumed daily.

In addition to the oxidation of amino acids to produce ATP, a second major metabolic use of protein during exercise is for gluconeogenesis, which is the production of glucose from a noncarbohydrate source. The glucose-alanine cycle illustrates how amino acids can be used to generate glucose (Figure 5.7). Exercising muscle may use carbohydrate (glucose and/or glycogen) for metabolism, and in this process some of the pyruvate produced by glycolysis is converted (via transamination) to the amino acid alanine. Alanine is not used by the skeletal muscle, but is released into the blood where it travels to the liver. In the liver, alanine (as well as other amino acids) can be converted to pyruvate to produce glucose, a process known as gluconeogenesis. The newly formed glucose can then be released into the blood where it circulates and is taken up and used by a variety of tissues, such as the brain, kidney, and muscle. Eighteen of the 20 amino acids can be converted into glucose (only leucine and lysine cannot), and this primarily takes place in the liver. Although various amino acids can provide the carbon skeleton for gluconeogenesis, during exercise lactate is most likely the major source.

The body is constantly breaking down proteins as well as building proteins.

The body is in a constant state of protein turnover. In other words, every day the body simultaneously degrades and synthesizes proteins. On average, the adult

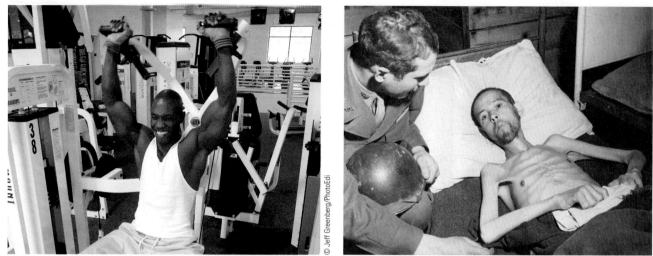

Athletes must be in positive nitrogen balance to increase skeletal muscle mass. Negative nitrogen balance, as a result of starvation, leads to substantially reduced skeletal muscle mass.

body turns over about 320 g of protein a day. It is estimated that 1–2 percent of the total protein in the body is degraded each day (that is, proteins are broken down to the amino acids that formed them). The source of most of the degraded protein is skeletal muscle, and these amino acids become part of the amino acid pool. About 80 percent of the amino acids that result from protein degradation are resynthesized into new proteins. However, at least 20 percent of the amino acids are not made into new proteins and are broken down to yield nitrogen and carbon skeletons. The nitrogen will be excreted in the urine and the carbon skeletons can be used for energy. Approximately 5 to 7 g of nitrogen, which is equivalent to 30 to 40 g of protein, is excreted via the urine

Cortisol: A glucocorticoid hormone that is secreted by the adrenal cortex that stimulates protein and fat breakdown and counters the effects of insulin.

Acute: Short-term.

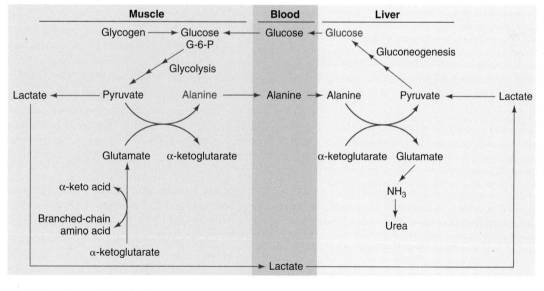

G-6-P = glucose-6-phosphate
NH$_3$ = ammonia

Figure 5.7 Glucose-alanine cycle

Muscle produces pyruvate as a result of using the glycolysis or glycogenolysis energy pathway. This pyruvate can be converted to alanine, which is then released into the blood and taken up by the liver. The liver can then convert the alanine to pyruvate for use in gluconeogenesis to produce glucose. The newly formed glucose can be released into the blood for distribution throughout the body and use by a variety of tissues.

Gallo Images/City Press / Alamy

AP Photo/Elaine Thompson

Both strength/power and endurance athletes need adequate protein daily.

each day. This fact alone makes it obvious that humans need to consume protein in their diets each day.

Protein metabolism is never static; rather, it is always changing. One way to measure and describe the changes is to determine nitrogen balance. Nitrogen balance is the difference between total nitrogen (protein) intake and total nitrogen loss (via the urine and feces), usually determined over several weeks. When intake is equal to loss, a state of nitrogen balance exists. When intake is greater than loss, a person is in positive nitrogen balance. Conversely, when loss is greater than intake, a state of negative nitrogen balance is present.

In addition to nitrogen balance, one must understand protein turnover and net protein balance. Net protein balance involves muscle protein synthesis (MPS) and muscle protein breakdown (MPB). Net protein balance is achieved when MPS = MPB. Positive net protein balance occurs when MPS > MPB, and negative net protein balance is when MPB > MPS (Tang and Phillips, 2009).

Under normal conditions most adults are in nitrogen balance. Their intake of nitrogen is in equilibrium with their nitrogen losses. There is also equilibrium in their protein turnover and so they are in net protein balance. In other words, the amount of protein being synthesized is equal to the amount of protein being degraded.

Protein turnover and nitrogen balance are not the same, but they are related. In a growth state, the body is in positive nitrogen balance and positive net protein balance. Protein synthesis is also outpacing protein degradation. Although the exact mechanisms for regulating protein balance and turnover are not entirely known, the amino acid pool does play a role. Unfortunately, protein turnover and nitrogen balance are difficult to study in humans, which is one reason why the amount of protein recommended for athletes, particularly strength athletes, is controversial.

Those adults who want to be in a growth state must achieve positive nitrogen balance and positive net protein balance. Adult growth states include pregnancy and substantial increases in skeletal muscle mass. Athletes who are trying to increase skeletal muscle size (**hypertrophy**) must be in positive nitrogen balance. To achieve positive nitrogen balance, both energy and protein intake must be sufficient. As athletes train hard and work toward their goal of increasing muscle size, there will be periods of protein imbalance. Resistance training results in muscle protein breakdown at the time of the exercise, but during recovery, rest and food intake stimulates muscle protein synthesis and results in positive net protein balance. More details about increasing skeletal muscle mass can be found in Chapter 11.

Negative nitrogen balance and negative net protein balance are generally not desirable, but they do occur in starvation and semistarvation states. For example, an athlete with an eating disorder, such as anorexia nervosa, would likely be in negative nitrogen balance and negative net protein balance. Athletes who severely restrict calories to lose weight are following a "starvation diet," and negative nitrogen and negative net protein balances may be an unintended consequence. Disease states such as fast-growing cancers and starvation due to famine are other conditions that can result in negative nitrogen balance and negative net protein balance.

Both bodybuilders and football players spend many hours lifting weights during training, but their performance demands are very different.

Because protein is always in a state of flux, the body needs several mechanisms by which it can make amino acids available for protein synthesis or metabolism. The amino acid pool has previously been described (Figure 5.3). In addition to the amino acid pool, the body has a **labile protein reserve**. This reserve of amino acids is found in liver and other organs, known as **visceral tissues**. Labile refers to something that readily or frequently undergoes change. The labile protein reserve allows the body to respond to very short-term changes in protein intake. For example, if on a given day a person fasted and consumed little food and protein, then the body can immediately tap the labile protein reserve and provide amino acids to the amino acid pool. Having the ability to quickly use liver and other visceral tissue proteins as a source of amino acids allows the body to protect the skeletal muscle from being used as an amino acid source for short-term emergencies (Institute of Medicine, 2002).

When faced with semistarvation or starvation, the body adapts by decreasing the rate of protein synthesis and increasing protein degradation. The rate of protein synthesis is decreased overall by at least 30 percent and to a greater degree in skeletal muscle than in other tissues. Earlier it was mentioned that visceral tissues (that is, liver and other organs) protect the skeletal muscle from being used as an amino acid source for short-term emergencies. During prolonged starvation the situation is reversed and the amino acids in skeletal muscle are used to protect the visceral tissues. The reason for this change is that visceral tissue proteins turn over very quickly and these visceral

proteins are critical for the body's survival. Skeletal muscle is sacrificed in long-term and extreme starvation states.

Skeletal muscle protein synthesis is complicated and is influenced by many factors.

The process of protein synthesis occurs by the stimulation of a specific gene within a cell, which then sets into motion a series of complex steps resulting in the assembly of a specific protein. A gene is a section of DNA in a chromosome of a cell that contains specific information to control hereditary characteristics. When stimulated, genes "express" these characteristics, usually through the synthesis of specific proteins. For example, the mechanical stress of force production by skeletal muscle that occurs as a result of strength training stimulates the genes that regulate the synthesis of actin and myosin. The target genes are signaled in a variety of ways beginning with the first step of the process, transcription. In transcription, the code for making each specific

Hypertrophy: An increase in size due to enlargement, not an increase in number; in relation to skeletal muscle, refers to an increase in the size of a muscle due to an increase in the size of individual muscle cells rather than an increase in the total number of muscle cells.

Labile protein reserve: Proteins in the liver and other organs that can be broken down quickly to provide amino acids.

Visceral tissue: Tissue of the major organs, such as the liver.

protein is copied to RNA. The next step is translation, during which RNA passes on the directions for manufacturing the protein to the ribosomes, cell organelles that assemble the amino acids from the amino acid pool into the correct sequence for that specific protein.

A simple analogy for this series of steps is the process of designing and building a house. The architect creates and keeps an original, detailed set of drawings and plans for the house. The architect does not build the house, however; this task is usually the responsibility of a building contractor. The contractor takes a copy of the architectural plans to the construction site and gives specific instructions to the construction workers to build the house as it appears in the plans. The original architectural plan is the gene (the specific DNA sequence in the cell), the "working copy" of the house construction plan is the RNA, and the construction site of cell protein synthesis is the ribosomes.

There are many factors that influence muscle protein synthesis and degradation, including one's genetic potential for cellular protein synthesis. The synthesis of skeletal muscle protein is strongly influenced by exercise, specifically strength training. The mechanical force that is developed by muscle during strength training stimulates both protein synthesis and protein breakdown. Protein synthesis is stimulated to a greater degree because one of the adaptations of skeletal muscle in response to strength training over time is hypertrophy, an increase in the amount of muscle tissue. The greatest increase is in **myofibrillar proteins**, the proteins that make up the force-producing elements of the muscle.

Feeding also promotes an anabolic state in the muscle, particularly if the meal contains adequate amino acids and is consumed immediately or within 2–3 hours after exercise. Therefore, it is important for athletes who want to increase skeletal muscle mass to consume an adequate amount of calories and protein daily and to correctly time their protein intake. Insulin has an important role in this anabolic process by stimulating protein synthesis and acting to inhibit protein degradation. Although genetics, resistance

Spotlight on...

Maximizing Skeletal Muscle Mass

Bodybuilders and other strength athletes focusing on developing maximal amounts of skeletal muscle mass are known to "push the envelope" of current research to find ways to achieve their goals. Some athletes use the following strategies in an effort to maximize gains in skeletal muscle mass, and they are reviewed here for their scientific merit and potential effects, both positive and negative.

Dietary Approaches

Large amounts of protein daily Many "hardcore" or "serious" male bodybuilders suggest that protein intake should be approximately 3.0 to 3.5 g/kg/d. Some of these individuals also suggest or imply the use of anabolic steroids. Based on this suggested range, an athlete weighing 220 pounds (100 kg) would consume between 300 and 350 g of protein daily. A critical question is whether such a high daily protein intake is safe and effective.

Part of normal amino acid metabolism involves the catabolism of amino acids, which takes place in the liver. As part of this process, ammonia is produced. Ammonia is toxic to the body and must be converted to urea for excretion. All humans have a maximum rate of urea synthesis, which is well above that needed to metabolize the amount of protein normally contained in the diet. However, the body cannot increase its maximum

Bodybuilders and other strength athletes may focus on developing maximal amounts of skeletal muscle mass.

SHANNON STAPLETON/Reuters/Corbis

rate of urea synthesis in response to an exceptionally high intake of protein. Therefore, the suggested maximum intake of protein is 2.5 g/kg/d. When protein intakes are above this level, there is a danger that the individual's maximum rate of urea synthesis will be exceeded. This would lead to an elevated blood

exercise, nutrients, and hormones are major factors, there are many other factors that influence protein synthesis including injury and disease.

Maximum versus optimum skeletal muscle protein synthesis. Most athletes engage in some form of strength training to maintain or increase skeletal muscle size and strength. It may be beneficial for novice athletes or those moving to more demanding levels of competition, such as high school to college or college to professional, to substantially increase skeletal muscle size and strength. In one sport, bodybuilding, the goal is to achieve the maximum *amount* of skeletal muscle mass that is genetically possible. However, most athletes are not focused solely on the maximum amount. Their goal is to achieve the optimal amount of muscle mass for their sport. Too much skeletal muscle can result in decreased speed or diminished performance in other ways. Both a bodybuilder and a linebacker on a football team will spend many hours strength training, but the performance demands of their sports are very different (Lambert et al., 2004).

Bodybuilders are not only looking to be in positive nitrogen balance; they also want to be in positive protein balance (that is, protein synthesis outpaces protein degradation). They hope to consume the amount of protein that correlates with *maximum* muscle protein synthesis. Such an amount has not yet been determined through scientific studies, but many bodybuilders (and some other strength athletes) may take large amounts of protein with the hope that it approximates the amount needed for maximum protein synthesis in skeletal muscle. Some of the protein-related techniques these athletes consider in their quest to maximize skeletal muscle mass are explained in the Spotlight on...Maximizing Skeletal Muscle Mass.

Myofibrillar proteins: The proteins that make up the force-producing elements of the muscle

ammonia level, a dangerous medical condition (Bilsborough and Mann, 2006). There is also no scientific evidence that such a high intake is more effective for increasing skeletal muscle size.

30–50 g of protein per feeding Many athletes wonder what is the maximum amount of protein that the body can absorb per feeding. A common belief among strength athletes is 30–50 g. This has resulted in diet plans that include multiple meals or snacks throughout the day, each containing 30–50 g of protein per feeding. The rationale is that providing the body the amount of protein that can be maximally absorbed will result in the maximum synthesis of skeletal muscle proteins.

Evidence suggests that this rationale is incorrect. When amino acids are rapidly absorbed, protein synthesis increases; however, amino acid oxidation also increases. Therefore, the net gain in protein is not greater. In general, the absorption of amino acids from protein is slow, about 5–8 g per hour. Whey protein isolate has the highest rate of absorption, ~8–10 g/h (Bilsborough and Mann, 2006). A combination of rapidly and slowly absorbed proteins at each meal or snack may be the best approach because "slow" proteins inhibit skeletal muscle breakdown whereas "fast" proteins increase protein synthesis. Protein-containing meals throughout the day are often encouraged, but not because of maximal amino acid absorption.

Consumption of leucine before bed and during the sleep cycle Those trying to build and maintain large amounts of muscle mass may consume supplements containing 3–4 g of leucine before and during the sleep cycle. The theory behind this practice is that sleep is a time of fasting and fasting results in protein degradation. It is assumed that skeletal protein breakdown can be minimized by consuming protein before sleep and midway through the sleep cycle (usually by setting an alarm clock and getting up to drink a protein supplement). The amino acid leucine is considered critical because leucine is the only amino acid that by itself can stimulate protein synthesis.

Leucine stimulates muscle protein synthesis and decreases the rate of protein degradation. Animal studies suggest that leucine activates mTOR (mammalian target of rapamycin), an important compound involved in skeletal muscle protein synthesis. Leucine also stimulates a temporary increase in insulin concentration. A small increase in circulating insulin level can also influence mTOR, although not directly (Little and Phillips, 2009). At the present time, there has not been enough research to suggest that leucine-rich feedings before or during the sleep cycle are associated with an increase in protein anabolism or a decrease in protein catabolism.

Adequate protein is necessary for the optimal function of the immune system.

Many different nutrients play a role in the support of the immune system, although protein is vital to optimal function and health. Low protein intake is associated with a compromised immune system, but reduced immunity is more complicated than the lack of a single nutrient. Typically, low protein intake is accompanied by a low caloric intake and a low intake of other nutrients, such as vitamins and minerals. Those who are chronically undernourished are more susceptible to infections. A proper diet provides the nutrients needed for optimal immune system function.

Impact of injury or infection on protein status. Even in an otherwise healthy individual, injury or infection has an effect on protein status. Injury or infection results in an increased loss of nitrogen from the body and negative nitrogen balance, which is intensified if a fever is present. Illness typically results in a lack of appetite, so individuals often consume less food (and less protein) when feeling ill. Because the body is in negative nitrogen balance, it must depend on its stores and reserves for amino acids. These amino acids are provided to a large degree by the breakdown of skeletal muscle tissue. Someone with good nutritional status prior to an injury or infection has an advantage over someone with poor nutritional status, but nitrogen loss and skeletal protein breakdown will still occur (Kurpad, 2006).

Of course, injury or infection should receive appropriate medical treatment, but the athlete should not overlook the importance of restoring protein status as a part of treatment. For bacterial infections, it takes approximately 2–3 times the duration of the infection to replete the protein that has been lost. In other words, if the bacterial infection lasts 3 days then it takes about 6 to 9 days for full restoration of protein status. As a rule of thumb, a protein intake of an additional 0.2–0.3 g/kg/day during the convalescent period would provide the additional protein needed for restoration (Kurpad, 2006). In practical terms, it means an athlete who consumes 1.2 g/kg/day might consume 1.4–1.5 g/kg daily while convalescing. This represents a slightly higher than normal intake of protein, but not an excessively high protein intake.

Effects of exercise on the immune system. One of the many benefits of moderate exercise is its positive effect on the immune system. Epidemiological studies suggest that those who exercise moderately (intensity and/or duration) on a regular basis have fewer infections than those who are sedentary. On the other hand, repeated strenuous exercise is a physiologic stress that may put some athletes at risk for suppression of

immune system functions and at risk for an increased susceptibility to infection and illness. Strenuous exercise is typically defined as 1.5 hours or more of moderate- to high-intensity exercise (55–75 percent VO_2max). Such a level of endurance exercise likely reduces the athlete's immune responses and can result in an increased number of infections such as upper respiratory tract infections (Gleeson, 2007).

A strenuous bout of exercise typically results in the transient or short-term impairment of some immune system cells for 3 to 24 hours after exercise. Thus a single strenuous bout of exercise temporarily suppresses immune function, but it returns to normal by the next day. However, many endurance and ultraendurance athletes rigorously train on a routine basis or have periods during their seasons that include repeated days of strenuous training. In such cases, immune suppression and dysfunction can become a chronic problem.

Prolonged strenuous exercise also results in some positive effects on the immune system, such as a reduction of **systemic inflammation**. Systemic

Key points

- The body is in a constant state of protein turnover.

- Skeletal muscle proteins are synthesized using amino acids from the amino acid pool, a process known as anabolism.

- Conversely, skeletal muscle can be broken down to provide amino acids back to the pool, a process known as catabolism.

- Some proteins are broken down and the nitrogen is excreted, one reason that some protein must be eaten daily.

- Athletes who want to build skeletal muscle tissue must be in positive nitrogen balance and positive net protein balance.

- Restricting energy intake ("dieting") favors breakdown of skeletal muscle mass.

- Proteins are synthesized through the process of gene expression, in which a specific area of a cell's DNA is stimulated to direct organelles to assemble amino acids into specific protein structures.

- Exercise, particularly strength training, stimulates muscle cells to increase protein synthesis.

- Adequate dietary intake of protein is necessary for the proper functioning of the immune system.

Why is the amino acid pool important?

Under what conditions are nitrogen and protein balance positive or negative?

How would you describe to a novice strength athlete the various factors that influence skeletal muscle protein synthesis?

Why is it important for someone who is injured or sick to consume an adequate amount of protein?

inflammation is related to an increase in certain cytokines, signaling proteins that can affect how a cell functions. Some cytokines induce inflammation, which is believed to be a major factor associated with many chronic diseases, such as cardiovascular disease. Strenuous exercise results in an increase in some of the anti-inflammatory cytokines. In this respect, prolonged strenuous exercise is a positive factor for good health (Gleeson, 2007).

The goal for the athlete who rigorously exercises is to try and offset any negative effects of strenuous exercise to the extent possible. Proper nutrition, including adequate protein intake, may help to reduce some of the negative effects. An adequate daily intake of calories and nutrients is important for good health, which allows the athlete to train and compete.

5.4 Protein Recommendations for Athletes

The Dietary Reference Intake (DRI) for adults is 0.8 g of protein/kg body weight daily (Institute of Medicine, 2002), although this amount may be too low to protect against the loss of skeletal muscle mass associated with aging (Elango et al., 2010). Athletes in training need more protein than nonathletes. There is a general consensus on the amount of protein needed daily by strength, endurance, and ultraendurance athletes, although controversies still exist and more research on highly trained athletes is needed. In addition to daily protein intake, athletes in training should also plan the timing of their protein intake, particularly immediately after exercise. An emerging area of research is the type of the protein consumed and its potential role in optimal protein synthesis.

Recommended ranges for protein intake by athletes are good guidelines but should be individualized for each athlete.

The precise amount of protein needed each day by athletes in training is not known. Relatively few studies have been conducted in athletic populations, and one of the traditional measures used in research, nitrogen balance, has known limitations for predicting requirements in well-trained athletes. Therefore, protein recommendations for athletes, especially those who wish to substantially increase skeletal muscle mass, are somewhat controversial.

As summarized in Table 5.3, protein recommendations usually fall between 1.0 and 2.0 grams of protein/kg body weight/day (g/kg/d). These recommendations make two assumptions: (1) total energy intake is

Table 5.3 General Protein Recommendations Based on Activity

Level of activity	Recommended daily protein intake
Sedentary adults	0.8 g/kg
Recreational athletes*	1.0 g/kg
Endurance athletes	1.2 to 1.4 g/kg
Ultraendurance athletes	1.2 to 2.0 g/kg
Strength athletes**	1.2 to 1.7 g/kg 1.5 to 2.0 g/kg

Legend: g/kg = grams per kilogram body weight
*Low to moderate training volume and intensity
**Consensus has not yet been reached among experts.

adequate, and (2) the quality of the protein is good. In the United States and other developed countries, protein quality is usually not a concern. However, it cannot be assumed that energy intake is adequate for all athletes.

Although 1.0 to 2.0 g/kg/d is the general protein guideline, more specific recommendations are made depending on the predominant type of training (that is, strength, endurance, or ultraendurance) as well the volume of training (for example, recreational versus elite athletes). Recreational athletes, who typically train at low to moderate intensities several times a week, may need only slightly more protein than nonathletes—1.0 g/kg/d. It is recommended that endurance athletes consume 1.2 to 1.4 g/kg of protein daily. Ultraendurance athletes may need even more protein (up to 2.0 g/kg/d) during certain times in their training cycles (Seebohar, 2011). Rigorous training, such as running the equivalent of 80 miles or more a week, results in an increase in amino acids being burned as fuel, so some additional protein is needed to replenish the protein lost through oxidation.

The American Dietetic Association, Dietitians of Canada, and the American College of Sports Medicine's (Rodriguez et al., 2009) recommended range for strength athletes is 1.2 to 1.7 g/kg of protein daily with the higher end of the range recommended for those who are trying to increase skeletal muscle mass. Some protein researchers suggest that strength athletes may need 1.5 to 2.0 g/kg/d (Kreider and Campbell, 2009). Their reasoning is that a higher protein intake is needed to maintain positive nitrogen balance and promote skeletal muscle synthesis (in the presence of resistance training and sufficient caloric intake). One counterargument is that one

Systemic inflammation: A chronic inflammation in many parts of the body, which may be a factor in some chronic diseases.

Establishing Dietary Protein Recommendations for Endurance and Strength Athletes

For years controversy and conflicting recommendations have existed for the amount of protein intake that would be both effective and safe for athletes. For a long time it was generally recommended that an athlete's protein intake remain near that of a sedentary adult—0.8 g/kg/day. Many nutrition professionals felt that was enough protein, even for athletes, and were concerned that it was potentially injurious to the kidneys to consume large amounts of excess protein for long periods of time. However, a belief that large amounts of protein are necessary for building and maintaining muscle resulted in the widespread promotion and practice of high-protein diets, often greater than 2.5 g/kg/day, particularly by bodybuilders and strength athletes. Little attention was given to the protein needs of endurance athletes as protein intake was mostly seen as a muscle-building issue, although prolonged endurance exercise increases protein breakdown. Clearly, more research was needed to answer the question: how much protein do athletes need?

This research study provided results that brought significant clarity to the issue of protein needs for both strength and endurance athletes: Tarnopolsky, M. A., MacDougall, J. D., & Atkinson, S. A. (1988). Influence of protein intake and training status on nitrogen balance and lean body mass. *Journal of Applied Physiology, 64*(1): 187–193.

What did the researchers do? These researchers designed their study to determine protein needs for bodybuilders and endurance athletes by carefully analyzing their nitrogen balance when consuming diets containing two different amounts of protein. As discussed on pages 166–169, nitrogen balance is a comparison of the amount of nitrogen (protein) consumed to the amount of nitrogen excreted (urine, feces, sweat) and gives an indication of the body's state of growth, maintenance, or breakdown of protein. An athlete in negative nitrogen balance would be suspected of not consuming enough protein, whereas positive nitrogen balance would indicate a sufficient level of protein intake to support the athlete's training.

The researchers tested three groups of subjects: sedentary control subjects, experienced bodybuilders who had been training for at least 3 years, and endurance athletes (runners or Nordic skiers) who had been training for at least 5 years. Each group had 6 subjects who completed two segments of the study: one in which they consumed a diet that was representative of a typical diet for their group and another in which the amount of protein was changed—increased for the sedentary control subjects and endurance athletes and decreased for the bodybuilders. The endurance athletes consumed 1.7 g/kg/day of protein in the first study segment, which then increased to 2.7 g/kg/day in the second segment, whereas bodybuilders began with 2.7 g/kg/day and reduced their protein intake to

1.0 g/kg/day in the second part of the study (see figure, next page). The bodybuilders consumed a larger amount of protein in the first segment because their "typical" daily diet was much higher in protein than was usual for the subjects in the other two groups. Note that these amounts of daily protein intake are higher than the amount recommended for sedentary adults, which is 0.8 g/kg/day.

In order to ensure good dietary control in the study, each diet segment was designed to have a 10-day adaptation period, which was followed immediately by a 3-day measurement period. During the 3-day measurement period, diet was strictly controlled and measurements of nitrogen balance were made by precisely determining the amount of nitrogen consumed in food and excreted from the body in urine, feces, and sweat. Although the amount of protein in the diet for the subjects was changed, the total amount of kilocalories was kept the same to make sure that only one variable—the amount of protein—was changed.

What did the researchers find? First, the results of the study indicated that bodybuilders do not need to consume very large amounts of protein. When the bodybuilder subjects consumed their "typical" diets (2.7 g/kg/day protein), they were very positive in their nitrogen balance, by 13.4 grams per day. When their protein intake was reduced approximately 64 percent, to 1.0 g/kg/day, nitrogen balance fell, but remained positive on average for the group at 1.1 g/kg/day. This indicates that 1.0 grams of protein per kilogram body weight per day is probably sufficient for many bodybuilders. However, two of the subjects were in negative nitrogen balance at this level of protein intake so it was not sufficient for all the bodybuilders.

The protein needs of endurance athletes had not been studied extensively in the past, and the results of this study showed that when these subjects consumed 1.7 g/kg/day of protein, they were in positive nitrogen balance by 1.8 grams per day. When their protein intake was increased approximately 60 percent, nitrogen balance became more positive, up to 7.1 grams per day (see figure). These results would indicate that although endurance athletes may not need large amounts of protein in their diets, an amount higher than the 0.8 g/kg/day would seem to be necessary to remain in positive nitrogen balance.

Tarnopolsky, MacDougall, and Atkinson then further analyzed their results to determine a minimum amount of dietary protein that would achieve a perfect balance between nitrogen (protein) intake and excretion—a nitrogen balance of zero. They did this by plotting the protein intake of the subjects and their resulting nitrogen balance on a graph and extrapolating or extending the line to the point where it intersected with a nitrogen balance of zero. Based upon this analysis, they concluded that a minimum protein intake of 0.82 g/kg/day by bodybuilders and

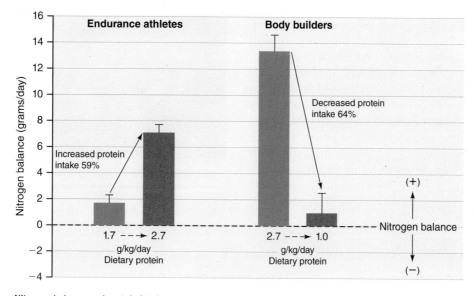

Nitrogen balance and protein intake

Endurance athletes consumed 1.7 g/kg/day of protein and were in positive nitrogen balance by 1.8 grams per day. When their protein intake was increased 59 percent, nitrogen balance became even more positive, up to 7.1 grams per day. When bodybuilders consumed their "typical" diets (2.7 g/kg/day protein), they were very positive in their nitrogen balance, by 13.4 grams per day. When their protein intake was reduced 64 percent to 1.0 g/kg/day, nitrogen balance fell, but remained positive on average for the group at 1.1 g/kg/day.

Drawn from data from: Tarnopolsky, M. A., MacDougall, J. D., & Atkinson, S. A. (1988)

1.37 g/kg/day for endurance athletes would be required for a zero nitrogen balance. Because of individual differences, inherent measurement error in the nitrogen balance method, and the desire to have a reasonable safety margin built in to the recommended amount of protein intake, the amounts suggested are slightly higher than these estimated minimum levels. By adding an amount equal to 1 standard deviation of the calculated minimum protein intake, the authors arrived at a recommended amount of protein intake for bodybuilders of 1.2 g/kg/day and for endurance athletes of 1.6 g/kg/day.

What was the significance of this research study? This research study made a significant contribution to our understanding of the protein intake requirements of athletes because the researchers considered the protein needs of endurance athletes as well as strength athletes, carefully controlled diet and exercise during the measurement period, and precisely measured both protein/nitrogen intake and all avenues of nitrogen loss/excretion, including sweat. They showed that the dietary protein requirements of athletes in training exceed the general recommendation for healthy young men. This study further demonstrated that although strength athletes such as bodybuilders do

need an increased protein intake over that of sedentary people, they do not need the very large amounts traditionally recommended and consumed by these groups. One of the more surprising and significant results of this study was the increased need for protein intake for endurance athletes, exceeding even that needed by strength athletes.

Answering the question: How much protein do athletes need? The 1988 Tarnopolsky, MacDougall, and Atkinson study was very important, but more research was needed. As other study results were published that confirmed the findings of Tarnopolsky, MacDougall, and Atkinson, it became clear that athletes in training needed more protein than sedentary individuals. Today, the American Dietetic Association, Dietitians of Canada, and the American College of Sports Medicine's (Rodriguez et al., 2009) recommended range for strength athletes is 1.2 to 1.7 g/kg of protein daily with the higher end of the range recommended for those who are trying to increase skeletal muscle mass. It is recommended that endurance athletes consume 1.2 to 1.4 g/kg of protein each day. Over the years, research has not shown that such intakes of protein are harmful to the kidneys, so these recommendations are considered both safe and effective.

of the adaptations to resistance training is nitrogen retention, which reduces the amount of protein needed daily to ~1.7 g/kg. More research is needed to fine-tune daily protein recommendations, particularly for strength athletes who are in an active phase of building skeletal muscle.

Although these ranges are good guidelines, recommendations must be individualized. Similar to recommendations for carbohydrate and fat consumption, protein intake may need to change as the training cycles change. For example, a female middle distance runner may typically consume ~1.2 g/kg on a daily basis, but may increase intake to 1.5 g/kg/d when training volume and intensity increases in preparation for her competitive season. Many athletes engage in both endurance and strength training to various degrees over the course of a year's training, and increasing

Juanmonino/iStockphoto

Vegans do not consume any products derived from animals.

Spotlight on…

Protein Intake Expressed as a Percentage of Total Calories Can Be Deceiving

Nutrient recommendations may be expressed on a gram per kilogram of body weight basis or as a percentage of total calories. In the case of protein, the recommendation for adults is 0.8 g/kg or 10–35 percent of total calorie (energy) intake. The typical American does not know his or her weight in kilograms, so a recommendation to consume 0.8 g of protein per kilogram body weight is not seen as being very practical.

This same practice has also been applied to athletes, in part because dietary analysis programs often calculate protein intake as a percentage of total energy intake. Protein recommendations for athletes are sometimes stated as 10–15 percent of total calories for endurance athletes, 15–20 percent for strength athletes, and 20–30 percent for strength athletes focusing on substantially increasing muscle mass. However, this approach is problematic for three reasons: (1) athletes may not be consuming an adequate amount of calories, (2) protein requirements increase when energy intake is deficient, and (3) the percentages of carbohydrate, protein, and fat needed will typically change as training volume increases.

The first problem with using this method is that percentages get distorted when caloric intake is too low. The example here is based on a 50 kg (110 lb) female endurance athlete. In both cases, protein represents 15 percent of total calories. However, when caloric intake is inadequate, 15 percent is not an adequate quantity of protein (in this example, 45 g). When protein is calculated on a g/kg basis the comparison is more telling (1.5 versus 0.9 g/kg). In this example, 15 percent of total calories from protein could be adequate, but ONLY if caloric intake is adequate.

A second problem is that protein requirements increase when energy intake is deficient. A daily diet containing 1.5 g of protein per kg of body weight may help athletes who are restricting calories to protect against loss of skeletal muscle mass. Continuing with this example, let's assume that this

	Consuming adequate calories	Consuming inadequate calories
Energy (kcal)	2,000	1,200
Protein (g)	75	45
% total calories from protein	15	15
g/kg protein	1.5	0.9

50 kg (110 lb) athlete reduces her usual intake from 2,000 kcal daily to 1,700 kcal daily in an effort to reduce body fat. To protect against the loss of skeletal muscle tissue, she would need to reduce her caloric intake but keep her daily protein intake the same. (This would likely be accomplished by restricting fat intake, although it should not be restricted too much.) As the chart below shows, there is a substantial difference between the g/kg/day for these two diets (1.1 versus 1.5 g/kg). However,

	Consuming adequate calories	Reduction of both calories and protein	Reduction of calories without restricting usual protein intake
Energy (kcal)	2,000	1,700	1,700
Protein (g)	75	55	75
% total calories from protein	15	13	~18
g/kg protein	1.5	1.1	1.5

or decreasing the amount of protein in their diets to match their training goals is appropriate.

Some experts suggest that a rule of thumb for maximum protein intake is 2.5 g/kg/d (Bilsborough and Mann, 2006). This level is based on the body's ability to metabolize the ammonia that is produced as a result of amino acid catabolism. Ammonia is toxic to the body and must be converted to urea for excretion. Protein intakes above 2.5 g/kg/d may put the individual at risk for elevated blood ammonia because the maximum rate of urea synthesis and excretion could be exceeded with an exceptionally high intake of protein.

The need for protein is influenced by energy intake. As a rule of thumb, when energy intake is deficient (for example, the athlete is "dieting"), protein intake should be ~1.5 g/kg/d to prevent the loss of skeletal muscle mass to the extent possible. Low protein intake when coupled with low energy intake over months of time will eventually reduce the body's "stores" of protein (that is, skeletal muscle mass), and training and performance will be negatively affected. Ultimately, the individual will be unable to engage in physical activity, but such cases in athletes are typically due to an eating disorder.

All of the protein recommendations for athletes mentioned so far have been expressed on a gram per kilogram body weight basis (g/kg). In other words, protein recommendations are stated as an absolute amount. It is considered inappropriate in the area of sports nutrition to express protein requirements as a percentage of energy intake (Mettler and Meyer, 2010). The Spotlight on...Protein Intake Expressed as a Percentage of Total Calories Can Be Deceiving discusses this issue in more depth.

if only percent of total calories was considered, the diet containing 13 percent protein would appear to be adequate.

Perhaps the best reason for not expressing carbohydrate, protein, or fat as a percentage of total calories is related to training volume. As shown in the charts below, as training volume increases, the need for protein on a g/kg basis increases for carbohydrate, protein, and fat but the percentage of energy intake from protein and fat actually decreases (Mettler and Meyer, 2010). It is for all these reasons that nutrient recommendations for athletes are expressed as g/kg and not as a percentage of total calories.

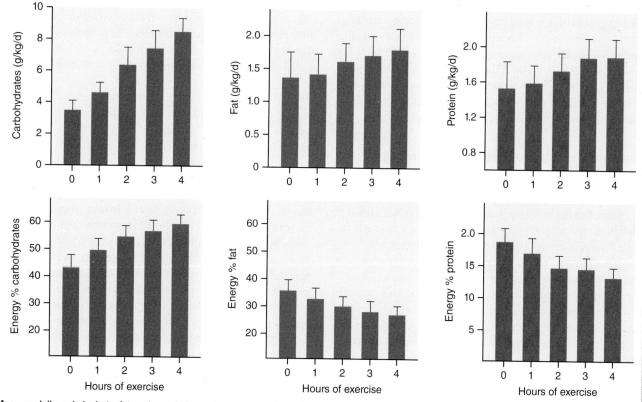

Average daily carbohydrate, fat, and protein intake based on training volume compared

As hours of exercise increases, the g/kg of protein increases but the percentage of energy intake from protein decreases. It is important to make recommendations to athletes on a g/kg basis.

Data from: Mettler, S., & Meyer, N.L., (2010). Food pyramids in sports nutrition. *SCAN'S PULSE, 29*(1), 12–18.

Recommended protein intake for vegetarian athletes. Vegetarians do not consume the flesh of animals, such as meat, fish or poultry, but may consume animal products, such as eggs, milk, or cheese. **Vegans** do not consume any product derived from animals. Due to these differences in food intake, it is advisable to distinguish vegans from vegetarians and nonvegetarians.

There have been no studies examining the amount of protein required by vegetarian or vegan athletes. A common recommendation for physically active vegetarians or vegans is to consume 10 percent more protein than the amount recommended for nonvegetarian athletes. This 10 percent figure is an adjustment for the lower digestibility of plant proteins when compared to animal proteins. Consuming this amount of protein is not difficult if energy intake is sufficient.

Vegetarians tend to consume less protein than nonvegetarians, and vegans tend to have lower protein intakes than nonvegans. Thus it is important for vegetarian and vegan athletes to make sure that their protein intakes are adequate in both quantity and quality. Vegetarians who consume animal products, such as eggs, milk or cheese, may consume an amount and quality of protein that is similar to a nonvegetarian diet.

Vegans should be aware that the greater restriction of protein foods could result in a lower quantity and quality of protein in their diets. It is recommended that vegan athletes emphasize more protein-rich vegetarian foods, such as beans, legumes, nuts, soy milk, and tofu, than vegetarian foods that are relatively low in protein, such as fruits and vegetables (Larson-Meyer, in press). Both vegan and vegetarians diets can support optimal performance if the diet contains the proper quantity and quality of protein.

Timing of protein intake is important, especially after exercise.

Obtaining an adequate amount of protein each day is fundamentally important, but athletes should not overlook the importance of the timing of protein intake throughout the day, especially after exercise.

The acute effect of exercise is to put the body into a catabolic state, breaking down certain tissues to provide the energy to sustain the exercise. For example, muscle glycogen is broken down to provide glucose, stored fats are broken down to mobilize fatty acids, and skeletal muscle proteins are broken down to provide amino acids that can be used as energy (for example, leucine). However, the catabolic state is followed by an anabolic state, an environment that allows for recovery from the acute effects of exercise and for skeletal muscle growth.

The first 1 to 2 hours after exercise is sometimes referred to as the "anabolic window." This is the time period when it is important to reverse the catabolic state and promote a hormonal and nutritional state

Inspirestock/Jupiter Images

Consuming protein after exercise is effective for support of skeletal muscle growth.

that favors replacement of energy stores and the synthesis of protein rather than its breakdown. As discussed in Chapter 4, the hours immediately after exercise are the most important for initiating optimal glycogen replacement. A similar "window" exists for protein to initiate skeletal muscle protein synthesis. Taking advantage of the favorable postexercise anabolic environment requires proper nutritional intake.

Protein consumption after resistance exercise. There is strong scientific evidence that postexercise protein intake is effective for support of skeletal muscle growth after resistance exercise (Tipton et al., 2004). However, protein intake immediately after exercise does not appear to be effective for increasing strength or power or changing body composition (Hoffman et al., 2009). Foods or supplements containing **intact proteins**, such as whey, casein, or soy, stimulate an increase in skeletal muscle mass. Particularly effective is the combination of whey and casein, which naturally occurs in milk but is also found in some protein supplements. Specific strategies for postexercise protein intake are explained below.

Timing: Immediately after exercise, when possible, but no later than 2 to 3 hours after exercise.

Although the optimal timing is not known, a reasonable time frame for the consumption of protein after resistance exercise is from immediately afterward to ~2 hours after exercise. The "anabolic window" may extend to 3 hours, but more research is needed (Kerksick et al., 2008).

Amount and quality: 10–20 g of high-quality protein.

Research suggests that 10–20 g of high-quality protein is an appropriate amount of protein to consume after exercise. The amount associated with maximum protein synthesis is 20 g. Taking more protein than this results in the amino acids being oxidized (burned for energy) because the skeletal muscle cannot utilize the excess protein (Moore et al., 2009). This is likely related to the saturation of the amino acid pool.

The protein consumed should contain a large amount of indispensable amino acids, one measure of protein quality. High-quality proteins include egg white, whey, casein, and soy. Although these proteins contain about the same amount of indispensable amino acids, they are digested and absorbed differently. Whey and soy are rapidly digested so blood amino acid concentration increases quickly but stays elevated for a relatively short period of time. Such proteins have become known as "fast acting" proteins. In contrast, ingestion of casein results in a slower but more sustained rise in blood amino acids. Casein is known as a "slow acting" protein.

Source of protein: Certain protein foods or properly formulated protein supplements

Both milk and soy provide about the same amount of indispensable amino acids, thus both will promote skeletal muscle synthesis. However, soy protein is a fast-acting protein whereas milk is a combination of fast-acting (whey) and slow-acting (casein) proteins. The ingestion of milk after resistance exercise appears to promote skeletal muscle growth to a greater extent in the short term than soy does. The rapid rise in blood amino acids that is seen with both soy and milk proteins promotes skeletal muscle growth. However, the slower, sustained rise due to the casein in milk also helps suppress skeletal muscle protein breakdown. Although either soy or milk intake after exercise is likely to be effective for promoting increased skeletal muscle mass over time, milk may accomplish this goal faster than soy (Wilkinson et al., 2007). Soy is an important source of postexercise protein for athletes who do not consume animal products.

Carbohydrate-protein consumption after exercise. As a practical matter, many athletes choose a postexercise food or drink that contains both carbohydrate and protein. As described in Chapter 4, carbohydrates, especially high-glycemic carbohydrates, consumed immediately after exercise are beneficial because they help restore muscle glycogen. The carbohydrates also stimulate the release of insulin. Although its primary role is cellular glucose uptake, insulin also increases amino acid uptake into muscle and inhibits the process of muscle degradation. The important point is to provide the body with the nutrients it needs immediately after exercise to begin resynthesis of tissue that has been catabolized during exercise. In this respect, consumption of carbohydrate and protein (both of which also provide energy) shortly after exercise ends is advantageous. Some popular choices for athletes to consume after exercise include low-fat chocolate milk, fruit-in-the-bottom yogurt, and sports beverages that have been formulated for consumption during recovery.

Though most of the postexercise protein consumption studies have been conducted in resistance-trained athletes, there are a number of studies conducted with endurance athletes. Such studies have shown no effect of the addition of protein on recovery from exercise-induced muscle injury or on next-day performance when compared to the consumption of carbohydrate alone (Green et al., 2008; Rowlands et al., 2007). The consumption of a food or beverage with both carbohydrate and protein is beneficial for restoring muscle glycogen, promoting protein synthesis, and providing energy (calories), but other benefits for endurance athletes have not been documented.

Application exercise

Scenario: *A competitive cyclist who is training for an upcoming multiday stage race feels he doesn't recover from long, hard training rides as well as his competitors. He is a male cyclist weighing 159 pounds (72 kg) who completes 2–3 hour training rides 4–5 times a week at a moderate to high intensity. His typical food intake after training rides is 12 oz (360 ml) water and two bananas.*

1. How much protein and carbohydrate does he consume now after training rides?
   - Determine how much protein and carbohydrate are in the water and bananas.

2. How much protein and carbohydrate should he be consuming right after training rides?
   - Determine the minimum amount of carbohydrate this athlete should consume in the 1–2 hours after his training rides.
   - Determine the amount of protein this athlete should consume in the 1–2 hours after his training rides.

3. What would be a good solution to his recovery problem?
   - Recommend a specific plan for consuming a proper amount of carbohydrate and protein—specific foods, amounts, and timing—for each of the following:
     a. A meal containing food and beverage
     b. A liquid-only meal
     c. A commercially available sports beverage/supplement product

---

Intact protein: A protein that has not been broken down (by a food processor) prior to ingestion.

Carbohydrate-protein consumption during exercise. In theory, the consumption of protein (along with carbohydrate) during exercise could help to offset some of the damage from rigorous resistance or endurance exercise. It would be advantageous for the athlete if the carbohydrate-protein beverage or food consumed during exercise could help to maximize the body's anabolic response to exercise. Studies to date have shown that carbohydrate and protein, when consumed together, may decrease muscle damage to some degree (Valentine et al., 2008; Saunders, Luden, and Herrick, 2007; Kerksick et al., 2008). However, not all studies have shown carbohydrate-protein beverages to be effective (Osterberg, Zachwieja, and Smith, 2008) so more research is needed.

Based on research studies that have shown an effect, the athlete would need to consume a beverage or gel that contains a carbohydrate-to-protein ratio of 3:1 or 4:1. For example, an athlete who consumes 60 grams of carbohydrate per hour would also consume 15–20 grams of protein per hour. Many gels and beverages sold for consumption during exercise are formulated using these ratios. Intake of protein along with carbohydrate has not been shown to improve endurance exercise performance over the consumption of carbohydrate alone (van Essen and Gibala, 2006).

Protein consumption before exercise. At the present time there is insufficient evidence to suggest that protein consumption prior to exercise is beneficial to performance or recovery. However, there is no evidence that such a practice is harmful. Some athletes may take in protein as part of their pre-event meal or snack, such as eating a turkey sandwich before their afternoon run, and this is not likely to hurt performance. Other athletes consume a food or beverage that contains protein in a deliberate attempt to prevent protein breakdown during exercise or stimulate protein synthesis after exercise. At the present time, there is a lack of evidence to suggest that manipulating protein intake in this way will positively affect performance or confer other advantages.

Most athletes consume a sufficient amount of protein, but some consume a low or excessive amount.

Table 5.4 summarizes the average daily protein and energy intakes of various athletes. The majority of athletes consume sufficient protein. However, some athletes consume protein that is considered excessively low or high. For example, Mullins and colleagues (2001) found that the average protein intake of the 19 female heptathletes studied was 1.4 g/kg or 95 g/d. However, protein intake ranged from a low of 0.7 g/kg (53 g/d) to a high of 3.1 g/kg (186 g/d). Total energy intake also varied considerably from 1,553 kcal to 5,276 kcal.

The majority of athletes consume sufficient protein.

Athletes with protein intakes at the extremes are at the greatest risk. Those who are at risk for low protein intake include those who restrict energy in an effort to keep body weight or body fat low, such as jockeys, ski jumpers, and females in sports where body appearance is part of the scoring. On the other end of the spectrum, some strength athletes consume large amounts of protein.

There are some practical problems associated with consuming an excessive amount of protein.

In the past, caution was raised regarding high-protein diets (typically defined as greater than 2.0 g/kg/d), especially for those athletes who took protein supplements. It was thought that excessive protein would stress healthy kidneys and liver in the short term. This does not seem to be the case for those with normal kidney function. A seven-day study of bodybuilders and other well-trained athletes detected no short-term harmful effects on renal function with protein intakes up to 2.8 g/kg/d (Poortmans and Dellalieux, 2000). All of the athletes in this study were healthy and had normal kidney and liver function. Athletes should be aware that they could experience problems with high-protein diets if they have latent (hidden) or known kidney or liver conditions. Athletes consuming large amounts of protein should monitor their health and contact a physician if problems appear.

Concern has also been expressed about the potential effects of a high-protein diet on bone health because high-protein diets increase urinary excretion of calcium. However, protein also improves calcium absorption, which appears to compensate for the increase in calcium lost in the urine. Though research

Table 5.4 Average Daily Protein and Energy Intakes of Athletes

Study	Sport and level	Age (range or average) and nationality	Subject characteristics	Average daily protein intake		Average daily energy intake
				g/kg	g	kcal
Havemann and Goedecke, 2008	Well-trained endurance cyclists	39 years	45 males	1.6	117	2,986
Aerenhouts et al., 2008	Trained indoor and outdoor track sprinters	15–17 years Flemish	29 girls	1.5	78	2,581
			31 boys	1.5	92	3,117
Soric, Misigoj-Durakovic, and Pedisic, 2008	National level aesthetic athletes	9–13 years Croatian	39 girls			
			Artistic gymnasts	2.3	68	1,941
			Rhythmic gymnasts	1.8	66	1,647
			Ballet dancers	2.1	69	1,731
Jonnalagadda, Ziegler, and Nelson, 2004	Elite figure skaters	19 years American	23 males	1.2	82	2,112
		15.5 years American	26 females	1.16	54	1,490
Nogueira and Da Costa, 2004	Well-trained triathletes	18–54 Brazilian	29 males	2.0	142	3,660
			9 females	1.6	88	2,300
Onywera et al., 2004	Elite runners (cross country and 1,500 m)	18–24 Kenyan	10 males	1.3	72	2,987
Leydon and Wall, 2002	Jockeys	23.5 years New Zealander	6 males	1.1	58	1,514
		24.5 years New Zealander	14 females	0.95	47	1,479
Poortmans and Dellalieux, 2000	Competitive bodybuilders	29 years Belgian	20 males	1.94	169	3,908
	Trained cyclists, judoka, and rowers	28 years Belgian	17 males	1.35	99	2,607

Legend: g/kg = grams per kilogram body weight; g = gram; kcal = kilocalorie

continues in this area, short-term high-protein diets do not appear to be detrimental to bone health. In fact, there are a number of epidemiological studies that suggest a protein intake of ~2.0 g/kg/day is associated with a high, not a low, bone mineral density (Kerstetter, 2009; Hunt, Johnson, and Fariba Roughead, 2009). High-protein diets appear to be safe over the short term, but more research is needed to clarify the long-term health effects (Bernstein et al., 2007).

From a practical perspective, athletes should be aware of potential training and performance problems that can be caused by a high intake of protein. The practical concerns are dehydration, low carbohydrate intake, and excessive caloric intake. Protein supplements make it easy to consume a lot of protein, but all of these concerns could also be associated with an excess of dietary protein.

Large amounts of protein can result in dehydration because additional water is needed to metabolize protein. Athletes who consume large amounts of protein should be aware of the need for adequate fluid intake. Urea, one of the by-products of protein metabolism, must be eliminated from the body by the kidneys via the urine. Urea is an osmotically active compound that draws more water into the collecting tubules of the kidneys, increasing the volume of the urine and resulting in the loss of water from the body. Water and fluid balance will be discussed further in Chapter 7.

Consuming large amounts of protein may come at the expense of the inclusion of enough carbohydrate foods. If protein *replaces* adequate carbohydrate intake, then the athlete will be consuming a high-protein, low-carbohydrate diet. This could result in lower muscle glycogen stores after several days of demanding training. If protein intake consistently *exceeds* usual intake, then caloric intake may be too high and body fat may increase over time. Maintaining macronutrient and energy balance is important.

Key points

- It is recommended that athletes consume 1.0 to 2.0 grams of protein/kg body weight/day, depending upon level and type of activity.

- The amount of protein intake should be determined on a gram per kilogram of body weight basis.

- Vegetarian and vegan athletes need to include high-quality protein in their diets and may want to increase protein intake approximately 10 percent over the general recommendation for athletes to account for the lower digestibility of plant proteins.

- Protein intake within 2–3 hours after resistance exercise is an important strategy for supporting skeletal muscle protein synthesis.

- Consumption of protein after endurance exercise may help slow down or reverse the process of muscle degradation and may help recovery, particularly when combined with carbohydrates.

- The consumption of proteins before or during endurance exercise has not been shown to improve performance over the consumption of carbohydrate alone.

- Short-term protein intake in excess of the recommended amount for athletes (>2.0 g/kg/day) does not appear to be detrimental, but is not associated with increased synthesis of muscle mass.

Why do endurance athletes have a need for protein that is higher than the DRI for adults?

What is the rationale for including protein (amino acids) in a sports drink intended to be consumed during exercise?

How is it possible for an athlete to consume 25 percent of his or her diet as protein, yet be deficient in protein intake?

The Internet café

Where do I find reliable information about protein, exercise, and health?

Athletes who search for information about protein on the Internet will often be directed to commercial sites that are selling protein supplements or high-protein weight loss diets. Unbiased information about protein needs for athletes is lacking. Some university or medical-related sites give general information about protein, such as that found at Harvard School of Public Health (**http://www.hsph.harvard.edu/nutrition-source/index.html**).

More information about vegetarianism can be found at the Vegetarian Resource Group, **http://www.vrg.org**, a non-profit organization dedicated to educating the public about vegetarianism.

5.5 Effect of Energy Intake on Protein Intake

The amount of protein required is related to energy intake. Under normal conditions, an adequate energy intake from either carbohydrate or fat spares amino acids from being used for energy and helps maintain nitrogen balance. Athletes need to be in nitrogen balance to maintain muscle mass and need to be in positive nitrogen balance to increase muscle mass.

Adjustments to protein intake will need to be made when energy intake is deficient. The amount of protein needed depends on the magnitude of the energy deficit and whether it is acute or chronic. The most serious situations are those athletes who self-impose starvation over several months or years. Some athletes maintain small chronic energy deficits, and they may need to slightly increase their protein intake as a compensatory measure.

Long-term, substantial energy deficits typically result in low protein intake.

When a chronic energy deficit is present, more protein is needed than when energy intake is sufficient. Nitrogen balance cannot be maintained if energy and protein intakes are too low. Studies of athletes with anorexia nervosa report low intakes of both energy and protein. In one study, the mean daily protein intake of athletes with anorexia nervosa was 0.7 g/kg body weight (range 0.5 to 1.0 g/kg/d) (Sundgot-Borgen, 1993). Athletes with eating disorders are struggling with psychological issues that interfere with the consumption of food and will need intense counseling from well-trained practitioners. Among the nutritional goals will be appropriate energy and protein intakes.

A study of female athletes with subclinical eating disorders (that is, the presence of some but not all the features of an eating disorder) reported that the mean daily energy intake was 1,989 kcal, or approximately 500 kcal/d less than estimated energy expenditure. Some athletes were consuming fewer than 1,700 kcal daily. Mean daily protein intake was 1.2 g/kg (Beals and Manore, 1998, 2000). These athletes could benefit from nutritional counseling that addresses disordered eating behaviors, determines appropriate energy intake, and evaluates protein intake in light of chronic energy restriction (see Chapter 13).

Long-term, small energy deficits are characteristic of a pattern of eating for some athletes.

Distance runners and female gymnasts and figure skaters are examples of athletes who commonly have small, but long-term energy deficits. They differ from those in their sports who have eating disorders

Some athletes have eating patterns that result in small daily energy (kcal) deficits that occur over long periods of time.

because caloric intake is restricted to a lesser degree, body image is accurate, and they do not have an excessive fear of weight gain. They do, however, want a body that is well matched for their sport. In the case of runners, they desire a lightweight body with a sufficient amount of skeletal muscle mass, and in the case of gymnasts and figure skaters, a strong but aesthetically pleasing body. These athletes try to achieve this by maintaining a low percentage of body fat without losing skeletal muscle mass.

One survey of female endurance and aesthetic sport athletes who did not have eating disorders reported an average daily consumption of approximately 2,400 kcal or about 100 kcal per day less than their estimated energy expenditure. Their average protein intake was approximately 1.5 g/kg (Beals and Manore, 1998). In other words, these athletes tended to slightly undereat and consume slightly more protein than that recommended for endurance athletes (that is, 1.2 to 1.4 g/kg/d). Such a diet could be nutritionally sound as long as carbohydrate intake is sufficient and a variety of nutrient-dense foods are consumed.

Intermediate-term, small-to-medium energy deficits ("dieting") may lead to loss of lean body mass.

Low-calorie diets can lead to a significant loss of lean body mass. There is much evidence that high-protein, low-energy diets can achieve positive nitrogen balance in rats. The body of literature in humans is smaller and not as well defined (Millward, 2001, 2004). Nevertheless, it is also recommended that when humans are energy deficient they should increase protein intake in an effort to maintain nitrogen balance.

Several studies in overweight and obese women have shown that the loss of lean body mass can be reduced (but not prevented entirely) with higher-protein diets. Evidence suggests that weight loss diets for the general (sedentary) population should be reduced in calories and total energy intake should be distributed as 35–50 percent carbohydrate, 25–30 percent protein, and 25–35 percent fat (Schoeller and Buchholz, 2005). However, subjects of these studies were not athletes, and substantially reducing carbohydrate intake will likely be detrimental to an athlete's training.

Because protection of lean body mass is critical to most athletes, a high-protein, low-calorie diet is often recommended to athletes who desire to lose weight primarily as body fat. They may alter the macronutrient balance of their current diets by slightly increasing protein intake, reducing carbohydrate and fat intakes, and eliminating alcohol. Although this recommendation may be prudent, scientific studies of effectiveness in athletes are lacking. The major concern about high-protein, calorie-reduced diets is that the carbohydrate content may not be sufficient to restore muscle glycogen and support training and performance. Research studies using trained athletes as subjects are needed to determine whether caloric-restricted, high-protein diets adversely affect long-term training and performance.

Short-term, substantial-energy deficits are used to "make weight," but such diets can have detrimental effects.

Many athletes periodically want to lose body fat quickly and will substantially reduce caloric intake for a short time (often a few weeks). When energy deficits are large and short term, such as when an athlete is "making weight," the goals are to lose body weight through fat and water loss but maintain as much skeletal muscle protein as possible. This is difficult to accomplish. Under semistarvation conditions, the loss of weight comes from several components, including water, glycogen, protein, and fat. In the first 10 days of fasting/starvation, only about one-third of the body weight lost is typically lost as body fat (Brownell, Steen, and Wilmore, 1987), whereas 6–16 percent is lost from protein stores. This type of severe energy restriction raises serious concerns about hydration status, the potential for heat illness, maintenance of skeletal muscle mass, ability to exercise due to depleted glycogen stores, hypoglycemia, and declines in resting metabolic rate.

Researchers are beginning to study the use of high-protein, low-calorie diets in athletes as a way to protect against the loss of lean body mass to the extent possible. Mettler, Mitchell, and Tipton (2010) found that in a group of resistance-trained athletes, a high-protein (~2.3 g/kg) weight loss diet for two weeks reduced the loss of lean body mass when compared to the control group (~1.0 g/kg). Both groups consumed 60 percent less than their usual caloric intake for 14 days and lost the same amount of fat mass (a little more than 1 kg or ~2.5 lb). However, the high-protein diet group lost significantly less lean body mass. This study used a diet containing protein at a high level, and it is not known if other levels of protein are equally effective. The authors note that the athletes in the high-protein diet group reported more fatigue and less feelings of well-being than the control group. These feelings did not impact their ability to maintain the required volume and intensity of training during the 14-day study, but it is not known if carbohydrate intake would be too low to support training if the diet were extended beyond two weeks. Much more research is needed.

There are problems associated with low protein and calorie intake.

Studies have shown that athletes who consume an inadequate amount of protein also usually consume an inadequate amount of energy (kcal), so the problems they face are numerous. The lack of protein is an especially important problem because of its critical role in building and maintaining muscle mass and supporting the immune system.

Inability to build or maintain skeletal muscle. Skeletal muscle hypertrophy (an increase in muscle size) cannot take place in the absence of food. Resistance exercise is a powerful stimulus for skeletal muscle protein synthesis, but it also results in skeletal muscle protein breakdown. Thus protein and carbohydrate ingestion, particularly in the two to three hours after exercise, is necessary for a net gain in skeletal muscle protein. To achieve positive net protein balance, a postexercise feeding is needed (Tang and Phillips, 2009; Koopman et al., 2007).

Protein synthesis and degradation is a constant process, and there must be a balance between protein intake and protein loss. The main problem with consuming a chronically low-protein, low-energy diet (that is, a semistarvation state) is that net protein balance cannot be maintained. The body has few intermediate-term options if dietary protein is not available over a period of weeks and months. It must reduce the synthesis of some body proteins and degrade skeletal muscle protein, thus returning the body to balance but at a lower functional level. Net protein balance will be achieved but at the expense of training, performance, and health. Severe starvation leads to the loss of homeostasis, general weakness, and, ultimately, death.

Under starvation conditions hormonal balance also changes. Food is not being consumed, so little insulin is being produced and the muscle and fat cells become resistant to the insulin that is present. Without the influence of insulin, protein synthesis is further reduced because insulin promotes the uptake of amino acids by the muscle cells. The hormonal mix in this case favors the breakdown of skeletal muscle protein to provide amino acids such as alanine, which can be used to make glucose.

Inability to support a fully functioning immune system. The immune system is highly dependent on protein because of the rapid turnover of immune system cells and the number of immune-related proteins, such as immunoglobulins, which act like antibodies. Low protein intake usually accompanies low energy intake and low nutrient intake, all of which can negatively affect the immune system. Though low protein intake and immune system function have not been studied directly in athletes, studies of people who restrict energy severely (that is, "dieters") have shown that some immune mechanisms are impaired. These studies are interpreted to mean that a low-protein intake is detrimental to the functionality of the immune system (Gleeson, Nieman, and Pedersen, 2004).

Key points

■ Most athletes consume a sufficient amount of protein daily.

■ High-protein diets are not associated with kidney dysfunction or excessive loss of calcium.

■ Low-protein, low-calorie diets can negatively affect performance and health.

■ Athletes who restrict calories should consume sufficient protein (~1.5 g/kg/d) to help preserve skeletal muscle mass.

If you looked across the weight room at the athletes' training center on a university campus, which athletes might be at risk for low caloric and protein intakes and which might be at risk for excessive protein intakes?

5.6 Translating Protein Intake Recommendations to Practical, Daily Food Choices

There are animal and plant sources of protein.

Proteins are found in both animal and plant foods. Animal sources include meat, fish, poultry, eggs, milk, and milk products. Beans, legumes, nuts, and seeds are popular plant protein sources. Grains and vegetables contain smaller amounts than other protein-containing foods, but because these foods may be eaten in large quantities, they often contribute a reasonable amount to total daily protein intake.

Table 5.5 Protein, Fat, and Carbohydrate Content of Selected Foods

Food	Amount	Protein (g)	Fat (g)	Carbohydrate (g)	Energy (kcal)
Animal sources of protein					
Beef (lean cuts such as flank steak)	3 oz	30	8	0	199
Chicken (dark meat, roasted)	3 oz	23	8	0	174
Chicken (white meat, roasted)	3 oz	26	3	0	140
Fish (nonoily such as cod, halibut, or roughy)	3 oz	19	0.5	0	89
Fish (oily such as salmon)	3 oz	20	4	0	118
Hamburger (regular ground beef)	3 oz	20	18	0	246
Hamburger (extra lean ground beef)	3 oz	24	13	0	225
Pork (tenderloin, roasted)	3 oz	24	4	0	139
T-bone steak	3 oz	21	16.5	0	238
Tuna (oil packed, drained)	3 oz	25	7	0	168
Tuna (water packed or fresh)	3 oz	23	0.5	0	106
Turkey (dark meat, roasted)	3 oz	24	6	0	159
Turkey (white meat, roasted)	3 oz	25.5	0.5	0	115
Veal (lean)	3 oz	27	6	0	167
Animal-related protein sources					
Cheese (cheddar)	1 oz	7	9	0	114
Cheese (fat-free)	1 slice	4	0	3	30
Egg (white and yolk)	1 large egg	6	5	1	77
Egg whites	¼ cup	6	0	1	30
Milk (whole, 3.3%)	8 oz	8	8	11	146
Milk (reduced fat, 2%)	8 oz	8	5	11	122
Milk (low-fat, 1%)	8 oz	8	2	12	102
Milk (nonfat)	8 oz	8	Trace	12	83
Yogurt (low-fat, fruit in the bottom)	8 oz	6	1.5	31	160
Yogurt (nonfat, artificially sweetened)	8 oz	11	Trace	19	122
Plant sources of protein					
Beans (such as navy or pinto)	½ cup cooked	8	0.5	24	129
Bread (white)	1 slice (25 g)	2	0.8	13	67
Bread (whole wheat)	1 slice (44 g)	2.5–3	~2.5	21.5	~119
Lentils	½ cup cooked	9	Trace	20	115
Nuts: Almonds (dry roasted)	¼ cup	8	18	7	206
Peanuts (oil roasted)	¼ cup	10	19	7	221
Peanut butter	2 T	8	16	6	188
Potato (baked)	1 large (~7 oz)	5	Trace	43	188
Rice (white or brown)	½ cup cooked	2–2.5	~1	22	108
Spaghetti noodles	½ cup cooked	3.5	0.5	19.5	95
Sunflower seeds (dry roasted)	¼ cup	6	16	8	186
Sunflower seeds (oil roasted)	¼ cup	6	17	4	178
Sweet potato	½ cup cooked	1	Trace	12	51

Legend: g = gram; oz = ounce; kcal = kilocalorie; T = tablespoon

Each macronutrient—carbohydrate, protein, fat—must be considered individually, but in reality most foods contain a mixture of macronutrients, and thus protein foods may not be chosen solely for the amount of protein contained. Table 5.5 lists the protein, fat, and carbohydrate contents in various protein-containing foods. For example, athletes who are looking for protein without much fat or carbohydrate could choose egg whites or very lean cuts of meat, fish, or poultry. Those athletes who want both proteins and carbohydrates but wish to limit fat intake could choose beans, legumes, and nonfat dairy products. Nuts provide heart-healthy fats as well as proteins and carbohydrates. Foods are also arranged by source (animal, animal products, and nonanimal). As Table 5.5 illustrates, there are many choices.

Vegetarians and vegans restrict their intake of certain protein foods.

In general, vegetarian diets have a positive effect on health, particularly the risk for developing chronic diseases. Vegetarians tend to have less heart disease, high blood pressure, diabetes, and obesity than non-vegetarians (Fraser, 2009). This is likely the result of a lower intake of saturated fat and cholesterol and a higher intake of fruits, vegetables, whole grains, nuts, and soy, which are nutrient-rich foods that are often high in fiber. Well-planned vegetarian diets are appropriate for all individuals, including athletes (Craig, Mangels, and American Dietetic Association, 2009).

The key is adequate planning. In the case of protein, both the quantity and quality of the amino acids must be accounted for. Although vegetarians

Spotlight on a real athlete

Lucas, a Cross Country Runner

As in the previous chapter, a 24-hour dietary intake was analyzed for Lucas, a collegiate cross country runner (see Figure a). Recall that due to the demands of his training, Lucas needs approximately 3,400 kcal (~54 kcal/kg) daily. His need for carbohydrate during preseason training that emphasizes longer distance runs (75 to 80 miles a week) is estimated to be 8 g/kg/d. The guideline for protein for endurance athletes is approximately 1.2 to 1.4 g/kg/d and may be as high as 2.0 g/kg/d for ultraendurance athletes during some phases of their training. In Lucas' case he would like to consume approximately 1.5 g/kg/d.

According to the dietary analysis, Lucas consumed 3,333 kcal and 532 g of carbohydrate (8.5 g/kg/d). These amounts are very close to his goals. Lucas's protein intake was 124 g or ~2 g/kg/d. He exceeded his goal for protein but he does not consume an excessive amount. In fact, his intake reflects that of many Americans and is in line with guidelines for ultraendurance athletes. His total energy intake was adequate and he met his goal for total carbohydrate intake. He consumed high-quality protein as both animal protein (for example, cheese and milk) and complementary proteins (for example, black beans and rice). His macronutrient and energy intakes are balanced.

Nutrient	DRI	Intake	0%	50%	100%
Energy					
Kilocalories	3365 kcal	3332.8 kcal			99%
Carbohydrate	379–547 g	532.15 g			
Fat, total	75–131 g	93.86 g			
Protein	84–294 g	124.09 g			

	Goal*	Actual	% Goal
Grains	10 oz. eq.	11.8 oz. eq.	118%
Vegetables	4 cup eq.	1.2 cup eq.	30%
Fruits	2.5 cup eq.	1.7 cup eq.	68%
Milk	3 cup eq.	5.4 cup eq.	180%
Meat & Beans	7 oz. eq.	8.8 oz. eq.	126%
Discretionary	648	1152	178%

MyPyramid.gov
STEPS TO A HEALTHIER YOU

*Your results are based on a 3200 calorie pattern, the maximum caloric intake used by MyPyramid.

Figure a Dietary analysis of a 24-hour diet of a male collegiate cross country runner

tend to consume less protein than nonvegetarians, the quantity of protein is typically sufficient if caloric intake is adequate. When animal foods are included, the likelihood is also high that the diet will provide protein of sufficient quality. When no animal proteins are consumed, protein quality should be assessed.

Plant proteins may lack either one or more of the indispensable amino acids or the proper concentrations of these amino acids. Of greatest concern are lysine, threonine, and the sulfur-containing amino acids cysteine and methionine. Legumes such as lentils and beans are low in methionine and cysteine.

However, grains, nuts, and seeds are high in these amino acids. Grains lack lysine but legumes have high levels. Vegetables lack threonine but grains have adequate amounts.

Consuming two plant proteins, each with a relatively high amount of the amino acid that the other lacks, can result in an adequate intake of all of the indispensable amino acids. This is the concept of complementary proteins. Beans and rice, lentils and rice, corn and beans, and peanut butter and bread consumed at the same meal or within the same day are examples of complementary proteins. Many cultures have a long history of combining plant proteins

From a performance perspective, the quantity and quality of Lucas's protein intake was appropriate.

Marcus, a Running Back (American Football)

Now consider the diet of a strength athlete such as a college football player. Marcus is a 6 ft (183 cm), 200 lb (91 kg) running back. Marcus's goals for the three months prior to fall practice include increasing body weight (as skeletal muscle mass) by 10 lb (4.5 kg), increasing muscle strength, and maintaining aerobic fitness. To achieve this he has developed a training plan that predominantly consists of strength training along with some speed and agility drills. For the hypertrophy or muscle-building phase, Marcus emphasizes an increased volume of strength training. To accomplish this, he works out four to six days a week in the weight room with multiple sets of each exercise. As

he gets closer to fall practice, he will reduce the number of repetitions but increase the resistance of each exercise to transition into a strength and power development phase. To support training, it is estimated that he needs approximately 4,000 kcal (44 kcal/kg), 600 g carbohydrate (6.6 g/kg), 155 g protein (1.7 g/kg), and 110 g of fat (1.2 g/kg) daily.

Marcus can easily meet his energy and nutrient needs as shown in Figure b. The composition of the diet shown is very close to the recommended guidelines, although these guidelines are just estimates and athletes' diets are not expected to match them exactly. His protein intake (2.0 g/kg/d) is slightly higher than recommended (1.7 g/kg/d), which is characteristic of many strength athletes. Marcus's diet as shown will provide the nutrients he needs to support his training and includes foods that are easy to prepare and readily available at the grocery store.

Breakfast: 2 cups oatmeal with ½ cup nonfat milk, ¼ cup raisins, one honey-wheat English muffin with 1 tablespoon each margarine and jelly, 8 ounces orange juice.

Lunch: 2 sandwiches, each including 2 slices of whole wheat bread, 3 ounces lean turkey, 1 slice Swiss cheese, sliced tomato, lettuce, mustard, and 1 tablespoon light mayonnaise; 1 apple, 8 ounces nonfat milk.

Dinner: 6 ounces grilled halibut, 1 large baked potato with ¼ cup fat-free sour cream, 1½ cups broccoli, 5 Fig Newtons, 1 cup frozen yogurt.

Snacks: 8 ounces reduced-fat chocolate milk (postexercise), 2 bananas, 1 peanut butter PowerBar.

Energy: 4,009 kcal (44 kcal/kg)

Carbohydrate: 589 g (6.5 g/kg)

Protein: 185 g (2.0 g/kg)

Fat: 112 g (1.2 g/kg)

Figure b Sample diet for a strength athlete

Legend: kcal = kilocalorie; g = gram; g/kg = grams per kilogram body weight

in traditional dishes or meals that result in the consumption of complementary proteins. New Orleans is famous for its red beans and rice, Mexican cuisine features corn tortillas and pinto beans, and Asian dishes often include stir-fried tofu (made from soy beans) and rice.

An important issue for vegans is that they consume a variety of plant proteins to ensure that they obtain all the indispensable amino acids in the proper quantities. Cereals are popular with vegans but cereals are low in lysine, which is an indispensable amino acid necessary for tissue growth. Vegans are typically encouraged to include beans and soy in their diets, especially if the quantity of protein in the vegan diet is low, to ensure an adequate intake of lysine (Craig, Mangels, and American Dietetic Association, 2009). Many vegans include soy protein isolate because it is usually considered comparable in quality to animal protein. Adequate energy intake is also important. The bottom line is that vegan and vegetarian athletes can meet their protein needs with an exclusive or primarily plant-based diet if the diet is well planned.

Protein supplements should be considered a part of an athlete's overall protein intake.

Protein supplements are popular among trained athletes, particularly bodybuilders and strength athletes. These supplements are marketed as powders, premixed drinks, and bars. They are heavily advertised to athletes who are building and maintaining large amounts of skeletal muscle mass. In the presence of resistance training and adequate calories, proteins from either food or supplement sources can contribute to increasing skeletal muscle size and strength. Protein supplements are neither more nor less effective than food proteins for skeletal muscle growth.

The purity of dietary supplements has become a critical issue for athletes as some supplements may be tainted with substances banned by their sports governing body. Some of these substances may be due to poor manufacturing practices, but there is also a high degree of suspicion that precursors to anabolic steroids may be added intentionally. A 2007 study of 52 supplements found that 25 percent contained a small amount of steroids (Judkins, Hall, and Hoffman, 2007). Athletes should be especially cautious of supplements described as "testosterone boosters."

As illustrated in the Spotlight on a real athlete, Marcus, a running back, and Lucas, a distance runner, both consumed more than enough protein from

Protein supplements are heavily advertised to strength/power athletes.

food alone. Obtaining protein from food is relatively easy and reasonably affordable. Obtaining protein from supplements may be more convenient because of portability and preparation (e.g., adding protein powder to water can be done quickly anywhere). For example, Marcus typically goes home after lifting weights and has a quick snack of chocolate milk and a banana. But he also keeps a protein powder handy for a number of reasons—he likes the variety, it is easy to bring to the gym if he is not going back home, it does not need to be refrigerated, and he sometimes comes home to find that his roommate has eaten his bananas and finished the carton of milk. The protein powder he buys has about the same number of calories as his usual postexercise snack but has more protein and less carbohydrate, so he must consider the differences in their nutrient content. Marcus may choose a protein supplement for a number of good reasons, but because he can get enough protein from food, protein supplements are not a required part of his diet. For those who choose to supplement, such supplements seem to be safe for healthy adults.

Protein supplement ingredients. Protein supplements typically contain whey, casein, egg, and soy proteins, the same proteins that are contained in milk (80 percent casein and 20 percent whey), egg whites, and tofu and edamame. Whey is a major ingredient in many protein supplements.

Both whey and casein are processed from milk. When milk is coagulated (thickened), whey is found in the liquid portion whereas casein is found in the semisolid portion known as curds. The whey can be processed further into whey protein isolate, whey protein concentrate, or whey powder. Whey protein

Table 5.6 Nutrient Content of Selected Protein Supplements

Protein supplement	Amount	Sources of protein	Energy (kcal)	Protein (g)	Fat (g)	Carbohydrate (g)
100% Whey protein fuel (powder)[1]	1 scoop (33 g) in 6 oz of water	Whey protein concentrate and isolate (from milk)	130	25	2	4
Myoplex Original (ready to drink)[2]	17 fl oz	Milk protein concentrate, calcium caseinate, whey protein isolate	300	42	7	20
Heavyweight Gainer 900[3]	4 scoops (154 g)	Beef protein, whey protein concentrate and hydrolysate, egg albumen	630	35	9.5	101
Protein Plus bar[4]	1 bar (85 g)	Milk protein concentrate, whey protein isolate and concencentrate, calcium caseinate, egg white, L-glutamine	310	32	9	32

Legend: kcal = kilocalorie; g = gram; oz = ounce; fl oz = fluid ounce

Note: All nutrition information was obtained from the Nutrition Facts label

[1]TwinLab, American Fork, UT; [2]EAS/Abbott Laboratories, Abbott Park, IL; [3]Champion Nutrition, Sunrise, FL; [4]Met-Rx, Ronkonkoma, NY

concentrate and whey powder contain lactose. Whey protein isolate, which is typically added to protein supplements and infant formulas, is a concentrated source of protein because both the carbohydrate (for example, lactose) and the fat are removed. The end product is high in indispensable amino acids, particularly the branched chain amino acids, leucine, isoleucine, and valine (Whey protein. Monograph, 2008). Casein has a different amino acid composition and is particularly high in glutamine, an amino acid that is considered conditionally indispensable under physiological stress such as endurance exercise.

When whey is compared to casein, the amino acids in whey are absorbed faster. This difference in absorption rate is not surprising because whey has a high percentage of indispensable amino acids, which are absorbed more rapidly than the dispensable amino acids. Whey is often referred to as a fast-acting protein whereas casein is described as a slow-acting protein.

Protein supplements must list the sources of the protein in the ingredient list on the label, but sometimes the label terms can be confusing. For example, what is the difference between whey protein isolate and whey protein concentrate? Isolate refers to the purest form, in this case the most protein because the majority of carbohydrate and fat have been removed. Concentrate does not contain as high a percentage of protein and is typically cheaper. Whey protein isolate is ~90 percent protein, whereas whey protein concentrate is ~80 percent protein.

Hydrolysate refers to enzymatic predigestion of the protein. Casein often appears as caseinate on the label. Egg white protein (or albumin) refers to dried, pasteurized egg whites.

Table 5.6 lists some popular products advertised as protein supplements. Protein powders may contain only protein but most also contain carbohydrate to make them more palatable and to provide some postexercise carbohydrate. Sometimes athletes will mix protein powders with sugary or artificially sweetened drinks. Protein bars vary in their protein, carbohydrate, and fat content, and many contain ~300 kcal. Careful label reading helps athletes determine the nutrient content of a product and how such a product may fit into their overall diet plan.

Effect of protein source on skeletal muscle growth. Supplements containing whey, casein, and soy have all been studied to determine their effects on protein synthesis in the presence of resistance training and sufficient calories. The amino acids found in whey protein supplements are rapidly absorbed, and several studies have shown that whey is superior to casein for stimulating skeletal muscle protein synthesis (Tang et al., 2009; Cribb et al. 2007; Kerksick et al., 2006). On the other hand, casein is absorbed more slowly, and the slow but sustained rise in amino acids appears to help suppress skeletal muscle protein breakdown (Tang et al., 2009). This is sometimes described as whey having an anabolic effect and casein having an anticatabolic effect. Many protein supplements contain both whey and casein to take advantage of their different absorption rates. Soy, which is a rapidly absorbed protein, also stimulates skeletal muscle protein synthesis (Wilkinson et al., 2007).

Key points

■ Protein is found in both plant and animal foods. Protein supplements are neither superior nor inferior to food proteins.

■ Vegetarian and vegan athletes can meet their protein needs if their diets are well planned and contain a variety and sufficient amount of plant proteins.

How can vegetarians or vegans ensure that they are consuming all of the indispensable amino acids in their diet?

5.7 Supplementation with Individual Amino Acids

Proteins are made up of amino acids, which have specific biochemical functions. For example, some amino acids can stimulate growth hormone release or muscle protein synthesis in the presence of resistance training, whereas others are integral to immune system function. A logical question is whether supplementation with individual amino acids can enhance specific biochemical functions and be an effective way to increase skeletal muscle size or strength or enhance immune system function, which in turn could lead to positive changes in body composition and performance.

For the amino acids discussed below, all are normally found in the body because each is either obtained from food or manufactured in the body. However, supplementation typically raises the concentration of the amino acid above the normal level because the supplement contains a concentrated dose. These supplements appear to be safe at recommended doses. However, many fall short when it comes to effectiveness. Some of the reasons for lack of effectiveness may be that enhanced function is not solely dependent on the amount of substrate present and that complicated processes, such as building skeletal muscle or supporting the immune system, cannot be affected to a large degree by a single factor.

Table 5.7 summarizes some of the popular amino acid-related supplements. Athletes should always be aware of the potential for contamination with anabolic steroids or other substances that may be banned by their sports governing body.

Spotlight on supplements

National Collegiate Athletic Association (NCAA) Bylaws and Nutritional Supplements

NCAA Bylaw 16.5.2 (g) states that only non-muscle-building nutritional supplements may be given to student athletes. Permissible supplements fall into four categories: vitamins and minerals, energy bars, calorie replacement drinks, and electrolyte replacement drinks. A permissible supplement can contain no more than 30 percent of its calories from protein. The following are impermissible for the institution or any of its staff members to provide:

- Amino acids (including amino acid chelates)
- Chondroitin (unless prescribed by a physician for a specific medical condition)
- Chrysin
- CLA (Conjugated Linoleic Acid)
- Creatine/compounds containing creatine
- Garcinia Cambogia (Hydroxycitric Acid)
- Ginkgo Biloba
- Ginseng
- Glucosamine (unless prescribed by a physician for a specific medical condition)
- Glutathione
- Glycerol (permissible if only being used as a binding agent)
- Green tea
- HMB (Hydroxy-methybutyrate)
- Melatonin
- MSM (Methylsulfonyl Methane)
- Protein powders
- St. John's Wort
- Tribulus
- Weight-gainers
- Yohimbe

Table 5.7 Summary of the Safety and Effectiveness of Amino Acid Supplements

Supplement	Safety	Effectiveness
Beta-alanine	Safe at recommended doses	Studies show promise of effectiveness for buffering muscle pH in athletes performing high-intensity (sprint) exercise
β-hydroxy-β-methylbutyrate (HMB)	Safe at recommended doses	Not effective in resistance-trained athletes; small to very small increases in overall, upper body, and lower body strength in untrained individuals
Branched chain amino acids (BCAAs)	Safe at recommended doses	Not effective for improving performance; studies show promise for immune system support and reduction of postexercise fatigue
Glucosamine/chondroitin sulfate	Safe at recommended doses	Generally not effective for reducing joint pain or increasing functionality in those with osteoarthritis, although individual responses vary
Glutamine	Safe at recommended doses	Not effective
Growth hormone releasers (arginine)	Safe at recommended doses	Effective for stimulating growth hormone release; if taken before exercise, decreases the effectiveness of exercise as a stimulator of growth hormone; not effective for increasing muscle mass or strength
Nitric oxide (NO)/arginine alpha-ketoglutarate (AAKG)	Safe at recommended doses	Effectiveness unknown due to lack of research

Beta-alanine may help to buffer muscle pH in high-intensity (sprint) exercise.

Beta-alanine is a dispensable amino acid. It is not the same as the amino acid alanine, which has a different structure. Beta-alanine is found in protein-containing foods such as meat, poultry, and fish as part of the dipeptide, carnosine. Carnosine is broken down in the gastrointestinal tract to beta-alanine and histidine. The liver can also manufacture beta-alanine.

Carnosine is manufactured in skeletal muscle but is found in greater concentrations in type II muscle fibers than in type I fibers. Of the two amino acids needed to manufacture carnosine, beta-alanine and histidine, beta-alanine is the rate-limiting factor. The enzyme carnosine synthase regulates the manufacture of carnosine. Two of the roles of carnosine in the muscles is to buffer acid and maintain muscle pH within an optimal range (Artioli et al., 2010). Enhanced buffering of muscle pH is particularly important during high-intensity exercise.

Research suggests that beta-alanine supplementation can substantially increase muscle carnosine concentration (Artioli et al., 2010; Harris et al., 2006). Studies have shown improved performance with beta-alanine supplementation during multiple bouts of high-intensity exercise as well as a single bout of exercise lasting greater than 60 seconds (Artioli et al., 2010; Stout et al., 2007; Hill et al., 2007). However, not all studies of beta-alanine supplementation in those performing repeated sprints have shown an effect (Sweeney et al., 2010).

Although most of the research has been conducted with athletes performing high-intensity exercise, a few

researchers are studying the effect of beta-alanine supplementation in endurance athletes. Endurance events often require a sprint to the finish line. van Thienen et al. (2009) found that beta-alanine supplementation improved sprint performance in endurance cyclists.

The doses used in research studies are ~3.2–6.4 g/day, with a higher initial dose followed by a lower maintenance dose. Many beta-alanine supplement manufacturers recommend an initial dose of 6 g/day for two weeks and then a daily dose of 4 g. Such doses appear to be safe.

β-Hydroxy-β-Methylbutyrate (HMB) does not appear to be effective for increasing skeletal muscle strength or reducing skeletal muscle damage after resistance exercise.

HMB is a metabolite of leucine, one of the indispensable amino acids. Leucine is one of the branched chain amino acids and has some anticatabolic properties. In theory, supplementation with HMB could minimize the protein breakdown that follows resistance exercise. Research has focused on whether HMB supplements could reduce muscle damage, increase muscle strength, and change body composition.

A 2009 meta-analysis (Rowlands and Thomson, 2009) analyzed the effects of HMB supplements on muscle strength, muscle damage, and body composition in trained and untrained lifters. In untrained lifters, HMB supplementation was effective for small increases in lower body and average strength and very small increases in upper body strength. In trained lifters, any gains were very small. Although there were reports of reduced muscle damage, there were not enough study results to be able to draw a conclusion. There were no substantial effects on body composition in either trained or untrained subjects. It appears that HMB supplementation holds the most promise for untrained individuals who begin resistance training, although the benefit over strength training alone is small.

The usual recommended dose is 3 g/day divided into three 1g doses. Recommended amounts appear to be safe.

Branched chain amino acids (BCAA) may help to support immune function in endurance athletes.

Leucine, isoleucine, and valine are the branched chain amino acids, so named because of their chemical structure. During prolonged endurance exercise when glycogen stores are low, skeletal muscle can metabolize these amino acids for energy. In addition to being used as an energy source, BCAAs compete with tryptophan, an amino acid associated with mental fatigue. BCAAs are also involved in the immune system (Newsholme and Blomstrand, 2006; Gleeson, 2005). Resistance-trained athletes use BCAA supplements in an effort to reduce skeletal muscle damage and muscular fatigue.

In theory, greater availability of BCAA late in prolonged exercise could provide a much-needed fuel source and help to delay mental and physical fatigue. However, studies of supplemental BCAA have not shown a delay fatigue or an improvement in endurance performance in elite athletes (Negro et al., 2008).

Although the trials are small, some positive effects have been reported in studies that examined immune response (Negro et al., 2008). Endurance-trained athletes, such as triathletes, may experience some suppression of the immune system, which could be a result of decreased glutamine. BCAA supplementation reverses the decline in glutamine because BCAAs are metabolized to glutamine in skeletal muscle. BCAA supplements may play a role in supporting immune function, but more research is needed.

Another area of research is the use of BCAA supplements to reduce exercise-induced muscle damage (Jackman et al., 2010; Greer et al., 2007; Shimomura et al., 2006). Preliminary studies with a small number of untrained subjects found that 5 g of BCAA prior to performing squat exercises reduced delayed-onset muscle soreness (DOMS) and muscle fatigue for several days after exercise. The mechanism is not known, but a possibility is that BCAA could reduce protein breakdown and stimulate protein synthesis in the exercised muscle. This has become a more active area of research.

The usual recommended dose for supplemental BCAA is 5–20 g/day in divided doses. Such amounts appear to be safe.

Glucosamine/chondroitin sulfate is generally not effective for reducing joint pain.

Glucosamine and chondroitin, which are usually sold together, is a dietary supplement marketed to relieve joint pain in those with osteoarthritis and prevent cartilage breakdown in athletes. Glucosamine is manufactured by the body from glucose and the amino acid glutamine, and is not related to dietary intake. It is part of glycosaminoglycan (an unbranched polysaccharide), which is found in the extracellular matrix of the joints. Because of its ability to attract water, it is referred to as a "joint lubricant." Chondroitin is also synthesized by the body and is part of a protein that aids in elasticity of cartilage.

Early studies of reduced joint pain in those with osteoarthritis were promising, but they were also controversial because they lacked scientific rigor. The National Institutes of Health funded a well-designed study, GAIT, to determine the effect of glucosamine and chondroitin on osteoarthritis of the knee (Clegg et al., 2006). The researchers found that 1,500 mg/day of glucosamine and 1,200 mg/day chondroitin sulfate, either alone or in combination, for 24 weeks did not reduce pain effectively in the overall group of patients with osteoarthritis of the knee. Further data analysis suggested that glucosamine/chondroitin sulfate supplements might be effective in the small subgroup of patients with moderate-to-severe knee pain. A continuation of the GAIT study found that glucosamine did not slow the narrowing of the knee joint space (Sawitzke et al., 2008). Nearly all the studies are conducted in subjects with osteoarthritis, so it is unknown if the breakdown of cartilage in athletes is reduced or prevented with supplementation.

The recommended dose of glucosamine and chondroitin is 1,500 mg/day and 1,200 mg/day, respectively. These doses are considered safe. Improvement of symptoms would not be expected for six to eight weeks. However, if improvement does not occur after eight weeks of continuous use, it is not likely that supplementation will help. Minimal adverse effects have been reported, especially compared with frequent use of nonsteroidal anti-inflammatory drugs (NSAIDs), such as ibuprofen.

Glutamine supplementation does not appear to be effective as a way to enhance the functioning of the immune system.

Under normal conditions glutamine is considered a dispensable amino acid and the body can manufacture a sufficient amount. Under physiologic stress, such as prolonged endurance exercise and illness, glutamine production may fall short of the amount needed.

Glutamine is a fuel source for immune system cells. If glutamine is not available, some immune cells are impaired, which increases the risk for infections. Plasma glutamine concentration can be decreased after strenuous endurance exercise, so the question of whether supplemental glutamine is necessary or beneficial has been raised. Some resistance-trained athletes use glutamine as a cell volumizer.

A review by Gleeson (2008) concluded that glutamine supplementation is not effective to counteract immunologic stress and does not decrease exercise-induced suppression of the immune system in endurance athletes. Study results suggest that glutamine supplementation by resistance-trained athletes does not improve recovery, decrease muscle catabolism, or increase skeletal muscle mass (Candow et al., 2001).

Recommendations made by manufacturers range from 5 to 10 g/day to more than 20 g/day. The estimated daily dietary intake of glutamine is ~3–6 g (based on a dietary protein intake of 0.8–1.6 g/kg body weight). Athletes who consume higher levels of protein may consume more. It is known that some endurance athletes consume low or marginal amounts of dietary protein, and increasing protein intake, and therefore glutamine intake, would likely be beneficial. Glutamine supplements at the usual recommended doses are considered safe (Gleeson, 2008).

Growth hormone releasers, particularly arginine, may be effective for stimulating the release of growth hormone.

Exercise is a potent stimulator of growth hormone. Certain amino acids, such as arginine, ornithine, and lysine, can also stimulate growth hormone release (Kanaley, 2008; Collier, Casey, and Kanaley, 2005; Collier, Collins, and Kanaley, 2006). Some athletes consume these amino acids before strength training to enhance the exercise-induced release of growth hormone and to promote greater gains in muscle mass and strength. These amino acids may be consumed individually or together and are often advertised as growth hormone releasers.

More research has been conducted with arginine than with ornithine or lysine. Oral ingestion of 5 to 9 g of arginine at rest results in a dose-dependent increase in growth hormone. A higher dose of 13 g causes considerable gastrointestinal distress without further raising growth hormone concentration (Collier, Casey, and Kanaley, 2005). Ingesting arginine alone increases resting growth hormone level at least 100 percent. By comparison, exercise increases growth hormone level by 300–500 percent (Kanaley, 2008). Ingesting arginine before exercise, however, *decreases* the growth hormone response to exercise to ~200 percent (Kanaley, 2008; Collier, Collins, and Kanaley, 2006). Since consuming arginine reduces the effect of exercise on growth hormone concentration, athletes should not take arginine supplements prior to exercise. Arginine supplements are effective for stimulating the release of more growth hormone but not effective for increasing muscle mass or strength. Recommended amounts appear to be safe.

The effectiveness of nitric oxide (NO)/arginine alpha-ketoglutarate (AAKG) is not known due to a lack of studies.

Nitric oxide supplements contain arginine alpha-ketoglutarate. Arginine is a dispensable amino acid that is needed to form the enzyme nitric oxide synthase. This enzyme catalyzes the oxidation of arginine to produce nitric oxide (a gas) and citrulline. Nitric oxide is a key signaling molecule in the cardiovascular system and promotes vasodilation (Gornik and Creager, 2004).

When vessels are dilated, blood flow and oxygen transport to tissues, such as skeletal muscle, are enhanced. The increased blood flow would also enhance delivery of nutrients and removal of wastes. NO supplements are marketed to resistance-trained athletes as a way to produce dramatic increases in muscle size and strength due to the increased delivery of oxygen and nutrients to skeletal muscle.

There is no evidence that AAKG supplements increase nitric oxide levels or blood flow to the muscles of healthy people who are sedentary (Gornik and Creager, 2004). Only a few studies have evaluated the effectiveness of supplements containing arginine alpha-ketoglutarate in resistance-trained adult men (Little et al., 2008; Campbell et al., 2006). Preliminary results suggest NO/AAKG supplements may increase muscle strength but not muscle size. However, more research is needed to determine effectiveness and safety.

Arginine alpha-ketoglutarate appears to be safe in doses up to 12 g/day for short-term use (one to eight weeks). High doses can cause gastrointestinal distress and diarrhea (Evans et al., 2004).

Key points

■ Although individual amino acid supplements seem safe, they generally are not effective for improving performance, body composition, or health.

What factors must a collegiate or professional athlete consider when deciding whether or not to consume a "muscle-building" supplement?

Keeping it in perspective

The Role of Protein for Athletes

Protein is an important nutrient, and it receives much attention because of its role in the growth and development of skeletal muscle. It deserves attention, but no nutrient should be the sole or predominant focus of the athlete's diet. Overemphasizing protein can mean losing sight of the broader dietary picture, which includes adequate energy, carbohydrate, and fat intakes.

Protein is no different from other nutrients in that both the "big" picture and the details are important. The big-picture issues include an adequate amount of protein (based on energy intake) and macronutrient balance (the amount of protein relative to the amount of carbohydrate and fat needed). There is no point to the strength athlete taking in large amounts of protein only to find that carbohydrate intake is too low to support a well-planned resistance-training program. Determining the appropriate energy intake and the amount of protein required are important first steps. Athletes need to assess their current dietary intake to determine if, and how much, more protein is needed. Once the daily protein need is determined, then the fine-tuning can take place. Athletes need to decide if they will obtain protein only from food or from a combination of food and supplements. Sources acceptable to the athlete—animal, animal derived, or plant—must be considered as well as any issues related to the speed of protein absorption. Protein is one of several nutrients that should be consumed in the postexercise recovery period, although the amount needed is small compared to the amounts of carbohydrate and fluid needed.

Some of the questions that athletes have do not yet have scientific answers. For example, how much protein is needed daily for maximum skeletal muscle protein synthesis? Could amino acid consumption before exercise help to offset muscle protein breakdown during exercise? The answers are not yet known. Part of keeping protein intake in perspective is to understand what is known, what is theorized, and what is pure conjecture. This is known: protein is one important aspect of the athlete's diet.

Post-Test — Reassessing Knowledge of Proteins

Now that you have more knowledge about protein, read the following statements and decide if each is true or false. The answers can be found in Appendix O.

1. Skeletal muscle is the primary site for protein metabolism and is the tissue that regulates protein breakdown and synthesis throughout the body.

2. In prolonged endurance exercise, approximately 3–5 percent of the total energy used is provided by amino acids.

3. To increase skeletal muscle mass, the body must be in positive nitrogen balance, which requires an adequate amount of protein and energy (calories).

4. Athletes who consume high-protein diets are at risk for developing kidney disease.

5. Strength athletes usually need protein supplements because it is difficult to obtain a sufficient amount of protein from food alone.

Summary

Proteins are made up of amino acids, which contain carbon, hydrogen, oxygen, and nitrogen. Protein is a critical nutrient, in part, because of the roles that it plays in the growth and development of tissues and the immune system. It has many other important roles including the synthesis of enzymes and hormones. Amino acids can also be catabolized to provide energy. Protein functions optimally when energy intake from carbohydrate and fat is sufficient. When energy intake is insufficient, such as with fasting or starvation, more amino acids are broken down to provide the carbon skeletons needed by the liver to manufacture glucose.

Athletes need more protein than nonathletes. The recommended daily protein intake for endurance athletes is 1.2 to 1.4 g/kg, although ultraendurance athletes may need as much as 2.0 g/kg/d during some rigorous phases of their training. Consensus has not yet been reached regarding daily protein requirements for strength athletes, with some experts recommending 1.2 to 1.7 g/kg and others suggesting a higher intake of 1.5 to 2.0 g/kg. Excessive amounts, greater than 2.5 g/kg/d, are generally not recommended.

To increase skeletal muscle mass, the athlete must engage in an appropriate resistance-training program and consume an adequate amount of calories and protein to achieve positive nitrogen balance and positive net protein balance. The timing of protein intake is important, and the consumption of a small amount of protein within 2 to 3 hours after exercise has been found to be beneficial for skeletal muscle protein synthesis. Most athletes consume an adequate amount of protein, but some consume too little protein, usually in conjunction with too few calories. Low protein and energy intakes have the potential to impair performance and health.

Proteins are found in both animal and plant foods. When animal foods are consumed, the likelihood is high that the diet will provide protein of sufficient quality. Vegans and vegetarians who do not consume any animal-derived proteins can also obtain sufficient protein quality by correctly combining different plant proteins. Athletes can consume enough food to meet their higher-than-average protein needs. Protein supplements are popular, particularly among bodybuilders and strength athletes, and seem to be safe but no more or less effective than proteins found in food. Individual amino acid supplements seem to be safe, but are generally not effective for improving performance, body composition, or health.

Self-Test

Multiple-Choice

The answers can be found in Appendix O.

1. The chemical composition of proteins differs from carbohydrate or fat because of the presence of:
   a. sodium.
   b. carbon.
   c. nitrogen.
   d. hydrogen.

2. The amino acids that are most rapidly absorbed from the intestine are:
   a. dispensable amino acids.
   b. indispensable amino acids.
   c. dietary supplements containing predigested proteins.
   d. dietary supplements containing casein.

3. Which athlete would use protein as an energy source to the greatest extent?
   a. recreational athlete
   b. strength athlete
   c. endurance athlete
   d. All of these athletes would use about the same amount.

4. Why do athletes who restrict their energy intake need more protein?
   a. More protein will be burned for energy.
   b. Protein is converted to muscle glycogen when carbohydrate intake is restricted.
   c. to maintain fluid balance
   d. to spare muscle glycogen from being used during exercise

5. How do whey and casein differ?
   a. Whey is a milk protein; casein is a plant protein.
   b. Each is comprised of different amino acids.
   c. Whey has fewer kcal.
   d. There is no difference.

Short Answer

1. Amino acids contain which chemical elements?

2. How do the structures of amino acids differ?

3. What is meant by protein quality?

4. Briefly describe the digestion, absorption, and transport of protein found in food.

5. What is the amino acid pool?

6. Describe anabolic and catabolic processes that involve amino acids.

7. What is nitrogen balance and protein balance?

8. Describe how skeletal muscle protein is synthesized.

9. What is the recommended daily intake of protein for adult nonathletes? Endurance athletes? Strength athletes? Ultraendurance athletes?

10. Why is the timing of protein intake important?

11. How are high- or low-protein diets detrimental to training, performance, or health?

12. Compare and contrast animal and plant proteins and why this information is important to vegans.

13. Compare and contrast whey, casein, and soy proteins.

14. Are protein and individual amino acid supplements safe and effective? What information is needed to draw conclusions about safety and effectiveness?

Critical Thinking

1. When is maximal not optimal? Give as many protein-related examples as possible.

2. Nitrogen and protein balance often increase or decrease in tandem. Explain how it might be possible to have nitrogen balance but negative net protein balance?

3. You are the sports nutritionist for a competitive bicycle racing team during a 2-week training camp. Toward the end of the camp, one of the cyclists tells you that his teammates have been complaining about his significant body odor, particularly the smell of ammonia. What are the potential causes for the ammonia-like body odor, and what potential solutions would you recommend to this athlete?

References

Aerenhouts, D., Hebbelinck, M., Poortmans, J. R., & Clarys, P. (2008). Nutritional habits of Flemish adolescent sprint athletes *International Journal of Sport Nutrition and Exercise Metabolism, 18*(5), 509–523.

Artioli, G. G., Gualano, B., Smith, A., Stout, J., & Lancha, H. A., Jr. (2010). The role of beta-alanine supplementation on muscle carnosine and exercise performance. *Medicine and Science in Sports and Exercise, 42*(6), 1162–1173.

Beals, K. A., & Manore, M. M. (1998). Nutritional status of female athletes with subclinical eating disorders. *Journal of the American Dietetic Association, 98*(4), 419–425.

Beals, K. A., & Manore, M. M. (2000). Behavioral, psychological, and physical characteristics of female athletes with subclinical eating disorders. *International Journal of Sport Nutrition and Exercise Metabolism, 10*(2), 128–143.

Bernstein, A. M., Treyzon, L., & Li, Z. (2007). Are high-protein, vegetable-based diets safe for kidney function? A review of the literature. *Journal of the American Dietetic Association, 107*(4), 644–650.

Bilsborough, S., & Mann, N. (2006). A review of issues of dietary protein intake in humans. *International Journal of Sport Nutrition and Exercise Metabolism, 16*(2), 129–152.

Brownell, K. D., Steen, S. N., & Wilmore, J. H. (1987). Weight regulation practices in athletes: Analysis of metabolic and health effects. *Medicine and Science in Sports and Exercise, 19*(6), 546–556.

Campbell, B., Roberts, M., Kerksick, C., Wilborn, C., Marcello, B., Taylor, L., et al. (2006). Pharmacokinetics, safety, and effects on exercise performance of L-arginine alpha-ketoglutarate in trained adult men. *Nutrition, 22*(9), 872–881.

Candow, D. G., Chilibeck, P. D., Burke, D. G., Davison, K. S., & Smith-Palmer, T. (2001). Effect of glutamine supplementation combined with resistance training in young adults. *European Journal of Applied Physiology, 86*(2), 142–149.

Clegg, D. O., Reda, D. J., Harris, C. L., Klein, M. A., O'Dell, J. R., Hooper, M. H., et al. (2006). Glucosamine chondroitin sulfate, and the two in combination for painful knee osteoarthritis. *New England Journal of Medicine, 354*(8), 795–808.

Collier, S. R, Casey, D. P., & Kanaley, J. A. (2005). Growth hormone responses to varying doses of oral arginine. *Growth Hormone & IGF Research, 15*(2), 136–139.

Collier, S. R., Collins, E., & Kanaley, J. A. (2006). Oral arginine attenuates the growth hormone response to resistance exercise. *Journal of Applied Physiology, 101*(3), 848–852.

Craig, W. J., Mangels, A. R., & American Dietetic Association. (2009). Position of the American Dietetic Association: Vegetarian diets. *Journal of the American Dietetic Association, 109*(7), 1266–1282.

Cribb, P. J., Williams, A. D., Stathis, C. G., Carey, M. F., & Hayes, M. (2007). Effects of whey isolate, creatine, and resistance training on muscle hypertrophy. *Medicine and Science in Sports and Exercise, 39*(2), 298–307.

Elango, R., Humayun, M. A., Ball, R. O., & Pencharz, P. B. (2010). Evidence that protein requirements have been significantly underestimated. *Current Opinion in Clinical Nutrition and Metabolic Care, 13*(1), 52–57.

Evans, R. W., Fernstrom, J. D., Thompson, J., Morris, S. M., Jr., & Kuller, L. H. (2004). Biochemical responses of healthy subjects during dietary supplementation with L-arginine. *Journal of Nutritional Biochemistry, 15*(9), 534–539.

Fraser, G. E. (2009). Vegetarian diets: What do we know of their effects on common chronic diseases? *American Journal of Clinical Nutrition, 89*(5), 1607S–1612S. Erratum in: *American Journal of Clinical Nutrition, 90*(1), 248.

Garrett, R. H., & Grisham, C. M. (2009). *Biochemistry.* Belmont, CA: Thomson/Wadsworth.

Gleeson, M. (2005). Interrelationship between physical activity and branched-chain amino acids. *Journal of Nutrition, 135*(6 Suppl.), 1591S–1595S.

Gleeson, M. (2007). Immune function in sport and exercise. *Journal of Applied Physiology, 103*(2), 693–699.

Gleeson, M. (2008). Dosing and efficacy of glutamine supplementation in human exercise and sport training. *Journal of Nutrition, 138*(10), 2045S–2049S.

Gleeson, M., Nieman, D. C., & Pedersen, B. K. (2004). Exercise, nutrition and immune function. *Journal of Sports Sciences, 22*(1), 115–125.

Gornik, H. L., & Creager, M. A. (2004). Arginine and endothelial and vascular health. *Journal of Nutrition, 134*(10 Suppl.), 2880S–2887S.

Green, M. S., Corona, B. T., Doyle, J. A., & Ingalls, C. P. (2008). Carbohydrate-protein drinks do not enhance recovery from exercise-induced muscle injury. *International Journal of Sport Nutrition and Exercise Metabolism, 18*(1), 1–18.

Greer, B. K., Woodard, J. L., White, J. P., Arguello, E. M., & Haymes, E. M. (2007). Branched-chain amino acid supplementation and indicators of muscle damage after endurance exercise. *International Journal of Sport Nutrition and Exercise Metabolism, 17*(6), 595–607.

Gropper, S. S., Smith, J. L., & Groff, J. L. (2009). *Advanced nutrition and human metabolism.* Belmont, CA: Thomson/Wadsworth.

Harris, R. C., Tallon, M. J., Dunnett, M., Boobis, L., Coakley, J., Kim, H. J., et al. (2006). The absorption of orally supplied beta-alanine and its effect on muscle carnosine synthesis in human vastus lateralis. *Amino Acids, 30*(3), 279–289.

Havemann, L., & Goedecke, J. H. (2008). Nutritional practices of male cyclists before and during an ultraendurance event. *International Journal of Sport Nutrition and Exercise Metabolism, 18*(6), 551–566.

Hill, C. A., Harris, R. C., Kim, H. J., Harris, B. D., Sale, C., Boobis, L. H., Kim, C. K., et al. (2007). Influence of beta-alanine supplementation on skeletal muscle carnosine concentrations and high intensity cycling capacity. Amino Acids, 32(2), 225–233.

Hoffman, J. R., Ratamess, N. A., Tranchina, C. P., Rashti, S. L., Kang, J., & Faigenbaum, A. D. (2009). Effect of protein-supplement timing on strength, power, and body-composition changes in resistance-trained men. *International Journal of Sport Nutrition and Exercise Metabolism, 19*(2), 172–185.

Howarth, K. R., Phillips, S. M., Macdonald, M. J., Richards, D., Moreau, N. A., & Gibala, M. J. (2010). Effect of glycogen availability on human skeletal muscle protein turnover during exercise and recovery. *Journal of Applied Physiology, 109*(2), 431–438

Hunt, J. R., Johnson, L. K., & Fariba Roughead, Z. K. (2009). Dietary protein and calcium interact to influence calcium retention: a controlled feeding study. *American Journal of Clinical Nutrition, 89*(5), 1357–1365.

Institute of Medicine. (2002). *Dietary Reference Intakes for energy, carbohydrate, fiber, fat, fatty acids, cholesterol, protein and amino acids* (Food and Nutrition Board). Washington, DC: National Academies Press.

Jackman, S. R., Witard, O. C., Jeukendrup, A. E., & Tipton, K. D. (2010). Branched-chain amino acid ingestion can ameliorate soreness from eccentric exercise. *Medicine and Science in Sports and Exercise, 42*(5), 962–970.

Jonnalagadda, S. S., Ziegler, P. J., & Nelson, J. A. (2004). Food preferences, dieting behaviors, and body image perceptions of elite figure skaters. *International Journal of Sport Nutrition and Exercise Metabolism, 14*(5), 594–606.

Judkins, C., Hall, D. & Hoffman, K. (2007). Investigation into supplement contamination levels in the US

market. HFL. Retrieved March 9, 2010, from http://www.usatoday.com/sports/hfl-supplement-research-report.pdf

Kanaley, J. A. (2008). Growth hormone, arginine and exercise. *Current Opinion in Clinical Nutrition and Metabolic Care, 11*(1), 50–54.

Kerksick, C., Harvey, T., Stout, J., Campbell, B., Wilborn, C., Kreider, R., et al. (2008). International Society of Sports Nutrition position stand: Nutrient timing. *Journal of the International Society of Sports Nutrition, 5*, 18.

Kerksick, C. M., Rasmussen, C. J., Lancaster, S. L., Magu, B., Smith, P., Melton, C., et al. (2006). The effects of protein and amino acid supplementation on performance and training adaptations during ten weeks of resistance training. *Journal of Strength and Conditioning Research, 20*(3), 643–653.

Kerstetter, J. E. (2009). Dietary protein and bone: A new approach to an old question. *American Journal of Clinical Nutrition, 90*(6), 1451–1452.

Koopman, R., Saris, W. H., Wagenmakers, A. J., & van Loon, L. J. (2007). Nutritional interventions to promote post-exercise muscle protein synthesis. *Sports Medicine, 37*(10), 895–906.

Kreider, R. B., & Campbell, B. (2009). Protein for exercise and recovery. *Physician and Sportsmedicine, 37*(2), 13–21.

Kurpad, A. V. (2006). The requirements of protein & amino acid during acute & chronic infections. *The Indian Journal of Medical Research, 124*(2), 129–148.

Lambert, C. P., Frank, L. L., & Evans, W. J. (2004). Macronutrient considerations for the sport of bodybuilding. *Sports Medicine, 34*(5), 317–327.

Larson-Meyer, D. E. (in press). Vegetarian athletes. In C. Rosenbloom (Ed.), *Sports nutrition: A practice manual for professionals.* Chicago: American Dietetic Association.

Leydon, M. A., & Wall, C. (2002). New Zealand jockeys' dietary habits and their potential impact on health. *International Journal of Sport Nutrition and Exercise Metabolism, 12*(2), 220–237.

Little, J. P., Forbes, S. C., Candow, D. G., Cornish, S. M., & Chilibeck, P. D. (2008). Creatine, arginine α-ketoglutarate, amino acids, and medium-chain triglycerides and endurance and performance. *International Journal of Sport Nutrition and Exercise Metabolism, 18*(5), 493–508.

Little, J. P., & Phillips, S. M. (2009). Resistance exercise and nutrition to counteract muscle wasting. *Applied Physiology, Nutrition, and Metabolism, 34*(5), 817–828.

Mettler, S., & Meyer, N. L. (2010). Food pyramids in sports nutrition. *SCAN'S PULSE, 29*(1), 12–18.

Mettler, S., Mitchell, N., & Tipton, K. D. (2010) Increased protein intake reduces lean body mass loss during weight loss in athletes. *Medicine and Science in Sports and Exercise, 42*(2), 326–337.

Millward, D. J. (2001). Protein and amino acid requirements of adults: Current controversies. *Canadian Journal of Applied Physiology, 26*(Suppl.), S130–S140.

Millward, D. J. (2004). Macronutrient intakes as determinants of dietary protein and amino acid adequacy. *Journal of Nutrition, 134*(6 Suppl.), 1588S–1596S.

Millward, D. J., Layman, D. K., Tomé, D., & Schaafsma, G. (2008). Protein quality assessment: impact of expanding understanding of protein and amino acid needs for optimal health. *American Journal of Clinical Nutrition, 87*(5), 1576S–1581S.

Moore, D. R., Robinson, M. J., Fry, J. L., Tang, J. E., Glover, E. I., Wilkinson, S. B., et al. (2009). Ingested protein dose response of muscle and albumin protein synthesis after resistance exercise in young men. *American Journal of Clinical Nutrition, 89*(1), 161–168.

Mullins, V. A., Houtkooper, L. B., Howell, W. H., Going, S. B., & Brown, C. H. (2001). Nutritional status of U.S. elite female heptathletes during training. *International Journal of Sport Nutrition and Exercise Metabolism, 11*(3), 299–314.

Negro, M., Giardina, S., Marzani, B., & Marzatico F. (2008). Branched-chain amino acid supplementation does not enhance athletic performance but affects muscle recovery and the immune system. *Journal of Sports Medicine and Physical Fitness, 48*(3), 347–351.

Newsholme, E. A., & Blomstrand, E. (2006). Branched-chain amino acids and central fatigue. *Journal of Nutrition, 136*(1 Suppl.), 274S–276S.

Nogueira, J. A., & Da Costa, T. H. (2004). Nutrient intake and eating habits of triathletes on a Brazilian diet. *International Journal of Sport Nutrition, 14*(6), 684–697.

Onywera, V. O., Kiplamai, F. K., Boit, M. K., & Pitsiladis, Y. P. (2004). Food and macronutrient intake of elite Kenyan distance runners. *International Journal of Sport Nutrition and Exercise Metabolism, 14*(6), 709–719.

Osterberg, K. L., Zachwieja, J. J., & Smith, J. W. (2008). Carbohydrate and carbohydrate + protein for cycling time-trial performance. *Journal of Sports Science, 26*(3), 227–233.

Poortmans, J. R., & Dellalieux, O. (2000). Do regular high protein diets have potential health risks on kidney function in athletes? *International Journal of Sport Nutrition and Exercise Metabolism, 10*(1), 28–38.

Rodriguez, N. R., DiMarco, N. M., Langley, S., American Dietetic Association, Dietetians of Canada, & the American College of Sports Medicine (2009). Position of the American Dietetic Association, Dietitians of Canada and the American College of Sports Medicine: Nutrition and athletic performance. *Journal of the American Dietetic Association, 109*(3), 509–527.

Roth, E. (2008). Nonnutritive effects of glutamine. *Journal of Nutrition, 138*(10), 2025S–2031S.

Rowlands, D. S., & Thomson, J. S. (2009). Effects of beta-hydroxy-beta-methylbutyrate supplementation during resistance training on strength, body composition, and muscle damage in trained and untrained young men: A meta-analysis. *Journal of Strength and Conditioning Research, 23*(3), 836–846.

Rowlands, D. S., Thorp, R. M., Rossler, K., Graham, D. F., & Rockell, M. J. (2007). Effect of protein-rich feeding on recovery after intense exercise. *International Journal of Sport Nutrition and Exercise Metabolism, 17*(6), 521–543.

Saunders, M. J., Luden, N. D., & Herrick, J. E. (2007). Consumption of an oral carbohydrate-protein gel improves cycling endurance and prevents postexercise muscle damage. *Journal of Strength and Conditioning Research, 21*(3), 678–684.

Sawitzke, A. D., Shi, H., Finco, M. F., Dunlop, D. D., Bingham, C.O., III, Harris, C. L., et al. (2008). The effect of glucosamine and/or chondroitin sulfate on the progression of knee osteoarthritis: A report from the glucosamine/chondroitin arthritis intervention trial. *Arthritis and Rheumatism, 58*(10), 3183–3191.

Schoeller, D. A., & Buchholz, A. C. (2005). Energetics of obesity and weight control: Does diet composition matter? *Journal of the American Dietetic Association, 105*(5, Suppl. 1), S24–S28.

Seebohar, B. (2011). *Nutrition periodization for endurance athletes: Taking traditional sports nutrition to the next level*. Boulder, CO: Bull.

Sherwood, L. (2010). *Human physiology: From cells to systems*. Belmont, CA: Thomson/Wadsworth.

Shimomura, Y., Yamamoto, Y., Bajotto, G., Sato, J., Murakami, T., Shimomura, N., et al (2006). Nutraceutical effects of branched-chain amino acids on skeletal muscle. *Journal of Nutrition, 136*(2), 529S–532S.

Soric, M., Misigoj-Durakovic, M., & Pedisic, Z. (2008). Dietary intake and body composition of prepubescent female aesthetic athletes. *International Journal of Sport Nutrition and Exercise Metabolism, 18*(3), 343–354.

Stout, J. R., Cramer, J. T., Zoeller, R. F., Torok, D., Costa, P., Hoffman, J. R., et al. (2007). Effects of beta-alanine supplementation on the onset of neuromuscular fatigue and ventilatory threshold in women. *Amino Acids, 32*(3), 381–386.

Sundgot-Borgen, J. (1993). Nutrient intake of female elite athletes suffering from eating disorders. *International Journal of Sport Nutrition, 3*(4), 431–442.

Sweeney, K. M., Wright, G. A., Brice, G. A., & Doberstein, S. T. (2010). The effect of beta-alanine supplementation on power performance during repeated sprint activity. *Journal of Strength and Conditioning Research, 24*(1), 79–87.

Tang, J. E., Moore, D. R., Kujbida, G. W., Tarnopolsky, M. A., & Phillips, S. M. (2009). Ingestion of whey hydrolysate, casein, or soy protein isolate: Effects on mixed muscle protein synthesis at rest and following resistance exercise in young men. *Journal of Applied Physiology, 107*(3), 987–992.

Tang, J.E. & Phillips, S.M. (2009). Maximizing muscle protein anabolism: the role of protein quality. *Current Opinion in Clinical Nutrition and Metabolic Care*, 12, 66-71.

Tarnopolsky, M. A., MacDougall, J. D., & Atkinson, S. A. (1988). Influence of protein intake and training status on nitrogen balance and lean body mass. *Journal of Applied Physiology, 64*(1): 187–193.

Tipton, K. D., Elliott, T. A., Cree, M. G., Wolf, S. E., Sanford, A. P., & Wolfe, R. R. (2004). Ingestion of casein and whey proteins result in muscle anabolism after resistance exercise. *Medicine and Science in Sports and Exercise, 36*(12), 2073–2081.

Valentine, R. J., Saunders, M. J., Todd, M. K., & St Laurent, T. G. (2008). Influence of carbohydrate-protein beverage on cycling endurance and indices of muscle disruption. *International Journal of Sport Nutrition, 18*(4), 363–378.

van Essen, M., & Gibala, M. J. (2006). Failure of protein to improve time trial performance when added to a sports drink. *Medicine and Science in Sports and Exercise, 38*(8), 1476–1483.

van Thienen, R., van Proeyen, K., Vanden Eynde, B., Puype, J., Lefere, T., & Hespel, P. (2009). Beta-alanine improves sprint performance in endurance cycling. *Medicine and Science in Sports and Exercise, 41*(4), 898–903.

Whey protein. Monograph. (2008). *Alternative Medicine Review, 13*(4), 341–347.

WHO Technical Report Series. (2007). Protein and amino acid requirements in human nutrition. Report of a Joint WHO/FAO/UNU Expert Consultation. Retrieved [date] from http://whqlibdoc.who.int/trs/WHO_TRS_935_eng.pdf

Wilkinson, S. B., Tarnopolsky, M. A., Macdonald, M. J., Macdonald, J. R., Armstrong, D., & Phillips, S. M. (2007). Consumption of fluid skim milk promotes greater muscle protein accretion after resistance exercise than does consumption of an isonitrogenous and isoenergetic soy-protein beverage. *American Journal of Clinical Nutrition, 85*(4), 1031–1040.

Wu, G. (2009). Amino acids: Metabolism, functions, and nutrition. *Amino Acids, 37*(1), 1–17.

6 Fats

Learning Plan

- **Classify** fats according to their chemical composition.
- **Distinguish** between saturated and unsaturated, monounsaturated and polyunsaturated, *cis* and *trans*, and omega-3, -6, and -9 fatty acids.
- **Describe** the digestion, absorption, transportation, and storage of fat.
- **Explain** the metabolism of fat, including mobilization, transportation, uptake, activation, translocation, and oxidation.
- **Explain** ketosis and the effect it may have on training.
- **Describe** how the body uses fat to fuel exercise.
- **State** fat recommendations for athletes and calculate the amount of fat needed daily.
- **Identify** sources of dietary fat.
- **Assess** an athlete's dietary fat intake.
- **Evaluate** dietary supplements related to fat metabolism.

Oils contain heart healthy fatty acids.

Pre-Test | Assessing Current Knowledge of Fats

Read the following statements about fats and decide if each is true or false.

1. Athletes typically need to follow a very low-fat diet.

2. At rest, the highest percentage of total energy expenditure is from fat, not carbohydrate.

3. To lose body fat, it is best to perform low-intensity exercise, which keeps one in the fat-burning zone.

4. To improve performance, endurance athletes should ingest caffeine because more free fatty acids are oxidized for energy and muscle glycogen is spared.

5. A low-calorie, low-carbohydrate diet that results in ketosis is dangerous for athletes because it leads to the medical condition known as ketoacidosis.

The word *fat* is used in many different ways. In nutrition, fats are energy-containing nutrients found in food. In medicine, fats are known as **lipids**, large fat-containing components in the blood. In physiology, a fat is a long chain of carbon molecules. In all of these disciplines, fat is also used to describe the body's long-term storage site for fats, although the precise term is adipose tissue. In everyday language *fat* is often used as an adjective, describing a body weight that is greater than desirable.

Fats are important nutrients for athletes because they are a primary energy source at rest and during low-intensity activity. Along with carbohydrates, fats are an important energy source for moderate-intensity exercise. Aerobic training can increase the body's ability to utilize fats. All humans need some fat in their diets to provide the essential fatty acids that the body cannot manufacture and for the absorption of the fat-soluble vitamins. Fats in foods also help to satisfy hunger.

Fats are the most concentrated form of energy, containing 9 kilocalories per gram (kcal/g). Athletes who expend a lot of energy during daily training need to include a sufficient amount of fat in their diets or they risk consuming too few calories to support their rigorous training. However, fat intake needs to be relative to carbohydrate and protein intakes, and many athletes are consuming too much fat and not enough carbohydrate and/or protein. Many athletes wish to lose body fat, and reducing dietary fat is a strategy frequently used to meet this goal. Lowering dietary fat intake can be appropriate, but too low of a fat intake can also be detrimental

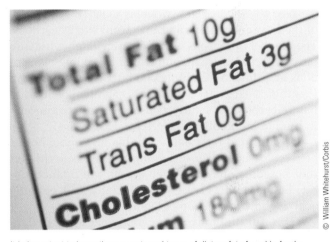

It is important to know the amounts and types of dietary fats found in foods.

© William Whitehurst/Corbis

to performance and health. Fat is an important nutrient to support training and performance, but because certain fats are associated with chronic diseases, notably cardiovascular diseases, their influence on health and wellness must be considered as well.

6.1 Fatty Acids, Sterols, and Phospholipids

Fats vary in their chemical composition. The predominant fats in food and in the body are triglycerides, which are made up of three fatty acids attached to a glycerol molecule. Sterols, such as cholesterol, and phospholipids, phosphate-containing fats, are also

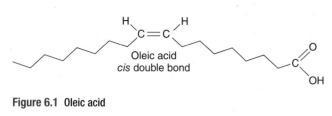

Figure 6.1　Oleic acid

Oleic acid is an 18-carbon fat that is found in plant and animal fats.

found in food and in the body. These three classes of fat compose the category known as lipids.

Fatty acids vary due to their chemical composition.

To understand the differences in the various fats, one must look closely at their chemical composition. This discussion begins with fatty acids, which are chains of carbon and hydrogen ending with a **carboxyl** group (a carbon with a **double bond** to oxygen and a single bond to an oxygen/hydrogen, written as COOH). The length of the fatty acid chain can range from 4 to 24 carbons. The number of carbons will be an even number because fatty acid chains are manufactured by adding two carbons at a time. An example of a fatty acid is shown in Figure 6.1. The fatty acid in this example, oleic acid, has 18 carbons. The fatty acids used most commonly in human metabolism have 16 or 18 carbons.

A saturated fatty acid contains no double bonds between carbons. The term *saturated* refers to the fact that no additional hydrogen atoms can be incorporated. An example of one saturated fatty acid, palmitic acid, is shown at the top of Figure 6.2.

Unsaturated fatty acids contain one or more double bonds between carbons, reducing the number of

Lipid: General medical term for fats found in the blood.

Cholesterol: A fatlike substance that is manufactured in the body and is found in animal foods.

Cardiovascular disease: Any of a number of diseases that are related to the heart or blood vessels.

Carboxyl group: Carbon with a double bond to oxygen and a single bond to oxygen/hydrogen.

Double bond: A chemical bond between two atoms that share two pairs of electrons.

Unsaturated fat: Fatty acids containing one or more double bonds between carbons.

Monosaturated fat: A fat containing only one double bond between carbons.

Polyunsaturated fat: A fatty acid with two or more double bonds between carbons.

Hydrogenated/hydrogenation: A chemical process that adds hydrogen. In food processing, used to make oils more solid.

Omega: The terminal carbon formed by the double bond between carbons that is counted from the last carbon in the chain (farthest from the carboxyl group carbon).

hydrogen atoms that can be bound to the structure. When only one double bond between carbons is present, it is referred to as a **monounsaturated** fatty acid (*mono* means "one"). When two or more double bonds are present, these fatty acids are referred to as **polyunsaturated** fatty acids (*poly* means "many"). Unsaturated fatty acids have 16 to 22 carbons and from one to six double bonds. Examples are shown in Figure 6.2.

When double bonds between carbons are present, as in the case of mono- and polyunsaturated fatty acids, the fatty acid can be in the **cis** or **trans** formation. *Cis* refers to groups that are on the same side of the double bond between carbons. *Trans*, which means "across" or "on the other side," refers to groups that are on opposite sides of the double bond between carbons. The vast majority of unsaturated fatty acids occur naturally in the *cis* form. The *cis* form allows fatty acids to "bend," which is an important feature when these fatty acids are incorporated into cell membranes. Although some *trans* fatty acids are found in nature, most are produced synthetically through the addition of hydrogen atoms to an unsaturated fatty acid. This results in the fatty acid chain being "straight." This **hydrogenation** process is used in commercial food processing to make liquid oils more solid (for example, soybean oil made into margarine) and to increase the shelf life of the product. Figure 6.3 illustrates unsaturated fatty acids in their *cis* and *trans* forms, and the Spotlight feature explains why *trans* fat intake should be as low as possible.

Polyunsaturated fatty acids can be further distinguished by their fatty acid series. This refers to the presence of a double bond between carbons that is counted from the last carbon in the chain (farthest from the carboxyl group carbon). This terminal carbon is referred to as **omega** or n-. The three fatty acid series are termed omega-3 (n-3), omega-6 (n-6), or omega-9 (n-9) families. Figure 6.4 shows an example of each. Omega-3 and -6 fatty acids are discussed in more detail in the next section.

Most fats in food are in the form of triglycerides.

Fatty acids are the building blocks of fat found in food. Nearly 95 percent of all the fat consumed in the diet is in the form of triglycerides. A triglyceride is composed of four parts—three fatty acids attached to a glycerol—as shown in Figure 6.5. The three fatty acids can all be the same, but a triglyceride usually contains a combination of different fatty acids. The nature of the individual fatty acids that make up a triglyceride influence the temperature at which the fat will melt. Those triglycerides with unsaturated fatty acids tend to be liquid at room temperature and are known as oils. Those triglycerides that contain primarily saturated fatty acids do not melt until the temperature is higher and are solid at room temperature. Although the term *triglyceride* is commonly used, it should be noted that the technically

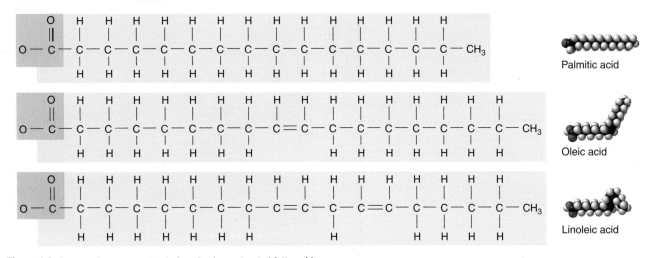

Figure 6.2 Saturated, monounsaturated, and polyunsaturated fatty acids

Palmitic acid (palmitate) is a 16-carbon saturated fat found in plant and animal fats. Oleic acid is an 18-carbon monounsaturated fat found in plant and animal fats. Linoleic acid is an 18-carbon polyunsaturated fat found in the seed and oil of plants such as corn, safflower, sunflower, soybean, and peanuts.

correct term is **triacylglycerol**, and this latter term is frequently used in the scientific literature.

The triglycerides found in food contain a combination of saturated, monounsaturated, and polyunsaturated fatty acids. Foods can be grouped according to the predominant fatty acid. For example, coconut oil contains 92 percent saturated fatty acids, 6 percent monounsaturated fatty acids, and 2 percent polyunsaturated fatty acids. Coconut oil is classified as a saturated fat because saturated fatty acids predominate. Other foods that contain predominantly saturated fatty acids include palm kernel oil, beef, and milk and milk products that contain fat (for example, whole milk, butter, ice cream).

Oils are predominantly, but not exclusively, polyunsaturated fatty acids. Most vegetable oils contain about 10–15 percent saturated fatty acids. Oils are usually distinguished by their predominant unsaturated fatty acid, which is either a mono- or polyunsaturated fat. Olive oil and canola oil contain primarily monounsaturated fatty acids. In other oils, such as safflower or corn oil, the polyunsaturated fatty acids predominate. Figure 6.6 illustrates the fatty acid distribution of some fat-containing foods.

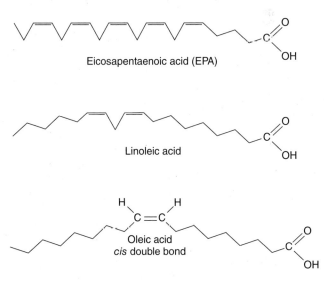

Figure 6.4 Omega-3, -6, and -9 fatty acids

Eicosapentaenoic acid (EPA) is an omega-3 fatty acid found in marine fish. Linoleic acid, an omega-6 fatty acid, is found in a variety of vegetable oils. Oleic acid, an omega-9 fatty acid, is found in animal and plant fats, particularly in olive oil.

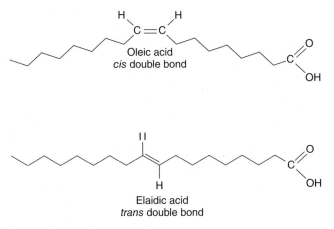

Figure 6.3 *cis* and *trans* formations

Oleic acid, which naturally occurs in the *cis* formation, is an 18-carbon monounsaturated fat that is widely found in plant and animal fats. Elaidic acid is the *trans* isomer of oleic acid and is found in hydrogenated vegetable oils such as margarine.

Triacylglycerol: The formal term for triglycerol.

Two essential fatty acids cannot be manufactured by the body.

Two 18-carbon fatty acids are essential fatty acids—**linoleic** and **alpha-linolenic** (α-linolenic). The body cannot manufacture these essential fatty acids, so they must be consumed in the diet. Fortunately, these two essential fatty acids are widely found in food. Linoleic, an omega-6 fatty acid, is in many vegetable oils such as corn, soy, safflower, and sunflower oils. Alpha-linolenic, a member of the omega-3 family, is found in soy, canola, and flaxseed oils. It is also found in leafy green vegetables, fatty fish, and fish oils.

Linoleic: An essential fatty acid.

Alpha-linolenic: An essential fatty acid.

Vegetable oils are excellent sources of monounsaturated and polyunsaturated fatty acids.

Spotlight on…

Trans Fatty Acids

In the early 1900s, many foods in the United States contained lard or butter, fats that are high in saturated fatty acids. By the middle of the century, health professionals were recommending a reduction in saturated fats so food manufacturers began to look for alternatives. This led to the hydrogenation of vegetable oils, a process that results in oils becoming more solid. These hydrogenated oils began to replace lard or butter in traditional products and were a prime ingredient in new "snack" products that were being introduced to consumers. Hydrogenation of vegetables oils give products a desirable texture and a longer shelf life but also produces *trans* fatty acids, a type of fat that naturally occurs in only a small number of foods. Prior to the widespread use of hydrogenated vegetable oils, *trans* fatty acid intake in the United States was low.

By the mid-1990s, it was becoming clear that *trans* fatty acids were associated with an increased risk for cardiovascular disease (see Chapter 13). *Trans* fatty acids result in an increase in the concentration of low-density lipoproteins, cholesterol transporters that have a tendency to deposit cholesterol in arteries. Health professionals now recommend that *trans* fatty acid intake be as low as possible. Specifically, the American Heart Association recommends that *trans* fat be limited to no more than 1 percent of total caloric intake. For example, a person consuming a 2,000 kcal diet should consume approximately 2 g or less of *trans* fatty acids (20 kcal). To meet this recommendation a person would need to avoid foods that contain or are prepared with partially hydrogenated oils.

There is no federal law in the United States that regulates the amount of *trans* fatty acids in foods. Instead, consumers are urged to read the Nutrition Facts label to determine *trans* fatty acid intake. However, this is nearly impossible for consumers to do because foods with less than 0.5 g *trans* fat can list the amount of *trans* fat on the label as 0. One clue to is to check the ingredient list for the words *hydrogenated* or *partially hydrogenated oil*, an indication that the product may contain *trans* fatty acids. The amount of *trans* fatty acids present in foods eaten away from home is very difficult to assess.

Some cities and states have enacted laws to limit the use of *trans* fats, and some countries, such as Denmark, have nearly eliminated *trans* fatty acids from commercial products. In the United States, most food manufacturers are exploring the use of alternative processes in an effort to reduce the amount of *trans* fat in their products.

The average intake of *trans* fat in the United States is not known, but has been estimated to be about 2 percent of energy intake or twice that recommended by the American Heart Association. As a rule of thumb, consumers can lower *trans* fatty acid consumption by eliminating or reducing intake of foods that contain hydrogenated vegetable oils such as chips, pastries, and candy bars or are cooked in such oils, such as French fries. It should be noted that not all *trans* fatty acids are unhealthy. For example, conjugated linoleic acid (CLA) is a naturally occurring *trans* fat found in beef, lamb, and dairy products that may confer some health benefits. However, most of the *trans* fatty acids consumed in the United States do not come from naturally occurring sources (Remig et al., 2010).

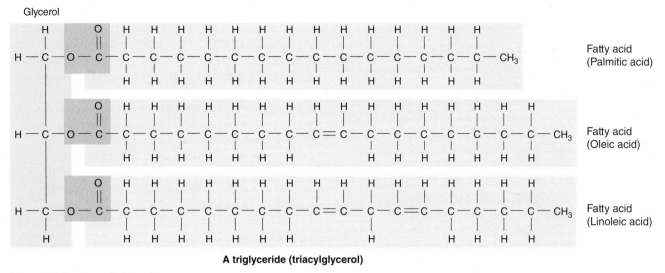

A triglyceride (triacylglycerol)

Figure 6.5 Structure of triglycerides

A triglyceride is composed of three fatty acids attached to a glycerol.

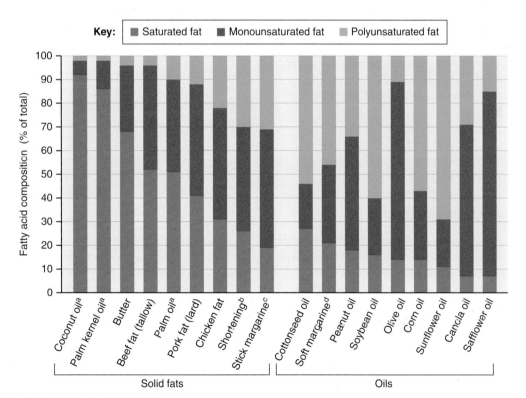

Figure 6.6 Fatty acid distribution in selected foods

a. Coconut oil, palm kernel oil, and palm oil are called oils because they come from plants. However, they are semi-solid at room temperature due to their high content of short-chain saturated fatty acids. They are considered solid fats for nutritional purposes. b. Partially hydrogenated vegetable oil shortening, which contains *trans* fats. c. Most stick margarines contain partially hydrogenated vegetable oil, a source of *trans* fats. d. The primary ingredient in soft margarine with no *trans* fats is liquid vegetable oil.

Source: U.S. Department of Agriculture, Agricultural Research Service, Nutrient Data Laboratory. USDA National Nutrient Database for Standard Reference, Release 22, 2009. Available at http://www.ars.usda.gov/ba/bhnrc/ndl. Accessed July 19, 2010.

Omega-3 fatty acids have many beneficial effects.

The omega series fatty acids include omega-3, -6, and -9 fatty acids, which are briefly summarized in Table 6.1. The best-known omega-3 fatty acids are alpha-linolenic acid (ALA), eicosapentaenoic acid (EPA), and docosahexaenoic acid (DHA). As noted on page 202, alpha-linolenic acid is an essential fatty acid and is widely distributed in food. The body can convert ALA into EPA and DHA, but the amount that can be converted is very limited. In the past 20 years, researchers have focused on the beneficial effects of EPA and DHA, which include their potential for reducing cardiovascular disease risk by having a positive effect on blood lipids (see Chapter 13), their impact on the immune system, and their roles in acute inflammation. The latter two are of particular interest for athletes who engage in prolonged and intense exercise, which results in an acute increase in inflammatory compounds and is a chronic physiologic stress on the immune system (see Spotlight on... Omega-3 Fatty Acids and Athletes).

Two physiologically important omega-6 fatty acids are linoleic acid, an essential fatty acid found in many vegetable oils, and arachidonic acid, which is found in animal fats. Arachidonic acid is a precursor to the prostaglandins and other related compounds, which are sometimes referred to as "local hormones." They act near the site where they are synthesized and generally have an effect at a low concentration. Excessive conversion of arachidonic acid to certain prostaglandins and related compounds has been associated with heart disease, acute and chronic inflammation, and immune dysfunction (Wall et al., 2010).

Though much emphasis is placed on the omega-3 fatty acids because of their positive effects on immune and inflammation responses, they do not operate independent of the omega-6 fatty acids. The omega-3 and -6 fatty acids are in competition for the same enzymes and they often have opposite effects on inflammatory processes. Therefore, it is recommended that the ratio of omega-6 to omega-3 fatty acids be approximately 4:1 or less. Typical Western diets have ratios of 15:1 or more due to high intake of animal fats and corn oil, which is rich in the omega-6 fatty acid, linoleic acid (Wall et al., 2010). To change the ratio, the high intake of corn, safflower, and soybean oils would need to be replaced with oils such as canola.

A frequently used term is *eicosanoids*. Eicosanoids are fatty acids that have 20 carbons. The omega-3 fatty acid, EPA, is an eicosanoid, as is the omega-6 fatty acid, arachidonic acid. Prostaglandins are also eicosanoids, and these hormone-like compounds can have potent biological effects.

Oleic acid is an omega-9 fatty acid. It is a monounsaturated fatty acid and can help reduce the risk for heart disease if it replaces saturated and *trans* fatty

Eicosanoid: Fatty acids that have 20 carbons produced in response to injury and inflammation

Cytokine: Signaling proteins secreted by lymphocytes (white blood cells)

Spotlight on...

Omega-3 Fatty Acids and Athletes

Athletes who engage in strenuous exercise are at risk for both acute inflammation and chronic suppression of the immune system (see Effects of Exercise on the Immune System in Chapter 5). Strenuous exercise is often defined as 1.5 hours or more of moderate- to high-intensity exercise (55–75 percent $\dot{V}O_{2max}$), thus many athletes in training fall into this category. After strenuous exercise there is an increase in many of the compounds associated with inflammation. This is a normal response to exercise and the muscle injury that occurs with exercise. There is also an increase in the body's anti-inflammatory compounds. Thus the goal for the athlete is to have the proper balance between the inflammatory and anti-inflammatory compounds.

As one way to offset the adverse effects of acute inflammation and chronic immune dysfunction, researchers have focused on the omega-3 fatty acids, alpha-linolenic acid (ALA), eicosapentaenoic acid (EPA), and docosahexaenoic acid (DHA).

ALA has not been shown to have the same benefits as EPA and DHA. To distinguish EPA and DHA from ALA, the term *marine omega-3* is sometimes used because ocean fish that feed on algae are excellent sources of EPA and DHA. Studies have shown that EPA and DHA can decrease the production of some of the inflammatory compounds, such as certain **eicosanoids** and **cytokines**. There are many different mechanisms involved including the replacement of arachidonic acid (a precursor to some inflammatory compounds) and a decrease in the release of certain cytokines (Wall et al., 2010).

As a rule of thumb, deep-water and "oily" fish have the highest amounts of EPA and DHA. Examples include salmon (~2 g in a 3 oz [100 g] portion), herring (2 g), mackerel (1.8 g), anchovies (1.5 g), sardines (1 g), and flounder and sole (0.5 g). The fish are high in EPA and DHA because of the algae that they eat, not because they naturally produce these fatty acids.

Table 6.1 Brief Summary of Omega-3, -6, and -9 Fatty Acids

	Name	Characteristics	Sources
Omega-3 fatty acids	α-linolenic acid (ALA)	Essential fatty acid that is widely distributed in food; can be used to form longer fatty acids	Abundant in flaxseed and rapeseed oils and green leafy vegetables; also found in soybean, canola, and linseed oils
	Eicosapentaenoic acid (EPA)	Eicosanoids made from EPA have anti-inflammatory effects and inhibit the formation of eicosanoids made from arachidonic acid, which have proinflammatory effects	Marine ("oily") fish; supplements
	Docosahexaenoic acid (DHA)	Major component of membrane phospholipids; eicosanoids made from DHA have anti-inflammatory effects and inhibit the formation of eicosanoids made from arachidonic acid, which have proinflammatory effects	Marine ("oily") fish; supplements
Omega-6 fatty acids	Linoleic acid (LA)	Essential fatty acid; cannot be synthesized by humans but can be metabolized into physiologically active compounds	Abundant in corn, safflower, and sunflower oils; also found in soybean, cottonseed, and peanut and peanut oils; cereals, whole grain bread, and animal fats
	Arachidonic acid (AA)	Cannot be synthesized by humans but can be metabolized into physiologically active compounds; eicosanoids made from AA have proinflammatory effects and are associated with chronic diseases, such as heart disease, diabetes, high blood pressure, and obesity	Small amounts in animal fats
Omega-9 fatty acid	Oleic acid	Can be made from unsaturated fats; cannot be used to make eicosanoids due to its chemical structure	Abundant in olive oil

Sources: Wall et al., 2010; Gropper, Smith, and Groff, 2009

It is estimated that the average North American intake of EPA and DHA is 130–150 mg/day (Denomme, Stark, and Holub, 2005).

Increasing the amount of EPA and DHA in the diet may not be feasible (for example, some people do not want to eat fish) or advisable (for example, some fish are contaminated with mercury and other toxins). Therefore, consumers may look to increase their EPA and DHA intake via fish oil supplements. It is common for fish oil capsules to contain 180 mg EPA and 120 mg DHA according to the label. Unfortunately, studies suggest that there is great variability in the actual content of the capsules. One study found that the mean content of the capsules tested was 82.4 percent of the content stated on the label for EPA and 90 percent of the labeled content for DHA. A laboratory analysis of 41 omega-3 supplements found that only 53 percent contained the amount of EPA claimed on the label (Abete et al., 2009). The Food and Drug Administration recommends that consumers not exceed a total of 3 g/d of EPA and DHA, with no more than 2 g/d from a dietary supplement (http://www.fda.gov).

Research studies examining the effect of EPA and DHA supplements on inflammation and immunity associated with exercise show mixed results. Those conducted on trained athletes have not shown definitively that these supplements counteract inflammation, positively affect the immune system, or improve performance (Nieman et al., 2009; Bloomer et al., 2009; Buckley et al., 2009; Raastad, Høstmark, and Strømme, 1997). It is theorized that the greatest benefit may be to those who have certain diseases (such as asthma) or are untrained. Trained athletes may not benefit because their anti-inflammatory responses may already be enhanced as a response to their training.

Cholesterol is found only in animal foods.

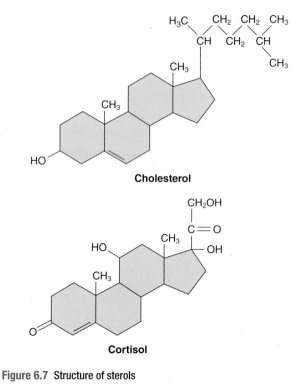

Figure 6.7 Structure of sterols

Note the four-ring nucleus (steroid) in these sterols.

acids in the diet (see Chapter 13). Olive oil is an excellent source of oleic acid.

The omega-3, -6, and -9 fatty acids continue to be active areas of research.

Sterols, such as cholesterol, and phospholipids are types of fat found in foods.

Almost 95 percent of the fat found in foods is in the form of triglycerides; the remaining fats in food are either sterols or phospholipids. Fatty acids are chains of carbon, but sterols have a different chemical composition. **Sterols** belong to a group of fats whose core structure is made up of four rings. This four-ringed nucleus is known as a steroid. Various side chains can be added to the steroid nucleus and many different compounds can be made, including cholesterol, vitamin D, and the steroid hormones, including the sex hormones (Figure 6.7). If the steroid-based compound has one or more **hydroxyl** (OH) groups attached and no **carbonyl** (=C=O) or carboxyl (COOH) groups, then the compound is known as a sterol.

The most common sterol found in food is cholesterol. Cholesterol is found only in animal foods such as meat, egg yolks, and milk and milk products in which the fat has not been removed. Cholesterol is an important component of human cell membranes. No plant foods contain cholesterol. However, plants do contain other sterols, known as phytosterols, including **estrogens**

similar to human sex hormones. Phytosterols are an important structural component of plant cell membranes.

Phospholipids are similar in structure to triglycerides but are distinguished by the inclusion of phosphate. They are a structural component of all living tissues, especially the cell membranes of animal cells. Lecithin is an example of a phospholipid. Although sterols and phospholipids are physiologically important, the primary nutritional focus in this chapter is the predominant fat found in foods and in the body—triglycerides.

Some fats lower the risk for heart disease.

Although the chemical differences between the various fats may be obvious, the impact of these differences on performance or health may not be immediately clear. From a performance perspective, fat intake is important because fat is metabolized to provide energy for low- to moderate-intensity exercise. The vast majority of fat used to provide this energy will come from triglycerides stored in either adipose tissue or muscle cells. The adipose and muscle triglycerides are manufactured from fatty acids originally found in food. From a purely metabolic perspective, the original source of the fat (saturated or unsaturated fatty acids found in food) is not important. What is important to the body is that it can metabolize the triglycerides for immediate energy or store them for future use as energy.

From a health perspective, there are certain fats that should be emphasized because they have been shown to lower the risk of cardiovascular disease.

Unsaturated fatty acids (that is, mono- and polyunsaturated fats) and omega-3 fatty acids are examples. Conversely, it is recommended that the consumption of saturated fatty acids be limited and that *trans* fatty acid intake be as low as possible, as excessive amounts of either may be detrimental to cardiovascular health. In some people, excess saturated fats and dietary cholesterol raise blood cholesterol concentration, and therefore, limiting these fats in the diet is recommended. The role that the various fats play in the development of cardiovascular and other chronic diseases is discussed in Chapter 13.

Because performance and health are both important, athletes first determine the total amount of fat appropriate in their daily diet. Then they can choose foods that contain certain kinds of fatty acids (for example, almonds are high in monounsaturated fat). The appropriate amount of fat intake for the athlete will depend on two factors: (1) overall energy (caloric) need and (2) macronutrient balance (that is, the proper proportions of carbohydrate, protein, and fat). These two factors will be explained later in this chapter.

Key points

- Differences in the fatty acids are due to their chemical composition.
- A triglyceride contains three fatty acids attached to a glycerol molecule.
- Triglycerides are the most abundant type of fat in both food and the body.
- The type of fat consumed can increase one's risk for heart disease, particularly the *trans* fatty acids found in snack-type foods.
- The type of fat consumed can influence inflammation and immune processes, such as the omega-3 and -6 fatty acids.

If cholesterol is needed for cell membranes, why is it often labeled as being "unhealthy"?

6.2 Digestion, Absorption, and Transportation of Fats

Digestion is the breakdown of foods into smaller parts. Absorption involves taking these smaller parts through the cells of the intestine where they will then be transported to other parts of the body. Fats present a particular digestion and absorption challenge because they are large molecules. Additionally, they do not mix well with water, the main component of blood, so transport in the blood requires most fat to be bound to protein.

Fat is primarily digested in the small intestine.

The process of fat digestion begins in the mouth and continues in the stomach to a small degree, but digestion of fat occurs predominantly in the small intestine. Undigested fat in the stomach has two effects: (1) it delays the emptying of the stomach contents (known as the gastric emptying rate) and (2) it results in a feeling of fullness (known as satiety). Athletes limit their fat intake temporarily in certain situations when they do not want gastric emptying delayed, such as during endurance events when rapid movement of carbohydrate-containing fluids into the intestine is beneficial. Conversely, athletes may include fat-containing foods in their meals when they want to avoid feeling hungry for several hours.

Fats are large and complex molecules that do not mix easily with water. Therefore, fats must be exposed to **bile** salts and digestive enzymes before they can cross the membranes of the intestinal cells (see figure 6.8). An important digestive enzyme is **pancreatic lipase**. As the name implies, this enzyme is secreted by the pancreas into the small intestine and helps to break down the large fatty acids into smaller components. Recall that all but 5 percent of the fat found in food is triglyceride (that is, triacylglycerol), which is composed of three fatty acids attached to a glycerol molecule. These three-unit fats are broken down by enzymes to two-unit fats known as **diglycerides** (that is, **diacylglycerols**), one-unit fats, **monoglycerides** (that is, **monoacylglycerols**), and free fatty acids. Phospholipids are involved in a similar digestive process, although the enzymes are different. Cholesterol is not broken down at this point. The process of digestion reduces the size of the fat particles and readies them for absorption by the mucosal cells of the intestine.

After being absorbed, the fatty acids are resynthesized into triglycerides.

The fat particles enter the mucosal cells by passive diffusion, a process in which molecules move from an area of higher concentration to an area of lower

Sterol: A fat whose core structure contains four rings.

Hydroxyl group: Formed when oxygen attaches to hydrogen (OH).

Carbonyl group: A group of atoms (=C=O).

Estrogen: A steroid hormone associated with the development of female sex characteristics.

Bile: Digestive fluid produced in the liver and stored in the gallbladder.

Pancreatic lipase: An enzyme secreted by the pancreas that helps to break down large fatty acids.

Diglyceride: A two-unit fat, known technically as a diacylglycerol.

Diacylglycerol: A two-unit fat.

Monoglyceride: A one-unit fat, known technically as a monoacylglycerol.

Monoacylglycerol: A one-unit fat.

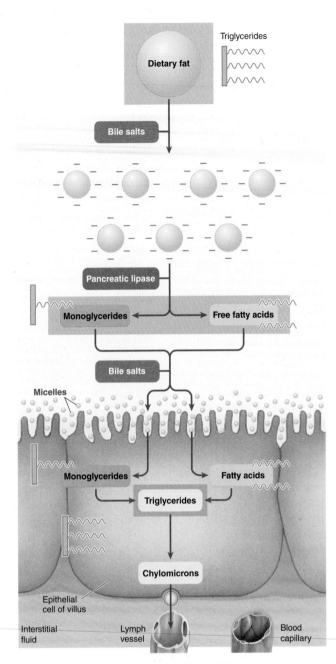

Figure 6.8 Digestion and absorption of fatty acids

Dietary fat, most of which is triglycerides, is broken down into smaller units so that intestinal cells can absorb it. Once absorbed, the monoglycerides and free fatty acids are resynthesized into triglycerides. The triglycerides are incorporated into chylomicrons, which transport fat via the lymphatic system.

concentration. Recall that the original triglyceride had 3 fatty acids and that each fatty acid could be a different length (from 4 to 24 carbons). Once in the mucosal cells, one-unit fats (monoglycerides) are resynthesized into triglycerides. The enzymes that assist in the resynthesis prefer to incorporate long-chain fatty acids (those with 12 to 18 carbons) into the newly synthesized triglycerides. The short- (4 carbons) and medium- (6 to 10 carbons) chain fatty acids will be

unchanged as they pass through the mucosal cell. The majority of dietary triglycerides are long-chained fatty acids (usually 16 to 18 carbons); thus the majority of fat eaten is broken down and then reincorporated into triglycerides by the mucosal cells.

The newly synthesized triglyceride then becomes part of a large protein and fat molecule known as a **chylomicron**. A chylomicron is one example of a **lipoprotein**. Lipoproteins transport fat throughout the body. Cholesterol and the partially digested phospholipids also become part of the chylomicron. The chylomicrons and the short- and medium-chain fatty acids are then ready to be transported out of the mucosal cells of the intestine.

The transportation of fats into the blood is a slow process.

To understand how fats are transported, one must understand the body's main transport fluids: blood and lymph. Blood consists of water, red and white blood cells, and many other constituents, including oxygen and nutrients. Blood enters tissues through arteries, leaves tissues through veins, and circulates within the tissues via the capillaries. Some components of blood are filtered out of the capillaries into the spaces of the tissue. This fluid is known as **interstitial fluid**. Most of the interstitial fluid is returned to the capillaries but some is not. That which is not returned is referred to as lymph. Lymph consists of white blood cells (which play an important role in immune function), proteins, fats, and other compounds. Lymph moves through its own set of vessels (the lymphatic system) that are separate from capillaries. Eventually, the lymph and blood vessels are joined near the heart.

The chylomicrons formed in the mucosal cells will be released slowly into lymphatic vessels. The release of the chylomicrons is an intentionally slow process that prevents a sudden increase in fat-containing compounds in the blood. Blood lipid (fat) levels are usually highest about 3 hours after fat consumption, but it may take as long as 6 hours for the dietary fat to be transported into the blood. The short- and medium-chain fatty acids in the mucosal cells that are not incorporated into chylomicrons will be released directly into the blood via the portal (liver) vein, where each will immediately be bound to a plasma protein, albumin.

The majority of the dietary fat consumed will have been incorporated into chylomicrons. This chapter will focus only on the transport and cellular absorption of the triglycerides contained in the chylomicrons, although fatty acids that are attached to albumin are absorbed in a similar manner. The transport and cellular absorption of the cholesterol found in the chylomicrons are more complicated, and an explanation is included in the discussion of the development of **atherosclerosis** in Chapter 13.

The chylomicrons circulate through all the tissues, but adipose, muscle, and liver tissues play very important roles in fat metabolism. The triglyceride portion of the chylomicron can be absorbed by adipose and muscle cells. As the chylomicron circulates in the blood, it comes in contact with **lipoprotein lipase** (LPL), an enzyme. LPL is found on the surface of small blood vessels and capillaries within the adipose and muscle tissues. This enzyme stimulates the release of the fatty acids from the triglyceride, which are then rapidly absorbed by the fat and muscle cells.

The absorption of fatty acids into the fat and muscle cells is of great importance to athletes and will be the focus of the metabolism section that follows. However, the liver also plays a substantial role that deserves mention here. In the liver, the triglyceride portion of the chylomicron is broken down and becomes part of the fatty acid pool, which will provide the fatty acids for the lipoproteins that the liver manufactures. Recall that a chylomicron is one example of a lipoprotein. Chylomicrons transport dietary (exogenous) fat. Other lipoproteins transport endogenous fats, that is, fats that are manufactured in the body by liver and other tissues. The functions of these lipoproteins, which include low- and high-density lipoproteins, are explained further in Chapter 13.

> **Key points**
>
> ■ The digestion and absorption of dietary fats involves several steps.
>
> ■ The transport of fats involves other compounds because fat and water do not mix.
>
> *Why might a high-fat meal 2 hours prior to a competition be a bad idea for an athlete?*

6.3 Metabolism of Fats

Once fats are digested, absorbed, and circulated, they are stored in the body largely in the form of triglycerides. The main sites of fat storage in the body are adipose tissue (in **adipocytes** [fat cells]), liver, muscle (as intramuscular triglycerides), and to a small degree in the blood. As has been previously discussed in Chapter 3, fats are metabolized for energy through oxidative phosphorylation, the aerobic energy system. In order to be metabolized, fats must be removed from storage, transported to cells, and taken up into mitochondria. There they are oxidized via the Krebs cycle, and ATP is produced via the electron transport chain.

Fat can be easily stored in the body.

The process of triglyceride formation is called **esterification**. The enzyme lipoprotein lipase exists in the walls of the capillaries that **perfuse** fat cells. When this enzyme is activated, it results in the breakdown

of circulating triglycerides from lipoproteins, freeing fatty acids for uptake into the fat cells. Once taken up into adipocytes, the fatty acids are re-formed into triglycerides for storage. The activity of LPL and the process of triglyceride formation are primarily stimulated by the hormone insulin. The pancreas secretes insulin in response to food consumption, particularly a meal containing carbohydrate. Therefore, in the hours after a meal (particularly a meal containing fat and carbohydrate), the body has the hormonal environment and the substrates that favor triglyceride formation and fat storage. An abundance of adipocytes can be found just beneath the skin (**subcutaneous fat**) and deep within the body surrounding the internal organs (**visceral fat**).

Muscle can also store fat, referred to as intramuscular triglycerides. Fat storage in muscle occurs primarily in muscle that is highly aerobic, for example, heart (myocardial) muscle and slow twitch (Type I) skeletal muscle. LPL in the capillary walls in muscle initiates this process in the same way that it does in adipocytes. Circulating lipoproteins are stimulated to break down the triglycerides they contain and release the fatty acids. The fatty acids are then taken up by muscle cells and reassembled into triglycerides for storage.

Fats are an excellent storage form of energy for several reasons. Compared to carbohydrates and proteins, fats contain more than twice the number of kcal per unit of weight—9 kcal/g for fat versus 4 kcal/g each for carbohydrate and protein. Therefore, fats are a very "energy dense" nutrient. When carbohydrate is stored in the form of glycogen, approximately 2 g of water are stored along with every gram of glycogen, increasing the weight without increasing the energy content. Fat is anhydrous, meaning it does not have water associated with it, making it an even more efficient storage form of energy on a per unit weight basis.

The importance of fat as a storage form of energy can be seen in comparison to carbohydrate storage. As discussed in Chapter 4, an average-size person stores approximately 500 g of carbohydrate in the form of muscle glycogen, liver glycogen, and blood glucose.

Chylomicron: A large protein and fat molecule that helps to transport fat.

Lipoprotein: A protein-based lipid (fat) transporter.

Atherosclerosis: Narrowing and hardening of the arteries.

Lipoprotein lipase: An enzyme that releases fatty acids from circulating triglycerides so the fatty acids can be absorbed by fat or muscle cells.

Adipocytes: Fat cells.

Esterification: The process of forming a triglyceride (triacylglycerol) from a glycerol molecule and three fatty acids.

Perfuse: To spread a liquid (for example, blood) into a tissue or organ.

Subcutaneous fat: Fat stored under the skin.

Visceral fat: Fat stored around major organs.

At 4 kcal/g, these carbohydrate reserves can provide approximately 2,000 kcal of energy. To illustrate this point, assume that a runner could use purely carbohydrate as a fuel source during an endurance run. These carbohydrate reserves would be essentially depleted in about 1½ hours. This same runner, however, has in excess of 100,000 kcal of energy stored as fat in adipocytes and as intramuscular triglycerides, enough energy to fuel more than 100 hours of running (assuming the runner could rely solely on fat metabolism).

Fat is an important source of energy for many athletes.

Fat is an excellent storage form of energy, and the metabolism of fat provides a high yield of ATP. However, there are a number of steps in the metabolism of fats that make the process complex and relatively slow. Fats must be mobilized from storage, transported to the appropriate tissues, taken up into those tissues, **translocated** (moved from one place to another) and taken up by mitochondria, and prepared for oxidation. At the onset of moderate-intensity, **steady-state** exercise, it may take 10–20 minutes for fat oxidation to reach its maximal rate of activity.

Fat mobilization, circulation, and uptake. In order for fats to be used in metabolism, triglycerides must first be taken out of storage. Triglycerides stored in adipocytes are broken down into the component parts—glycerol and the fatty acid chains. This process of **lipolysis** is catalyzed by an enzyme found in fat cells, **hormone-sensitive lipase** (HSL). Hormone-sensitive lipase is stimulated by catecholamines (epinephrine and norepinephrine), growth hormone, glucocorticoids (cortisol), and thyroid-stimulating hormone (TSH), and is inhibited by insulin. Therefore, mobilization of stored fat is inhibited after meals have been consumed when insulin is high, and is encouraged during the postabsorptive state (the period from approximately three to four hours after eating until food is eaten again), fasting, starvation, and when stressed. Stress, such as exercise, stimulates the sympathetic nervous system, which releases epinephrine and glucocorticoids such as cortisol from the adrenal glands, and growth hormone from the anterior pituitary gland into the blood. These hormones then interact with fat cells to promote lipolysis. Norepinephrine is released by nerve endings of sympathetic nervous system cells and also contributes to the activation of HSL and the mobilization of fat.

Following lipolysis, the fatty acid chains and glycerol circulate in the blood. Glycerol, a sugar alcohol, is water soluble and is easily carried in the blood. The liver contains the enzymes necessary to metabolize glycerol, so the liver takes up most of the glycerol that enters the circulation. There it

can be converted to glucose via gluconeogenesis or eventually reassembled into triglycerides. The fatty acid chains liberated in lipolysis are not water soluble, however, and must be attached to a carrier to be transported in the blood. The carrier is typically a plasma protein, the most common of which is albumin. When mobilized and circulated, these fatty acid chains are often referred to as free fatty acids (FFA), which is somewhat of a misnomer, as very few of them exist "free" in the circulation. Transport in the blood by albumin is crucial to the metabolism of fat. Disruption of the ability of albumin to transport fatty acids impairs fat metabolism, particularly during higher-intensity exercise, and will be discussed later in the chapter. The mobilization and transport of stored triglyceride is illustrated in Figure 6.9.

Once in the blood, fatty acids can be distributed to other tissues throughout the body for use in metabolism. A key element for the utilization of fat as a fuel is the delivery of adequate amounts of fatty acids to the tissue. Some fat-utilizing tissues have a greater number of receptor sites and transport mechanisms embedded in the cell wall that are unique for fatty acids and facilitate the movement of fatty acids from the blood into the cells. For example, heart muscle cells have a higher capacity for fat utilization than slow-twitch muscle fibers, which in turn have a higher capacity than fast-twitch muscle fibers. Certain tissues also have a more extensive network of capillaries that allow a more effective delivery of fatty acids, and therefore enhanced fat oxidation. Examples again include heart muscle and slow-twitch muscle fibers, particularly compared to fast-twitch muscle fibers, which have a less dense capillary network.

Activation and translocation within the cell. After mobilization, circulation, and uptake into the cells, the fatty acids must go through an activation step before being transported into the mitochondria. Similar to the beginning of glycolysis where two ATP are used in the first 3 steps to initiate the process, one ATP is used along with coenzyme A (CoA) to convert the fatty acid chain to a compound called fatty acyl-CoA, and one ATP is used to ensure that the reaction is irreversible. This process takes place in the outer mitochondrial membrane prior to the translocation of the fatty acid into the mitochondrial matrix. As with the beginning steps of glycolysis, this initial investment of energy is recouped by the subsequent production of a relatively large amount of ATP.

Once the fatty acid chain has been converted to fatty acyl-CoA in the mitochondrial membrane, it must be transported into the mitochondria where it goes through the process of beta-oxidation (β-oxidation) and is eventually metabolized aerobically. Fatty acyl-CoA is translocated into the mitochondria by a carnitine transport mechanism, catalyzed by a group of enzymes, known collectively as carnitine acyl

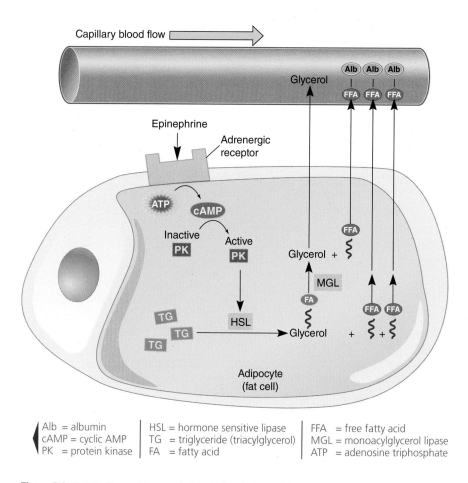

Alb = albumin
cAMP = cyclic AMP
PK = protein kinase

HSL = hormone sensitive lipase
TG = triglyceride (triacylglycerol)
FA = fatty acid

FFA = free fatty acid
MGL = monoacylglycerol lipase
ATP = adenosine triphosphate

Figure 6.9 Mobilization and transportation of stored triglyceride

Within an adipocyte, hormone-sensitive lipase will break down a stored triglyceride into glycerol and three fatty acids, which diffuse into the capillary circulation. The fatty acids are bound to albumin for transport. This process is initiated with the stimulation of adrenergic receptors on the cell membrane by stress hormones such as epinephrine and norepinephrine.

transferases (for example, CAT I and CAT II). This process involves removing the CoA, translocating the long fatty acid chain that is attached to carnitine into the mitochondria, then removing the carnitine and replacing the CoA (Figure 6.10). Each fatty acid chain has a specific carnitine acyl transferase. Palmitate, for example, is translocated into the mitochondria through the use of carnitine palmitate transferase (CPT1). Because this important transportation step involves carnitine, one approach that has been used to attempt to increase fat metabolism to improve endurance exercise performance has been the use of dietary carnitine supplements, which are discussed later in the chapter.

Because fatty acids are ultimately metabolized inside mitochondria, translocating the fatty acids into the mitochondria is a critical and potentially limiting step. Cells that have a large number of well-developed mitochondria, such as slow-twitch muscle fibers, have an increased capacity to take up and metabolize fat. Athletes with a genetic disposition for a greater number of slow-twitch muscle fibers have an increased

ability to metabolize fat. Regular endurance exercise training also results in an increase in the number and size of mitochondria in the trained muscles, which further enhances an athlete's ability to metabolize fat.

Beta-oxidation. There is an additional series of steps that must be completed before fatty acids can be metabolized to produce ATP—beta-oxidation. Beta-oxidation is a series of 4 chemical steps during which two-carbon segments are cleaved off the fatty acid chain, and converted to acetyl CoA. In a fatty acid

Translocation: Moving from one place to another.

Steady-state: Exercise or activity at an intensity that is unchanging for a period of time.

Lipolysis: The breakdown of a triglyceride (triacylglycerol) releasing a glycerol molecule and three fatty acids.

Hormone-sensitive lipase: An enzyme found in fat cells that helps to mobilize the fat stored there.

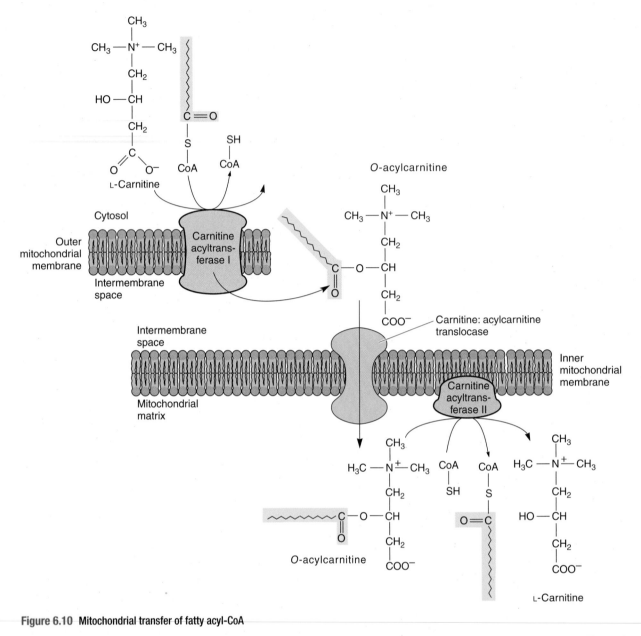

Figure 6.10 Mitochondrial transfer of fatty acyl-CoA

In the outer mitochondrial membrane, the CoA of fatty acyl-CoA is removed and joined to carnitine via carnitine acyltransferase. The acylcarnitine is then translocated into the mitochondria where the fatty acid is separated from the carnitine and is rejoined with a CoA.

chain, the first carbon is labeled alpha (α) and the second is labeled beta (β)—the carbons are clipped off the chain at the location of the second carbon, therefore the name β-oxidation. Once the two-carbon segment has been converted to acetyl CoA, it can enter the Krebs cycle as previously explained in Chapter 3. Each acetyl CoA that is derived from a fatty acid chain can be oxidized to eventually form 12 ATP. Each fatty acid that is metabolized contains an even number of carbons, so a series of acetyl CoA can be formed by β-oxidation. The final two carbons

of the fatty acid chain are already formed as acetyl CoA, so they do not have to go through the process of β-oxidation and can proceed directly to the Krebs cycle. In addition to forming acetyl CoA that can be oxidized, the process of β-oxidation involves two oxidation-reduction reactions, during which one NAD (nicotinamide adenine dinucleotide) and one FAD (flavin adenine dinucleotide) pick up electrons that can be shuttled through the electron transport chain to produce ATP. A detailed figure of β-oxidation can be found in Appendix K.

The value of fat metabolism can be seen in the number of acetyl CoA that are formed and oxidized, leading to a very large amount of ATP produced. Palmitate, a 16-carbon fatty acid, is used as an example. Metabolism of palmitate results in a total of eight acetyl CoA available for oxidation. After accounting for the ATP utilized in the activation steps, the complete oxidation of the acetyl CoA, and the additional electrons from α-oxidation, the final ATP production from the complete metabolism of palmitate is 129 ATP (calculated using the traditional 3 ATP yield from NADH and 2 ATP yield from FADH). When compared to the 2 ATP obtained from glucose by anaerobic glycolysis or even the 36 ATP from glucose by aerobic metabolism, fat metabolism has a substantial advantage in the provision of energy. The major disadvantages of fat metabolism are the numerous steps involved and the necessity to consume additional oxygen.

Ketosis. Fat is primarily metabolized as described above, but acetyl CoA may also be catabolized to produce ketone bodies—acetoacetate, β-hydroxybutyrate, and acetone. This is a normal metabolic pathway that is sometimes referred to as an "overflow" pathway. It is estimated that the liver can produce as much as 185 g of ketones daily and that after an overnight fast, ketones supply ~2–6 percent of the body's total energy needs. The normal blood ketone concentration is less than 0.05 mmol/L, with the highest concentration being present after an overnight fast. The normal urine ketone concentration is typically zero (VanItallie and Nufert, 2003).

Ketone production is increased when fatty acid oxidation is accelerated. This can occur when carbohydrate intake is low due to self-restriction or involuntary starvation. It can also occur when carbohydrate metabolism is impaired, which is the case for those with diabetes mellitus. For medical purposes, ketosis is defined as an abnormal increase in ketone bodies or a blood ketone concentration >0.06 mmol/L. In someone with diabetes, ketosis is a potentially dangerous complication because it can result in ketoacidosis, a condition in which the pH of the blood is more acidic than the body tissues. Ketoacidosis can result in diabetic coma and death. However, ketosis in those without diabetes rarely leads to ketoacidosis. In discussions of ketosis it is very important to distinguish between individuals who have diabetes and those who do not have diabetes.

To understand how ketones are formed, a brief review of carbohydrate metabolism is needed (see Chapter 4 for details). Glycolysis is the process that converts glucose to pyruvate. Pyruvate is transported into the mitochondria where it is metabolized to acetyl CoA. Acetyl CoA joins with oxaloacetate, the first step in the Krebs cycle. With a low carbohydrate intake, the body must find other sources of acetyl CoA, namely fatty acids and some amino acids. As more fatty acids are broken down to provide acetyl CoA, it begins to accumulate because of a low supply of oxaloacetate. In response to the accumulating acetyl CoA, ketone bodies are produced. The ketones become especially important as a source of energy for the brain because its usual energy source, glucose, is declining and fatty acids cannot cross the blood-brain barrier.

In the first two to three days of carbohydrate and energy restriction, the body produces glucose from alternative sources, such as lactate and amino acids provided by muscle. At this point, approximately two-thirds of the fuel used by the brain is glucose with the remaining one-third provided by ketones. Sustaining this alternative metabolic pathway would be untenable because too much muscle protein would need to be degraded to provide fuel for the central nervous system. As carbohydrate and energy restriction continues past a few days (referred to as starvation), more metabolic adaptations take place. Glucose will be primarily manufactured from glycerol (obtained from the breakdown of fatty acids), and ketones will become the primary source of fuel for the brain. After six weeks of starvation, about 70 percent of the brain's energy sources will be ketones, with less than 30 percent provided by glucose.

Starvation also produces changes in skeletal muscle. Restricted carbohydrate intake results in the muscle cells using a fuel source other than glucose, which is now in short supply because muscle glycogen has been depleted and dietary carbohydrate is not providing glucose for glycogen resynthesis.

A low-carbohydrate, calorie-restricted diet that results in ketosis is one popular method for weight loss used by overweight and obese individuals (see Chapter 13). Athletes may wonder if such a diet plan is appropriate for them. Ketosis is a result of both restricted energy (caloric) and carbohydrate intakes, dietary manipulations that would likely have a negative effect on an athlete's training. Some loss of muscle protein would occur, although a high dietary protein intake may help to **attenuate** muscle degradation. Although ketosis in a nondiabetic athlete will not likely result in ketoacidosis, the

Attenuate: To reduce the size or strength of.

disadvantages (for example, low glycogen stores, inability to sustain training and/or train at higher intensities) seem to outweigh the advantages (for example, loss of weight as fat) when viewed from the perspective of optimal performance.

Key points

- Fats are stored as triglycerides formed from a glycerol and three fatty acids.

- Fat is energy dense, containing 9 kcal per gram.

- In order to be used, fats must be translocated, or broken down, from their storage form and transported to the site of usage.

- In order to be metabolized aerobically, fatty acids must be converted to fatty acyl-CoA, transported into the mitochondria, and converted to acetyl CoA via β-oxidation.

- Fats are more readily metabolized by tissues with a higher aerobic capacity.

- Ketone bodies can be produced as a result of fat metabolism, and an overreliance on fat metabolism can result in ketosis.

Why does it take a number of minutes for fat metabolism to increase during aerobic exercise?

Why might a bodybuilder hear from other bodybuilders that ketosis is beneficial?

6.4 Fats as a Source of Energy during Exercise

Fat is an important fuel source for energy production at rest and during exercise. As mentioned in previous chapters, carbohydrate and fat are the two major fuel sources, with protein playing a much smaller role. The degree to which either of these fuel sources may contribute to the body's energy needs is dependent upon a variety of factors that will be discussed in this section, with an emphasis on the factors that influence fat oxidation (Jeukendrup, Saris, and Wagenmakers, 1998a, 1998b).

The use of fat as a fuel has a number of important advantages: fat is abundant in the food supply, energy dense (high caloric content on a per unit weight basis), stored in substantial amounts in adipose tissue, and when metabolized, provides a large amount of ATP. Disadvantages, however, include the many steps and the time involved in metabolizing fat. In addition, it should be recalled that fat can be used only in aerobic metabolism. Its use as a fuel is therefore limited to activities and exercise that can be supported by oxidative phosphorylation (for example, low- to moderate-intensity). This is in contrast to carbohydrate metabolism—glucose and glycogen can either be metabolized via oxidative phosphorylation or used anaerobically through anaerobic glycolysis to support higher-intensity activities. As shown in the example with palmitate, the complete metabolism of one molecule of a fatty acid results in the rephosphorylation of a large amount of ATP compared to the metabolism of glucose. The metabolism of fatty acids requires significantly more oxygen, however. When the ATP yield is analyzed relative to the amount of oxygen consumed, fat metabolism is less efficient than carbohydrate because it requires more oxygen for each ATP replenished.

It is important to know the relative (percentage) and absolute amount of fat utilized as a fuel.

To understand the use of fat as a fuel at rest and during exercise, it is important to understand how the utilization of this fuel is determined and expressed. Recall from Chapter 3 that the most common way of determining fuel utilization during aerobic exercise or activity is through indirect calorimetry and the use of the respiratory exchange ratio (RER). The RER is the ratio of carbon dioxide produced to the amount of oxygen consumed. Metabolism of carbohydrate requires the utilization of an amount of oxygen equal to the amount of carbon dioxide produced to give a higher RER, approaching or equal to 1.0. Because oxidation of fat requires a larger consumption of oxygen than the carbon dioxide produced, the RER during fat metabolism is lower, approaching 0.70.

The respiratory exchange ratio that is determined during a steady-state aerobic activity gives an indication of the percentage of the energy expenditure that is derived from fat oxidation relative to the percentage derived from carbohydrate oxidation. This is an expression of the fuel sources in a relative fashion, as a percentage of the total energy expenditure. Because only nonprotein sources of fuel are typically considered, fat and carbohydrate can be expressed as a percentage relative to the other. Sometimes it is important to know the percentage of energy expenditure provided by fat and carbohydrate oxidation (see case study on the next page).

The expression of fuel utilization in a relative fashion is useful, but does not provide any information about the actual amount of energy being expended, in other words, the total amount of energy that is being provided by fat and carbohydrate. An expression of the total amount of energy expenditure is termed absolute. A common way of expressing energy expenditure is the use of rate, the number of kilocalories (kcal) expended each minute (min).

Energy expenditure and fuel utilization during running can be measured.

SimSearch/Masterfile

If one knows both the absolute energy expenditure (kcal/min) and the relative contribution of each fuel (percentage from fat and carbohydrate), the absolute caloric expenditure (kcal/min) from fat and carbohydrate can be determined. The case study presented here will illustrate a number of important concepts about fat metabolism during exercise and the relationship of fat and carbohydrate as fuel sources.

Consider the case of a 49-year-old male who is in the process of training for his first marathon to celebrate his 50th birthday. He had some previous recreational running experience and had run several 10-kilometer races and two half-marathons. At 5'6" (168 cm) and 182 lb (82.5 kg), he realizes that losing weight and body fat will help him accomplish his goal of running a marathon. He participated in an indirect calorimetry study at rest and while running on a treadmill to learn more about his energy expenditure and fuel utilization at different running paces (Table 6.2).

First, observe the results from the metabolic study at rest. Heart rate is low (72 bpm), RER is low (0.77), and total energy expenditure is low (1.4 kcal/min). The low RER indicates that a large proportion (78.2 percent) of energy at rest is provided by fat oxidation, and a relatively small percentage (21.8 percent) is provided by metabolizing carbohydrate. Of the total energy expenditure of 1.4 kcal/min, 1.1 kcal/min is provided by fat. This is a typical metabolic response at rest, particularly if it has been several hours since the last meal—long enough for any food to be completely digested and absorbed (that is, postabsorptive state).

Fat oxidation during exercise. Next, observe the metabolic response when he begins running at a steady pace of 9 minutes and 30 seconds (9:30) per mile, a modest exercise intensity for this athlete. As one would expect, heart rate increases above resting (127 bpm) and total energy expenditure increases substantially to 11.5 kcal/min. In other words, he burns 11.5 kcal every minute he runs at this pace. The RER rises to 0.88, indicating the percentage of energy derived from fat has dropped to 40.8 percent whereas that provided by carbohydrate has increased to 59.2 percent. Even at this fairly modest exercise intensity, fat is no longer the predominant source of energy for running. However, even though the percentage of energy from fat has declined, the absolute number of kcal from fat metabolism has increased dramatically over what was seen at rest. This makes perfect sense—although the percentage is less (40.8 percent compared to 78.2 percent at rest), it is a smaller percentage of a much larger number—the total energy expenditure has increased 8-fold, from 1.4 kcal/min at rest to 11.5 kcal/min during exercise.

Finally, observe the metabolic response as exercise intensity continues to increase as the running pace gets faster, to 9:00, 8:30, and finally 8:00 minutes per mile. Again, as expected with increasing exercise

Table 6.2 Energy Expenditure and Fuel Utilization

Run pace (min/mile)	Heart rate (bpm)	RER	Percent energy from fat	Percent energy from CHO	Total energy expenditure (kcal/min)	Fat (kcal/min)	CHO (kcal/min)
Rest	72	0.77	78.2	21.8	1.4	1.1	0.3
9:30	127	0.88	40.8	59.2	11.5	4.7	6.8
9:00	138	0.89	37.4	62.6	13.4	5.0	8.4
8:30	144	0.91	30.6	69.4	14.2	4.4	9.8
8:00	153	0.92	27.2	72.8	15.3	4.2	11.1

Heart rate, RER, and relative and absolute energy expenditure from fat and carbohydrate at rest and at four different running paces for a 49-year-old male marathon runner.

Legend: min = minutes; bpm = beats per minute; RER = respiratory exchange ratio; CHO = carbohydrate; kcal = kilocalories

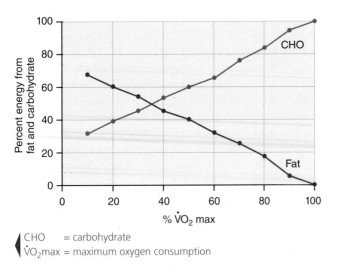

CHO   = carbohydrate
$\dot{V}O_2max$ = maximum oxygen consumption

Figure 6.11 Percentage of fat and carbohydrate used as exercise intensity increases

As exercise intensity increases (as a percentage of $\dot{V}O_{2max}$), the percentage of energy provided by fat metabolism decreases and the percentage of energy from carbohydrate metabolism increases.

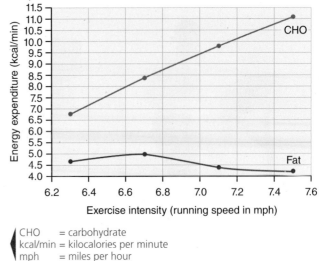

CHO   = carbohydrate
kcal/min = kilocalories per minute
mph   = miles per hour

Figure 6.12 Absolute fat and carbohydrate oxidation

Absolute energy expenditure from carbohydrate oxidation increases as exercise intensity increases and is higher than fat oxidation at all running paces. Absolute fat oxidation increases as the exercise intensity increases from a running speed of 6.3 to 6.7 mph, but then declines as the exercise intensity continues to increase.

intensity, the heart rate increases and the total energy expenditure increases. The RER also increases with each successive increase in exercise intensity, indicating a continuing decline in the percentage of energy that is supplied by fat metabolism and a continuing increase in that provided by carbohydrate metabolism. The pattern of response is also illustrated in Figure 6.11. There is a point in exercise intensity when carbohydrate increases and becomes the predominant source of energy and the utilization of fat declines to a lesser percentage. This point has been described as the "crossover concept" by Brooks, Fahey, and Baldwin (2005).

The *relative* expression of fuel utilization provides only a portion of the true picture, however. To complete the story, one must examine the *absolute* energy expenditure from fat and carbohydrate. As the running pace increases from 9:30 to 9:00 minutes per mile, there is an increase in absolute fat oxidation, from 4.7 to 5.0 kcal/min (Table 6.2). Even though the percentage of energy from fat has declined, the larger total energy expenditure indicates that the total absolute amount of fat being metabolized each minute has actually increased. There is a maximum point of absolute fat oxidation, however, and as speed continues to increase beyond this point, the absolute amount of fat metabolism goes down (in this case, from 5.0 to 4.4 then 4.2 kcal/min) along with a consistent decline in the percentage of energy derived from fat. The patterns of response of absolute fat and carbohydrate oxidation

by the marathon runner to increasing exercise intensity are shown in Figure 6.12.

If fat is such a good energy source, why does fat metabolism decline as exercise intensity increases? Again, understand that there is an initial increase in absolute fat oxidation with lower-intensity exercise. The body responds to the exercise by stimulating lipolysis through sympathetic nervous system stimulation of hormones such as epinephrine, norepinephrine, and growth hormone. The free fatty acids that are mobilized are delivered to exercising muscle fibers by the increase in blood flow that occurs with exercise activity. Increases in oxygen consumption by muscle are made possible by increases in breathing and circulation of oxygen-laden blood to the exercising muscle, so aerobic metabolism increases, including an increase in fat oxidation.

As exercise intensity continues to increase, however, a number of changes occur that may reduce the body's ability to metabolize fat. Increased exercise intensity means an increased reliance on fast-twitch muscle fibers. These fibers rely more on anaerobic energy systems, particularly anaerobic glycolysis, and have a relatively poor ability to oxidize fat. As these fibers use anaerobic glycolysis, they produce lactate. Lactate in the blood is known to inhibit the ability of the plasma protein albumin to bind and carry free fatty acids in the blood. Without this carrier mechanism available to transport the fatty acids, they essentially become "stuck" in the fat cells, unable to be transported anywhere else in the body. In addition, when the exercise intensity is higher, a successively larger proportion of the body's blood flow is diverted to exercising muscle, and may not be as available to circulate through adipose tissue to pick up fatty acids. Because fat oxidation requires greater oxygen consumption than carbohydrate oxidation to produce the same amount of ATP, it is more energetically efficient to oxidize carbohydrate under conditions when oxygen consumption is very high, that is, during higher-intensity exercise.

Fat oxidation during prolonged steady-state exercise. The subject in this case study had a goal to run a marathon, a prolonged distance run of 26.2 miles. The data from the metabolic study show what happens to energy expenditure and fuel utilization if the athlete runs at different intensities, but what metabolic response occurs when he runs at a steady pace for several hours? This runner did indeed complete a marathon in celebration of his 50th birthday. His final time for the marathon was 3 h, 55 min, 48 sec, an average pace of 9:00 minutes per mile. Reviewing the metabolic study data, at a pace of 9:00 minutes per mile this runner was obtaining approximately 37.4 percent of his energy from fat metabolism and 62.6 percent from carbohydrates. As a runner continues to exercise for a prolonged period of time, however, carbohydrate stores are reduced significantly, eventually leading to muscle glycogen depletion. As the available carbohydrate stores are diminished, the body has no choice but to rely more on fat oxidation. Therefore, a very common metabolic response to prolonged exercise is a slight, gradual decline in the RER, indicating a reduced reliance on carbohydrate and an increased dependence on fat oxidation. This metabolic response to exercise duration is illustrated in Figure 6.13.

Do you have to burn fat to lose fat? Like many people, the subject of this case study wanted to lose body fat by exercising. Because fat can be used as a fuel source during exercise, it seems logical to focus on fat "burning" as a primary way to reduce the body's fat stores. This idea has unfortunately led to erroneous recommendations by some people in the fitness industry. A common recommendation is to exercise at a "fat-burning" intensity. Commercial exercise equipment such as treadmills even come programmed with "fat-burning zones." This recommendation to "burn fat to lose fat" is faulty on at least two levels.

The recommended "fat-burning zone" is typically lower-intensity aerobic exercise, with the rationale that people burn more fat at lower exercise intensities. It is true that RER is typically lower during lower-intensity exercise (Table 6.2), but this indicates only that fat burning is higher *as a percentage of the total*

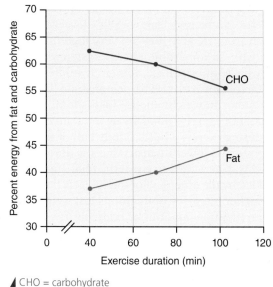

CHO = carbohydrate
min = minutes

Figure 6.13 Percentage of fat and carbohydrate used as exercise duration increases

As steady-state exercise continues for a prolonged period of time, the percentage of energy derived from fat may increase slightly as the percentage of energy derived from carbohydrate oxidation declines.

energy expenditure. If having the highest possible *percentage* of fat burning were the key factor in losing body fat, the best strategy would be to lie on the couch all day! A person typically has the lowest RER and highest *percentage* of fat utilization when they are at rest. The marathon runner in the case study was getting nearly 80 percent of his energy from fat metabolism while resting, but total energy expenditure (1.4 kcal/min) and the absolute number of fat calories (1.1 kcal/min) being burned were very low.

To find the exercise intensity that results in the highest *amount* of fat being burned, look again at the metabolic study of the marathon runner. The highest rate of absolute fat oxidation (5 kcal/min) was reached at a running pace of 9:00 min/mile, even though as a percentage, the relative contribution of fat had dropped to approximately 37 percent. If this subject runs for an hour at this pace, he would burn 804 kcal, 300 kcal of which are from fat. If he runs for an hour at a lower intensity (for example, slows down to 9:30 minutes per mile) he would burn only 690 kcal, 282 kcal of which are from fat. It is clear that a low exercise intensity does not necessarily result in a greater amount of "fat burning."

The second level of faulty reasoning behind these recommendations is based on the incorrect assumption that in order to lose body fat, a person must burn fat *during* exercise. Casual observation of athletes such as sprinters, weight lifters, and bodybuilders reveals that these athletes can be very lean with low levels of body fat, yet their exercise activities are very high intensity, relying mostly on anaerobic energy systems during which very little fat is utilized. Research studies, such as that by Grediagin et al. (1995), demonstrate the most important factor in weight and fat loss is total caloric expenditure, not exercise intensity or the source of the fuel used during exercise. In this study half of the subjects exercised at a lower intensity, but the duration of the exercise was increased to equal the total amount of kcal expended by the other half of the subjects that exercised at a higher intensity. Both groups lost weight and both lost an identical amount of body fat. If exercise

time (duration) is limited, the most appropriate strategy for weight/fat loss is to not worry about the source of the fuel, but to exercise in a way that maximizes caloric expenditure during the allotted time.

Returning to the example of the marathon runner, if he were to increase his running pace from 9:00 to 8:00 min/mile for his 1-hour run, he would expend 918 kcal instead of 804 kcal. The number of kcal expended from fat would decline from 300 to 250, but the total energy expenditure during the same amount of time is over 100 kcal greater. To summarize and answer the question—Do you have to burn fat to lose fat?—no, you do not need to burn fat during a short-term exercise session to accomplish long-term body fat loss.

Focus on research

Determining the Effect of High-Fat Diets on Fat Metabolism during Exercise and Endurance Exercise Performance

Because carbohydrate is an important fuel source during endurance exercise and is limited in supply, numerous attempts have been made to spare carbohydrates and improve performance by enhancing the use of fat as a fuel. Various experimental strategies to manipulate fat oxidation have been employed, including fasting, caffeine ingestion, and high-fat diets. Until the following study was conducted, few studies employed a high-fat diet for a time period long enough to allow for a stable metabolic adaptation to occur:

Phinney, S. D., Bistrian, B. R., Evans, W. J., Gervino, E., & Blackburn, G. L. (1983). The human metabolic response to chronic ketosis without caloric restriction: Preservation of submaximal exercise capability with reduced carbohydrate oxidation. *Metabolism*, *32*(8), 769–776.

What did the researchers do? This research paper was actually the second part of a research study with the overall purpose of examining the metabolic adaptations of healthy, lean adults to a long-term carbohydrate-restricted diet that resulted in ketosis. The researchers used a subset of the subjects in the overall research project in a "study within a study" to determine the effect of the chronic high-fat diet on endurance exercise capability. Five of the subjects were well-trained cyclists who, in addition to consuming the prescribed high-fat diet, continued their exercise training for the duration and completed an endurance exercise test at the beginning and end of the study.

In this well-designed and conducted study, subjects first consumed a standardized, balanced diet for a week before switching to a high-fat diet. The high-fat diet provided the same amount of kilocalories (eucaloric) but restricted carbohydrate to less than 20 grams per day, and was composed of

approximately 85 percent fat. Sufficient protein, vitamins, and minerals were included to ensure good health, as the high-fat diet resulted in ketosis and was followed for 4 weeks. The study design provided for good dietary and lifestyle control because meals were prepared and served to the subjects, and although they could go about their normal daily activities, they lived in a clinical research center during the study. Before the high-fat diet was initiated and at the end of 4 weeks of adaptation to the high-fat diet, a test of endurance exercise ability was performed. This test, often referred to as a "time to exhaustion" test, consisted of the subject pedaling on a cycle ergometer for as long as possible at a power output that corresponded to 60–65 percent of his or her $\dot{V}O_{2max}$. The amount of time the subject could pedal at this intensity was recorded as his or her endurance exercise capability.

Blood samples were taken regularly throughout the study to examine various metabolites and hormones in blood such as glucose, lactate, free fatty acids, 3-hydroxybutyrate, amino acids, insulin, and so on. Measurements were made of glucose oxidation rates using a radioactive isotope technique, and muscle biopsies were obtained to determine the use of muscle glycogen during the endurance exercise tests.

What did the researchers find? First, Phinney et al. (1983) observed that all subjects tolerated the carbohydrate-restricted, high-fat diet for 4 weeks without adverse effects and underwent substantial metabolic adaptation to the long-term diet. The subjects developed ketosis and resting blood glucose oxidation rates were significantly reduced, indicating an enhanced reliance on fat metabolism at rest. However,

The body adapts to endurance exercise training by improving its ability to metabolize fat.

Endurance exercise training results in an enhanced ability to oxidize fat. This is potentially advantageous to endurance athletes—if they can rely more on fat metabolism during an endurance event, they may be able to "spare" the body's limited carbohydrate stores (that is, muscle glycogen) for use later in the event and improve their performance.

The regular stimulus of chronic exercise training taxes the oxidative phosphorylation energy system and the oxidative pathways of fat and carbohydrate metabolism. Over time a number of physiological adaptations occur that enhance the body's fat oxidation capability. Fatty acids are mobilized from adipocytes more easily and are taken up into muscle cells more readily. Cardiovascular adaptations include an increase in the capillary network in muscle, which allows for an enhanced delivery of fatty acids. One of the most important adaptations to endurance training that aids fat oxidation is an increase in mitochondrial mass in the muscle due to an increase in both the number and size of these important oxidative organelles. Mitochondrial mass can double in response to endurance training and results in an overall increase in activity of oxidative enzymes, once again enhancing the body's ability to metabolize fat.

---

fasting blood glucose levels did not fall below normal limits. For the cycling endurance exercise test, after 4 weeks on a carbohydrate-restricted, high-fat diet the total endurance exercise time increased from 147 to 151 minutes; however, this 4-minute increase was not statistically significant. After 4 weeks on the high-fat diet, the muscle glycogen levels of the subjects before the exercise test were substantially lower, but they finished with the same low levels as they did on the first exercise test, indicating a reduced reliance on muscle glycogen as a fuel source. Measurements of the glucose oxidation rate during the first 2 hours of the exercise test showed a fourfold decline in metabolism of blood glucose. The respiratory exchange ratio (RER) during the exercise test was dramatically lower, indicating a doubling in the rate of fat metabolism.

What was the significance of this research study? This research study was well designed and conducted and provided significant insight into the metabolic adaptations to a carbohydrate-restricted, high-fat diet by lean, healthy adults. It demonstrated that a longer time period was needed for these adaptations to take place and that the diet was well tolerated and did not have significant adverse effects on the subjects. The main adaptations were a significant shift in metabolism away from carbohydrate to enhanced fat oxidation. The authors were careful to conclude that "maximal oxygen uptake and aerobic endurance capacity … are not compromised."

Answering the question: Do high-fat diets enhance fat metabolism and improve endurance exercise performance? Whereas this study yielded very important information about the metabolic adaptations of endurance athletes to high-fat diets, the results of Phinney et al. (1983) cannot be used to definitively answer this question. Although this study is cited frequently to support the use of low-carbohydrate, high-fat diets by endurance athletes, it is often done so erroneously as there are a number of important limitations revealed by a close examination of the design, methods, and results. People who cite this study often claim that the high-fat diet actually resulted in improved exercise performance because the exercise time to exhaustion increased by 4 minutes. The authors carefully note, however, that this difference was not a statistically significant increase. In fact, they use these results to conclude that the high-fat diet did not result in the exercise capability declining, which is what many would predict. There is also an important distinction to be made based upon the type of exercise test used. The "time to exhaustion" tests do not mimic real competitive events very well and are therefore often referred to as tests of endurance capacity or capability. Additionally, the intensity of the test was set at 60–65 percent of $\dot{V}O_{2max}$, an exercise intensity well below what a competitive cyclist would be performing at during a race. Although adaptation to the high-fat diet resulted in a dramatic shift toward increasing fat oxidation during exercise, it is likely that increased carbohydrate oxidation would be seen if exercise intensity was increased to a more competition-like level. The authors recognized this and concluded, "There are indications…that the price paid for such extreme conservation of carbohydrate… appears to be a limitation on the intensity of exercise that can be performed." Ultimately, it is important to understand that even well-designed and conducted studies have their limitations, and scientists and practitioners must take care to avoid applying the results of research studies to situations that are not warranted.

The enhanced ability to metabolize fat during exercise can be seen at the same absolute exercise intensity and to a lesser degree at the same exercise intensity relative to the athlete's maximum. In other words, an athlete running at the exact same pace after months of training will be able to oxidize more fat due to the increase in fat oxidation capability. This is the same absolute exercise intensity, but because the athlete's maximal exercise capacity has increased, it now represents a lower percentage of his or her maximum. For example, if a 9:00 min/mile running pace was 75 percent of the runner's $\dot{V}O_{2max}$ before training, it may now be only 65 percent of the new, higher $\dot{V}O_{2max}$. To a certain degree, the ability to oxidize fat increases at the same relative exercise intensity in response to endurance training. If this runner increases the running pace to the point that it represents 75 percent of the new, higher $\dot{V}O_{2max}$, fat oxidation will be slightly higher after training than at the same percentage of maximum before training. This increased ability to metabolize fat during exercise may be beneficial to the endurance athlete during training and during some periods of lesser effort during competition; however, during a competitive event the intensity of exercise at "race pace" means that the predominant fuel source will be carbohydrate.

Dietary manipulations to enhance fat metabolism. Long-term consumption of a diet that is high in fat will result in a metabolic adaptation to favor fat oxidation at rest and during exercise of certain intensities. As a strategy to increase fat metabolism and potentially spare carbohydrate usage, the use of high-fat diets ("fat loading") has been studied to determine if endurance performance can be improved.

The study of Phinney, Bistrian, Evans, Gervino, and Blackburn (1983) is often referenced in support of this strategy. In this study a group of cyclists were fed a high-fat ketogenic diet for 4 weeks that provided less than 20 g of carbohydrate each day. Endurance ability was tested before and after the high-fat diet period by having the cyclists ride as long as they could on a cycle ergometer in the lab at an intensity of approximately 60 percent of $\dot{V}O_{2max}$. The RER was significantly lower, indicating an enhancement in fat oxidation, and the "ride-to-exhaustion" time was 4 minutes longer on average after a month on the high-fat diet. The results of this study are sometimes used erroneously to suggest high-fat diets improve performance, although the 4-minute difference was not statistically significant. This study is also sometimes used to suggest that adaptation to a high-fat diet at least does not hurt endurance performance. However, this study used a small sample size, only five cyclists, and in the posttest of endurance only three of the subjects increased their endurance time whereas two decreased their time. Most relevant to the question of performance for the endurance athlete was

the intensity of the performance task. The cyclists were asked to ride at a relatively low percentage of their maximum for as long as they could—a measure of endurance time that fails to mimic the demands of an endurance event. The exercise intensity was also one that was far below that of an athlete actually competing in an endurance race event.

Subsequent research studies (Carey et al., 2001) and reviews (Helge, 2000) show that consumption of a high-fat diet can indeed alter the metabolic response at rest and during light- to moderate-intensity exercise in favor of fat oxidation. However, there does not appear to be any practical benefit when it comes to endurance performance.

A newer strategy by endurance athletes to try and achieve greater fat oxidation is the use of an acute, or short-term carbohydrate-restricted diet. Such a diet would result in low muscle glycogen stores. When carbohydrate is not available due to dietary restriction, one of the adaptations the body makes is to increase fat availability, specifically by increasing free fatty acids in the blood and/or increasing muscle triglycerides (Hawley and Burke, 2010). This approach, sometimes referred to as "train low," is explained in Chapter 4.

Effect of caffeine on fat usage. In the past, caffeine was thought to improve endurance performance by enhancing free fatty acid release, which in turn would spare

---

The Internet café

What about Wikipedia?

Wikipedia (**http://www.wikipedia.org**) is an open-access, free, collaborative encyclopedia that is written and edited by volunteers. Known as "the free encyclopedia that anyone can edit," it is currently available in more than 240 languages and is one of the top 10 most frequently visited websites. Some of the guiding principles include accuracy, verifiability (referencing of reliable sources), and an unbiased (neutral) point of view. Articles are rated for quality as follows (low to high): Stub, Start, C-class, B-class, Good Article, A-class, and Featured Articles. Articles classified as Good, A-class, or Featured are considered high quality.

The English-language Wikipedia is a frequently used source for health-related information that has been found to contain no more mistakes or inaccuracies than more traditional sources of information such as *Encyclopedia Britannica* (Laurent & Vickers, 2009). The number of Featured Articles is small compared to the total number of articles, and much of the information on Wikipedia could be more complete. However, Wikipedia provides freely accessible and generally reliable information, although the information is always in flux and its accuracy depends on the willingness of knowledgeable people to edit Wikipedia articles. As with other sources of information on the Internet, Wikipedia should not be relied upon as a sole source of information without verification from other reliable sources.

muscle glycogen. Caffeine has now been well studied, and the consensus opinion is that caffeine may enhance free fatty acid mobilization during endurance exercise; however, fat oxidation is not significantly increased, nor is muscle glycogen spared. The benefit of caffeine is likely due to its role as a central nervous system stimulant, resulting in a heightened sense of awareness and a decreased perception of effort (Burke, 2008; Davis and Green, 2009; Ganio et al., 2009). Caffeine is discussed further in the dietary supplement section.

Key points

■ Fat utilization can be determined by the respiratory exchange ratio (RER), which is the ratio of carbon dioxide produced to oxygen consumed.

■ The percentage of energy derived from fat metabolism decreases as activity and exercise intensity increases.

■ As exercise intensity increases, the total amount of energy derived from fat metabolism increases, reaches a peak, and then begins to decrease.

■ If exercise continues at the same intensity for long periods of time (for example, hours), the relative amount of energy derived from fat metabolism increases slightly (if the athlete does not consume any carbohydrate during exercise).

■ Regular aerobic exercise training increases the body's ability to metabolize fat, particularly during exercise at the same absolute exercise intensity.

■ Increasing the amount of fat in the diet increases reliance on fat as an energy source at rest and during lower intensities of exercise, but does not result in improved performance at higher exercise intensities.

■ Caffeine may increase the mobilization of fatty acids during exercise, but does not appear to improve performance due to increased fat metabolism.

Why does fat metabolism increase slightly after several hours of exercising at the same intensity?

Why is it erroneous to recommend a low "fat-burning" exercise intensity to help someone lose body fat?

6.5 Fat-Related Dietary Supplements

Supplements that are involved in "fat burning" are marketed to both athletes and nonathletes. Those sold as ways to enhance the metabolism of fat, reduce inflammation, enhance the immune system, or improve performance are summarized in Table 6.3. Supplements sold as an adjunct to weight loss are discussed in Chapter 11.

Caffeine is a central nervous stimulant that helps to delay fatigue.

Athletes in various sports use caffeine to improve performance and to delay fatigue. The strongest scientific evidence shows that caffeine can enhance endurance performance in distance runners, cyclists, and cross country skiers. There is also evidence that caffeine can improve performance for those engaged in high-intensity activities lasting 1–20 minutes, including runners, cyclists, swimmers, and rowers. Athletes in other sports have been studied, but the evidence for improving performance is not conclusive (Burke, 2008; Davis and Green, 2009; Ganio et al., 2009).

Caffeine improves endurance performance because it is a central nervous system stimulant that results in a heightened sense of awareness and a decreased perception of effort. Although caffeine may enhance free fatty acid mobilization during endurance exercise, more fat is not oxidized and reliance on muscle glycogen is not reduced. Many athletes engaged in resistance training use caffeine to delay fatigue while training. Caffeine may enhance contractile force in skeletal muscle during submaximal contractions. It may also increase the athlete's threshold for pain or perceived exertion, which could result in longer training sessions (Tarnopolsky, 2008).

Table 6.3 Summary of the Safety and Effectiveness of Fat-Related Supplements

Supplement	Safety	Effectiveness
Caffeine	Safe at recommended doses	Effective as a central nervous system stimulant; effective for improving endurance performance and high-intensity activities lasting up to 20 minutes
Carnitine	Safe at recommended doses	Not effective
Medium-chain triglycerides (MCT)	Safe at recommended doses, but unknown effect on blood lipids	Not effective
Omega-3 fatty acids	The Food and Drug Administration recommends intake of EPA and DHA not to exceed 3 g/d with no more than 2 g/d from supplement sources. The amount in the supplement may not match the amount listed on the label.	Not effective to reduce inflammation, enhance the immune system, or improve performance; shows promise for use in athletes with exercise-induced bronchoconstriction due to asthma

To achieve the desired effects, an athlete must have a significant amount of caffeine in the blood. In the past, the recommended dose for endurance athletes was 5–6 mg/kg of body weight. Newer studies in endurance athletes suggest that a more moderate dose is effective—2–3 mg/kg of body weight (Burke, 2008). Athletes should use larger doses with caution because caffeine has side effects, such as an increase in heart rate, gastrointestinal distress, and interference with sleep. The optimal dosage will need to be established by trial and error, and 2 mg/kg of body weight is a reasonable starting point. For the 110-pound (50 kg) person, a reasonable starting dose would be 100 mg of caffeine. The optimal timing of caffeine intake is being studied, but results are not yet conclusive (Burke, 2008).

Caffeine could be consumed in a variety of ways, including caffeine-containing pills (1 tablet = 100 mg), energy bars with caffeine (1 bar = 50 or 100 mg), caffeinated gels (1 oz = ~50 mg), strongly brewed coffee (8 oz = ~85 mg), or caffeine-containing soft drinks (12 oz = ~36 mg). Pills, energy bars, and gels are popular because they contain a standardized concentrated dose, whereas the caffeine content of brewed coffee can vary considerably.

Caffeine is legal and socially acceptable throughout the world. At certain concentrations caffeine is a banned substance by some sports-governing bodies, such as the National Collegiate Athletic Association (NCAA). In a postcompetition urine analysis, a urinary caffeine level exceeding 15 mcg/ml would subject an NCAA athlete to disqualification. However, such a level would be difficult to reach (that is, the equivalent of ~6 to 8 cups of caffeinated coffee 2 to 3 hours prior to competition) and would likely impair performance in other ways, such as shaking, rapid heartbeat, or nausea. However, some athletes have reached this level and were disqualified. The International Olympic Committee (IOC) monitors caffeine use in athletes but it does not disqualify anyone based on urinary caffeine level. The IOC has not found evidence of caffeine abuse by athletes, and not specifying a urinary caffeine threshold allows athletes to consume caffeinated beverages and use caffeine-containing cold medications without risk for disqualification (Burke, 2008).

Caffeine is considered safe at recommended dosages. However, it is addictive and sudden withdrawal results in severe headaches. There is no evidence that consuming caffeine results in dehydration or electrolyte imbalances in athletes. Caffeine recommendations are intended for adults, and there is concern about caffeine use by adolescents and children (Burke, 2008). Caffeine intoxication is possible but is rare. More information on caffeine is found in Chapters 7, 10, and 11.

Carnitine. Carnitine is essential to transport fatty acids into the mitochondria where they can be broken down for energy. Whenever a substance is known to have a direct role in metabolism, an intriguing question is raised: Would a concentrated amount, such as that found in a supplement, enhance the normal metabolic process?

Carnitine is found in food and can be synthesized in the body from the amino acid lysine. Deficiencies have been reported in humans but they are rare. As would be expected, in carnitine-deficient individuals supplementation normalizes long-chain fatty acid metabolism. Healthy adults are not carnitine deficient and exercise does not result in the loss of carnitine in the muscle. Studies have shown that carnitine supplements do not increase the carnitine content of muscles, probably because the transport of carnitine into the muscle is well controlled and only very small amounts are needed for proper fatty acid metabolism (Brass, 2004; Muller et al., 2002).

There is no compelling scientific evidence that carnitine supplements alter fat metabolism or improve performance (Kraemer, Volek, and Dunn-Lewis, 2008; Broad, Maughan, and Galloway, 2008). Supplement manufacturers recommend a dose of 2 to 4 g/day, which is the dose used in most scientific studies. Carnitine supplementation appears to be safe at these doses and there is no evidence that carnitine is detrimental to performance.

Medium-chain triglycerides. As described earlier in the chapter, medium-chain triglycerides contain 6 to 10 carbon atoms. They are rapidly absorbed via the portal vein and are easily transported into the mitochondria. For these reasons, MCT are sometimes advertised as being an energy source that is as readily available as carbohydrate. It is important to know if the use of MCT by endurance athletes could increase fat oxidation during moderate- to high-intensity exercise, reduce reliance on muscle glycogen stores, or enhance performance (Hawley, 2002; Horowitz and Klein, 2000).

Several studies of well-trained endurance athletes have found that MCT ingestion does not alter fat metabolism, spare muscle glycogen, or improve performance (Goedecke et al., 2005; Misell et al., 2001; Horowitz et al., 2000; Angus et al., 2000). In fact, Goedecke et al. (2005) found that ingestion of MCT by ultraendurance cyclists compromised sprint performance—high-intensity, short-duration cycling bouts required as part of some ultraendurance competitions. The negative effect on sprint performance may have been due to gastrointestinal upset from the MCT solution. This is an example of a supplement that not only fails to improve performance but also may actually impair it.

Omega-3 fatty acid supplements. Omega-3 fatty acid supplements (fish oil supplements) are marketed to athletes as a way to reduce inflammation, reduce the effects of oxidative stress, and counteract immune dysfunction associated with strenuous exercise. Typically, omega-3 fatty acid supplements contain eicosapentaenoic acid (EPA) and docosahexaenoic acid (DHA), although some may also contain alpha-linolenic acid (ALA). EPA is usually found in the greatest proportion. The dosages used in research studies are generally greater than 2.4 g/day of omega-3 fatty acids, with various mixtures of EPA, DHA, and, sometimes, ALA.

In studies of trained athletes, omega-3 fatty acid supplements do not appear to positively affect inflammation or immune responses or improve performance (Nieman et al., 2009; Bloomer et al., 2009; Buckley et al., 2009; Raastad, Høstmark, and Strømme, 1997). The greatest benefit of omega-3 fatty acid supplements containing EPA and DHA may be as a protective effect for those athletes who have exercise-induced bronchoconstriction due to asthma (Mickleborough et al., 2006). Malaguti et al. (2008) found damage to red blood cell membranes in the study group that received 3 g/d fish oil supplements, suggesting that excessive amounts may negatively affect some of the mechanisms the body has to counteract oxidative stress.

Ads targeting athletes often suggest supplementing with 1–2 g of omega-3 fatty acids daily. Recommendations for higher doses for trained athletes, particularly endurance athletes, have been made (Simopoulus, 2008); however, there is a lack of evidence to support such recommendations and there are concerns about high-dose supplementation with omega-3 fatty acids. No Dietary Reference Intake (DRI) has been established to date; however, the Food and Drug Administration recommends that consumers not exceed a total of 3 g/d EPA and DHA from all sources, with no more than 2 g/d from a dietary supplement (http://www.fda.gov).

Key points

- Caffeine is an effective central nervous system stimulant and can improve endurance performance and high-intensity activities lasting up to 20 minutes.

- Carnitine, MCT, and omega-3 fatty acids are not effective in healthy athletes.

- Research studies suggest that the use of omega-3 fatty acids in athletes with exercise-induced bronchoconstriction due to asthma is promising.

What might be the downside to an endurance athlete using caffeine as a performance enhancer?

6.6 Fat Recommendations for Athletes

The appropriate amount of dietary fat for the athlete will depend on two factors—overall energy (caloric) need and macronutrient balance. Recall that the four energy-containing nutrients are carbohydrate, fat, protein, and alcohol. Though each nutrient can be considered separately, the relationships among them are also important. Typically only carbohydrate, fat, and protein are included in macronutrient balance discussions. Alcohol is usually not included in general recommendations for athletes because it contains no essential nutrients and many athletes cannot legally consume alcohol due to age restrictions.

Total daily fat intake depends on total energy, carbohydrate, and protein intakes.

To determine the amount of dietary fat needed, one must also know how much carbohydrate, protein, and total energy (kcal) the athlete needs. This discussion assumes that the athlete is in energy balance. In other words, the athlete does not want to change body weight or composition and energy intake is equal to energy output. In such cases, how much dietary fat does such an athlete need to consume?

In some respects this is a mathematical problem. Because the athlete wishes to remain in energy balance, the daily number of kcal needed to match energy expenditure must be determined. The amount of carbohydrate necessary to support the demands of the sport and the goals of the training cycle must be established. The daily protein goal is also important information. Once those figures are obtained, the amount of fat can be calculated. The recommended fat intake will be determined within the context of energy and macronutrient balance as explained in the example below.

Emily, a 140 lb (~64 kg) elite 800 m runner, needs approximately 2,700 kcal (~42 kcal/kg) daily to maintain energy balance. To restore the glycogen used during her most demanding training cycle, she needs a carbohydrate intake of 7 g/kg (~445 g daily). Her goal for protein intake is 1.4 g/kg (~89 g daily). Together, carbohydrate and protein provide approximately 2,136 kcal. With an estimated total energy need of 2,700 kcal, fat needs to provide about 564 kcal or approximately 63 grams of fat daily.

The carbohydrate and protein recommendations are expressed on a gram per kilogram body weight basis (g/kg). In other words, carbohydrate and protein recommendations are stated as an absolute amount. They are not stated as a relative amount such as a percentage of total energy intake (for example, 60 percent of total calories as carbohydrate). Although carbohydrate and protein recommendations are commonly expressed on an absolute basis in the planning of athletes' diets, fat recommendations have not usually been expressed this way. It is common to express fat as a percentage of total energy intake, even though this method is inconsistent with carbohydrate and protein recommendations. However, that is changing as some sports dietitians and sports nutrition researchers are beginning to state fat recommendations on a g/kg basis.

A very general guideline for daily fat intake by athletes is approximately 1.0 g/kg. This figure is also consistent with the range recommended by the Dietary Reference Intakes (DRI) and the Dietary Guidelines, which is 20 to 35 percent of total caloric intake. Endurance athletes may need up to 2.0 g/kg to adequately replace intramuscular triglycerides (Horvath, Eagen, Fisher et al., 2000). When energy needs are exceptionally high, as is the case with ultraendurance athletes, the amount of fat in the diet may be as high as 3.0 g/kg (Seebohar, 2011). These g/kg guidelines are based on observations of the amount of fat trained athletes consume when they are in energy and macronutrient balance, not on research studies that have examined optimal dietary fat intake. In the example above, Emily's fat intake was approximately 1.0 g/kg or 21 percent of total energy intake, which puts her at the lower end of the Dietary Guideline recommendations for fat intake. If well-trained athletes consume the amount of carbohydrate and protein recommended, then the diet is relatively low or moderate in fat because the intake of carbohydrate and protein is relatively high. However, fat intake typically falls within the DRI and the Dietary Guidelines.

Unfortunately, many well-trained athletes do not meet minimum carbohydrate and protein guidelines. Heaney et al. (2010) assessed the intake of elite Australian female athletes in the sports of cycling and triathlon, netball, softball, track and field, volleyball, and water polo. Of the 57 athletes in this study, 65 percent did not meet the minimum recommendation for carbohydrate intake (5 g/kg/d) and 30 percent did not meet the minimum recommendation for protein intake (1.2 g/kg/d). Their mean macronutrient intake was 46 percent carbohydrate, 18 percent protein, and 31 percent fat. If these athletes do not want to increase caloric intake, then fat intake would need to be reduced if carbohydrate or protein is increased. In fact, the only groups of athletes in this study who did meet the carbohydrate and protein recommendations were the cyclists and triathletes. Their average macronutrient intake was 50.3 percent carbohydrate, 18.7 percent protein, 26.8 percent fat, and little or no alcohol. Their average fat intake was 87 g per day.

Although fat may be somewhat limited to accommodate higher carbohydrate and protein needs, athletes' diets do not need to be overly restrictive or devoid of fat. One study has shown that male and female distance runners (minimum 35 miles/wk) who consumed low-fat diets (16 percent of total energy intake as fat) for four weeks also consumed significantly fewer kilocalories when compared to medium- to high-fat diets of 31 percent and 44 percent of total energy, respectively. Interestingly, the low-fat diet that resulted in the consumption of 19 percent fewer kilocalories was also associated with a statistically significant decrease in endurance performance (Horvath, Eagen, Fisher et al., 2000).

In general, people believe that a high intake of dietary fat results in increases in body fat. Because of this belief, fat is perceived negatively and fat intake is often restricted (Wenk, 2004). However, increases in body fat are due to the excess consumption of total energy (kcal). The overconsumption of any of the energy-containing compounds—carbohydrate, fat, protein, and/or alcohol—can result in increased body fat. Dietary fat intake is not the sole determinant of the extent of body fat stores. Athletes should not consider fat a forbidden nutrient or be misled by the erroneous belief that all fat is unhealthy (see Spotlight on...Must an Athlete's Diet Be a "Low-Fat" Diet?).

Reducing caloric intake by reducing dietary fat intake over several weeks or months may help athletes achieve a loss of body fat.

Many athletes do not wish to be in energy balance; instead they wish to maintain an energy deficit for a period of time. The reason for the deficit is to force the body to use stored body fat for energy. The athlete's goal is to attain a lower percentage of body fat, which presumably will translate to a performance advantage. A low percentage of body fat may be advantageous to performance depending on the sport. For example, there is a potential performance advantage for a 10,000 m (6.2-mile) runner to attain a low percentage of body fat because there is less body mass (weight) to be moved. Excess body fat is "dead weight" as it does not produce any force to help with the exercise task and is extra weight that must be carried. However, performance and health can be negatively affected when body fat stores are too low or when dietary fat intake is severely limited. Given that a loss of body fat would be advantageous, how does the athlete best achieve an energy deficit?

Again, the athlete must consider the four energy-containing nutrients—carbohydrate, fat, protein, and alcohol. If the athlete is consuming all four of these nutrients, then the obvious nutrient to reduce is alcohol because its intake is not essential and may be counterproductive. If alcohol is not a part of the athlete's diet, then only the 3 remaining energy-containing nutrients could be adjusted. Sufficient carbohydrate is needed to adequately replenish glycogen stores, and sufficient protein is needed to maintain muscle mass. Since both of these nutrients are directly or indirectly related to performance, fat often becomes the nutrient to reduce by default.

Although athletes in many different sports may want to maintain an energy deficit over several weeks or months to produce a change in body fat, the sport of bodybuilding provides an excellent illustration. Six to 12 weeks prior to a contest, bodybuilders change their diets so they are deficient in energy on a daily basis. Protein intake is kept at a relatively high level, since they want to maintain the maximum amount of muscle mass possible, and weeks of low energy intake will result in some protein being used as an energy source. Carbohydrate intake must be adequate to replenish glycogen used during weight lifting and for aerobic activities. By design, precontest diets are low in fat, which results in an energy deficit when compared to energy intake in previous months. Adding or increasing aerobic exercise achieves a further energy deficit (Lambert, Frank, and Evans, 2004).

Consider the case of Kevin, a 220 lb (100 kg) bodybuilder who routinely consumes about 5,500 kcal (55 kcal/kg). His usual fat intake is about 1.5 g/kg or 150 g daily (~25 percent of total energy intake). Two to three months before a big contest he will reduce his energy intake to approximately 3,500 kcal. In addition to reducing caloric intake, he will slightly increase protein intake, slightly reduce carbohydrate intake, and substantially reduce fat intake. His precontest fat intake will likely be about 65 g daily or 0.65 g/kg, less than half of his usual fat consumption. During the 7 days before the contest, he will reduce fat intake even more in an effort to lose as much body fat as possible without sacrificing muscle mass or the ability to maintain training. After the contest he will return to his usual intake of 5,500 kcal, which includes a routine fat intake of about 1.5 g/kg daily.

If done correctly, a reduction in dietary fat intake can be a strategy that results in the safe loss of body fat. However, health professionals have concerns about athletes who maintain fat and energy deficits that last several months or years. Male and female distance runners and jockeys and female figure skaters, gymnasts, and dancers are examples of athletes who often exhibit chronic fat and energy deficits to attain or maintain a low percentage of body fat. The ultimate degree of leanness possible is largely determined by genetics, and forcing the body to maintain a very low percentage of body fat that is not biologically

Kevin Dodge/Masterfile

The fat intake of a bodybuilder will vary depending on the training cycle.

comfortable may be achievable only with a long-term semistarvation diet. In some cases, chronic or severe fat restriction is a result of a fat phobia—an irrational fear that dietary fat will become body fat. Dietary and psychological counseling can be beneficial in such cases. Athletes need to consider the effects on training, performance, and physical and mental health any time that fat and kilocalories are severely restricted.

Inadequate fat intake can negatively affect training, performance, and health.

Chronic inadequate fat intake, and the energy restriction that usually accompanies it, has the potential to negatively affect training, performance, and health. These effects may include: (1) inadequate replenishment of intramuscular fat stores, (2) inability to manufacture sex-related hormones, (3) alterations in the ratio of high- and low-density lipoproteins (HDL:LDL), and (4) inadequate fat-soluble vitamin intakes.

Intramuscular fat stores, such as muscle triglycerides, are reduced after endurance exercise and even more so after ultraendurance exercise. These muscular fat stores must be replenished. A routine very low-fat diet may not be sufficient for the resynthesis of muscle triglycerides, just as a routine low-carbohydrate diet does not provide enough carbohydrates for the re-synthesis of muscle glycogen (Pendergast, Leddy, and Venkatraman, 2000).

Chronic fat restriction may negatively impact the manufacture of sex-related hormones such as **testosterone**. Some studies of healthy men have shown that a low-fat diet (between 18 and 25 percent of total calories) with a high ratio of polyunsaturated to saturated fat lowered testosterone concentration (Dorgan et al., 1996; Hamalainen et al., 1984). There have also been reports of lowered testosterone concentration in wrestlers who consumed fat- and energy-restricted diets (Strauss, Lanese, and Malarkey, 1985). These studies did not examine the effect that low testosterone concentration may have on muscle mass, but they have been interpreted to mean that chronic and severe fat restriction is not desirable for male athletes because of the potential effect on testosterone production (Lambert, Frank, and Evans, 2004).

The effect of chronic fat restriction by females on the manufacture of sex-related hormones such as estrogen has been hard to ascertain. Female athletes with exercise-related menstrual irregularities tend to have both low-fat and low-energy intakes (De Cree, 1998). For some, their fat and energy restrictions are aspects of an eating disorder, **anorexia athletica** (Sudi et al., 2004). Because there are several interrelated factors, it is not known if and how one factor, the chronic low intake of dietary fat, influences the low estrogen concentrations that are observed. However, increasing both dietary fat and caloric intake is one facet of the treatment for those with anorexia athletica.

As part of the Dietary Reference Intakes for fat intake, it is recommended that daily dietary fat intake not be below 20 percent of total energy intake. Studies have shown that very low-fat diets in healthy individuals can result in declines in high-density lipoprotein (HDL) concentrations. HDL is a lipid carrier that tends to remove cholesterol from the surface of arteries and transports it back to the liver where the cholesterol can be metabolized. Low HDL concentrations are a risk factor for cardiovascular disease as explained in Chapter 13 (Institute of Medicine, 2002).

Inadequate fat-soluble vitamin intake is a concern when dietary fat intake is low (that is, <0.75 g/kg). Four vitamins are fat-soluble: vitamins A, D, E, and K. These vitamins are found in foods that contain fat, and a small amount of fat must be present for their proper absorption. Surveys of endurance and ultra-endurance athletes suggest that vitamin E intake is low in both males and females as a result of low fat and energy intakes (Tomten and Høstmark, 2009; Machefer et al., 2007).

Linoleic and alpha-linolenic are essential fatty acids that cannot be manufactured by the body.

Spotlight on...

Must an Athlete's Diet Be a "Low-Fat" Diet?

Many athletes wonder if they must routinely consume a low-fat diet. Before answering this question, the term *low-fat* must be clarified. For the general population, a low-fat diet has been defined historically as one that contains less than 30 percent of total calories as fat. This definition was developed in the mid-1960s when the average fat intake in the U.S. diet was 40–42 percent of total calories, a level associated with health risks (Chanmugam et al., 2003). Using this typical definition, athletes often consume a "low-fat" diet because fat intake is often less than 30 percent of total calories.

Because Americans are normally very sedentary, the typical diet consumed in the United States is too high in fat and calories. Today people eat more grams of fat, on average 76 g daily, and more calories than they did in the 1960s. The public health message is still the same—reduce intake of dietary fat, particularly saturated and *trans* fats.

If athletes consume sufficient calories to match their high caloric output due to exercise, they will likely find that their diets are moderate in fat rather than "low" in fat. However, some athletes follow a low-fat diet because they want to limit caloric intake in an effort to decrease body fat.

Although a low-fat diet can be appropriate for athletes, they should be aware that a very low-fat diet could be detrimental to training, performance, and health. A very low fat diet is typically defined as less than 15–20 percent of total calories as fat. In most cases, such diets are also too low in energy over the long term to support training and can result in impaired performance. A low-fat diet can result in a lower caloric intake, which can lead to a lower intake of other nutrients (Horvath, Eagen, Ryder-Calvin et al., 2000).

The reality for athletes is that their diets are usually lower in fat than those of the general population who are not attempting to lose weight. However, they do not need to be exceptionally low in fat. In fact, very low-fat diets may be detrimental for athletes.

Essential fatty acid deficiencies have been reported in humans that have diseases that result in fat malabsorption. However, in healthy adults essential fatty acid deficiencies are unlikely even among people who chronically consume low-fat diets. The reason is that about 10 percent of the fat stored in adipose cells is linoleic acid. This protects the adult body against essential fatty acid deficiencies that are a result of long-term fat and energy deficits (Institute of Medicine, 2002). Although essential fatty acid deficiencies are not likely, for the other reasons stated above, chronically low fat intake can be detrimental to performance and health.

Key points

- To determine the appropriate fat intake, the athlete must consider total caloric, carbohydrate, and protein intakes.
- Some athletes consume too much fat relative to the amount of carbohydrate and protein needed. They need to slightly increase carbohydrate and protein intakes and reduce fat intake to a degree.
- Some athletes chronically consume too little dietary fat as part of a semistarvation diet, and training, performance, and health may suffer.
- If done appropriately, reducing fat intake can be an effective strategy for body fat loss.

What concerns might you have if an athlete told you that she considers all fatty foods "bad"?

6.7 Translating Daily Fat Recommendations to Food Choices

Many athletes fail to consume an appropriate amount of fat. For some, fat intake is too high and for others fat consumption is too low. In addition to the total amount of fat needed, athletes should also be aware of the kinds of fatty acids that are associated with good health. Certain unsaturated fatty acids may help to reduce heart disease risk, whereas *trans* fatty acids are associated with an increased risk for cardiovascular disease. Excess saturated fat intake should be avoided. Therefore, it is important for athletes to know both the amounts and types of fat found in food.

The amount and type of fat in foods varies.

The Nutrition Facts label is a good starting point for determining the amount and type of fat found in food

Nutrition Facts

Serving Size 1 cup (228g)
Servings Per Container 2

Amount Per Serving	
Calories 250	Calories from Fat 110
	% Daily Value*
Total Fat 12g	**18%**
Saturated Fat 3g	**15%**
Trans Fat 3g	
Cholesterol 30mg	**10%**
Sodium 470mg	**20%**
Potassium 700mg	**20%**
Total Carbohydrate 31g	**10%**
Dietary Fiber 0g	**0%**
Sugars 5g	
Protein 5g	
Vitamin A	**4%**
Vitamin C	**2%**
Calcium	**20%**
Iron	**4%**

* Percent Daily Values are based on a 2,000 calorie diet. Your Daily Values may be higher or lower depending on your calorie needs.

	Calories	2,000	2,500
Total Fat	Less than	65g	80g
Sat Fat	Less than	20g	25g
Cholesterol	Less than	300mg	300mg
Sodium	Less than	2,400mg	2,400mg
Total Carbohydrate		300g	375g
Dietary Fiber		25g	30g

Figure 6.14 Nutrition Facts label

The fat-related information found on a food label includes Calories from fat, total, saturated and *trans* fat, cholesterol, and % Daily Value.

(Figure 6.14). The label is required to show the amount of total fat, saturated fat, *trans* fat, and cholesterol. The amount of monounsaturated or polyunsaturated fat may be included, but listing these values is voluntary. The number of Calories from fat is near the top of the label. Remember that all of these figures are based on the serving size listed, which may be smaller than the amount eaten. The % Daily Value is also listed, but this is based on a 2,000 kcal diet, which is lower than the amount of calories needed by many athletes in training.

Table 6.4 lists foods that are 100 percent fat or nearly 100 percent fat, such as oils and margarine, and the amount of fat in one serving. The predominant fat contained (for example, monounsaturated, polyunsaturated, or saturated) is noted. Most of these fats are added to foods or used for food preparation. Table 6.5 lists fat-containing foods that also

Testosterone: A steroid hormone associated with the development of male sex characteristics.

Anorexia athletica: An eating disorder unique to athletes. May include some elements of anorexia nervosa and bulimia and excessive exercise.

Table 6.4 Fats and Oils

Food*	Amount	Energy (kcal)	Fat (g)	Predominant type of fat**
Olives, black	6 medium	30	3	Monounsaturated
Olives, green	4 medium, stuffed	40	3	Monounsaturated
Olive oil	1 T	120	13.5	Monounsaturated
Canola oil	1 T	120	13.5	Monounsaturated
Peanut oil	1 T	120	13.5	Monounsaturated
Safflower oil, ~70% oleic	1 T	120	13.5	Monounsaturated
Safflower oil, ~70% linoleic	1 T	120	13.5	Polyunsaturated
Corn oil	1 T	120	13.5	Polyunsaturated
Soybean oil	1 T	120	13.5	Polyunsaturated
Flaxseed oil	1 T	115	13	Polyunsaturated
Margarine, liquid (squeezable)	1 T	100	11	Polyunsaturated
Margarine, soft (tub)	1 T	100	11	Polyunsaturated/ monounsaturated
Margarine, hard (stick)	1 T	100	11	Saturated
Mayonnaise	1 T	100	11	Polyunsaturated
Salad dressing, oil and vinegar	1 T	85	8	Depends on the type of oil used
Salad dressing, Ranch type	1 T	73	8	Polyunsaturated
Coconut oil	1 T	117	13.5	Saturated
Bacon grease	1 T	112	12	Saturated/monounsaturated
Butter, stick	1 T	108	12	Saturated
Butter, whipped	1 T	82	9	Saturated
Coconut oil	1 T	120	13.5	Saturated
Cream, half and half	1 T	20	1.5	Saturated
Lard	1 T	114	12.5	Monounsaturated/saturated
Shortening	1 T	110	12	No one type is predominant

Examples of foods that are 100 percent or nearly 100 percent fat and the predominant type of fat they contain.

Legend: kcal = kilocalorie; g = gram; T = tablespoon; oz = ounce

*All foods listed are either 100 percent fat or nearly 100 percent (contain <1 g of protein and carbohydrate).

**When two fats are listed, both are found in approximately equal amounts.

have some carbohydrate and/or protein, such as nuts and seeds. The amount of energy, fat, carbohydrate, and protein are listed. The predominant type of fat is also noted. Notice that there are substantial differences in the foods collectively called fats and oils. For example, some oils contain a high percentage of monounsaturated fatty acids (olive oil has the highest concentration), whereas others are predominantly polyunsaturated. The fatty acid content of margarines differs, depending on whether the margarine is liquid, soft, or hard. Although it won't affect athletic performance, the type of fat may affect long-term health.

The amounts of carbohydrate and protein required by athletes in training are substantial; thus fat intake is comparatively lower, especially if an energy (caloric) deficit is desirable. Low-fat or nonfat versions of foods may better fit into an athlete's diet plan. In particular, lean meat, fish, and poultry and modified-fat dairy products are often chosen. Table 6.6 compares high-, medium-, and low-fat versions of similar foods.

What may not be readily apparent is the amount or type of fat that is in a processed food. This is sometimes referred to as "hidden fat." Reading food labels is necessary if athletes want to discover a processed food's nutrient composition. Table 6.7 lists the amount of total and saturated fat found in some snack foods.

The typical American diet is usually too high in fat for an athlete in training.

The typical American diet is characterized by a high intake of red meat, processed meat, high-fat dairy products, French fries, refined grains, sweets, and desserts. Large portions of these foods are frequently served in restaurants and at home. This dietary pattern is associated with the overconsumption of calories, total fat, saturated and *trans* fats, cholesterol, refined carbohydrates, and sodium. Such a pattern tends to be too high in calories and fat and too low

in carbohydrate and certain vitamins and minerals to support training and performance. Additionally, the typical American diet has been shown to contribute to chronic diseases such as cardiovascular disease, type 2 diabetes, and metabolic syndrome as explained in Chapter 13.

Traditional Mediterranean, Asian, and Latin American diets differ from the typical American diet in that meats, sweets, and eggs are used sparingly (often on a monthly or weekly basis) and fruits, vegetables, fish, beans, whole grains, and oils are eaten daily. Portion sizes are also smaller. There is strong evidence that a diet containing fruits, vegetables, whole grains, and nuts can reduce the risk for developing these chronic diseases (Hu, 2009). Such a dietary pattern can also support training, so some athletes look to these traditional dietary patterns as a way to achieve both good health and good performance.

Some athletes adopt a vegetarian or vegan diet and in doing so may lower their intake of fat, particularly saturated and *trans* fats and cholesterol. However, meat and animal products do not have to be excluded.

Table 6.5 High-Fat Foods That Also Contain Carbohydrate and Protein

Food	Amount	Energy (kcal)	Fat (g)	Predominant type of fat*	CHO (g)	Protein (g)
Avocado	One (173 g)	306	30	Monounsaturated	12	3.5
Peanuts	¼ c, oil roasted	213	18	Monounsaturated	6	10
Almonds	¼ c, dry roasted	206	18	Monounsaturated	7	8
Hazelnuts (filberts)	¼ c, dry roasted	183	18	Monounsaturated	5	4
Pecans	¼ c, dry roasted	187	19	Monounsaturated	4	2.5
Pistachios	¼ c, dry roasted	183	15	Monounsaturated	9	7
Walnuts	¼ c	196	19.5	Polyunsaturated	4	4.5
Sesame seeds	1 T	51	4	Polyunsaturated/monounsaturated	2	1.5
Tahini (sesame seed paste)	1 T	89	8	Polyunsaturated/monounsaturated	3	2.5
Sunflower seeds	¼ c, oil roasted	208	19	Polyunsaturated	5	7
Pumpkin seeds	¼ c, oil roasted	296	24	Polyunsaturated	8	19
Flax seeds	1 T	59	4	Polyunsaturated	4	2
Bacon	2 slices	70	6	Monounsaturated/saturated	0	4
Canadian-style bacon (pork sirloin)	2 slices	50	1.5	Saturated	0	8
Coconut, sweetened, shredded	2 T	58	4	Saturated	5.5	0
Coconut milk	¼ c	138	14	Saturated	3	1

Legend: kcal = kilocalorie; g = gram; CHO = carbohydrate; c = cup; T = tablespoon

*When two fats are listed, both are found in approximately equal amounts.

Table 6.6 High-, Medium-, and Low-Fat Meat, Fish, Poultry, and Dairy Products

Food	Preparation method	Amount	Fat (g)
Ground beef, regular	Broiled	3 oz	17.5
Ground beef, lean	Broiled	3 oz	16
Ground beef, extra lean	Broiled	3 oz	14
Tuna salad	Mayonnaise added to tuna	¾ c	20
Light tuna, canned in oil	Drained	2 oz	7
Light tuna, canned in water	Drained	2 oz	0.5
Chicken wing (meat and skin), flour coated	Fried	3 oz	19
Chicken wing (meat and skin)	Roasted	3 oz	16.5
Chicken leg (dark meat)	Roasted	3 oz	7
Chicken breast (white meat)	Roasted	3 oz	3
Whole milk (3.3% butterfat)		8 oz	8
Reduced fat milk (2% butterfat)		8 oz	5
Low-fat milk (1% butterfat)		8 oz	2
Nonfat (skim) milk		8 oz	0.2
Creamed cottage cheese (4% butterfat)		½ c	5
Low-fat cottage cheese (2% butterfat)		½ c	2
Low-fat cottage cheese (1% butterfat)		½ c	1
Dry curd cottage cheese (0.4% or less butterfat)		½ c	~0.5

Legend: g = gram; oz = ounce, c = cup

Table 6.7 Amount of Total and Saturated Fat in Selected Snack Foods

Food	Amount	Energy (kcal)	Total fat (g)	Saturated fat (g)
Oreo cookie ice cream	5 oz	355	20.5	12
Glazed donut	1	350	19	5
Reese's Peanut Butter Cups	2 pieces	250	14	5
Oreo cookies	6 cookies	318	14	3
Nestlé plain milk chocolate candy bar	1 (1.45 oz)	210	13	8
Trail mix	¼ c	173	11	2
Milky Way candy bar	1	270	10	5
Fritos	1 oz	160	10	1.5
Brownie	1 piece (1.5 oz)	170	8	1.5
Cheese Whiz	2 T	90	7	5
Wheat Thins	16 crackers (1 oz)	140	6	1
Hostess Twinkie	1	150	5	2
Pop-Tart (frosted toaster pastry)	1	200	5	1

These foods may contain *trans* fat.

Legend: kcal = kilocalorie; oz = ounce; c = cup; g = gram

Hu (2003) noted that a dietary pattern that includes poultry and low-fat dairy products is also consistent with a reduced risk for cardiovascular disease. There are various ways to modify the typical American diet to one that supports training, performance, and long-term health.

Figure 6.15 illustrates the macronutrient content of a diet containing foods associated with the typical American diet—bacon and eggs for breakfast; a ham and cheese sandwich, potato chips, and cookies for lunch; a super-size cheeseburger and fries at

Spotlight on a real athlete

Lucas, a Cross Country Runner

As in previous chapters, a one-day dietary intake of Lucas, a collegiate cross country runner, is analyzed. Recall that the appropriate amount of dietary fat for an athlete depends on two factors—overall energy (caloric) need and macronutrient balance. Lucas's need for energy is approximately 3,400 kcal (~54 kcal/kg) daily. Due to the demands of his training (running 75 to 80 miles per week), his daily goal for carbohydrate is 8 g/kg. His daily protein goal is 1.5 g/kg. It is in this context that his fat intake should be evaluated. An analysis of the one-day diet that Lucas consumed is shown in the figure below.

According to the dietary analysis, Lucas consumed approximately 3,333 kcal, 532 g of carbohydrate (~8.5 g/kg), and 124 g of protein (~2 g/kg). His fat intake was 94 g or 1.5 g/kg (~25 percent of total energy intake). His total energy intake was neither too high nor too low. He exceeded his goals for carbohydrate and protein intake. His fat intake allowed him to achieve macronutrient and energy balance. From a performance perspective, Lucas's fat intake was appropriate. An evaluation of Lucas's diet from a health perspective is included in Chapter 13.

It should be pointed out that Lucas included a number of nonfat and low-fat foods in his diet. For example, the smoothie that he had for breakfast was made from nonfat milk and low-fat yogurt. The lunchtime burritos were made from low-fat tortillas and low-fat refried black beans. For dinner he drank nonfat milk and chose white meat turkey for his sandwich. Because Lucas began running cross country in high school, he had already made adjustments to his diet. Had he eaten full-fat versions of these foods he would have exceeded his fat and energy goals.

Nutrient	DRI	Intake	0%	50%	100%
Energy					
Kilocalories	3365 kcal	3332.8 kcal			99%
Carbohydrate	379–547 g	532.15 g			
Fat, total	75–131 g	93.86 g			
Protein	84–294 g	124.09 g			
Fat					
Saturated fat	<10%	36.63 g			
Monounsaturated fat	no rec	19.45 g			
Polyunsaturated fat	no rec	11.65 g			
Cholesterol	300 mg	195.79 mg			65%
Essental fatty acids (efa)					
Omega-6 linoleic	17 g	1.9 g			11%
Omega-3 linolenic	1.6 g	0.47 g			30%

	Goal*	Actual	% Goal
Grains	10 oz. eq.	11.8 oz. eq.	118%
Vegetables	4 cup eq.	1.2 cup eq.	30%
Fruits	2.5 cup eq.	1.7 cup eq.	68%
Milk	3 cup eq.	5.4 cup eq.	180%
Meat & Beans	7 oz. eq.	8.8 oz. eq.	126%
Discretionary	648	1152	178%

MyPyramid.gov
STEPS TO A HEALTHIER YOU

*Your results are based on a 3200 calorie pattern, the maximum caloric intake used by MyPyramid.

Dietary analysis of a 24-hour diet of a male collegiate cross country runner

Nutrient	DRI	Intake	0%	50%	100%
Energy					
Kilocalories	2427 kcal	3540.98 kcal			146%
Carbohydrate	273–394 g	406.75 g			
Fat, total	54–94 g	176.99 g			
Protein	61–212 g	99.2 g			
Fat					
Saturated fat	<10%	60.03 g			
Monounsaturated fat	no rec	34.5 g			
Polyunsaturated fat	no rec	13.36 g			
Cholesterol	300 mg	666.58 mg			222%
Carbs					
Dietary fiber, total	38 g	20.01 g			53%
Sugar, total	no rec	179.14 g			

MyPyramid.gov
STEPS TO A HEALTHIER YOU

	Goal*	Actual	% Goal
Grains	8 oz. eq.	8.5 oz. eq.	106%
Vegetables	3 cup eq.	1.3 cup eq.	43%
Fruits	2 cup eq.	0 cup eq.	0%
Milk	3 cup eq.	2.4 cup eq.	80%
Meat & Beans	6.5 oz. eq.	6.2 oz. eq.	95%
Discretionary	362	1816.7	502%

*Your results are based on 2427 calorie pattern (sedentary male)

Figure 6.15 Dietary analysis of a typical American diet (see text below for foods included)

Source: USDA (U.S. Department of Agriculture)

a fast-food restaurant for dinner; and chocolate ice cream for dessert. This diet provides more than 3,500 kcal and is high in fat, saturated fat, and cholesterol. Fiber and vegetable intake are low and fruit intake is zero! One can easily see how many Americans consume high-fat, high-calorie diets.

For illustration purposes, now assume that this dietary analysis represents Lucas's alter ego and former high school teammate, Luke. Luke has never much cared about his diet because he didn't think it was all that important. In high school he ate whatever was convenient, which included fast foods most days, and he was still the best cross country runner in his region. He also figured that his diet was fine as long as he didn't gain any body fat. But a dietary analysis reveals some interesting facts. Luke's diet has an excess of energy by about 140 kcal (3,541 kcal consumed compared to 3,400 kcal needed). His carbohydrate intake is ~6.5 g/kg, 1.5 g/kg lower than his daily goal of ~8 g/kg. His fat intake is 177 g or ~2.8 g/kg. He does not meet recommended goals in three crucial areas: energy, carbohydrates, and macronutrient balance. If this one-day diet reflects his usual intake, over time he could fail to meet his training and performance goals. His carbohydrate intake is insufficient to adequately restore muscle glycogen levels depleted by prolonged

endurance training, and he could expect to gain 15 lb (~7 kg) of body fat over the course of the year if he consumes an excess of 140 kcal daily. From a health perspective, his intake of total fat, saturated fat, and cholesterol exceeds the recommendations, and his intake of fruits, vegetables, whole grains, and nuts is low or nonexistent. This example clearly shows why the typical American diet is not consistent with many athletes' training, performance, and health goals.

There are ways to modify the typical American diet so it is lower in fat.

Sports dietitians frequently counsel athletes about ways to modify their current diet to lower fat intake. Strategies include reducing portion sizes, preparing and buying foods with less fat, adding less fat to food, choosing lower-fat meat and poultry, consuming low-fat or non-fat dairy products, and substituting high-fat refined grains and desserts with fresh fruits and vegetables. There is no one strategy that must be used. Instead, sports dietitians help athletes create an individualized diet plan utilizing some or all of these strategies. In many cases, the amount of fat is too high relative to the amount of carbohydrate or protein consumed,

Table 6.8 Burger Meals Compared

Meal	Energy (kcal)	Fat (g)
Big Mac®	540	29
Large French fries (5.4 oz)	500	25
Total	**1,040**	**54**
Quarter Pounder® (no cheese)	410	19
Medium French fries (4.1 oz)	380	19
Total	**790**	**38**
Cheeseburger	300	12
Small French fries (2.5 oz)	230	11
Total	**530**	**23**

Ordering a smaller-size meal results in lower total fat and energy intakes.

Legend: kcal = kilocalorie; g = gram; oz = ounce

Used with permission from McDonald's Corporation.

The Americanization of traditional ethnic meals often results in the addition of fat.

Grilling and steaming are preparation methods that do not require additional fat.

so foods high in fat may need to be "replaced" with foods that help athletes to achieve the proper macronutrient balance.

- **Reduce portion size.** Reducing the portion size of high-fat foods is one obvious way to reduce fat and caloric intake. Table 6.8 compares small, medium, and large meals containing a cheeseburger and French fries sold at McDonald's. These meals range from a high of 54 g of fat and 1,040 kcal to a low of 23 g of fat and 530 kcal. When high-fat foods are the only option, one strategy to reduce fat and energy intake is to simply order a smaller size.

- **Prepare foods with less fat.** Much fat can be added to food when it is prepared. Ways to prepare food that minimize the amount of fat used for cooking include grilling, roasting, broiling, baking, steaming, or poaching. Deep-fat frying adds substantial amounts of fat to the original food. For example, a 3 oz piece of fish that has been coated with flour and fried has about 11 g of fat. The same fish prepared by steaming or baking has about 1 g of fat.

- **Add less fat to foods.** Some people add substantial amounts of fat to food, such as butter or margarine on bread or potatoes. Potatoes naturally contain very little fat so a baked potato contributes less than 1 g of fat when eaten plain. But what about the addition of butter, sour cream, or cheese sauce? All of these can be sources of substantial amounts of fat, depending on how much is added. A baked potato with sour cream and chives, which is available at several fast-food

restaurants, has approximately 14 g of fat, nearly all coming from the sour cream. Most baked potatoes with cheese sauce sold at fast-food restaurants have about 28 g of fat. The baked potato prepared at home can be equally fat laden, depending on how much fat is added.

- **Order carefully at restaurants.** Most restaurants have items on the menu that are lower in fat and calories, but in many cases the number of items is limited. Regular items on the menu can be modified, such as asking that sauces be left off or by substituting high-fat items with lower-fat ones. Ethnic restaurants may have "Americanized" their menus based on consumer demand, so it may be necessary to ask for a more traditional version of the dish. The key is to actually order the lower-fat versions, something that people may have difficulty doing.

Beverages may be a source of fat.

© Susan Van Etten/PhotoEdit

- **Be aware of "hidden fats."** The fat content of some foods is surprising. A Caesar salad served in a restaurant can have 30 to 40 g of fat. A Mrs. Fields peanut butter cookie has 16 g of fat. Beverages as sources of fat should not be overlooked. A Double Chocolaty Chip Frappuccino® Blended beverage (16 oz) from Starbucks contains 20 g of fat. Many processed snack-type foods such as crackers and chips and desserts such as pastries contain much more fat than one might realize. Reading nutrition labels and looking up nutrition information on commercial websites are ways of finding out how much fat is in a particular food.
- **Consume lower-fat cuts of meat or poultry and low-fat or nonfat dairy products.** White meat chicken, water-packed tuna, flank steak, skim milk, and low-fat cottage cheese are examples of foods that often become the staples of many athletes' diets because these foods are low in fat. Eating the lower-fat version of a product is a strategy frequently used by athletes because these foods taste good and are widely available.
- **Choose lower-fat versions of high-fat processed foods.** There are at least 5,000 foods on supermarket shelves that are lower-fat versions of full-fat foods. Consumers have tremendous choice when it comes to buying foods with a lower fat content. A word of caution: these lower-fat foods may not be lower in kilocalories (energy) than the full-fat version. This is because the fat content of the food may be replaced with carbohydrate, often in the form of sugar.
- **Substitute fruits and vegetables for fat-containing snack foods.** One strategy that athletes may find helpful is to substitute a food that is naturally low in fat, such as most fruits and vegetables, for a high-fat food. This may also result in a lower energy intake and greater carbohydrate intake, which are goals for some athletes.

Application exercise

In some respects, athletes need to be food detectives, so they know if the foods they are eating are providing the nutrients they are intending to eat. For the scenarios listed below, determine which food is the right choice. Note: if a product doesn't have a Nutrition Facts label, nutrient information is generally available on the manufacturer's website.

1. A female athlete is trying to lose weight as body fat by reducing the amount of fat and calories she consumes. She goes with friends to a fast-food restaurant. To meet her goals, which is the better meal to order, the salad or the small burger and small fries?

2. This same athlete stops for a dish of soft-serve yogurt while her friends get ice cream. Is yogurt a better choice for her than ice cream?

3. A baseball player is trying to reduce his intake of sunflower seeds during games because he wants to reduce his intake of fat and calories. Would banana chips be a better choice?

4. A body builder is trying to keep his fat intake low but finds himself craving chocolate cake. He sees fat-free chocolate cake in the grocery store. Would this be a good choice?

Some foods are made with fat substitutes.

Fat imparts some of the most appealing flavors and textures found in foods. It is also the most concentrated source of energy in the diet at 9 kcal/g. Because dietary fat is so desirable, there have been many efforts made to find acceptable **fat substitutes**.

Fat substitutes. During the 1990s, when the rates of overweight and obesity were steadily climbing in the

United States, the government encouraged the food industry to create more than 5,000 reduced-fat foods. At one point more than a thousand such products were introduced each year (Wylie-Rosett, 2002). Some of these foods simply had the fat removed (for example, skim milk), but many of the products involved fat substitutes.

Fat substitutes are most often made from carbohydrate sources, although some are derived from fat and at least one is protein based. Carbohydrates are used because fibers and starches retain water, and they add body and texture to the food when the fat is removed. Such foods have much less fat but often a similar amount of kilocalories when compared to the original product.

Protein-based fat substitutes are egg or milk proteins that have been ultracentrifuged to produce extremely small particles. The microparticles roll over each other in the mouth so the product has the same feel and texture as the full-fat product. Protein-based fat substitutes are used in dairy products, such as ice cream and baked goods.

Many reduced-fat bakery products contain monoglycerides and diglycerides. These fats are derived from vegetable oils and have been emulsified with water. Although all fats contain 9 kcal/g, a gram of monoglyceride or diglyceride contains a considerable amount of water, so the total amount of fat is reduced when these compounds replace other fats on a per weight basis.

Caprenin and salatrim are fatty acids that are used in baked goods and dairy products because of their similar textural properties to cocoa butter. These fatty acids essentially contain 5 kcal/g because they are only partially digested and absorbed. Substituting these for traditional fats reduces the fat content by almost half. However, gastrointestinal symptoms, such as nausea and cramps, have been reported, especially as the amount consumed increases.

In a class by itself, both structurally and for regulatory purposes, is **Olestra** (Olean®). Olestra is a nonabsorbable fat substitute composed of sucrose polyester, a compound made up of sucrose with 6 to 8 fatty acids attached. Because of its chemical structure, Olestra is resistant to pancreatic lipase and remains unabsorbed in the gastrointestinal tract. When first approved, all products containing Olestra were required to carry a warning label: "This Product Contains Olestra. Olestra may cause abdominal cramping and loose stools. Olestra inhibits the absorption of some vitamins and other nutrients. Vitamins A, D, E, and K have been added." In 2003, the Food and Drug Administration removed the warning label requirement.

Similar to artificial sweeteners, fat substitutes are not the weight loss panacea that both industry and

Fat substitute: Compounds that replace the fat that would be found naturally in a food. Most are made from proteins or carbohydrates.

Olestra: A fat substitute that cannot be absorbed by the body.

Keeping it in perspective

Fat is for Fuel and Fun

Angelina looked solemnly at the piece of cake that she had been served. To the others in the room it was simply a small piece of birthday cake, but to her it was 10 g of fat. Even though she had run 5 miles that morning and had a figure that others said they envied, Angelina couldn't bring herself to eat the cake. She feared that "a minute on the lips" meant "a lifetime on the hips." So she said that she was a bit nauseated and didn't eat it, even though it was her birthday.

Fat is a calorie-dense nutrient and there are problems associated with excessive intake. However, it is very important for athletes to be able to keep their fat intake in perspective. On the fat intake continuum, athletes should stay centered and not restrict fat too severely or consume fat excessively. They should view fat as a nutrient, not as something that is inherently "bad."

At times they must limit fat intake, such as during a training cycle when energy and fat is restricted in an effort to lose weight. But they also need to be flexible with fat intake, such as eating and enjoying a piece of cake on their birthday.

Fat is an excellent fuel source. When energy, carbohydrate, protein, and fat intakes are balanced there is no need to fret about fat. Fat also imparts a wonderful flavor and texture to food, and there is a joy and satisfaction that comes with eating fat-containing foods. Backpackers often stop at the top of a steep pass and enjoy a chocolate bar. Getting to the top of the mountain gives them perspective because they can look at the world from afar. That same kind of perspective is necessary for athletes to evaluate their fat intake and discover that fat, at least in part, is for fuel and fun.

consumers had hoped they would be. Many "fat-free" products do not have fewer kilocalories than the full-fat product, just more sugar. Controlled randomized trials have shown that individuals who choose foods with fat substitutes consume less total fat and less saturated fat but only to a small degree. In one study of Olestra, total dietary fat intake was reduced by 2.7 percent and saturated fat intake was reduced by 1.1 percent (Wylie-Rosett, 2002).

Key points

■ The amount and type of fat in food varies.

■ Athletes often choose lower fat versions of food.

■ Many "fat-free" products are not low in calories.

How can the Nutrition Facts label be used to determine if the product has "hidden" fat?

Post-Test　Reassessing Knowledge of Fats

Now that you have more knowledge about fats, read the following statements and decide if each is true or false. The answers can be found in Appendix O.

1. Athletes typically need to follow a very low-fat diet.

2. At rest, the highest percentage of total energy expenditure is from fat and not carbohydrate.

3. To lose body fat, it is best to perform low-intensity exercise, which keeps one in the fat-burning zone.

4. To improve performance, endurance athletes should ingest caffeine because more free fatty acids are oxidized for energy and muscle glycogen is spared.

5. A low-calorie, low-carbohydrate diet that results in ketosis is dangerous for athletes because it leads to the medical condition known as ketoacidosis.

Summary

Fat is the most energy-dense nutrient found in food. The predominant fat in food and in the body is the triglyceride (triacylglycerol), which is made up of three fatty acids attached to a glycerol molecule. The fatty acids, which are chains of carbon and hydrogen, vary in length and chemical composition. Fats are large molecules, and their digestion, absorption, and transport are complicated because they must be broken down into smaller components.

Once absorbed, triglycerides must be re-formed. The main sites of fat storage are adipocytes (fat cells), liver, and muscle cells. After a meal, particularly a meal containing fats and carbohydrates, the hormonal environment is such that triglyceride formation and fat storage are favored. To be used in exercise metabolism, fats must be taken out of storage. An enzyme, hormone-sensitive lipase, catalyzes lipolysis of triglycerides stored in adipocytes.

Fat is the primary energy source at rest and during low-intensity activity. The advantages of fat include its abundance in food, energy density, ability to be easily stored, and ability to produce a large amount of ATP. However, fat usage is limited to rest and low-intensity activity because of the time it takes to metabolize fatty acids and the oxygen necessary for metabolism.

Because the need for carbohydrate and protein is relatively high, athletes find that their diets tend to be relatively lower in fat than the typical American diet. Heart-healthy fats, oils such as olive, canola, or flaxseed, nuts, and fatty fish or fish oils should be emphasized. Athletes can reduce the amount of fat in their diets by reducing portion sizes, preparing and buying foods with less fat, adding less fat to food, ordering carefully at restaurants, choosing lower-fat meat and poultry, consuming low-fat or nonfat dairy products, and substituting high-fat refined grains (for example, snack foods and desserts) with fresh fruits and vegetables. Caution should be used when restricting fat because athletes can reduce the fat in their diets too much.

Self-Test

Multiple-Choice

The answers can be found in Appendix O.

1. The predominant fat in food and the body is:
   a. cholesterol.
   b. triglyceride.
   c. lipoprotein.
   d. lecithin.

2. The appropriate amount of fat for training and performance depends on:
   a. overall energy need.
   b. macronutrient balance.
   c. proportion of saturated to unsaturated fat.
   d. both overall energy need and macronutrient balance.
   e. all of the above.

3. Why would an endurance athlete limit the amount of fat in a precompetition meal?
   a. contributes to faster dehydration
   b. delays gastric emptying
   c. slows the metabolism of carbohydrate
   d. all of the above

4. When the body is stressed by exercise, fat stored in adipocytes is:
   a. mobilized.
   b. demobilized.
   c. unaffected.

5. The mechanism by which caffeine most likely improves endurance performance is that it:
   a. spares muscle glycogen.
   b. enhances the release of muscle and liver glycogen.
   c. increases the oxidation of fatty acids during exercise.
   d. decreases perception of exertion.

Short Answer

1. What might people mean when they say "too much fat"?

2. What are the differences between saturated, monounsaturated, and polyunsaturated fatty acids? Name foods in which each type predominates.

3. What is the chemical difference between *cis*- and *trans* fatty acids? Why might people be advised to eat whole (less processed) foods?

4. What is a triglyceride? Where are triglycerides found in food? In the body?

5. Explain the two major factors that dictate the appropriate amount of dietary fat intake for an athlete.

6. Briefly explain the digestion, intestinal absorption, and transportation of fats.

7. Briefly explain the major steps in the metabolism of fatty acids in the body.

8. What happens to the use of fat as a fuel as exercise intensity increases? What happens to the use of fat as a fuel when steady-state exercise progresses in duration?

9. Explain relative and absolute fat oxidation.

10. Why do athletes acutely and chronically restrict fat intake? Why is consuming enough dietary fat important for athletes?

11. In what ways may severe, chronic fat and energy restriction affect performance and health?

12. What are the recommended guidelines for fat intake for athletes?

13. Describe some strategies that athletes could use to reduce the fat content of their diets.

14. What are fat substitutes?

Critical Thinking

1. Does a person need to burn fat during exercise in order to lose body fat?

2. An athlete figures that if his or her diet has the right amount of calories and body composition is stable, then the amount of fat consumed is also the right amount. Why might this be erroneous thinking?

References

Abete, P., Testa, G., Galizia, G., Della-Morte, D., Cacciatore, F., & Rengo, F. (2009). PUFA for human health: Diet or supplementation? *Current Pharmaceutical Design, 15*(36): 4186–4190.

Angus, D. J., Hargreaves, M., Dancey, J., & Febbraio, M. A. (2000). Effect of carbohydrate or carbohydrate plus medium-chain triglyceride ingestion on cycling time trial performance. *Journal of Applied Physiology, 88*(1), 113–119.

Bloomer, R. J., Larson, D. E., Fisher-Wellman, K. H., Galpin, A. J., & Schilling, B. K. (2009). Effect of eicosapentaenoic and docosahexaenoic acid on resting and exercise-induced inflammatory and oxidative stress biomarkers: A randomized, placebo controlled, cross-over study. *Lipids in Health and Disease, 19*(8), 36.

Brass, E. P. (2004). Carnitine and sports medicine: Use or abuse? *Annals of the New York Academy of Sciences, 1033,* 67–78.

Broad, E. M., Maughan, R. J., & Galloway, S. D. (2008). Carbohydrate, protein, and fat metabolism during exercise after oral carnitine supplementation in humans. *International Journal of Sport Nutrition and Exercise Metabolism, 18,* 567–584.

Brooks, G. A., Fahey, T. D., & Baldwin, K. M. (2005). *Exercise physiology: Human bioenergetics and its applications* (4th ed.). New York: McGraw-Hill.

Buckley, J. D., Burgess, S., Murphy, K. J., & Howe, P. R. (2009). DHA-rich fish oil lowers heart rate during submaximal exercise in elite Australian Rules footballers. *Journal of Science and Medicine in Sport, 12*(4), 503–507.

Burke, L. M. (2008). Caffeine and sports performance. *Applied Physiology, Nutrition, and Metabolism, 33,* 1319–1334.

Carey, A. L., Staudacher, H. M., Cummings, N. K., Stepto, N. K., Nikolopoulos, V., Burke, L. M., et al. (2001). Effects of fat adaptation and carbohydrate restoration on prolonged endurance exercise. *Journal of Applied Physiology, 91*(1), 115–122.

Chanmugam, P., Guthrie, J. F., Cecilio, S., Morton, J. F., Basiotis, P. P., & Anand, R. (2003). Did fat intake in the United States really decline between 1989–1991 and 1994–1996? *Journal of the American Dietetic Association, 103*(7), 867–872.

Davis, J. K., & Green, J. M. (2009). Caffeine and anaerobic performance: Ergogenic value and mechanisms of action. *Sports Medicine, 39,* 813–832.

De Cree, C. (1998). Sex steroid metabolism and menstrual irregularities in the exercising female. A review. *Sports Medicine, 25*(6), 369–406.

Denomme, J., Stark, K. D., & Holub, B. J. (2005). Directly quantitated dietary (n-3) fatty acid intakes of pregnant Canadian women are lower than current dietary recommendations. *Journal of Nutrition, 135*(2), 206–211.

Dorgan, J. F., Judd, J. T., Longcope, C., Brown, C., Schatzkin, A., Clevidence, B. A., et al. (1996). Effects of dietary fat and fiber on plasma and urine androgens and estrogens in men: A controlled feeding study. *American Journal of Clinical Nutrition, 64*(6), 850–855.

Ganio, M. S., Klau, J. F, Casa, D. J, Armstrong, L. E., & Maresh, C. M. (2009). Effect of caffeine on sport-specific endurance performance: A systematic review. *Journal of Strength and Conditioning Research, 23,* 315–324.

Goedecke, J. H., Clark, V. R., Noakes, T. D., & Lambert, E. V. (2005). The effects of medium-chain triacylglycerol and carbohydrate ingestion on ultra-endurance exercise performance. *International Journal of Sport Nutrition and Exercise Metabolism, 15*(1), 15–27.

Grediagin, A., Cody, M., Rupp, J., Benardot, D., & Shern, R. (1995). Exercise intensity does not effect body composition change in untrained, moderately overfat women. *Journal of the American Dietetic Association, 95*(6), 661–665.

Gropper, S. S., Smith, J. L., & Groff, J. L. (2009). *Advanced nutrition and human metabolism* (5th ed.). Belmont, CA: Wadsworth, Cengage Learning.

Hamalainen, E., Adlercreutz, H., Puska, P., & Pietinen, P. (1984). Diet and serum sex hormones in healthy men. *Journal of Steroid Biochemistry, 20*(1), 459–464.

Hawley, J. A. (2002). Effect of increased fat availability on metabolism and exercise capacity. *Medicine and Science in Sports and Exercise, 34*(9), 1485–1491.

Hawley, J. A., & Burke, L. M. (2010). Carbohydrate availability and training adaptation: Effects on cell metabolism. *Exercise and Sport Sciences Reviews, 38*(4), 153–160.

Heaney, S., O'Connor, H., Gifford, J., & Naughton, G. (2010). Comparison of strategies for assessing nutritional adequacy in elite female athletes' dietary intake. *International Journal of Sport Nutrition and Exercise Metabolism, 20*(3), 245–256.

Helge, J. W. (2000). Adaptation to a fat-rich diet: Effects on endurance performance in humans. *Sports Medicine, 30*(5), 347–357.

Horowitz, J. F., & Klein, S. (2000). Lipid metabolism during endurance exercise. *American Journal of Clinical Nutrition, 72*(2 Suppl.), 558S–563S.

Horowitz, J. F., Mora-Rodriguez, R., Byerley, L. O., & Coyle, E. F. (2000). Preexercise medium-chain triglyceride ingestion does not alter muscle glycogen use during exercise. *Journal of Applied Physiology, 88*(1), 219–225.

Horvath, P. J., Eagen, C. K., Fisher, N. M., Leddy, J. J., & Pendergast, D. R. (2000). The effects of varying dietary fat on performance and metabolism in trained male and female runners. *Journal of the American College of Nutrition, 19*(1), 52–60.

Horvath, P. J., Eagen, C. K., Ryer-Calvin, S. D., & Pendergast, D. R. (2000). The effects of varying dietary fat on the nutrient intake in male and female runners. *Journal of the American College of Nutrition, 19*(1), 42–51.

Hu, F. B. (2003). Plant-based foods and prevention of cardiovascular disease: An overview. *American Journal of Clinical Nutrition, 78*(3 Suppl.), 544S–551S.

Hu, F. B. (2009). Diet and lifestyle influences on risk of coronary heart disease. *Current Atherosclerosis Reports, 11*(4), 257–263.

Institute of Medicine. (2002). Dietary Reference Intakes for energy, carbohydrate, fiber, fat, fatty acids, cholesterol, protein and amino acids (Food and Nutrition Board). Washington, DC: National Academies Press.

Jeukendrup, A. E., Saris, W. H. M., & Wagenmakers, A. J. M. (1998a). Fat metabolism during exercise: A review. Part I: fatty acid mobilization and muscle metabolism. *International Journal of Sports Medicine, 19*(4), 231–244.

Jeukendrup, A. E., Saris, W. H. M., & Wagenmakers, A. J. M. (1998b). Fat metabolism during exercise: A review. Part II: regulation of metabolism and the effects of training. *International Journal of Sports Medicine, 19*(5), 293–302.

Kraemer, W. J., Volek, J. S., & Dunn-Lewis, C. (2008). L-carnitine supplementation: influence upon physiological function. *Current Sports Medicine Reports, 7,* 218–223.

Lambert, C. P., Frank, L. L., & Evans, W. J. (2004). Macronutrient considerations for the sport of bodybuilding. *Sports Medicine, 34*(5), 317–327.

Laurent, M. R., & Vickers, T. J. (2009). Seeking health information online: Does Wikipedia matter? *Journal of the American Medical Informatics Association, 16*(4), 471–479.

Machefer, G., Groussard, C., Zouhal, H., Vincent, S., Youssef, H., Faure, H., et al. (2007). Nutritional and plasmatic antioxidant vitamins status of ultra endurance athletes. *Journal of the American College of Nutrition, 26*(4), 311–316.

Malaguti, M., Baldini, M., Angeloni, C., Biagi, P., & Hrelia, S. (2008). High-protein-PUFA supplementation, red blood cell membranes, and plasma antioxidant activity in volleyball athletes. *International Journal of Sport Nutrition and Exercise Metabolism,* 18(3), 301–312.

Mickleborough, T. D., Lindley, M. R., Ionescu, A. A., & Fly, A. D. (2006). Protective effect of fish oil supplementation on exercise-induced bronchoconstriction in asthma. *Chest, 129*(1), 39–49.

Misell, L. M., Lagomarcino, N. D., Schuster, V., & Kern, M. (2001). Chronic medium-chain triacylglycerol consumption and endurance performance in trained runners. *Journal of Sports Medicine and Physical Fitness, 41*(2), 210–215.

Muller, D. M., Seim, H., Kiess, W., Loster, H., & Richter, T. (2002). Effects of oral L-carnitine supplementation on in vivo long-chain fatty acid oxidation in healthy adults. *Metabolism, 51*(11), 1389–1391.

Nieman, D. C., Henson, D. A., McAnulty, S. R., Jin, F., & Maxwell, K. R. (2009). n-3 polyunsaturated fatty acids do not alter immune and inflammation measures in endurance athletes. *International Journal of Sport Nutrition and Exercise Metabolism, 19*(5), 536–546.

Pendergast, D. R., Leddy, J. J., & Venkatraman, J. T. (2000). A perspective on fat intake in athletes. *Journal of the American College of Nutrition, 19*(3), 345–350.

Phinney, S. D., Bistrian, B. R., Evans, W. J., Gervino, E., & Blackburn, G. L. (1983). The human metabolic response to chronic ketosis without caloric restriction: Preservation of submaximal exercise capability with reduced carbohydrate oxidation. *Metabolism, 32*(8), 769–776.

Raastad, T., Høstmark, A. T., & Strømme, S. B. (1997). Omega-3 fatty acid supplementation does not improve maximal aerobic power, anaerobic threshold and running performance in well-trained soccer players. *Scandinavian Journal of Medicine & Science in Sports, 7*(1), 25–31.

Remig, V., Franklin, B., Margolis, S., Kostas, G., Nece, T., & Street, J.C. (2010). *Trans* fats in America: A review of their use, consumption, health implications, and regulation. *Journal of the American Dietetic Association, 110*(4), 585–592.

Seebohar, B. (2011). *Nutrition periodization for athletes* (2nd ed.). Boulder, CO: Bull.

Simopoulos, A. P. (2008). Omega-3 fatty acids, exercise, physical activity and athletics. *World Review of Nutrition and Dietetics, 98,* 23–50.

Strauss, R. H., Lanese, R. R., & Malarkey, W. B. (1985). Weight loss in amateur wrestlers and its effect on serum testosterone levels. *Journal of the American Medical Association, 254*(23), 3337–3338.

Sudi, K., Ottl, K., Payerl, D., Baumgartl, P., Tauschmann, K., & Muller, W. (2004). Anorexia athletica. *Nutrition, 20*(7–8), 657–661.

Tarnopolsky, M. A. (2008). Effect of caffeine on the neuromuscular system—potential as an ergogenic aid. *Applied Physiology, Nutrition, and Metabolism, 33,* 1284–1289.

Tomten, S. E., & Høstmark, A. T. (2009). Serum vitamin E concentration and osmotic fragility in female long-distance runners. *Journal of Sports Science, 27*(1), 69–76.

VanItallie, T. B., & Nufert, T. H. (2003). Ketones: Metabolism's ugly duckling. *Nutrition Reviews, 61*(10), 327–341.

U.S. Department of Agriculture, Agricultural Research Service, Nutrient Data Laboratory. USDA National Nutrient Database for Standard Reference, Released 22, 2009. Available at http://www.ars.usda.gov/ba/bhnrc/ndl. Accessed July 19, 2010.

Wall, R., Ross, R. P., Fitzgerald, G. F., & Stanton, C. (2010). Fatty acids from fish: the anti-inflammatory potential of long-chain omega-3 fatty acids. *Nutrition Reviews, 68*(5), 280–289.

Wenk, C. (2004). Implications of dietary fat for nutrition and energy balance. *Physiology & Behavior, 83*(4), 565–571.

Wylie-Rosett, J. (2002). Fat substitutes and health: An advisory from the nutrition committee of the American Heart Association. *Circulation, 105*(23), 2800–2804.

7 Water and Electrolytes

Learning Plan

- **Describe** the approximate amount, distribution, and roles of body water.
- **Discuss** the processes by which water movements occur between compartments in the body.
- **Define** hypohydration, euhydration, hyperhydration, and dehydration.
- **Identify** avenues of water and sodium loss and intake.
- **Discuss** the effect of exercise on fluid balance, and outline strategies for maintaining fluid balance before, during, and after exercise.
- **Identify** the role fluid plays in body temperature regulation during exercise and on performance and health.
- **Discuss** the effect of caffeine on hydration status.
- **Explain** the phenomenon of hyponatremia, and outline a strategy for prevention in endurance and ultraendurance athletes.

Maintenance of body fluid balance during exercise is critical to performance.

Pre-Test   Assessing Current Knowledge of Water and Electrolytes

Read the following statements about water and electrolytes and decide if each is true or false.

1. The two major aspects of fluid balance are the volume of water and the concentration of the substances in the water.

2. Now that sports beverages are precisely formulated, it is rare that water would be a better choice than a sports beverage for a trained athlete.

3. Athletes should avoid caffeinated drinks because caffeine is a potent diuretic.

4. A rule of thumb for endurance athletes is to drink as much water as possible.

5. Under most circumstances, athletes will not voluntarily drink enough fluid to account for all the water lost during exercise.

Water is often considered the most important nutrient. Failure to consume other nutrients may result in harmful deficiencies over a span of weeks, months, or years, but humans can live for only a few days without water. It is the most abundant substance in the body, comprising approximately 60 percent of an average person's body weight. Water provides the **aqueous** medium for chemical reactions and other processes within cells, transports substances throughout the body, facilitates **thermoregulation** (maintenance of body temperature), and is critical to most other physiological processes. Because of the additional physiological stress generated by physical activity, exercise, and sport, fluid balance is an important consideration for athletes and active people.

Loss of body water can be detrimental to both performance and health, as can excessive water consumption. In extreme cases, these situations can be fatal. The challenge for athletes, especially those exercising in hot and humid environments, is to adequately replenish water that is lost during (if possible) and after exercise. Some simple postexercise assessments such as scale weight, urine color, and degree of thirst, can help athletes monitor if fluid has been adequately restored. The loss and intake of electrolytes (for example, sodium) must also be balanced.

Each athlete needs an individualized plan for fluid and/or electrolyte intake before, during, and after exercise under normal training conditions. The plan will need to be adjusted to reflect changing environmental conditions, such as increasing temperature or the stress of competition. The amount and timing of fluid and/or electrolyte intake are critical elements of the

Water may be part of the athlete's individualized plan for fluid and/or electrolyte intake before, during, and after exercise.

© Steve Cole/iStockphoto.com

plan. For many athletes, carbohydrate intake will also be critical, and the amount, type, and concentration of carbohydrate are included as part of the plan because fluid is often used as a vehicle for carbohydrate delivery.

7.1 Overview of Water and Electrolytes

There are two major aspects to fluid balance that must be considered and understood: (1) water volume and (2) the concentration of **solutes** in body fluid. First, the body must have an adequate volume of water to meet physiological demands, a condition referred to as **euhydration**. An excess amount of water is generally a temporary condition in healthy people and is called

Table 7.1 Electrolytes Involved in Fluid Balance

Cations	Anions
Sodium (Na$^+$)	Chloride (Cl$^-$)
Potassium (K$^+$)	Bicarbonate (HCO$_3^-$)
Calcium (Ca^{2+})	Phosphate (PO$_4^{3-}$)
Magnesium (Mg^{2+})	Protein

Cations are positively charged electrolytes, and anions are negatively charged electrolytes.

hyperhydration, whereas an insufficient volume of water in the body is termed hypohydration. The term **dehydration** refers to the process of losing body water and moving from a state of euhydration to hypohydration. Dehydration is often used interchangeably with hypohydration, but these terms have different meanings.

Second, because of the potential for water to move from one area to another by **osmosis** due to concentration differences, the overall concentration of substances dissolved in body water must be considered as well. This is the concept of **tonicity**; body fluids are considered to be **hypotonic**, **isotonic**, or **hypertonic** if they have a concentration of solutes that is less than, the same as, or greater than the concentration of solutes in the cells respectively. Although a number of substances are osmotically active, the tonicity of body fluids is due largely to the concentration of electrolytes, electrically charged **cations** such as sodium (Na$^+$) and potassium (K$^+$), and **anions** such as chloride (Cl$^-$) and phosphate (PO$_4^{3-}$). Table 7.1 lists the cations and anions involved in fluid balance.

The amount of water in the body depends on many factors.

The amount of water in the body depends on a variety of factors, including body size, gender, age, and body composition. In general, larger people have more body water compared to those of smaller stature, and males have more water than females because men typically have more muscle mass and less body fat than women. Body water percentage has an inverse relationship with both age and body fatness; it declines with advancing age and increasing body fatness. On average, an adult's body is approximately 60 percent water by weight, but individuals may range from 40 to 80 percent. An average 70 kg (154 lb) male has approximately 42 liters (L) of total body water, and the average female approximately 30 L. Expressed as a nonmetric measurement, the average female and male have approximately 8 and 11 gallons of body water, respectively.

Different body tissues contain varying proportions of water. For example, blood plasma is largely fluid

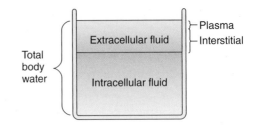

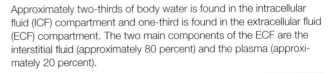

Figure 7.1 Body water compartments

Approximately two-thirds of body water is found in the intracellular fluid (ICF) compartment and one-third is found in the extracellular fluid (ECF) compartment. The two main components of the ECF are the interstitial fluid (approximately 80 percent) and the plasma (approximately 20 percent).

and consists of approximately 90 percent water. Muscle and other organ tissue can range from 70 to 80 percent water, whereas bone contains much less water, about 22 percent. Lipids are **anhydrous**; therefore, fat tissue contains very little water—approximately 10 percent.

Body water is distributed as intracellular or extracellular fluid.

Water is distributed throughout the body. This distribution is often separated into two major compartments: intracellular fluid (ICF) and extracellular fluid (ECF). Intracellular fluid consists of all the water contained within the trillions of cells in the body. Some cells have higher concentrations of water than others. All of these cells maintain their integrity because of their cell membranes, which separates the fluid inside the cells from the extracellular fluid. The ECF is further divided into subcompartments. One subcompartment is the plasma, the watery portion of the blood. Another is the **interstitial fluid**, the fluid that is found between the cells. Approximately two-thirds of total body water is in the ICF, leaving approximately one-third in the ECF (Figure 7.1).

Intracellular fluid. The body contains trillions of cells, many with different structures, composition, and functions. The water content of cells may vary dramatically. For example, myocytes, or muscle cells, may contain as much as 75–80 percent water by weight, whereas osteocytes, or bone cells, may consist of as little as 22 percent water. Collectively, though, the cells of the body contain approximately two-thirds of the body's total water volume, and all cells together are considered the intracellular fluid compartment. If an average male has 42 L (~11 gal) of body water, approximately 28 L (~7.5 gal) is contained in the cells or about 40 percent of total body weight.

Extracellular fluid. All of the water in the body not contained inside cells is considered a part of the extracellular fluid compartment. This is approximately one-third of the total body water, or about 14 L (~3.5 gal) in the average

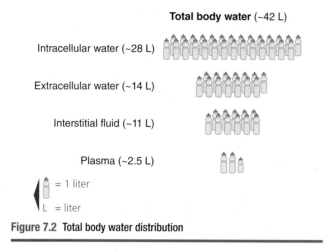

Figure 7.2 Total body water distribution

male. As shown in Figure 7.1, the ECF can be further subdivided into discrete areas such as the plasma (~2.8 L or ~3 qt) and the interstitial fluid (11.2 L or 3 gal).

The plasma serves as the fluid transportation medium to transport red blood cells, gases, nutrients, hormones, and other substances throughout the body. It is a major reservoir of fluid and plays a critical role in thermoregulation, particularly for activity in hot and humid environments. Of the total ECF volume, plasma accounts for approximately one-fifth, or 20 percent. The plasma is contained within the vascular system and is separated from the interstitial fluid by the walls of the blood vessels. The thicker walls of the larger blood vessels (for example, arteries and veins) provide a substantial barrier to fluid movement, whereas the smaller and thinner walls of the capillaries are very permeable and can allow considerable movement of water between compartments.

The major component of the ECF compartment (approximately 80 percent) is interstitial fluid. The interstitial fluid surrounds the cells and provides protection and an avenue for exchange with the cells of the body. The remaining ECF compartments are considered negligible compared to plasma and interstitial fluid because they are so small. Lymph is fluid contained in the lymphatic system, which returns fluid from the interstitial space to the blood. Transcellular fluids are found in specialized cells such as the brain and spinal column (cerebrospinal fluid), joints (synovial fluid), areas surrounding the internal organs, heart, and lungs (peritoneal, pericardial, and intrapleural fluids, respectively), eyes (intraocular fluid), and digestive juices. Although these fluids play critical functional roles, the total amount of water contained in these fluids is small and generally stable, so these fluid compartments are usually not included in discussions of body fluid balance. Figure 7.2 illustrates total body water distribution.

Water movement between compartments. The water that is found in ICF and ECF is not static. Water can be added to or removed from these compartments,

and although there are cell membrane barriers separating the various compartments, fluid moves between compartments relatively easily. Because intracellular fluid exists in isolated cells, water must pass through the extracellular fluid compartment in order to reach cells. The ECF therefore acts as a gateway for water entry into the body, initially through the plasma.

All cells are freely permeable to water so water can move through cell membranes easily, but there must be some force that stimulates the movement of water. The two major forces that result in the movement of water are fluid (hydrostatic) pressure and osmotic pressure.

Fluid, or hydrostatic pressure, is created when there is a difference in fluid pressure between two areas. For example, the cardiovascular system uses hydrostatic pressure to move blood throughout the body. When the heart contracts, it squeezes the blood that fills it, increasing blood pressure. This increase in blood pressure creates the driving force to propel blood through the blood vessels in the pulmonary and **systemic circulation**. This type of pressure can also result in water moving from the area of higher pressure (blood plasma), to areas of lower pressure (interstitial spaces). In another example of the fluid shifts resulting from hydrostatic pressure, feet and ankles

Aqueous: Consisting mostly of water.

Thermoregulation: Maintenance of body temperature in the normal range.

Solute: A substance dissolved in a solution.

Euhydration: "Good" hydration (eu = good); a normal or adequate amount of water for proper physiological function.

Hyperhydration: A temporary excess of water; beyond the normal state of hydration.

Dehydration: The process of going from a state of euhydration to hypohydration.

Osmosis: Fluid movement through a semipermeable membrane from a greater concentration to a lesser concentration so the concentrations will equalize.

Tonicity: The ability of a solution to cause water movement.

Hypotonic: Having a lower osmotic pressure than another fluid.

Isotonic: Having an equal osmotic pressure to another fluid.

Hypertonic: Having a higher osmotic pressure than another fluid.

Cation: A positively charged ion.

Anion: A negatively charged ion.

Anhydrous: Containing no water.

Interstitial fluid: Fluid found between cells, tissues, or parts of an organ.

Systemic circulation: Circulation of blood to all parts of the body other than the lungs.

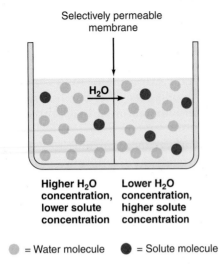

Selectively permeable membrane

H_2O

Higher H_2O concentration, lower solute concentration

Lower H_2O concentration, higher solute concentration

● = Water molecule   ● = Solute molecule

Figure 7.3 Osmosis

Water will move by osmosis across a selectively permeable membrane from an area of lower solute concentration to an area of higher solute concentration.

may swell if a person stands for long periods of time. When a person stands upright, blood rushes towards the feet due to gravity. This increase in hydrostatic pressure inside the blood vessels in the lower extremities results in more movement of water to the interstitial spaces in the feet and lower legs, resulting in swelling, or **edema**.

The second cause of water movement is due to osmosis, the tendency of water to move from areas of high solute concentration to areas of lower solute concentration. This would not be a factor if the fluid in the various compartments contained only water. However, fluids in the body contain a wide variety of solutes dissolved in the water. Figure 7.3 illustrates the net movement of water by osmosis from an area of lower to higher solute concentration across a selectively permeable membrane. The membranes of cells are selectively permeable, allowing free movement of water but often maintaining differences in the distribution of solutes.

Osmotic pressure is measured in milliosmoles (mOsm). When the number of particles (solute) is measured per *kilogram* of **solvent**, the correct term is **osmolality**; when measured per *liter* of solvent, the correct term is **osmolarity**. In nutrition and medicine, *osmolarity* is the standard term, whereas in exercise physiology the term *osmolality* is more commonly used because osmolality is not affected by temperature. In humans, there is little difference between the values when the two methods are calculated and compared (Gropper, Smith, and Groff, 2009). In this textbook, the term *osmolarity* will be used.

The two major subcompartments of the ECF—plasma and interstitial fluid—have a nearly identical composition and distribution of electrolytes (Figure 7.4). In extracellular fluid the major cation

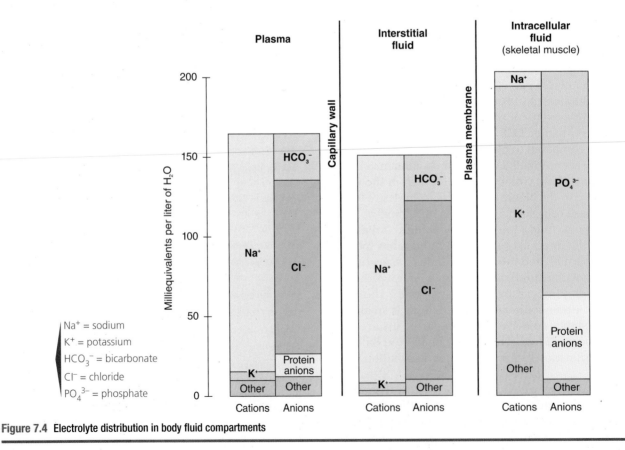

Na^+ = sodium
K^+ = potassium
HCO_3^- = bicarbonate
Cl^- = chloride
PO_4^{3-} = phosphate

Figure 7.4 Electrolyte distribution in body fluid compartments

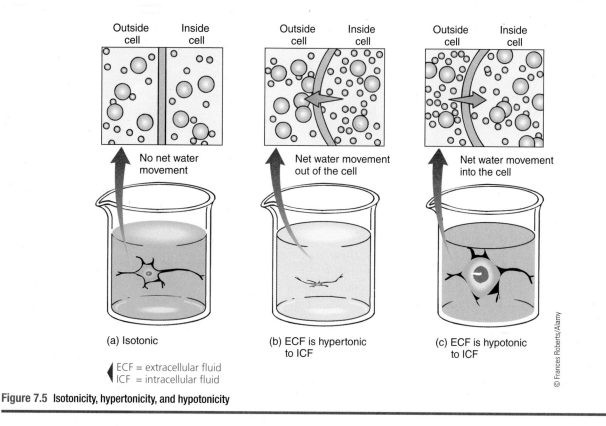

Figure 7.5 Isotonicity, hypertonicity, and hypotonicity

(positively charged ion) is sodium. The amount and concentration of potassium and other cations are much smaller. The major anions (negatively charged ions) are chloride and bicarbonate. In addition, there are some negatively charged proteins in plasma (for example, albumin) that are typically not found in the interstitial fluid, but the amount is small compared to chloride and bicarbonate.

The composition and distribution of electrolytes inside the cells are quite different from those found in the extracellular fluid (Figure 7.4). Potassium is the primary cation in the ICF, with sodium being present but at a much lower concentration. This distribution is opposite that of the ECF, and represents an important concentration differential for each ion. Because of these concentration differences, there is constant pressure for sodium to leak into cells and for potassium to leak out of cells. Normal intracellular and extracellular concentrations are maintained by the action of the sodium-potassium pumps located in the cell membranes, which constantly pump sodium ions out of the cells, while simultaneously pumping potassium ions back into the cells. The major anions inside the cells are phosphate and the negatively charged intracellular proteins.

It is important to note that although the ionic composition differs between the ICF and the ECF, the osmolarity or total concentration of all solutes in those compartments is generally the same. Shifts in fluid between the ECF and ICF occur solely due to osmosis, the movement of water from an area of lower concentration to an area of higher concentration. Under normal homeostatic conditions, the osmolarities of the ECF and ICF are the same, and there is no net movement of water. However, if the concentration in either compartment changes, a fluid shift may occur. If sodium increases in concentration in the extracellular fluid, water would move by osmosis out of the cells and into the ECF in an attempt to dilute the extracellular fluid and restore balance. For example, heavy sweating can cause a large loss of plasma volume due to water loss, resulting in an increased concentration of sodium in the plasma. This stimulates movement of water out of the cells (that is, ICF) and into the plasma (that is, ECF), causing the cells to shrink. Conversely, if the concentration of sodium in the extracellular fluid is decreased, the osmolarity of the ECF would be less than that in the cells, and water would move by osmosis into the cells in an attempt to correct the concentration imbalance. The resulting movement of water into the cells would cause the cells to expand.

Edema: An abnormal buildup of fluid between cells.

Solvent: A substance (usually a liquid) in which other substances are dissolved.

Osmolality: Osmoles of solute per kilogram of solvent.

Osmolarity: Osmoles of solute per liter of solution.

The concentration of solutes, or osmolarity, in a particular fluid is not static, and it can sometimes change relatively quickly. The concept of tonicity describes this change in solute concentration. Recall that isotonicity refers to a concentration of solutes that are equal to each other. Water will pass in and out of the cell, but the net movement of water will be zero. When a fluid has a higher concentration of solutes compared to another fluid, it is said to be hypertonic. For example, when a person sweats heavily, there is a loss of water from the extracellular fluid and the ECF becomes hypertonic in relation to the intracellular fluid. The opposite concentration difference is called hypotonic, when a fluid has a lower osmolarity than the reference fluid. If a person consumes a large amount of water very quickly, this water is absorbed into the extracellular fluid, resulting in a dilution of the ECF and making it hypotonic compared to the intracellular fluid (Figure 7.5).

If one understands how water moves by osmosis, one can easily understand how fluid shifts in the body. The movement of fluid occurs as a result of controlling the amount of water in the ECF and the osmolarity of the ECF. The amount of water and the osmolarity in the ECF is controlled by water intake and loss, sodium intake and loss, and by compensatory regulatory mechanisms in the kidney and gastrointestinal (GI) tract.

Key points

- An adult's body is approximately 60 percent water.

- Water is distributed between two major compartments in the body, intracellular (ICF) and extracellular (ECF) water.

- Water may move from an area of higher pressure to an area of lower pressure (hydrostatic pressure).

- Water may move from an area of higher solute concentration to an area of lower solute concentration (osmotic pressure).

Where in the body is the majority of water found?

An athlete consumes a supplement that results in an increase in solute concentration in muscle cells. Describe any subsequent movement of water in the body (assume the athlete continues to consume an adequate amount of water).

7.2 Water Loss, Intake, Balance, and Imbalance

Fluid balance in the body is partially a result of the daily balance of water intake and water loss. There are a limited number of avenues for water intake, but a wider variety of ways that water may be lost from the body (Figure 7.6). There are constant changes in body water resulting in temporary water imbalances.

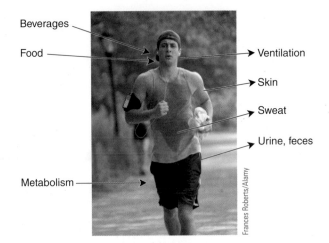

Figure 7.6 Avenues of water gain and loss

Water can be gained by the body through the consumption of food and/or beverages, and as a result of metabolism. Water can be lost via ventilation, insensible loss through the skin, sweating, and urine and fecal losses.

Water is lost in a variety of ways.

Water loss from the body is generally categorized as either **insensible** or **sensible**. Insensible refers to avenues of loss that are not normally noticed by the individual, including water lost through ventilation and through nonsweat diffusion through the skin. With each breath, the inspired (inhaled) air is humidified to protect delicate lung tissues from drying. The water vapor that is added is then lost with the subsequent expiration, because the water is not recaptured before the air is exhaled from the body. This water loss is typically not noticed except on cold days when we can "see our breath." When warm, humid air from the lungs is rapidly cooled by the cold air outside, water vapor condenses into water droplets that can be readily seen. Water losses by this route can increase in environments in which the air is colder and drier, as more water vapor needs to be added to the inspired air. Increased levels of ventilation as a result of exercise may also cause an increased insensible loss of water.

Skin must be kept moist to prevent drying and cracking, and some of the water that diffuses into the skin is lost from the body. Water loss by this mechanism may also increase in dry environments with low humidity. Total insensible water losses average approximately 1,000 milliliters (ml) or ~36 oz (~4.5 cups) per day for the average person.

The three major areas of sensible water loss are the fluid lost in urine, feces, and sweat (Figure 7.6). The amount of water in feces is variable, but daily loss by this route averages approximately 100 ml (~3.5 oz) per day. This is not typically a major avenue of water loss

One avenue of water loss is through ventilation.

by the body, unless an individual has a disease such as dysentery that results in large volumes of watery diarrhea.

The renal system provides the major physiological mechanism for controlling fluid balance in the body via the production and excretion of urine. Urine output can vary dramatically, but for the average person under homeostatic conditions it is approximately 1,500 ml per day (~54 oz or 6–7 cups). The amount of renal water loss is highly variable and can be influenced by the amount of fluid and salt intake, renal function, the action of various hormones, and the consumption of compounds that have a **diuretic** effect.

A diuretic is a substance that increases urine output. All fluids can have a diuretic effect. Water is a diuretic if a person is in fluid balance and ingests a large volume of water. However, the focus is generally on substances that exert a diuretic effect by a mechanism in addition to an increased volume of fluid. For example, alcohol has a mild diuretic effect because it inhibits the production of the antidiuretic hormone (ADH). Nearly 200 herbs are known to have a diuretic effect. Some diuretics are prescription medications (for example, Lasix™) and can block the reabsorption of fluid and/or electrolytes from the renal tubules. These substances may have a substantial effect on hydration status, and athletes are cautioned about their use especially if they are already hypohydrated.

Caffeine (found in caffeinated coffee and some soft drinks) and theophylline (found in tea) increase urine output by increasing the blood flow in the kidneys and increasing sodium and chloride excretion. In the past, athletes were cautioned about consuming these beverages because they are mild diuretics, but avoidance is no longer recommended. Armstrong et al. (2007) diligently reviewed studies of caffeine ingestion and the effect on fluid and electrolyte imbalance. Based on the current body of scientific literature, a daily caffeine intake for adults of ~450 mg or less does not have a

negative effect on fluid or electrolyte balance. The caffeine content of brewed coffee can vary considerably, but a rule of thumb measurement in the United States is that an 8 oz (240 ml) cup of caffeinated coffee contains approximately 80–150 mg of caffeine. Thus a little more than 2 cups of strongly brewed caffeinated coffee daily (or the equivalent from other beverages) as part of a normal diet is not considered detrimental to euhydration, and it is no longer recommended that caffeine be completely avoided by adult athletes. More information about caffeine and alcohol can be found in Chapter 10.

Another highly variable avenue of water loss is sweating, one of the body's major thermoregulatory mechanisms. Sweat is water that is secreted from sweat glands onto the surface of the skin. When this water evaporates (that is, changes from water to water vapor), it results not only in a loss of water but also in a transfer of heat away from the body. If an individual is not in a hot environment, sweating is minimal and water loss by this mechanism averages approximately 100 ml (~3.5 oz) daily. As environmental temperatures rise and/or activity increases, thermoregulatory demands increase, and water loss via sweating increases. It would not be uncommon for active adults in hot and/or humid environments to lose several liters of water each day through sweating. The effect of exercise and activity on water loss will be discussed in greater detail later in this chapter.

Water is added to the body primarily through the intake of beverages and foods.

The addition of water to the body is primarily accomplished through the fluid content of beverages and foods consumed each day, and secondarily through metabolism (Figure 7.6). On average, an adult will take in approximately 2,350 ml (~84 oz or 10.5 cups) of water each day from beverage and food sources, but there can be considerable variation depending on how much an individual eats and drinks. Adults in the United States consume ~20–25 percent of their total daily water intake via food, 35–40 percent from tap or bottled water, and the remainder from other beverages (Sharp, 2007).

The water content of foods can vary dramatically, as can be illustrated by comparing watermelon (~90 percent water) and bran cereal (~2 percent water). Beverages also vary in their water content as shown in

Insensible: Imperceptible, typically not noticeable.

Sensible: Perceptible.

Diuretic: Causing an increased output of urine.

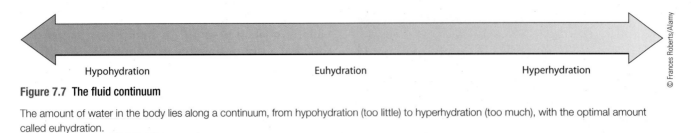

© Frances Roberts/Alamy

Figure 7.7 The fluid continuum

The amount of water in the body lies along a continuum, from hypohydration (too little) to hyperhydration (too much), with the optimal amount called euhydration.

Table 7.2. All beverages are predominantly water, but each can have a different effect on fluid balance and osmolarity because of the other substances contained.

Another source of water is a result of aerobic metabolism. As explained in Chapter 3, in the metabolic pathway of oxidative phosphorylation, oxygen is required in the mitochondria to be the final electron acceptor at the end of the electron transport chain. Oxygen (O) molecules pick up the electrons in the form of hydrogen (H) molecules, and are thus converted to water (H_2O). Aerobic metabolism contributes approximately 350 ml (~12.5 oz) of water each day. For the average adult in homeostasis, approximately 2,700 ml (~96 oz or 12 cups) of water are consumed or produced by the body each day via food and beverage intake and metabolism.

There are constant changes in body water, resulting in temporary water imbalances.

The amount of water in the body is constantly changing, with fluid being added and removed through the mechanisms discussed above. The body is said to be in fluid balance if a sufficient amount of fluid is present that allows for the body to function normally. Although the body has sensitive mechanisms to maintain fluid balance over time, it is possible to exceed the capabilities of these mechanisms in either the short or long term. In general, the amount of fluid consumed is not carefully matched to the body's daily needs. The usual overall strategy is to drink an excess of water, retain what is needed to maintain fluid balance, and excrete the excess. Deviations from this basic strategy can result in imbalances.

Hypohydration, euhydration, and hyperhydration. The amount of water in the body lies along a continuum, from too little to too much. Hypohydration, euhydration, and hyperhydration are terms that describe the status of body water, although these terms refer to a general condition rather than to a specific amount of water (Figure 7.7). Because of the difficulties of measuring total body water and the variability between individuals, specific dividing lines between these conditions on this continuum have not been identified. The discussion that follows assumes that a person is healthy and has normal kidney function.

Euhydration refers to a "normal" amount of water to support fluid balance and to easily meet required physiological functions. Enough water is present to maintain osmolarity of the extracellular fluid, prevent large water shifts between compartments, and support critical processes such as cardiovascular function and thermoregulation. As previously discussed, this optimal level of hydration is typically achieved by consuming fluids in excess of need and allowing the renal system to excrete the unneeded amount.

Hyperhydration refers to body water above that considered normal and is typically a short-term condition. Acute hyperhydration can be achieved by consuming excess fluids, but the renal system usually acts quickly (within minutes to hours) to excrete the excess water by forming greater volumes of urine. For example, a person who drinks 2 cups (480 ml) of water with lunch will usually need to urinate by the middle of the afternoon.

Consumption of certain osmotic substances such as glycerol along with excess water may slightly prolong the state of hyperhydration. This induced

Table 7.2 Water Content of Various Beverages

Food or beverage	% Water	Predominant substances
Water	99	Some minerals depending on the hardness of the water and traces of other compounds
Coffee or tea (black)	99	Some minerals depending on the hardness of the water and traces of other compounds such as caffeine
Sports beverages	90–94	Sugars
Fruit juice	88	Sugars
Milk	88–91 (depending on the fat content)	Protein, minerals, milk sugar, possibly fat

hyperhydration is a strategy that may be employed by athletes to aid thermoregulation when competing in hot environments, and is discussed later in this chapter.

Consumption of large amounts of water very quickly (for example, 3,000 ml or ~104 oz or ~13 cups in 4 hours, as some slow marathon runners have done) can result in a state of hyperhydration that can actually be dangerous to one's health, and can even result in death (Almond et al., 2005). The excess water dilutes the concentration of solutes in the extracellular fluid. Before the kidneys have a chance to excrete the extra water, the reduced osmolarity in the ECF provokes a shift of water from the ECF into the cells, causing them to swell. Nerve cells, especially those in the brain, are particularly sensitive to this swelling and may cease to function properly, resulting in impaired brain function, coma, or even death. This condition can occur during certain types of endurance exercise and is discussed later in this chapter (see section on hyponatremia).

Hypohydration is the term used to describe body fluid levels that are below normal or optimal. It is the result of either an inadequate intake of water, excessive loss of water, or a combination of both. Hypohydration is a relative term (compared to euhydration) and is not defined as a certain amount of body water below euhydration. Hypohydration occurs initially in the ECF, which results in an increased concentration of solutes in the ECF due to a relative lack of water. Because the osmolarity of the ECF increases, water shifts from the cells to the ECF in an attempt to balance the solute concentrations between these two fluid compartments. When water levels decline in the cells, the cells shrink and this shrinkage may impair cellular function. Hypohydration may have severe adverse effects on exercise performance and thermoregulation, so athletes need to be especially conscious of ways to prevent excessive hypohydration. The term dehydration is commonly used interchangeably with hypohydration, but it is more accurate to use the term dehydration to describe the *process* of moving from a state of euhydration to a state of hypohydration.

Electrolytes and water distribution. Distribution of water throughout the body is regulated by both the volume of water and the osmolarity of the extracellular fluid. Osmolarity is influenced by the concentration of solutes, particularly certain electrolytes. Sodium is the most important electrolyte in the extracellular fluid because it exists in the largest amount and therefore has a large and direct affect on osmolarity. The body must respond to changes in the amount of sodium in the ECF by adjusting water volume. An increase in sodium in the ECF will increase the

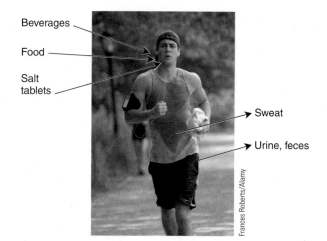

Figure 7.8 **Routes of sodium intake and excretion**

Sodium intake occurs with the consumption of sodium-containing foods and beverages, or rarely, salt tablets. Sodium can be lost via sweat, urine, and feces.

volume of water and a decrease in sodium will result in a decrease in ECF water volume. Other electrolytes that may influence water distribution are potassium, calcium, magnesium, chloride, and phosphate (see Table 7.3).

Sodium intake and excretion. The only route of intake for sodium is by ingestion, through either foods, fluids, or rarely, salt tablets (sodium chloride). Similar to water, sodium is generally consumed in excess of the body's requirements and the body relies on the renal system to excrete what is not needed. Many factors influence the amount of sodium in the diet, but the largest factor in industrialized countries is the consumption of processed foods at home and at restaurants. For the average American adult, processed food is the source of 77 percent of daily sodium intake. The addition of table salt (that is, sodium chloride) to foods accounts for approximately 11 percent of daily sodium intake. The remaining sodium (~12 percent) occurs naturally in water and in foods such as milk, vegetables, and grains (Institute of Medicine, 2004).

There is considerable variation in sodium intakes among men and women in the United States and around the world. Consumption can be reported as either sodium or salt intake, which often causes confusion because these terms are not the same. Intakes are routinely reported in both milligrams and grams, which adds to the confusion.

The highest reported consumption worldwide is in northern Japan where intake of sodium is estimated to be more than 10,000 milligrams (mg), or 10 g, daily due to the salting and pickling of foods. The average daily intake of sodium in the United States is approximately 4,200 mg (4.2 g) for men and 3,300 mg (3.3 g) for women.

Table 7.3 Intake and Output of Electrolytes Involved in Fluid Balance

	Intake	Output
Sodium (Na$^+$)	Food, beverages, use of salt shaker; salt tablets; intake can be excessive	Sweat, urine, feces; sodium loss in sweat is generally low but can be substantial for some people ("salty sweaters"); sodium output in urine is typically high due to tight homeostatic control by the renal system; fecal losses of sodium are small
Potassium (K$^+$)	Food, beverages; occasionally the use of salt substitute; intake tends to be low due to lack of fruits, vegetables, and minimally processed foods	Small losses in sweat or feces; losses in urine tightly controlled by the renal system; atypical losses such as vomiting and potassium-depleting diuretics can be dangerous
Calcium (Ca^{2+})	Food (both naturally occurring and fortified), beverages, supplements; many people consume less than is recommended	Small losses in sweat; larger losses in feces (unabsorbed calcium) and urine; blood calcium tightly regulated by several hormones, which affect absorption, retention, and excretion
Magnesium (Mg^{2+})	Food, beverages, supplements	Small losses in sweat and feces; most magnesium is lost via the urine; some diuretics increase urinary excretion
Chloride (Cl$^-$)	Found as sodium chloride in food, beverages, salt	See sodium above

For most people in the United States, sodium is consumed in the form of sodium chloride (salt added to food) and this sometimes leads to the reporting of salt intake rather than sodium intake. One gram (that is, 1,000 mg) of salt contains ~40 percent sodium or ~400 mg of sodium. One-fourth teaspoon is the equivalent of 1.5 g of table salt, thus this amount of salt contains 590 mg of sodium. It is easy to see how adding salt to food results in a high sodium intake. Reported salt intake by U.S. men is approximately 10 g per day, but for some salt intake may be as high as 25 g daily.

The Dietary Reference Intake for sodium for adults under the age of 50 is 1,500 mg (1.5 g) daily. The Tolerable Upper Intake Level is 2,300 mg (2.3 g) daily. These recommendations are made to the general population to reduce the prevalence of high blood pressure associated with aging. However, the Institute of Medicine (2004) clearly states that these recommendations do not apply to highly active individuals who lose large amounts of sweat daily. Sodium intake and hypertension are discussed in Chapter 13.

Sodium intake by athletes varies, based on caloric intake and choice of foods. Low caloric intake may mean lower sodium intake. For example, the daily energy intake of 19 New Zealand jockeys was approximately 1,500 kcal whereas the average sodium intake was approximately 1,900 mg (1.9 g) (Leydon and Wall, 2002). As caloric intake increases, sodium levels rise, so even an athlete who is careful about avoiding high-sodium foods (for example, soy sauce, fast foods, salty-tasting snacks) may find that sodium intake is above 3,000 mg (3 g) daily. Some athletes consume fairly high levels of sodium (for example, ≥5,000 mg or 5 g), especially if they add table salt to their food.

Although sodium intake may not be carefully matched to need, urinary sodium excretion is precisely controlled to maintain proper ECF osmolarity. Sodium is regulated by the kidneys and can be either reabsorbed or excreted in the urine. Because sodium intake usually far exceeds daily needs, a large amount of sodium is excreted in the urine to maintain sodium balance. Average urinary sodium excretion is approximately 1 to 5 g per day. More detailed information about renal function can be found in a human physiology text such as Sherwood (2010).

A small amount of sodium is lost each day in sweat and feces. This obligatory loss amounts to approximately 0.5 g (500 mg) per day. Sodium loss can increase substantially if there is heavy sweating and will be discussed later in this chapter.

Popcorn, pretzels, nuts, and chips typically have salt (sodium chloride) added, although low-salt varieties are available.

Potassium intake and excretion. Potassium, the primary intracellular cation, is consumed via foods and beverages or occasionally, through the use of a salt substitute (potassium chloride). Potassium is abundant is unprocessed foods such as fruits, vegetables, whole grains, beans, and milk. Low dietary potassium intake in the United States is a reflection of a low daily fruit and vegetable intake and the high intake of processed foods; processing results in substantial potassium losses. Examples of foods with a high potassium content include bananas, orange juice, and avocadoes. Less than 3 percent of U.S. adults are likely to receive an adequate amount of potassium from their diets (Institute of Medicine, 2004). Potassium found in food is easily absorbed from the gastrointestinal tract (greater than 90 percent absorption).

An Adequate Intake (AI) for adults is 4,700 mg daily. The average daily intake in the United States is approximately 2,200 to 2,400 mg for women and 2,800 to 3,300 mg for men. With such low average intakes, it is likely that some adults in the United States will have a moderate potassium deficiency. Such a deficiency is associated with increased blood pressure (see Chapter 13) and increased bone turnover due to an increase in urinary calcium excretion. However, a moderate potassium deficiency is hard to detect with laboratory tests because blood potassium concentration will remain within the normal range (3.5 to 5.0 mmol/L) (Institute of Medicine, 2004).

Athletes who are concerned about the potential for a moderate potassium deficiency should focus on consuming a variety of fruits and vegetables daily. Self-prescribed potassium supplements are not recommended because of the potential for hyperkalemia (that is, elevated blood potassium concentration), which has been occasionally reported in bodybuilders (Perazella, 2000; Appleby, Fisher, and Martin, 1994; Sturmi and Rutecki, 1995).

Hypokalemia, a severe deficiency state, is defined as a blood potassium concentration less than 3.5 mmol/L and would be a rare occurrence in an otherwise healthy individual. Symptoms include muscle weakness and cardiac arrhythmias, which can be fatal. Hypokalemia is *not* a result of low dietary intake of potassium because the amount consumed in food is high enough to prevent this condition. Rather, hypokalemia is a result of substantial potassium loss, usually through severe and prolonged vomiting or diarrhea or the use of potassium-depleting diuretic drugs without adequately replenishing potassium.

Under normal conditions the primary pathway for potassium loss is urinary excretion, although small amounts are lost in the feces and sweat. The kidneys precisely control the amount of potassium either reabsorbed or excreted. Problems occur when potassium is lost atypically. For example, an athlete who

is struggling with an eating disorder may frequently self-induce vomiting, which over time could result in hypokalemia, cardiac arrhythmias, and death.

Other electrolytes. Although there are several electrolytes involved in fluid and electrolyte balance, the initial dietary focus tends to be on sodium and potassium. Two other cations—calcium and magnesium—are discussed in Chapter 9. The corresponding anions, such as chloride and phosphate, receive little dietary attention. The chloride content of the diet can be reasonably well predicted from salt intake and phosphorus is widely found in food. A dietary deficiency of either would be extremely rare.

Key points

- Water is lost from the body by noticeable means (for example, sweating, urination) and by less noticed avenues such as ventilation.
- Diuretics are substances that increase water loss through an increase in urine output.
- Water loss from sweat may range from 100 ml to several liters per day.
- The main avenue of water intake is the consumption of fluids and food.
- Having an appropriate amount of body water to support fluid balance and body functions is called euhydration.
- A state of too little body water is called hypohydration, and too much body water is called hyperhydration.
- Excessive hypohydration or hyperhydration can adversely affect physical performance and can potentially be dangerous to one's health.
- The average daily intake of sodium in the United States is over 3,000 mg for women and over 4,000 mg for men.
- Approximately 40 percent of salt intake is sodium.
- The majority of excess sodium consumed is excreted in the urine.
- A low daily intake of fruits and vegetables may result in a mild potassium deficiency.

Why do heavy sweaters increase their sodium intake by salting food?

Why is taking potassium supplements not a good idea?

7.3 Effect of Exercise on Fluid Balance

Under normal conditions, most healthy sedentary individuals regulate their fluid balance (that is, achieve homeostasis) relatively easily. Thirst and hunger mechanisms usually lead people to consume water in excess of the body's daily needs, with the excess being excreted. If there is a shortfall in fluid consumption, the body can respond in the short-term by reducing urine excretion, which conserves water and maintains fluid balance.

Exercise challenges fluid homeostasis because of the critical role that body fluids play in thermoregulation (maintaining an appropriate body temperature). Exercise causes an increase in body temperature, and a major mechanism for lowering body temperature is the evaporation of sweat. The loss of fluid through sweat may have a large impact on fluid balance, in both the short and long term. Physical activity or exercise, especially in hot and humid conditions, represents a substantial challenge for the regulation of body temperature, fluid homeostasis, and, subsequently, performance (Armstrong and Epstein, 1999). In contrast to sedentary individuals who can regulate their fluid balance easily, athletes may have a difficult time preventing dehydration and severe hypohydration. These conditions can negatively impact training, performance, and health (Hargreaves and Febbraio, 1998).

Exercise can have dramatic effects on water loss, particularly due to sweating.

Exercise can result in water shifting between compartments within the body and in accelerated loss of water from the body. As the cardiovascular system adjusts to the demands of exercise by increasing blood flow and oxygen delivery, the increased pressure in the blood vessels results in some of the fluid leaking into the surrounding interstitial space. This fluid is lost from the plasma and plasma volume therefore declines slightly as the water shifts within the extracellular fluid compartments. This decrease in plasma volume occurs within the first few minutes after exercise begins with the amount being largely dependent upon the exercise intensity. The decline may be up to approximately 5 percent of plasma volume if the exercise is intense. As explained earlier, an average adult male has approximately 2.8 L of plasma. A 5 percent loss would be approximately 140 ml (5 oz) of water, an amount that would not likely impair exercise performance. However, plasma volume losses up to 10–20 percent can occur with prolonged exercise, and losses of this magnitude may compromise cardiovascular function and ultimately reduce exercise performance (Convertino; 1987).

Exercise can cause substantial dehydration through increased sweating. The increase in energy expenditure associated with physical activity and exercise results in an increase in heat production. To prevent this heat load from causing an excessive increase in body temperature, heat loss mechanisms must be activated, one of which is sweating. As the internal (core) temperature of the body rises, sweat glands are

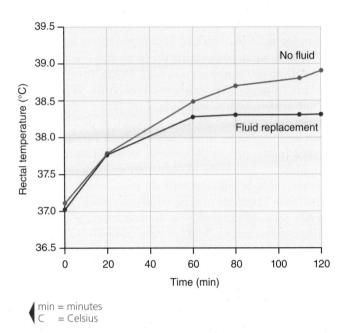

min = minutes
C = Celsius

Figure 7.9 Rise in rectal temperature during exercise

Body temperature rises with exercise, but replacing fluid to prevent dehydration prevents the excessive rise in body temperature that occurs when no fluid is consumed.

stimulated to secrete sweat to reduce the heat load. A review of thermoregulation may be beneficial at this point (see Sherwood, 2010), but in short, when sweat evaporates from the surface of the skin, heat is transported away from the body. It is not uncommon for body temperatures to rise to 39° or 40°C (102°F to 104°F) during exercise, even in temperate climates (Figure 7.9) (Hamilton et al., 1991).

Factors affecting sweat rate. The rate at which the body loses water through sweating depends upon a variety of factors. Exercise intensity can influence sweat rate through its independent effect on core temperature. At any given room temperature, increased exercise intensity results in higher body temperatures, and higher body temperatures will stimulate higher sweat rates. Increased exercise intensity requires higher energy expenditure, which results in a greater amount of metabolic energy being converted to heat.

Environmental conditions can also dramatically influence sweat rate. Independent of exercise intensity, higher ambient temperatures result in higher core temperatures and higher sweat rates. The relative humidity, or amount of water vapor in the air, can also influence sweat rate and can make fluid loss by sweating more visible. Increased water vapor in the air reduces the ability of sweat secreted onto the skin to be evaporated, and it is the *evaporation* of sweat, not just the process of sweating that leads to heat transfer.

Elevated humidity may be found in certain geographical areas (such as southeastern United States) or may be encountered in indoor environments where people are exercising and there is insufficient ventilation. When conditions are more humid, sweating is a less effective means of thermoregulation and body temperatures may be higher, resulting in a stimulus for an even higher sweat rate.

Clothing, uniforms, and protective gear may further influence the rate of sweating by providing a barrier to heat loss. This type of clothing may provide an insulating effect, trapping more heat in the body, or it may adversely affect the evaporation of sweat by reducing the surface area of the skin that is exposed. Athletes in sports with certain uniform traditions or requirements may be at greater risk for heat injury and fluid imbalances. Football uniforms provide an excellent example. They generally cover most of the skin and have thick padding in many places that can have an insulating effect. Protective gear such as helmets, arm wrappings, padding, and gloves add to the thermoregulatory challenge and may result in greater fluid loss through sweat. Other examples include the fire protection gear worn by firefighters, hazardous materials suits worn by public safety workers, and chemical warfare protection suits worn by military personnel.

The training status of the athlete and the degree of acclimation to the heat are also important factors affecting sweat rate. Because of regular training and the increased demands for thermoregulation, trained athletes typically have higher sweat rates than do sedentary individuals (Armstrong, Costill, and Fink, 1987). Trained athletes will start sweating sooner as body temperature begins to increase and will sweat at a higher rate, so they have a greater potential for water loss through sweating. The advantage, of course, is that they have a more effective mechanism for controlling their body temperature. Exposure to higher ambient temperatures during training (for example, a hotter climate) results in a number of adaptations that improve thermoregulation, one of which is an increased sweat rate.

Amount of sweat lost during exercise. Obviously, there are a variety of factors that can influence sweat rate, either alone or in conjunction with one another. Just how much fluid can a person lose through sweating? As discussed earlier, an adult who is performing usual occupational activities throughout the day in a temperate (normal room temperature) environment may lose approximately 0.1 L (100 ml or ~3.5 oz) of water each day as sweat. If that person performs those same daily activities in a hotter, more desertlike environment (high temperature but low relative humidity),

Firemen work in high-temperature environments and wear protective clothing that adversely affects the evaporation of sweat.

Pharaoh Richard/PhotoLibrary

sweat loss may increase to a rate greater than 1 L (1,000 ml or ~36 oz or ~4.5 cups) per *hour*. The addition of protective clothing such as the firefighter's suit can result in sweat rates of 1.0 to 2.0 L (1,000–2,000 ml or 36–71 oz or ~4.5–9 cups) per hour. Sweat rates in excess of 2.5 L (2,500 ml or ~89 oz or ~11 cups) per hour have been observed in athletes competing in team sports (for example, soccer) and in individual athletes engaged in prolonged endurance sports (for example, marathon running) in hot environments (Armstrong and Maresh, 1998).

In many athletes the loss of water through sweating is very visible, but in others it may not be as noticeable. Athletes that compete and train in the water, such as swimmers and water polo players, may sweat as part of their thermoregulatory response, but this avenue of water loss is not readily apparent. One study of elite male water polo players showed an average sweat rate of 287 ml per hour during training and 786 ml per hour during competitive matches. The same study showed elite male and female swimmers having an average sweat rate during training of 138 ml and 107 ml per kilometer swum, respectively (Cox et al., 2002). The hydration needs of these athletes should not be overlooked.

Although much attention is paid to exercising in hot and humid environments, substantial fluid loss can occur with exercise in cold climates as well (for example, winter sports, mountain climbing). Water loss from ventilation is usually greater in cold environments because the air is dry (low relative humidity) and the body must add more water vapor to the cold, dry air that is inspired. If the individual's ventilation is elevated during exercise, a large portion of this water is lost through exhalation. Activities requiring a high energy expenditure will result in elevations in body

Nick Laham/Getty Images Sport/Getty Images

This heavily favored Olympic marathoner was unable to finish the race, which was held in hot and humid conditions.

temperature even in cold temperatures. Participants need to carefully consider the clothing they wear. Clothing that is sufficiently warm at rest may prove to be too warm during exercise or activity, resulting in increased sweating and fluid loss.

Consider a world-class female runner competing in the Olympic Marathon, which is contested during the Summer Games, often in hot and humid conditions. She will run for a little over 2 hours and could lose approximately 2 L of sweat per hour or a total of 4 L of water! This may approach 10 percent of total body water or almost 10 pounds of water weight loss in a 100-pound runner if she were to consume no fluids. With such a sweat rate, it is easy to see how large amounts of fluid can be lost very quickly under certain exercise or environmental conditions. Because fluid homeostasis is a balance of water intake and loss, fluid intake is essential. But can an athlete prevent overall fluid loss when sweating heavily by drinking water or other fluids?

Unfortunately, it is very difficult to match water intake with water loss when exercising in hot and humid conditions for several reasons. Most people will experience "voluntary dehydration" and will not consume enough fluid during the activity to match the amount of fluid being lost. Voluntary dehydration occurs because

Focus on research

How Often and How Does Hyponatremia Occur during Ultraendurance Events?

Participation in ultraendurance events such as Ironman™-length triathlons, in which athletes swim 2.4 miles (3.8 km), bike 112 miles (180 km), and run 26.2 miles (42.2 km), tests the limits of human endurance performance. These events are typically contested in hot and humid environments, which provide a substantial challenge for thermoregulation and fluid balance that may have serious health consequences as well as performance implications. Observations of athletes seeking medical attention during and after ultraendurance events revealed a surprising percentage of cases of hyponatremia, a lowering of sodium concentration in the blood plasma. This is surprising because this is the opposite response that is expected in athletes who are sweating heavily and becoming dehydrated. Sodium is lost in sweat, but water is lost at a faster rate, leaving a higher concentration of sodium in the blood. It is vital for physicians and clinicians to understand this phenomenon because failure to recognize hyponatremia may result in treatment that may make the condition worse, and this electrolyte disorder is potentially fatal.

How does exertional hyponatremia occur and how frequently does it occur among participants in ultraendurance events? A large-scale, field-based research study provided important information that helped define and characterize this phenomenon:

Speedy, D. B., Noakes, T. D., Rogers, I. R., Thompson, J. M. D., Campbell, R. G. D., Kuttner, J. A., Boswell, D. R., Wright, S., & Hamlin, M. (1999). Hyponatremia in ultraendurance triathletes. *Medicine and Science in Sports and Exercise, 31*(6), 809–815.

What did the researchers do? Prior to this study being conducted by Speedy et al., a number of published studies noted hyponatremia in ultraendurance participants. These studies typically involved only participants who sought medical care at the races and were typically limited in the number of observations, leaving our understanding of the extent and reason for the hyponatremia unknown. It is often necessary for scientists to step out of the laboratory in order to study individuals' responses as they occur in a real athletic event, and the Speedy et al. study is an excellent example of a large-scale, field-based study.

The purpose of the study was to obtain plasma sodium measurements from athletes who completed an Ironman™-length triathlon, determine the incidence of hyponatremia in these participants, and see if there was a relationship with change in body weight during the race. Subjects had their body weight measured before and after the race to determine weight change, which in a race of this type would be mostly due to fluid (and some food) ingestion. They also had a blood sample taken after the race to determine plasma sodium levels. All registrants of an Ironman™-length triathlon were invited to participate (n = 660); a large percentage agreed, and when the study was completed, a total of 330 subjects had complete measurements for analysis.

A field-based study has the advantage of realism, studying athletes before, during, and after a real athletic event, but it has its difficulties and limitations as well. The main limitation is the loss of control over certain conditions and ability to make

the human thirst mechanism responds too slowly and does not have enough precision to stimulate drinking enough fluids to compensate for the high rate of fluid loss. If fluid loss exceeds fluid intake, athletes will gradually dehydrate, but they do not feel thirsty initially.

Even if athletes are encouraged to consume more fluid than they would likely take in of their own volition, the rate at which the water can be emptied from the stomach and absorbed from the gastrointestinal tract may be less than the rate of fluid loss due to sweating. The maximal rate of gastric emptying (how fast the fluid leaves the stomach) appears to be less than the maximal rate of absorption from the intestines, and therefore limits the overall rate of fluid intake (Sawka and Coyle, 1999). The maximal rate of gastric emptying for an average-size adult male is approximately 1.0–1.5 L per hour. This rate may be slightly reduced with high-intensity exercise and with an increased heat load.

Some level of activity or movement may actually aid gastric emptying through agitation of the stomach contents, and exercise up to approximately 70 percent of an athlete's maximal oxygen consumption generally does not have an adverse effect on fluid uptake. More intense exercise may impede gastric emptying, though, and delay the replacement of water and carbohydrate. Compared to bouts of low-intensity exercise (walking), athletes that performed intermittent, high-intensity exercise that mimicked the intensity of a soccer match showed a significant decline in the gastric emptying of a beverage that contained carbohydrate as well as a noncarbohydrate beverage (Leiper et al., 2005).

Therefore, even if athletes are encouraged or even forced to consume fluids in an attempt to match fluid loss, there is usually a gradual fluid loss because intake and absorption cannot keep pace with loss. It is not uncommon for athletes exercising under these conditions to lose 2–6 percent of their body weight during the activity. It is important to attempt to match the amount of fluid lost during exercise with appropriate fluid intake strategies. It is also very important to pay attention to rehydration strategies after the activity to ensure a return to euhydration.

certain measurements. For example, obtaining body weight can be done precisely in the laboratory, but picture trying to do so quickly and accurately when hundreds of subjects and volunteers are milling around on the morning of a race. The researchers therefore weighed many subjects during the registration period 2 days before the race and reweighed as many people as possible immediately before the race. They found this to be an acceptable limitation as prerace body weight varied little over those 2 days (~0.1 kg). Athletes were weighed again immediately after the race either at the finish line or in the medical tent. Similarly, the blood samples taken after the race were analyzed somewhat differently because the athletes whose blood samples were taken in the medical tent were analyzed immediately because of their medical condition whereas others' were analyzed the following day. However, additional analyses were performed so that comparisons of blood samples were accurate and valid.

What did the researchers find? Based upon the postrace blood samples, the researchers reported that 17 percent of participating subjects had plasma sodium levels that reached or exceed the defined levels for hyponatremia. They further found a significant inverse relationship between blood levels of sodium and change in weight over the course of the triathlon. The average weight loss was 4.1 percent of body weight, but the vast majority (72 percent) of athletes who either maintained their weight or gained weight were hyponatremic. The athletes who maintained or gained weight during the triathlon apparently did so by consuming large amounts of fluids, potentially leading to the dilution of the sodium levels in the body. The study also revealed tremendous individual differences, with wide ranges of body weight changes, large variability in plasma sodium levels, and different symptomatic responses. For example, some people with sodium levels that were clinically hyponatremic showed no symptoms whereas others with normal levels were symptomatic and sought medical treatment.

What is the significance of the research? The results confirmed the incidence of hyponatremia seen in previous studies, provided insight into the reasons it occurred, and served as the basis for additional research and specific fluid and electrolyte consumption recommendations for ultraendurance athletes. As triathletes sweat for long periods of time they lose water and sodium, typically leading to acute body weight loss, dehydration, and an increase concentration of plasma sodium. If athletes consume a large amount of fluid that is hypotonic, for example, water, the fluid is replaced but not the sodium, so plasma sodium levels become diluted resulting in hyponatremia. Subsequent practice among triathletes has been to consume fluids with sodium during triathlons, although some "salty sweaters" may need additional sodium supplementation. Because of the tremendous individual variability, each athlete needs to determine the fluid and electrolyte replenishment plan that works to optimize race performance while minimizing health concerns.

Core temperature is affected by hydration status.

Dehydration, or moving to a state of hypohydration as a result of fluid loss, can have an adverse impact on core temperature and ultimately on exercise performance (Sawka et al., 1998). When there is a loss of body water, the majority of the water comes initially from the ECF, specifically from the plasma. Therefore, as body water is lost through heavy sweating, there is a gradual loss of blood volume, a condition known as **hypovolemia**. Because sweat is hypotonic in relation to blood, fewer electrolytes and other solutes are lost in sweat in proportion to the amount of water lost. In this case, the plasma that remains is more concentrated and its osmolarity increases. Both of these conditions may adversely affect thermoregulation and exercise performance.

Spotlight on...

Intentional, Rapid Dehydration

Athletes who must meet certain weight classifications and have their weight certified before competition include wrestlers, boxers, and lightweight rowers. There is a long history in these sports of using dehydration as a rapid weight loss method. In 1990, Steen and Brownell surveyed college (n = 63) and high school (n = 368) wrestlers to assess weight loss practices and concluded that traditional practices used to "make weight" were still being employed. Some of the common practices were dehydration, fasting, and restriction of food intake. Practices that were less common but used by some wrestlers included vomiting and the use of laxatives and diuretics. The survey documented what had long been suspected: "weight" loss was rapid, large, and frequent, and after the target weight was met, rapid weight gain followed. Forty-one percent of the collegiate wrestlers reported weight fluctuations of ~11 to 20 pounds (~5 to 9 kg) each *week* throughout the season. For high school wrestlers the weekly losses were smaller, ~6 to 10 pounds (2.7 to 4.5 kg), and reported by fewer athletes (23 percent), but suggested that such practices had their roots early in the athletes' competitive careers.

In 1994, Scott, Horswill, and Dick found that collegiate wrestlers competing in the season-ending NCAA tournament gained a substantial amount of weight between weigh-in and competition (approximately 20 hours apart). In this study wrestlers gained an average of 4.9 percent of body weight, with the wrestlers in the lower weight categories gaining the most weight. In 1996, the American College of Sports Medicine (Oppliger et al., 1996) published a Position Stand on weight loss in wrestlers, in which concern was expressed about rapid and large weight losses and the methods used to achieve them and called for rule changes that would help to limit these losses.

During a 1-month period in 1997, three collegiate wrestlers, ages 19, 21, and 22, died from hyperthermia. The accounts of their precompetition preparation were remarkably and tragically similar. They restricted food and fluid intake, wore vapor-impermeable suits under their warm-up clothing to promote sweating, and engaged in excessive exercise. In the case of the 19-year-old, his preseason weight was 233 lb (105.9 kg) and he was attempting to "make weight" to compete in the 195-pound weight class. He needed to lose 15 lb (6.8 kg) over a 12-hour period. The 22-year-old wrestler's preseason weight was 178 lb (80.9 kg). He had lost 8 lb (3.6 kg) in the 4 previous days but was attempting to lose 4 additional pounds (1.8 kg) in 4 hours. Rectal temperature at the time of death was 108°F (42°C). The third victim had a preseason weight of 180 lb (81.8 kg) and had lost 11 lb (5 kg) over a 3-day period. He was attempting to lose 6 additional pounds (2.7 kg) in 3 hours so he could wrestle in the 153-pound weight class. When he began his attempt he weighed 159 lb (72.2 kg). After 90 minutes of exercise he had lost 2.3 lb (1 kg) and lost an additional 2 lb after another 75 minutes of exercise. After a short rest he began exercising again in an effort to lose the remaining weight, 1.7 lb. He collapsed and died from cardiorespiratory arrest. A postmortem laboratory analysis indicated that blood sodium concentration was 159 mmol/L, far above the normal range (136 to 146 mmol/L) and a sign of severe dehydration (Centers for Disease Control and Prevention, 1998).

These deaths were the first documented collegiate wrestling deaths, and they did for the sport what no amount of research studies, position stands, or previously expressed concerns could do—raise awareness at all levels of the sport and promote change in wrestling rules. For example, the NCAA instituted new rules for collegiate wrestlers in the 1997–98 season, and high schools have since adopted rules to prevent rapid, large weight (water) losses and large weight fluctuations.

The National Wrestling Coaches Association and the National Federation of State High School Associations have established weight management programs for high school wrestlers. A minimum wrestling weight is established for each wrestler by first measuring the wrestler's specific gravity in urine (an indication of hydration status) and then, if properly hydrated, his or her body composition. By monitoring weight and percent body fat changes over the course of the season, these sports-governing bodies are making good on their promises to try and stop the use of large and rapid fluid losses as a method of weight loss. These new rules are reducing the prevalence of rapid and large weight losses (Shriver, Betts, and Payton, 2009; Oppliger et al., 2006).

The main function of blood flow is to deliver oxygen-laden blood to tissues, and the need for oxygen delivery is greatly increased during exercise, particularly in the exercising muscles. Blood flow also helps to control body temperature, and this thermoregulatory function is used to a greater extent during exercise in the heat. A finite and relatively small amount of blood is available to fulfill both these functions, and exercise in the heat sets up a competition for this limited resource. The situation gradually becomes worse as the athlete dehydrates because total blood volume continues to decline.

A state of hypohydration will result in an increased core temperature during exercise, in the heat and even in normal ambient conditions. A loss of only 1 percent body weight as water can result in measurable increases in body temperature, and the greater the loss of body water, the greater the increase in temperature. It is estimated that for every 1 percent of body weight that is lost as water, core temperature will be elevated 0.1°C–0.23°C (Sawka and Coyle, 1999). Conversely, as demonstrated in Figure 7.9, replenishment of fluid lost during exercise results in body temperature rising more slowly and reaching a steady state at a lower body temperature.

It is critical for performance that athletes remain well hydrated, because the adverse effects of hypohydration may offset the benefits gained by having a higher aerobic fitness level or by becoming acclimated to the heat. When hypohydrated, heat dissipation ability is impaired, resulting in higher body temperatures. At the same body temperature (that is, same level of stimulus) the hypohydrated athlete will experience a decrease in whole-body sweating rate and a decrease in skin blood flow, which can substantially narrow the two main avenues of heat loss.

In addition to the declining blood volume, fluid loss may also adversely affect body temperature through the increase in osmolarity of the blood. The portion of the brain that controls the body's thermoregulatory responses is in the hypothalamus. The blood becoming hypertonic may have a direct effect on the cells in this region of the brain and their ability to maintain body temperature. The worst-case scenario of increasing core temperature and deteriorating thermoregulatory control is **hyperthermia** (abnormally high body temperature), which may lead to coma and death (for example, heat stroke).

Some athletes manipulate fluid intake and fluid balance in an effort to "make weight." For example, wrestlers may restrict fluid and food intake while engaging in excessive exercise, wearing clothing while exercising that increases body temperature and sweating, and taking diuretics. Although the goal may be weight loss through dehydration, the result may be hyperthermia. Tragically, some wrestlers have died from hyperthermia because of these practices

(Centers for Disease Control and Prevention, 1998). (See Spotlight on...Intentional, Rapid Dehydration.)

Hypohydration clearly affects body temperature, but what effect does it have on exercise performance? In general, a loss of more than 2 percent of body weight can be detrimental, especially when exercising in the heat, because physical and mental performance is reduced (Table 7.4). However, there are individual variations that result in some people being more or less tolerant to changes in hydration status (Sawka et al., 2007).

Maximal aerobic power ($\dot{V}O_{2max}$) and endurance performance are reduced when an athlete is hypohydrated. With a loss of 3 percent or more of body weight as water, measurable reductions in $\dot{V}O_{2max}$ can be observed. Losses of body water less than 3 percent of body weight are not typically associated with decreased $\dot{V}O_{2max}$ when the athlete is tested in normal ambient temperatures. When tested in the heat, however, modest fluid losses of 2–4 percent of body weight can result in significant declines in maximal oxygen consumption (Sawka, Montain, and Latzka, 2001).

In addition to maximal aerobic performance, endurance exercise performance is impaired when an athlete is hypohydrated (Cheuvront, Carter, and Sawka, 2003; Von Duvillard et al., 2004). The decline in performance is larger with increased distance or duration of the endurance activity. For example, performance in a marathon will decline more than in a 10-kilometer race when a runner is hypohydrated. Hypohydration can also impair performance of lesser-intensity activities such as walking or hiking in the heat. A hypohydrated athlete exercising in the heat becomes fatigued more quickly compared to an athlete exercising in a euhydrated state. Body temperatures will rise in both athletes, but the hypohydrated athlete will fatigue at a lower body temperature. When hypohydrated, thermoregulation is impaired and the athlete cannot tolerate the same amount of increase in body temperature without fatiguing.

There is good evidence to suggest, however, that a loss of 3–5 percent of body weight does not reduce anaerobic performance or muscular strength (Sawka et al., 2007). However, this magnitude of body water loss does affect thermoregulation and increases the risk for potentially fatal heat illnesses such as heat stroke.

Electrolyte loss, particularly sodium loss, during exercise can be substantial.

Fluid loss through sweating is the major concern for the exercising athlete, but sweat is composed of more

---

Hypovolemia: Less than the normal volume.

Hyperthermia: Abnormally high body temperature.

Table 7.4 Effects of Dehydration on Aerobic Performance

Sport	Degree of dehydration*	Effect on performance
Basketball	2%	Movement slows, shooting is less accurate, and attention declines; performance declines continue as the degree of dehydration increases
Distance cycling	2.5%	On a 2-hour ride with hills, power output for hill climbing declines
1,500 m run	2%	Running time increases by ~3%
5,000 or 10,000 m run	2%	Running time increases by ~5%
Soccer	1.5–2%	Playing ability and fitness declines; perceived exertion increases

*Percent loss of body weight as water

Sources: Armstrong, Costill, and Fink, 1985; Baker, Conroy, and Kenney, 2007; Baker, Dougherty, et al., 2007; Ebert et al., 2007; Edwards et al., 2007

than water. Sweat contains the electrolytes sodium, potassium, and chloride; small amounts of minerals such as iron, calcium, and magnesium; and trace amounts of urea, uric acid, ammonia, and lactate. Of these, sodium is present in the largest amount. During light sweating, sodium and chloride are reabsorbed from the tubule of the sweat gland and are not lost in large amounts. During heavy sweating, however, the sweat moves through the tubule at a rate that is too fast for substantial reabsorption, so sodium and chloride losses are proportionally greater.

The sodium content of sweat ranges from 10 to 70 mEq per liter with the average concentration of approximately 35 mEq/L (Sawka et al., 2007). One mEq of sodium is equal to 23 mg of sodium, thus the upper end of the range, 70 mEq/L, is equivalent to 1,610 mg or 1.6 g of sodium per liter. This rate and amount of sodium loss typically does not pose a problem for the athlete if the exercise duration is not over an hour or two. During exercise lasting less than 2 hours, the athlete would need to pay more attention to fluid replacement to address the water loss through sweating than to sodium replacement. Water works well as a fluid replacement beverage under these conditions. The duration is short enough that excessive electrolyte loss is not likely, and therefore electrolyte replacement is not a priority. The duration of the activity is also within the range that endogenous carbohydrate (that is, muscle and liver glycogen) stores are not likely to be depleted, so the inclusion of carbohydrate in the beverage for exercise of this duration is likely not a necessity. Consumption of beverages that do contain carbohydrate and electrolytes are not likely to pose any problems for the athlete.

More than 2 hours of continuous exercise typically marks the transition when the use of fluids that contain sodium and carbohydrate becomes appropriate. As explained earlier, it is possible for athletes engaged in prolonged moderate- to high-intensity exercise in hot environments to sustain sweat rates of 1 L to over 2 L per hour, resulting in substantial sodium loss (~1.5–3.0 g per hour). In such cases, sodium replacement should begin during exercise by consuming fluids containing sodium or eating salty-tasting foods. The addition of sodium may also encourage greater voluntary drinking and may aid in the uptake of water from the small intestine.

Athletes who train and compete at these distances and durations should experiment during training to determine the need for sodium and carbohydrate replacement during exercise. For example, an athlete who is a heavy sodium excreter can look for accumulation of salt on skin or clothing after training and experiment with various replacement strategies during and after exercise (Maughan and Shirreffs, 2008). Although the focus of this chapter is water and electrolytes, the intensity and length of time of activities such as marathon running and Olympic distance triathlons also result in substantial utilization of carbohydrate stores, and the ingestion of carbohydrates to maintain the availability and use of this fuel is also important (Sawka et al., 2007).

Very long duration activities (that is, >4 hours), such as ultramarathon running or Ironman™-length triathlons, may also result in substantial sodium loss even in athletes who are not "heavy" sweaters (Rehrer, 2001). Much lesser amounts of potassium, magnesium, calcium, and chloride are lost, even at high sweat rates (up to 0.5, 0.02, 0.04, and 2.1 g per liter, respectively). More careful consideration of both fluid and electrolyte replacement (particularly sodium) must be made by the athlete during prolonged activities. Sweat is hypotonic in relation to blood, so as water and electrolytes are lost in sweat, there is a proportionally greater loss of water. Therefore, as the athlete dehydrates, the blood becomes more hypertonic. The combination of dehydration and hypertonicity may significantly impair performance, so the athlete must have a strategy to prevent or delay these, and the fatigue that can accompany them, from occurring.

One strategy for sodium replacement during prolonged exercise when there are large sodium losses through sweating is the use of sports drinks that contain more sodium, often called "endurance formula" beverages. As an example, the Gatorade G Series Pro 02 Perform beverage contains 200 mg of sodium per 8 ounces (240 ml), nearly double the sodium found in the original

Table 7.5 Sodium-Containing Products Marketed to Athletes

Product	Serving size	Sodium	Other electrolytes
Caution: Do not exceed the amount or timing recommended on the label.			
Gatorade G Series Pro, 02 Perform	8 oz (240 ml) ready to drink	200 mg	Potassium: 90 mg
Gatorlytes	1 pouch (3.4 g) in 20 oz water*	780 mg	Potassium: 400 mg Magnesium: 40 mg
Lava Salts™	1 capsule	166 mg**	Sodium chloride: 255 mg Sodium bicarbonate: 80 mg Sodium citrate: 75 mg Sodium phosphate: 50 mg Potassium chloride: 30 mg Magnesium stearate: 10 mg
Saltstick™	1 capsule	215 mg	Potassium: 63 mg Calcium: 22 mg Magnesium: 11 mg
Succeed!™	1 capsule (1 g)	341 mg	Potassium: 21 mg
Succeed Ultra Energy Drink	1 packet (38 g) in 20 oz water	189 mg	Potassium: 42 mg

*The manufacturer suggests adding this product to Gatorade G Series Pro, 02 Perform.

**Approximate value calculated from product label.

Gatorade formula. The use of sodium supplements in the form of salt tablets is another strategy to dramatically increase the intake of sodium during prolonged exercise. Triathletes competing in Ironman™-length events may be exercising vigorously in a thermally challenging environment for 8 to 12 hours or more, experiencing substantial fluid and sodium loss. As these athletes have become aware of the dangers of hyponatremia that may occur in these events (see section on hyponatremia), some have attempted to counter the sodium loss by consuming it in the more concentrated form of salt tablets (Speedy et al., 2003). A number of commercial products are marketed for this purpose as shown in Table 7.5. Research on the effectiveness of this strategy for preventing the occurrence of hyponatremia is not extensive, although the results of two studies (Hew-Butler et al., 2006; Speedy et al., 2003) suggest athletes supplementing with salt tablets during an Ironman-length triathlon do not have significantly different serum sodium concentrations than athletes that consumed a placebo. It bears repeating that this high level of sodium intake during exercise is for those events and conditions where there is large sweat loss for prolonged periods of time, that is, several hours or more.

Exercise-related muscle cramping, often associated with dehydration or electrolyte loss, may have other causes.

Athletes may experience painful muscle cramps during or immediately after exercise, known as exercise-associated muscle cramping (EAMC). It has long been believed that such cramping is due to dehydration, changes in electrolyte concentrations, or both. Therefore, athletes have been advised to hydrate properly and eat more foods rich in potassium such as bananas. In addition, potassium, calcium, and magnesium supplements are advertised to athletes as a way to prevent or recover from muscle cramps. The ingestion of sodium via 1–2 oz (30–60 ml) of pickle juice has also been recommended.

Despite a widespread belief that EAMC in all athletes is caused by dehydration and electrolyte loss, the scientific evidence that supports this theory is rather limited (Schwellnus, 2009). Although EAMC often occurs in conjunction with heavy sweating, dehydration, and electrolyte losses, there is not strong evidence that athletes who experience cramping have significantly different levels of dehydration or electrolytes than non-cramping athletes. Experimental intervention studies also do not provide strong evidence for this approach. For example, limited studies of pickle juice have failed to show its effectiveness in relieving EAMC (Miller et al., 2010; Miller, Mack, and Knight, 2009). The American College of Sports Medicine (ACSM) (Sawka et al., 2007) position paper notes that recommendations to avoid dehydration and sodium deficits to prevent muscle cramps is based on consensus and usual practice (that is, Evidence category C) not experimental evidence (that is, Evidence categories A and B).

At present, scientists theorize that exercise-related muscle cramping may be due to altered neuromuscular control (Schwellnus, 2009). In this theory, muscle cramping is more related to the development of muscle fatigue than to dehydration and electrolyte imbalances. Muscle fatigue can result from repetitive contractions over a long period of time (for example, marathon running), and may be further influenced by the level of conditioning of the athlete, the intensity of the exercise, stressful environmental conditions, and so on. When reaching a certain state of fatigue, the sensory feedback from the muscle and the resultant nervous system response may, in effect, malfunction resulting in a failure to send appropriate relaxation signals and the sending of excess contraction signals, causing the muscle to cramp.

Some athletes experience exertional heat cramps, which is total-body cramping when exercising in the heat. This type of cramping can involve muscles all over the body, not just the ones directly involved in the exercise, and does appear to be caused by sodium depletion and dehydration as well as muscle fatigue. Case studies of tennis and football players suggest that exertional heat cramps may be the result of rapid and large losses of fluid and sodium. Those who fall into this group, known as "salty sweaters," benefit from sodium-containing beverages during exercise and the consumption of an adequate amount of sodium and water after exercise (Bergeron, 2007). Adding salt to food, eating salty foods, and consuming beverages with sodium are all ways to

replenish sodium before the next exercise session. Salty sweaters must closely match salt (sodium chloride) and fluid intake to sodium, chloride, and fluid losses on a daily basis. As there are many nonnutritional causes of cramping (for example, lack of stretching), each athlete should determine the likely causes of their cramping and, through trial and error, institute strategies that are known to prevent the causative factors.

7.4 Strategies to Replenish Water and Electrolytes

Exercise and physical activity can have a substantial impact on fluid balance. Water and electrolyte loss must be compensated for to maintain long-term fluid homeostasis. Single events or bouts of exercise may cause large disruptions in fluid balance and need immediate attention. However, many athletes experience small deficits on a daily basis, and these small cumulative deficits become more pronounced deficits over several days or weeks. Athletes should develop a practical approach to monitoring their hydration status and a strategy for water and electrolyte replacement to maintain a status of euhydration and appropriate osmolarity.

Hydration status should be assessed and monitored.

There are a variety of ways to monitor the body's hydration status. These methods have varying degrees of accuracy, difficulty, and expense. Typically, the methods that are the most accurate are also the most difficult and time-consuming to perform and the most expensive. These methods are best suited for use in research studies and are rarely practical for day-to-day use by athletes or those who exercise or compete recreationally. Daily monitoring requires an approach that is practical, and easy to administer and understand (Armstrong, 2005).

To precisely determine an individual's hydration status, the amount of total body water and the osmolarity of the plasma must be known. Accurate measures of total body water are often determined using isotope dilution, most commonly deuterium oxide (also discussed in Chapter 2). When a known volume of water having a known radioactivity level is consumed, it is diluted as it is absorbed and distributed throughout the water compartments of the body. After distribution and equilibration, a sample of body water can be taken and analyzed for its radioactivity. Total body water volume can then be determined from the degree of dilution of radioactivity. This method, though accurate, is expensive and time-consuming, requires trained personnel, and is not suitable for daily monitoring.

Key points

- Exercise can result in large fluid losses, primarily from sweating.

- Sweat rate can be affected by the ambient temperature, the exercise intensity, the athlete's training status, acclimation to the heat, and clothing, uniforms, and protective gear.

- Sweat rates may be as high as 2.5 L per hour in hot, humid conditions.

- It is difficult for athletes to consume enough fluid to offset losses when sweating heavily due to an inability to drink and/or absorb enough fluid during exercise.

- Fluid loss during exercise resulting in hypohydration results in a greater rise in core temperature and a decline in exercise performance.

- Heavy sweating results in sodium loss that should be replaced during exercise if the duration of the activity exceeds 1–2 hours.

- Muscle cramping may be associated with dehydration, electrolytes losses, and/or altered nervous system control of the muscle due to fatigue.

Why is exercise in a tropical environment (hot and humid) more challenging than exercise in a desertlike environment (hot and dry)?

A second method of testing fluid balance involves measuring plasma solutes. The plasma osmolarity level associated with euhydration is 285 mOsm/kg (Institute of Medicine, 2004). In order to determine this value, however, a blood sample is needed, as well as access to a clinical laboratory for analysis. This method has been shown to provide an accurate assessment of hydration at a single point in time, however, it is not very practical or desirable for frequent monitoring because of the time, expense, and necessity for drawing blood, but it is often used in research settings (Cheuvront et al., 2010).

Because the renal system is the major physiological mechanism for regulating fluid balance (Figure 7.10), an analysis of urine can also provide information about hydration status. When an individual becomes hypohydrated, the amount of urine produced is often lower than usual due to water conservation by the body. A 24-hour urine collection can easily measure volume, but this can be a time-consuming and unpleasant task and one that is not practical for an athlete to perform on a regular basis. However, in the hypohydrated individual the urine also has a higher specific gravity (concentration of particles), higher osmolarity, and a darker color. Urine specific gravity of ≤1.020 and urine osmolality of ≤700 mOsm/kg are considered consistent with euhydration (Sawka et al., 2007). In the past, such measurements required specific equipment or a clinical laboratory, but the availability of inexpensive urine testing strips now allows athletes to easily test urine for specific gravity as well as the presence of other compounds (for example, glucose, protein).

Observation of urine color is more subjective and may not be as precise as laboratory measures, but is a more practical approach that can be easily conducted whenever the athlete urinates (Armstrong et al., 1998). It is typically recommended that athletes observe the color of the first void (urination) of the day after awakening from a night's sleep. Armstrong (2000) has suggested the use of a urine color scale (Figure 7.11) to estimate the degree of hypohydration. Although it lacks precision, urine color may be an easy, practical marker that an athlete can use in conjunction with other markers to assess hydration status. Urine color may be affected by diet (for example, consumption of beets), dietary supplements, or medications, and these reduce the accuracy of a color test.

Weight loss that occurs as a result of a single exercise bout is likely due to fluid loss; thus changes in body weight can be used as a marker of short-term fluid loss and as a benchmark for subsequent rehydration. Changes in body fat or lean body mass over time complicate the use of this marker, but daily weight loss or gain over the course of a single workout can be used to determine water loss, and therefore hydration

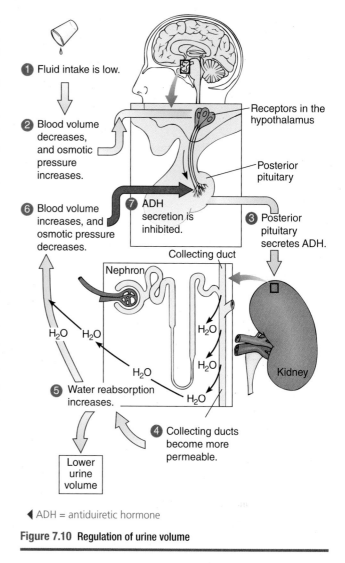

ADH = antiduiretic hormone

Figure 7.10 Regulation of urine volume

status, with reasonable accuracy (Cheuvront et al., 2010). One liter of water weighs approximately 1 kg (2.2 lb). If an athlete completes a hard workout lasting approximately 1 hour and loses 2 kg (4.4 lb) of body weight, it can be assumed that approximately 2 L of fluid have been lost (2 kg × 1 L/kg). If this athlete takes a scale weight the next morning and has gained back only 1 kg (2.2 lb), it is likely that not enough fluids were consumed over the intervening day. Weight is easy to track, and when used in conjunction with other markers may provide a reasonably accurate and practical method for monitoring hydration status. However, checking body weight every day may be detrimental if an athlete is struggling with disordered eating, an eating disorder, or anxiety related to degree of body fatness. These athletes may misinterpret a 1 kg (2.2 lb) weight gain as a gain in body fat and not an increase in body water. In such cases, taking daily weights would be discouraged and urine color may be the most appropriate measure of hydration status.

Figure 7.11 Urine color chart

A urine color chart can be used as a general assessment of hydration status. A urine sample collected the first thing in the morning can be viewed against a white background in good light and the color compared to the chart. A lighter urine color in the 1, 2, or 3 range, can be considered well hydrated, whereas a darker urine color in the 6, 7, or 8 range can be considered hypohydrated. Other factors, such as the use of medications or vitamins, may also affect urine color; therefore, this method should be used cautiously and in conjunction with other methods such as acute changes in body weight.

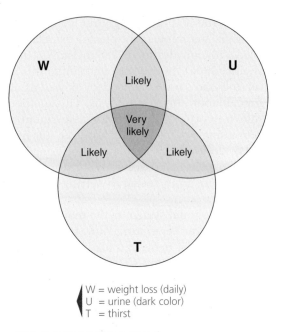

W = weight loss (daily)
U = urine (dark color)
T = thirst

Figure 7.12 Hydration assessment tool

Each morning, athletes need to evaluate whether: (1) they are thirsty, (2) their urine is dark yellow, and (3) their body weight is noticeably lower than the previous morning. If one of these conditions is present, then they *may* be hypohydrated; if two conditions are present, then it is *likely* they are hypohydrated; and if all three conditions are present, then it is *very likely* they are hypohydrated.

Cheuvront and Sawka (2005) have suggested a simple, logical approach using a combination of factors: thirst, body weight loss, and dark urine color (Figure 7.12). The presence of any one of these factors may not provide sufficient evidence of inadequate hydration, but there is increasing probability with the overlap of two or all three factors. Each morning, athletes need to evaluate whether: (1) they are thirsty, (2) their urine is dark yellow, and (3) their body weight is noticeably lower than the previous morning. If one of these conditions is present, then they *may* be hypohydrated; if two conditions are present, then it is *likely* they are hypohydrated; and if all three conditions are present, then it is *very likely* they are hypohydrated. Rehydration strategies can then be pursued accordingly.

General guidelines have been developed for the type, timing, and amount of fluids and electrolytes consumed before, during, and after exercise.

Each athlete should have an individualized plan for consuming water and/or electrolytes before, during, and after exercise (Maughan and Shirreffs, 2008; Shirreffs, Armstrong, and Cheuvront, 2004). Many athletes also need to consume carbohydrate during these periods, and having carbohydrate in a beverage makes doing so convenient. It is critical that each athlete plan a strategy for obtaining the nutrients needed and then test that plan during training under various environmental conditions. Because heat and humidity are substantial factors influencing fluid loss, athletes need to determine usual losses and successful rehydration strategies under different environmental conditions. A plan for adjusting fluid intake is especially important if competition is held in an environment with higher heat and humidity than the usual training setting. A basic plan, with adjustments for changing environmental conditions and the stress of competition, helps the athlete to be proactive in preventing or delaying dehydration and other nutrient-related problems. Trial and error is important for determining the ways in which the athlete will meet the ultimate goal: proper consumption of fluids, electrolytes, and/or carbohydrates.

Intake prior to training and performance. The type and amount of fluid consumed prior to exercise depends on the athlete's goals and tolerances. Typical goals include:

- Being fully hydrated prior to exercise
- If not fully hydrated, rehydrating to the extent possible
- Avoiding gastrointestinal upset
- Consuming carbohydrate, if appropriate

Athletes should be conscious of being adequately hydrated at the beginning of an exercise session or competitive event by consuming adequate fluids throughout the previous day. Pre-exercise fluid consumption should begin at least 4 hours prior to training or performance, if possible. Assuming that the athlete has adequately rehydrated from the previous day's exercise, the slow intake of fluids is recommended. Although the amount will depend on the individual, a rule of thumb for fluid intake is ~5–7 ml/kg at least 4 hours prior to exercise (Sawka et al., 2007). For example, a 50 kg (110 lb) female may establish a goal of consuming 250–350 ml (~8–12 oz or 1–1½ cups) of fluid before exercise. This amount of fluid 4 hours prior should be sufficient to maintain euhydration and allow for urination prior to training or competition. When this same rule of thumb measure is applied to larger athletes, such as a 260 lb (118 kg) male, an appropriate goal may be ~600–800 ml (~20–27 oz), highlighting the difficulty in making general recommendations for athletes.

The pre-exercise fluid intake strategy is different if the athlete is not euhydrated. Entering exercise in a state of hypohydration may be due to inadequate restoration of fluid from the previous day's exercise, multiple exercise bouts in hot and/or humid conditions, or the voluntary restriction of fluid to reduce body weight prior to weight certification. In such cases a more aggressive approach to pre-exercise fluid intake is needed. In addition to the rule-of-thumb recommendation outlined above, ~3–5 ml/kg 2 hours prior is recommended. Using the previous example, a 50 kg (110 lb) female might want to consume an additional 150–250 ml (5–12 oz), whereas the 260 lb (118 kg) male may need ~350–600 ml (~12–20 oz) more.

Table 7.6 summarizes the composition of various pre-exercise beverages. In most cases, water is

Table 7.6 Composition of Various Pre-Exercise Beverages

Product (8 oz or 240 ml unless otherwise noted)	Sodium (mg)	Potassium (mg)	CHO (g)	CHO (%)	Caffeine (mg)	Other	Energy (kcal)
Water	trace	trace	0	0	0		0
Black coffee	trace	trace	0	0	~80–150		0
Decaf coffee (black)	trace	trace	0	0	3–12		0
Tea (plain)	trace	trace	0	0	~40–120	Some antioxidants	0
Decaf tea (plain)	trace	trace	0	0	5	Some antioxidants	0
Gatorade G Series, O_2 Perform (formerly Original Thirst Quencher)	110	30	14	6	0		50
Gatorade G Series, O_2 Perform, G2-Low Cal	110	30	5	2	0		20
Gatorade G Series Pro, O_2 Perform, Endurance Formula	200	90	14	6	0		50
Gatorade G Series Pro, O_2 Perform, Gatorlytes (3.4 g to be added to another beverage)	780	400	0	0	0	For cramp-prone athletes; to be added to Endurance Formula or water	0
Hydrade	86	75	9	4	0	Contains glycerol (5.1%) and 24 mg vitamin C	40
All Sport Body Quencher	55	60	16	6.5	0	Contains 24 mg vitamin C	60
Powerade ION4	100	25	14	6	0	Contains B_3, B_6, and B_{12}	50
Powerade ZERO	100	25	0	0	0	Contains B_3, B_6, and B_{12}	0
Accelerade Hydro (1 scoop [31 g] in 12 oz water)	180	55	10	4	0	2.5 g protein; 20 mg Ca; 140 mg Mg	50

Legend: oz = ounce; ml = milliliter; CHO = carbohydrate; g = gram; kcal = kilocalorie; Ca = calcium; Mg = magnesium

sufficient. However, sodium does help to stimulate thirst, retain body water, and encourage people to drink more, so a pre-exercise source of sodium may be beneficial. Also, if the athlete has been sweating heavily in the preceding days, care must be taken to consume adequate electrolytes, particularly sodium and potassium. If obtained in a beverage, the recommended amount is 20–50 mEq/L or 460–1,150 mg/L of sodium. If the athlete is also attempting to optimize carbohydrate stores, the ingestion of a carbohydrate-containing beverage or a small carbohydrate meal may be beneficial.

The amount and timing of the fluid ingestion prior to exercise should be determined on an individual basis during training by trial and error. Too little fluid intake may leave the athlete with a suboptimal hydration status and may allow for a greater degree of dehydration during the activity, predisposing the athlete to potentially poorer performance. Ingesting too much fluid may pose problems as well. The athlete may feel too full or bloated, potentially leading to gastrointestinal disturbances during exercise. An overabundance of fluid intake may also lead to excess urine production, discomfort, or interruption of training or competition. Each athlete's situation is really an experiment of one, and he or she needs to determine the most appropriate amount and timing of fluid consumption before exercise. The stress of competition may result in a need to alter one's usual pattern of intake before exercise.

Intake during training and performance. Many athletes benefit from fluid consumption during exercise. The intake of electrolytes and carbohydrate may also be beneficial. Typical goals include:

- Replacing lost body water to the extent possible
- Delaying dehydration to the extent possible
- Avoiding the overconsumption of water
- Replacing sodium if losses are large or rapid
- Consuming carbohydrate if appropriate
- Avoiding gastrointestinal upset

As athletes lose fluid during exercise they should attempt to replace these losses to maintain fluid balance (Sawka and Montain, 2000). This is the easy and logical answer, but replacing fluids equal to those lost during exercise is not always possible, at least during the activity. Voluntary fluid replacement often falls short of fluid losses, in part due to "voluntary dehydration." Athletes may not be able to, or may not want to take time from their sporting activity to drink fluids. Certain activities, such as cycling, provide the means to carry more fluids and greater opportunity to consume them during the activity. Contrast this with activities such as running, during which fluid and/or

food consumption is difficult. Some team sports may have time-outs, substitutions, or other breaks in the action that provide an opportunity to drink, but often voluntary fluid intake does not compensate for large fluid losses.

There is also a physiological reason that makes it difficult to avoid hypohydration—the inability to empty water from the stomach and absorb it into the blood as fast as it is being lost. Although the maximal rate of gastric emptying and fluid uptake in adults approximates 1 L (~36 oz or ~4.5 cups) per hour, water may be lost in sweat at rates twice that (over 2 L of sweat per hour) with very heavy sweating. In some situations it is not physically possible to take in enough fluid to match what is lost. A recognized goal is to prevent *excessive* dehydration (that is, >2 percent of body water loss) and *excessive* changes in electrolyte balance (Sawka et al., 2007).

In the past, a recommended guideline for fluid replacement during exercise was to consume 150 to 350 ml (~6–12 ounces) of fluid at 15- to 20-minute intervals, beginning at the onset of exercise (Convertino et al., 1996). The ACSM (Sawka et al., 2007) and the American Dietetic Association and Dietitians of Canada (Rodriquez, DiMarco, and Langley, 2009) now recommend a customized plan that considers sweat rate, sweat composition, duration of exercise, clothing, and environmental conditions. Recognizing that the lack of recommendations regarding specific amounts to be consumed make it difficult for many athletes to determine the amount of fluid needed, the ACSM recommends as a starting point that marathon runners consume 0.4–0.8 L of fluid per hour. To avoid overconsumption of water and hyponatremia, it is further recommended that the lower end of the range would be more appropriate for slower-paced, lighter-weight runners competing in cooler environmental temperatures whereas the upper end of the range would be more appropriate for faster-paced, heavier-weight athletes running in hotter, more humid conditions. This range is only a guideline, and it is well recognized that the lower or upper ends of the range could result in over or under consumption of fluid by a particular runner. The need for an individualized plan for fluid replacement cannot be overemphasized.

Guidelines have not been developed for sports other than marathon runners but the principles used to establish the guideline for marathon runners do have application for other athletes. For example, an interior lineman in the National Football League may need the equivalent of 0.8 L of fluid per hour because of a large body size, a high sweat rate, clothing that prevents the evaporation of sweat, and environmental conditions at the competition site that are hotter and more humid than the athlete is accustomed to.

From a practical perspective, the amount of fluid consumed will vary depending upon the rate of fluid

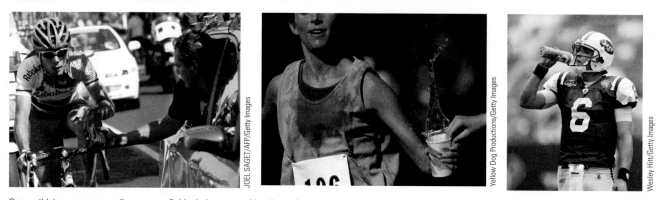

Some athletes can more easily consume fluids during competition than others.

loss and the individual's tolerance for fluid intake during the activity. Liquids that are cool are generally better tolerated than those that are warmer (although their coldness has no noticeable effect on body temperature). People can also generally better tolerate small amounts of fluid consumed more frequently than large amounts consumed less frequently. Frequent drinking of small amounts of fluid will maintain an amount of fluid in the stomach that will stimulate gastric emptying and fluid uptake by the body. However, each athlete must find his or her "gastric tolerance." Again, the frequency and volume of fluid replacement should be determined during training so as not to introduce a new, unfamiliar process during competition.

If the athlete consumes a typical amount of dietary sodium, there is little need for the fluid replacement beverage to contain sodium unless the exercise duration is more than 2 hours and there is a very high

sweat rate. In such cases, it is recommended that 1 g of sodium per hour be consumed. This recommendation is made for endurance athletes who sweat heavily and whose sweat contains a large amount of sodium (that is, "salty sweaters") (Murray, in press). However, a sodium content of 20–30 mEq/L (460–690 mg/L) may be beneficial for endurance athletes exercising in the heat who do not sweat heavily because the sodium stimulates thirst, stimulates voluntary consumption of more fluid, increases the palatability of a beverage, and promotes body water retention (Sawka et al., 2007).

Similarly, if the athlete has adequate carbohydrate stores, there is little need for the fluid to contain carbohydrate if the exercise duration is less than 2 hours. If exercise is in excess of 2 hours, however, it is recommended that carbohydrate be consumed as well as sodium. Table 7.7 summarizes the nutrient content of typical beverages consumed during exercise.

Table 7.7 Composition of Various Beverages Consumed during Exercise

Product (8 oz or 240 ml)	Sodium (mg)	Potassium (mg)	CHO (mg)	CHO (%)	Caffeine (mg)	Energy (kcal)
Gatorade G Series, 02 Perform (formerly Original Thirst Quencher)	110	30	14	6	0	50
Gatorade G Series Pro, 02 Perform, Endurance Formula	200	90	14	6	0	50
Gatorade G Series, 02 Perform, G2-Low Cal	110	30	5	2	0	20
All Sport Body Quencher	55	60	16	6.5	0	60
Powerade ION4	100	25	14	6	0	50
Powerade ZERO	100	25	0	0	0	0
Accelerade (1 scoop [31 g])	190	65	21	8	0	120
Accelerade Hydro (1 scoop [31 g] in 12 oz water)	180	55	10	4	0	60

Legend: oz = ounce; ml = milliliter; mg = milligram; CHO = carbohydrate; kcal = kilocalorie; g = gram

Table 7.8 Composition of Various Beverages Consumed Postexercise

Product (8 oz or 240 ml)	Sodium (mg)	Potassium (mg)	CHO (g)	Protein (g)	Fat (g)	Caffeine (mg)	Other	Energy (kcal)
Note: Although these beverages provide some fluid, the emphasis is on the carbohydrate and protein content.								
Gatorade G Series, 03 Recover	120	45	7	8	0	0	40 mg Ca	60
Gatorade G Series Pro, 03 Recover, Protein Recovery Shake	190	270	33	15	1	0	300 mg Ca	200
Endurox R⁴ (2 rounded scoops [75 g])	210	120	52	13	1	0	470 mg vitamin C, 400 IU vitamin E, 100 mg Ca, 240 mg Mg, 420 mg L-glutamine	270
Low-fat chocolate milk	153	425	26	8	3	0	288 mg Ca, other vitamins and minerals	170
High-protein boost	170	380	33	15	6	0	21 vitamins and minerals	240
Boost	130	400	41	10	4	0	26 vitamins and minerals	240

Legend: oz = ounce; ml = milliliter; mg = milligram; CHO = carbohydrate; g = gram; kcal = kilocalorie; Ca = calcium; Mg = magnesium

If consuming carbohydrate, the carbohydrate content of the beverage should be less than 10 percent (≤10 g of carbohydrate in 100 ml of water) for more effective fluid replacement and thermoregulation. Many sports beverages contain 6 to 8 percent carbohydrate, a concentration that does not induce gastric distress for many people. However, some athletes find that they need more dilute solutions based on their individual tolerances. Carbohydrate concentrations greater than 10 percent may slow gastric emptying and fluid uptake if consumed during exercise. However, there may be times that the carbohydrate concentration of the beverage may be greater than 10 percent. For example, ultraendurance athletes may benefit from large amounts of glucose during competition because the need for carbohydrate is so great. These athletes will need to experiment with more concentrated carbohydrate solutions during training. Trial-and-error experimentation will help ultraendurance athletes test their gastric tolerance for beverages containing more than 10 percent carbohydrate. More information about carbohydrate intake during exercise is found in Chapter 4.

Replenishment after training and performance. The recovery period begins immediately after exercise. Rehydration is a focus for all athletes, but the recovery period also involves the intake of other nutrients. Typical goals include:

- Restoring lost body water to achieve euhydration
- Replacing sodium and other electrolytes lost
- Consuming adequate carbohydrate to fully restore muscle glycogen
- Consuming adequate protein to build and repair skeletal muscle
- Avoiding gastrointestinal upset

Fluid balance is typically compromised during exercise training or performance, so athletes must pay particular attention to rehydration strategies after exercise is complete (Maughan and Shirreffs, 2008; Shirreffs and Maughan, 2000; Shirreffs, 2001). It is recommended that athletes drink approximately 1.5 L (~50 oz or ~6 cups) of fluid per kg body weight lost, beginning as soon after exercise as is practical. In other words, a 2.2 lb loss of scale weight requires ~1.5 L fluid intake (Sawka et al., 2007). In the past, a rule of thumb for rehydration was "a pint, a pound," a catchy way of reminding athletes to consume 1 pint (2 cups) of fluid for each pound of scale weight lost. However, a pint is ~480 ml and is not enough to restore a 1 lb water loss, which requires ~700 ml to replenish. "A pint, a pound" is typically not enough, especially if the athlete began the exercise session in a mild state of hypohydration. Water is not as effective in achieving euhydration as a beverage that contains some sodium, because sodium increases the body's drive to drink and results in a temporary decrease in urine output (Murray, in press). Table 7.8 lists the nutrient content of some typical postexercise beverages. Note that although these beverages provide some fluid, the emphasis is on the carbohydrate and protein content. The sodium in beverages and foods consumed after exercise help to replenish electrolytes lost during exercise. The sodium content of beverages can vary considerably. If the amount of sodium is too high to be palatable, then the athlete may not voluntarily consume a large volume due to taste and GI intolerance. It is recommended that athletes who lose large amounts of sodium in sweat use a combination of strategies to replenish sodium, such as drinking beverages that contain sodium, salting their food, and/or consuming salty-tasting snacks after exercise. The amount of sodium an athlete should consume daily must be individualized. Consultation with a physician and a sports dietitian is recommended since excessive sodium intake results in elevated blood pressure in some individuals (see Chapter 13).

Each athlete should develop an individualized plan for choosing foods and beverages that meet fluid and electrolyte needs.

One of the challenges for athletes and professionals who work with them is to translate scientifically based recommendations into practice. Athletes have many questions about water, sports beverages, and foods that may be used to replenish fluid, electrolytes, and carbohydrates. No single strategy or product is "best" for all athletes; each athlete needs a personalized plan. Development of an individualized plan involves four steps: assessment, goals, actions, and reassessment (Macedonio and Dunford, 2009).

Assessment. Successful planning begins with assessment of hydration and electrolyte status. For practical reasons, most athletes self-assess their hydration status by using thirst, urine color, and/or body weight to roughly determine if they are adequately hydrated. Self-assessment of electrolytes is more difficult and is often based on observing salt crystals on skin or clothing, tasting salty sweat, or experiencing conditions known to result in the loss of electrolytes such as vomiting or diarrhea. Any athlete who has experienced electrolyte-related conditions, such as hyponatremia or exertional heat cramps, should consult with medical professionals, such as a sports medicine physician or a sports dietitian.

Is the Athlete Generally Euhydrated?
- If yes, 24-hour fluid intake appears to be appropriate.
- If no, total daily fluid intake should be evaluated and adjusted.

What Are the Athlete's Goals Prior to Exercise?
- If euhydration only, water intake is sufficient.
- If carbohydrate is needed, choose between a sports beverage, carbohydrate gel and water, or carbohydrate-containing food and water.

What Are the Athlete's Goals during Exercise?
- If only fluid is needed, water may be sufficient.
- If carbohydrate is also needed, choose a sports beverage or carbohydrate gel and water, or carbohydrate-containing food and water.
- If sodium is also needed, choose a sodium-containing sports beverage or sodium-containing food and water.

What Are the Athlete's Goals after Exercise?
- Replenish water, carbohydrate, and sodium as needed with a combination of foods and beverages.

Figure 7.13 Assessment and establishment of a fluid, electrolyte, and carbohydrate replenishment plan

Goals. Once the athlete's current status is known, then pre-, during, and postexercise goals are used to help develop a personalized plan for appropriately timed fluid, electrolyte, and carbohydrate intake. Some evaluation questions are listed in Figure 7.13. For example, is the athlete generally euhydrated? If yes, an appropriate plan for fluid replenishment may already be established and only fine-tuning is needed as conditions change. In many cases, however, athletes are in a routine state of hypohydration and overall daily fluid intake needs to be increased. One simple adjustment may be to drink more water during the course of a day. Some substantial adjustments may also need to be made.

Prior to exercise, some athletes are concerned only about maintaining euhydration and not about consuming additional carbohydrate. In such cases water is an appropriate pre-exercise beverage. As an example, a sprinter may drink only water during the warm-up period and while waiting for the competition to begin. Athletes engaged in prolonged exercise may choose a carbohydrate-containing beverage, because both carbohydrate and fluid are needed during exercise. These athletes will experiment during training with various carbohydrate concentrations, finding the sports beverage or food-and-water combination that provides sufficient carbohydrate without creating gastrointestinal distress. Similarly, a sports beverage can also provide sodium, a nutrient that may be needed during prolonged or ultraendurance exercise. Postexercise intake should reflect the water and nutrients lost during exercise, and beverages play an important role.

Advantages of Water

Noncaloric
Refreshing taste
Widely available (bottled, drinking fountains, hoses)
Depending on hardness or softness, may provide some electrolytes

Advantages of Sports Beverages

Provide carbohydrate
Sweet taste
Contain electrolytes in known quantities
Rapid rate of absorption due to sugar and sodium content
Convenient

Advantages of Fruit Juices

Provide carbohydrate
Sweet taste
Often high in potassium
May contain vitamins, minerals, and phytochemicals

Advantages of Soft Drinks

Provide carbohydrate
Sweet taste
Widely available
Provide stimulatory effect if caffeinated

Disadvantages of Water

Provides no carbohydrate
Electrolyte content of unbottled water not known and variable

Disadvantages of Sports Beverages

Could provide unwanted calories if overconsumed

Disadvantages of Fruit Juices

High concentration of carbohydrate
May cause some gastrointestinal distress
Could provide unwanted calories if overconsumed
In children, may displace milk intake

Disadvantages of Soft Drinks

High concentration of carbohydrate
Carbonation may contribute to gastrointestinal distress
Low nutrient density
A source of excess calories for many adults and children
In children, may displace milk intake
Provide unwanted stimulatory effect if caffeinated

Figure 7.14 Advantages and disadvantages of popular beverages

Because each athlete has different requirements, it is important to determine the amount of fluid and other nutrients needed and then choose foods and beverages accordingly.

Action plan. Fluid, electrolyte, and carbohydrate guidelines are often given as ranges and athletes use these ranges to create and fine-tune an individualized plan through trial and error. The nutrient composition of food and beverages is important information needed to make wise decisions about the specific foods and beverages to include and the timing of those foods and beverages. All beverages provide water, but athletes must check the label to determine carbohydrate (amount and source), sodium, potassium, and energy (kcal) content. Additionally, athletes may need information about caffeine content, carbohydrate concentration, or glycemic response, which is not required on the label but may be stated in promotional materials. Tables 7.6, 7.7, and 7.8 list the nutrient content for beverages commonly used by athletes before, during, and after exercise. These nutrients can also be supplied by meals and snacks.

Many types of fluids are available to an athlete: water, sports beverages, energy drinks, fruit juices, and soft drinks. The advantages and disadvantages of each are listed in Figure 7.14. An important point is that athletes need to choose the beverages that provide water and nutrients in the necessary quantities while making sure they do not contain unwanted or excess substances. Equally important is that any food or beverage chosen is done so because it fits into the athlete's personalized plan. Trial and error helps athletes to determine the appropriate combinations for successful training and performance.

For illustration purposes, consider two marathon runners who have the same fluid and carbohydrate goals during the race but achieve them in different ways. The first runner prefers to consume only a sports beverage throughout the race. This preference is based on convenience (for example, always available along the course, easy to consume while running), taste, and predictability (for example, known nutrient content, no history of gastrointestinal distress, no disruption to mental routine). The second runner prefers to consume a variety of carbohydrate and fluid sources—sports beverages, sports gels, bananas, and water (Figure 7.15). This runner experiences taste fatigue and voluntary dehydration if limited to just sports beverages and prefers a variety of tastes and textures during the marathon. She also wants to be able to match food and beverage intake to how she is feeling during the run (for example, blood glucose concentration and degree of gastrointestinal stress), and increasing or decreasing fluid, semisolid, and solid food intake is part of her racing strategy. Each of these athletes has developed

	Product	Water	Carbohydrate (g)	Calories (kcal)	Sodium (mg)	Potassium (mg)
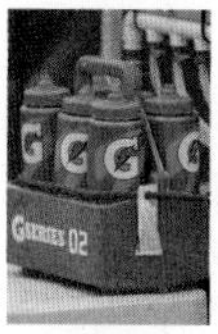 AP PHoto/Paul Abell	8 oz (240 ml) 6% carbohydrate beverage	~7.5 oz	14	50	110	30
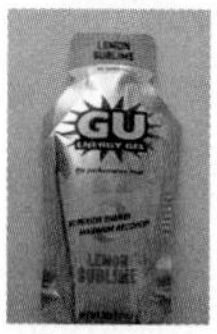 Christina Micek	Gel + 8 oz (240 ml) water	8 oz	25	100	40	35
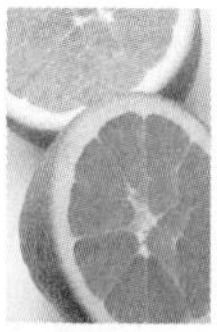 © Peter Scholey/almay	Large orange	~5 oz	22	86	0	333
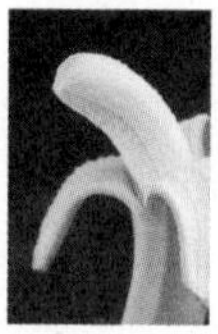 ABW Photography/Purestock/ SuperStock	7- to 8-inch banana + 6 oz water	~8 oz	27	105	5	422

Legend: kcal = kilocalorie; g = gram; mg = milligram; ml = milliliter

Figure 7.15 Comparison of selected food/water combinations to a sports beverage

an individualized plan that appropriately meets her fluid and carbohydrate goals, but the plans are very different even though they are in the same sport.

Reassessment. Athletes sometimes overlook the reassessment process, especially if everything appears to be going well. However, periodic evaluation allows the athlete to determine if goals are being met, if progress is too fast or too slow, and if the plan has unintended longer-term consequences, such as weight gain or caffeine dependency. Although there is no set schedule, there are logical time points for athletes to reassess such as early in the preseason, when environmental temperatures change, and when the duration and intensity of exercise training substantially increases or decreases.

In the process of replenishing fluids and electrolytes, athletes may be consuming other nutrients.

Although some beverages may contain only fluids and electrolytes, many beverages marketed to athletes contain other nutrients such as carbohydrates, caffeine, or vitamins. In some cases, carbohydrate is very appropriate because some athletes need additional sources of glucose during endurance and ultraendurance events (see Chapter 4). However, some beverages may be a source of excess sugars, calories, caffeine or vitamins that are not needed, and in some cases, are detrimental to the athlete's long-term goals or general health.

Sugars in sport beverages. Sport beverages and energy drinks fall along the sweetened-beverage continuum that runs from flavored water to highly concentrated carbohydrate solutions. Such drinks can be highly appropriate for an athlete under certain circumstances. However, many of these drinks contain only sugars and have a low nutrient density. A 3- or 7-day dietary analysis can reveal if the athlete's sports beverage intake is appropriate or if it is displacing other foods or drinks that provide needed nutrients, such as vitamins and minerals.

In some cases, the drinks are a source of excess kilocalories, and water may be a better choice because it is noncaloric. This may be the case for those participating in low-energy-expenditure sports and includes

Table 7.9 Caloric, Carbohydrate, and Caffeine Content of Selected Energy Drinks

Beverage	Serving size (oz)	Calories (kcal)	Carbohydrate (g)	Caffeine (mg)
AMP Energy	8	110	29	71
Red Bull	8.3	110	28	80
Rockstar Energy Drink	8	140	31	80 (+25 mg guarana)
SoBe Adrenaline Rush	8.3	140	37	86
Venom Energy	8	120	29	80

Legend: oz = ounce; kcal = kilocalorie; g = gram; mg = milligram

athletes (for example, pinch hitter in baseball), occasional exercisers, and children (for example, T-ball). Not everyone engaged in a "sport" needs a "sports beverage." Although sports beverages may be less sweet and have fewer calories than soft drinks, some recreational and trained athletes consume too much added sugar daily and need to be cautious about excessive intake from sports beverages.

Caffeinated "energy" drinks. "Energy" drinks are marketed to individuals who are mentally or physically fatigued, particularly adolescents and young adults. These drinks are popular among high school and collegiate athletes and triathletes because they mask fatigue for about 60 to 90 minutes (Duchan, Patel, and Feucht, 2010). The immediate feeling of "energy" is most likely due to caffeine, a neurological stimulant, as well as a rise in blood glucose.

Table 7.9 includes the energy, carbohydrate, and caffeine content of some popular energy drinks. Most provide about 28 to 37 g of carbohydrate and 110 to 140 kcal in an 8 oz serving. The caffeine content is about 80 mg per 8 oz (240 ml) serving, but many are sold in 16 oz (480 ml) containers. Note that these products are not recommended for children, pregnant and nursing women, and those sensitive to caffeine. The source of caffeine may be from herbs, such as guarana, kola nuts, or maté. These drinks often contain some B vitamins, taurine (a sulfur-containing amino acid), and herbs such as ginseng or ginkgo. Typically, the amounts of these compounds are generally considered safe, although there have been some case reports of seizures, cardiac arrest, and erosion of dental enamel. The Food and Drug Administration does not regulate energy drinks. Athletes should also be aware that these drinks are not designed to be fluid replacement beverages (Duchan, Patel, and Feucht, 2010).

For those who are caffeine-naïve (never or rare users) or caffeine-sensitive, a large dose of caffeine can cause a jittery, nervous response. They feel "overstimulated" rather than "energized," which may be detrimental to training, performance, or sleep. In some individuals, caffeine can cause an irregular heartbeat.

Athletes should be aware of their caffeine threshold for adverse side effects and cognizant of their daily intake. A 12 oz caffeinated soft drink provides approximately 35 to 55 mg of caffeine, less caffeine than an "energy" beverage. The caffeine content of an 8 oz cup of black caffeinated coffee brewed at home can vary considerably but is typically 80–100 mg. Many U.S. consumers buy "grande"-size (16 oz) coffee, which typically has 250 to 330 mg of caffeine (Center for Science in the Public Interest, 2008).

Athletes routinely struggle with fatigue, so it is very tempting to buy a drink that promises more "energy." However, an athlete's fatigue may be caused by insufficient caloric (energy) or carbohydrate intake, hypohydration, iron-deficiency anemia, lack of sleep, overtraining, increased body temperature, or a combination of these factors. Caffeine may positively impact performance because of central nervous system stimulation that alters perception of fatigue (Burke, 2008). But altering perception will not offset substantial contributors to fatigue such as low muscle glycogen stores or iron-deficiency anemia in endurance athletes. Athletes who mask fatigue with stimulants are encouraged to determine and address the fundamental causes of their fatigue.

Application exercise

Choose an athlete in a sport that you are familiar with and devise a plan for fluid and/or electrolyte intake before, during, and after exercise. What are the important considerations given the demands of the sport and the environmental conditions under which training and performance occur? Why might an athlete not like the plan that you created?

Hyponatremia, or plasma sodium being too low, is a serious electrolyte disturbance that can be fatal.

A potentially serious medical complication that may occur in endurance athletes during prolonged exercise such as ultramarathons or triathlons is hyponatremia (Noakes et al., 1985; Armstrong et al., 1993). Clinically, hyponatremia occurs when plasma sodium concentration falls below 135 mmol/L (from a typical level of 140 mmol/L). Exercise-associated hyponatremia is often characterized by a rapid drop to 130 mmol/L or below and is particularly serious when it drops rapidly and remains low. Because of the important role sodium has in maintaining osmotic and fluid balance between body water compartments and in the electrochemical gradient necessary for the transmission of nerve impulses, large disruptions in plasma sodium concentration can have serious physiological and medical consequences. Low sodium concentration in the extracellular fluid will stimulate the movement of water by osmosis from the plasma into the intracellular spaces, causing cells to swell. If nerve cells swell too much they cease to function properly, which can result in symptoms of dizziness, confusion, seizure, coma, and even death (see Spotlight on a real athlete: Hyponatremia in a Boston Marathon Runner).

Although rare in occurrence in shorter events, hyponatremia has been reported in up to 10 percent of runners in certain ultraendurance running events and in as many as 29 percent of triathletes in the Ironman™ Triathlon (Speedy, Rogers, Noakes, Thompson, et al., 2000). Note that these events involve endurance exercise lasting over 7 hours in duration, often in conditions of high heat and humidity, resulting in significant sweat loss. The physiological mechanisms of this exertional hyponatremia are not completely understood, but a leading hypothesis is that it is due to a combination of loss of sodium through heavy sweating and an overconsumption of hypotonic fluids, particularly water (Noakes et al., 2005; Noakes, 1992, Speedy, Rogers, Noakes, Wright, et al., 2000). When dehydration is prevented by copious consumption of water, sodium lost in sweat is not replaced, leading to a dilution of the sodium in the extracellular fluid. Hyponatremia can also occur in slow marathon runners, who are on the course for 5 or more hours, all the while consuming water or other beverages with a low sodium content (Almond et al., 2005).

The strategy to prevent hyponatremia is twofold: replacement of sodium and prevention of fluid overload or overdrinking. If exercise is to extend beyond 2 or 3 hours, sodium replacement should be considered along with fluid replacement, either in a beverage, as a supplement, or with salty-tasting foods. The recommended amount is 0.5 to 0.7 g of sodium per liter of fluid (Rodriquez, DiMarco, and Langley, 2009). Athletes may simply add salt to traditional sports beverages. For example if an athlete adds ¼ teaspoon of table salt to 32 ounces (~960 ml or 4 cups) of a sports beverage, 590 mg of sodium will be added. The additional sodium is ~0.6 g of sodium per liter of fluid, the midpoint of the guideline.

Athletes should be encouraged to consume enough fluid to match fluid loss and prevent performance-attenuating hypohydration, while not exceeding the amount of fluid lost. A substantial reduction in the incidence of hyponatremia in an ultradistance triathlon was observed after participants were educated about appropriate fluid intake, and access to fluids during the race was decreased slightly by reducing the number of fluid stations and increasing the distance between them (Speedy, Rogers, Noakes, Thompson, et al., 2000) Because the body's fluid balance mechanisms can be temporarily overwhelmed, an athlete's fluid and sodium intake is an important part of fluid homeostasis.

Increasing fluid levels above normal is hyperhydration.

Dehydration can have an adverse effect on training and performance, thus athletes have attempted to

Spotlight on a real athlete

Hyponatremia in a Boston Marathon Runner

The *Wall Street Journal* (2005) recounted the harrowing story of a 27-year-old male running his first Boston Marathon. The projected temperature for the 2004 race was 90°F, and his goal was to finish the race in less than 4 hours. The runner knew that he sweated heavily in warm weather and was concerned that he would dehydrate quickly. He drank more than a gallon of water prior to the race and water at every rest stop. He was on pace at mile 19 when he developed nausea and leg cramps. By mile 23 he was unable to run but walked to the finish line. After finishing he drank approximately 2 quarts of water but felt worse and experienced vomiting and diarrhea. On the subway ride home, the vomiting continued so he continued to drink water and a carbohydrate-electrolyte beverage. At home he became unconscious and fell (breaking his shoulder). Relatives called 911 and he was transported to the hospital, during which time he was given IV fluids as he was incorrectly diagnosed as being hypohydrated. He lapsed into a coma and was placed on life support for 4 days. In the hospital he was correctly diagnosed as having hyponatremia. Happily, this condition was reversed and he recovered.

Keeping it in perspective

Fluid and Electrolyte Balance Is Critical

Athletes must keep many aspects of nutrition in perspective, including their intake of energy, carbohydrates, proteins, fats, alcohol, vitamins, and minerals. What makes the fluid and electrolyte perspective different is that balance can change to imbalance quickly and the impact on health can be immediate and potentially fatal. There is no one-size-fits-all approach to fluid balance, so each athlete must consider fluids and electrolytes from an individual perspective. Well-meaning advice—"drink as much as you can"—does not consider the case of the slow-paced endurance athlete. "Reduce sodium intake" does not consider the athlete who loses large amounts of sodium in sweat. "Drink a pint, a pound" does not consider the athlete who was hypohydrated before exercise began.

"Just sweat it out and you'll make weight" does not consider the athlete's body temperature or degree of dehydration.

The proper perspective is to match fluid and electrolyte intake with fluid and electrolyte losses, although an exact "match" is often not possible or necessary. Part of maintaining a proper perspective is to recognize changing environmental conditions and needs and adjust accordingly. There can be a lot of trial and error involved; however, severe errors (for example, hyponatremia, elevated core temperature) can be fatal. That makes the fluid and electrolyte perspective more time-critical and an everyday concern for the athlete who is training and competing and the professionals who work with athletes.

manipulate body fluid levels prior to exercise by hyperhydrating. The idea is to increase the amount of body water prior to exercise so when fluid is lost during exercise, a critical level of hypohydration is not reached as quickly. Theoretically, this could prevent or delay a decline in performance.

Short-term hyperhydration can be achieved relatively easily by fluid overload, the consumption of excess fluids in the hours before exercise. This overconsumption results in an increase in total body water, an increase in plasma volume, and a potential improvement in thermoregulation and exercise performance in the heat. Because the kidneys react quickly to an overload of fluid, urine production is increased and the resulting full bladder and need to urinate may be an interfering factor for the upcoming exercise. Consumption of large quantities of fluid may also result in gastric discomfort. Hyponatremia may also be a concern if very large volumes of hypotonic fluid are consumed.

Glycerol loading. Another strategy for hyperhydrating involves the ingestion of an osmotically active, water-retaining molecule along with the increased amounts of fluid prior to exercise. One such compound is glycerol. Glycerol is easily absorbed and distributed throughout fluid compartments in the

body where it exerts an osmotic force to attract water.

A typical glycerol-loading regimen is to consume 1.0–1.2 g of glycerol per kg body weight along with 25–35 ml of water per kg body weight. For a 150 lb (~68 kg) runner this would be approximately 80 g of glycerol and 2 L (2,000 ml or ~71 oz or ~9 cups) of water. This approach generally results in fluid retention and hyperhydration of approximately 500 ml (~18 oz or ~2¼ cups), more than with fluid overload alone. However, it is unclear if this additional water "storage" prior to the onset of exercise improves performance. Theoretically, the additional body water would provide extra fluid for sweating and maintaining blood volume to enhance the control of body temperature without unduly compromising the cardiovascular function needed to sustain exercise performance. Although a small number of studies have shown this to be the case, the scientific literature lacks a sufficient number of studies with similar results to support a strong consensus opinion (Burke, 2001).

Hyperhydrating strategies do not show a clear benefit for improving performance over euhydration, and create the potential for needing to urinate during exercise. Therefore, hyperhydrating before competition, particularly prolonged endurance exercise, is not typically encouraged (Rodriquez, DiMarco, and Langley, 2009).

The potential adverse effects that may accompany glycerol loading must be considered, particularly the weight gain that results from hyperhydration. Excess body weight may negatively affect performance, particularly in weight-bearing activities such as running. Consumption of glycerol may also result in gastrointestinal upset and nausea, as might the ingestion of large amounts of fluid. This hyperhydration strategy may have some benefit, but should be attempted by an athlete only under the supervision of a sports medicine professional. This practice should first be instituted during training to ascertain the potential effects, both positive and negative, on performance.

The Internet café

Where do I find reliable information about water and electrolytes?

Several organizations have published position papers on the replenishment of water and electrolytes by athletes. These position papers outline general recommendations for fluid and electrolyte intake before, during, and after exercise. However, since body size varies tremendously among athletes, these guidelines must be individualized so that the amount of fluid consumed is well matched to the amount of fluid lost.

American College of Sports Medicine Position Stand, "Exercise and Fluid Replacement," **http://www.acsm-msse.org/**

American College of Sports Medicine, American Dietetic Association, and Dietitians of Canada Position Statement, "Nutrition and Athletic Performance," **http://www.acsm-msse.org/**

USA Track and Field, "Proper Hydration for Distance Running—Identifying Individual Fluid Needs," **http://www.usatf.org/**

National Athletic Trainers' Association, "Position Statement: Fluid Replacement for Athletes," **http://nata.org/sites/default/files/FluidReplacementsForAthletes.pdf**

The Gatorade Sports Science Institute (**http://www.gssiweb.com/**) produces a large amount of excellent materials intended for professional audiences such as exercise physiologists, sports dietitians, strength and conditioning coaches, and athletic trainers. This is a commercial website and features products made by Gatorade.

Key points

- Athletes need to monitor their hydration status in a practical, easy, and understandable way.

- Urine color and changes in body weight are not as precise as some laboratory methods but may be practical methods for athletes to assess their hydration on a daily basis, particularly when combined with their perception of thirst.

- Every athlete should develop an individual plan for fluid and electrolyte consumption before, during, and after exercise with the goal of preventing or delaying dehydration that will adversely affect performance and/or health.

- The amount of fluid consumed before exercise should achieve and maintain euhydration without stimulating excessive urine production.

- The amount of fluid consumed during exercise should match water losses from sweat as closely as possible, but overconsumption of water should be avoided during prolonged exercise.

- Fluid should be consumed after exercise to replenish water losses and return the athlete to euhydration.

How does hyponatremia occur during exercise?

If an athlete has lost 5 pounds during a long endurance run in the heat, how much fluid (water) should he or she consume after exercise?

Post-Test   Reassessing Knowledge of Water and Electrolytes

Now that you have more knowledge about water and electrolytes, read the statements and decide if each is true or false. The answers can be found in Appendix O.

1. The two major aspects of fluid balance are the volume of water and the concentration of the substances in the water.

2. Now that sports beverages are precisely formulated, it is rare that water would be a better choice than a sports beverage for a trained athlete.

3. Athletes should avoid caffeinated drinks because caffeine is a potent diuretic.

4. A rule of thumb for endurance athletes is to drink as much water as possible.

5. Under most circumstances, athletes will not voluntarily drink enough fluid to account for all the water lost during exercise.

Summary

Water is critical to the normal physiological functioning of the body. Optimal fluid balance is of further importance to the athlete, as exercise can place severe demands on the body to maintain fluid homeostasis. Substantial water losses can occur during exercise and activity, particularly if the environmental conditions are severe. Water losses can compromise the body's ability to regulate body temperature, which may in turn impair training, performance, and health. Extreme losses of water and electrolytes may be dangerous or fatal due to severe hypohydration and an inability to keep body temperature from rising. Athletes should be conscious of water losses and employ strategies to ensure adequate hydration and electrolyte replacement before,

during, and after exercise. Water in excess of need can also be problematic, as in the case of hyponatremia.

Recommendations for fluid and electrolyte consumption before, during, and after exercise are good guidelines for intake but must be individualized. Many sports beverages provide a convenient way to consume water and sodium, as well as carbohydrate, although food and water combinations are also used. Trial and error during training helps athletes determine the most appropriate fluid and electrolyte intake before and during competition. The athlete's usual rehydration strategy may need adjustment when environmental conditions change, especially increases in heat and humidity.

Self-Test

Multiple-Choice

The answers can be found in Appendix O.

1. In which compartment is the largest amount of body fluid stored?
   a. blood plasma
   b. interstitial fluid
   c. extracellular fluid
   d. intracellular fluid

2. The primary cation in extracellular fluid is:
   a. sodium.
   b. potassium.
   c. calcium.
   d. chloride.

3. Wearing uniforms and protective gear while exercising in the heat typically:
   a. keeps the sun off the skin and prevents body temperature from rising.
   b. restricts the evaporation of sweat and results in body temperature rising.
   c. has little effect on body temperature.

4. Water only is an appropriate pre-exercise beverage for:
   a. exercise lasting less than 60 minutes.
   b. exercise lasting from 1 to 4 hours.
   c. exercise lasting greater than 4 hours.
   d. exercise lasting from 1 to 4 hours or greater than 4 hours.

5. Two factors that may be associated with hyponatremia include:
   a. large losses of sodium in sweat and low water intake.
   b. large losses of sodium in sweat and excessive water intake.
   c. intake of salt tablets and low water intake.
   d. intake of salt tablets and excessive water intake.

Short Answer

1. Why is water so critical to athletic performance?

2. How is water distributed throughout the body? How does the body maintain fluid balance between the extracellular and intracellular compartments?

3. Compare and contrast euhydration, hypohydration, and hyperhydration. Discuss the reasons athletes might attempt to be hypohydrated or hyperhydrated and the ways in which they achieve these states. What are the dangers associated with hypohydration and hyperhydration?

4. Compare and contrast the intake, physiological roles, and excretion of sodium and potassium.

5. What effect does exercise have on fluid balance? What effect does hypohydration have on exercise performance? On health?

6. Outline the ways that hydration status can be monitored. Which ways are easy and practical for athletes to use?

7. What is the current recommendation for caffeine intake by adult athletes?

8. What is hyponatremia? In which sports are athletes at risk for developing hyponatremia? How can this condition be prevented?

Critical Thinking

1. Why is the extracellular fluid considered the "gateway" for fluid and electrolyte movement between compartments?

2. Describe and explain hyperhydration in athletes as (a) a normal condition, (b) a precompetition strategy, and (c) a dangerous medical condition.

3. Outline a fluid and electrolyte replenishment plan for an athlete competing in the Ironman™ Triathlon in Hawaii. What are the specific considerations that must be taken into account when developing this plan?

4. Outline a fluid and electrolyte replenishment plan for an interior lineman (offensive or defensive lineman) for preseason training and for the first game of the season in Texas.

References

Almond, C. S., Shin, A. Y., Fortescue, E. B., Mannix, R. C., Wypij, D., Binstadt, B. A., et al. (2005). Hyponatremia among runners in the Boston Marathon. *New England Journal of Medicine, 352*(15), 1550–1556.

Appleby, M., Fisher, M., & Martin, M. (1994). Myocardial infarction, hyperkalaemia and ventricular tachycardia in a young male body-builder. *International Journal of Cardiology, 44*(2), 171–174.

Armstrong, L. E. (2000). *Performing in extreme environments.* Champaign, IL: Human Kinetics.

Armstrong, L. E. (2005). Hydration assessment techniques. *Nutrition Reviews, 63*(6, Pt. 2), S40–S54.

Armstrong, L. E., Casa, D. J., Maresh, C. M., & Ganio, M. S. (2007). Caffeine, fluid-electrolyte balance, temperature regulation, and exercise-heat tolerance. *Exercise and Sport Sciences Reviews, 35*(3), 135–40.

Armstrong, L. E., Costill, D. L., & Fink, W. J. (1985). Influence of diuretic-induced dehydration on competitive running performance. *Medicine and Science in Sports and Exercise, 17*(4), 456–461.

Armstrong, L. E., Costill, D. L., & Fink, W. J. (1987). Changes in body water and electrolytes during heat acclimation: Effects of dietary sodium. *Aviation, Space and Environmental Medicine, 5*(2), 143–148.

Armstrong, L. E., Curtis, W. C., Hubbard, R. W., Francesconi, R. P., Moore, R., & Askew, E. W. (1993). Symptomatic hyponatremia during prolonged exercise in heat. *Medicine and Science in Sports and Exercise, 25*(5), 543–549.

Armstrong, L. E., & Epstein, Y. (1999). Fluid-electrolyte balance during labor and exercise: Concepts and misconceptions. *International Journal of Sport Nutrition, 9*(1), 1–12.

Armstrong, L. E., & Maresh, C. M. (1998). Effects of training, environment, and host factors on the sweating response to exercise. *International Journal of Sports Medicine, 19*(Suppl. 2), S103–S105.

Armstrong, L. E., Soto, J. A., Hacker, F.T., Jr., Casa, D. J., Kavouras, S. A., & Maresh, C. M. (1998). Urinary indices during dehydration, exercise, and rehydration. *International Journal of Sport Nutrition, 8*(4), 345–355.

Baker, L. B., Conroy, D. E., & Kenney, W. L. (2007). Dehydration impairs vigilance-related attention in male basketball players. *Medicine and Science in Sports and Exercise, 39*(6), 976–983.

Baker, L. B., Dougherty, K. A., Chow, M., & Kenney, W. L. (2007). Progressive dehydration causes a progressive decline in basketball skill performance. *Medicine and Science in Sports and Exercise, 39*(7), 1114–1123.

Bergeron, M. F. (2007). Exertional heat cramps: Recovery and return to play. *Journal of Sport Rehabilitation, 16*(3), 190–196.

Burke, L. M. (2001). Nutrition needs for exercise in the heat. *Comparative Biochemistry and Physiology. Part A, Molecular & Integrative Physiology, 128*(4), 735–748.

Burke, L. M. (2008). Caffeine and sports performance. *Applied Physiology, Nutrition, and Metabolism, 33,* 1319–1334.

Center for Science in the Public Interest. (2008 with periodic updates). How much is that caffeine in the window? Retrieved December, 14, 2010 from http://www.cspinet.org/nah/02_08/caffeine.pdf

Centers for Disease Control and Prevention (CDC). (1998). Hyperthermia and dehydration-related deaths associated with intentional rapid weight loss in three collegiate wrestlers—North Carolina, Wisconsin, and Michigan, November–December 1997. *MMWR Morbidity and Mortality Weekly Report, 47*(6), 105–108.

Cheuvront, S. N., Carter, R., III, & Sawka, M. N. (2003). Fluid balance and endurance exercise performance. *Current Sports Medicine Reports, 2*(24), 202–208.

Cheuvront, S. N., Ely, B. R., Kenfick, R. W., & Sawka, M. N. (2010). Biological variation and diagnostic accuracy of dehydration assessment markers. *American Journal of Clinical Nutrition, 92,* 565–573.

Cheuvront, S. N., & Sawka, M.N. (2005). Hydration assessment of athletes. *Sports Science Exchange, 97*(18), 2 [Suppl.].

Convertino, V. A. (1987). Fluid shifts and hydration state: Effects of long-term exercise. *Canadian Journal of Sport Sciences, 12*(Suppl. 1), 136S–139S.

Convertino, V. A., Armstrong, L. E., Coyle, E. F., Mack, G. W., Sawka, M. N., Senay, L. C., Jr., et al. (1996). Position Stand on Exercise and Fluid Replacement. *Medicine and Science in Sports and Exercise, 28*(1), i–vii.

Cox, G. R., Broad, E. M., Riley, M. D., & Burke, L. M. (2002). Body mass changes and voluntary fluid intakes of elite level water polo players and swimmers. *Journal of Science and Medicine in Sport, 5*(3), 183–193.

Duchan, E., Patel, N. D., & Feucht, C. (2010). Energy drinks: A review of use and safety for athletes. *The Physician and Sportsmedicine, 38*(2), 171–179.

Ebert, T. R., Martin, D. T., Bullock, N., Mujika, I., Quod, M. J., Farthing, L. A., et al. (2007). Influence of hydration status on thermoregulation and cycling hill climbing. *Medicine and Science in Sports and Exercise, 39*(2), 323–329.

Edwards, A. M., Mann, M. E., Marfell-Jones, M. J., Rankin, D. M., Noakes, T. D., & Shillington, D. P. (2007). Influence of moderate dehydration on soccer performance: Physiological responses to 45 min of outdoor match-play and the immediate subsequent performance of sport-specific and mental concentration tests. *British Journal of Sports Medicine, 41*(6), 385–391.

Gropper, S. S., Smith, J. L., & Groff, J. L. (2009). *Advanced nutrition and human metabolism* (5th ed.). Belmont, CA: Thomson/Wadsworth.

Hamilton, M. T., Gonzalez-Alonso, J., Montain, S. J., & Coyle, E. F. (1991). Fluid replacement and glucose infusion during exercise prevent cardiovascular drift. *Journal of Applied Physiology, 71*(3), 871–877.

Hargreaves, M., & Febbraio, M. (1998). Limits to exercise performance in the heat. *International Journal of Sports Medicine, 19*(Suppl. 2), S115–S116.

Hew-Butler, T. D., Sharwood, K., Collins, M., Speedy, D., & Noakes, T. (2006). Sodium supplementation is not required to maintain serum sodium concentrations during an Ironman™ triathlon. *British Journal of Sports Medicine, 40*(3), 255–259.

Institute of Medicine. (2004). *Dietary Reference Intakes for water, potassium, sodium, chloride, and sulfate* (Food and Nutrition Board). Washington, DC: National Academies Press.

Leiper, J. B., Nicholas, C. W., Ali, A., Williams, C., & Maughan, R. J. (2005). The effect of intermittent high-intensity running on gastric emptying of fluids in man. *Medicine and Science in Sports and Exercise, 37*(2), 240–247.

Leydon, M. A., & Wall, C. (2002). New Zealand jockeys' dietary habits and their potential impact on health. *International Journal of Sport Nutrition and Exercise Metabolism, 12*(2), 220–237.

Macedonio, M.A., & Dunford, M. (2009). *The Athlete's Guide to Making Weight: Optimal Weight for Optimal Performance.* Champaign, IL: Human Kinetics.

Maughan, R. J., & Shirreffs, S. M. (2008). Development of individual hydration strategies for athletes. *International Journal of Sport Nutrition and Exercise Metabolism, 18*(5), 457–472.

Miller, K. C., Mack, G., & Knight, K. L. (2009). Electrolyte and plasma changes after ingestion of pickle juice, water, and a common carbohydrate-electrolyte solution. *Journal of Athletic Training, 44*(5), 454–461.

Miller, K. C., Mack, G. W., Knight, K. L., Hopkins, J. T., Draper, T. O., Fields, P. J., et al. (2010). Reflex inhibition of electrically induced muscle cramps in hypohydrated humans. *Medicine and Science in Sports and Exercise, 42*(5), 953–961.

Murray, B. (in press). Fluid, electrolytes and exercise. In C. Rosenbloom(Ed.), *Sports nutrition: A practice manual for professionals* (5th ed., Chicago: American Dietetic Association.

Noakes, T. D. (1992). The hyponatremia of exercise. *International Journal of Sport Nutrition, 2*(3), 205–228.

Noakes, T. D., Goodwin, N., Rayner, B. L., Branken, T., & Taylor, R. K. N. (1985). Water intoxication: A possible complication during endurance exercise. *Medicine and Science in Sports and Exercise, 17*(3), 370–375.

Noakes, T. D., Sharwood, K., Speedy, D., Hew, T., Reid, S., Dugas, J., et al. (2005). Three independent biological mechanisms cause exercise-associated hyponatremia: Evidence from 2,135 weighed competitive athletic performances. *Proceedings of the National Academy of Sciences, 102*, 18550–18555.

Oppliger, R. A., Case, H. S., Horswill, C. A., Landry, G. L., Shelter, A. C., & American College of Sports Medicine. (1996). Position stand on weight loss in wrestlers. *Medicine and Science in Sports and Exercise, 28*(6), ix–xii.

Oppliger, R. A., Utter, A. C., Scott, J. R., Dick, R. W., & Klossner, D. (2006). NCAA rule change improves weight loss among national championship wrestlers. *Medicine and Science in Sports and Exercise, 38*(5), 963–970.

Perazella, M. A. (2000). Drug-induced hyperkalemia: Old culprits and new offenders. *American Journal of Medicine, 109*(4), 307–314.

Rehrer, N. J. (2001). Fluid and electrolyte balance in ultra-endurance sport. *Sports Medicine, 31*(10), 701–715.

Rodriquez, N. R., DiMarco, N. M., & Langley, S. (2009). American College of Sports Medicine position stand. Nutrition and athletic performance. *Medicine and Science in Sports and Exercise, 41*(3), 709–731.

Sawka, M. N., Burke, L. M., Eichner, E. R., Maughan, R. J., Montain, S. J., Stachenfeld, N. S., et al. (2007). Position stand on exercise and fluid replacement. *Medicine and Science in Sports and Exercise, 39*(2), 377–390.

Sawka, M. N., & Coyle, E. F. (1999). Influence of body water and blood volume on thermoregulation and exercise performance in the heat. *Exercise and Sport Science Reviews, 27*, 167–218.

Sawka, M. N., Latzka, W. A., Matott, R. P., & Montain, S. J. (1998). Hydration effects on temperature regulation. *International Journal of Sports Medicine, 19*(Suppl. 2), S108–S110.

Sawka, M. N., & Montain, S. J. (2000). Fluid and electrolyte supplementation for exercise heat stress. *American Journal of Clinical Nutrition, 72*(2 Suppl.), 564S–572S.

Sawka, M. N., Montain, S. J., & Latzka, W. A. (2001). Hydration effects on thermoregulation and performance in the heat. *Comparative Biochemistry and Physiology. Part A, Molecular & Integrative Physiology, 128*(4), 679–690.

Schwellnus, M. P. (2009). Cause of exercise associated muscle cramping (EAMC)—altered neuromuscular control, dehydration or electrolyte depletion? *British Journal of Sports Medicine, 43*(6), 401–408.

Scott, J. R., Horswill, C. A., & Dick, R. W. (1994). Acute weight gain in collegiate wrestlers following a tournament weigh-in. *Medicine and Science in Sports and Exercise, 26*(9), 1181–1185.

Sharp, R. L. (2007). Role of whole foods in promoting hydration after exercise in humans. *Journal of the American College of Nutrition, 26*(5 Suppl.), 592S–596S.

Sherwood, L. (2010). *Human physiology: From cells to systems* (7th ed.). Belmont, CA: Thomson Brooks/Cole.

Shirreffs, S. M. (2001). Restoration of fluid and electrolyte balance after exercise. *Canadian Journal of Applied Physiology, 26*(Suppl.), S228–S235.

Shirreffs, S. M., Armstrong, L. E., & Cheuvront, S. N. (2004). Fluid and electrolyte needs for preparation and recovery from training and competition. *Journal of Sports Sciences, 22*(1), 57–63.

Shirreffs, S. M., & Maughan, R. J. (2000). Rehydration and recovery of fluid balance after exercise. *Exercise and Sport Science Reviews, 28*(1), 27–32.

Shriver, L. H., Betts, N. M., & Payton, M. E. (2009). Changes in body weight, body composition, and eating attitudes in high school wrestlers. *International Journal of Sport Nutrition and Exercise Metabolism, 19*(4), 424–432.

Speedy, D. B., Noakes, T. D., Rogers, I. R., Thompson, J. M. D., Campbell, R. G. D, Kuttner, J. A., et al. (1999). Hyponatremia in ultraendurance triathletes. *Medicine and Science in Sports and Exercise, 31*(6), 809–815.

Speedy, D. B., Rogers, I. R., Noakes, T. D., Thompson, J. M. D., Guirey, J., Safih, S., et al. (2000). Diagnosis and prevention of hyponatremia at an ultradistance triathlon. *Clinical Journal of Sport Medicine, 10*(1), 52–58.

Speedy, D. B., Rogers, I. R., Noakes, T. D., Wright, S., Thompson, J. M., Campbell, R., et al. (2000). Exercise-induced hyponatremia in ultradistance triathletes is caused by inappropriate fluid retention. *Clinical Journal of Sport Medicine, 10*(4), 272–278.

Speedy, D. B., Thompson, J. M., Rodgers, I., Collins, M., Sharwood, K., & Noakes, T. D. (2003). Oral salt supplementation during ultradistance exercise. *Clinical Journal of Sport Medicine, 12*(5), 279–284. Erratum in: *Clinical Journal of Sport Medicine,* 2003, *13*(1), 67.

Steen, S. N., & Brownell, K. D. (1990). Patterns of weight loss and regain in wrestlers: Has the tradition changed? *Medicine and Science in Sports and Exercise, 22*(6), 762–768.

Sturmi, J. E., & Rutecki, G. W. (1995). When competitive bodybuilders collapse. A result of hyperkalemia? *Physician and Sportsmedicine, 23*, 49–53.

Von Duvillard, S. P., Braun, W. A., Markofski, M., Beneke, R., & Leithauser, R. (2004). Fluids and hydration in prolonged endurance performance, *Nutrition, 20*(7–8), 651–656.

8 Vitamins

Learning Plan

- **Classify** vitamins and describe their general and specific roles.

- **Explain** how vitamin inadequacies and excesses can occur and why either might be detrimental to performance and health.

- **Explain** how the Dietary Reference Intakes (DRI) and the Tolerable Upper Intake Levels (UL) should be interpreted.

- **Describe** if, and how, exercise increases the need for or accelerates the loss of a particular vitamin.

- **Compare** and contrast the average intake of vitamins by sedentary adults in the United States and by athletes, particularly those who restrict energy intake.

- **Differentiate** between a clinical and subclinical vitamin deficiency.

- **Explain** how vitamins are involved in energy metabolism and blood formation and summarize the results of studies conducted with athletes.

- **Compare** and contrast vitamins A, C, and E, particularly their antioxidant functions.

- **Compare** and contrast vitamins based on their source—naturally occurring in food, added to foods during processing, and found in supplements.

- **Evaluate** the need for vitamin supplements based on food intake.

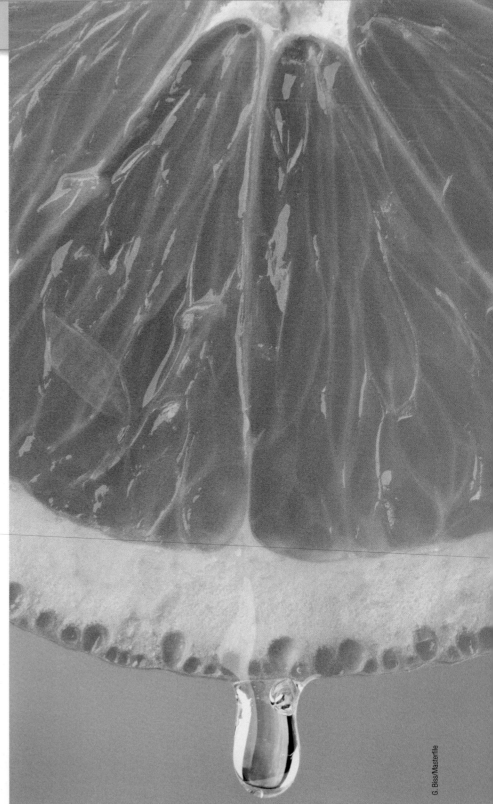

Vitamins are provided by a balanced diet containing a variety of foods.

G. Biss/Masterfile

Pre-Test   Assessing Current Knowledge of Vitamins

Read the following statements and decide if each is true or false.

1. Exercise increases the usage of vitamins, so most athletes need more vitamins than sedentary people.
2. Vitamins provide energy.
3. The amount of vitamins an athlete consumes is generally related to caloric intake.
4. When antioxidant vitamins are consumed in excess, they act like pro-oxidants instead of antioxidants.
5. Vitamin supplements are better regulated than other dietary supplements because the U.S. Food and Drug Administration sets a maximum dose (amount) for each vitamin.

Vitamins are essential nutrients needed for the proper functioning of the body. They are fundamental to many of the metabolic processes involved with physical activity, growth of tissue, and general health. Some vitamins are involved in energy metabolism, red blood cell production, and antioxidant functions, so it is natural that athletes would look at the functionality of these vitamins and their potential to improve exercise-related processes. Athletes should develop strategies for consuming enough of all the vitamins without consuming too much.

Endurance and ultraendurance exercise may increase the need for some vitamins and may result in small losses of others via urine or sweat. An increase in exercise also requires that more energy be consumed to maintain energy balance and body composition. If the additional foods eaten to support exercise and training are nutrient dense, then vitamin-rich foods can easily provide the additional vitamins needed. But many athletes do not consume vitamin-rich foods such as fruits, vegetables, and whole grains. Athletes may also be limiting caloric intake in an effort to reduce body fat. In such cases vitamin intake may be inadequate unless the diet is planned with nutrient density in mind.

A dietary analysis can help determine if the diet is adequate in vitamins. If there are inadequacies, the athlete is typically advised to make dietary changes, since vitamin inadequacies are usually found along with other nutrient deficiencies, all of which could hamper training, performance, or recovery. In some cases, a vitamin supplement might be needed.

8.1 Classification of Vitamins

Vitamins are essential nutrients needed in small quantities for the proper functioning of the body. Table 8.1 lists 13 vitamins that have been identified as essential. Vitamins are often classified based on their solubility. Those vitamins that are fat soluble include vitamins A, D, E, and K. All of the B vitamins (thiamin, riboflavin, niacin, pantothenic acid, biotin, folate, B_6, and B_{12}) and vitamin C are water soluble. Some of the characteristics associated with vitamins are related to their solubility.

The fat-soluble vitamins are absorbed and transported in the same way as fat (see Chapter 6). Absorption may take several hours and transport in the blood requires that they be bound to a carrier. Fat-soluble vitamins are stored in liver and adipose (fat) cells. Although each fat-soluble vitamin has a recommended daily intake, the ability to store these vitamins means that daily intake can vary without immediate risk for deficiency. For example, on days when vitamin E intake is lower than usual, the body has a ready store of vitamin E for use. On days when vitamin E intake is adequate, stores that have been reduced can be increased. This ability to store fat-soluble vitamins helps the body to guard against deficiencies, but it also means that toxicities can occur if excessive amounts are consumed over long periods of time. These toxicities, although rare, can cause substantial health problems, especially in a major organ such as the liver. Optimal intake—not too little, not too much—is an important goal.

In contrast to fat-soluble vitamins, water-soluble vitamins are easily absorbed and circulate in the

Table 8.1 Fat- and Water-Soluble Vitamins

Fat soluble	Water soluble
Vitamin A	Vitamin B_1 (Thiamin)
Vitamin D	Vitamin B_2 (Riboflavin)
Vitamin E	Vitamin B_3 (Niacin)
Vitamin K	Pantothenic acid
	Biotin
	Folate (Folic acid, folacin)
	Vitamin B_6
	Vitamin B_{12}
	Vitamin C

Choline is sometimes listed as a water-soluble vitamin, but it is technically an amine (a derivative of ammonia). It is included with the B vitamins as part of the Dietary Reference Intakes (DRI).

blood without the need for a carrier. There is no designated storage site. Instead, tissues can become saturated with the vitamin, and when the saturation point is reached, the excess is excreted via the urine. For some water-soluble vitamins, such as B_1 (thiamin), B_2 (riboflavin), B_{12}, pantothenic acid, or biotin, this saturation/excretion system works exceptionally well and no **toxicity** symptoms have been reported. However, toxicity symptoms have been reported for some other water-soluble vitamins when excessive amounts are consumed. For example, excessive amounts of vitamin B_6 can damage the nervous system and result in headaches, difficulty with reflexes or walking, and numbness. A moderate intake helps prevent both deficiencies and toxicities.

Solubility is one way of classifying vitamins, but another classification method uses physiological function. In many cases several vitamins are needed for a physiological process to occur. Some common classifications include (1) vitamins related to energy metabolism, (2) vitamins needed for red blood cell formation, (3) vitamins associated with **antioxidant** functions, and (4) growth and development. When classified this way, water- and fat-soluble vitamins may be in the same category. For example, vitamins

Vitamin: Essential organic (carbon-containing) compound necessary in very small quantities for proper physiological function.

Toxicity: State or relative degree of being poisonous.

Beta-carotene: One form of carotene, a precursor to vitamin A.

Antioxidant: Substance that inhibits oxidative reactions and protects cells and tissues from damage.

with antioxidant properties include the fat-soluble vitamins A (as **beta-carotene**) and E as well as the water-soluble vitamin C.

In many cases the vitamins are not classified together but are addressed individually. For each of the 13 vitamins listed in Table 8.1, there is a tremendous amount of information known. This information has been summarized in Table 8.2. For each vitamin the following appears: common and alternative names, major physiological functions, solubility, the deficiency and/or toxicity disease, symptoms associated with deficiency and/or toxicity, association with disease prevention, food sources, and miscellaneous information. More detailed information on vitamins can also be found in basic nutrition textbooks or at the Food and Nutrition Information Center at http://www.nal.usda.gov/fnic.

A recommended daily intake has been established for each vitamin.

Vitamins play an important role in overall health. Because each vitamin plays a specific role that cannot be replaced or substituted by another vitamin, it is important to consume an adequate amount of each vitamin. Consumption of excessive amounts of vitamins should be avoided since toxicities, even of certain water-soluble vitamins, can occur. Two sets of guidelines have been created that help quantify adequate but not excessive amounts. As discussed in Chapter 1, the Dietary Reference Intakes is a set of values that helps answer the question, "How much [of a nutrient] is needed each day?" The Tolerable Upper Intake Levels (UL) help address the question, "How much is too much?"

Table 8.3 is a quick and easy reference when considering these questions. The complete DRI and UL for vitamins are listed in the gatefold located in the front of this textbook, but the values for adult males and adult, nonpregnant females have been repeated in Table 8.3. Notice that the DRI are the same for adult males and females of any age for some of the vitamins. However, the recommended intake of several vitamins is higher based on male gender (for example, vitamins A, C) or increasing age (for example, vitamins D, B_6). Observe that upper intake levels have been established for only 8 of the 14 vitamins listed; UL have not been established for the other 6 vitamins because of a lack of scientific data. Excessive amounts of these vitamins may cause adverse effects; however, at the present time the amounts at which adverse effects may occur is not known.

The Dietary Reference Intakes are the current standard used to determine nutrient goals for individuals. The standard was developed using scientific

Table 8.2 Summary of Fat- and Water-Soluble Vitamin Characteristics

	Vitamin A
Names	In animal sources: retinol, retinal, and retinoic acid (also known as preformed vitamin A) In plant sources: carotenoids (precursors to vitamin A) including beta-carotene
Major physiological functions	Overall health of cells and membranes resulting in proper vision, reproduction, bone and tooth development, immune system function; carotenoids are antioxidants
Solubility	Fat soluble
Deficiency disease	Hypovitaminosis A
Symptoms of deficiency	Night blindness, permanent blindness, more frequent and severe infections, lack of growth, inability to reproduce
Toxicity disease	Hypervitaminosis A (from preformed vitamin A)
Symptoms of toxicity	Blurred vision, lack of growth, birth defects, hemorrhaging, liver failure; can be fatal
Health promotion and disease prevention	Adequate intake daily is necessary for good health; lack of evidence that vitamin A or beta-carotene supplements prevent disease; some evidence that mortality is increased with these supplements
Food sources	Animal (preformed vitamin A): liver, fish oil, milk and milk products (fortified); plant (provitamin A): dark-green leafy vegetables (for example, spinach); orange fruits and vegetables (for example, carrots, cantaloupe, tomatoes)
Other	Beta-carotene supplements are not recommended and may promote tumor growth in smokers and those exposed to asbestos; the yellowing of the skin that occurs is thought to be harmless but is indicative of a high level of carotenoid intake.
	Vitamin D
Names	Calciferol, cholecalciferol
Major physiological functions	Regulates bone mineralization, cardiac and skeletal muscle, and growth of normal and cancerous cells
Solubility	Fat soluble
Deficiency disease	Rickets, osteomalacia
Symptoms of deficiency	Bowing of the legs, demineralization of bones, joint pain, muscle spasms
Toxicity disease	Hypervitaminosis D
Symptoms of toxicity	Calcification of tissues including blood vessels; kidney stones; general gastrointestinal and nervous system complaints
Health promotion and disease prevention	Adequate intake daily is necessary for good health; vitamin D and calcium supplements together reduce the loss of bone mass and help prevent bone fractures; vitamin D supplements help prevent falls in older people
Food sources	Fish oil; some fish such as salmon, mackerel, tuna, and shrimp; milk (fortified); margarine (fortified)
Other	Ultraviolet light (sunshine) can activate a vitamin D precursor in the skin
	Vitamin E
Names	Tocopherol (for example, alpha-tocopherol, beta-tocopherol)
Major physiological functions	Antioxidant; proper red blood cell formation
Solubility	Fat soluble
Deficiency disease	Deficiencies are rare
Symptoms of deficiency	Anemia, muscle weakness
Toxicity disease	None; toxicities are rare
Symptoms of toxicity	General symptoms such as fatigue or nausea
Health promotion and disease prevention	Adequate intake daily is necessary for good health; if dietary intake is low, increasing dietary sources of vitamin E may be beneficial for preventing coronary heart disease; lack of evidence that vitamin E is effective in the treatment of Alzheimer's disease or mild cognitive impairment; some evidence that mortality is increased with vitamin E supplements in people with heart disease or diabetes
Food sources	Oil; soybeans; almonds and other nuts; sunflower seeds; wheat germ
Other	Powerful antioxidant

(Continued)

Table 8.2 Summary of Fat- and Water-Soluble Vitamin Characteristics (Continued)

	Vitamin K
Names	Phylloquinone
Major physiological functions	Normal blood clotting; role in bone mineralization
Solubility	Fat soluble
Deficiency disease	Vitamin K deficiency
Symptoms of deficiency	Hemorrhaging; poor bone mineralization
Toxicity disease	None
Symptoms of toxicity	Not known
Food sources	Green leafy vegetables
Other	Synthesized by bacteria in the intestine; vitamin K supplements are prescription only since excessive vitamin K could interfere with medications that prevent clotting of the blood
	Vitamin B₁
Names	Thiamin
Major physiological functions	Release of energy from carbohydrates, proteins, and fats via thiamin-containing enzymes; normal nervous system function
Solubility	Water soluble
Deficiency disease	Beriberi
Symptoms of deficiency	Muscle wasting, weight loss, cardiovascular problems
Toxicity disease	None
Symptoms of toxicity	None known
Health promotion and disease prevention	Adequate intake daily is necessary for good health; at the present time there is not enough research to adequately determine if B_1 supplementation has a beneficial effect on Alzheimer's, Parkinson's, or related diseases
Food sources	Whole grain breads and cereals; bread and cereals made from processed grains or flour (fortified); dried beans, pork
Other	Deficiencies seen in the United States are usually due to alcohol abuse or gastric bypass surgery
	Vitamin B₂
Names	Riboflavin
Major physiological functions	Release of energy from carbohydrates, proteins, and fats via riboflavin-containing enzymes; normal skin development
Solubility	Water soluble
Deficiency disease	Riboflavin deficiency disease
Symptoms of deficiency	Changes to the mouth, lips, and tongue; skin rash
Toxicity disease	None
Symptoms of toxicity	None known
Health promotion and disease prevention	Adequate intake daily is necessary for good health
Food sources	Milk; leafy green vegetables; whole grain breads and cereals
Other	Exercise increases the need for riboflavin, although most athletes consume enough

Table 8.2 Summary of Fat- and Water-Soluble Vitamin Characteristics (Continued)

	Vitamin B$_3$
Names	Niacin, nicotinic acid, nicotinamide
Major physiological functions	Release of energy from carbohydrates, proteins, and fats via niacin-containing enzymes
Solubility	Water soluble
Deficiency disease	Pellagra
Symptoms of deficiency	Diarrhea, mental changes, skin rash
Toxicity disease	Not named; usually referred to as niacin toxicity
Symptoms of toxicity	Flushing, itching, rash, sweating
Health promotion and disease prevention	Adequate daily intake is necessary for good health; high doses of niacin may be prescribed by a physician to help raise high-density lipoprotein cholesterol ("good cholesterol") to prevent cardiovascular events such as heart attack
Food sources	Meat, fish, poultry, and eggs; milk; nuts; whole grain breads and cereals; bread and cereals made from processed grains or flour (fortified)
Other	Tryptophan, an amino acid found in foods, is a precursor to niacin. Niacin rush (flushing, itching, rash) may be a result of a high intake of supplemental vitamin B$_3$ in a short period of time
	Vitamin B$_6$
Names	Pyridoxine
Major physiological functions	Release of energy stored in muscle glycogen; role in gluconeogenic processes (for example, manufacture of glucose from protein fragments); red blood cell formation
Solubility	Water soluble
Deficiency disease	Not named
Symptoms of deficiency	Microcytic (small cell) anemia
Toxicity disease	Not named; usually referred to as vitamin B$_6$ toxicity
Symptoms of toxicity	Nervous system impairment including fatigue, difficulty walking, numbness, depression
Health promotion and disease prevention	Adequate daily intake is necessary for good health; lack of evidence that supplementation of vitamin B$_6$ alone or in combination with vitamin B$_{12}$ and folic acid helps to prevent cardiovascular events such as heart attack or stroke
Food sources	Whole grain breads and cereals; dried beans; leafy green vegetables; bananas; meat, fish, and poultry
Other	Not all symptoms of vitamin B$_6$ toxicity are reversed after supplementation is withdrawn; exercise increases turnover and lose of vitamin B$_6$
	Vitamin B$_{12}$
Names	Cobalamin, cyanocobalamin
Major physiological functions	Synthesis of new cells; nervous system; red blood cell formation; activation of folate
Solubility	Water soluble
Deficiency disease	Not named
Symptoms of deficiency	Fatigue; nerve cell degeneration; numbness; lack of vitamin B$_{12}$ absorption results in pernicious anemia
Toxicity disease	None
Symptoms of toxicity	None known
Health promotion and disease prevention	Adequate intake daily is necessary for good health; lack of evidence that supplementation of vitamin B$_{12}$ alone or in combination with vitamin B$_6$ and folic acid helps to prevent cardiovascular events such as heart attack or stroke

(Continued)

Table 8.2 Summary of Fat- and Water-Soluble Vitamin Characteristics (Continued)

Vitamin B₁₂	
Food sources	Animal foods only; specially formulated yeast or other fortified foods
Other	Intrinsic factor (IF), which is produced in the stomach, is needed for vitamin B_{12} absorption in the intestine. Lack of IF may require vitamin B_{12} injections; injections do not boost energy in athletes
Folate	
Names	Folate (when found naturally in food) or folic acid (when added to fortified foods or in supplements)
Major physiological functions	Synthesis of new cells; red blood cell formation
Solubility	Water soluble
Deficiency disease	Not named
Symptoms of deficiency	Megaloblastic (large cell) anemia; depression; in pregnancy, increased risk for neural tube defects
Toxicity disease	None
Symptoms of toxicity	None known
Health promotion and disease prevention	Adequate intake daily is necessary for good health and to reduce the prevalence of neural tube defects in utero; lack of evidence that supplementation of folic acid alone or in combination with vitamins B_6 and B_{12} helps to prevent cardiovascular events such as heart attack or stroke
Food sources	Leafy green vegetables; whole grain breads and cereals; dried beans; bread and cereals made from processed grains or flour (fortified); orange juice
Other	Folate supplementation can mask the symptoms of vitamin B_{12} deficiency and can delay diagnosis
Pantothenic acid	
Names	No other name
Major physiological functions	Release of energy from carbohydrates, proteins, and fats via acetyl CoA
Solubility	Water soluble
Deficiency disease	Not named; rare
Symptoms of deficiency	Fatigue
Toxicity disease	None
Symptoms of toxicity	None known
Food sources	Widely distributed in food
Biotin	
Names	No other name
Major physiological functions	Release of energy from carbohydrates, proteins, and fats
Solubility	Water soluble
Deficiency disease	Not named; rare
Symptoms of deficiency	Fatigue; loss of appetite
Toxicity disease	None
Symptoms of toxicity	None known
Food sources	Widely distributed in food
Vitamin C	
Names	Ascorbic acid
Major physiological functions	Collagen synthesis; antioxidant; immune function; aids absorption of iron
Solubility	Water soluble

Table 8.2 Summary of Fat- and Water-Soluble Vitamin Characteristics (Continued)

	Vitamin C
Deficiency disease	Scurvy
Symptoms of deficiency	Poor wound healing; bleeding gums, small blood vessel hemorrhages
Toxicity disease	Not named
Symptoms of toxicity	Diarrhea, fatigue, kidney stones in some people
Health promotion and disease prevention	Adequate intake daily is necessary for good health; if dietary intake is low, increasing dietary sources of vitamin C may be beneficial for preventing coronary heart disease; vitamin C supplementation does not prevent colds in the general population but does reduce the duration and severity of the cold; some evidence that supplementation may prevent colds in those who engage in rigorous exercise, especially in cold temperatures
Food sources	Oranges, grapefruit, and other citrus fruits; strawberries, cabbage, broccoli, peppers, tomatoes
Other	Antioxidant that works independently of and in conjuction with vitamin E

Source: Gropper, Smith, and Groff (2009a).

Table 8.3 DRI and UL for Adult Males and Adult, Nonpregnant Females

	Dietary Reference Intakes (DRI)/d	Tolerable Upper Intake Levels (UL)/d
Vitamin A	700 mcg (females) 900 mcg (males)	3,000 mcg**
Vitamin D	5 mcg (ages 19 to 50) 10 mcg (ages 51 to 70) 15 mcg (over age 70)	50 mcg
Vitamin E	15 mg	1,000 mg***
Vitamin K*	90 mcg	Not established
Thiamin	1.1 mg (females) 1.2 mg (males)	Not established
Riboflavin	1.1 mg (females) 1.3 mg (males)	Not established
Niacin	14 mg (females) 16 mg (males)	35 mg
Vitamin B_6	1.3 mg (ages 19 to 50) 1.5 mg (females over 50) 1.7 mg (males over 50)	100 mg
Vitamin B_{12}	2.4 mcg	Not established
Folate	400 mcg	1,000 mcg
Pantothenic acid*	5 mg	Not established
Biotin*	30 mcg	Not established
Choline*	425 mg (females) 550 mg (males)	3,500 mg
Vitamin C	75 mg (females)**** 90 mg (males)****	2,000 mg

Legend: d = day; mg = milligrams; mcg = micrograms

*These values are based on Adequate Intake (AI). The remaining DRI are based on the Recommended Dietary Allowances (RDA). See Chapter 1 for further explanation.

**Refers to preformed vitamin A, not beta-carotene.

***Via supplements or fortified foods.

****Values are increased for smokers. Female smoker, 110 mg; male smoker, 125 mg.

studies of healthy people who are moderately active. A reasonable question raised by athletes is whether exercise that is greater than moderate intensity and/or duration, especially endurance and ultraendurance exercise, substantially changes vitamin needs. For example, do athletes need more of a particular vitamin to meet the demands of moderate or rigorous training? Conversely, does training bring about metabolic adaptations that result in more efficient use of vitamins? Do athletes lose more vitamins through sweat or urine than sedentary individuals? Unfortunately, these questions are not easily answered because the body of literature is small, and in some cases, nonexistent.

Moderate to rigorous exercise may increase the need for some vitamins, but the increase is small.

There are a number of ways in which exercise could alter vitamin requirements. These include (1) decreased absorption from the gastrointestinal tract, (2) increased loss via sweat or urine, (3) increased utilization due to the stress of exercise, or (4) increased need associated with large gains and maintenance of skeletal muscle mass. Alternatively, there are a number of adaptations the body can make to the stress of exercise that might preserve vitamins. For example, exercise may cause the body to decrease excretion or effectively recycle vitamins. Because the body has so many adaptive mechanisms in response to exercise, an increase in utilization does not necessarily mean an increase in dietary need.

Research in this area has been limited because it is difficult to conduct. To determine the effect of exercise on vitamin requirements both trained and untrained subjects are needed. These subjects would need to consume sufficient energy and controlled amounts of the vitamin being studied. Extensive measurements are required to determine the effect of training on vitamin absorption, excretion, metabolism, and utilization. It is not surprising that the body of scientific literature in this area is small (Akabas and Dolins, 2005).

The B vitamins, particularly vitamin B_6, thiamin, and riboflavin, have been the focus of much of the research in active populations. These vitamins are involved in a number of the chemical reactions of the energy systems in the body. A smaller number of studies have examined folate and vitamin B_{12}, vitamins involved in the synthesis of red blood cells. Woolf and Manore (2006) concluded that exercise appears to increase the requirements for vitamin B_6 and riboflavin, but not thiamin.

Specifically, exercise appears to increase the turnover and loss of vitamin B_6 and increase the need for riboflavin. The riboflavin studies were conducted in people who exercised for fitness (2.5 to 5 hours per

Athletes can improve their vitamin status by increasing their intake of fruits, vegetables, and whole grains.

week) rather than trained athletes, but a reasonable assumption is that strenuous exercise also increases the need for riboflavin. There have been too few studies of folate or vitamin B_{12} to draw conclusions about the effect of exercise on these vitamins.

Vitamins E, C, and A (as beta-carotene) act as antioxidants. Endurance and ultraendurance exercise increases the amount of oxidants (free radicals) produced. To reduce tissue damage from oxidation, more antioxidants are needed. A critical question is whether the requirements for antioxidant vitamins are increased as exercise duration increases. Of the three antioxidant vitamins, vitamin E is the most studied in athletes. Exercise reduces the utilization of vitamin E and also reduces vitamin E stores, but there is no evidence that athletes have deficiencies of vitamin E. This would suggest that stores are sufficient to meet the increased utilization of vitamin E in trained athletes (Manore, Meyer, and Thompson, 2009).

Although there is some evidence that exercise increases the demand for some vitamins, the demands can likely be met by consuming the amounts recommended by the Dietary Reference Intakes (DRI). When the DRI is established for each vitamin there is a built-in margin of safety. This margin of safety likely exceeds the increased demand associated with the exercise. It is not difficult for athletes, including endurance and ultraendurance athletes, to meet the DRI if they are consuming sufficient calories. However, caloric restriction is a barrier to meeting vitamin needs. In those consuming sufficient calories, low intake of vitamins can result from a low consumption of vitamin-rich foods, such as fruits, vegetables, and whole grains (Manore, Meyer, and Thompson, 2009; Volpe 2007). The bottom line is that exercise may increase need slightly, but dietary intake is a much greater influence on vitamin status.

Poor food choices by athletes and sedentary people often lead to low vitamin intake.

At the present time the effect that exercise has on vitamin requirements is presumed to be relatively small, and any small additional demand is adequately covered by the requirements set forth for sedentary humans. Perhaps the most striking feature that stands out in a review of the scientific literature in this area is not evidence of an increased demand for vitamins imposed by the stress of exercise, but the marginal dietary intake of vitamins that exists for some athletes as well as by sedentary adults.

Figure 8.1 shows the probability of adequate vitamin intake from food by Americans. For example, only 7 percent are likely to receive an adequate amount of vitamin E from their diets. Other vitamins of concern are vitamins A and C, which are likely to be adequate for 56 and 69 percent of Americans, respectively. An adequate amount of some vitamins, such as niacin and riboflavin, are consumed by the vast majority of Americans (Dietary Guidelines Advisory Committee, 2010). A dietary analysis would need to be performed for an individual to know the approximate amount of vitamins he or she consumes daily, but vitamins A, C, and E are typically of greatest concern. Low intake of vitamins A and C often reflect a low fruit and vegetable intake. Low intake of vitamin E is typically a result of a low consumption of oils and nuts. Athletes may also consume such foods in low amounts, so it is not unusual for athletes to fall short of some of the same nutrients as the sedentary population.

An important factor in obtaining excellent nutritional status is adequate energy intake. Many studies have shown that for both male and female athletes, inadequate energy intake is associated with inadequate nutrient intake (Jonnalagadda, Ziegler, and Nelson, 2004; Leydon and Wall, 2002; Papadopoulou, Papadopoulou, and Gallos, 2002; Ziegler et al., 2002; Ziegler, Nelson, and Jonnalagadda, 1999; Lukaski, 2004). Energy-restricted athletes such as female gymnasts, ballet dancers, wrestlers, and jockeys often consume low dietary intakes of one or more vitamins. Low vitamin E intake may reflect a low-fat diet, a common dietary pattern for athletes such as distance runners who must emphasize higher carbohydrate and protein consumption because of the demands of training and competition (Horvath et al., 2000).

Although energy intake is often associated with vitamin intake, it is not always a predictor. Sometimes energy intake is adequate, or even excessive, but vitamin intake is low. This is most likely a result of poor food choices. Soric, Misigoj-Durakovic, and Pedisic (2008) found that energy intake was adequate in the 9- to 13-year-old aesthetic athletes they studied, but that intake was low for some vitamins, particularly vitamin A and niacin. The best approach is to conduct a

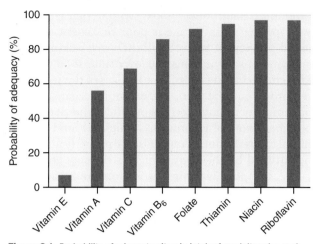

Figure 8.1 Probability of adequate vitamin intake for adult males and females

Many adults likely consume inadequate amounts of several vitamins, and in particular, low amounts of vitamins A, C, and E.

dietary analysis and determine the usual caloric and vitamin intake of the individual athlete.

It is important to guard against both vitamin deficiencies and toxicities.

Vitamin deficiencies do not occur overnight, especially in previously well-nourished adults. Any vitamin deficiencies will progress through stages—at first mild, then moderate, and ultimately, severe. Severe deficiencies are termed clinical deficiencies whereas mild and moderate deficiencies are called subclinical deficiencies. These terms describe indistinct points on a continuum. There are no clear-cut divisions between mild and moderate and moderate and severe deficiencies. The stages are outlined in Figure 8.2.

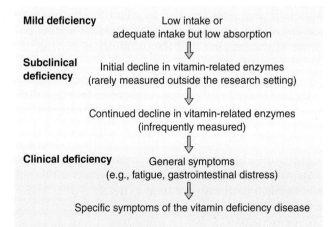

Figure 8.2 Stages associated with vitamin deficiency

Vitamin deficiencies progress from mild to moderate (subclinical) to severe (clinical). The progression through the mild- and moderate-deficiency stages is difficult to monitor and recognize.

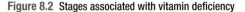

Developing mild deficiencies. Mild deficiencies can develop if vitamin intake is poor or absent. As discussed previously, many adult diets are lacking sufficient amounts of vitamins and, over time, mild vitamin deficiencies can develop. In a few cases the intake of a vitamin could be zero. For example, vegans do not consume any animal-derived products and their diets could be devoid of vitamin B_{12}, which is found only in animal foods. Vitamins must also be properly absorbed and utilized by the body, which is the case for most people, but poor vitamin absorption can be a consequence of some gastrointestinal diseases. Inadequate vitamin intake from the diet is one of the few factors associated with vitamin deficiencies that can be easily documented. The first step in determining a potential vitamin deficiency is an assessment of usual vitamin intake from food.

Developing subclinical deficiencies. If poor intake or absorption is not reversed or resolved, a subclinical deficiency can develop over time. With a subclinical vitamin deficiency, the person shows no medical signs of a disease. However, the lack of a particular vitamin can begin to affect functionality. For example, a lack of vitamin A ultimately results in blindness, a sign of a clinical vitamin A deficiency. However, long before blindness appears, problems with normal vision are present, such as difficulty in being able to see in the dark or adjusting vision to low-light conditions. Any subtle changes are hard to detect early, and in most cases there are few reliable tests to determine whether a subclinical vitamin deficiency exists (Tanumihardjo, 2004).

Vitamin status can be assessed through biochemical measurements of the vitamins or their metabolites, which are by-products of the vitamin's metabolism. These tests involve the analysis of blood or urine. Although such tests are used in scientific studies, they are not often performed outside a research setting because of the expense and the difficulty in getting accurate measurements. The amount of a vitamin in the blood does not reflect the amount in storage in the body and is typically a poor measure of functionality. Functionality can be determined by measuring an enzyme that contains a certain vitamin, but no such tests have been developed for several vitamins (Manore, Meyer, and Thompson, 2009). Even when laboratory tests are available, small changes in vitamin-related biochemical pathways are hard to detect. For all these reasons, biochemical assessment of vitamins is not often practical.

An important question is how prevalent are subclinical vitamin deficiencies? That is a difficult question to answer because of a lack of studies, particularly of athletes. Nutrition scientists suspect that some people in the general population have subclinical vitamin deficiencies because vitamin intakes are well below recommendations. However, subclinical vitamin deficiencies have been well documented for only a few vitamins.

Researchers have only scratched the surface in identifying subclinical vitamin deficiencies in athletes. For example, athletes who substantially restrict caloric intake over time are at risk for developing subclinical deficiencies of vitamin B_6, thiamin, and riboflavin (Manore, Meyer, and Thompson, 2009). The few studies that have been conducted in athletes suggest that the prevalence of a subclinical vitamin D deficiency may be as high as 40 percent (Willis, Peterson, and Larson-Meyer, 2008). Until more research is conducted, it is hard to know how many athletes may have subclinical vitamin deficiencies.

Developing clinical deficiencies. As has been shown throughout human history, the lack of a particular vitamin can lead to a vitamin deficiency disease. Several vitamin deficiency diseases were common and widespread throughout the world, including the United States, until the 1950s. For example, beriberi is a disease of the nervous system caused by a deficiency of thiamin that has essentially been eliminated in the United States. Rickets, a bone malformation disease in children caused by a lack of vitamin D, has been reported throughout history and is on the rise again in the United States and other industrialized countries.

With a clinical vitamin deficiency, the person shows medical signs of a disease. For example, someone suffering from pellagra, which is due to a lack of the B vitamin niacin, would have changes in skin similar to sunburn, a red and swollen tongue, diarrhea, and mental confusion. In the United States and other developed nations, clinical vitamin deficiency diseases are rare because of an abundant food supply, vitamin fortification of foods, and use of vitamin supplements. The chances of young or middle-aged, active U.S. adults manifesting a clinical vitamin deficiency are slim. One exception, which may apply to some athletes, is an individual with a severe eating disorder (see Chapter 12). A clinical vitamin deficiency that is due to poor intake can be treated and reversed by administering the missing vitamin, usually in the form of a supplement.

Developing toxicities. Vitamin deficiencies do not occur overnight; neither do vitamin toxicities. Most vitamin toxicities take months or years to develop. Initially, the symptoms of vitamin toxicity are vague—a general feeling of **lethargy**, also known as **malaise**. With continued exposure to high amounts, more specific symptoms emerge, often related to major organ systems such as the liver or nervous system. Vitamin toxicities are rare but they can occur and have been reported for vitamins A, D, and B$_6$ (Ramanathan et al., 2010; Heaney, 2008; Bendich, 2000).

Vitamin A toxicity, known as **hypervitaminosis A**, provides an excellent example of the many issues related to vitamin overdose. Myhre et al. (2003) conducted a meta-analysis of all vitamin A toxicities reported in the medical literature between 1944 and 2000. Recall from Chapter 1 that a meta-analysis is a powerful statistical method used to compare similar research studies and review the collective body of scientific research on a particular topic. These researchers found that there were 259 reported cases of vitamin A overdose worldwide. The largest number of cases, 105, was from the United States. This meta-analysis confirms that although the number of cases is extremely small, vitamin A toxicity does occur.

The researchers found evidence of hypervitaminosis A with both acute (short-term) and chronic (long-term) excessive intake of **retinol**, a type of preformed vitamin A. Those who developed an acute toxicity did so within a few weeks. In contrast, the chronic toxicity was the result of months or years of excessive ingestion. The key finding was the form of the supplement consumed. Those who developed an acute toxicity took water-**miscible**, **emulsified**, and solid preparations, whereas those who developed a chronic toxicity consumed an oil-based supplement. Retinol in water-miscible, emulsified, and solid preparations is very readily absorbed; faster absorption led to acute toxicity. Therefore, the form of the vitamin supplement can be important as well as the dose consumed.

The widespread use of multivitamin supplements in the United States raises the question of whether the consumption of such supplements results in toxicities. The accumulated data from U.S. Poison Control Centers indicate that the ingestion of multivitamins (including some that contained the mineral iron) represents ~2.8 percent of the total number of human exposures reported. Many of these cases were in children under the age of 6 and it is very important that supplements of any kind are kept away from children. Vitamin toxicities do occur, although serious health consequences, especially in adults, are rare.

Once vitamin toxicity is diagnosed, the usual treatment is to curtail the use of that supplement. When the body is no longer exposed to high doses, tissue concentrations decrease over time and symptoms generally subside. The best prevention of vitamin toxicities is for individuals to not consume doses higher than the Tolerable Upper Intake Levels, which are discussed in more detail later in this chapter.

Key points

- The Dietary Reference Intakes (DRI) are a guideline for the amount of vitamins that people need each day.
- The Tolerable Upper Intake Level (UL) is a guideline for avoiding excessive intake.
- Moderate and rigorous training increases the need for only a few vitamins (for example, vitamin B$_6$ and riboflavin).
- The increased need for vitamins with exercise training is small and likely met by consuming the DRI, which includes a margin of safety.
- Vitamin deficiencies and toxicities are typically slow to develop.
- Subclinical deficiencies are hard to identify.
- Avoid the extremes—too little or too much causes problems.

What tools do athletes and those who work with them have to determine if an athlete may be at risk for vitamin deficiencies?

8.2 The Roles of Vitamins in the Body

Each vitamin has a unique chemical composition as well as specific biochemical roles. Many vitamins are involved with enzymatic activity, particularly as part of a coenzyme. Enzymes are proteins that regulate metabolic reactions. Some enzymes depend solely on their protein structure to function, but many require a cofactor, which helps the enzyme to be more stable. The cofactor may be a vitamin or a mineral ion. When the enzyme contains a cofactor it is referred to as a coenzyme. Coenzymes are usually involved in speeding up a reaction, typically by transferring a functional group as substrates are converted to other compounds.

All B vitamins act as coenzymes, although each is involved in very specific reactions. For example, riboflavin (vitamin B$_2$) is part of a coenzyme that transfers hydrogen whereas vitamin B$_{12}$ helps to transfer amino groups. Many of the coenzymes are involved with energy metabolism, thus the B vitamins are often

Lethargy: Physically slow or mentally dull.

Malaise: A general feeling of sickness but a lack of any specific symptoms.

Hypervitaminosis: Excessive intake of one or more vitamins.

Retinol: Preformed vitamin A.

Miscible: Two or more liquids that can be mixed together.

Emulsified: Suspending small droplets of one liquid in another liquid, resulting in a mixture of two liquids that normally tend to separate, for example, oil and water.

associated with the production of energy. Vitamin C also acts as a coenzyme but primarily in reactions that involve synthesis. For example, three reactions necessary for collagen synthesis involve vitamin C.

Although many vitamins function as cofactors, vitamins are not only associated with coenzymes. Some have chemical properties that make them important compounds in complex processes. For example, vitamin E, vitamin C, and beta-carotene have antioxidant properties. These vitamins can interact with compounds that have oxidant activity and reverse the oxidation. At the biochemical level, vitamins often have chemical properties that are necessary for reactions to continue or for complex processes to work optimally.

Several vitamins are involved in the differentiation, growth, development, and maintenance of cells. For example, vitamin A is needed for proper cell differentiation, the process by which an immature cell becomes a specific type of cell. Folate and vitamin B_{12} are needed for normal cell division, and vitamin D is necessary for proper bone formation. These growth and development functions are one reason that vitamin deficiencies result in specific diseases.

It is easy to focus on the individual roles that vitamins play but vitamins also interact with each other, allowing another vitamin or mineral to function optimally. Folate and vitamin B_{12} are both involved in the metabolism of the amino acid methionine. Vitamin C interacts with the minerals iron and copper, allowing for their optimal absorption.

Vitamin functionality is such a large topic that it is hard to generalize. In fact, biochemistry texts typically do not have a separate chapter for vitamins because they are so intimately involved in the biochemical reactions that take place in the body. However, grouping together vitamins that have similar functions helps to make a big topic more manageable and understandable. To that end, the discussion of vitamins that follows focuses on their roles in energy metabolism, as antioxidants, in red blood cell formation, and in growth and development.

Nutrient-dense carbohydrate-containing foods, such as breads, cereals, and grains, are excellent sources of thiamin.

Some of the B-complex vitamins are associated with energy metabolism.

Thiamin (B_1), riboflavin (B_2), niacin (B_3), vitamin B_6, pantothenic acid, and biotin are often referred to as the B-complex vitamins. These vitamins are primarily involved in the production of ATP as they are part of the enzymes that regulate these reactions. Table 8.4 lists some of the vitamins and their associated coenzymes and biochemical pathways, whereas Figure 8.3 highlights the vitamin-containing compounds involved in energy metabolism.

Thiamin. Thiamin (B_1) is part of a coenzyme involved in the release of energy from carbohydrates, proteins, and fats. The majority of the thiamin in the body (~80 percent) is found as thiamin diphosphate (TDP), which is also known as thiamin pyrophosphate (TPP). TDP **catalyzes** reactions involving pyruvate and a-ketoglutarate (Figure 8.3) and the branched chain amino acids (leucine, isoleucine, and valine). TDP is one of several enzymes needed in these biochemical

Table 8.4 Vitamins and Associated Coenzymes

Vitamin	Coenzyme	Biochemical pathway
Thiamin (B_1)	Thiamin pyrophosphate (TPP); also known as thiamin diphosphate (TDP)	Decarboxylation (removal of –COOH group to form CO_2) of pyruvate and alpha-ketoglutarate
Riboflavin (B_2)	Flavin mononucleotide (FMN); flavin adenine dinucleotide (FAD)	Numerous oxidation-reduction reactions (can accept and release a pair of hydrogen atoms)
Niacin (B_3)	Nicotinamide adenine dinucleotide (NAD and NADH); nicotinamide adenine dinucleotide phosphate (NADP and NADPH)	NAD and NADH transfer electrons in the electron transport chain; NADP and NADPH are involved in reduction reactions in many parts of the cell
Vitamin B_6	Pyroxidoxal phosphate (PLP)	Needed for amino acid metabolism, including transamination, (pyridoxine) transferring and removing sulfur, cleavage, and synthesis
Pantothenic acid	Component of acetyl CoA	Critical intermediate compound in energy production (carbohydrates, proteins, and fats)

Source: Gropper, Smith, & Groff, (2009a)

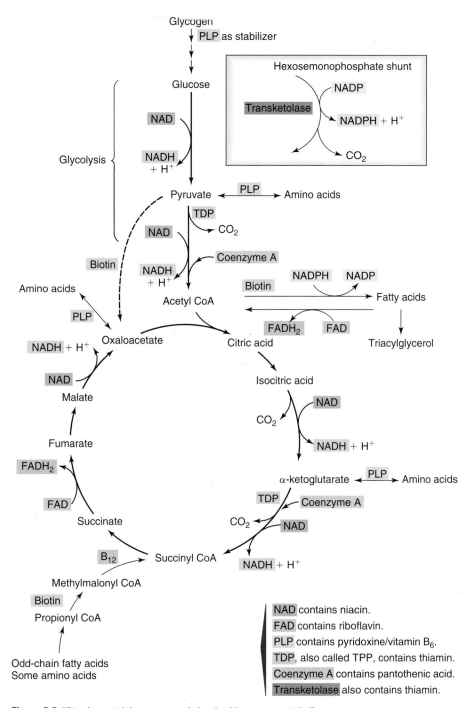

Figure 8.3 Vitamin-containing compounds involved in energy metabolism

reactions. Half of the thiamin in the body is found in skeletal muscle, where many of these reactions occur.

Given the link between thiamin and ATP production, a logical question is whether an increase in thiamin intake above that which is normally needed could result in an increase in TDP and energy production. An athlete might ask, "Does taking more thiamin give me more energy?" The answer is no. Consuming the recommended amount of thiamin likely results in tissues being saturated with thiamin. Enzymes and coenzymes function at maximum velocity when saturated with

a substrate. After the point of saturation, more substrate will not result in a greater number of enzymes or greater speed. When enzymes and tissues are saturated with water-soluble vitamins, any excess is excreted in urine. Energy production is not increased with excess thiamin intake because the extra thiamin is not utilized and is excreted when it exceeds the point of saturation.

Catalyze: Increase the rate of, such as speeding up a chemical reaction.

Riboflavin is found in a wide variety of foods, with dairy products being excellent sources.

Green leafy vegetables and protein-containing foods are excellent sources of vitamin B_6.

Excellent sources of niacin include protein-containing foods such as chicken, tuna, and pork.

Athletes in training produce more ATP than non-athletes, so another logical question is whether athletes need more thiamin than sedentary individuals. Although the number of studies is limited, it does not appear that exercise increases the need for thiamin. Dietary intake studies have found that most athletes consume a sufficient amount of thiamin. When energy expenditure is high, many athletes focus on consuming high-quality carbohydrates, such as breads, cereals, and grains, to ensure that muscle glycogen is adequately resynthesized. In doing so, they consume sufficient thiamin and meet or exceed the DRI for this vitamin. Low thiamin intake in athletes is typically associated with caloric restriction and consumption of low-nutrient-dense carbohydrates, such as foods or beverages high in sugar and low in fiber and vitamins (Woolf and Manore, 2006).

Riboflavin. Riboflavin is part of two coenzymes involved in ATP production, flavin mononucleotide (FMN) and flavin adenine dinucleotide (FAD). These coenzymes are necessary for the numerous oxidation-reduction reactions that occur because they can accept or release hydrogen atoms (Figure 8.3). The synthesis of these enzymes is under hormonal control. When tissues are saturated with riboflavin the excess is excreted in the urine. Exercise increases the need for riboflavin (Woolf and Manore, 2006). Most athletes consume sufficient riboflavin, although there have been reports of low intake by athletes who consume too few kilocalories. Riboflavin is found in a wide variety of foods such as breads and cereals, vegetables, meat, and dairy products such as milk, and athletes consuming a sufficient amount of energy would not likely be deficient.

Niacin. Niacin is part of two coenzymes, nicotinamide adenine dinucleotide (NAD) and nicotinamide adenine dinucleotide phosphate (NADP). These coenzymes are involved in numerous reactions and are part of almost 200 enzymes in the body. NAD/NADH play critical roles in the production of ATP (Figure 8.3), including the transfer of electrons in the electron transport chain. In addition, NADP/NADPH is needed in a variety of reactions such as synthesis of fatty acids and the oxidation of the amino acid glutamate. Many cellular reactions involve niacin as a cofactor.

Severe or prolonged deficiencies could occur, but evidence of subclinical deficiencies affecting performance is lacking. Athletes who consume sufficient energy also consume sufficient niacin (Lukaski, 2004). Excellent sources of niacin include protein-containing foods such as chicken, tuna, and pork. Cereal, because it has niacin added, is also high in niacin.

Vitamin B_6. Vitamin B_6 (pyridoxine) is involved in amino acid metabolism and the release of glucose from glycogen (Figure 8.3). Pyroxidoxal phosphate (PLP) is one of the coenzymes that catalyze reactions that transform amino acids, including amino acid cleavage and transamination. For example, the synthesis of the conditionally indispensable amino acid cysteine from the indispensable amino acid methionine requires PLP. PLP is also part of the enzyme that helps release glucose from glycogen stores, and most of the vitamin B_6 in skeletal muscle is found as PLP.

Exercise increases the need for vitamin B_6 because both the turnover and the loss of vitamin B_6 are increased. The current DRI for vitamin B_6 is 1.3 mg for those ages 19 to 50 and 1.5 or 1.7 mg for females and males older than 50, respectively. Some vitamin B_6 researchers suggest athletes in training need 2.0 to 3.0 mg daily. Limited studies suggest that a vitamin B_6 deficiency may impair exercise capacity (Woolf and Manore, 2006).

Vitamin B_6 is found in both plant and animal sources. Not surprisingly, given the vitamin's role in amino acid metabolism, protein-containing foods such as meat, fish, and poultry are good sources of vitamin B_6. Other good sources include whole grains, bananas, and green leafy vegetables. Dietary intake studies suggest that some athletes have low vitamin B_6 intake, especially those with low caloric intake. However, vitamin B_6 intake may be low even in those with sufficient caloric intake if certain food groups are not consumed (Woolf and Manore, 2006). For example, some athletes eliminate meat and others have a very low leafy green vegetable intake. Vitamin B_6 differs from thiamin, riboflavin, and niacin in that it is not one of the vitamins included in the fortification of grain products. The lack of intake and the increased need are two reasons that vitamin B_6 status may be compromised in some trained athletes, despite an adequate caloric intake.

Pantothenic acid and biotin. Many people are somewhat familiar with thiamin, riboflavin, niacin, and vitamin B_6 but may be unfamiliar with the remaining B-complex vitamins, pantothenic acid and biotin. Although these vitamins may be mentioned less often, they are almost always included in vitamin B-complex supplements. Pantothenic acid is part of coenzyme A (CoA), an important compound in aerobic metabolism. Like other enzymes, acetyl CoA is made up of two parts, acetate (also known as acetic acid) and coenzyme A. Pantothenic acid is intimately involved in energy metabolism as a part of this compound, but large amounts of pantothenic acid do not increase the rate of energy reactions. Biotin is also involved in a number of energy-related reactions (Figure 8.3).

Not surprisingly, pantothenic acid and biotin are widely found in food. If vitamins that are so universally involved in energy metabolism were concentrated in a few select foods, it is likely that deficiencies would occur more frequently. However, pantothenic acid and biotin deficiencies in humans are extremely rare because nearly all foods contain some of these compounds.

To summarize, the B-complex vitamins are important nutrients because they are part of the enzymes that catalyze the biochemical reactions associated with energy (Table 8.4 and Figure 8.3). They play a critical role in energy metabolism but, by themselves, do not provide "energy" (see Spotlight on...Vitamins and "Energy"). Excessive amounts of the B vitamins

Spotlight on...

Vitamins and "Energy"

Vitamins play critical roles in energy metabolism but they are *indirect* roles. The biological energy that is needed to perform work is provided by carbohydrates, proteins, fats, and alcohol and this energy is measured in kilocalories. Vitamins contain no energy but facilitate the production of energy. Why then do people say "vitamins give me energy"? Why do ads for vitamin supplements proclaim "vitamins for energy"? One reason is that a *clinical* vitamin deficiency of one or more of the B-complex vitamins results in physical fatigue. Theoretically, a subclinical deficiency of any one of the B-complex vitamins could also result in fatigue, although proof of this is lacking for several vitamins. Broad statements linking vitamins to energy are correct, but it needs to be understood that the roles vitamins play in energy metabolism are not simple, direct, or independent; rather, they are complicated, indirect, and in partnership with other compounds. The vitamin/energy claims are typically overstated because they are out of context for the athlete who is unlikely to have a clinical vitamin deficiency.

Athletes can be fatigued for a number of nutrient-related reasons, including inadequate energy (caloric) intake, lack of

sufficient carbohydrates, hypohydration (insufficient volume of body water), and subclinical B-complex vitamin deficiencies. If the reason is a lack of B-complex vitamins as part of a low caloric intake, then consuming more nutritious foods will likely resolve both the low energy and vitamin consumption. If caloric intake is sufficient but B-complex vitamin intake is low (a scenario that is not as likely but could exist due to the high intake of foods that are calorie dense but not nutrient dense), increasing consumption of foods containing B vitamins is a logical first step. B-complex vitamin supplements could also be considered. If an athlete's vitamin status has been compromised (for example, poor long-term vitamin intake due to an eating disorder) and a subclinical deficiency is suspected, consumption of appropriate amounts of the deficient vitamins will restore normal levels and will address vitamin-related issues of fatigue. The consumption of B-complex vitamins above and beyond recommended amounts in an athlete who is not deficient does not provide "extra" energy.

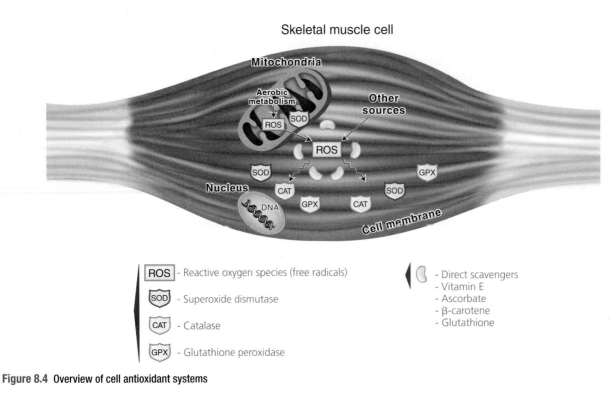

Figure 8.4 **Overview of cell antioxidant systems**

Reactive oxygen species (ROS) or free radicals formed in skeletal muscle cells by aerobic metabolism and from other sources have the potential to damage DNA, cell membranes, and other cellular structures. Antioxidant enzymes such as superoxide dismutase (SOD), catalase (CAT), and glutathione peroxidase (GPX) interact chemically to neutralize the free radicals and protect cellular structures. Direct free radical scavengers such as vitamin E, ascorbate, beta-carotene, and glutathione are also part of the cell's protective, antioxidant system.

do not result in more enzymes or greater enzymatic activity. A clear theme has emerged—consuming sufficient energy (kilocalories) and eating a variety of nutrient-dense foods usually ensures that enough of the vitamins needed to metabolize the energy will be ingested. Athletes who are at risk for low intake and potential subclinical deficiencies are those who restrict their caloric intake, eliminate certain food groups, and consume foods that are nutrient poor (for example, high in sugar and low in vitamins).

Some vitamins have antioxidant properties that help protect cells from damage.

Oxygen is needed to produce ATP from carbohydrates, fats, or proteins by our aerobic energy pathways, that is, oxidative phosphorylation. The majority of oxygen used in oxidative phosphorylation reactions is reduced to water but a small percentage (~4–5 percent) is not. Instead, free radicals are produced. Free radical is a broad term that includes **reactive oxygen species** (ROS), such as ozone and superoxide radicals, and **reactive nitrogen species** (RNS), such as nitric oxide. Free radicals are unstable chemical compounds that can destroy cells by damaging cellular membranes, proteins, and DNA. These compounds

will always be present because they are part of normal physiological processes, including exercise. In fact, some free radicals help to destroy bacteria and other foreign particles, so they can have beneficial effects. However, in excessive amounts free radicals have many detrimental effects and the body has several mechanisms to counteract them. The key issue is that there is a balance between rate of production and rate of clearance. When the balance favors the overproduction of free radicals then **oxidative stress** occurs, which can lead to damaged cells, tissues, and organs.

The formation of free radicals is directly related to exercise intensity and duration. As exercise intensity and duration increases, the body makes greater use of the aerobic energy system (oxidative phosphorylation), and as a result, more free radicals are formed. Therefore, endurance and ultraendurance exercise has the potential to produce more tissue-damaging free radicals. A positive result of aerobic exercise training, however, is a buildup of the body's natural defenses against free radicals, both the enzymatic and nonenzymatic antioxidants (Jackson et al., 2004).

Each cell has many antioxidant systems located predominantly in the cell membrane, cytoplasm, and mitochondria as shown in Figure 8.4. These antioxidant systems rely on sufficient amounts and types of vitamins, such as vitamins E, C, and A as beta-carotene. They also depend on mineral, including selenium, zinc, iron, and copper. One protective mechanism is the conversion of a free radical to a compound that can be disposed of safely. This is one way that vitamin C helps to counteract reactive oxygen species. Another mechanism is the activity of antioxidant enzymes. These enzymes contain mineral cofactors. Both vitamin E and beta-carotene can interact directly with free radicals and break the chain of reactions that lead to cellular damage. Thus the body has various antioxidant mechanisms, and many of them depend directly or indirectly on vitamins and minerals.

In the process of acting as an antioxidant, the original antioxidant donates an electron in an effort to stabilize the original free radical. The original antioxidant becomes a new free radical because it now has an unpaired electron. Therefore, antioxidants must also be regenerated. Vitamin C helps to regenerate vitamin E and niacin helps to regenerate vitamin C (Gropper, Smith, and Groff, 2009b). This underscores a point made earlier about balance. This balance is not only between the antioxidants and the free radicals but also among the antioxidant vitamins themselves.

Endurance and ultraendurance athletes engage in rigorous training. One concern is that the production of free radicals may outstrip the body's ability to defend against them, leading to oxidative stress (Williams et al., 2006; Evans, 2000; Powers and Lennon, 1999). Another concern is a low intake of antioxidant vitamins, which diminishes the body's ability to counteract the harmful effects of oxidation or repair the damage. For these reasons, many endurance and ultraendurance athletes consider taking supplements containing vitamin A (as beta-carotene), C, and/or E.

McGinley, Shafat, and Donnelly (2009) conducted a review of the literature to answer the question, does antioxidant vitamin supplementation protect against muscle damage? Their review was limited to vitamin C and E supplements. They found little evidence that these supplements protect against muscle damage and some emerging evidence of interference with some of the beneficial signaling functions associated with reactive oxygen species. Ristow and colleagues (2009) found that vitamin C and E supplementation blocked the positive effects that exercise had on insulin sensitivity. More studies are needed to determine positive and/or negative effects of antioxidant supplements in those who exercise.

Williams et al. (2006) reviewed the results of 41 studies of antioxidant supplements (primarily vitamins E, C, or A as beta-carotene) conducted in endurance athletes prior to 2005. Of the 47 trials (some studies tested more than one supplement), 20 found that exercise-induced oxidative stress was decreased with supplementation of antioxidant vitamins, whereas such supplements had no effect in 23 trials. Four trials reported an increase in oxidative stress. Half of the 20 trials that showed a decrease in oxidative stress involved vitamin E supplements. The authors of both reviews concluded that the results of antioxidant supplement studies in endurance athletes are **equivocal** (difficult to interpret) and state that there is currently insufficient evidence to recommend antioxidant supplements to endurance athletes. However, there is one area of scientific agreement—all athletes should consume a diet that is rich in antioxidant-containing foods, particularly fruits, vegetables, and whole grains (McGinley, Shafat, and Donnelly, 2009; Williams et al., 2006).

One of the more intriguing aspects of this type of research is the reported increase in oxidative stress in some athletes who use antioxidant supplements. Scientists have been trying to determine why antioxidant supplements might produce detrimental effects. One theory is that antioxidant vitamins in high concentrations act as **pro-oxidants**. Pro-oxidants increase the formation of free radicals and enhance oxidative damage. In the case of vitamin C, pro-oxidant activity can occur at high concentrations but the effect is indirect since the vitamin C reacts with copper and iron, which then interact with other compounds to form free radicals. The concentration of vitamin C in the blood is one of at least three factors that determines if the vitamin C acts as a pro-oxidant or an antioxidant (Li and Schellhorn, 2007).

Viewed on a continuum, both low and high amounts of antioxidant vitamins may be detrimental, but for different reasons. Insufficient or excessive amounts likely upset the balance of antioxidants. Until scientists can better clarify the complex mechanisms, a prudent approach is to obtain a sufficient amount of all the antioxidant vitamins from food sources (see Spotlight on...Antioxidant Vitamins and Health).

Reactive oxygen species: Oxygen ions, free radicals, and peroxides that are highly reactive because of the presence of unpaired electrons.

Reactive nitrogen species: Free radicals (and some nonradicals) that contain nitrogen and are highly reactive.

Oxidative stress: Damage to cells, organs, or tissues due to reactive oxygen or nitrogen species.

Carotenoid: A precursor to vitamin A, characterized by an orange or red pigment.

Equivocal: Open to more than one interpretation; difficult to interpret or understand.

Pro-oxidant: Compound that increases the formation of reactive oxygen species or free radicals.

Exploring Free Radical Production during Exercise, Muscle Damage, and Antioxidant Supplementation

Oxidative stress, or an increase in the production of free radicals, occurs with an increase in exercise intensity and duration. Because of the potentially damaging effect free radicals may have, muscle damage may be higher under conditions where oxidative stress is increased and more free radicals are produced. Because of the dramatic increase in the use of the oxidative phosphorylation energy system during aerobic exercise, this type of exercise has typically been studied in relation to oxidative stress and muscle damage. Prior to this study, little attention was given to other types of exercise, such as shorter-duration, high-intensity exercise, for example, strength or resistance training, and their effect on free radical formation and muscle damage.

A response of the body to the stimulus of increased oxidative stress is the enhancement of the body's antioxidants defense systems. There are several vitamins that have antioxidant properties, but vitamin E is particularly powerful. Vitamin E protects cells from damage by protecting cell membranes. It is not clear if damage caused by free radicals can be further reduced by dietary supplementation of antioxidant vitamins, such as vitamin E.

A research study that was the first to explore the effect of strength training (resistance exercise) on free radical production and the potential antioxidant effects of vitamin E supplementation was:

McBride, J. M., Kraemer, W. J., Triplett-McBride, T., & Sebastianelli, W. (1998). Effect of resistance exercise on free radical production. *Medicine & Science in Sports & Exercise*, *30*(1), 67–72.

The purpose of this research study was twofold: (1) to determine if high-intensity strength training increased the production of free radicals, and (2) to see if dietary supplementation with an antioxidant vitamin (vitamin E) would reduce free radical formation and associated muscle damage. Therefore, the study was designed to have subjects complete an intense strength-training workout, before and after which measurements were made to determine free radical formation and muscle damage. In studies such as these it is not possible to measure free radical formation in muscle cells directly, so the researchers measured the amount of malondialdehyde (MDA) in the blood, which is a marker or indicator of free radicals in the muscle. They also measured the amount of creatine kinase in the blood, which is a marker for muscle cell damage, particularly for disruption of the muscle cell membrane.

What did the researchers do? The subjects recruited for this study were young males who had at least 1 year of strength-training experience and no excessive vitamin E intake in the previous 6 months. After warming up, all subjects completed an intense strength-training session consisting of eight exercises that involved muscle groups in both the upper and lower body. Ten repetitions of each exercise were performed and each exercise was repeated three times in a circuit-training fashion with a minimal amount of rest between each set. Blood samples were taken before, immediately after the strength training session, and 6, 24, and 48 hours after the exercise.

In the 2 weeks before the strength-training session, the researchers had the subjects complete a diet history questionnaire to make sure that subjects did not have an unusual intake of vitamin E. They kept a diet diary for 3 days, which was analyzed to more accurately quantify average vitamin E intake. Subjects were randomly assigned to either a vitamin E supplementation group or a placebo group, with six subjects in each group. The 2-week dietary supplementation was 992 mg/day of vitamin E, which is an amount that approaches the Tolerable Upper Intake Level for this vitamin.

What did the researchers find? As seen in the first figure, the results indicate significantly elevated MDA levels in the blood immediately, 6, and 24 hours after the intense strength-training session. The elevated malondialdehyde levels indicate free radical production in excess of the body's ability to remove them. The second figure illustrates an increase in creatine kinase in the blood following the resistance exercise bout, which is an indicator that some muscle damage resulted from the intense strength-training session. This figure also illustrates an important difference found for the subjects taking the vitamin E supplement compared to the placebo group. The amount of creatine kinase in the blood 24 hours after the exercise was significantly lower for the vitamin E group than for the placebo group, which is seen as being an indicator of less muscle damage.

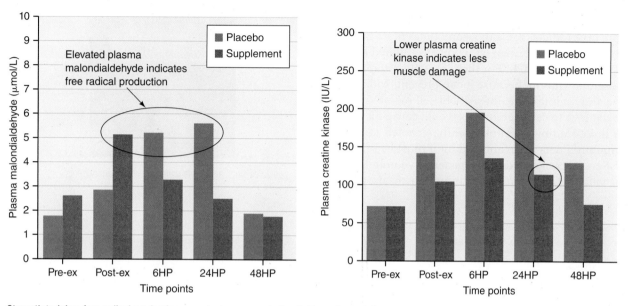

Strength training, free radical production, muscle damage, and vitamin E supplementation

Malondialdehyde (MDA) in the blood was elevated immediately, 6 hours and 24 hours after an intense strength-training session, which is a marker for increased free radical production. Subjects who consumed a vitamin E supplement for 2 weeks prior to the strength-training session had lower creatine kinase levels in the blood 24 hours after the exercise, which is an indication of a lesser degree of muscle damage.

Redrawn from McBride et al. (1998) with permission.

What was the significance of this research study? The McBride et al. (1998) research study was the first to show evidence of increased free radical production with shorter-duration, high-intensity exercise, which had primarily been studied during aerobic exercise. Whereas free radical production in aerobic exercise is thought to be a result of greater use of the oxidative phosphorylation energy system, these researchers theorized that in this study it was due to ischemia-reperfusion—the interruption and return of blood flow to muscle cells that occurs with the repetitive high-intensity muscle contractions used in resistance training. This study also provided evidence that the body's defense against free radicals may be enhanced, reducing muscle cell membrane damage with dietary vitamin E supplementation.

Answering the questions: Does exercise cause an increase in free radical production? Do athletes need to take vitamin E supplements? It is apparent from a number of research studies that free radical production increases with a variety of types of exercise, and McBride et al. (1998) demonstrated this for strength training. The main factor in oxidative stress appears to be the intensity and duration rather than the specific type of exercise. The increase in oxidative stress may not automatically result in an increase in muscle damage, as antioxidant defenses against free radicals are enhanced to some extent with exercise training. Although the results of the McBride et al. (1998) study suggested that less muscle damage occurred as a result of taking a vitamin E supplement, an analysis of a number of similar studies completed since that time suggests that there is not enough scientific evidence to definitively recommend vitamin E supplementation for this purpose. However, there is no question athletes should consume a diet that contains plenty of fruits, vegetables, and whole grains—all antioxidant-containing foods.

Vitamin E. Vitamin E is the primary antioxidant found on or near cell membranes. Cell membranes contain a high proportion of polyunsaturated fatty acids, which can be oxidized by a variety of free radicals. The destruction of the lipid in the cell membrane is a chain reaction requiring several steps. Vitamin E's role is to break the chain reaction. Free radicals are a thousand times more likely to react with vitamin E than with polyunsaturated fatty acids (Viitala et al., 2004). For these reasons, vitamin E is an essential vitamin because it can prevent oxidative damage and maintain the integrity of cell membranes. Severe damage to cell membranes would cause the cell to leak fluid, leading to cellular death. As noted earlier, once vitamin E acts as an antioxidant it must be regenerated. Several compounds can regenerate vitamin E including vitamin C. The ratio of vitamin E to unsaturated

Fruits, vegetables, nuts, and whole grains are rich in antioxidants.

Spotlight on…

Antioxidant Vitamins and Health

A frequent health recommendation is to consume at least five servings of fruits and vegetables daily. However, more than 75 percent of the adult population in the United States fails to meet this minimum recommendation (2010 Dietary Guidelines, http://www.cnpp.usda.gov/Publications/DietaryGuidelines/2010/DGAC/Report/D-2-NutrientAdequacy.pdf [pp. D2–D16]). In some cases, people turn to supplements, particularly to the antioxidant vitamins A, C, and E. Chun et al. (2010) estimated the intake of antioxidant vitamins in U.S. adults between 1999 and 2002. Average intake of vitamin C was 208 mg daily. Of this, 46 percent (96 mg) was obtained from food and 54 percent (112 mg) was in the form of a supplement. Vitamin E intake was 20 mg with the majority (64 percent) coming from supplements. In contrast, the majority of beta-carotene came from food (86 percent) not supplements.

Do foods that contain antioxidants, such as fruits and vegetables, improve health? Table 8.5 summarizes some of the meta-analyses published regarding the role of fruit and vegetable intake on health. The strongest association between a fruit and vegetable intake of at least 5 servings per day and disease prevention is the reduced risk of coronary heart disease and stroke. Although it was once thought that fruit and vegetable intake was strongly associated with reducing colon cancer risk, more recent research suggests that there is not a strong correlation. At present, it does not appear that the risk for developing type 2 diabetes is lowered with fruit and vegetable intake. None of these studies have shown that fruit and vegetable intake is detrimental.

Do antioxidant supplements improve health? Table 8.6 summarizes the effects of antioxidant

Table 8.5 Effect of Fruit and Vegetable Intake on Health

Study	Condition	Effect of five servings of fruits and vegetables
Koushik et al., 2008	Cancer	Not strongly associated with lowering overall colon cancer risk but may lower risk of cancer in the distal (lowest part) colon
He et al., 2007	Coronary heart disease	Risk is reduced especially when a low intake (less than three servings/day) is increased to five servings/day
He, Nowson, and MacGregor, 2006	Stroke	Strong association between intake of five or more servings daily and major reduction in the incidence of stroke
Hamer and Chida, 2007	Development of type 2 diabetes	Three or more servings/day not associated with reduced risk of type 2 diabetes

Note: All studies are meta-analyses, which combine the results of similar studies for analysis thereby improving the statistical strength of the research.

fatty acids in cell membranes is estimated at 9:1,000–2,000. Therefore, a very important part of the body's defense mechanisms is the regeneration of vitamin E (Gropper, Smith, and Groff, 2009b).

The Dietary Reference Intake for Vitamin E for adults is 15 mg daily. Excellent dietary sources of vitamin E include oils, seeds, nuts, whole grains, and vegetables. Deficiencies are rare and are associated with fat malabsorption syndromes. Limited surveys of endurance and ultraendurance athletes suggest that vitamin E intake is low in both males and females. This is a result of a low caloric intake and a low fat intake (Tomten and Høstmark, 2009; Machefer et al., 2007).

Vitamin E is a popular dietary supplement among athletes, particularly endurance athletes (de Silva et al., 2010; Morrison, Gizis, and Shorter, 2004). Theoretically,

Excellent sources of vitamin E include oils, seeds, nuts, whole grains, and vegetables.

© Bill Aron/PhotoEdit

supplements on health. Most studies have not found an association between antioxidant supplements and reduced risk for disease. Some studies have found no effect and others have found detrimental effects. Bjelakovic et al. (2008) found that supplemental vitamin A, beta-carotene, and vitamin E, consumed singly or combined, did not prevent heart disease or reduce mortality. In fact, this 2008 analysis indicated that antioxidant supplement use was associated with a higher rate of premature mortality.

These studies highlight some important points. First, it cannot be assumed that a beneficial component found in food will be equally beneficial if concentrated in a pill and taken as a supplement. A likely explanation is that the component in food is found in low concentrations and works in conjunction with other compounds also found in the food. These features are lost when a single vitamin is consumed as a supplement. Second, apparently benign compounds can act in unpredictable ways in the body. Although many supplements are safe, it cannot be assumed that all compounds sold as supplements are safe.

Table 8.6 Effect of Antioxidant Supplements on Health

Study	Focus of the study	Conclusion
Bardia et al., 2008	Beta carotene supplements and cancer	Increases cancer incidence and death in smokers
	Vitamin E supplements and cancer	No effect on cancer incidence and mortality
Chong et al., 2007	Vitamins A, C, E, and beta-carotene supplements and age-related macular degeneration (AMD)	No effect of antioxidant supplements on AMD
Huang et al., 2006	Multivitamin and mineral supplements to prevent cancer, heart disease, cataracts, or AMD	Insufficient evidence to prove the presence or absence of benefits with multivitamin and mineral supplements for chronic disease prevention
Bjelakovic et al., 2008	Vitamin A, E, and beta-carotene supplements and gastrointestinal (GI) cancers	Does not prevent GI cancers and likely increases mortality
Bjelakovic et al., 2008	Vitamins A, C, E, and beta-carotene supplements and preventing and delaying chronic disease and increasing survival	Vitamin A, beta-carotene, and vitamin E supplements increase mortality
Sesso et al., 2008	Vitamin E and C supplements to prevent heart disease	Neither vitamin E or C prevented heart disease

endurance athletes would be at greater risk for oxidative damage due to their prolonged aerobic training, which may take place under environmental conditions that generate free radical production, such as high altitude and exposure to ozone and ultraviolet light. It has also been postulated that resistance exercise could be associated with oxidative damage. Resistance exercise can damage muscle tissue that leads to inflammation, a process associated with an increase in ROS (Viitala and Newhouse, 2004). At the present time, antioxidant supplements for athletes are generally not recommended because the study results are inconclusive and there is some evidence that oxidative stress could be increased with antioxidant supplementation (Williams et al., 2006; Lukaski, 2004; Viitala and Newhouse, 2004; Viitala et al., 2004).

Although the current body of research does not support using vitamin E supplementation as a way to enhance exercise training or performance, researchers acknowledge that not all scientists agree. Vitamin E remains a popular supplement because of the theoretical *potential* to decrease exercise-induced oxidative damage in muscle tissue. Vitamin E supplements are commonly sold in dosages of 100 or 200 mg daily, amounts that are not achievable through food alone and well above the DRI for adults, 15 mg daily. In those studies in which an increase in oxidative damage was reported, the amount of vitamin E consumed was greater than 200 mg (Williams et al., 2006). It should be noted that on supplement labels vitamin E may be listed in International Units (IU), an old unit of measurement. To receive 15 mg of supplemental vitamin E, one must choose a supplement that contains 22 IU of natural vitamin E or 33 IU of synthetic vitamin E. Some of the vitamin E found in synthetic forms is inactive, thus more of the synthetic vitamin E is needed when compared to more active forms. The Tolerable Upper Intake Level is 1,000 mg daily, which is equivalent to 1,500 IU of natural vitamin E or 1,100 IU of synthetic vitamin E (Institute of Medicine, 2000). The Spotlight on supplements: Applying Critical Thinking Skills to Evaluating Dietary Supplements uses vitamin E supplementation as an example.

Vitamin C. Vitamin C is an antioxidant vitamin that works both independently of and in conjunction with vitamin E. Vitamin C is a water-soluble vitamin, and its antioxidant activity occurs primarily in extracellular tissue. In these ways vitamin C is very different from vitamin E. However, vitamin E also depends on vitamin C. When vitamin E interacts with a free radical, it becomes oxidized; vitamin C can interact with the oxidized vitamin E, reducing the compound and regenerating the vitamin E. These vitamins work both separately and together, which is the reason many researchers include both vitamins in their study

Vitamin C-containing fruits and vegetables include strawberries, citrus fruits, and peppers.

protocols and athletes may hear recommendations to take these supplements together.

It is very hard to study vitamin C as an antioxidant during exercise. The number of studies in athletes is small and the results are conflicting (Moreira et al., 2007). Some researchers suggest that athletes have a greater need for vitamin C than the general population but this has not been proved. Those who recommend supplementation generally do so because they believe that it will help athletes counteract the free radicals that result from exercise. Others do not recommend supplementing and are concerned that vitamin C supplements could upset the balance of antioxidants and reduce the effectiveness of the body's antioxidant response system. At present, this debate continues, but there is widespread agreement that athletes should consume vitamin C-containing fruits and vegetables daily.

The DRI for nonsmoking females is 75 mg/d and 90 mg daily for nonsmoking males. Limited studies suggest that some athletes may not be consuming enough vitamin C. A study conducted in France found that one-third of trained male marathon runners did not meet the French DRI, which is 100 mg/d (Machefer et al., 2007). However, the remainder did, with ~40 percent consuming more than 150 percent of the French DRI. This range is characteristic of reported intake in other sports (Teixeria et al., 2009). Due to widespread use of vitamin C supplements and the known variations in dietary intake, the best approach is to determine vitamin C intake on an individual basis. Once usual intake is known, appropriate recommendations can be made.

Of the three antioxidant vitamins, vitamin C is the most crucial to consume on a daily basis because it is water soluble. Athletes, particularly endurance and ultraendurance athletes, are highly encouraged to meet the DRI daily. For all practical purposes this means consuming vitamin C-containing fruits (for example, oranges, grapefruit, strawberries) and vegetables (for example, cabbage, broccoli, peppers, tomatoes). Vitamin C supplementation and its relationship to colds, asthma, and the immune system are discussed in Spotlight on supplements: Vitamin C and Colds.

Vitamin A as carotenoids. Vitamin A is a broad term and includes both preformed vitamin A (for example, retinol) and vitamin A precursors, known as carotenoids. Carotenoids, which are found in the red, orange, and yellow pigments in plants, have some antioxidant properties. Beta-carotene is the best known and most studied carotenoid but it is a less powerful antioxidant than some of the other carotenoids, such as **lycopene** and **lutein**. Compared to vitamin E, carotenoids are weak antioxidants. However, they do interact with some reactive oxygen species (for example, singlet oxygen molecules) and inactivate them. Due to their chemical properties, the carotenoids do not need to be regenerated, as is the case for vitamin E and C.

Preliminary studies have been conducted in athletes to determine the role that carotenoids play as part of the body's antioxidant system. In general, exercise increases the body's antioxidant capacity as long as there is a sufficient amount of antioxidants

© David Young-Wolff/PhotoEdit

Dark green, orange, and red-colored fruits and vegetables are excellent sources of vitamin A.

Lycopene: One of the carotenoids with a red pigment.

Lutein: One of the carotenoids with an orange pigment. Also found in some animal fats such as egg yolk.

Quercetin

Phytochemicals are defined as nonnutritive compounds that have biological activity. There are tens of thousands of phytochemicals in plants, including a group of ~8,000 known as polyphenols. One of the largest subgroups of polyphenols is the flavonoids. Quercetin is a flavonoid that has received attention and study because it is a powerful antioxidant. Reported benefits, such as anti-inflammatory effects and improved performance, have resulted in quercetin becoming a popular supplement among athletes. The usual dose is 1,000 mg/d, which is thought to be safe.

Studies of tissue cultures and mice are promising, especially since quercetin has been shown to be five times more potent an antioxidant than vitamin C (Davis et al., 2010). However, there is no convincing evidence that quercetin supplementation prevents oxidative tissue damage in humans that exercise (Utter et al., 2009; Quindry et al., 2008).

There is some evidence that supplementation can help to counter inflammation in well-trained athletes. In a study of trained cyclists, quercetin supplementation of 1,000

mg/d of for 2 weeks was an effective anti-inflammatory agent after 3 days of heavy training (Nieman et al., 2009). However, other studies of cyclists have not found a similar effect (McAnulty et al., 2008), so more research is needed.

Similarly, some studies in untrained men showed improved performance (Davis et al., 2010; Nieman et al., 2010) whereas others have not (Cureton et al., 2009; Cheuvront et al., 2009). There is no strong evidence for a beneficial performance effect in trained athletes. Dumke and colleagues (2009) found no effect of 1,000 mg/day of quercetin on cycling efficiency. In ultraendurance runners, quercetin supplementation did not reduce perceived exertion during a 160 km (100 mile) run (Utter et al., 2009).

Quercetin is an active area of study, but at the present time there is not enough evidence in trained athletes to suggest that supplementation will improve performance or reduce oxidative damage. It may have an anti-inflammatory effect, although more study is needed.

consumed and blood levels of antioxidants are decreased. Markers of inflammation are increased if dietary antioxidants are restricted (Plunkett et al., 2010; Carlsohn et al., 2010).

Some athletes do not consume an adequate amount of vitamin A and fail to meet the DRI, which is 700 mcg for adult nonpregnant females and 900 mcg for adult males. Surveys of athletes suggest that low beta-carotene intake is due to poor food choices (Soric, Misigoj-Durakovic, and Pedisic, 2008; Machefer et al., 2007). Low intake can easily be rectified with greater consumption of "colorful" fruits and vegetables.

Vitamin B$_{12}$ and folate are two vitamins associated with red blood cell function.

The human body contains approximately 30 trillion red blood cells (erythrocytes) that are primarily responsible for transporting oxygen as well as carbon dioxide. With an average life span of only 120 days, the body must replace the erythrocytes as rapidly as they die—about 2 to 3 million cells per second. Each red blood cell (RBC) contains more than 250 million molecules of hemoglobin, an iron-containing compound that is found only in RBC. For these reasons the mineral iron is at the center of discussions about red blood cells. Iron, its role in red blood cell formation, and the impact of iron-deficiency anemia are discussed in depth in Chapter 9.

Several vitamins are necessary to produce red blood cells, but **erythropoiesis** could not proceed without vitamin B$_{12}$ and folate. These two vitamins are needed alone and in combination with each other since they activate one another. Deficiencies of these vitamins can result in anemia. Vitamin-related anemia is different from the more common mineral-related iron-deficiency anemia.

Vitamin B$_{12}$. Vitamin B$_{12}$ (cobalamin) is a water-soluble vitamin but some is stored in the liver. It works in conjunction with folate to form a coenzyme that is needed to produce red blood cells. Other cells also require vitamin B$_{12}$, but deficiencies are first recognized in RBC because these cells are produced so rapidly. A deficiency of vitamin B$_{12}$ may be a result of

Spotlight on supplements

Applying Critical Thinking Skills to Evaluating Dietary Supplements

Listen to an endurance athlete explain why he is taking a vitamin E supplement.

Most people don't realize it but oxygen has both beneficial and harmful effects. I try to counter the harmful effects of free radicals by taking a vitamin E supplement. Most studies have shown that vitamin E supplementation has a positive effect on endurance performance. Besides, studies have shown that there is a health benefit to vitamin E supplements. I'm concerned that my vitamin E intake may be low since I purposefully keep my caloric and fat intakes low so I don't gain body fat. Someone told me that when you take high levels of vitamin E it actually has an opposite effect and it no longer works as an antioxidant, but I don't think that is true. That just doesn't sound right. Anyway, I've read that no one gets a vitamin E toxicity.

This athlete has drawn both correct and incorrect conclusions. Oxygen does have both beneficial and harmful effects. His vitamin E intake may be low since he restricts his intake of both energy (kilocalories) and fats. Vitamin E is a supplement that could theoretically play a role in enhancing performance. Toxicity from vitamin E supplementation would be extremely rare. This athlete has correctly concluded some things about vitamin E supplements.

However, some incorrect conclusions have also been drawn. The majority of studies have *not* shown that vitamin E supplementation has a positive effect on endurance performance. In the past, some studies have shown that there is a health benefit to vitamin E supplementation, but more recent studies have not duplicated earlier findings and there is increasing concern that such supplements may be detrimental to health. Although it may sound illogical that an antioxidant vitamin taken in large doses could produce the opposite effect (pro-oxidant activity), this is one theory that is being investigated. There is evidence to suggest that detrimental effects can occur from vitamin E supplements, at levels well below that associated with toxicity. This athlete has also overlooked the need for balance among antioxidants.

An appropriate course of action for this athlete is to have a dietary analysis performed by a sports dietitian to determine if energy and nutrient intake is sufficient. If intake of vitamin E is insufficient, dietary recommendations can be made and the athlete could change his food intake and consume the recommended amount. If supplementation is determined to be appropriate, the source and amount of supplementation could be discussed. Vitamin supplements are not well regulated and are heavily advertised, sometimes in ways that are misleading. Supplementation requires an evaluation of dietary intake and good critical thinking skills to judge the safety, effectiveness, risk for contamination, and potential advantages and disadvantages of use.

Vitamin B$_{12}$ is naturally found only in foods of animal origin.

poor dietary intake or poor absorption, a condition associated with older adults.

The absorption of vitamin B$_{12}$ requires a compound known as intrinsic factor (IF). IF is produced in the stomach but does not bind with vitamin B$_{12}$ until both reach the small intestine. The IF-vitamin B$_{12}$ complex must travel to the ileum (the lowest portion of the small intestine) where it is absorbed over the next 3 to 4 hours. As people age, IF production declines. Many people over the age of 50 do not produce IF and are at risk for developing **pernicious anemia**.

Those who lack IF need vitamin B$_{12}$ injections. Those who have a dietary deficiency (but normal IF production) can reverse the deficiency by consuming more vitamin B$_{12}$-containing foods (naturally occurring or fortified) or by taking vitamin B$_{12}$ supplements orally. Some athletes who are not vitamin B$_{12}$ deficient and have normal IF secretion inject vitamin B$_{12}$ based on anecdotal evidence (that is, personal accounts) that these injections "boost energy." There is no scientific evidence that vitamin B$_{12}$ injections will influence energy metabolism, decrease fatigue, or affect red blood cell production in athletes without a vitamin B$_{12}$ deficiency (Lukaski, 2004). Vitamin B$_{12}$ injections are not recommended for athletes because there is no apparent benefit and there are some risks associated with injections, such as soreness, abscesses, needle contamination, and the presence of banned substances.

Vitamin B$_{12}$ is found only in foods of animal origin. Examples include meat, fish, poultry, eggs, and milk and milk products. The vitamin B$_{12}$ in milk and fish has high bioavailability (Vogiatzoglou et al., 2009). Vegans, who eliminate all sources of animal products from their diets, are at risk for a subclinical vitamin B$_{12}$ deficiency. In adults who ate animal products before becoming vegan, a deficiency could take decades to develop due to stores of vitamin B$_{12}$ in the liver.

Some vegetarians are also at risk for a subclinical vitamin B$_{12}$ deficiency. Although vegetarians obtain some vitamin B$_{12}$ from their diets, their intake over time may be low (Elmadfa and Singer, 2009). The Dietary

Erythropoiesis: The production of red blood cells.

Pernicious anemia: Anemia caused by a lack of intrinsic factor, which is needed to absorb vitamin B$_{12}$.

Spotlight on supplements

Vitamin C and Colds

During infection there is a rapid decline in the concentration of vitamin C in the plasma and white blood cells. This knowledge has led to the use of vitamin C supplements to self-treat upper respiratory infections such as the common cold. It is also popular for people to routinely consume vitamin C supplements in an effort to prevent colds.

A meta-analysis of 30 studies conducted between 1966 and 2006 found that the incidence (the number of new cases) of colds was not reduced with routine vitamin C supplementation greater than 200 mg daily. However, once a cold was present, the duration (number of sick days) was reduced by 8 percent in adults and the severity (extent of symptoms and confinement) was also reduced. Interestingly, in the studies in which vitamin C supplements were introduced as a treatment in those who did not routinely consume them, vitamin C supplements up to 4 g (4,000 mg) were not beneficial (Douglas et al., 2007).

In endurance athletes, there is some evidence that supplemental vitamin C may help prevent colds. A small number of studies have suggested that routine vitamin C supplementation to prevent respiratory infections might be beneficial for marathon runners, skiers, and soldiers exposed to low environmental temperatures (Douglas et al., 2007). However, other studies have not found any benefit (Gleeson, Nieman, and Pedersen, 2004).

The bottom line: Vitamin C supplements do not prevent colds in the general population. There is some evidence that supplementation may prevent colds in those who engage in rigorous exercise, especially In cold temperatures. Once a cold is present, supplementation helps to reduce the duration and severity of the cold.

Reference Intake for vitamin B_{12} for nonpregnant females and male adults is 2.4 mcg daily. No Tolerable Upper Intake Level has been established for vitamin B_{12} (Institute of Medicine, 1998).

The effect of exercise on vitamin B_{12} metabolism has not been well studied. In general, studies of athletes have not found B_{12} deficiencies (Woolf and Manore, 2006). A prudent approach is to determine the vitamin B_{12} intake of vegan or vegetarian athletes. It is recommended that vegetarians, particularly vegans, consume vitamin B_{12}-fortified foods or supplemental vitamin B_{12} (Elmadfa and Singer, 2009).

Folate. Folate is an important B-complex vitamin involved with DNA synthesis and amino acid metabolism. Folate is the term used when referring to this vitamin in the body or when it naturally occurs in food. Folic acid refers to the form used in fortified foods or supplements. Folate works in conjunction with vitamin B_{12} to form a coenzyme that is needed to produce red blood cells. Other cells also require folate to divide, but deficiencies are usually detected first in the rapidly produced RBC. At conception, folate is critical to support rapid cell division and prevent neural tube defects such as spina bifida (incomplete closure of spinal cord tissue). The addition of folic acid to cereals and other grain products in the United States began in 1998 in an effort to eliminate neural tube defects.

The Dietary Reference Intake for adults is 400 mcg of dietary folate equivalents daily. An equivalency measure is necessary because naturally occurring folate is not as bioavailable as the synthetic form. The Tolerable Upper Intake Level is 1,000 mcg from fortified foods and supplements, sources that have highly absorbable forms of folic acid (Institute of Medicine, 1998).

A dietary deficiency of folate will eventually result in **megaloblastic anemia**. A clinical deficiency will be preceded by subclinical deficiencies—a drop in folate concentration in the blood and a decrease in folate concentration in the red blood cells. Folic acid supplements will reverse a clinical deficiency but supplementation will not improve performance in those with subclinical deficiencies or normal blood folate concentrations (Woolf and Manore, 2006; Lukaski, 2004).

As is the case with vitamin B_{12}, the effect of exercise on folate metabolism has not been well studied. Information about dietary intake in athletes is also sparse (Woolf and Manore, 2006). In general, male athletes consume a sufficient amount due to their high caloric intake, which includes fortified cereals and grains, leafy green vegetables, oranges, bananas, and nuts (Lukaski, 2004). Female athletes who restrict caloric intake may have low folate intake. However, some may meet the DRI because they are consuming a multivitamin that contains the standard dose of folic acid. Individual assessment is critical, as folic acid supplementation may be warranted.

Christina Micek

Breads, cereals, beans, oranges, and bananas provide folic acid.

Interactions between vitamin B_{12} and folate. Vitamin B_{12} and folate work in concert with each other. Therefore, a deficiency or an excess of either one can affect normal metabolism and health. Under normal conditions, both are present in sufficient quantities. A deficiency of either vitamin B_{12} or folate can result in megaloblastic anemia. If the vitamin B_{12} deficiency is caused by a lack of intrinsic factor, it is a subtype of megaloblastic anemia referred to as pernicious anemia. In addition to the anemia, a deficiency of vitamin B_{12} also results in neurological symptoms. These symptoms include tingling in the arms and legs and loss of concentration and memory.

The majority of individuals with pernicious anemia are elderly, because intrinsic factor production declines with age. If those individuals take folic acid supplements, the pernicious anemia is resolved but the neurological symptoms are not. In other words, folic acid masks the vitamin B_{12} deficiency, and the neurological symptoms, which cannot be reversed with folic acid, get progressively worse. This scenario illustrates three important vitamin-related points: (1) each vitamin is essential because it has some unique functions, (2) intake of all the vitamins needs to be adequate, and (3) because vitamins interact there needs to be balance among the vitamins.

Many vitamins are associated with growth and development, including vitamins A and D.

Vitamins A and D are associated with growth and development in various ways. Although both are fat soluble, these two vitamins have very different chemical structures. Vitamin D has four rings, one of which is broken. This structure allows it to easily bind with proteins. The basic structure of vitamin A is a ring and a polyunsaturated side chain, but because the side chain can be different there are a number of closely related compounds that have vitamin A activity. These compounds can interact with or be converted to other compounds, so vitamin A has many different functions in the body.

Table 8.7 Role of Vitamin D in the Prevention and Treatment of Disease

Study	Condition	Effect of vitamin D
Pittas et al., 2010	Cardiovascular outcomes or blood pressure	No clinically significant effect of vitamin D supplementation in studies through 2009
DIPART Group, 2010; Stránský and Rysavá, 2009	Prevention of bone fractures	Supplemental vitamin D is effective to reduce the number of hip and total fractures if taken with calcium. Not effective if taken alone.
Stránský and Rysavá, 2009	Osteoporosis	Calcium and vitamin D supplements together reduce loss of bone mass
Bischoff-Ferrari et al., 2009	Risk of falling in the elderly	Supplemental vitamin D of 700–1,000 IU may help to decrease the risk of falling
Yin et al., 2009a, 2009b	Cancer	Low blood concentration of vitamin D is associated with an increased risk of colon cancer; most studies show no association with prostate cancer
Ferguson and Chang, 2009	Cystic fibrosis	No evidence of benefit or harm from vitamin D supplements

Note: All studies are meta-analyses, which combine the results of similar studies for analysis thereby improving the statistical strength of the research.

Vitamin D. Scientists began to study vitamin D in the early 1900s in response to a high prevalence of rickets in children, which results in improper bone development. A form of vitamin D, **calcitriol**, helps to regulate blood calcium levels. The regulation of calcium is intricate and involves intestinal, bone, and kidney cells. Vitamin D targets those cells to help maintain calcium balance in the blood and ensure proper bone development. This is further explained in Chapter 9, "Minerals."

Bone growth and development continues to be a major focus of vitamin D research. However, it is now widely recognized that vitamin D regulates many other tissues, such as cardiac and skeletal muscle, and is intimately involved with cellular growth of both normal and cancerous cells (Kulie et al., 2009). Table 8.7 summarizes the role of vitamin D, typically administered via supplementation, in the prevention or treatment of various diseases.

Vitamin D is obtained in three ways: (1) consumption of vitamin D-containing foods, (2) ingestion of supplemental vitamin D, and (3) exposure to ultraviolet (UV) light. Naturally occurring food sources are typically of animal origin, such as beef or fatty fish. Milk and milk products are fortified with vitamin D, as is some orange juice. Most people do not obtain the required amount of vitamin D from food (Stroud et al., 2008). Supplemental vitamin D is widely available.

Exposure to UV light can vary considerably. It will depend on environmental factors such as latitude, season of the year, and time of day. Time spent outdoors, degree of clothing, and use of sunscreen are very substantial behavioral factors. Skin color and degree of conversion in the skin are influential physiological characteristics. Table 8.8 summarizes the many factors

Calcitriol: 1,25-dihydroxyvitamin D_3, a hormonally active form of vitamin D.

Megaloblastic anemia: A type of anemia characterized by large red blood cells.

Table 8.8 Factors Affecting UV Light Conversion of Vitamin D

Factor	Explanation
Latitude	Living at >37 degrees N or S reduces exposure especially during the winter months
Season	Highest in summer; lowest in winter
Time of day	Highest between 10 a.m. and 3 p.m.
Amount of exposure	Greater time increases exposure and conversion to vitamin D as well as sunburn, risk for skin cancer and risk for eye damage
Lack of sun exposure	Housebound or institutionalized are at the greatest risk for deficiency
Clothing	Head-to-toe clothing reduces exposure to UV light
Use of sunscreen	Sunscreen reduces conversion of vitamin D via UV light by at least 90%
Skin color	Darker skin color requires more exposure than lighter skin color for conversion of vitamin D
Aging (elderly)	Decreased rate of conversion of the vitamin D precursor to the active form of vitamin D, absorption from food is reduced, food intake and UV exposure often low

that can affect the amount of vitamin D available due to exposure to UV light.

It is hard to estimate vitamin D status because many food databases have missing values and assessing UV exposure is difficult (Willis, Peterson, and Larson-Meyer, 2008). Vitamin D status can be determined with a laboratory test that measures the main form of vitamin D in the blood, 25-OH D_3 (also known as 25-hydroxy vitamin D or 25-hydroxycholecalciferol). A concentration of <10–11 ng/ml (25–27.5 nmol/L) is considered a clinical deficiency because this range is associated with rickets in children and the equivalent disease in adults, osteomalacia. A blood level of <10–15 ng/mL (<25–37.5 nmol/L) is reflective of a subclinical deficiency due to its association with inadequate bone health and overall health. However, there is a controversy about the concentration that defines insufficiency. A concentration >15 ng/mL (>37.5 nmol/L) is generally considered adequate, but some vitamin D researchers feel that an insufficiency is present unless the 25-OH D_3 concentration is >30 ng/mL (>75 nmol/L) (http://ods.od.nih.gov/factsheets/vitamind.asp).

It is estimated that the vast majority of people in the United States have a blood vitamin D level of <30 ng/mL (<75 nmol/L). This includes 90 percent of the Asian, Black, and Hispanic populations and 75 percent of the Caucasian population in the United States. Such wide-scale insufficiencies have led many experts to recommend vitamin D supplements for most adults (Adams and Hewison, 2010).

The DRI for vitamin D was revised in 2010 (Institute of Medicine, 2010). The current DRI is 15 µg (600 IU) for ages 1 to 70 years. After age 70, the DRI is 20 µg (800 IU). The UL is 100 µg (4000 IU). Although vitamin D toxicities from supplements can occur, they are rare. There is a physiological limit to how much vitamin D can be converted via UV light, so prolonged UV exposure is not a source of vitamin D toxicity.

Vitamin D and athletes. The vitamin D status of athletes has received more attention as the prevalence of low vitamin D intake and the use of sunscreen to block UV light exposure has increased among people of all ages. Very limited data in athletes would suggest that some athletes are vitamin D deficient and that more have an insufficiency. An athlete could determine status by having 25-OH D_3 measured. Recommendations for athletes are speculative at this time. One recommendation is 20–50 µg (800–2000 IU) daily, an amount unlikely to be obtained from food alone. Safe UV sun exposure, which is defined as 5–30 minutes between 10 a.m. and 3 p.m. twice a week based on latitude, season of the year, and skin pigment, or vitamin D supplementation is recommended (Willis, Peterson, and Larson-Meyer, 2008).

Vitamin A. Vitamin A is a broad term that includes preformed vitamin A and provitamin A. Preformed vitamin A, also known as retinoids, is obtained from animal sources, such as fish-liver oils, liver, egg yolks, and butter. Provitamin A refers to compounds that can be converted to vitamin A in the body. Some carotenoids, which are red and yellow pigments found primarily in fruits and vegetables, have provitamin A activity. For example, beta-carotene is converted to an active form of vitamin A. Some carotenoids, such as lutein and lycopene, are not converted to vitamin A, although these compounds support good health in other ways.

Vitamin A affects the growth and development of many different tissues. Preformed vitamin A is an integral part of normal vision, and those with a severe deficiency experience blindness or difficulty seeing in low-light conditions. Preformed vitamin A is also needed for proper cell differentiation, and many aspects of growth can be affected if vitamin A is lacking.

One of the carotenoids, beta-carotene, has important antioxidant functions, as explained earlier in this chapter. This area has been the focus of most of the studies involving athletes. The consumption of fruits and vegetables, which contain many different carotenoids, has been associated with good health and prevention of heart disease and strokes (He et al., 2007; He, Newson, and MacGregor, 2006). However, most people do not consume a sufficient amount of fruits and vegetables daily and some choose to supplement with compounds found in these foods, such as beta-carotene. Studies have shown that beta-carotene supplements do not prevent heart disease, cancer, or macular degeneration, an eye disease associated with aging (Table 8.6). In fact, several studies found that beta-carotene supplements *increased* the likelihood of dying from chronic disease, especially in people who smoke. The best advice is to consume at least five servings of a variety of brightly colored fruits and vegetables daily.

The DRI for vitamin A for adults is 700 mcg for females and 900 mcg for males. These values are given as retinol activity equivalents (RAE) because both animal and plant sources can contribute vitamin A. Vitamin A deficiency is uncommon in North America, Europe, and Australia, but many children and adults in Asia, Central and South America, and Africa lack sufficient vitamin A.

Vitamin A toxicity is rare, but it can occur. The UL is 3,000 mcg of preformed vitamin A. Excess preformed vitamin A can cause liver damage. Vitamin A toxicity is associated with birth defects, so pregnant women are cautioned to supplement with a safe dose of vitamin A as prescribed by their physician. Vitamin A toxicity from beta-carotene is highly unlikely because of its relatively low rate of absorption and the amount that can be converted to preformed vitamin A is well regulated by the body.

Key points

- Some vitamins are involved in energy reactions, but vitamins are not metabolized for or do not directly provide energy.

- Enzymes and coenzymes need certain vitamins to function properly, but consuming an excessive amount of those vitamins does not result in better or faster functionality.

- Cellular damage from oxidation results when free radical production outpaces clearance.

- Vitamins E, C, and A (as carotenoids) are antioxidants that can counteract free radicals.

- Balance is the key, not only between production and clearance of free radicals (oxidants) but also among the antioxidants.

- Athletes should consume antioxidant-rich food daily, such as fruits, vegetables, whole grains, and oils.

- Vitamin B_{12} and folate deficiencies can affect various types of cells, but red blood cells are affected quickly because they are rapidly produced.

- Vitamin B_{12} and folate interact, so an adequate amount of both is necessary.

- There is no evidence that vitamin B_{12} injections boost energy in athletes.

Which vitamins play a role (1) as coenzymes, (2) in cell growth or development, (3) in energy metabolism, and (4) as antioxidants?

8.3 Sources of Vitamins

Most vitamins are obtained from three sources: they exist naturally in food, they are added to foods during processing, and they are manufactured as dietary supplements. An orange is a food that naturally contains approximately 70 mg of vitamin C. A fruit punch drink with vitamin C added has about 50 mg. Vitamin C supplements come in 100, 300, 500, and 1,000 mg tablets. The vitamin C found in each is the same chemical compound but the dose can vary considerably.

Food has always been a way for people to obtain nutrients. Before the discovery and isolation of nutrients in the laboratory, food stood on the merit of the nutrients it contained naturally. The word *naturally* is used to describe nutrients put in food by nature, such as the vitamin C in an orange. For each of the vitamins discussed in this chapter, there are excellent naturally occurring food sources as shown in Table 8.2.

The nutrient content of foods began to change when food processing became widespread. When raw foods are processed some nutrients are lost. For example, if an orange is juiced, the ½ cup of juice that it yields has about 60 to 65 mg of vitamin C. This is a small loss of vitamin C since the orange originally had about 70 mg. But some foods lose a considerable amount of nutrients when processed. When whole grains are processed into white flour, many of the vitamins are lost, some almost completely (for example, vitamin B_6). Modern processing may also add nutrients to foods. Sometimes a vitamin that is lost is added back (enrichment), and sometimes vitamins that were never present before processing are added (fortification). For example, flour is enriched with thiamin, riboflavin, and niacin but milk is fortified with vitamin D. Some foods become an excellent source of a particular vitamin because it has been added by food processors.

Vitamin supplements are concentrated sources of vitamins found naturally in food, thus the amount consumed as a supplement is often more than would be consumed from food alone. In some cases the vitamin supplement is sold in a form that has high bioavailability. Bioavailability refers to the degree to which a compound can be absorbed and utilized by the body, and it is not necessarily greater in naturally occurring sources. Folate found naturally in foods has about 50 percent of the bioavailability of the folic acid added to food or found in most supplements.

Although there are three sources of vitamins—naturally occurring in food, added to foods during processing, and supplements—the lines between them are blurry. Oranges, fruit punch with vitamin C added, and vitamin C supplements are fairly clear points on a continuum. But additional vitamin C is being added to orange juice to enhance the amount naturally found, and cereals are fortified with 100 percent of the DRI for many nutrients including vitamin C. Many new food and beverage products have added vitamins because this is a way to attract consumers. Several popular brands of cereal advertise that the cereal contains 100 percent of the Daily Value (DV) for many vitamins and minerals. The differences between the amounts found in foods, added to foods, and contained in supplements are not as distinct as in the past.

Each person must decide the best ways to obtain an adequate amount of vitamins.

Athletes, like other consumers, have many questions about vitamins. Can athletes meet their daily vitamin needs by eating food alone? Are there advantages to getting vitamins from food instead of from supplements? Are there advantages to using vitamin supplements? Will taking a multivitamin every day do the trick? As is often the case in a complicated subject like nutrition, many factors must be considered before such questions can be answered.

A strong argument can be made for obtaining vitamins from foods in which they naturally occur.

Throughout history, people have survived by only eating food. There is an amazing array of vitamins that occur naturally in food. An orange, which has a substantial amount of vitamin C, also contains small amounts of thiamin, riboflavin, niacin, vitamin B_6, folate, and vitamin A. As is often the case, a certain food may be an excellent source of one vitamin but also contribute to the overall intake of many vitamins. A diet that includes a variety of nutrient-dense foods can provide all of the vitamins that people need. There is historical precedent for obtaining vitamins from "food first."

Nutrition scientists who analyze the nutrient content of food have found that nutrients that occur naturally in foods are in the right proportion to other nutrients, a factor that probably helps nutrients to be absorbed. Fruits and vegetables, foods that contain numerous vitamins, also contain other biologically active compounds, such as the carotenoids that are not converted to vitamin A. The interactions between vitamins and these compounds may be necessary for proper biological function and protection from chronic diseases. Whole grains and beans are excellent sources of many vitamins, and they come "packaged" with carbohydrates, proteins, and fiber, nutrients that are important for athletes both for training and for health. There is also scientific backing for obtaining vitamins from "food first."

Although it is true that people have survived with only the nutrients found naturally in foods, it is also true that some people who did so suffered from vitamin deficiencies. Obtaining an adequate amount of vitamins only from foods that naturally contain them requires that a wide variety of nutrient-dense foods be eaten. In reality, most people in the United States do not consume a wide variety of vitamin-rich foods.

Many of the foods consumed in the United States are processed foods. Food processing can strip vitamins from the food in which they naturally occurred. For example, the processing of wild rice into polished white rice or whole grains into bleached white flour reduces the vitamin content of the processed food substantially. In the United States in the 1930s and 1940s, vitamin deficiencies were widespread. These deficiencies reflected changing dietary habits, specifically the use of white bread instead of whole wheat bread. White bread had become so widely accepted that it was a staple in the diet of many Americans. To address the vitamin deficiencies that were occurring, a law was passed requiring that three vitamins lost in the processing of whole grains—thiamin, riboflavin, and niacin—be added back. The enrichment of flour products has greatly improved the nutrient status of Americans.

The addition of nutrients to food during processing is really a form of supplementation. Most Americans consume foods that contain added vitamins. Those that are required by law to be added to flour include thiamin,

riboflavin, niacin, and folic acid (as well as the mineral iron). Fortification of milk with vitamin D is also a law. Such laws have helped to prevent widespread vitamin deficiencies, but the addition of selected nutrients does not necessarily make the highly processed product as nutritious as the original product. In the case of whole grain flour, some vitamins (for example, vitamin B_6), minerals (for example, zinc, magnesium), and fiber lost in the processing are not restored.

The natural outcome of required fortification is the voluntary addition of nutrients to foods. Some products are fortified with vitamins that would be lacking in the diet. For example, some vegetarian products are fortified with vitamin B_{12}, which is found naturally only in animal products. Most ready-to-eat breakfast cereals contain many added vitamins and minerals. They are often advertised to highlight the addition of these nutrients, most of which are added to a level that will provide 100 percent of the Daily Value. **Daily Value (DV)** is a term that is used on food and supplement labels and is an estimate of the amount of a nutrient needed each day; it is similar to but not as specific as the DRI. Consuming a highly fortified cereal is really the same as eating a food that contains a multivitamin and mineral supplement, as shown in Table 8.9.

For many years, ready-to-eat cereal was one of only a handful of foods that essentially had a multivitamin and mineral supplement added. Today, vitamins (and minerals) are being added to many foods. Energy and sports bars, popular products for athletes, have many added nutrients, and manufacturers highlight the vitamin content of these products as a prominent

Table 8.9 Fortified Cereal and Multivitamin Supplement Compared

Vitamin	Daily Value (DV)	Fortified cereal (% DV)	Multivitamin supplement (% DV)
Thiamin	1.5 mg	100	100
Riboflavin	1.7 mg	100	100
Niacin	20 mg	100	100
Vitamin B_6	2 mg	100	100
Vitamin B_{12}	6 mcg	100	100
Folate	400 mcg	100	100
Vitamin C	60 mg	100	100
Vitamin A	900 mcg	15	100
Vitamin E	20 a-TE or 30 IU	100	100
Vitamin D	400 IU or 10 mcg	10	100

Legend: α-TE = alpha-tocopherol equivalents; IU = International Units; mg = milligram; mcg = microgram

part of their advertising. Even some bottled water has added vitamins.

Vitamin supplements are also widely available. Supplements may be limited to one vitamin (for example, vitamin B_1) or may contain several vitamins (for example, multivitamin or B-complex supplements). The amount contained in each tablet or capsule can vary. Supplements that have been tested for purity and potency may be verified. Verification programs assure consumers that good manufacturing practices have been followed during production and that the amount listed on the label is accurate. One of the best protections athletes have is to buy supplements that have been verified (see The Internet café).

Although many supplements contain 100 percent of the Daily Value of a particular vitamin, some contain more. A supplement with a high amount is sometimes referred to as a "megadose." *Megadose* does not have a standard definition but it is used to describe a supplement containing several times the DRI. Some supplements on the market contain an amount that meets or exceeds the Tolerable Upper Intake Level. Consumers should *always* be aware of and monitor the dose of any supplements consumed. In this respect, supplements are more similar to over-the-counter medications than to naturally occurring, vitamin-rich foods.

It is time to return to the original questions raised at the beginning of this section. Can eating food alone meet the daily vitamin needs of an athlete? The answer is yes. However, the athlete must consume enough food (that is, adequate kilocalories) with sufficient variety, and the foods chosen must be nutrient dense. This is a total-diet approach that takes both time and effort. Such a diet supports training, has the potential to improve performance, and helps to prevent chronic diseases. In reality, many athletes do not meet the daily recommendations for all the vitamins through food alone.

Are there advantages to getting vitamins from food? The answer is yes. Vitamins are found naturally in food with water, minerals, and other biologically active compounds and they provide needed calories in the form of carbohydrate, protein, and/or fat. In other words, when you get vitamin-rich foods you get more than just the vitamins. Vitamin supplements are singular in focus—they provide only vitamins. As mentioned previously, vitamins are required for energy metabolism but they do not provide energy-containing compounds—carbohydrates, proteins, or fats.

Are there advantages to using vitamin supplements? The answer is yes. However, the well-documented advantages are typically related to reversing or preventing vitamin deficiencies that result from a poor diet. Putting a Band-Aid on a cut may help the cut to heal, but the best strategy is not to get cut in the first place. A sound approach would be for athletes to adopt a "food first, supplements second" philosophy in which they try to obtain a sufficient amount of vitamins via their diet but take a vitamin supplement to cover any vitamins that may be lacking when they do not eat vitamin-rich foods.

Will taking a multivitamin every day do the trick? A multivitamin supplement does provide insurance against low vitamin intake. However, any underlying nutritional problems not related to vitamins would not be covered by such an insurance policy. Such problems include low or excessive caloric intake, or low carbohydrate, protein, fat, and fiber intakes. In a study of British athletes, nearly 73 percent took multivitamin supplements. The predominant reason was to avoid

The Internet café

Where do I find information about vitamins and exercise?

Finding noncommercial factual information about vitamins and vitamin supplementation for athletes on the Internet can be difficult. Here are some links that may be helpful:

- Fortify Your Knowledge about Vitamins (general information)

  http://www.fda.gov/ForConsumers/ConsumerUpdates/ucm118079.htm

- The Australian Institute of Sport fact sheets on vitamin and multivitamin supplements

  http://www.ausport.gov.au/ais/nutrition

- Natural Medicines Comprehensive Database (subscription required)

  http://www.naturaldatabase.com

- SupplementWatch (information about various supplements)

  http://www.supplementwatch.com

- Supplement verification websites:

  ConsumerLab.com (Athletic Banned Substances Screened Products) **http://www.consumerlab.com**

  NSF (Certified for Sport™) **http://www.nsf.org**

  United States Pharmacopeia **http://www.usp.org/USPVerified**

  Informed-Choice **http://www.informed-choice.org**

Daily Value (DV): A term used on food labels; estimates the amount of certain nutrients needed each day. Not as specific as Dietary Reference Intakes.

2 fried eggs
2 pieces of white toast w/1 T butter
2 slices bacon
8 oz coffee w/1 T each cream, sugar
Ham & cheese sandwich
1 oz bag potato chips
16 oz soft drink
3 Oreo cookies
Cheeseburger
Medium fries
16 oz soft drink
½ cup chocolate ice cream

Figure 8.5 An example of a high-fat, high-sugar, low-fiber diet

Legend: T = tablespoon; oz = ounce

sickness. Other reasons mentioned less frequently were advice from a physician, no time to prepare meals, and to overcome injury (Petróczi et al., 2007).

The vitamin content of a diet can vary tremendously based on the amounts and types of food consumed.

In a previous section, the answer to the question "Can eating food alone meet the daily vitamin needs of an athlete?" was yes. But each athlete must ask another question, "Does *my* diet provide the vitamins I need?" The answer to that question depends on the amount and types of foods consumed. One way to estimate the nutrient content of one's diet is to record food intake

1 cup oatmeal
1 slice whole wheat toast
1 T peanut butter
8 oz nonfat milk
8 oz orange juice
8 oz black coffee
1.5 cups lentil soup
Large (1.5 cups) green salad
1 oz avocado (~⅕ of an avocado)
2 T oil & vinegar dressing
2 whole wheat rolls
8 oz nonfat milk
1 banana
6 oz chicken breast
1½ cups brown rice
1 cup broccoli
1 slice whole wheat bread w/1 tsp *trans free* margarine
1 cup strawberries
Water
1 cup nonfat frozen yogurt
¼ cup dry-roasted sunflower seeds

Figure 8.6 An example of nutrient-dense, whole-foods diet

Legend: T = tablespoon; oz = ounce; tsp = teaspoon

Table 8.10 Vitamin Intake of a High-Saturated-Fat, High-Sugar, Low-Fiber Diet Compared to the DRI

Vitamin	Intake	DRI*	% of DRI
Thiamin	1.1 mg	1.2 mg	92
Riboflavin	2.1 mg	1.3 mg	162
Niacin	13 mg	16 mg	81
Vitamin B$_6$	0.41 mg	1.3 mg	32
Vitamin B$_{12}$	2.1 mcg	2.4 mcg	88
Folate	210 mcg	400 mcg	53
Vitamin C	20 mg	90 mg	22
Vitamin A	633 mcg	900 mcg	70
Vitamin E	1.9 mg	15 mg	13

Legend: DRI = Dietary Reference Intakes; mg = milligrams; mcg = micrograms

*20-year-old male

for 3 to 7 days by using a food diary (see Appendix E). All foods and beverages consumed can then be entered into a computer program that estimates the nutrient content of the diet and compares it to the DRI. In the case of vitamins, estimated intake can also be compared to the Tolerable Upper Intake Levels.

For comparison purposes, two very different dietary patterns, each containing sufficient caloric intake, are examined here, and the vitamin content of these diets is compared to the DRI and to each other. Both diets contain the same amount of energy, about 2,500 kcal. Figure 8.5 shows a diet that consists of three meals that are typical of those consumed away from home—a bacon and fried egg breakfast; a lunch

Table 8.11 Vitamin Intake of a Nutrient-Dense, Whole-Foods Diet Compared to the DRI

Vitamin	Intake	DRI*	% of DRI
Thiamin	1.9 mg	1.2 mg	160
Riboflavin	2.2 mg	1.3 mg	171
Niacin	42 mg	16 mg	264
Vitamin B$_6$	3.3 mg	1.3 mg	255
Vitamin B$_{12}$	4 mcg	2.4 mcg	168
Folate	446 mcg	400 mcg	111
Vitamin C	316 mg	90 mg	351
Vitamin A	3,410 mcg	900 mcg	379
Vitamin E	22 mg	15 mg	148

Legend: DRI = Dietary Reference Intakes; mg = milligrams; mcg = micrograms

*20-year-old male

consisting of a sandwich, chips, cookies, and a soda; and dinner at a fast-food restaurant.

This high-saturated-fat, high-sugar, low-fiber dietary pattern is low in all vitamins except riboflavin (Table 8.10). The diet came closest to meeting the recommendations for thiamin (92 percent of the DRI), vitamin B_{12} (88 percent), niacin (81 percent), and vitamin A (70 percent). Of great concern is the low intake of vitamin E (13 percent), vitamin C (22 percent), vitamin B_6 (32 percent), and folate (53 percent). This is a good example of how a diet can provide sufficient (or excess) kilocalories but is lacking in many vitamins. This diet is also high in saturated fats and sugar and low in fiber, a dietary pattern that may increase the risk for some chronic diseases as discussed in Chapter 13.

The dietary pattern featured in Figure 8.6 is nutrient dense. This diet meets the recommended daily intakes for all of the vitamins as shown in Table 8.11. Notice that the diet emphasizes fruits, vegetables, legumes, nuts, and whole grains, naturally occurring foods that are unprocessed or minimally processed. *Whole food* is a term that is used to describe such foods. Many of these foods are easy to prepare, although some planning is needed so that the refrigerator and pantry are well stocked. The colored bars in Figure 8.7 illustrate the obvious differences in the vitamin content of the two diets. The caloric intake is the same, approximately 2,500 kcal, but the vitamin intake is substantially different.

Figure 8.8 illustrates a diet that is very low in energy (~800 kcal), macronutrients, and most vitamins. It represents the intake of a female athlete who is trying to rapidly lose weight and who is severely restricting caloric intake for a very short period of

This diet consists of one 4 oz bagel with 1 T of low-fat cream cheese, 2 slices of wheat bread, 2 oz of sliced turkey with 2 tsp mustard, 1 medium banana, 16 oz of diet cola, 1 cup green salad with 2 T low-calorie dressing, 16 oz artificially sweetened iced tea, and 1 package of animal crackers. The approximate nutrient content is shown below:

Calories: 828 kcal (36% of recommended) or 17 kcal/kg
CHO: 144 g or 3.0 g/kg
Protein: 34 g or 0.7 g/kg
Fat: 14 g or 0.29 g/kg
Fiber: 13 g (52%)
Vitamins (% DRI)
Thiamin: 94
Riboflavin: 75
Niacin: 73
Vitamin B_6: 55
Vitamin B_{12}: 4
Folate: 100
Vitamin C: 78
Vitamin A: 110
Vitamin E: 8

When the caloric content of the diet is very low, it is difficult to obtain the recommended amounts of vitamins. In this example, seven of the nine vitamins analyzed were below the DRI.

Figure 8.8 Vitamin content of a low-energy diet

Legend: oz = ounce; T = tablespoon; tsp = teaspoon; kcal = kilocalorie; kcal/kg = kilocalorie per kilogram body weight; CHO = carbohydrate; g = gram; g/kg = gram per kilogram body weight; DRI = Dietary Reference Intakes

time. In this case the athlete is eating some foods that are "healthy" such as lean turkey, green salad, and banana but drastically limiting the amount of food consumed. Because energy intake is so low, the diet lacks the recommended amounts of most vitamins with the exception of folate and vitamin A. This illustrates how caloric intake and vitamin intake are closely tied. Athletes who are restricting calories are usually advised to take a multivitamin supplement. With supplementation, a low-energy (calorie) diet can meet the DRI for each vitamin that was previously lacking.

Vitamins are added to many foods marketed to athletes.

Adding vitamins to food began in the United States as a way to prevent vitamin deficiencies, which were widespread in the 1930s and 1940s. In the 2000s, manufacturers began adding vitamins to foods such as meal replacement bars, energy bars, and energy beverages, products that are highly marketed to athletes. To gain market share in the highly competitive new product market, manufacturers are adding "nutrient horsepower" to foods. In other words, they are adding many vitamins in high doses so consumers perceive their products as being more nutritious than other similar products.

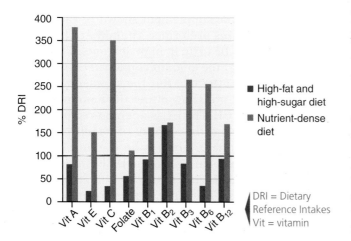
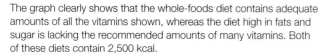

Figure 8.7 Vitamin content of two diets compared

The graph clearly shows that the whole-foods diet contains adequate amounts of all the vitamins shown, whereas the diet high in fats and sugar is lacking the recommended amounts of many vitamins. Both of these diets contain 2,500 kcal.

Table 8.12 High Vitamin Content of Diet When Highly Fortified Foods Are Consumed

Vitamin	Intake	DRI*	% of DRI	UL
Thiamin	4.9 mg	1.2 mg	406	Not established
Riboflavin	5.4 mg	1.3 mg	412	Not established
Niacin	63 mg	16 mg	394	35 mg**
Vitamin B$_6$	6.4 mg	1.3 mg	492	100 mg
Vitamin B$_{12}$	19 mcg	2.4 mcg	788	Not established
Folate	1,352 mcg	400 mcg	338	1,000 mcg**
Vitamin C	240 mg	90 mg	267	2,000 mg
Vitamin A	758 mcg	900 mcg	84	3,000 mcg
Vitamin E	64 mg	15 mg	428	1,000 mg**

Legend: DRI = Dietary Reference Intakes; UL = Tolerable Upper Intake Level; mg = milligrams; mcg = micrograms

With the exception of one vitamin, the consumption of three highly fortified foods (cereal, energy bar, and meal replacement beverage) provides two to seven times the DRI for the nine vitamins analyzed.

*26-year-old male

**The UL for these nutrients is based on the intake of fortified food and supplements only. For the remaining vitamins, the UL is based on all food (fortified and nonfortified), water, and supplements.

Table 8.12 illustrates the amount of vitamins in 1 cup of fortified breakfast cereal, one fortified energy bar, and one meal replacement beverage. These foods were chosen because they represent foods that could realistically be consumed in 1 day by the average athlete. As this table is limited to the vitamins found in the three fortified foods, keep in mind that the total daily vitamin intake would be much higher if all the food and beverages consumed in a 24-hour period were analyzed.

The table includes the Tolerable Upper Intake Levels. The UL for vitamins is based on the intake of all food (fortified and nonfortified), water, and supplements except for three vitamins. For vitamin E, niacin, and folate, the UL is based on fortified food and supplements only. The UL cannot be established for some vitamins because of a lack of scientific data. This example clearly illustrates that the intake of many highly fortified foods provides vitamins at levels several times higher than the DRI and, in the case of two vitamins, higher than the UL. As mentioned previously, the line between fortified foods and supplements is beginning to blur.

The dose and potency of a vitamin supplement can vary substantially from brand to brand.

It behooves athletes to look carefully at the dose of any multivitamin supplement that they are considering taking. The amounts shown in Table 8.13 and Table 8.14 are from actual supplements available for purchase. Table 8.13 is typical of the amounts shown in a one-a-day type multivitamin supplement, whereas Table 8.14 is typical of a high-potency vitamin supplement. The UL is provided as a basis of comparison but is not required to appear on the supplement label. Notice that the amounts contained in the high-potency vitamin supplement are very high when compared to the Daily Value. Despite the high levels, such vitamin preparations are legal to sell.

The amount and form of vitamin A is a particular concern for the supplement shown in Table 8.14 because it contains preformed vitamin A. This form of vitamin A can be toxic in high amounts. The chart clearly shows that the amount contained (3,003 mcg) is comparable to the UL. But the supplement label (Figure 8.9) will not show either the amount in micrograms or the UL. The UL is not required by law to be on the label, and the amount of vitamin A listed on many supplement labels is an old unit of measure, IU (International Unit). The Daily Value is also based on an older, higher figure for vitamin A (5,000 IU or 1,500 mcg). The current recommended daily intake of vitamin A is lower than in the past (Penniston and Tanumihardjo, 2003). An athlete who looks at this supplement will see that it contains 200 percent of the DV and may conclude that 10,000 IU is not an excessive amount. The consumer is unlikely to be able to convert IU to micrograms, compare the amount contained to the Tolerable Upper Intake

Table 8.13 Vitamin Content of a One-A-Day Vitamin Supplement Compared to the Daily Value (DV)

Vitamin	Intake	% of DV	UL
Thiamin	25 mg	1,666	Not established
Riboflavin	25 mg	1,470	Not established
Niacin	100 mg	500	35 mg*
Vitamin B_6	25 mg	1,250	100 mg
Vitamin B_{12}	100 mcg	1,666	Not established
Folate	400 mcg	100	1,000 mcg*
Vitamin C	150 mg	250	2,000 mg
Vitamin A (beta-carotene)	10,000 IU or 1,000 mcg	200	3,000 mcg
Vitamin E	100 mg	333	1,000 mg*

Legend: DV = Daily Value; UL = Tolerable Upper Intake Level; mg = milligrams; mcg = micrograms; IU = International Units

The vitamin content of a typical one-a-day type vitamin supplement ranges from 100 percent to over 1,600 percent of the Daily Value (DV).

*The UL for these nutrients is based on the intake of fortified food and supplements only. For the remaining vitamins, the UL is based on all food (fortified and nonfortified), water, and supplements.

Table 8.14 Vitamin Content of a High-Potency Vitamin Supplement Compared to the Daily Value (DV)

Vitamin	Intake	% of DV	UL
Thiamin	50 mg	3,333	Not established
Riboflavin	50 mg	2,941	Not established
Niacin	50 mg	250	35 mg*
Vitamin B_6	50 mg	2,500	100 mg
Vitamin B_{12}	50 mcg	833	Not established
Folate	400 mcg	100	1,000 mcg*
Vitamin C	500 mg	833	2,000 mg
Vitamin A as fish liver oil, (preformed)	10,000 IU or 3,003 mcg	200	3,000 mcg
Vitamin E	400 mg	1,333	1,000 mg*

Legend: DV = Daily Value; UL = Tolerable Upper Intake Level; mg = milligrams; mcg = micrograms; IU = International Units

The vitamin content of this high-potency vitamin supplement ranges from 100 percent to over 3,300 percent of the Daily Value (DV).

*The UL for these nutrients is based on the intake of fortified food and supplements only. For the remaining vitamins, the UL is based on all food (fortified and nonfortified), water, and supplements.

Level, or recognize that the amount contained is potentially toxic.

The UL represents decades of research and allows scientists and consumers a yardstick by which to measure adequate (that is, DRI) and relatively safe (that is, UL) intakes of vitamins. The UL has become particularly important as more processed foods have added vitamins and vitamin supplements usage has increased. Some nutrition scientists are beginning to question whether there are effects associated with chronic intake of highly fortified foods and routine use of vitamin supplements. For example, researchers

Keeping it in perspective

The Need for an Adequate but Not Excessive Amount of Vitamins

Moderation is the process of limiting the extremes, an important principle that can be applied to many areas, including vitamins. To obtain and maintain excellent health, both vitamin deficiencies and toxicities must be avoided. Vitamins may function as antioxidants at moderate levels in part because they are properly balanced. A high intake of one antioxidant may upset the balance needed to respond optimally to oxidative stress. An adequate vitamin intake is typically associated with adequate caloric intake, whereas the caloric extremes are often associated with poor vitamin intake. These are just a few examples of why vitamin intake should be moderate and not extreme.

There is no moderation when it comes to information about vitamins and vitamin supplements. A comprehensive Internet search of the word *vitamin* yields about 200 million matches, of which millions are for vitamin supplements. With the amount of information and advertisement about vitamins, it may be hard for consumers, including athletes, to keep vitamin needs and intake in perspective.

SUPPLEMENT FACTS

Serving size: 1 tablet	
Servings per container: 60	
Amount per serving	% Daily Value
Thiamin 50 mg	3,333%
Riboflavin 50 mg	2,941%
Niacin 50 mg	250%
Vitamin B$_6$ 50 mg	2,500%
Vitamin B$_{12}$ 50 mcg	833%
Folate 400 mcg	100%
Vitamin C 500 mg	833%
Vitamin A 10,000 IU	200%
Vitamin E 400 mg	1,333%

Ingredients: Vitamin A as fish liver oil (preformed), asorbic acid, di-alpha tocopherol acetate, thiamin mononitrate, riboflavin, niacinamide, pyridoxine HCl, folic acid, cyanocobalamin.
Storage: Keep in a cool dry place, tightly closed.
Suggested Use: As a dietary supplement, take one tablet daily.

Keep out of reach of children
Expiration date: Dec 2010

Figure 8.9 High-potency vitamin supplement label

Although the amount and percent of the Daily Value are required to appear on the label, this information may be difficult for consumers to interpret. In this example, consumers would need to convert the amount stated in IU to micrograms and compare it to the UL (Tolerable Upper Intake Level).

have found that folate intake from all sources (naturally occurring, fortified foods, and supplements) was low in some groups, such as women of childbearing age, but high in other groups, such as those who are older than 50. In fact, 5 percent of those older than 50 years of age exceeded the UL for folate intake (Bailey et al., 2010). Although this study was not conducted in athletes, it brings to light an important lesson about dosage and total vitamin intake from all sources.

Athletes must develop a plan for getting enough of each vitamin without getting too much. It is ironic that it may take less time and effort to consume too much than to consume an appropriate amount by eating a variety of nutrient-dense foods. Food and supplements are not mutually exclusive; rather, they are complementary. Athletes should first determine the amount of vitamins normally obtained from food and then evaluate if they need to supplement. If vitamin supplements are necessary, then decisions must be made regarding dosage. Sports dietitians, physicians, and pharmacists are professionals who can help the athlete decide whether supplements are necessary and, if so, the proper dose and form of a vitamin supplement.

Application exercise

Scenario: *A high school-age competitive female gymnast knows she does not eat as well as she should. Her mother has decided that she needs to take a multivitamin supplement.* She enters multivitamin supplement *into her favorite search engine to see what might be available.*

1. Enter the term *multivitamin supplement* into your favorite search engine (for example, Google, Bing, and so on).
   - How many results/matches are there?
   - What are the differences in the search results between the Sponsored Links (ads) and the list of other search results?
   - How would you decide which entry to click on?

2. Click on one of the links that sells a multivitamin supplement and print the product information. Answer the following questions about that multivitamin supplement.
   - Does the supplement contain more than 100 percent of the Daily Value (DV) for any vitamin?
   - Is the Tolerable Upper Intake Level (UL) exceeded for any vitamin?
   - Is the vitamin supplement high-potency?
   - Does the advertising information appear to be factual? Give examples.
   - Is any of the information factual but taken out of context or appear to be overstated? Give examples.

3. Decision making
   - Using the multivitamin supplement product you selected, outline the specific factors that you think would influence the decision to purchase/use this supplement or to not purchase/use this supplement.
   - If you were a registered dietitian, would you recommend this supplement? Why or why not?

Key points

- Vitamins can be obtained naturally in food, from fortified foods, and in vitamin supplements.
- Low caloric intake usually means low vitamin intake from food.

Don't guess how much you might be taking in—conduct a dietary analysis to determine whether usual vitamin intake is inadequate, adequate, or excessive.

How do the DRI, UL, or DV help an athlete make decisions about vitamin supplements?

Post-Test   Reassessing Knowledge of Vitamins

Now that you have more information about vitamins, read the following statements and decide if each is true or false. The answers can be found in Appendix O.

1. Exercise increases the usage of vitamins, so most athletes need more vitamins than sedentary people.

2. Vitamins provide energy.

3. The amount of vitamins an athlete consumes is generally related to caloric intake.

4. When antioxidant vitamins are consumed in excess, they act like pro-oxidants instead of antioxidants.

5. Vitamin supplements are better regulated than other dietary supplements because the U.S. Food and Drug Administration sets a maximum dose (amount) for each vitamin.

Summary

Vitamins are essential nutrients needed for the proper functioning of the body, particularly energy metabolism, red blood cell production, antioxidant functions, and growth and development. Exercise increases the need for some vitamins, however the increase is small and likely covered by meeting the amounts recommended by the Dietary Reference Intakes (DRI). Unfortunately, many athletes do not consume a sufficient amount of vitamins daily, usually due to low caloric intake or low intake of vitamin-rich foods. The best advice to athletes is to consume vitamins in the amounts recommended by the Dietary Reference Intakes and to not exceed the Tolerable Upper Intake Level. Presuming that absorption and utilization are normal, this strategy should prevent subclinical deficiencies and vitamin toxicities.

Vitamins perform specific biochemical functions and often interact with each other. The B-complex vitamins, including thiamin, riboflavin, and niacin, are involved in energy metabolism because they are part of enzymes that catalyze energy reactions. The antioxidant vitamins—vitamin E, C, and beta-carotene (a form of vitamin A)—are a key part of the body's antioxidant mechanisms. The body's antioxidant system depends on balance, not only between oxidants and antioxidants but also among the vitamins themselves. Vitamins are also required for proper red blood cell production and growth and development.

Adequate amounts of all the vitamins are needed to support training, performance, and health. A "food first, supplement second" philosophy can serve athletes well. Obtaining the DRI for vitamins from food requires that athletes consume adequate energy and eat vitamin-rich foods such as fruits, vegetables, whole grains, legumes, and nuts. Many athletes take a multivitamin supplement. Such a supplement can provide vitamins missing from the diet but does not provide carbohydrates, proteins, fats, or fiber, important nutrients to support training, performance, and health.

Before choosing to supplement with any vitamin, an athlete should evaluate dietary intake and determine if any vitamin deficiencies likely exist. If so, the athlete should develop a strategy for consuming an adequate amount of vitamins daily. An initial approach could be changing food intake. In cases where supplementation is warranted, the supplement dose and ingredients should be evaluated so safety, effectiveness, and potential advantages and disadvantages of use can be judged appropriately. Athletes should read vitamin supplement labels carefully and recognize that some supplements may contain high doses that could be potentially harmful.

Self-Test

Multiple-Choice

The answers can be found in Appendix O.

1. According to the current body of scientific research, what effect does exercise have on an athlete's vitamin requirements?
   a. increases the need substantially, making vitamin supplementation necessary
   b. increases the need substantially but supplementation not necessary
   c. increases the need slightly but supplementation not necessary
   d. increases the need slightly, making vitamin supplementation necessary

2. The intake of which vitamin is likely to be low in both sedentary and athletic populations?
   a. thiamin
   b. riboflavin
   c. niacin
   d. vitamin E

3. The amount of energy (kcal) provided by the B-complex vitamins is:
   a. 4 kcal/g.
   b. 7 kcal/g.
   c. 9 kcal/g.
   d. none.

4. Which of the following is true regarding athletes and the consumption of an adequate amount of vitamin A from food?
   a. Vitamin A is a difficult vitamin to obtain from food.
   b. Low vitamin A intake from food is rare.
   c. Vitamin A intake generally reflects fruit and vegetable intake.
   d. Vitamin A intake generally reflects energy intake.

5. A female figure skater consumes a low-calorie diet consisting of approximately 1,200 kcal daily. She says that she is taking a multivitamin supplement as an insurance policy. What is the multivitamin supplement insuring her against?
   a. low energy intake
   b. low vitamin intake
   c. low carbohydrate intake
   d. both low energy intake and low vitamin intake

Short Answer

1. Compare and contrast fat- and water-soluble vitamins.
2. Explain the physiological and performance effects of vitamin intake that is too low, adequate, and excessive.
3. Do vitamins provide energy? Explain.
4. Explain the effects of exercise on vitamin need. Are the requirements for vitamins increased for athletes?
5. Why is it important for athletes to determine their intake of vitamins from food?
6. Do sedentary adults consume enough vitamins? Do athletes? Explain.
7. Describe how a low vitamin intake could become a subclinical vitamin deficiency. Describe how excessive vitamin intake could become a toxicity.
8. Explain the differences between a clinical and subclinical vitamin deficiency.
9. List and explain the roles of vitamins associated with antioxidant activity, energy metabolism, red blood cell formation, and growth and development.
10. Why do athletes inject vitamin B_{12}? Is this practice safe? Is it effective?
11. Are vitamin supplements helpful to athletes? If so, in which ways? Are vitamin supplements harmful to athletes? If so, in which ways?
12. Outline a process an athlete could follow to determine if a vitamin supplement is needed.

Critical Thinking

1. Suppose that both of these high school athletes have vitamin deficiencies: a basketball player from an economically disadvantaged area with little parental support and a cross country runner who appears to have every economic and social advantage. Why might each be vitamin deficient and what solutions might you propose?
2. Studies have shown that people who eat nutritiously tend to take vitamin supplements whereas people who don't eat well do not supplement. Why might this be the case?

References

Adams, J. S., & Hewison, M. (2010). Update in vitamin D. *Journal of Clinical Endocrinology and Metabolism, 95*(2), 471–478.

Akabas, S. R., & Dolins, K. R. (2005). Micronutrient requirements of physically active women: What can we learn from iron? *American Journal of Clinical Nutrition, 81*(5), 1246S–1251S.

Bailey, R. L., Dodd, K. W., Gahche, J. J., Dwyer, J. T., McDowell, M. A., Yetley, E. A., et al. (2010). Total folate and folic acid intake from foods and dietary supplements in the United States: 2003–2006. *American Journal of Clinical Nutrition*, 91(1), 231–237.

Bardia, A., Tleyjeh, I. M., Cerhan, J. R., Sood, A. K., Limburg, P. J., Erwin, P. J., et al. (2008). Efficacy of antioxidant supplementation in reducing primary cancer incidence and mortality: Systematic review and meta-analysis. *Mayo Clinic Proceedings*, 83(1), 23–34.

Bendich, A. (2000). The potential for dietary supplements to reduce premenstrual syndrome (PMS) symptoms. *Journal of the American College of Nutrition*, 19(1), 3–12.

Bischoff-Ferrari, H. A., Dawson-Hughes, B., Staehelin, H. B., Orav, J. E., Stuck, A. E., Theiler, R., et al. (2009). Fall prevention with supplemental and active forms of vitamin D: A meta-analysis of randomised controlled trials. *British Medical Journal*, 339, b3692. doi: 10.1136/bmj.b3692.

Bjelakovic, G., Nikolova, D., Gluud, L. L., Simonetti, R. G., & Gluud, C. (2008). Antioxidant supplements for prevention of mortality in healthy participants and patients with various diseases. Cochrane Database of Systematic Reviews. Apr 16;(2):CD007176.

Carlsohn, A., Rohn, S., Mayer, F., & Schweigert, F. J. (2010). Physical activity, antioxidant status and protein modification in adolescent athletes. *Medicine and Science in Sports and Exercise*, 42(6), 1131–1139.

Cheuvront, S. N., Ely, B. R., Kenefick, R. W., Michniak-Kohn, B. B., Rood, J. C., & Sawka, M. N. (2009). No effect of nutritional adenosine receptor antagonists on exercise performance in the heat. *American Journal of Physiology. Regulatory, Integrative and Comparative Physiology*, 296, R394–R401.

Chong, E. W., Wong, T. Y., Kreis, A. J., Simpson, J. A., & Guymer, R. H. (2007). Dietary antioxidants and primary prevention of age related macular degeneration: Systematic review and meta-analysis. *British Journal of Medicine*, 335(7623), 755.

Chun, O. K., Floegel, A., Chung, S. J., Chung, C. E., Song, W. O., & Koo, S. I. (2010). Estimation of antioxidant intakes from diet and supplements in U.S. adults. *Journal of Nutrition*, 140(2), 317–324. Erratum in: *Journal of Nutrition*, 2010, 140(5), 1062.

Cureton, K. J., Tomporowski, P. D., Singhal, A., Pasley, J. D., Bigelman, K. A., Lambourne, K., et al. (2009). Dietary quercetin supplementation is not ergogenic in untrained men. *Journal of Applied Physiology*, 107(4), 1095–1104.

Davis, J. M., Carlstedt, C. J., Chen, S., Carmichael, M. D., & Murphy, E. A. (2010). The dietary flavonoid quercetin increases VO($_2$max) and endurance capacity. *International Journal of Sport Nutrition and Exercise Metabolism*, 20(1), 56–62.

de Silva, A., Samarasinghe, Y., Senanayake, D., & Lanerolle, P. (2010). Dietary supplement intake in national-level Sri Lankan athletes. *International Journal of Sport Nutrition and Exercise Metabolism*, 20(1), 15–20.

Dietary Guidelines Advisory Committee (2010). Report on the Dietary Guidelines Committee on Dietary Guidelines for Americans, 2010, to the Secretary of Agriculture and the Secretary of Health and Human Services. U.S. Department of Agriculture, Agricultural Research Service, Washington, D.C.

DIPART (Vitamin D Individual Patient Analysis of Randomized Trials) Group. (2010). Patient level pooled analysis of 68, 500 patients from seven major vitamin D fracture trials in US and Europe. *British Medical Journal*, 12;340:b5463. doi:10.1136/bmj.b5463.

Douglas, R. M., Hemila, H., Chalker, E. B., & Treacy, D. (2007). Vitamin C for preventing and treating the common cold. *Cochrane Database of Systematic Reviews*, 3, CD000980. Update of 2004 (4):CD000980.

Dumke, C. L., Nieman, D. C., Utter, A. C., Rigby, M. D., Quindry, J. C., Triplett, N. T., et al. (2009). Quercetin's effect on cycling efficiency and substrate utilization. *Applied Physiology, Nutrition, and Metabolism*, 34(6), 993–1000.

Elmadfa, I., & Singer, I. (2009). Vitamin B$_{12}$ and homocysteine status among vegetarians: A global perspective. *American Journal of Clinical Nutrition*, 89(5), 1693S–1698S.

Evans, W. J. (2000). Vitamin E, vitamin C, and exercise. *American Journal of Clinical Nutrition*, 72(2 Suppl.), 647S–652S.

Ferguson, J. H., & Chang, A. B. (2009). Vitamin D supplementation for cystic fibrosis. Cochrane Database of Systematic Reviews. Oct 7;(4):CD007298.

Gleeson, M., Nieman, D. C., & Pedersen, B. K. (2004). Exercise, nutrition and immune function. *Journal of Sports Sciences*, 22(1), 115–125.

Gropper, S. S., Smith, J. L., & Groff, J. L. (2009a). *Advanced Nutrition and Human Metabolism* (5th ed.). Belmont, CA: Wadsworth, Cengage Learning.

Gropper, S. S., Smith, J. L., & Groff, J. L. (2009b). Perspective: The antioxidant nutrients, reactive species, and disease. In *Advanced nutrition and human metabolism* (5th ed., pp. 417–427). Belmont, CA: Wadsworth, Cengage Learning.

Hamer, M., & Chida, Y. (2007). Intake of fruit, vegetables, and antioxidants and risk of type 2 diabetes: Systematic review and meta-analysis. *Journal of Hypertension*, 25(12), 2361–2369.

He, F. J., Nowson, C. A., & MacGregor, G. A. (2006). Fruit and vegetable consumption and stroke: Meta-analysis of cohort studies. *Lancet*, 367(9507), 320–326.

He, F. J., Nowson, C. A., Lucas, M., & MacGregor, G. A. (2007). Increased consumption of fruit and vegetables is related to a reduced risk of coronary heart disease: Meta-analysis of cohort studies. *Journal of Human Hypertension*, 21(9), 717–728.

Heaney, R. P. (2008). Vitamin D: Criteria for safety and efficacy. *Nutrition Reviews*, 66(10, Suppl. 2), S178–S181.

Horvath, P. J., Eagen, C. K., Ryer-Calvin, S. D., & Pendergast, D. R. (2000). The effects of varying dietary fat on the nutrient intake in male and female runners. *Journal of the American College of Nutrition*, 19(1), 42–51.

Huang, H.Y., Caballero, B., Chang, S., Alberg, A., Semba, R., Schneyer, C., et al. (2006). Multivitamin/mineral supplements and prevention of chronic disease.

Evidence Report/Technology Assessment (Full Rep), May; (139), 1–117.

Institute of Medicine. (1998). Dietary Reference Intakes for thiamin, riboflavin, niacin, vitamin B_6, folate, vitamin B_{12}, pantothenic acid, biotin and choline (Food and Nutrition Board). Washington, DC: National Academies Press.

Institute of Medicine. (2000). Dietary Reference Intakes for vitamin C, vitamin E, selenium and carotenoids (Food and Nutrition Board). Washington, DC: National Academies Press.

Institute of Medicine (2010). Dietary Reference Intakes for calcium and vitamin D. Food and Nutrition Board. Washington, DC: National Academies Press.

Jackson, M. J., Khassaf, M., Vasilaki, A., McArdle, F., & McArdle, A. (2004). Vitamin E and the oxidative stress of exercise. *Annals of the New York Academy of Sciences*, *1031*, 158–168.

Jonnalagadda, S. S., Ziegler, P. J., & Nelson, J. A. (2004). Food preferences, dieting behaviors, and body image perceptions of elite figure skaters. *International Journal of Sport Nutrition and Exercise Metabolism*, *14*(5), 594–606.

Koushik, A., Hunter, D. J., Spiegelman, D., Beeson, W. L., van den Brandt, P. A., Buring, J. E., et al. (2007). Fruits, vegetables, and colon cancer risk in a pooled analysis of 14 cohort studies. *Journal of the National Cancer Institute*, *99*(19), 1471–1483.

Kulie, T., Groff, A., Redmer, J., Hounshell, J., & Schrager, S. (2009). Vitamin D: An evidence-based review. *Journal of the American Board of Family Medicine*, *22*(6), 698–706. Erratum in: *Journal of the American Board of Family Medicine*, *23*(1), 138.

Leydon, M. A., & Wall, C. (2002). New Zealand jockeys' dietary habits and their potential impact on health. *International Journal of Sport Nutrition and Exercise Metabolism*, *12*(2), 220–237.

Li, Y., & Schellhorn, H. E. (2007). New developments and novel therapeutic perspectives for vitamin C. *Journal of Nutrition*, *137*(10), 2171–2184.

Lukaski, H. C. (2004). Vitamin and mineral status: Effects on physical performance. *Nutrition*, *20*(7–8), 632–644.

Machefer, G., Groussard, C., Zouhal, H., Vincent, S., Youssef, H., Faure, H., et al. (2007). Nutritional and plasmatic antioxidant vitamins status of ultra endurance athletes. *Journal of the American College of Nutrition*, *26*(4), 311–316.

Manore, M. M., Meyer, N. L., & Thompson, J. (2009). *Sport Nutrition for Health and Performance* (2nd ed.). Champaign, IL: Human Kinetics.

McAnulty, S. R., McAnulty, L. S., Nieman, D. C., Quindry, J. C., Hosick, P. A., Hudson, M. H., et al. (2008). Chronic quercetin ingestion and exercise-induced oxidative damage and inflammation. *Applied Physiology, Nutrition, and Metabolism*, *33*(2), 254–262.

McBride, J. M., Kraemer, W. J., Triplett-McBride, T., & Sebastianelli, W. (1998). Effect of resistance exercise on free radical production. *Medicine & Science in Sports & Exercise*, 30, 1, 67–72.

McGinley, C., Shafat, A., & Donnelly, A. E. (2009). Does antioxidant vitamin supplementation protect against muscle damage? *Sports Medicine*, *39*(12), 1011–1132.

Miller, E. R., III, Pastor-Barriuso, R., Dalal, D., Riemersma, R. A., Appel, L. J., & Guallar, E. (2005). Meta-analysis: High-dosage vitamin E supplementation may increase all-cause mortality. *Annals of Internal Medicine*, *142*(1), 37–46.

Moreira, A., Kekkonen, R. A., Delgado, L., Fonseca, J., Korpela, R., & Haahtela, T. (2007). Nutritional modulation of exercise-induced immunodepression in athletes: A systematic review and meta-analysis. *European Journal of Clinical Nutrition*, *61*(4), 443–460.

Morrison, L. J., Gizis, F., & Shorter, B. (2004). Prevalent use of dietary supplements among people who exercise at a commercial gym. *International Journal of Sport Nutrition and Exercise Metabolism*, *14*(4), 481–492.

Myhre, A. M., Carlsen, M. H., Bohn, S. K., Wold, H. L., Laake, P., & Blomhoff, R. (2003). Water-miscible, emulsified, and solid forms of retinol supplements are more toxic than oil-based preparations. *American Journal of Clinical Nutrition*, *78*(6), 1152–1159.

Nieman, D. C., Henson D. A., Maxwell K. R., Williams A. S., McAnulty S. R., Shanely, R. A., et al. (2009). Effects of quercetin and EGCG on mitochondrial biogenesis and immunity. *Medicine and Science in Sports and Exercise*, *41*(7), 1467–1475.

Nieman, D. C., Williams, A. S., Shanely, R. A., Jin, F., McAnulty, S. R., Triplett, N. T., et al. (2010). Quercetin's influence on exercise performance and muscle mitochondrial biogenesis. *Medicine and Science in Sports and Exercise*, *42*(4), 338–345.

Papadopoulou, S. K., Papadopoulou, S. D., & Gallos, G. K. (2002). Macro- and micro-nutrient intake of adolescent Greek female volleyball players. *International Journal of Sport Nutrition and Exercise Metabolism*, *12*(1), 73–80.

Penniston, K. L., & Tanumihardjo, S. A. (2003). Vitamin A in dietary supplements and fortified foods: Too much of a good thing? *Journal of the American Dietetic Association*, *103*(9), 1185–1187.

Petróczi, A., Naughton, D. P., Mazanov, J., Holloway, A., & Bingham, J. (2007). Limited agreement exists between rationale and practice in athletes' supplement use for maintenance of health: A retrospective study. *Nutrition Journal*, 6, 34.

Pittas, A.G., Chung, M., Trikalinos, T., Mitri, J., Brendel, M., Patel, K., et al. (2010). Systematic review: Vitamin D and cardiometabolic outcomes. *Annals of Internal Medicine*, *152*(5), 307–314.

Plunkett, B. A., Callister, R., Watson, T. A., & Garg, M. L. (2010). Dietary antioxidant restriction affects the inflammatory response in athletes. *British Journal of Nutrition*, *103*(8), 1179–1184.

Powers, S.K. & Lennon, S.L. (1999). Analysis of cellular responses to free radicals: Focus on exercise and skeletal muscle. *The Proceedings of the Nutrition Society*, *58*(4), 1025–1033.

Quindry, J. C., McAnulty, S. R., Hudson, M. B., Hosick, P., Dumke, C., McAnulty, L. S., et al. (2008). Oral quercetin supplementation and blood oxidative capacity in response to ultramarathon competition. *International Journal of Sport Nutrition and Exercise Metabolism*, *18*(6), 601–616.

Ramanathan, V. S., Hensley, G., French, S., Eysselein, V., Chung, D, Reicher, S., et al. (2010). Hypervitaminosis A inducing intra-hepatic cholestasis—a rare case report. *Experimental and Molecular Pathology*, *88*(2), 324–325.

Ristow, M., Zarse, K., Oberbach, A., Klöting, N., Birringer, M., Kiehntopf, M., et al. (2009). Antioxidants prevent health-promoting effects of physical exercise in humans. *Proceedings of the National Academy of Sciences of the United States of America*, *106*(21), 8665–8670.

Sesso, H. D., Buring, J. E., Christen, W. G., Kurth, T., Belanger, C., MacFadyen, J., et al. (2008). Vitamins E and C in the prevention of cardiovascular disease in men: The Physicians' Health Study II randomized controlled trial. *Journal of the American Medical Association*, *300*(18), 2123–2133.

Sherwood, L. (2007). *Human Physiology: From Cells to Systems* (6th ed.). Belmont, CA: Thomson Brooks/Cole.

Soric, M., Misigoj-Durakovic, M., & Pedisic, Z. (2008). Dietary intake and body composition of prepubescent female aesthetic athletes. *International Journal of Sport Nutrition and Exercise Metabolism*, *18*(3), 343–354.

Stránský, M., & Rysavá, L. (2009). Nutrition as prevention and treatment of osteoporosis. *Physiological Research*, *1*, S7–S11.

Stroud, M. L., Stilgoe, S., Stott, V. E., Alhabian, O., & Salman, K. (2008). Vitamin D—a review. *Australian Family Physician*, *37*(12),1002–1005.

Tanumihardjo, S.A. (2004). Assessing vitamin A status: Past, present and future. *Journal of Nutrition*, *134*(1), 290S–293S.

Teixeira, V., Valente, H., Casal, S., Marques, F., & Moreira, P. (2009). Antioxidant status, oxidative stress, and damage in elite trained kayakers and canoeists and sedentary controls. *International Journal of Sport Nutrition and Exercise Metabolism*, *19*(5), 443–456.

Tomten, S. E., & Høstmark, A. T. (2009). Serum vitamin E concentration and osmotic fragility in female long-distance runners. *Journal of Sports Science*, *27*(1), 69–76.

Utter, A. C., Nieman, D. C., Kang, J., Dumke, C. L., Quindry, J. C., McAnulty, S. R., et al. (2009). Quercetin does not affect rating of perceived exertion in athletes during the Western States endurance run. *Research in Sports Medicine*, *17*(2), 71–83.

Viitala, P., & Newhouse, I. J. (2004). Vitamin E supplementation, exercise and lipid peroxidation in human participants. *European Journal of Applied Physiology*, *93*(1–2), 108–115.

Viitala, P. E., Newhouse, I. J., LaVoie, N., & Gottardo, C. (2004). The effects of antioxidant vitamin supplementation on resistance exercise induced lipid peroxidation in trained and untrained participants. *Lipids in Health and Disease*, *3*, 14.

Vogiatzoglou, A., Smith, A. D., Nurk, E., Berstad, P., Drevon, C. A., Ueland, P. M., et al. (2009). Dietary sources of vitamin B-12 and their association with plasma vitamin B-12 concentrations in the general population: The Hordaland Homocysteine Study. *American Journal of Clinical Nutrition*, *89*(4), 1078–1087.

Volpe, S. L. (2007). Micronutrient requirements for athletes. *Clinics in Sports Medicine*, *26*(1), 119–130.

Williams, S. L., Strobel, N. A., Lexis, L. A., & Coombes, J. S. (2006). Antioxidant requirements of endurance athletes: Implications for health. *Nutrition Reviews*, *64*(3), 93–108.

Willis, K. S., Peterson, N. J., & Larson-Meyer, D. E. (2008). Should we be concerned about the vitamin D status of athletes? *International Journal of Sport Nutrition and Exercise Metabolism*, *18*(2), 204–224.

Woolf, K., & Manore, M. M. (2006). B-vitamins and exercise: Does exercise alter requirements? *International Journal of Sport Nutrition and Exercise Metabolism*, *16*(5), 453–484.

Yin, L., Grandi, N., Raum, E., Haug, U., Arndt, V., & Brenner, H. (2009a). Meta-analysis: Longitudinal studies of serum vitamin D and colorectal cancer risk. *Alimentary Pharmacology & Therapeutics*, *30*(2), 113–125.

Yin, L., Grandi, N., Raum, E., Haug, U., Arndt, V., & Brenner, H. (2009b). Meta-analysis of longitudinal studies: Serum vitamin D and prostate cancer risk. *Cancer Epidemiology*, *33*(6), 435–445.

Ziegler, P. J., Nelson, J. A., & Jonnalagadda, S. S. (1999). Nutritional and physiological status of U.S. national figure skaters. *International Journal of Sport Nutrition and Exercise Metabolism*, *9*(4), 345–360.

Ziegler, P., Sharp, R., Hughes, V., Evans, W., & Khoo, C.S. (2002). Nutritional status of teenage female competitive figure skaters. *Journal of the American Dietetic Association*, *102*(3), 374–379.

9 Minerals

Learning Plan

- **Classify** minerals and describe their general roles.
- **Explain** how mineral inadequacies and excesses can occur and why each might be detrimental to performance and health.
- **Explain** how the Dietary Reference Intakes (DRI) and the Tolerable Upper Intake Levels (UL) should be interpreted.
- **Describe** if, and how, exercise increases the need for or accelerates the loss of a particular mineral.
- **Compare** and contrast the average intake of minerals by sedentary adults and athletes in the United States.
- **Explain** the differences between a clinical and subclinical deficiency for calcium, iron, and zinc.
- **Discuss** the minerals associated with bone formation, red blood cell production, and the immune system, and explain how low intake affects performance and health.
- **Compare** and contrast minerals based on their source—naturally occurring in food, added to foods during processing, and found in supplements.
- **Evaluate** the need for mineral supplements based on food intake and the safety and effectiveness of mineral supplements.

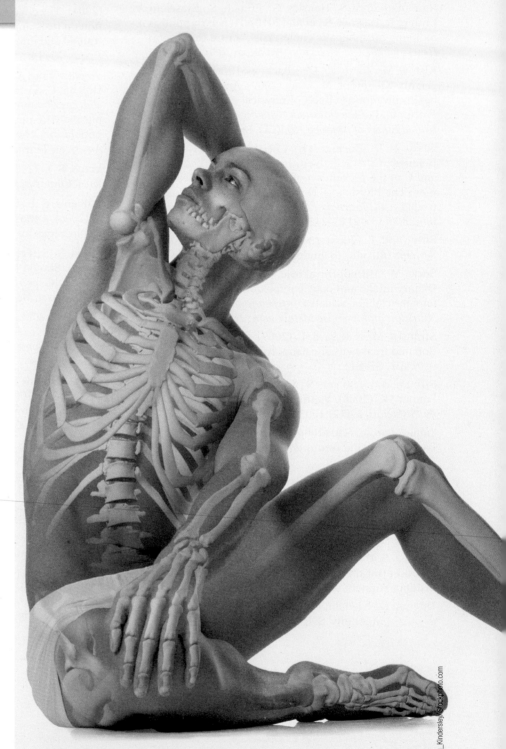

Minerals are nutrients essential to structure and function of the body.

Pre-Test | Assessing Current Knowledge of Minerals

Read the following statements about minerals and decide if each is true or false.

1. The basic functions of minerals include building body tissues, regulating physiological processes, and providing energy.

2. In general, the body absorbs a high percentage of the minerals consumed.

3. In most cases, exercise does not increase mineral requirements above what is recommended for healthy, lightly active individuals.

4. If dietary calcium is inadequate over a long period of time, the body maintains its blood calcium concentration by reabsorbing skeletal calcium.

5. Iron deficiency anemia has a negative impact on performance.

Vitamins and minerals are often mentioned in the same breath. Although there are some similarities between these classes of nutrients, **minerals** differ from vitamins, especially water-soluble vitamins, in several ways. The chemistry, absorption, metabolism, and excretion of minerals are generally very different when compared to vitamins. In the case of two minerals, calcium and iron, there are medical tests that can help quantify the amount in the body, providing valuable information for detecting and treating low bone density and iron deficiency.

Vitamins and minerals are found in a wide variety of foods. Both are added to foods and sold as dietary supplements, either singly or in combination with other nutrients. Both are essential to good health and performance, but there are more differences between vitamins and minerals than there are similarities. This chapter reviews information about minerals needed for the proper functioning of the skeleton, blood, and immune systems with an emphasis on studies conducted with athletes.

9.1 Classification of Minerals

Twenty-one minerals have been identified as essential as shown in Table 9.1. Minerals are often divided into two categories based on the amount found in the body. Those found in relatively large amounts (about 5 g in a 60 kg [132 lb] person) are termed macrominerals and include calcium, phosphorous, magnesium, sodium, potassium, chloride, and sulfur. Microminerals, also known as trace minerals, are found in comparatively smaller amounts in the body. These include well-known minerals, such as iron, as well as lesser-known ones, such as manganese and molybdenum.

Another classification method for minerals is based on function. Minerals critical to proper bone formation include calcium, phosphorus, magnesium, and fluoride. Several minerals are electrolytes and have either a positive or negative charge. Sodium, potassium, and chloride are prime examples of minerals that function as electrolytes and help to maintain

Table 9.1 Minerals

Macrominerals	Microminerals (trace minerals)
Calcium	Iron
Phosphorous (phosphorus)	Zinc
Magnesium	Copper
Sodium	Fluorine (fluoride)
Potassium	Iodine
Chloride	Chromium
Sulfur	Selenium
	Manganese
	Molybdenum
	Cobalt
	Silicon
	Boron
	Nickel
	Vanadium

Macrominerals are found in relatively large amounts in the body, whereas microminerals are found in trace amounts.

body fluid balance. Iron is central to proper red blood cell formation. Many enzymes contain minerals such as zinc, selenium, or copper, and some of these minerals are necessary for the proper functioning of the immune system.

Although some minerals have been studied more than others, basic information about each mineral is known. This information is summarized in Table 9.2.

For each mineral the following appears: chemical symbol, major physiological functions, symptoms associated with a deficiency or toxicity, association with disease prevention, and food sources. More detailed information on minerals can also be found in basic and advanced nutrition textbooks or at the Food and Nutrition Information Center at http://www.nal.usda.gov/fnic.

Table 9.2 Summary of Mineral Characteristics

Boron	
Chemical symbol	B
Major physiological functions	Needed for normal calcium metabolism; most likely functions as a reaction catalyst or regulator
Symptoms of deficiency	Abnormal growth, low sperm count
Symptoms of toxicity	Gastrointestinal symptoms (nausea, vomiting, diarrhea), loss of appetite, fatigue
Health promotion and disease prevention	Recommended daily intake has not been established; further study is needed to better define its roles in bone health and as an anti-inflammatory
Food sources	Green leafy vegetables; some fruits
Other	Little is known about this mineral

Calcium	
Chemical symbol	Ca^{2+} (divalent cation)
Major physiological functions	Mineralization of bones and teeth, muscle contraction, nerve conduction, secretion of hormones and enzymes
Symptoms of deficiency	A dietary calcium deficiency is not associated with any signs or symptoms as bone mineral density declines; spine, wrist, or hip fractures are often the first symptoms; low calcium intake can contribute to hypertension (high blood pressure); low blood calcium concentration, a sign of deficiency, is associated with disease states such as renal (kidney) failure
Symptoms of toxicity	Elevated blood calcium levels, impaired renal function, decreased absorption of other minerals; toxicity typically caused by a disease state but could be caused by excessive intake of either calcium or vitamin D
Health promotion and disease prevention	Adequate intake daily is necessary for good health and is associated with a decreased risk for osteoporosis, high blood pressure, colon cancer, and obesity
Food sources	Milk and milk products; green leafy vegetables; fish with bones such as salmon or sardines; calcium-fortified products such as soy milk or orange juice

Chloride	
Chemical symbol	Cl⁻ (anion)
Major physiological functions	Helps to maintain fluid balance; component of hydrochloric acid (HCl) found in the stomach
Symptoms of deficiency	Failure to thrive (infants)
Symptoms of toxicity	No toxicities from chloride in food reported; rarely toxicities can occur from extremely high ingestion of sodium chloride (table salt); symptoms would include vomiting, muscle weakness, severe dehydration, acidosis
Health promotion and disease prevention	Adequate intake daily is necessary for good health
Food sources	Table salt (sodium chloride, abbreviated NaCl); fish; meat; milk; eggs

Table 9.2 Summary of Mineral Characteristics (Continued)

Chromium	
Chemical symbol	Cr^{3+} (trivalent cation)
Major physiological functions	Enhances insulin sensitivity; involved in carbohydrate, protein, and fat metabolism
Symptoms of deficiency	Rare but likely to be glucose intolerance and weight loss
Symptoms of toxicity	Few side effects reported from excess chromium from food
Health promotion and disease prevention	Adequate intake daily is necessary for good health; chromium may improve glycemic control in those with type 2 diabetes
Food sources	Whole grain breads and cereals; mushrooms; beer
Cobalt	
Chemical symbol	Co
Major physiological functions	Part of vitamin B_{12} (cobalamin)
Symptoms of deficiency	Anemia
Symptoms of toxicity	None known due to food consumption
Food sources	Same as vitamin B_{12} (animal products)
Copper	
Chemical symbol	Cu^{2+} (divalent cation)
Major physiological functions	Part of copper-containing enzymes; role in normal hemoglobin synthesis
Symptoms of deficiency	Anemia, demineralization of bone
Symptoms of toxicity	Gastrointestinal distress; liver damage
Health promotion and disease prevention	Adequate intake daily is necessary for good health
Food sources	Seafood; nuts; seeds; whole grains; dried beans; some green leafy vegetables
Fluorine (fluoride)	
Chemical symbol	F^-
Major physiological functions	Component of bones and teeth. Strengthens bone crystal and resists tooth decay when taken in proper but not excessive amounts
Symptoms of deficiency	Dental caries, osteoporosis
Symptoms of toxicity	Mottled teeth, fragile bones, joint pain
Health promotion and disease prevention	Associated with less dental decay and greater bone strength
Food sources	Fluoridated water; fluoridated vitamins (infants and children)
Iodine (Iodide)	
Chemical symbol	I^-
Major physiological functions	Synthesis of thyroid hormones
Symptoms of deficiency	Mental retardation, impaired growth and development, goiter (enlargement of the thyroid gland), inadequate thyroid hormone production
Symptoms of toxicity	Thyroid-related medical problems
Health promotion and disease prevention	Adequate intake daily is necessary for good health
Food sources	Iodized salt; salt-water fish; mushrooms; eggs

(Continued)

Table 9.2 Summary of Mineral Characteristics (Continued)

Iron	
Chemical symbol	Fe^{2+} (divalent cation; ferrous) or Fe^{3+} (trivalent cation; ferric)
Major physiological functions	Component of hemoglobin, which is necessary for oxygen and carbon dioxide transport; component of enzymes necessary for cellular use of oxygen; immune system functions
Symptoms of deficiency	Fatigue, loss of appetite, reduced resistance to infection
Symptoms of toxicity	Gastrointestinal distress; in those who overabsorb iron, excess iron storage in liver and subsequent liver dysfunction (hemochromatosis)
Health promotion and disease prevention	Adequate intake daily is necessary for good health; always keep iron supplements away from children to prevent accidental overdose; never self-prescribe iron supplements due to risk for iron overload
Food sources	Well-absorbed (heme) sources: clams, oysters, liver, meat, fish, poultry; lesser-absorbed (nonheme) sources: dried beans and legumes, green leafy vegetables, dried fruit, iron-fortified grains
Other	Fe^{2+} = ferrous (storage) form; Fe^{3+} = ferric (transport) form
Magnesium	
Chemical symbol	Mg^{2+} (divalent cation)
Major physiological functions	Bone formation, component of more than 300 enzymes
Symptoms of deficiency	Muscle weakness, confusion, loss of appetite
Symptoms of toxicity	Diarrhea
Food sources	Green leafy vegetables; nuts and seeds; dried beans and legumes
Manganese	
Chemical symbol	Mn^{2+} (divalent cation) or Mn^{3+} (trivalent cation)
Major physiological functions	Role in bone formation; necessary for proper carbohydrate, protein, and fat metabolism
Symptoms of deficiency	Not likely but could result in impaired growth and metabolism
Symptoms of toxicity	Elevated blood manganese; nervous tissue toxicity
Food sources	Nuts; dried beans; whole grains; some vegetables
Molybdenum	
Chemical symbol	Mo^{4+} or Mo^{6+}
Major physiological functions	Component of three enzymes
Symptoms of deficiency	No reports in humans
Symptoms of toxicity	Minimal effects in humans; impaired growth and weight loss in laboratory animals
Food sources	Legumes; whole grains; nuts
Nickel	
Chemical symbol	Ni
Major physiological functions	May be a component of some enzymes and may enhance iron absorption, but few studies have been conducted in humans
Symptoms of deficiency	Not known
Symptoms of toxicity	Low toxicity in humans; gastrointestinal distress associated with nickel poisoning from accidental ingestion of large doses
Food sources	Legumes; nuts; whole grains

Table 9.2 Summary of Mineral Characteristics (Continued)

Phosphorous (phosphorus)	
Chemical symbol	P
Major physiological functions	Component of bone, component of phospholipids (cell membranes), helps to maintain normal pH, part of ATP, involved in cellular metabolism
Symptoms of deficiency	Impaired growth, bone pain, muscle weakness
Symptoms of toxicity	Impaired calcium regulation, calcification of the kidney, possible reduction of calcium absorption
Food sources	Widely found in foods, especially animal foods

Potassium	
Chemical symbol	K^+ (cation)
Major physiological functions	Intracellular cation, proper cellular function
Symptoms of deficiency	Severe deficiency: hypokalemia (low blood potassium) resulting in cardiac arrhythmias and muscle weakness Moderate deficiency: contributes to hypertension and calcium loss from bone
Symptoms of toxicity	Most people readily excrete potassium in urine so toxicity is associated with impaired potassium excretion, which results in hyperkalemia (high blood potassium) and risk for cardiac arrhythmias
Health promotion and disease prevention	Adequate intake daily is necessary for good health; low intake is associated with high blood pressure, glucose intolerance, and an increased risk for kidney stones
Food sources	Vegetables, especially green leafy vegetables; dried beans and peas; orange juice; bananas; melons; potatoes; milk and yogurt; nuts

Selenium	
Chemical symbol	Se (can be an anion or cation: Se^{2-} or Se^{4+} or Se^{6+})
Major physiological functions	Part of antioxidant enzymes
Symptoms of deficiency	Depressed immune function; Keshan disease, which results in cardiac problems, and Keshan-Beck disease, which affects cartilage, have been reported in Asia
Symptoms of toxicity	Selenosis resulting in brittle hair and nails, gastrointestinal distress, fatigue, impaired nervous system
Health promotion and disease prevention	Adequate intake daily is necessary for good health; large-scale studies do not show an association between selenium intake (from food or supplements) and risk for prostate cancer
Food sources	Found in plants; amount varies depending on the selenium content of the soil in which they are grown

Silicon	
Chemical symbol	Si
Major physiological functions	Not known but probably involved with bone formation
Symptoms of deficiency	Not known
Symptoms of toxicity	None known due to food consumption
Food sources	Water; grains; vegetables

(Continued)

Table 9.2 Summary of Mineral Characteristics (Continued)

Sodium	
Chemical symbol	Na^+ (cation)
Major physiological functions	Extracellular cation, helps to maintain fluid balance
Symptoms of deficiency	Not likely since sodium is widely found in foods and the body has a remarkable capacity to conserve sodium by limiting loss in urine and sweat
Symptoms of toxicity	Elevated blood pressure (depends on genetic predisposition to sodium sensitivity)
Health promotion and disease prevention	High intake of sodium is associated with high blood pressure, especially in those who are salt sensitive; as blood pressure increases, risk for stroke and cardiovascular disease also increases
Food sources	Table salt (NaCl); addition of sodium to processed foods
Sulfur	
Chemical symbol	S
Major physiological functions	Needed for the synthesis of sulfur-containing compounds, which are essential for the synthesis of many compounds in the body
Symptoms of deficiency	Not likely in the United States but deficiencies could result in stunted growth
Symptoms of toxicity	Diarrhea
Food sources	Protein-containing foods and water; the majority of sulfur is provided by the breakdown of body protein and the reuse of the sulfur found in the sulfur-containing amino acids
Vanadium	
Chemical symbol	V^{2+} to V^{5+}
Major physiological functions	Not known
Symptoms of deficiency	Not known
Symptoms of toxicity	None known due to food consumption
Food sources	Mushrooms; shellfish; black pepper; parsley; dill
Other	Use vanadium with caution; there is no justification for adding vanadium to food
Zinc	
Chemical symbol	Zn^{2+} (divalent cation)
Major physiological functions	Component of hundreds of enzymes; needed for proper cellular function and proper immune system function
Symptoms of deficiency	Impaired growth, poor immunity
Symptoms of toxicity	Immunosuppression; decrease in high-density lipoproteins (HDL); impaired copper metabolism
Health promotion and disease prevention	Adequate daily intake is necessary for good health; mild zinc deficiency is associated with immune system dysfunction
Food sources	Animal foods (for example, meat and milk); whole grains

Source: (1) Institute of Medicine (1997, 2000, 2001, 2004, 2010), Dietary Reference Intakes, National Academies Press, (2) Gropper, S. S., Smith, J. L., and Groff, J. L. (2009). *Advanced Nutrition and Human Metabolism* (5th ed.). Belmont, CA: Wadsworth, Cengage Learning, and (3) Office of Dietary Supplements, National Institutes of Health, http://dietary-supplments.info.nih.gov/Health_Information/Information_About_Individual_Dietary_Supplements.aspx

A recommended daily intake has been established for many minerals.

Dietary Reference Intakes have been established for 15 of the 21 minerals (referred to as elements). Some are needed in relatively large amounts daily, such as calcium and potassium, whereas others are needed in much smaller amounts, as is the case for iron or zinc. A few are needed in very small amounts, such as chromium. Although the amount needed may vary considerably, each mineral is

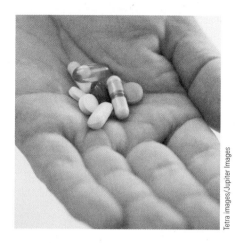

Minerals are found in a wide variety of foods, but are also sold as supplements.

important for proper physiological functioning and good health.

Absorption of minerals is generally low and well regulated but toxicities can occur. Consequently, it is important to address the question, "How much is too much?" A Tolerable Upper Intake Level (UL) has been established for adults for 14 minerals. There is a lack of data about adverse effects for the other minerals, and there are occasional reports of problems associated with excessive intake.

Table 9.3 lists the DRI and UL for adult males and adult, nonpregnant females. The complete DRI and UL for minerals (elements) for all ages and both genders are listed in the gatefold located at the front of this textbook. Notice that recommended daily intake is the same for both genders of the same age for several minerals including sodium, chloride, and potassium. For some minerals, such as zinc and fluoride, values differ for men and women due to differences in average body size. For females, iron recommendations reflect life cycle stage; those between the ages of 19 and 50 need more because of menstruation and the potentially large losses of iron in blood. In the case of calcium, adult needs vary based on age and gender (Institute of Medicine, 1997, 2000, 2001, 2004, 2010).

The Dietary Reference Intakes are established using scientific studies of healthy people who are moderately active. Athletes use these same standards. However, exercise does influence the body's mineral status, and there are athletes who lose substantial amounts of certain minerals due to a high level of training, which results in larger-than-normal losses in sweat, blood, or urine (see section below). Although some athletes have exceptional losses, most do not, and the DRI remains the standard that athletes use to judge the adequacy of their dietary mineral intake.

The Internet café

Where do I find reliable information about minerals?

The Office of Dietary Supplements, a part of the National Institutes of Health (NIH), was established to "strengthen knowledge and understanding of dietary supplements by evaluating scientific information, stimulating and supporting research, disseminating research results, and educating the public to foster an enhanced quality of life and health for the U.S. population." Fact sheets have been created for several minerals including calcium, chromium, iron, magnesium, selenium, and zinc. Each fact sheet gives an overview of the mineral, the foods that provide it, the Dietary Reference Intakes, information about deficiencies and toxicities, information about supplementation, a thorough discussion of current issues and controversies, and a list of references. The fact sheets are written by registered dietitians at the clinical research hospital at the NIH and reviewed by expert scientific reviewers. These fact sheets, as well as a plethora of other information, can be found at **http://ods.od.nih.gov/**.

Moderate to rigorous exercise increases the loss of some minerals.

The influence of exercise on minerals has been difficult to study. Sweat and urine represent the two most likely sources of larger-than-normal losses, with sweat being the bigger concern because some athletes lose a substantial amount of sweat during training or competition. Sweat contains several macrominerals such as sodium, chloride, and potassium; losses of these electrolytes, particularly the substantial loss of sodium by "salty" sweaters who exercise for many hours, are discussed in detail

Mineral: An inorganic (noncarbon-containing) compound essential for human health.

Table 9.3 DRI and UL for Adult Males and Adult, Nonpregnant Females

Mineral	Dietary Reference Intakes (DRI)/d	Tolerable Upper Intake Levels (UL)/d
Calcium	1,000 mg (females, ages 19 to 50, and males, ages 19 to 70) 1,200 mg (females, ages 51 and above, and males, over age 71)	2,500 mg (ages 19 to 50) 2,000 mg (ages 51 and above)
Chromium*	25 mcg (females, ages 19 to 50) 20 mcg (females, ages 51 and above) 35 mcg (males, ages 19 to 50) 30 mcg (males, ages 51 and above)	Not established
Copper	900 mcg	10,000 mcg
Fluoride*	3 mg (females) 4 mg (males)	10 mg
Iodine	150 mcg	1,100 mcg
Iron	8 mg (males; females 51 and above) 18 mg (females, ages 19 to 50)	45 mg
Magnesium	310 mg (females, ages 19 to 30) 320 mg (females, ages 31 and above) 400 mg (males, ages 19 to 30) 420 mg (males, ages 31 and above)	350 mg (supplement sources only)
Manganese*	1.8 mg (females) 2.3 mg (males)	11 mg
Molybdenum	45 mcg	2,000 mcg
Phosphorus	700 mg	4,000 mg (ages 19 to 70) 3,000 mg (ages 71 and above)
Selenium	55 mcg	400 mcg
Zinc	8 mg (females) 11 mg (males)	40 mg
Potassium*	4.7 g	Not established
Sodium*	1.5 g (ages 19 to 50) 1.3 g (ages 51 to 70) 1.2 g (ages 70 and above)	2.3 g
Chloride*	2.3 g (ages 19 to 50) 2.0 g (ages 51 to 70) 1.8 g (ages 70 and above)	3.6 g
Boron	Not established	20 mg
Nickel	Not established	1 mg
Vanadium	Not established	1.8 mg

Legend: d = day; mg = milligram; mcg = microgram; g = gram

*These values are based on Adequate Intake (AI). The remaining DRI are based on the Recommended Dietary Allowances (RDA). See Chapter 1 for further explanation.

Source: Institute of Medicine (1997, 2000, 2001, 2004, 2010)

in Chapter 7. The loss of sodium and chloride in sweat varies considerably on an individual basis.

Although other minerals such as calcium, magnesium, iron, copper, and zinc are also lost in sweat, there is no evidence that substantial losses of these minerals occur even with many hours of exercise with sustained sweating. In fact, one of the adaptations made by athletes as they acclimate to the heat is that these minerals are conserved and not as much is secreted in sweat (Chinevere et al., 2008; Montain et al., 2007).

Minerals may also be lost in urine. For example, acute exercise results in increased postexercise zinc blood concentration, as zinc moves from muscle cells into the **extracellular fluid**. Some of this zinc may then be excreted via the urine. Similarly, more iron may be lost in the urine of athletes than in sedentary individuals (Gleeson, Nieman, and Pedersen, 2004; Gleeson, Lancaster, and Bishop, 2001). Although it is recognized that mineral loss can be greater in athletes, it is also known that the consumption of excess minerals, such as zinc and iron, can be detrimental to the athlete by compromising the immune system (Gleeson, Nieman, and Pedersen, 2004). Therefore, thought must be given to the best way to compensate for larger-than-normal losses due to exercise. Moderate losses of minerals via sweat or urine can be offset by adequate mineral intake from food. Athletes who have substantial losses may need to increase their dietary intake or supplement the diet with the lost mineral(s). The best approach should be determined on an individual basis.

Poor food choices by athletes and sedentary people often lead to low mineral intake.

Table 9.4 shows the level of adequate mineral intakes from food for U.S. adults. Less than 3 percent of Americans are likely to receive an adequate amount of potassium from their diets, typically due to low fruit and vegetable intake. Calcium intake from food is also low for many adults with approximately 36 percent of Americans meeting recommended guidelines. However, some of these individuals take calcium supplements daily. The mineral that is most likely to be consumed in adequate amounts in adult diets is phosphorus, which is widely found in food (Dietary Guidelines Advisory Committee, 2010).

It is difficult to draw broad conclusions about mineral intake by athletes because individual intake can differ substantially among teammates in the same sport. In general, athletes with a low energy intake are likely to be deficient in one or more of the following minerals: calcium, iron, zinc, selenium, magnesium, and copper. Energy-restricted athletes known to be at risk include distance runners, female gymnasts, ballet dancers, teenage synchronized skaters, wrestlers, and jockeys. Some studies have shown substantial deficits, particularly of iron and zinc. Some of the trace minerals have not been assessed, but it is likely that athletes who are deficient in iron and zinc are also deficient in some of the other trace minerals. These athletes are also likely to have low vitamin intakes (Ziegler et al., 1999, 2002, 2005; Jonnalagadda, Ziegler, and Nelson, 2004; Leydon and Wall, 2002; Venkatraman and Pendergast, 2002). However, it is possible for athletes who are restricting food intake to consume a sufficient amount of some minerals if the foods chosen are highly fortified (for example, calcium and iron added). Some of these athletes also routinely take mineral supplements.

There is a greater chance that mineral intake will be adequate when caloric intake is adequate. However, mineral adequacy typically is a reflection of consumption of a variety of nutrient dense foods, some of which are

Table 9.4 Level of Adequate Mineral Intake from Food for Americans

Nutrient	% adequacy
Potassium	<3
Calcium	36
Magnesium	52
Zinc	88
Iron	95
Copper	95
Phosphorus	95

Source: http://www.cnpp.usda.gov/DGAs2010-DGACReport.htm

fortified. Studies that report an adequate mineral intake when data are pooled also find that there are individual athletes who do not consume an adequate amount of iron, calcium, and magnesium (S. Heaney et al., 2010). The best approach for determining mineral intake is a personalized dietary assessment for each athlete.

Key points

- There are 21 essential minerals.
- A DRI and UL have been established for most minerals.
- Moderate to rigorous exercise may increase the loss of some minerals, which typically can be replaced with properly chosen foods.
- In the United States, dietary intake of several minerals is less than recommended.

Why do many athletes and nonathletes fail to consume an adequate amount of minerals?

9.2 Mineral Deficiencies and Toxicities

Survival requires that the body be in a state of **homeostasis**, and the status of minerals is no exception to this rule. One of the major ways the body maintains its mineral balance is by altering either the amount absorbed, the amount excreted, or both. In general, absorption is low or moderate for most minerals, in part, because **excretion** is normally low.

Extracellular fluid: All fluids found outside cells. Includes Interstitial fluid and plasma.

Homeostasis: A state of equilibrium.

Excretion: The process of eliminating compounds from the body, typically in reference to urine and feces.

Most minerals can be stored in tissues, and high or low storage levels alter the amount absorbed or excreted. For example, when iron storage is high, absorption may drop to a low level, whereas if iron storage is low the body can increase absorption to a small degree. High storage levels may also affect excretion. When the body needs to limit iron intake, it may leave iron stranded in the mucosal cells, which are sloughed off the intestinal villa and excreted. Mineral homeostasis is maintained through the interplay of storage levels, absorption, and excretion (Gropper, Smith, and Groff, 2009).

However, mineral metabolism is much more complicated than simply adjusting intake and output. Calcium is an example of a mineral under substantial hormonal control. Its metabolism is regulated by several hormones that influence not only calcium absorption and excretion but also its deposition and **resorption** from bone. Bone is a substantial storage site for several minerals, including calcium and sodium. Absorption and excretion may be the bookends, but in between there are other substantial influences such as hormones, altered metabolism, or storage capacity that help the body maintain mineral homeostasis.

Many factors influence mineral absorption.

It is not possible to know exactly how much of any nutrient is absorbed from the gastrointestinal (GI) tract. A number of factors can affect mineral absorption, including age, gender, stage in the life cycle, genetics, general health, and the health of the GI tract. This section will discuss factors that directly affect absorption such as the presence of a deficiency state, the amount consumed, competition from other nutrients, the presence or absence of food in the intestinal tract, and any compounds that interfere with or enhance absorption. Table 9.5 summarizes the factors that are known to influence mineral absorption.

The amount of any nutrient absorbed from food depends on whether the body is in a state of deficiency. For example, under normal conditions 70 percent of the phosphorus consumed is absorbed. When the body is deficient in phosphorus, as is the case of some elderly women, absorption can increase to about 90 percent. Notice that the body tries to compensate for a deficiency by increasing absorption but that it cannot increase absorption to 100 percent. There are limits to the body's ability to adapt. This is one reason mineral deficiencies occur.

The amount consumed on a daily basis is important to prevent eventual mineral deficiencies. The percentage absorbed decreases as the amount consumed increases. However, the total amount absorbed is greater when more is consumed. For example, when 5 mg of zinc is consumed, maximum absorption is approximately 40 percent or 2 mg. When 10 mg of zinc is consumed,

Table 9.5 Factors Influencing Mineral Absorption

Factor	Influence on absorption
Age	Generally decreases with age
Gender	Varies with the mineral
Life cycle stage	Growth states generally increase absorption; growth states include infancy and childhood growth, puberty, and pregnancy
Genetics	Varies with the individual; absorption could be low, normal, high, or excessive
General and gastrointestinal (GI) health	Poor health, especially poor gastrointestinal health, generally results in poorer absorption
Presence of a deficiency state	Generally results in increased absorption
Amount consumed	In food, higher intakes usually result in greater absorption
Presence of other minerals	In food, reduces absorption to a small degree; large amounts found in supplements may reduce absorption of competing minerals to a large degree
Presence of food in the GI tract	Generally enhances absorption
Compounds found in food	Phytic acid, oxalate, and insoluble fiber are known to inhibit absorption; soluble fiber enhances absorption; some compounds enhance absorption (for example, lactose aids calcium absorption, vitamin C aids nonheme iron absorption)
Chemical form of the mineral	Most minerals have a chemical form that results in greater absorption; some supplements and fortified foods contain highly absorbable forms

absorption declines to about 30 percent, but the total amount absorbed is higher, about 3 mg (10 mg × 0.30). The maximum amount that can be absorbed is ~7 mg daily (Hambidge et al., 2010). The best way to prevent mineral deficiencies is to consume enough of each mineral daily, in other words, the amount recommended in the Dietary Reference Intakes.

Minerals compete with each other for absorption due to chemical similarities. Minerals that are divalent cations (that is, two positively charged atoms), such as calcium, iron, zinc, copper, and magnesium, use the same binding agents and cellular receptor sites. The major factor influencing the competition appears to be the amount of each mineral consumed.

Problems can occur when intake of one mineral is excessive, such as the consumption of a large amount through supplementation. Copper, iron, and calcium absorption are all reduced when zinc intake

The vitamin C in oranges increases the absorption of iron found in plant foods, such as the beans used in chili.

is excessive. In the case of magnesium or calcium, if either is taken in excess, then the absorption of the other is substantially decreased. Calcium and iron are also competitors. These interactions raise questions about both the benefits and potential problems associated with supplementation of minerals, either singly or as part of a multimineral supplement. Obtaining minerals from food may be more beneficial than obtaining them from supplements that provide excessive amounts because of the effects on absorption.

Minerals taken in adequate but not excessive amounts seem to circumvent the problem of substantially reduced or favored absorption. Competition is still present because minerals normally share the same absorption mechanisms and pathways. However, when all the minerals are present in reasonable amounts, no one mineral has an overwhelming absorption advantage. In nature, calcium-containing foods, such as milk, are poor sources of iron, and iron-containing foods, such as meat, are poor sources of calcium. When eaten apart, the competing nutrient is absent. When both meat and milk are eaten as part of the same meal, there will be some competition for absorption but one will not be absorbed to the exclusion of the other.

Minerals are better absorbed when food is present in the GI tract. This may be due to the slower movement of undigested food (increased transit time), the presence of enzymes that are secreted as part of the normal digestive process, or a favorable pH (Sabatier et al., 2002). Some mineral supplements, such as calcium carbonate, are also better absorbed when consumed with food. However, large doses of single supplements, such as iron, may be better if taken on an empty stomach so that they do not interfere with absorption of other minerals.

A number of compounds found naturally in food are known to interfere with or enhance absorption. Those that interfere include phytic acid, oxalate,

insoluble fiber, and fat. Phytic acid (phytates) and oxalates are known to inhibit the absorption of iron, zinc, calcium, and manganese by binding with the mineral and blocking absorption. These compounds are found in spinach, Swiss chard, seeds, nuts, beans, and legumes (Hambidge et al., 2010; Hurrell, 2003). Insoluble fiber, such as wheat, decreases calcium, magnesium, manganese, and zinc absorption because of a decreased transit time. In other words, the insoluble fiber causes the contents to move through the GI tract more quickly (decreased transit time), resulting in less contact time with the mucosal cells and less opportunity for the minerals to be absorbed (Greger, 1999).

Sugars such as lactose (milk sugar) may increase calcium absorption as well as magnesium and zinc. The mechanism is not entirely known but it may be due to the effect the sugar has on increasing the permeability of the GI tract. Soluble fibers, such as pectins or gums, may also have a favorable effect (Greger, 1999). Several factors can enhance iron absorption, including ascorbic acid (vitamin C) and the presence of the meat, fish, poultry (MFP) factor (Hurrell and Egli, 2010).

Although numerous factors have been identified as either increasing or decreasing absorption, their influences are probably subtle when considered in the larger context of daily, weekly, and yearly intake. Most reports of clinical mineral deficiencies are from poor countries where the food supply is severely limited. Eating a variety of nutrient-dense foods may be the best approach for minimizing competition and maximizing mineral absorption.

Resorb, Resorption: To break down and assimilate something that was previously formed.

It is important to guard against mineral deficiencies.

Despite the body's adaptive mechanisms, mineral deficiencies can occur if intakes are too low over time. Any mineral deficiency will be progressive—at first mild and difficult to detect, then moderate, and, ultimately, severe, if not detected and treated. As with vitamin deficiencies, mild and moderate mineral deficiencies, also called subclinical deficiencies, often progress over a long period of time with no visible signs or symptoms. When signs or symptoms do appear, they are usually subtle and nonspecific.

Zinc deficiency is a good example. A mild zinc deficiency forces the body to prioritize the functions of its zinc-containing enzymes to ensure that the available zinc is incorporated into higher-priority enzymes. Initially, a small reduction in the number and type of zinc-containing enzymes would not be noticeable and, because of limitations in measuring such activity, not likely to be detected with laboratory tests. As a mild deficiency became a moderate one, signs and symptoms would emerge but they would be nonspecific—reduced appetite, poor wound healing, or dermatitis (inflammation of the skin)—making a diagnosis of zinc deficiency difficult since many medical conditions have these same general symptoms. As a zinc deficiency became more severe (that is, a clinical deficiency) specific signs would emerge such as impaired taste or night blindness.

Prevalence of subclinical mineral deficiencies. In the early stages of a subclinical mineral deficiency a person would show few or no outward signs that mineral status is declining. A subclinical iron deficiency is known as iron deficiency without anemia. Such deficiencies are prevalent in the United States, with estimates ranging from 2–5 percent of males (highest prevalence in 12- to 15-year-olds), 9–16 percent of female adolescents, and 12 percent of nonpregnant adult females (Centers for Disease Control and Prevention, 2002).

A subclinical calcium deficiency is known as **osteopenia** or low bone mineral density (BMD). It is estimated that 30 percent of men (11.8 million) and 49 percent of women (22.7 million) in the United States age 50 or older have osteopenia (Looker et al., 2010). Estimates of low bone mineral density in athletes range from ~11 to 22 percent (Hoch et al., 2009).

The prevalence of those in the United States who have a subclinical zinc deficiency is not known. However, a handful of studies of U.S. children and adults have shown that some people benefit from zinc supplementation, suggesting that subclinical deficiencies do exist in the United States. A lack of zinc could result in a depressed immune system and a higher incidence of colds (Hambidge et al., 2010; Hambidge, 2000).

In addition to iron, calcium, and zinc, it is reasonable to assume that some people could have subclinical deficiencies of selenium and magnesium because surveys suggest that dietary intake is below recommended levels (Ford and Mokdad, 2003). These data must be interpreted cautiously, however, because prolonged low dietary intake is suggestive but not predictive of a subclinical deficiency. Increased absorption or decreased excretion may offset low intake. For several of the trace minerals, the amount contained in food is not known; thus dietary intake cannot be estimated and no conclusions about potential subclinical deficiencies can be made solely from dietary analysis.

Prevalence of clinical mineral deficiencies. In industrialized countries where food is abundant and iodine is added to salt, clinical mineral deficiencies are typically limited to iron and calcium. In individuals with cirrhosis (scar tissue in the liver), a zinc deficiency may be present.

A clinical iron deficiency results in iron deficiency anemia. Iron deficiency anemia results in fatigue and impairs performance by reducing aerobic capacity and endurance.

It is estimated that 3 percent of women ages 12 to 49 have iron deficiency anemia (Centers for Disease Control and Prevention, 2002). The percentage of female athletes who have a clinical iron deficiency is not known, but it is not likely to be less than 3 percent. Reports of iron deficiency anemia in female distance runners suggest that the prevalence of iron deficiency anemia is considerably higher than the general population in some sports (Malczewska, Raczynski, and Stupnicki, 2000). Some males, including a few endurance athletes, may manifest iron deficiency anemia but the prevalence is very low.

Osteoporosis is a clinical calcium deficiency and normally develops over many decades. Based on 2002 figures, 8 million women and 2 million men in the United States over the age of 50 have osteoporosis (National Osteoporosis Foundation, 2010). Loss of calcium from bone is exacerbated in women when **estrogen** production declines substantially. For most women this estrogen decline is a result of **menopause**, but for some female athletes, low circulating estrogen is a result of a prolonged low caloric intake concurrent with the high energy expenditure of intense training (see Chapter 12). Low energy availability can result in **amenorrhea**, the cessation of menstruation. In two studies of amenorrheic female distance runners between the ages of 20 and 30, 10–13 percent were diagnosed with osteoporosis (Khan et al., 2002).

Clinical mineral deficiencies negatively affect performance and undermine the athlete's health. Subclinical deficiencies are never desirable because they could theoretically impair an athlete's ability to train or perform and put the athlete's health at risk. As is often said in sports, "The best defense is a good offense." The athlete's best defense against mineral deficiencies is an adequate intake of minerals daily through the consumption of nutrient-dense foods in sufficient quantities to meet caloric needs.

Mineral toxicities are rare but possible.

For most of human history food was the only source of minerals. Very small amounts of trace minerals are found in food, so there was little risk of consuming toxic amounts from the diet. As medical research advanced, mineral deficiencies could be detected. When a deficiency was diagnosed, it was easily reversed with either a change in food intake or short-term mineral supplementation. Supplementation was monitored as part of the medical treatment and toxic levels could be avoided. The scientific focus was on one of two points on the continuum—deficiency or toxicity.

In the last few decades, especially in the United States, mineral supplementation by generally well-nourished individuals has increased. This has occurred because of research that has shown a relationship between various minerals and chronic diseases, as well as increased advertising for mineral supplements. More foods are now fortified with minerals, and more people are consuming supplements of a single mineral in amounts higher than would occur naturally from the consumption of food only. The scientific focus is now on three points on the continuum—deficiency, health benefits to well-nourished individuals, and toxicity (Fraga, 2005).

The distance between deficiency and toxicity has been fairly well defined; however, the difference between the amounts needed to prevent chronic diseases in the well-nourished individual and those that result in toxicity is poorly understood and hard to measure. The best advice for those who supplement with minerals, especially trace minerals, is to supplement carefully and monitor dosages to avoid potential toxicities. Fraga (2005) suggests that self-prescribed, poorly monitored intake of trace mineral supplements will put some people on the borderline of toxicity.

Table 9.6 lists the Tolerable Upper Intake Level that has been established for 14 of the minerals and the potential adverse effects associated with excess intake. For all of the nutrients except magnesium, the UL includes all foods, water, and supplements as mineral sources. In the case of magnesium, the established level is based on supplements only. To properly evaluate if an individual is approaching or exceeding the UL, a nutritional analysis of the usual diet should be conducted to determine approximate mineral intake from food. The amounts consumed as supplements should be added to these estimates and the total intake compared to the UL. High-potency multimineral supplements can provide surprisingly high amounts of several minerals (see Spotlight on supplements: Evaluating a High-Potency Multimineral Supplement Advertised to Athletes).

Key points

- An adequate amount of each mineral is needed for proper biological function, but an excessive amount can be detrimental.
- Moderate to rigorous exercise increases the loss of some minerals, but with a few exceptions (for example, sodium) the losses are small and easily replenished with food.
- Most athletes can obtain sufficient amounts of minerals from food if the diet is well balanced.
- Low calcium and iron intakes, which are more prevalent in females, may result in subclinical and clinical deficiencies. Supplementation is appropriate for some individuals.
- Mineral toxicities are rare but possible.

Why is it recommended that people check with their doctor before taking a mineral supplement?

9.3 The Roles of Minerals in Bone Formation

At least eight minerals are involved in bone formation. Eighty to 90 percent of the mineral content of bone consists of calcium and phosphorus incorporated into **hydroxyapatite** crystals, $Ca_5(PO_4)_3OH$. Fluoride is also incorporated, increasing the size of the crystal and making it less fragile. However, too much fluoride makes the crystal too large and brittle, and fragility is increased. In addition to the structural minerals, several minerals play indirect roles. Magnesium sits on the surface of the hydroxyapatite crystal and helps to regulate bone metabolism. Iron, zinc, and copper are part of various enzymes that are needed to synthesize collagen (Tucker, 2009).

Osteopenia: Low bone mineral density. A risk factor for osteoporosis.

Osteoporosis: Disease of the skeletal system characterized by low bone mineral density and deterioration of the bone's microarchitecture.

Estrogen: Female sex hormone.

Menopause: Period of time when menstruation diminishes and then ceases. Typically occurs around the age of 50.

Amenorrhea: Absence or suppression of menstruation.

Hydroxyapatite: The principal storage form of calcium and phosphorus in the bone.

Table 9.6 Tolerable Upper Intake Level and Potential Adverse Effects of Minerals

Nutrient	Tolerable Upper Intake Level for adult males and nonpregnant females (UL)/d	Potential adverse effects
Calcium	2,500 mg (ages 19 to 50) 2,000 mg (ages 51 and above)	Kidney stone formation; hypercalcemia (high blood calcium) and renal (kidney) insufficiency; decreased absorption of other minerals
Phosphorus	4,000 mg (ages 19 to 70) 3,000 mg (ages 71 and above)	Impaired calcium regulation; calcification of the kidney; possible reduction of calcium absorption
Fluoride	10 mg	Fluorosis of the teeth and bones resulting in mottled teeth, fragile bones, and joint pain
Magnesium	350 mg (supplement sources only)	Diarrhea
Iron	45 mg	Gastrointestinal distress
Copper	10,000 mcg	Gastrointestinal distress; liver damage
Zinc	40 mg	Immunosuppression; decrease in high-density lipoproteins (HDL); impaired copper metabolism
Sodium	2.3 g	Blood pressure tends to rise with increased sodium intake, especially in those with high blood pressure, diabetes, and chronic kidney disease; those who are older or African American tend to be more sensitive to the effects of sodium
Chloride	3.6 g	As sodium chloride (table salt), increased blood pressure
Selenium	400 mcg	Selenosis resulting in brittle hair and nails, gastrointestinal distress, fatigue, impaired nervous system
Boron	20 mg	Low toxicity in humans; gastrointestinal distress associated with boron poisoning from accidental ingestion of large doses
Iodine	1,100 mcg	Thyroid-related medical problems
Manganese	11 mg	Elevated blood manganese levels; nervous tissue toxicity
Molybdenum	2,000 mcg	Minimal effects in humans; impaired growth and weight loss in laboratory animals
Nickel	1 mg	Low toxicity in humans; gastrointestinal distress associated with nickel poisoning from accidental ingestion of large doses
Vanadium	1.8 mg	Use vanadium with caution. There is no justification for adding vanadium to food

Legend: d = day; mg = milligram; mcg = microgram; g = gram

Source: Institute of Medicine (1997, 2000, 2001, 2004, 2010)

Nutrients other than minerals are also critical to proper bone development. Vitamin D, a general term that includes several different chemical forms, is an important influence because one form functions as a hormone and helps to regulate calcium absorption and excretion. Vitamin K assists with incorporation of calcium into the hydroxyapatite crystal whereas vitamin C is necessary for proper collagen formation. In addition to minerals and vitamins, protein also plays a role in proper bone formation and maintenance. Low protein intake negatively affects bone health, which is a particular problem for the elderly (Genaro and Martini, 2010).

Although all the nutrients involved in bone formation are important, calcium and vitamin D tend to receive the most attention. Calcium is emphasized because it makes up a large proportion of bone structure as the hydroxyapatite crystal; vitamin D ensures proper intestinal absorption of calcium and proper mineralization of bone. However, the other nutrients should not be overlooked because all are needed for optimal bone health (Cashman, 2007).

Bones have both structural and metabolic functions.

The more than 200 bones in the body obviously are involved in skeletal support, movement, and protection of vital organs. However, bones play other important roles

Spotlight on supplements

Evaluating a High-Potency Multimineral Supplement Advertised to Athletes

The following multimineral supplement is advertised as a "high absorbance formula for stressed people, athletes and bodybuilders, and cancer, AIDS, and HIV patients." The manufacturer recommends one tablet daily. Table 9.7 lists the dose found in the supplement for each mineral and compares it to the Dietary Reference Intake for a 22-year-old male. Information about the UL is also included.

This mineral supplement contains several minerals in amounts that are more than twice the DRI. Many male athletes who consume a sufficient amount of energy (kilocalories) are already obtaining the recommended amounts from food, and these doses will bring daily mineral intake several times higher than recommended. The supplement is advertised as being highly absorbable. Excessive or preferential absorption could result, although it is impossible to know how much of any mineral will be absorbed.

Five minerals contained in this supplement are of particular concern—magnesium, zinc, manganese, vanadium, and silica. The amount of magnesium exceeds the Tolerable Upper Intake Level established. The amount of zinc contained (22.5 mg) is higher than many nutrition professionals recommend. No more than 15 mg of supplemental zinc is generally recommended because high doses of zinc can interfere with iron and copper absorption and can suppress the immune system. This supplement provides 10 mg of manganese, 91 percent of the UL. When added to the amount consumed in food, intake of this nutrient could be high. It is not illegal to have high dosages in a supplement, even those that exceed the UL.

Although there is no UL for vanadium, the committee that established the UL noted that vanadium supplements should be used with caution (Institute of Medicine, 2001). The European Food Safety Authority (2004) concluded that vanadium is not an essential mineral for humans and noted that adverse effects on kidneys and reproduction have been reported in rats and mice. The silica (silicon dioxide) in this supplement is extracted from horsetail herb and represents 20–50 percent of the estimated daily dietary intake of silicon. Although a UL has not been established, the Institute of Medicine also noted that silicon should not be added to food or supplements.

The label states that 2 mg of copper are contained in the supplement, but the DRI and UL are listed in micrograms, which makes it hard for the consumer to compare. Comparisons are unlikely, though, because mineral supplement labels do not require that the UL be listed.

It is curious that a supplement recommended to cancer, AIDS, and HIV-positive patients would also be recommended to athletes. Athletes are typically healthy individuals whose need for minerals is likely met by consuming amounts recommended by the DRI. Cancer and AIDS patients have wasting diseases that substantially affect nutrient intake and nutrient requirements. The use of the DRI is not appropriate with people who are ill because the DRI is established using data from healthy individuals. This supplement's advertising message was "more is better," but that conclusion is not necessarily true.

Table 9.7 Example of a High-Potency Multimineral Supplement

Mineral	Dose	DRI*	Dose compared to DRI
Calcium	1,000 mg	1,000 mg	Equal
Iron	18 mg	8 mg	2.25 times greater
Iodine	150 mcg	150 mg	Equal
Magnesium	500 mg	400 mg	1.25 times greater; exceeds the UL (350 mg)
Zinc	22.5 mg	11 mg	2 times greater
Selenium	50 mcg	55 mcg	10% less
Copper	2 mg**	900 mcg	2.2 times greater
Manganese	10 mg	2.3 mg	4 times greater
Chromium	100 mcg	35 mcg	2.8 times greater
Molybdenum	10 mcg	45 mcg	Less than 25%
Potassium	99 mg	4,700 mg	Less than 2%
Vanadium	2 mcg	Not established	UL not established; use vanadium supplements with caution
Silica (horsetail herb)	10 mg	Not established	UL not established but report states that silicon should not be added to food or supplements

Legend: DRI = Dietary Reference Intakes; mg = milligram; mcg = microgram; UL = Tolerable Upper Intake Level
*22-year-old male
**2 mg = 2,000 mcg

such as maintaining mineral homeostasis and acid-base balance. The strength and hardness of bones and lack of dimensional growth in adults may lead people to believe that bone has "finished growing" after adolescence and is not metabolically active in adulthood. To the contrary, bone is a dynamic tissue that is biologically active from birth throughout the entire life span.

Three major bone-related processes are growth, modeling, and remodeling. Bones grow both longitudinally (in length) and **radially** (in thickness), as is evidenced by children and adolescents who grow taller and whose bones get thicker. Modeling is the process by which bones are reshaped, often in response to mechanical force. Mechanical stress is placed on bones by our everyday weight-bearing activities resisting the force of gravity and by the actions of muscles on bones during exercise. Modeling occurs in children and adolescents and to a smaller degree in adults. For adults, the most common of the three processes is remodeling. Remodeling is important because "old" bone that has been microdamaged is replaced by "new" bone, which helps to maintain bone strength. The remodeling process is also vital to maintaining proper metabolism, such as calcium balance in the blood (Clarke, 2008).

Skeletal mass consists of both cortical and trabecular bone. Approximately 80 percent of the skeleton is cortical bone, which is found in the shafts of the long bones and on the surface of the bones. In contrast, about 20 percent is trabecular bone, which is found at the ends of the long bones and below the surface. Cortical bone is compact and is laid down in concentric circles around Haversian systems (canals) that contain blood vessels, nerves, and other tissues. In contrast, trabecular bone is a series of interconnecting plates housing the bone marrow that is the source of blood cells. Trabecular bone has much greater surface volume and a higher rate of metabolic activity and turnover than cortical bone (Sherwood, 2010). One to 2 percent of the entire skeletal mass in adults is being remodeled at any given time, but 20 percent of the trabecular bone is being remodeled, underscoring its high turnover. In the adult skeleton, approximately 10,000 to 20,000 new remodeling sites become active each day. At any given time, approximately 1 million sites are being actively remodeled. Over 10 years' time an adult's entire skeleton will have been remodeled (Sherwood, 2010).

Once the bone remodeling process begins at a site, the time to completion is a function of age. In children the remodeling takes several weeks, whereas in young adults it takes about 3 months. In older adults the time between bone breakdown and complete restoration can be anywhere from 6 to 18 months (R. P. Heaney, 2001). Most of the remodeling time is spent in bone formation, not bone breakdown (Compston, 2001).

Throughout life bone is constantly remodeled as existing bone is resorbed (broken down) and new bone is formed. This process, also known as bone turnover, involves **osteoclasts** (cells that resorb bone) and **osteoblasts** (cells that form bone). Osteoclastic activity is triggered by many factors, including mechanical force (physical activity), microfractures, and changes in concentration of hormones, such as **parathyroid hormone** (PTH) and calcitriol (1,25-dihydroxyvitamin D_3). In children and adolescents deposition outpaces resorption. In young adults the amount of bone formed typically equals the amount of bone resorbed, but as people age, osteoblastic activity does not equal osteoclastic activity. The result is a net loss in bone density (Kenny and Prestwood, 2000).

Achieving peak bone mineral density is critical to long-term health.

Peak mineral density (PMD) or peak bone mass refers to the highest bone mineral density achieved during one's lifetime. The largest amount of bone mineral is added during childhood and adolescence. By the end of adolescence, 95 percent of the adult skeleton has been formed; only 5 percent of bone density is accumulated between ages 20 and 35 (Rizzoli et al., 2010). Peak bone density of trabecular bone is achieved between the ages of 20 and 30, whereas cortical bone density peaks later, usually between ages 30 and 35.

Figure 9.1 shows four major influences on peak bone mass. Approximately 60 percent of peak bone density is genetically determined, although the exact genes are not known. To reach one's genetic potential for PMD, nutrient intake must be adequate. Although many nutrients are important in bone formation, the emphasis is put on consuming an adequate amount of calcium, vitamin D, and protein. Two substantial mechanical factors are physical activity and body weight. Hormones, such as estrogen and testosterone, also influence peak bone mass.

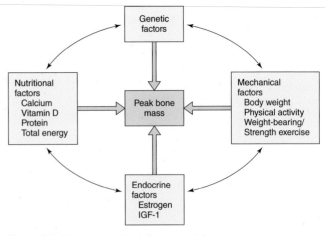

Figure 9.1 Major factors that influence peak bone mass

Legend: IGF-1 = Insulin-like growth factor 1

Nutritional factors affecting peak bone density. Calcium, vitamin D, and protein are major influences on bone health. A low calcium intake during childhood and adolescence can reduce peak bone mineral density by as much as 5–10 percent. Looking at this same issue from a positive perspective, a 10 percent increase in peak bone density is associated with up to a 50 percent reduction in fracture risk in adulthood (Rizzoli et al., 2010). As average life expectancy in the United States is approximately 78 years, obtaining peak bone mineral density is critical to long-term health (Centers for Disease Control and Prevention, 2007).

A severe vitamin D deficiency in children (defined as a serum level of 25-(OH)D$_3$ <10–11 ng/ml [<25–27.5 nmol/L]) can cause rickets, a disease associated with bone deformities and an increased risk for fractures. Studies of children with low or marginal vitamin D status (<15 ng/mL [<37.5 nmol/L]) have reported impaired bone growth and reduced bone mineralization at certain sites (Rizzoli et al., 2010).

Adequate protein in children is associated with bone growth and peak bone density. The amino acids are necessary to build the matrix around the bone. Protein is also necessary for the secretion of a powerful bone-building hormone, IGF-I. In the elderly, a low-protein diet results in bone demineralization (Rizzoli et al., 2010).

Mechanical factors affecting peak bone density. Physical activity or participation in sports or exercise programs has an important impact on the development of peak bone mineral density. Weight-bearing exercise or activity that exposes the body to repeated stress in excess of the normal effects of gravity is needed. This weight-bearing stress stimulates bone to increase bone mineral content over time. People who are physically active generally have greater bone mineral density than those who are sedentary. Exercise-related factors that influence peak bone density include the type, intensity, and frequency of exercise; age at which the activity is begun; and number of years that the exercise continues (Beck and Snow, 2003).

Studies in children have shown that jumping, hopping, and similar high-impact activities increase bone mineral content in the hip and spine and are safe for children to perform (Fuchs, Bauer, and Snow, 2001). A subsequent study showed that gains in bone density were sustained in the hip even after 7 months of detraining (Fuchs and Snow, 2002). Childhood and adolescence are critical times for increasing bone mass and high-impact activities (for example, jumping rope) and sports (for example, gymnastics, volleyball, basketball) should be encouraged.

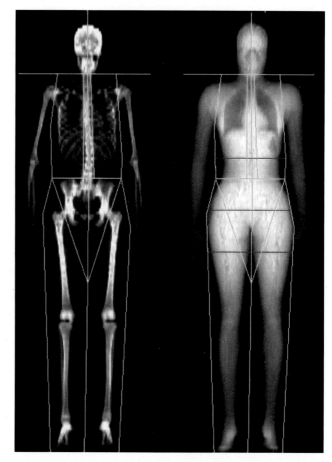

Bone mineral content and density can be determined by a dual-energy X-ray absorptiometry (DEXA) scan.

In young adults, high-impact activities and strength training such as power lifting increase bone mineral density to the greatest degree. Weight-bearing exercise at intensities of greater than 60 percent of $\dot{V}O_{2max}$ are more beneficial than weight-bearing activities performed at lower intensities. In contrast to children, bone-stimulating activities must be continued to maintain the gains made in bone density (Winters and Snow, 2000). Although good for cardiovascular fitness, nonweight-bearing activities such as cycling or swimming do not stimulate gains in bone mineral density. Sedentary adults have lower bone

Radially: Outward from the center.

Osteoblast: Bone-forming cell.

Osteoclast: Bone-removing cell.

Parathyroid hormone: A hormone produced in the parathyroid glands that helps to raise blood calcium by stimulating bone calcium resorption.

mineral density than adults who are physically active. Although high-impact activities have the greatest effect, weight-bearing activity, especially at higher intensities, is also beneficial. Predictably, more frequent activity has a greater effect than less frequent activity (Beck and Snow, 2003).

For the person who is overweight or obese, excess body weight contributes to greater bone mineral density. However, the negative health effects of excess body fat can have a considerable adverse impact on overall health and well-being.

Bone loss is associated with aging.

Achieving peak bone density is critically important because bone loss is a natural consequence of aging. Estimates of yearly bone loss for women between the ages of 18 and 50 years range from 0.25 to 1 percent per year (Vondracek, Hansen, and McDermott, 2009). With the onset of menopause, estrogen deficiency results in a yearly bone loss of 1–2 percent, initially much of it from the vertebrae. In the decade after menopause, women can lose a total of 20–30 percent of bone density from trabecular bone and up to 5–10 percent from cortical bone (Ilich and Kerstetter, 2000). Older men lose bone density at a fairly constant rate of about 1 percent per year (Kenny and Prestwood, 2000).

One reason that bone mineral is lost is that bone resorption increases with both age and estrogen deficiency. The most important mechanism appears to be the increase in the number of units that undergo remodeling. A second mechanism is the incomplete bone formation that occurs when bone resorption outpaces bone formation. Once the remodeling process is complete at a remodeling site, further modifications cannot be made. Thus if a remodeling cycle is completed and formation did not equal resorption, then the bone loss in that remodeling unit is irreversible (Compston, 2001).

Another reason that bone mineral is lost is that bone serves as an available source of calcium to achieve calcium homeostasis. Bones have both structural and metabolic functions, and when calcium intake is adequate both functions can be met. However, when dietary calcium intake is low, the body must tap its calcium reserves in bones to meet the metabolic demand for calcium. Over time, this undermines the structural functions.

Calcium may be taken from bone to maintain calcium homeostasis.

Calcium metabolism is hormonally controlled both on the micro level (for example, the amount of calcium in the blood) and on the macro level (for example, the amount of calcium absorbed, distributed, and excreted). Two critical hormones are parathyroid hormone (PTH) and calcitriol (a form of vitamin D). The regulation of calcium in the blood and extracellular fluid (ECF) is referred to as calcium homeostasis and is primarily under the control of PTH. Calcium balance describes the body's total absorption, distribution, and excretion of calcium and is regulated by PTH and calcitriol. Calcium homeostasis and calcium balance describe different aspects of calcium metabolism, although the two functions are related.

Normal blood calcium concentration is between 8.5 and 10.5 mg/dl, a small physiological range. This range is tightly regulated because calcium is a cellular messenger and enzyme regulator. Approximately half of the calcium in the blood is bound to proteins and cannot leave the blood plasma. The remaining half is unbound and can diffuse into the extracellular fluid. On average, the extracellular fluid contains approximately 1,000 mg of calcium. The amount of calcium in the ECF directly affects the function of the nerve and muscle cells, so it must be well controlled.

Homeostasis of plasma calcium occurs because calcium can be quickly moved into the ECF from a pool of calcium in the bone fluid. Bone fluid surrounds the membranes that connect osteoblasts with **osteocytes**. When blood calcium concentration decreases, PTH activates calcium pumps located in the membranes surrounding the bone fluid to quickly move calcium into the blood and restore homeostasis. Known as fast exchange, the calcium is coming from bone fluid, not from mineralized bone. PTH also stimulates the kidney to resorb more calcium (so it is not lost in urine) and activates calcitriol, a vitamin D-related hormone that increases calcium absorption in the intestine. Thus PTH is maintaining homeostasis via fast exchange but it is also influencing calcium balance by decreasing the loss of calcium in the urine and increasing the supply of calcium from food (Figure 9.2).

Under normal conditions, a sufficient amount of calcium is consumed daily and calcium balance is maintained by the interplay of calcium absorption and excretion. Calcium absorption and excretion can be increased or decreased to maintain balance. Part of calcium balance also involves bone. Normally, calcium is exchanged with bone at an equal rate. Of the nearly 1,000,000 mg (1,000 g) of calcium found in bone, approximately 550 mg is exchanged daily. Obviously, bone contains a large amount of calcium, which can be used to offset a long-term, low calcium intake.

Figure 9.3 illustrates the numerous tissues and hormones associated with calcium regulation. In adults, the average absorption is approximately 30 percent of the calcium entering the GI tract via food. Thus of a 1,000 mg daily intake of calcium, only about 300 mg is absorbed. The amount absorbed is regulated by vitamin D and does vary. Absorption

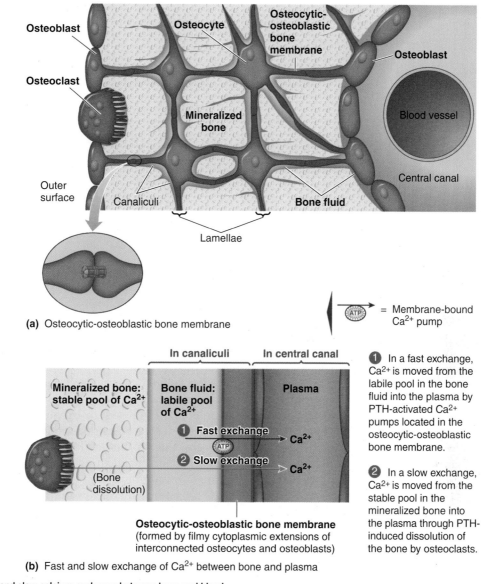

(a) Osteocytic-osteoblastic bone membrane

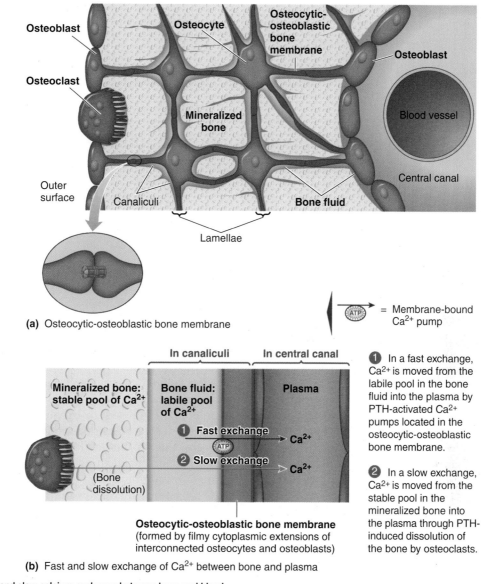

= Membrane-bound Ca²⁺ pump

① In a fast exchange, Ca²⁺ is moved from the labile pool in the bone fluid into the plasma by PTH-activated Ca²⁺ pumps located in the osteocytic-osteoblastic bone membrane.

② In a slow exchange, Ca²⁺ is moved from the stable pool in the mineralized bone into the plasma through PTH-induced dissolution of the bone by osteoclasts.

Osteocytic-osteoblastic bone membrane (formed by filmy cytoplasmic extensions of interconnected osteocytes and osteoblasts)

(b) Fast and slow exchange of Ca²⁺ between bone and plasma

Figure 9.2 Fast and slow calcium exchange between bone and blood

of calcium in adults in estimated to be between 10 and 50 percent and as high as 75 percent in children (Gropper, Smith, and Groff, 2009; R. P. Heancy, 2001). Absorption from calcium supplements also varies depending on the chemical composition (for example, calcium carbonate, calcium citrate), but typically ranges from 25 to 40 percent. The body cannot compensate for low calcium intake by increasing absorption to 100 percent.

On average, approximately 300 mg of calcium is also lost through the urine and **secretions** from the extracellular fluid into the GI tract. When the amount of calcium consumed is adequate, then the amount of calcium excreted is equal to the amount absorbed and the body can maintain both calcium

homeostasis and calcium balance (Gropper, Smith, and Groff, 2009).

Unfortunately, many people do not consume a sufficient amount of calcium daily, which affects calcium balance. The body must maintain calcium homeostasis, but faced with low calcium intake over extended periods of time, it cannot do so by using only fast exchange, the PTH-stimulated calcium pumping

Osteocyte: An osteoblast that has become embedded within the bone matrix.

Secretion: The process of releasing a substance to the cell's exterior.

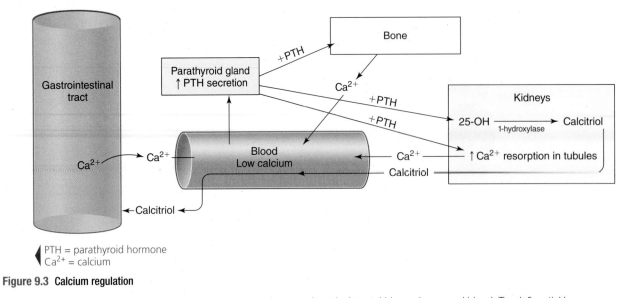

Figure 9.3 Calcium regulation

Calcium homeostasis and balance are complex and involve the gastrointestinal tract, kidneys, bone, and blood. Two influential hormones are parathyroid hormone and calcitriol (a form of vitamin D).

mechanism. A second mechanism, known as slow exchange, must be relied upon to make available the calcium needed. In slow exchange, PTH stimulates the dissolution of bone by increasing osteoclast activity and inhibiting osteoblast activity. The bone crystal, $Ca_5(PO_4)_3OH$, dissolves, releasing both Ca^{2+} (calcium) and PO_4^{3-} (phosphate). The calcium is used to maintain blood calcium within the normal range, and the phosphate is typically excreted in the urine. Over time, this process undermines the integrity of the bone's structure because the mineral density of the bone is decreased (Sherwood, 2010).

Bone loss is associated with lack of estrogen.

In addition to aging and low calcium intake, estrogen deficiency is a powerful factor in bone loss. Estrogen affects osteoclast number, activity, and life span with a decrease in estrogen being associated with an increase in osteoclast proliferation and activity. Estrogen deficiency may also be associated with increased erosion depth in trabecular bone, which negatively affects bone strength (Seeman, 2002; Compston, 2001).

Estrogen deficiency is generally associated with menopause when estrogen production naturally and substantially declines. In the United States, the average age of the onset of menopause is 50 years. However, estrogen deficiency can also be present in adolescent and young adult female athletes with amenorrhea or oligomenorrhea (absent or irregular menstruation, respectively). These conditions may result from consistently low energy (caloric) intake coupled with high energy expenditure (known as energy drain), inadequate nutrient intake, and the increase in exposure to stress hormones such as **epinephrine**, **norepinephrine**, and cortisol that accompanies intense training. When both energy and estrogen deficiencies are present in exercising women, bone formation is suppressed and bone resorption is increased (De Souza et al., 2008). This predisposes these young women to a failure to achieve peak bone mineral density, the loss of calcium from bone, a greater incidence of stress fractures, osteopenia (low bone mineral density), and osteoporosis (De Souza and Williams, 2005). Therein lies the danger of assuming that estrogen deficiency is associated only with age.

Distance runners, ballerinas, and gymnasts are at greatest risk for delaying the onset of menstruation, experiencing irregular menstruation, or developing amenorrhea (Gordon, 2000). Some, but not all, of these athletes have eating disorders. Low energy intake, amenorrhea, and low bone mineral density are three interrelated factors known as the Female Athlete Triad. Each of these factors is discussed in depth in Chapter 12.

Distance runners often have significantly lower bone mineral density when compared to age-matched controls (see Chapter 12 for a review of specific studies). Gymnasts, however, have above-normal bone

Epinephrine: Adrenaline. Primary blood pressure-raising hormone.

Norepinephrine: Noradrenaline. Hormone and neurotransmitter.

mineral density even when estrogen concentration and menstrual status are abnormal. The high-impact training associated with gymnastics seems to offset the negative effects on bone density associated with amenorrhea. Although amenorrhea is not a normal or desirable condition, the differences in bone density between highly trained distance runners and gymnasts highlight the substantial effect of high-impact exercise on bone mineral density (Beck and Snow, 2003).

The roles of calcium and exercise in preventing or reducing bone loss associated with aging have not been fully established.

During the period when peak bone mineral density can be attained, the emphasis is on adequate calcium intake and weight-bearing exercise. After this physiological period is complete, the goal is to prevent and then slow the loss of calcium from bone. This is also accomplished with adequate calcium intake and

Focus on research

Does the Disruption of the Menstrual Cycle That Occurs in Some Athletes Have Health Implications?

The number of women participating in sports began to increase dramatically in the 1970s and 1980s, both in organized school sports programs and independently (road races, triathlons, and so on). As the number of women participating in sports and exercise programs increased, so did the observation by physicians and clinicians of increased incidence of certain health conditions unique to female athletes, such as amenorrhea, or disruption of the menstrual cycle. In addition to the initial concerns for the reproductive health of these athletes, scientists began to determine if there were other associated health concerns such as decreased bone mineral content.

Female athletes involved in endurance sports, such as distance running, have a higher incidence of amenorrhea than in other sports. As a weight-bearing exercise, running is typically thought to provide a strong stimulus for maintaining bone density. However, the following research study provided important early evidence that amenorrheic athletes may also have bone mineral declines that represent a serious health issue: Drinkwater, B. L, Nilson, K., Chesnut, C. H., Bremner, W. J., Shainholtz, S., & Southworth, M. B. (1984). Bone mineral content of amenorrheic and eumenorrheic athletes. *The New England Journal of Medicine, 311*, 277–281.

What did the researchers do? Drinkwater and colleagues studied 28 female athletes, half of whom were amenorrheic, having no more than one menstrual cycle in the previous year. The 14 eumenorrheic subjects were selected to closely match the amenorrheic athletes for sport, age, weight, height, and frequency and duration of exercise training. Subjects completed a questionnaire detailing their athletic and menstrual history and kept a 3-day diary of food intake to determine nutritional intake. Body composition (percent body fat) was determined by underwater weighing. Bone mineral density and bone mineral content were determined in two areas of the body by DEXA, a relatively new technology at the time described as "single-photon and dual-photon absorptiometry." The subjects came to the laboratory once a week for 4 weeks to have blood samples taken, which were analyzed for estradiol, progesterone, prolactin, and testosterone.

What did the researchers find? The major finding of this classic research study was that the amenorrheic athletes had vertebral bone mineral density that was significantly lower than the normally menstruating subjects. Although these subjects had an average age of 25 years, their bone density was at a level typically seen in postmenopausal (>50 years old) women. The reduced bone mineral density was seen in these subjects despite their having similar amounts of calcium in their diet and running significantly more miles in training each week. Blood levels of estradiol, progesterone, and prolactin in the amenorrheic athletes were significantly lower than the eumenorrheic athletes, confirming their lack of menstruation and ovulation. Otherwise, the athletes were similar in body weight, body composition (percent body fat), and other characteristics.

What was the significance of the study? This study provided important early information about the incidence of amenorrhea and reduced bone mineral density in female athletes. This reduced bone mineral density occurred despite participation in substantial amounts of weight-bearing exercise, which was thought to protect against bone loss, but in the case of these amenorrheic runners the bone loss was to a degree that would be considered a future health risk. Although not designed as a cause-and-effect study, the results yielded important information for future research to determine the mechanisms in female athletes associated with low estrogen, disruption of the menstrual cycle, and low bone mineral density. For example, low body weight and extreme leanness was thought to be a cause, but the body weight and percent body fat of the amenorrheic and eumenorrheic athletes in this study were not different. Although not statistically significant, there was a very strong trend in the Drinkwater study for a lower intake of total kilocalories and kcal from fat by the amenorrheic athletes. Along with the significantly increased energy expenditure through training, the potentially decreased caloric intake results from this study provided evidence that insufficient energy intake and negative energy balance may be part of the cause of the amenorrhea and decreased bone mineral density, a theory that is widely accepted today.

weight-bearing exercise. How effective is calcium intake in preventing loss of calcium from bone? Between the period of the attainment of peak bone mineral density and menopause, adequate calcium intake through diet or supplements helps to slow the loss of calcium by about 1 percent a year. Recall that bone loss during this period ranges from about 0.25 percent to 1 percent. A reasonable conclusion is that bone calcium loss is reduced or prevented in many middle-age women and men if calcium intake is adequate (Bonura, 2009).

Studies suggest that an adequate calcium intake in older women and men is important because it helps to protect against loss of calcium from bone (Tucker, 2009). There is some evidence that calcium and vitamin D, when supplemented together, reduce the risk for fracture in the elderly (DIPART, 2010). However, the body of evidence that calcium supplementation helps to reduce fracture risk is weak (Seeman, 2010; North American Menopause Society, 2010; Tucker, 2009). This may seem counterintuitive since one often hears about the benefit of calcium supplementation for preventing fractures, but when only the strongest studies are considered (that is, well-controlled, randomized, prospective cohort studies) calcium supplementation does not significantly reduce the risk for hip fracture in older women and men (Tucker, 2009; Bischoff-Ferrari et al., 2007).

Despite the weak evidence for reduction in the risk of fracture, the consensus opinion is that calcium supplementation for postmenopausal women, particularly those over the age of 70, and older men is beneficial because average dietary calcium intake is not adequate and this age group is at risk for calcium deficiency. Calcium deficiency, in both older men and women, can result in hyperparathyroidism. Recall that when blood calcium is low, parathyroid hormone is elevated, which stimulates bone resorption and normalizes the blood calcium concentration. It is common for women over 70 to have low calcium intakes, reduced vitamin D absorption and/or conversion, and increased parathyroid hormone concentrations. Calcium and vitamin D supplementation resolves the hyperparathyroidism, thus preventing the bone resorption that accompanies this medical condition (Morgan, 2001; Kenny and Prestwood, 2000).

Adequate calcium intake via calcium-containing foods or supplements after age 35 is a prudent approach and may help to slow the loss of calcium from bone in some cases, but it is limited in its effectiveness by the other powerful effects of aging on bone remodeling. Calcium intake is not the sole influence on bone mineral density; thus, food or supplemental calcium cannot be expected to offset all of the various factors that cause a decline in BMD. The important point is the need for prevention; after age 35 even a sufficient calcium intake cannot compensate for a relatively low amount of calcium in bone since peak bone mineral density is achieved in childhood, adolescence, and young adulthood.

Results of exercise studies have been mixed, but high-intensity weight-bearing activity and resistance training generally maintains or increases bone mineral density in postmenopausal women. However, these activities must be safe for older women to perform, and those who have been diagnosed with osteoporosis should discuss with their physicians the most appropriate types of exercise. Unfortunately, low-intensity weight-bearing activities, such as walking, do not put enough stress on the bone to positively impact bone density, although there are many other benefits to walking. Any exercise by older women that increases muscle strength and stability is beneficial for the prevention of falls, which cause ~90 percent of hip fractures and ~50 percent of vertebral fractures (Beck and Snow, 2003).

Recommended dietary intakes of calcium and vitamin D have been revised.

Dietary intake of calcium throughout the life cycle is critical. In 2010, the Dietary Reference Intakes for calcium and vitamin D were updated (see Table 9.8). The recommended amount of calcium intake for infants and children changes to meet increased needs during high growth periods. Between the ages of 9 and 18 recommended calcium intake is at its highest—1,300 mg daily. Recommended intakes remain relatively high throughout the life cycle. Nonpregnant females between the ages of 19 and 50 and males up to age 70 need 1,000 mg daily. Calcium recommendations are

Table 9.8 Dietary Reference Intakes for Calcium and Vitamin D

Age group (yr)	Calcium (mg/d)	Vitamin D (IU/d)
0 to 0.5	*	*
0.5 to 1	**	**
1 to 3	700	600
4 to 8	1,000	600
9 to 18	1,300	600
19 to 50	1,000	600
51 to 70 (males)	1,000	600
51 to 70 (females)	1,200	
71 and above	1,200	800

Legend: yr = year; mg = milligram; d = day; IU = International Units

*Adequate Intake (AI) of calcium is 200 mg/d and 400 IU of vitamin D.

**Adequate Intake (AI) of calcium is 260 mg/d and 400 IU of vitamin D.

Note: Recommendations for those greater than 1 year are based on the Recommended Dietary Allowance (RDA). RDA is the average daily dietary intake sufficient to meet the nutrient requirements of nearly all healthy individuals. When the RDA cannot be determined, the AI is used. The AI is an estimate or approximation. See Chapter 1.

Source: Institute of Medicine (2010). Dietary Reference Intakes for calcium and vitamin D. Food and Nutrition Board. Washington, DC: National Academies Press.

Table 9.9 Average Daily Calcium Intakes from Food and Supplements in the United States, 2000–2006

Age (years)	Calcium			
	Males		Females	
	Daily average intake (mg)	Adequate intake (%)	Daily average intake (mg)	Adequate intake (%)
1–3	1008	96	977	97
4–8	1087	83	974	67
9–13	1093	23	988	15
14–18	1296	42	918	13
19–30	1259	65	945	38
31–50	1220	64	1055	44
51–70	1092	32	1186	39
71 and older	1087	31	1139	39

Adapted from: Bailey, R. L., Dodd, K. W., Goldman, J. A., Gahche, J. J., Dwyer, J. T., et al. (2010). Estimation of total usual calcium and vitamin D intakes in the United States. *Journal of Nutrition, 140*, 817–822.

increased to 1,200 mg daily for females 51 years and older and males older than 71 to help offset the loss of calcium from bone. The Tolerable Upper Intake Level for calcium includes intake from food, water, and supplements. For adults ages 50 and younger, the UL is 2,500 mg daily. However, the UL is 2,000 mg daily for those 51 and older. These amounts are not likely to be attained without supplement use; intake above the UL may increase the risk for the development of kidney stones (Institute of Medicine, 2010).

Vitamin D is a powerful regulator of calcium and phosphorus metabolism, so an adequate calcium intake should be accompanied by an adequate vitamin D intake. Calcium absorption by the cells of the small intestine is enhanced by calcitriol, a form of vitamin D (Figure 9.3). More information about vitamin D can be found in Chapter 8.

Many people consume an inadequate amount of calcium daily.

To accurately report the amount of calcium consumed daily, it is necessary to record calcium intake from both food and supplements. It is estimated that 43 percent of Americans consume calcium supplements, representing a mean intake of 331 mg of supplemental calcium per day. This figure includes children, adolescents, and adults. In some cases, such as elderly women, calcium supplements provide the amount of calcium recommended daily, but in most cases the calcium supplements contribute only some of the calcium needed daily (Bailey et al., 2010). Some postmenopausal women may be consuming excessive amounts of calcium supplements, which increase their risk for developing kidney stones (Institute of Medicine, 2010).

Despite the use of supplements, some people have low intakes of calcium as shown in Table 9.9. Of particular concern is the mean intake of calcium of teenage females; only 15 percent or less of this age group have an adequate intake even with the use of dietary calcium supplements. Many adults are falling short of the recommended amounts of calcium, but others receive an adequate amount via food and/or supplements (Bailey et al., 2010; Institute of Medicine, 2010).

Low dietary calcium intake results in lower contributions of calcium into the pool. For example, the average daily calcium intake from food for adult females is estimated at ~650 mg. Assuming absorption of 30 percent, a 650 mg daily intake would likely result in 195 mg of calcium being absorbed; or 65 percent of that absorbed if the recommended calcium intake of 1,000 mg was consumed (195 mg ÷ 300 mg = 65 percent). However, in adult women with low vitamin D concentrations (due to poor intake, absorption, and/or conversion to an active form), calcium absorption may be very low—approximately 10 percent—for both food and supplements.

R. P. Heaney (2001) estimates that of a 750 mg calcium supplement, only 75 mg will be absorbed. Of that 75 mg, approximately 36 mg are retained (much is lost in the urine) and available to offset bone resorption. Although calcium absorption does vary and absorption can increase to some degree when an individual is deficient, it is easy to see how low calcium and vitamin D intakes negatively affect bone health.

Long-term calcium deficiency is a risk factor for osteoporosis, a disease of the skeletal system characterized by low bone mineral density and deterioration of the bone's microarchitecture. Osteoporotic bones

Milk and milk products are excellent sources of calcium.

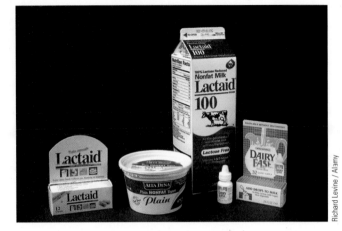

Those with lactose intolerance may use some of the products shown, which allows them to include calcium-dense dairy foods in their diets.

are fragile and more prone to fractures, particularly in the spine (vertebrae), wrist, and hip. Assessing bone density and preventing and treating osteoporosis are discussed in detail in Chapter 13.

There are numerous strategies for increasing dietary calcium consumption.

Calcium is a critical nutrient for all people of all ages. There is no one food that must be consumed or dietary pattern that must be followed in order for calcium needs to be met. However, each person needs to consume an adequate amount, and some or all of the following may be consumed as a strategy to obtain sufficient calcium daily:

- milk and milk products (dairy products)
- lactase-treated products or lactase tablets
- fermented milk products such as yogurt or aged cheeses
- calcium-containing vegetables such as cabbage, broccoli, greens
- calcium-fortified products such as orange juice, soy milk, cereal, sports bars
- calcium supplements

Milk and milk products are excellent sources of calcium. An 8 oz (240 ml) glass of milk contains approximately 300 mg, so three to four glasses a day would nearly meet the recommended calcium intake for children and adults of all ages. One 8 oz cup of yogurt also contains approximately 300 mg of calcium. Athletes who are trying to limit fat and/or caloric intake often choose nonfat versions of these foods. Other dairy products, such as 1½ oz of cheese, contain the equivalent amount of calcium found in 8 oz of nonfat milk but their fat contents are higher.

Dairy products are particularly valuable because in addition to being a concentrated calcium source, these products also contain protein and are fortified with vitamin D (R. P. Heaney, 2009). Milk was once accepted as the ideal beverage for children and adolescents, but milk consumption is declining in those ages 2 to 18 due, in part, to increased soda and juice consumption (Popkin, 2010; Bowman, 2002). This trend is particularly worrisome because this time period is critical for the development of peak bone mineral density.

Many adolescents and adults cannot digest lactose (milk sugar) because they lack sufficient lactase, the enzyme needed for lactose breakdown. They may wish to use lactase tablets, which provide the enzyme needed to break down lactose, or consume lactase-treated milk and milk products, which predigest the lactose. Some people have moderately reduced lactase production and can tolerate milk products that have been fermented, which reduces the lactose content. Examples of such products include yogurt and aged cheeses such as Parmesan, cheddar, and Gouda. These strategies help lactose-intolerant individuals to include some calcium-dense dairy foods in their diets.

Although milk and milk products contain a relatively large amount of calcium, calcium is not found exclusively in dairy foods. It is more difficult, but not impossible, to consume enough naturally occurring, nondairy calcium-containing foods. The difficulty is a result of a lack of concentrated sources of calcium and the need to consume a wide variety of foods. For example, 1 cup of cooked broccoli has about 100 mg of calcium, one-third the amount of an 8 oz glass of milk. Other vegetable sources of calcium include Brussels sprouts, collard greens, green cabbage, kale, kohlrabi, mustard greens, and turnip greens. Calcium is also found in some fish, beans, grains, and nuts in varying amounts. Table 9.10 lists the calcium content of some

Dark green vegetables are good nondairy sources of calcium.

Soy milk and rice drinks are often fortified with calcium.

nondairy sources. Eaten over the course of a day, the foods in the amounts listed in the table would total 1,051 mg of calcium.

Some foods are calcium fortified. Soy milk, rice milk, and orange juice are beverages that may be fortified with calcium. Even when calcium is added the amount can vary among brands, so the label must be checked to determine the calcium content. A common level of fortification for soy milk is 300 to 350 mg in an 8 oz glass, approximately the same amount of calcium found in dairy milk. However, the form of calcium added may not be as well absorbed, so 300 mg

of calcium from soy milk may only be the equivalent of 225 mg of calcium found in cow's milk (R. P. Heaney et al., 2000).

Tofu (soy bean curd) can be processed with either calcium sulfate or calcium chloride. Tofu that has been processed using these compounds contains approximately 110 mg of calcium in a ½-cup (3 oz) serving. Many other foods have calcium added, including breakfast cereals and sports bars. It may not be immediately clear from the label how many milligrams of calcium are contained because calcium may be listed as a percentage. This percentage is calculated using the Daily Value (DV) for calcium, 1,000 mg. For example, if the label on an energy bar states that it contains 2 percent of the DV for calcium, there are 20 mg of calcium in that bar (1,000 mg × 0.02 = 20 mg).

Table 9.10 Nondairy Sources of Calcium

Food and amount	Calcium (mg)
2 pancakes	176
1 cup blackberries	46
1 whole wheat English muffin	175
2 T peanut butter	12
2 figs	54
8 oz San Pellegrino mineral water	50
4 oz canned salmon	240
1 cup broccoli	94
1 oz almonds	74
½ cup cooked acorn squash	54
1 mixed grain roll	24
1 medium orange	52

Legend: mg = milligram; T = tablespoon; oz = ounce

Together these foods total 1,051 mg of calcium.

Phosphorus, fluoride, and magnesium are also involved with bone health.

Calcium receives most of the attention for bone formation, but bone crystal is calcium phosphate. Phosphorus is widely found in food, and 85 percent of women and almost 95 percent of men consume an adequate amount. The percentage absorbed is so high that deficiencies are unlikely to occur. The women with low phosphorus intakes are usually elderly. If these women take high-dose calcium supplements, the calcium may bind with the small amount of phosphorus found in the food (R. P. Heaney, 2004).

Concern has been raised about high dietary phosphorus intake due to the consumption of carbonated soft drinks, which contain phosphoric acid. At the present time the scientific research does not support an association between high phosphorus intake and osteoporosis (R. P. Heaney, 2004).

Fluoride is typically obtained by infants and children through the use of fluoridated vitamins and by adults through the consumption of fluoridated water. The fluoride content of tap water is available to consumers by contacting their local water agency. Magnesium is found in green leafy vegetables, nuts, seeds, and beans and legumes, such as soybeans, kidney beans, pinto beans, and lentils. It is found naturally in whole grains (for example, wheat germ, brown rice) but is lost in the processing when grains are highly refined (for example, white bread, white rice). Drinking water is described as "hard" if it contains minerals; one of the minerals found in hard water is magnesium.

Key points

- Adequate nutrient intake, in particular calcium, vitamin D, and protein, and weight-bearing exercise are necessary throughout life for bone health.

- The key to preventing osteoporosis is achieving peak bone mineral density.

- Calcium intake is particularly low by teenage females, raising serious concerns about future bone health.

- Calcium supplementation in mid- and later life has some benefit, but cannot completely offset the calcium loss from bone that accompanies aging.

- Athletic amenorrhea is detrimental to bone health.

In what ways does society promote poor calcium intake and a sedentary lifestyle?

9.4 The Roles of Minerals in Blood Formation

A favorite adage among endurance athletes is "oxygen is everything." It is no wonder that athletes, particularly endurance athletes, look at training and nutrition strategies that result in optimal oxygen delivery. The nutrient most associated with oxygen is iron.

Iron is an integral part of hemoglobin.

Blood consists of three types of cells—**erythrocytes** (red blood cells), **leukocytes** (white blood cells), and **platelets**. This section will focus only on erythrocytes, the blood cells whose primary function is the transport of oxygen. Secondary functions include carbon dioxide and nitric oxide transport. Simply stated, oxygen must be picked up from the lungs and transported to cells. Conversely, carbon dioxide must be picked up from cells and transported to the lungs. Both of these processes depend on hemoglobin, which transports approximately

98 percent of the oxygen (~2 percent is dissolved in the blood plasma) and 30 percent of the carbon dioxide (60 percent is transported as bicarbonate and 10 percent is dissolved in blood plasma) (Sherwood, 2010).

Hemoglobin (**heme** = iron, globin = protein) is an iron-containing protein found in the red blood cells that can bind oxygen (Figure 9.4). At the center of the heme portion of the molecule is iron (Fe). This iron atom forms six bonds, four with nitrogen (to maintain the molecule's ring structure), one with the amino acids in the protein portion of the molecule (globin), and one with oxygen. There are four heme molecules in each molecule of hemoglobin; thus a fully saturated hemoglobin molecule can carry four molecules of oxygen. Each red blood cell (RBC) contains more than 250 million molecules of hemoglobin. There are approximately 30 trillion red blood cells, so it is easy to see that the body has a phenomenal capacity for oxygen transport (Sherwood, 2010).

To transport oxygen throughout the body in the necessary amounts, blood must have a carrier or binding mechanism. Oxygen must bind to this carrier easily and rapidly in the lungs, remain bound as it is distributed throughout the body, yet release easily from the carrier so oxygen can be removed from the blood at sites in the body where the oxygen is needed. Iron-containing hemoglobin has unique properties that allow it to be an ideal oxygen carrier in the blood.

In areas of the body where oxygen levels are high (high **partial pressures** of oxygen), such as the lungs, hemoglobin has a high affinity for oxygen and is able to bind it readily. This is important for the fast and complete diffusion of oxygen from the lungs into the blood flowing through the pulmonary circulation. Under most circumstances, both at rest and for most people during exercise, there is nearly 100 percent saturation of oxygen on the hemoglobin molecules in the blood that passes through the pulmonary circulation. This is reflected in a common clinical test, the O_2-Hb saturation percentage, usually measured with a pulse oximeter or by the more invasive blood gas analysis.

Hemoglobin molecules have a unique ability to change their binding affinity for oxygen in areas of the body that contain less oxygen, in other words, when the partial pressure of oxygen is reduced. When the oxygen-laden blood arrives at tissues in the body that need oxygen, such as exercising muscle, the environment that has a lower partial pressure of oxygen results in hemoglobin reducing its binding affinity for oxygen, which allows the red blood cells to release oxygen molecules for transport into the cells. Iron-containing hemoglobin is critical to the effective uptake and delivery of oxygen to all tissues in the body. This process must continue constantly at rest, but is put under particular stress during exercise when oxygen demands can be dramatically increased.

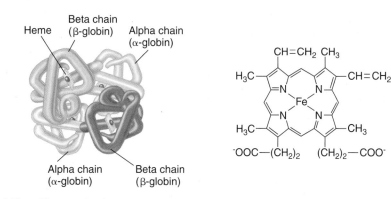

Figure 9.4 Simplified hemoglobin and heme molecules

Hemoglobin consists of four polypeptide chains and four heme molecules. The iron in each heme molecule forms six bonds—four with nitrogen, one with globin (protein chain), and one with oxygen.

Sufficient oxygen-carrying capacity is critical for all athletes, but especially endurance athletes. One measure of normal oxygen-carrying capacity is **hematocrit**, which is the amount of red blood cells expressed as a percentage of the total volume of blood plasma (the liquid portion of the blood). Normal hematocrit is approximately 42 percent for women and 45 percent for men. The general term *anemia* refers to a reduced oxygen-carrying capacity and is reflected in a hematocrit of approximately 30 percent. Another measure is the concentration of hemoglobin in the blood, which averages about 15 g/dl (dl = deciliter, or 100 milliliters) in males and slightly less, 14 g/dl, in females. Anemias can be caused by nutritional or nonnutritional factors as shown in Table 9.11.

Nutritional anemias are a result of a nutrient deficiency due to low intake or poor absorption. Some are vitamin related (for example, vitamin B_{12} or folic acid), but the most prevalent nutritional anemia is due to iron deficiency. Having sufficient iron stores by consuming an adequate amount of iron daily can prevent iron deficiency anemia. Approximately 25 percent of the body's iron is found in storage in the liver, spleen, and bone marrow, and this stored iron can be released and transported for incorporation into hemoglobin. On average, a well-nourished adult male has about 800 to 1,000 mg of stored iron. A well-nourished adult female has considerably less but still has sufficient stores to support normal red blood cell formation. With near-maximum iron stores, an adult male could sustain normal hemoglobin synthesis for about 2 years while consuming an iron-poor diet (Shah, 2004).

However, not all adults have excellent iron stores. Some may also experience higher-than-normal iron losses (for example, large losses of blood via menstruation, small daily losses as a result of GI bleeding), low iron absorption, and low iron intake. These factors, singly or in combination, may tax an already low supply of stored iron. When iron stores are depleted and

Table 9.11 Nutritional and Nonnutritional Anemias

Nutritional anemias	Nonnutritional anemias
Iron deficiency anemia	Aplastic anemia (RBC production depressed)
Vitamin B_{12} deficiency anemia	Hemolytic anemia (RBCs are destroyed)
Folate deficiency anemia	Sickle cell anemia (RBCs are abnormally shaped)
Anemia can result from a deficiency of any nutrient needed for RBC production (for example, zinc, copper)	

Legend: RBC = red blood cell

iron intake is low, then red blood cell production is negatively affected. Iron deficiency anemia results in a decreased number of red blood cells, smaller cells, and a lower concentration of hemoglobin per cell. This results in less hemoglobin being available to transport oxygen. Not surprisingly, a common symptom associated with iron deficiency anemia is fatigue.

Normal hemoglobin synthesis is also dependent on copper, another of the microminerals. Iron is stored in its ferrous form (Fe^{2+}) but must be converted to its ferric form (Fe^{3+}) to be transported in the blood. This conversion requires a copper-containing enzyme,

Erythrocyte: Red blood cell.

Leukocyte: White blood cell.

Platelet: A cell found in the blood that assists with blood clotting.

Heme: Iron. Also refers to a form of iron that is well absorbed.

Partial pressure: The pressure exerted by one gas within a mixture of gases.

Hematocrit: The percentage of the volume of blood that is composed of red blood cells.

ceruloplasmin. Humans can develop an anemia that is a result of a long-term copper deficiency. Excessive zinc interferes with copper absorption and can be one of the causes of this kind of anemia.

Myoglobin is an iron-containing protein found in muscle fibers that functions very similarly to hemoglobin, only in skeletal muscle. Myoglobin binds small amounts of oxygen within the muscle to provide a small reservoir for rapid increases in oxygen utilization. The body doesn't "store" oxygen per se, but myoglobin acts as an oxygen buffer when demand is increased until oxygen delivery from the blood can be increased. Myoglobin also facilitates the transfer of oxygen molecules from the red blood cells in the blood, through the muscle cells, and into the mitochondria. Highly aerobic tissues such as slow-twitch (type I) and intermediate (IIa) muscle fibers contain a higher concentration of myoglobin. Because aerobic exercise training stimulates the oxidative energy system, a common adaptation is an increase in myoglobin concentration in muscle, which helps increase the body's aerobic capacity.

Blood tests can help detect iron deficiency.

Iron is unique among minerals in that there are a variety of blood tests that can detect normal and reduced stores and physiological function. These tests measure the amount of iron in red blood cells and estimate the level of iron storage. The most common measures are hemoglobin, hematocrit, and **serum** ferritin. When these values are within the recommended ranges, iron status is considered to be normal and red blood cell production is adequate. Normal laboratory values are listed in Table 9.12.

When iron intake is poor, iron status declines over time. Iron stores become depleted, and this can be detected by measuring ferritin. **Ferritin** is the protein that stores iron in tissues (predominantly in the liver). The amount of ferritin circulating in the blood reflects the amount stored, so this blood test is a good indicator of iron storage. The normal values range from 12 to 300 ng/ml for males and 12 to 150 ng/ml for females. As the amount of storage iron declines, the amount of ferritin in the blood declines; so although the absolute value is important, repeated blood tests over time (for example, every 6 months) also indicate if stores are declining. For example, an endurance athlete may have a complete blood count (CBC) two times a year for 2 years. The four consecutive ferritin tests are all within the normal range—120, 110, 85, and 63 ng/ml—but these "normal" values also suggest that iron stores are declining. In a moderate subclinical iron deficiency, hemoglobin and hematocrit values are in the normal range because iron deficiency anemia develops slowly even after iron depletion. This subclinical deficiency is referred to as iron deficiency without anemia.

Hemoglobin measures the iron-containing protein found in red blood cells. Values below the normal range may indicate iron deficiency anemia, a recognized disease, or **false (runner's) anemia**, which is not a true anemia. Normal hemoglobin values are 13.8 to 17.2 g/dl for males and 12.1 to 15.1 g/dl for females. Those who live or train at higher altitudes generally have hemoglobin concentrations nearer the upper end of the normal range. Values below 13.8 and 12.1 g/dl, for males and females, respectively, are usually indicative of iron deficiency anemia, but false anemia should be ruled out. In iron deficiency

Table 9.12 Iron-Related Blood Tests

Blood test	This test measures:	Normal values*	In those with iron deficiency anemia, values tend to be:
Hematocrit	The proportion of red blood cells in blood plasma. Values vary with altitude.	40.7 to 50.3% (males) 36.1 to 44.3% (females)	Below normal
Hemoglobin	The iron-containing protein in red blood cells. Values vary with altitude.	13.8 to 17.2 g/dl (males) 12.1 to 15.1 g/dl (females)	Below normal
Ferritin	The amount of iron stored.	12 to 300 ng/ml (males) 12 to 150 ng/ml (females)	Below normal
Transferrin	The amount of iron being transported.	204 to 360 mg/dl	Above normal
Serum iron	The amount of iron in transferrin.	60 to 170 mcg/dl	Below normal
Transferrin saturation	The proportion of transferrin carrying iron.	20 to 50%	Below normal
Total iron binding capacity (TIBC)	Capacity to bind iron with transferrin. Indirect measure of transferrin.	240 to 450 mcg/dl	Above normal

Legend: g/dl = grams per deciliter; ng/ml = nanograms per milliliter; mcg/dl = micrograms per deciliter

*Normal values may vary by laboratory.

anemia, the body lacks the iron it needs to produce a normal amount of red blood cells. Hematocrit is also reduced because it is a measure of the proportion of red blood cells in blood plasma. In false (runner's) anemia, the slightly decreased hemoglobin value is due to plasma volume expansion associated with endurance training.

Iron deficiency anemia negatively affects performance.

It is known that iron deficiency anemia impairs performance. When iron deficiency anemia is present, $\dot{V}O_{2max}$ (that is, aerobic capacity) declines and subsequently increases with iron supplementation. The reduction in $\dot{V}O_{2max}$ is a result of impaired oxygen transport. Studies have documented that $\dot{V}O_{2max}$ can decline between 10 and 50 percent and reflects the severity of the anemia. Aerobic capacity does not seem to decline in people with an iron deficiency that has not progressed to anemia because oxygen transport is normal (Haas and Brownlie, 2001). Recall that those with iron deficiency without anemia (that is, subclinical iron deficiency) have normal hemoglobin and hematocrit concentrations.

Iron deficiency anemia also affects endurance capacity or the length of time until exhaustion at a given workload. This is different from aerobic capacity because endurance capacity depends on both oxygen transport and oxygen utilization. The role of iron in oxygen transport has been explained, but iron also plays a role in oxygen utilization because some oxidative enzymes contain iron.

Energy is produced as either heat or ATP from the flow of hydrogen ions down the electron transport chain located in cell mitochondria. As part of this metabolic pathway, iron is oxidized and reduced so that the transfer of electrons can proceed. A decrease in iron-containing compounds would negatively affect oxygen utilization. Studies have documented that iron-deficiency anemia reduces endurance capacity. Studies in animals suggest that iron deficiency without anemia also affects endurance capacity, but this has not been documented in human studies. A prudent approach is to maintain normal iron status to avoid any potential problems with endurance capacity (Haas and Brownlie, 2001).

The prevalence of iron deficiency and iron deficiency anemia in female athletes is likely higher than in the general population.

The prevalence of iron deficiency and iron deficiency anemia in athletes is hard to determine. In the United States, the prevalence of iron deficiency is estimated at 2 percent in adult males under the age of 70, 5 percent

```
---------------- BLOOD CELL COUNTS ----------------

TEST NAME    WBC    RBC    HGB    HCT   MCV   MCHC
UNITS       K/MM3  M/MM3  GM/DL   %    CUM    %
 DATE TIME
04/04R1127   4.7   4.76   15.8   44.2   93   35.7

------ ANEMIAS / HEMOGLOBINOPATHIES / MISCELLANEOUS -----

TEST
NAME         SERUM  TRANS-  TIBC   IRON
             IRON   FERRIN ------  SAT.
UNITS       MCG/DL  MG/DL  MCG/DL   %
 DATE TIME
04/04R1127   118    253     288    41

-------------------- COAGULATION --------------------

TEST         PLAT
NAME         COUNT
             K/MM3
 DATE TIME
04/04R1127   168
```

Laboratory blood tests are a tool physicians use to assess iron status.

Andy Doyle

of males ages 12 to 15, 9–16 percent of female adolescents, and 12 percent of women ages 20 to 49 (Centers for Disease Control and Prevention, 2002). Premenopausal women who engage in regular physical activity are at risk for developing an iron deficiency because exercise appears to have a negative effect on iron storage. One study reported that 26 percent of the female endurance athletes had a subclinical iron deficiency whereas another found that 28 percent of female recreational runners in the Zurich marathon had iron depletion (Malczewska, Raczynski, and Stupnicki, 2000; Mettler and Zimmermann, 2010). Studies of female military recruits suggest that about 15 percent enter training with an iron deficiency and ~25 percent will experience iron deficiency after 6 months of training (Merkel et al., 2009).

Iron deficiency anemia is estimated to occur in 2 percent of adolescent females and 4 percent of women ages 20 to 49 (Centers for Disease Control and Prevention, 2002). The prevalence of iron deficiency anemia is low in the male population because most adult males can easily meet their need for iron (8 mg daily) through diet. It is occasionally seen in male endurance runners and other male athletes who routinely take medications that relieve pain but also induce bleeding (for example, **aspirin**, **Ibuprofen**).

Serum: The fluid that separates from clotted blood. Similar to plasma but without the clotting agents.

Ferritin: Iron-containing storage protein.

False anemia: A reduced hematocrit and hemoglobin concentration associated with exercise training that does not represent a true anemia. Also known as runner's anemia or dilutional anemia.

Aspirin: Medication used to relieve pain, reduce inflammation, and lower fever. Active ingredient is salicylic acid.

Ibuprofen: Nonsteriodal anti-inflammatory drug (NSAID) used to relieve pain, reduce inflammation, and lower fever.

Some adolescent male athletes may experience iron deficiency anemia if they lack both sufficient kilocalories and iron-containing foods in their diet. The demand for iron resulting from rapid growth during adolescence, including an expanding blood volume, can outstrip the intake of iron. Still, the prevalence is low partly because very demanding periods of physiological growth favor iron absorption.

Female athletes who are menstruating are at risk for manifesting iron deficiency anemia. The loss of menstrual blood, thus the loss of iron, requires that iron be resupplied. This *can* be accomplished through diet alone, although the need for iron is relatively high (18 mg). Many factors may result in a poor dietary iron intake. Low caloric intake, low or absent animal protein intake, or high intake of iron inhibitors (for example, fiber) can result in an insufficient supply of dietary iron (Beard and Tobin, 2000). Some female athletes may also have an appreciable loss of iron in sweat, feces, and urine due to the stress of heavy training.

Although any female athlete may be at risk, iron deficiency anemia is most prevalent in female distance runners, gymnasts, and other athletes who have a restricted eating style. These athletes tend to have a low caloric intake compared to their energy output and therefore a low iron intake. The energy imbalance may be substantial and is due to a high energy expenditure from a heavy training volume and a commensurate inadequate intake of energy, often in the belief that losing weight and being a very low percentage of body fat will aid performance. In addition to inadequate total energy intake, these athletes often choose high-carbohydrate, low-fat, low-protein diets that are also low in iron. Some prefer to be vegetarians, and iron found in plant sources is not as well absorbed as that found in animal foods (Beard and Tobin, 2000).

Regardless of the estimated prevalence in the general athletic population or the specific sport, each athlete should have iron status periodically assessed. This is easily accomplished with a CBC and serum ferritin as part of a yearly physical. Those who are known to be at greatest risk—female, vegetarian, low-energy intake athletes—should have their iron status assessed more frequently, often two times a year. All athletes can be proactive in avoiding declines in iron status by consuming sufficient dietary iron daily.

Athletes should consume a variety of iron-containing foods.

The Dietary Reference Intakes for iron are shown in Table 9.13. Adult males and females 51 years and above

have only small amounts of iron loss daily, thus 8 mg daily is recommended. The DRI is substantially higher for females ages 19 to 50 (18 mg/d) and exceptionally high during pregnancy (27 mg/d). The Tolerable Upper Intake Level is 45 mg because large amounts of iron cause gastrointestinal distress. Iron supplements typically contain large amounts (for example, 65 mg) and a common complaint is nausea. Iron supplementation should always be taken under physician supervision because some people are at risk for iron overload.

Sufficient dietary iron intake is correlated with adequate energy intake. Although there are a few foods that are excellent sources of iron (for example, oysters, clams) and some foods that contain moderate amounts (for example, meat, dried beans, green leafy vegetables), many foods contain only small amounts. This wide distribution of small amounts of iron in food means that iron intake generally increases as caloric intake increases.

The average adult diet in the United States contains 6 to 7 mg of iron for every 1,000 kcal consumed (Beard and Tobin, 2000). Most males, especially male athletes, consume over 2,000 kcal daily. Even without emphasizing iron-rich foods in their diets, males are likely to consume at least 12 mg daily, more than the 8 mg recommended. Adequate dietary intake is one reason that males rarely manifest iron deficiency anemia.

Table 9.13 Dietary Reference Intakes for Iron

Age group (yr)	Males—iron (mg/d)	Females—iron (mg/d)
0 to 0.5	*	*
0.5 to 1	11	11
1 to 3	7	7
4 to 8	10	10
9 to 13	8	8
14 to 18	11	15*
19 to 50	8	18*
51 to 70	8	8
Greater than 70	8	8

Legend: yr = year; mg = milligram; d = day

Pregnant females: 27 mg

*Adequate Intake (AI) of iron is 0.27 mg/d

Note: Recommendations for those greater than 6 months are based on the Recommended Dietary Allowance (RDA). RDA is the average daily dietary intake sufficient to meet the nutrient requirement of nearly all healthy individuals. When the RDA cannot be determined, the AI is used. The AI is an estimate or approximation. See Chapter 1.

Source: Institute of Medicine. (2001). *Dietary Reference Intakes for vitamin A, vitamin K, arsenic, boron, chromium, copper, iodine, iron, manganese, molybdenum, nickel, silicon, vanadium, and zinc* (Food and Nutrition Board). Washington, DC: National Academies Press.

Iron is found in many foods, but often in small amounts. Adequate dietary iron intake is associated with adequate energy intake.

It is easy to see why women are at much greater risk for iron deficiency. If a woman consumed 2,000 kcal daily, then her likely intake of 12 to 14 mg of iron would be less than the recommended 18 mg. But many females, including female athletes, restrict caloric intake. A female athlete who is trying to attain or maintain a low percentage of body fat and restricts energy intake to 1,500 kcal daily would be expected to consume only 9 mg of iron daily, half the recommended amount. Females who prefer a dairy-based diet and consume high levels of milk, yogurt, and cheese also tend to have a low iron intake since milk and milk products are relatively poor sources of iron compared to meat, poultry, fish, beans, grains, nuts, and some vegetables.

In addition to the amount of iron consumed, the form of iron is a factor because it affects absorption. Iron is found in one of two forms in food—heme or **nonheme**. Heme iron, which is found in animal foods, is well absorbed. Absorption is estimated to be 15–35 percent of the heme iron consumed. In contrast, nonheme iron, which is found in plant foods (as well as animal tissues), has much lower absorption, in the range of 2–20 percent. The presence of ascorbic acid (vitamin C) enhances the absorption of nonheme iron and is one way for vegetarians to optimize their iron absorption. A compound found in meat, fish, and poultry, known as MFP factor, also enhances nonheme iron absorption if the animal food is eaten at the same time as the plant food. For example, vegetarian chili is made from beans and tomatoes. The iron in the beans is better absorbed due to the vitamin C contained in the tomatoes. However, nonheme iron absorption from chili also containing meat is greater than vegetarian chili due to the influence of the MFP factor.

Nonheme sources of iron (for example, beans and legumes, grains, vegetables) may contain iron absorption inhibitors such as phytates and oxalates. In vegetarian diets, phytate is the primary inhibitor of iron absorption. Black tea is also known to inhibit the absorption of iron (Hurrell and Egli, 2010).

How can athletes best use this information to plan a diet that is adequate in iron? Athletes should focus first on obtaining sufficient energy (kilocalories). Those who have low or marginal energy intakes should focus on including iron-rich foods in their diets. Increasing intake of iron-dense foods helps to provide more iron while not substantially increasing caloric intake. Vegetarian athletes, who do not consume the more absorbable heme iron or MFP factor, need to focus on the quantity of iron consumed as well as ensuring that a variety of plant foods are eaten. The focus on variety helps to modulate the intake of compounds that are known to inhibit iron absorption. Consuming vitamin C-containing foods at the same meal can also be beneficial.

Table 9.14 lists iron-containing foods. The table has been subdivided into heme and nonheme sources. Foods are then listed in descending order according to the amount of iron contained. Choosing foods nearer the top of each category is recommended for athletes who are at risk for iron deficiency anemia.

Although it receives less attention, an adequate copper intake is also important for proper red blood cell formation. Excellent sources of copper include seafood, nuts, and seeds. Copper is also found in whole grains, dried beans, and some green leafy vegetables. Most nutrient analysis programs do not include the copper content of foods in the database, so it is difficult to assess dietary intake. The DRI is 900 mcg for adults, and the average intake by U.S. adults is more than 1,000 mcg.

Key points

- Iron is necessary for proper red blood cell formation.
- Low iron intake may result in iron deficiency or iron deficiency anemia.
- Iron deficiency anemia negatively affects performance.
- Premenopausal female athletes have the highest risk of developing iron deficiency and iron deficiency anemia, particularly if caloric intake is restricted and the diet lacks iron-rich foods.
- Iron supplementation may be beneficial for some female athletes, but it should not be self-prescribed.

Why might a female distance athlete need a blood test every 6 months?

Nonheme: A form of iron with a lower rate of absorption (see heme).

Table 9.14 Iron-Containing Foods

Food	Iron (mg)
Heme sources	
3 oz steamed clams	26.6
5 steamed oysters	5 to 16.5
3 oz beef liver	5.3
3 oz beef	3
1 cup tuna fish	2.4
3 oz dark meat chicken	1.3
3 oz light meat chicken	1.1
3 oz halibut	0.9
3 oz pork loin	0.8
Nonheme sources	
1 cup cereal (iron added, amount depends on the brand)	2 to 16
1 cup instant oatmeal (iron added)	10
1 cup soy beans	8.8
1 cup lentils	6.6
1 cup pinto beans	4.2
½ cup spinach, cooked from fresh	3.2
½ cup spinach, cooked from frozen	1.4
1 cup soy milk	1.4
½ cup tofu	1.3
1½ oz raisins	0.87
1 slice bread (iron added)	0.8
1 egg, white and yolk	0.72

Legend: mg = milligram; oz = ounce

9.5 The Roles of Minerals in the Immune System

Intense training and prolonged exercise are immuno-suppressive. In other words, heavy training, especially for endurance athletes, depresses the immune system. Many endurance athletes have frequent infections, particularly upper respiratory infections, during periods of intense or prolonged training, which then undermines their ability to maintain training and negatively affects performance. The two nutritional factors most associated with proper immune system function are adequate protein and total energy intakes. However, several minerals are also involved, such as zinc, magnesium, selenium, and inadequate intake of these minerals impairs immune response. Conversely, excessive levels of some minerals, such as zinc and iron, can impair the immune system.

The immune system protects the body from disease.

The immune system is the body's defense against disease. Effective immunity relies upon a system of tissues and organs that are supported by adequate nutrition; nearly every tissue is involved. The skin serves as a physical barrier, hairlike projections known as cilia and mucosal secretions guard against invasion via the respiratory tract, and the GI tract can kill microorganisms with hydrochloric acid, digestive enzymes, and other secretions. These are examples of nonspecific immunity, the body's first line of defense against potentially harmful microorganisms. Another form of nonspecific resistance is inflammation, which involves the release of chemicals from phagocytic cells that destroy microbes.

Although nonspecific immunity is important, the body must have more specific ways to fight disease. This is known as acquired immunity, a special immune system response that forms antibodies and activates lymphocytes (white blood cells). Acquired immunity cannot occur until a specific microorganism has invaded the body (now called an antigen) and the immune system forms antibodies in response. Each antigen (microorganism) has a unique protein or large polysaccharide that the body can detect and remember. When reexposed to that specific antigen, the antibody response is rapid (within a few hours) and potent (circulating for months rather than weeks). All antibodies are immunoglobulins, compounds that are made of several polypeptide (protein) chains.

Regulation of the immune system is the responsibility of the cytokines. Cytokines are protein-containing compounds that are released in response to chronic inflammation, infections, and other disease processes. Interleukins (IL) and interferon (INF) are examples of regulatory cytokines. Detailed information about the immune system can be found in any physiology or medical physiology textbook (Sherwood, 2010; Guyton and Hall, 2011).

The effect of exercise on the immune system follows the "too much of a good thing may be harmful" theory. Considering just one outcome of immune function, an upper respiratory tract infection (URTI), the results of a number of studies show that moderate exercise bolsters immune function as evidenced by lower risk for or incidence of URTIs. However, people involved in more rigorous or prolonged exercise, such as marathon running or prolonged running training, have a greatly increased risk for upper respiratory tract infections (Nieman, 2008; Gleeson, Nieman, and Pedersen, 2004).

Zinc. Zinc is necessary for proper cellular function because of its role as a constituent/cofactor in more than 200 enzyme systems. It has wide-reaching effects on DNA, RNA, and cellular functions; thus many different types of immune system cells are affected when zinc is deficient or excessive. Zinc deficiency results in damaged skin and gastrointestinal cells, both of which are involved with nonspecific immunity. Examples of ways that zinc deficiency negatively affects specific immunity include decreased production and function of lymphocytes. An excessive zinc intake also decreases lymphocyte response and inhibits copper absorption. A copper deficiency is immunosuppressive (Shankar and Prasad, 1998).

The Dietary Reference Intake for zinc is 8 mg for females and 11 mg for males. The Tolerable Upper Intake Level is 40 mg. The median (50th percentile) intake for adults in the United States is 9 mg for women and 13 mg for men, so clearly zinc requirements can be met through diet alone. Those who meet the DRI for zinc tend to consume sufficient kilocalories and animal foods such as red meat and milk (Institute of Medicine, 2001).

Unfortunately, it is estimated that up to 90 percent of endurance athletes do not meet the DRI for zinc. Many endurance athletes limit red meat intake and consume high-carbohydrate, relatively low-protein diets (Micheletti, Rossi, and Rufini, 2001). This puts them at risk for subclinical zinc deficiencies and makes zinc supplementation attractive. But supplemental zinc can interfere with the absorption of other nutrients, especially iron and copper, and can suppress the immune system. Since caution is warranted, no more than 15 mg of supplemental zinc daily is recommended.

Recurring infections are the bane of many endurance athletes. Researchers have studied the effect of zinc supplements on prevention of upper respiratory tract infections and treatment of colds. The study results have been mixed, and the evidence in support of zinc supplement use to prevent colds and infections is limited. Gleeson, Nieman, and Pedersen (2004) recommend that endurance athletes monitor their zinc status. This can be done by assessing dietary zinc intake and through a specialized blood test. Given poor intake, higher-than-normal losses in sweat and urine, and evidence of altered zinc status, low-dose zinc supplementation would be appropriate.

Selenium. At least 20 selenium-containing proteins are involved in cellular metabolism; thus selenium is an essential nutrient. Its exact roles in the immune system are not known, but depressed immunity is associated with selenium deficiency. When selenium is deficient there is less proliferation of lymphocytes, immunoglobin production is decreased, and the ability to kill pathogens is reduced (Arthur, McKenzie, and Beckett, 2003). It does not appear that athletes are selenium deficient (Speich, Pineau, and Ballereau, 2001). Food sources of selenium include meat, fish, poultry, whole grains, and nuts. Acute toxicity is possible; such was the case when 200 people in the United States took a liquid dietary supplement that contained 200 times the selenium concentration listed on the label (MacFarquhar et al., 2010).

Iron. Iron has already been discussed in detail due to its role in oxygen transport. Iron also plays a role in the proper functioning of the immune system. Most studies suggest that individuals with iron deficiency are at a greater risk for infection. Iron is necessary for proper cellular function, and a deficiency can affect the iron-containing enzymes of immune cells. Iron also helps to regulate the cytokines (Beard, 2001). Excess iron can impair immune system function (Gleeson, Nieman, and Pedersen, 2004).

Key points

- Rigorous training and prolonged exercise suppress the immune system.
- Inadequate intake of zinc, magnesium, or selenium impairs the immune system, as does excessive intake of zinc or iron.

How can athletes determine if their intake of minerals is above inadequate but below excessive?

9.6 The Adequate Intake of All Minerals

At first glance, it may seem as if obtaining the proper intake of the 21 known essential minerals would be difficult. Fortunately, it is not necessary to measure and track the intake of all the macro- and microminerals in the diet. Because minerals are found in small amounts in many different foods, the best way to ensure adequate mineral status is to: (1) consume adequate kilocalories daily, (2) eat a variety of nutritious foods that have been minimally processed, and (3) consume an adequate amount of calcium and iron from food. Calcium and iron, when obtained from naturally occurring food sources, are fairly good predictors of the intake of the other minerals. In other words, when calcium and iron intake from food is adequate then the intake of the remaining 19 essential minerals is likely to be adequate.

In a nutshell, a nutritious diet contains a variety of fruits, vegetables, whole grains, beans and legumes, lean sources of proteins, heart-healthy fats, and a sufficient, but not excessive amount of energy. This dietary

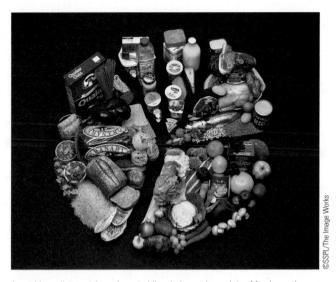

A nutritious diet contains adequate kilocalories and a variety of foods, such as those shown here. This dietary pattern is likely to provide sufficient carbohydrates, proteins, fats, vitamins, and minerals.

©SSPL/The Image Works

pattern is consistent with adequate mineral intake. This same pattern also provides athletes the carbohydrates, proteins, and fats needed to support training and competition.

The key to obtaining all the minerals needed from food is to consume a nutrient-dense, whole-foods diet.

Two diets that were discussed in Chapter 8 are repeated here in Figures 9.5 and 9.6. One is a high-fat, high-sugar, low-fiber dietary pattern that is low in most, but not all, vitamins for a 20-year-old male. The other diet

emphasizes fruits and vegetables, legumes, nuts, and whole grain, less processed foods. This whole-foods diet meets the recommended daily intakes for all of the vitamins. How well do these diets fare when mineral content is analyzed? Tables 9.15 and 9.16 reveal the answers.

Four minerals were analyzed—calcium, iron, magnesium, and zinc. Other than sodium and potassium (covered in Chapter 7), these four minerals are the only ones included in most nutrient analysis programs. The high-fat, high-sugar, low-fiber diet contains ~2,500 kcal, yet three of the four minerals assessed are low because of a lack of nutrient-dense foods. Magnesium and zinc intake are particularly low, 15 percent and 32 percent, respectively, and calcium intake is approximately 50 percent of the amount recommended for a 20-year-old.

Contrast that with the nutrient-dense, whole-foods diet. This diet has approximately the same amount of energy (kilocalories), but substantially more minerals. The nutrient-dense diet provides more than 100 percent of the DRI for calcium, iron, magnesium, and zinc. In the case of iron, the diet provides 20 mg, which meets the DRI for both adult males (8 mg) and

2 fried eggs

2 pieces of white toast w/ 1 T butter

2 slices bacon

8 oz coffee w/ 1 T each cream, sugar

Ham & cheese sandwich

1 oz bag potato chips

16 oz soft drink

3 Oreo cookies

Cheeseburger

Medium fries

16 oz soft drink

½ cup chocolate ice cream

Figure 9.5 An example of a high-fat, high-sugar, low-fiber diet

Legend: T = tablespoon; oz = ounce

1 cup oatmeal

1 slice whole wheat toast

1 T peanut butter

8 oz nonfat milk

8 oz orange juice

8 oz black coffee

1.5 cups lentil soup

Large (1.5 cups) green salad

1 oz avocado (~⅕ of an avocado)

2 T oil & vinegar dressing

2 whole wheat rolls

8 oz nonfat milk

1 banana

6 oz chicken breast

1½ cups brown rice

1 cup broccoli

1 slice whole wheat bread w/ 1 tsp *trans free* margarine

1 cup strawberries

Water

1 cup nonfat frozen yogurt

¼ cup dry-roasted sunflower seeds

Figure 9.6 An example of a nutrient-dense, whole-foods diet

Legend: T = tablespoon; oz = ounce

Table 9.15 Mineral Intake of a High-Saturated-Fat, High-Sugar, Low-Fiber Diet Compared to the DRI

Mineral	Intake	DRI*	% of DRI
Calcium	509 mg	1,000 mg	51
Iron	12.25 mg	8 mg	156
Magnesium	61 mg	400 mg	15
Zinc	3.5 mg	11 mg	32

Legend: DRI = Dietary Reference Intakes; mg = milligram

*20-year-old male

Table 9.16 Mineral Intake of a Nutrient-Dense, Whole-Foods Diet Compared to the DRI

Mineral	Intake	DRI*	% of DRI
Calcium	1,145 mg	1,000 mg	114
Iron	20.39 mg	8 mg	255
Magnesium	629 mg	400 mg	157
Zinc	16 mg	11 mg	148

Legend: DRI = Dietary Reference Intakes; mg = milligram

*20-year-old male

Table 9.17 Example of a Multimineral Supplement

Mineral	Amount	% of DV	UL*
Calcium	1,000 mg	100	2,500 mg (2,000 mg for females older than 51 and males older than 71)
Magnesium	500 mg	125	350 mg
Zinc	30 mg	200	40 mg
Manganese	10 mg	500	11 mg
Iron	10 mg	55	45 mg
Copper	2 mg	100	10 mg
Iodine	150 mcg	100	1,100 mcg
Selenium	200 mcg	285	400 mcg
Chromium	200 mcg	167	Not established
Molybdenum	500 mcg	667	2,000 mcg

Legend: DV = Daily Value; UL = Tolerable Upper Intake Level; mg = milligram; mcg = microgram

*The UL for minerals is based on the intake of all food (fortified and nonfortified), water, and supplements except for magnesium, which is based on supplements only.

19- to 50-year-old females (18 mg). The food with the highest iron content is lentil soup, which contains 4 mg. The remainder of the iron is gathered from the variety of foods included in this diet. With the exception of margarine and oil, all of the foods contributed iron, although no one food except the soup had more than 2 mg. An athlete consuming this dietary pattern could reasonably assume that the other minerals not analyzed are also adequate.

The dose and potency of a mineral supplement can vary substantially from brand to brand.

Adequate mineral intake can be achieved from food alone. However, this requires that a variety of nutrient-dense foods are consumed and, realistically, many people simply do not eat that way. They may wish to consume foods that have been highly fortified or enriched with minerals. Cereal is one product that has minerals added, and more mineral-fortified foods, such as energy bars, are being manufactured, in part, because people are known to consume mineral-poor diets.

Multimineral supplements are also sold. Earlier in the chapter, a supplement advertised as "high potency" was evaluated (see Spotlight on supplements: Evaluating a High-Potency Multimineral Supplement Advertised to Athletes). But what about a mineral supplement that is not "high potency"? Another example of a product that is advertised to athletes is shown in Table 9.17. This supplement is advertised as a way to obtain the various minerals that may be missing from an athlete's usual diet. Notice that one mineral, magnesium, exceeds the Tolerable Upper Intake Level. The amount of zinc contained is 30 mg, twice as high as the 15 mg dose that is often recommended. Given that the average male consumes 13 mg of zinc daily, total zinc intake would likely exceed the UL if a male athlete consumed this supplement. This supplement also contains iron. Iron supplementation is generally not recommended for males because the risk of iron overload is greater than the risk of iron deficiency anemia. About 5 of every 1,000 people in the United States, mostly Caucasian males,

carry two copies of the abnormal gene responsible for overabsorption of iron, which can lead to excessive iron storage in tissues. These individuals should never take iron supplements. Before consuming any multimineral supplement, consumers should carefully check the dose of all the minerals contained.

The biggest question about multiple minerals being added to food or taken as supplements is whether the minerals are well absorbed. Multimineral supplements contain too many minerals to test so this remains an unanswered question. A 2005 review article of iron and zinc supplementation together found that zinc reduced, but did not completely block, iron absorption (Fischer Walker et al., 2005). More studies are needed to determine both the potential benefits and harms associated with multiple mineral supplements.

Supplementing with individual minerals. Many people adopt a "food first, supplements second" philosophy. They try to obtain all of their nutrients from food if possible, but consume supplements when dietary intake is habitually low. The most frequently supplemented individual minerals are calcium and iron. Females are more likely than males to have insufficient habitual intake, increased avenues of loss, or both, potentially necessitating supplements. If both iron and calcium are supplemented, these supplements should be taken at different times if possible to reduce competition for absorption.

If a physician diagnoses iron deficiency anemia, the appropriate treatment is iron supplementation. High doses of ferrous sulfate (325 mg of which 65 mg is iron) are often recommended. Because this form and quantity can cause GI upset, adjustments may need to be made in the supplement's dose or type. Despite high supplemental doses, the restoration of depleted iron stores is a slow process that takes 3 to 6 months. Iron supplementation should be discontinued (or started) only after consultation with a physician.

Obtaining calcium from food sources is advised, but supplemental calcium is generally absorbed as well as calcium from milk. The best absorption occurs when supplemental calcium is consumed as a 400 to 500 mg dose. The Tolerable Upper Intake Level is 2,000–2,500 mg depending on age, so supplemental calcium of up to 1,000 mg daily is probably safe for most people, although the exact amount will depend on dietary intake.

Athletes may self-prescribe other individual mineral supplements (for example, zinc, chromium, magnesium) because they are concerned that dietary intake may be low. Although a single supplement may provide more than enough of that particular mineral, there are concerns that a single supplement, especially one containing a high dose, may be detrimental to the absorption of other nutrients. A prudent approach is to take no more than the DRI in supplement form. However, low-dose single mineral supplements are often hard to find so individual mineral supplements may need to be taken every other day or every third day to reduce exposure.

Sometimes supplements are advertised as having a higher **bioavailability** than food or another brand of supplements. The term *bioavailability* is often used interchangeably with the term *absorption*, but technically, bioavailability refers to absorption, utilization, and retention. Bioavailability should be thought of as a point on a continuum. The goal should be adequate bioavailability, not poor or excess bioavailability. If bioavailability is too low, absorption or utilization would likely be poor. But excess bioavailability can be a problem, too, due to excess absorption, a high blood concentration of that mineral, and substantially reduced absorption of other minerals that use the same absorption pathways.

An example is supplemental chromium, which may be sold as chromium picolinate. The picolinate makes the compound very stable and the chromium is absorbed in much higher amounts than would be absorbed from food. Excess chromium could cause damage to cells, so highly bioavailable chromium is not as beneficial as it may sound. Chromium supplements, which are popular with athletes, are featured in the Spotlight on supplements: How Beneficial Is Chromium Supplementation for Athletes?

Key points

- Consumption of a variety of nutrient-dense foods and sufficient caloric intake are associated with an adequate intake of minerals.

- Supplementation, particularly iron and calcium, may be beneficial to provide minerals missing from the diet.

- Iron supplements should not be self-prescribed.

When it comes to minerals, what is the danger for athletes who think that "more" is always "better"?

Keeping it in perspective

Minerals as Building Blocks

Remember the story of the Three Little Pigs? The first pig built his house out of straw, the second pig out of sticks, and the third pig out of bricks. The straw and stick houses were built quickly, and the pigs had time to play while the third pig built his house of bricks. But the houses made of straw and sticks were not well constructed and could not withstand the huffing and puffing of the big, bad wolf.

Consuming minerals in proper balance is like building a house of bricks. The most obvious analogy is with calcium and the other bone-forming nutrients. These minerals help the bones to be dense and strong and to better withstand the loss of bone calcium that accompanies aging. Many minerals are also involved in the proper function of the immune system, which must withstand pathological microorganisms and other sources of harm on a daily basis over a lifetime. A low or unbalanced mineral diet over a long period of time is like a house of straw or sticks. Consuming enough minerals is like building and living in a house of bricks. You will be glad you did when the big, bad wolf (for example, aging and disease) comes huffing and puffing at your door.

Application exercise

Scenario: *A high school-age competitive female gymnast knows she does not eat as well as she should. Her mother has decided that she needs to take a supplement containing iron, calcium, and other minerals. She enters* mineral supplement *into her favorite search engine to see what might be available.*

1. Enter the term *mineral supplement* into your favorite search engine (for example, Google, Bing, and so forth).
   • How many results/matches are there?
   • What are the differences in the search results between the Sponsored Links (ads) and the list of other search results?
   • How would you decide which entry to click on?
2. Click on one of the links that sells a multimineral supplement and print the product information. Answer the following questions about that multimineral supplement.

• Does the supplement contain more than 100 percent of the Daily Value (DV) for any mineral?
• Is the Tolerable Upper Intake Level (UL) exceeded for any mineral?
• Is the mineral supplement advertised as high potency or high bioavailability?
• Does the advertising information appear to be factual? Give examples.
• Is any of the information factual but taken out of context or appear to be overstated? Give examples.

3. Decision making:
   • Using the multimineral supplement product you selected, outline the specific factors that you think would influence the decision to purchase/use this supplement or to not purchase/use this supplement.
   • If you were a registered dietitian, would you recommend this supplement? Why or why not?

Bioavailability: The degree to which a substance is absorbed, utilized, and retained in the body.

Spotlight on supplements

How Beneficial is Chromium Supplementation for Athletes?

Chromium is a mineral supplement sometimes taken by athletes for the purpose of increasing muscle mass and decreasing body fat. Bodybuilders and other strength athletes are frequent users. The most common supplement form is chromium picolinate. The picolinate, which makes the compound extremely stable, increases gastrointestinal absorption of the chromium. An important point is that chromium picolinate is not absorbed in the same manner as dietary chromium. It appears that chromium picolinate supplements can result in an increase in free radical (oxidant) production. More research is needed to determine if there is a long-term effect, but athletes should be aware of this possibility (Vincent, 2000).

Chromium enhances insulin sensitivity and improves glucose utilization by increasing the number of insulin receptors (Volpe, 2008; Speich, Pineau, and Ballereau, 2001). Although chromium supplementation can help to improve blood glucose levels in those with diabetes, there is no evidence that such an effect occurs in individuals who are not diabetic (Balk et al., 2007). Insulin also has anabolic properties, and enhanced sensitivity could promote amino acid uptake into muscle cells. Uptake of amino acids is known to stimulate protein synthesis.

Animal studies and early studies in humans suggested that muscle mass was increased and body fat was decreased with chromium supplementation. In humans, the changes were small but significant, ~1.8 kg (~4 lb) increase in muscle mass and ~3.4 kg (~7.5 lb) decrease in body fat. Subsequent studies, with stricter methodology including better measurements of body composition, did not replicate these early results. Most studies use daily doses of chromium supplements of 200 to 400 mcg (Lukaski, 1999).

The Dietary Reference Intakes for adults range from 20 to 35 mcg daily depending on age and gender. The average daily dietary intake of chromium is approximately 25 mcg for adult females and 33 mcg for adult males. If sufficient kilocalories (energy) are consumed, a chromium deficiency would not be expected. No Tolerable Upper Intake Level has been established. Chromium supplementation of 50 to 200 micrograms daily seems to be safe. The effects of higher daily intakes are not known, but there is some suspicion that iron absorption may be decreased and that accumulated chromium in the body can damage DNA (Lukaski, 2000).

Post-Test   **Reassessing Knowledge of Minerals**

Now that you have more knowledge about minerals, read the following statements and decide if each is true or false. The answers can be found in Appendix O.

1. The basic fusnctions of minerals include building body tissues, regulating physiological processes, and providing energy.

2. In general, the body absorbs a high percentage of the minerals consumed.

3. In most cases, exercise does not increase mineral requirements above what is recommended for healthy, lightly active individuals.

4. If dietary calcium is inadequate over a long period of time, the body maintains its blood calcium concentration by reabsorbing skeletal calcium.

5. Iron deficiency anemia has a negative impact on performance.

Summary

More than 20 minerals are needed for the proper functioning of the body. Dietary calcium, potassium, and magnesium intakes are low for some men and women in the general population. Of great concern is the very low calcium intake of teenage females. Dietary iron intake may also be low for women. Athletes' intakes of these, as well as other minerals, are likely to be low if energy intake is restricted or a well-balanced diet is not consumed.

Adequate calcium, vitamin D, and protein intakes and weight-bearing exercise are needed throughout life so that peak bone mineral density is achieved and loss of calcium from bone is prevented or slowed. Bone loss is accelerated by a lack of estrogen. A lack of estrogen is usually associated with menopause but may also be present in some adolescent and young adult female athletes, putting them at great risk for low bone mineral density and fractures. Milk and milk products are concentrated sources of calcium, but calcium is found in a variety of nondairy products, although typically in lower amounts. Many people consume calcium supplements, which help to protect against loss of calcium from bone. However, an excessive amount of supplemental calcium can increase the risk of developing kidney stones.

Iron is closely linked to athletic performance because depleted iron stores lead to iron deficiency anemia, resulting in a reduced oxygen-carrying capacity of the red blood cells. Oxidative capacity at the cellular level is reduced and performance is impaired, particularly for the endurance athlete. With such clear impacts on performance, preventing iron deficiency anemia is important for athletes. Several minerals are also involved in maintaining a healthy immune system, including iron, zinc, and selenium.

The best ways to ensure adequate mineral status is to consume enough kilocalories daily and to eat a variety of nutritious foods that have been minimally processed. Although all minerals are important, there is an emphasis on obtaining adequate amounts of calcium and iron in the diet. When dietary intake of these two minerals is adequate, the intake of other minerals is likely to be adequate.

Because dietary mineral intake is low for many people, including athletes, there is much interest in taking mineral supplements. Dose is very important because large amounts of one mineral can hamper the absorption of other minerals. Large doses can also increase the risk for toxicities. Supplementation with minerals should be done thoughtfully to avoid potential problems. Mineral deficiency diseases, such as iron-deficiency anemia, are successfully treated with mineral supplementation.

Self-Test

Multiple-Choice

The answers can be found in Appendix O.

1. A mineral that is under substantial hormonal control is:

   a. zinc.   c. calcium.
   b. iron.   d. selenium.

2. In the absence of a deficiency, as mineral consumption increases, the amount absorbed:

   a. increases.
   b. decreases.
   c. stays the same.

3. The two factors that best explain the increased risk for osteoporosis in elite female distance runners are:
   a. age and poor diet.
   b. African-American heritage and poor diet.
   c. positive family history and low estrogen.
   d. low energy intake and low estrogen.

4. What effect does iron deficiency anemia have on performance?
   a. decline in aerobic capacity
   b. decline in endurance capacity
   c. decline in oxygen utilization
   d. all of the above

5. What is the potential problem with consuming excess supplemental zinc?
   a. anorexic (appetite-suppressing) effect
   b. increase in urinary tract infections
   c. interference with iron absorption
   d. excessive bruising

Short Answer

1. Explain the physiological and performance effects of mineral intakes that are too low or excessive.

2. Explain absorption problems associated with minerals found in food and supplements.

3. Are the requirements for minerals increased for athletes?

4. Do sedentary adults consume enough minerals? Do athletes? Explain.

5. Describe some of the potential adverse effects associated with consumption of minerals in excess of the Tolerable Upper Intake Level.

6. Explain the differences between a clinical and subclinical mineral deficiency, using either calcium or iron as an example.

7. Name the bone-forming minerals and explain why each is important for normal physiological function.

8. Briefly describe the roles of minerals and hormones in maintaining calcium homeostasis and calcium balance, including their roles in absorption, excretion, deposition, and resorption.

9. What effect does exercise have on attaining peak bone density and maintaining bone density across the life cycle?

10. How does iron deficiency, with and without anemia, affect athletic performance?

11. Explain the roles of minerals in proper immune function. Describe problems associated with both deficiencies and excesses.

12. How do athletes know if they are getting enough of all the minerals? What general dietary principles are associated with adequate mineral intake?

13. Are mineral supplements helpful or harmful to athletes? If so, in which ways?

14. What are the pros and cons of consuming a multi-mineral supplement daily?

Critical Thinking

1. Consider the information in both the vitamin and mineral chapters (Chapters 8 and 9). What similarities do vitamins and minerals share? How are these two classes of nutrients different?

2. Write a compelling argument for obtaining all minerals from naturally occurring food sources. Write an equally compelling argument for the use of mineral supplements. Describe an evaluation process that an athlete could use to determine a strategy for adequate mineral intake.

References

Arthur, J. R., McKenzie, R. C., & Beckett, G. J. (2003). Selenium in the immune system. *Journal of Nutrition*, *133*(5, Suppl. 1), 1457S–1459S.

Bailey, R. L., Dodd, K. W., Goldman, J. A., Gahche, J. J., Dwyer, J. T., Mishfegh, A. J., et al. (2010). Estimation of total usual calcium and vitamin D intakes in the United States. *Journal of Nutrition*, *140*, 817–822.

Balk, E. M., Tatsioni, A., Lichtenstein, A. H., Lau, J., & Pittas, A .G. (2007). Effect of chromium supplementation on glucose metabolism and lipids: A systematic review of randomized controlled trials. *Diabetes Care*, *30*, 2154–2163.

Beard, J. L. (2001). Iron biology in immune function, muscle metabolism and neuronal functioning. *Journal of Nutrition*, *131*(2S–2), 568S–579S.

Beard, J., & Tobin, B. (2000). Iron status and exercise. *American Journal of Clinical Nutrition*, *72*(2 Suppl.), 594S–597S.

Beck, B. R., & Snow, C. M. (2003). Bone health across the lifespan—exercising our options. *Exercise and Sport Sciences Reviews*, *31*(3), 117–122.

Bischoff-Ferrari, H. A., Dawson-Hughes, B., Baron, J. A., Burckhardt, P., Li, R., Spiegelman, D., et al. (2007). Calcium intake and hip fracture risk in men and women: a meta-analysis of prospective cohort studies and randomized controlled trials. *American Journal of Clinical Nutrition, 86*(6), 1780–1790.

Bonura, F. (2009). Prevention, screening, and management of osteoporosis: An overview of the current strategies. *Postgraduate Medicine, 121*(4), 5–17.

Bowman, S. A. (2002). Beverage choices of young females: Changes and impact on nutrient intakes. *Journal of the American Dietetic Association, 102*(9), 1234–1239.

Cashman, K. D. (2007). Diet, nutrition, and bone health. *Journal of Nutrition, 137*(11 Suppl.), 2507S–2512S.

Centers for Disease Control and Prevention. (2002). Iron deficiency—United States, 1999–2000. *Morbidity and Mortality Weekly Report, 51*(40), 897–899.

Centers for Disease Control and Prevention. (2007). Deaths: Preliminary data for 2007. Retrieved October 25, 2010, from http://www.cdc.gov/nchs/fastats/lifexpec.htm

Chinevere, T. D., Kenefick, R. W., Cheuvront, S. N., Lukaski, H. C., & Sawka, M. N. (2008). Effect of heat acclimation on sweat minerals. *Medicine and Science in Sports and Exercise, 40*(5), 886–891.

Clarke, B. (2008). Normal bone anatomy and physiology. *Clinical Journal of the American Society of Nephrology, 3*(Suppl. 3), S131–S139.

Compston, J. E. (2001). Sex steroids and bone. *Physiological Reviews, 81*(1), 419–447.

De Souza, M. J., West, S. L., Jamal, S. A., Hawker, G., Gundberg, C., & Williams, N. I. (2008). The presence of both an energy deficiency and estrogen deficiency exacerbate alterations of bone metabolism in exercising women. *Bone, 43*(1), 140–148.

De Souza, M. J., & Williams, N. I. (2005). Beyond hypoestrogenism in amenorrheic athletes: Energy deficiency as a contributing factor for bone loss. *Current Sports Medicine Reports, 4*(1), 38–44.

Dietary Guidelines Advisory Committee (2010). Report on the Dietary Guidelines Advisory Committee on the Dietary Guidelines for Americans, 2010, to the Secretary of Agriculture and the Secretary of Health and Human Services. U.S. Department of Agriculture, Agricultural Research Service, Washington, D.C.

DIPART (Vitamin D Individual Patient Analysis of Randomized Trials) Group. (2010). Patient level pooled analysis of 68,500 patients from seven major vitamin D fracture trials in US and Europe. *British Medical Journal,* 340:b5463 doi: 10.1136/bmj.b5463.

Drinkwater, B. L., Nilson, K., Chesnut, C. H., Bremner, W. J., Shainholtz, S., & Southworth, M. B. (1984). Bone mineral content of amenorrheic and eumenorrheic athletes. *The New England Journal of Medicine, 311,* 277–281.

European Food Safety Authority. (2004). *Opinions of the Scientific Panel on Dietetic Products, Nutrition and Allergies.* Retrieved [December 21, 2010] from http://ww.efsa.europa.eu/

Fischer Walker, C., Kordas, K., Stoltzfus, R. J., & Black, R. E. (2005). Interactive effects of iron and zinc on

biochemical and functional outcomes in supplementation trials. *American Journal of Clinical Nutrition, 82*(1), 5–12.

Ford, E. S., & Mokdad, A. H. (2003). Dietary magnesium intake in a national sample of US adults. *Journal of Nutrition, 133*(9), 2879–2882.

Fraga, C. G. (2005). Relevance, essentiality and toxicity of trace elements in human health. *Molecular Aspects of Medicine, 26*(4–5), 235–244.

Fuchs, R. K., Bauer, J. J., & Snow, C. M. (2001). Jumping improves hip and lumbar spine bone mass in prepubescent children: A randomized controlled trial. *Journal of Bone Mineral Research, 16*(1), 148–156.

Fuchs, R. K., & Snow, C. M. (2002). Gains in hip bone mass from high-impact training are maintained: A randomized controlled trial in children. *Journal of Pediatrics, 141*(3), 357–362.

Genaro, P deS., & Martini, L. A. (2010). Effect of protein intake on bone and muscle mass in the elderly. *Nutrition Reviews, 68*(10), 616–623.

Gleeson, M., Lancaster, G. I., & Bishop, N. C. (2001). Nutritional strategies to minimise exercise-induced immunosupression in athletes. *Canadian Journal of Applied Physiology, 26*(Suppl.), S23–S35.

Gleeson, M., Nieman, D. C., & Pedersen, B. K. (2004). Exercise, nutrition and immune function. *Journal of Sports Sciences, 22*(1), 115–125.

Gordon, C. M. (2000). Bone density issues in the adolescent gynecology patient. *Journal of Pediatric and Adolescent Gynecology, 13*(4), 157–161.

Greger, J. L. (1999). Nondigestible carbohydrates and mineral bioavailability. *Journal of Nutrition, 129*(7 Suppl.), 1434S–1435S.

Gropper, S. S., Smith, J. L., & Groff, J. L. (2009). *Advanced nutrition and human metabolism* (5th ed.). Belmont, CA: Thomson/Wadsworth.

Guyton, A. C., & Hall, J. E. (2011). *Textbook of medical physiology* (12th ed.). Philadelphia: Saunders.

Haas, J. D., & Brownlie, T., IV. (2001). Iron deficiency and reduced work capacity: A critical review of the research to determine a causal relationship. *Journal of Nutrition, 131*(2S–2), 676S–688S; discussion 688S–690S.

Hambidge, M. (2000). Human zinc deficiency. *Journal of Nutrition, 130*(5S Suppl.), 1344S–1349S.

Hambidge, K. M., Miller, L. V., Westcott, J. E., Sheng, X., & Krebs, N. F. (2010). Zinc bioavailability and homeostasis. *American Journal of Clinical Nutrition, 91*(5):1478S–1483S.

Heaney, R. P. (2001). The bone remodeling transient: Interpreting interventions involving bone-related nutrients. *Nutrition Reviews, 59*(10), 327–333.

Heaney, R. P. (2004). Phosphorus nutrition and the treatment of osteoporosis. *Mayo Clinic Proceedings, 79*(1), 91–97.

Heaney, R. P. (2009). Dairy and bone health. *Journal of the American College of Nutrition, 28*(Suppl. 1), 82S–90S.

Heaney, R. P., Dowell, M. S., Rafferty, K., & Bierman, J. (2000). Bioavailability of the calcium in fortified soy imitation milk, with some observations on method. *American Journal of Clinical Nutrition, 71*(5), 1166–1169.

Heaney, S., O'Connor, H., Gifford, J., & Naughton, G. (2010). Comparison of strategies for assessing nutritional adequacy in elite female athletes' dietary intake. *International Journal of Sport Nutrition and Exercise Metabolism, 20*(3), 245–256.

Hoch, A. Z., Pajewski, N. M., Moraski, L., Carrera, G .F., Wilson, C. R., Hoffmann, R. G., et al. (2009). Prevalence of the female athlete triad in high school athletes and sedentary students. *Clinical Journal of Sport Medicine, 19*(5), 421–428.

Hurrell, R. F. (2003). Influence of vegetable protein sources on trace element and mineral bioavailability. *Journal of Nutrition, 133*(9), 2973S–2977S.

Hurrell, R. F., & Egli, I. (2010). Iron bioavailability and dietary reference values. *American Journal of Clinical Nutrition, 91*(5), 1461S–1467S.

Ilich, J. Z., & Kerstetter, J. E. (2000). Nutrition in bone health revisited: A story beyond calcium. *Journal of the American College of Nutrition, 19*(6), 715–737.

Institute of Medicine. (1997). *Dietary Reference Intakes for calcium, phosphorus, magnesium, vitamin D and fluoride* (Food and Nutrition Board). Washington, DC: National Academies Press.

Institute of Medicine. (2000). *Dietary Reference Intakes for vitamin C, vitamin E, selenium and carotenoids* (Food and Nutrition Board). Washington, DC: National Academies Press.

Institute of Medicine. (2001). *Dietary Reference Intakes for vitamin A, vitamin K, arsenic, boron, chromium, copper, iodine, iron, manganese, molybdenum, nickel, silicon, vanadium, and zinc* (Food and Nutrition Board). Washington, DC: National Academies Press.

Institute of Medicine. (2004). *Dietary Reference Intakes for water, potassium, sodium, chloride, and sulfate.* (Food and Nutrition Board). Washington, DC: National Academies Press.

Institute of Medicine. (2010). *Dietary Reference Intakes for calcium and vitamin D* (Food and Nutrition Board). Washington, DC: National Academies Press.

Jonnalagadda, S. S., Ziegler, P. J., & Nelson, J. A. (2004). Food preferences, dieting behaviors, and body image perceptions of elite figure skaters. *International Journal of Sport Nutrition and Exercise Metabolism, 14*(5), 594–606.

Kenny, A. M., & Prestwood, K. M. (2000). Osteoporosis: Pathogenesis, diagnosis, and treatment in older adults. *Rheumatic Diseases Clinics of North America, 26*(3), 569–591.

Khan, K. M., Liu-Ambrose, T., Sran, M. M., Ashe, M. C., Donaldson, M. G., & Wark, J. D. (2002). New criteria for female athlete triad syndrome? As osteoporosis is rare, should osteopenia be among the criteria for defining the female athlete triad syndrome? *British Journal of Sports Medicine, 36*(1), 10–13.

Leydon, M. A., & Wall, C. (2002). New Zealand jockeys' dietary habits and their potential impact on health.

International Journal of Sport Nutrition and Exercise Metabolism, 12(2), 220–237.

Looker, A. C., Melton, L. J., III, Harris, T. B., Borrud, L. G., & Shepherd, J. A. (2010). Prevalence and trends in low femur bone density among older US adults: NHANES 2005–2006 compared with NHANES III. *Journal of Bone Mineral Research, 25*(1), 64–71.

Lukaski, H. (1999). Chromium as a supplement. *Annual Review of Nutrition, 19*, 279–302.

Lukaski, H. (2000). Magnesium, zinc, and chromium nutriture and physical activity. *American Journal of Clinical Nutrition, 72*(Suppl.), 585S–593S.

MacFarquhar, J. K., Broussard, D. L., Melstrom, P., Hutchinson, R., Wolkin, A., Martin, C., et al. (2010) Acute selenium toxicity associated with a dietary supplement. *Archives of Internal Medicine, 170*(3), 256–261.

Malczewska, J., Raczynski, G., & Stupnicki, R. (2000). Iron status in female endurance athletes and in nonathletes. *International Journal of Sport Nutrition and Exercise Metabolism, 10*(3), 260–276.

Merkel, D., Huerta, M., Grotto, I., Blum, D., Rachmilewitz, E., Fibach, E., et al. (2009). Incidence of anemia and iron deficiency in strenuously trained adolescents: Results of a longitudinal follow-up study. *Journal of Adolescent Health, 45*(3), 286–291.

Mettler, S., & Zimmermann, M. B. (2010). Iron excess in recreational marathon runners. *European Journal of Clinical Nutrition, 64*(5), 490–494.

Micheletti, A., Rossi, R., & Rufini, S. (2001). Zinc status in athletes: Relation to diet and exercise, *Sports Medicine, 31*(8), 577–582.

Montain, S. J., Cheuvront, S. N., & Lukaski, H. C. (2007). Sweat mineral-element responses during 7 h of exercise-heat stress. *International Journal of Sport Nutrition and Exercise Metabolism, 17*(6), 574–582.

Morgan, S. L. (2001). Calcium and vitamin D in osteoporosis. *Rheumatic Diseases Clinics of North America, 27*(1), 101–130.

National Osteoporosis Foundation. (2010). *America's bone health: The state of osteoporosis and low bone mass.* Washington, DC: Author.

Nieman, D. C. (2008). Immunonutrition support for athletes. *Nutrition Reviews, 66*(6), 310–320.

North American Menopause Society. (2010). Management of osteoporosis in postmenopausal women: 2010 position statement of the North American Menopause Society. *Menopause, 7*(1), 25–54.

Popkin, B. M. (2010). Patterns of beverage use across the lifecycle. *Physiology & Behavior, 100*(1), 4–9.

Rizzoli, R., Bianchi, M. L., Garabedian, M., McKay, H. A., & Moreno, L. A. (2010). Maximizing bone mineral mass gain during growth for the prevention of fractures in the adolescents and elderly. *Bone, 46*(2), 294–305.

Sabatier, M., Arnaud, M. J., Kastenmayer, P., Kastenmayer, P., Rytz, A., & Barclay, D. V. (2002). Meal effect on magnesium bioavailability from mineral water in healthy women. *American Journal of Clinical Nutrition, 75*(1), 65–71.

Seeman, E. (2002). Pathogenesis of bone fragility in women and men. *Lancet, 359*(9320), 1841–1850.

Seeman, E. (2010). Evidence that calcium supplements reduce fracture risk is lacking. *Clinical Journal of the American Society of Nephrology, 5*(Suppl. 1), S3–S11.

Shah, A. (2004). Iron deficiency anemia. Part I. *Indian Journal of Medical Sciences, 58*(2), 79–81.

Shankar, A. H., & Prasad, A. S. (1998). Zinc and immune function: The biological basis of altered resistance to infection. *American Journal of Clinical Nutrition, 69*(2 Suppl.), 447S–463S.

Sherwood, L. (2010). *Human physiology: From cells to systems* (7th ed.). Belmont, CA: Thomson Brooks/Cole.

Speich, M., Pineau, A., & Ballereau, F. (2001). Minerals, trace elements and related biological variables in athletes and during physical activity. *Clinica Chimica Acta; International Journal of Clinical Chemistry, 312*(1–2), 1–11.

Tucker, K. L. (2009). Osteoporosis prevention and nutrition. *Current Osteoporosis Reports, 7*(4), 111–117.

Venkatraman, J. T., & Pendergast, D. R. (2002). Effect of dietary intake on immune function in athletes. *Sports Medicine, 32*(5), 323–340.

Vincent, J. (2000). The biochemistry of chromium. *Journal of Nutrition, 130*(4), 715–718.

Volpe, S. L. (2008). Minerals as ergogenic aids. *Current Sports Medicine Reports, 7*, 224–229.

Vondracek, S. F., Hansen, L. B., & McDermott, M. T. (2009). Osteoporosis risk in premenopausal women. *Pharmacotherapy, 29*(3), 305–317.

Winters, K. M., & Snow, C. M. (2000). Detraining reverses positive effects of exercise on the musculoskeletal system in premenopausal women. *Journal of Bone Mineral Research, 15*(12), 2495–2503.

Ziegler, P. J., Kannan, S., Jonnalagadda, S. S., Krishnakumar, A., Taksali, S .E., & Nelson, J. A. (2005). Dietary intake, body image perceptions, and weight concerns of female US international synchronized figure skating teams. *International Journal of Sport Nutrition and Exercise Metabolism, 15*(5), 550–566.

Ziegler, P. J., Nelson, J. A., & Jonnalagadda, S. S. (1999). Nutritional and physiological status of U.S. national figure skaters. *International Journal of Sport Nutrition, 9*(4), 345–360.

Ziegler, P., Sharp, R., Hughes, V., Evans, W., & Khoo, C.S. (2002). Nutritional status of teenage female competitive figure skaters. *Journal of the American Dietetic Association, 102*(3), 374–379.

Diet Planning: Food First, Supplements Second

Learning Plan

- **Define** the word *diet*.
- **Explain** how energy intake and nutrient density are fundamental to diet planning.
- **Create** a 1-day diet plan for an athlete based on MyPyramid.
- **Translate** sports nutrition recommendations to food choices.
- **Make** practical suggestions for food and fluid intake prior to, during, and after exercise.
- **Discuss** the safety and effectiveness of caffeine and alcohol.
- **Outline** the information included on the Supplement Facts label.
- **Discuss** the role of supplementation in an athlete's diet.

Athletes are encouraged to take a total diet approach, obtaining the nutrients they need from a variety of nutrient-dense foods.

© Profimedia International s.r.o./Alamy

Pre-Test   **Assessing Current Knowledge of Diet Planning for Athletes**

Read the following statements and decide if each is true or false.

1. When planning dietary intake, athletes should first consider how much dietary fat would be needed.

2. The key to eating nutritiously without consuming excess kilocalories is to choose foods that have a high nutrient density.

3. There is no room in the athlete's diet for fast foods.

4. Processed foods tend to be more nutrient dense than unprocessed foods.

5. Most dietary supplements are effective for improving training and performance.

The word _diet_ is used in two distinct ways in the English language. By definition, a **diet** consists of the food and drink that a person normally consumes. Thus each person is always "on a diet." But the word _diet_ is also used to describe a restricted intake of food and drink, usually for the purpose of weight loss. Thus it is common to hear a person say "I need to go on a diet," "I'm on a diet," or "I've gone off my diet."

A diet is a pattern of eating. Sometimes people change their usual pattern of eating by adopting a weight-loss diet in an effort to change body composition. This is true for athletes, who wish to fine-tune body composition, and for sedentary people, many of whom are overweight or obese. One's usual diet may also need to change due to a medical condition, such as diabetes or heart disease. Changing one's usual pattern of eating is never easy, especially if it involves restriction.

Athletes need to match their dietary intake to their training. Diet planning often begins with the establishment of an energy goal—energy balance or imbalance for the purpose of weight loss or gain. Periodic adjustments of carbohydrate, protein, and/or fat intakes are necessary to support training, performance, and recovery. An athlete's diet may need to change abruptly in response to an injury. An athlete is always on a "diet," a pattern of eating, but that pattern may not always be the same.

Many nutrient recommendations are given on a g/kg basis. These calculations must then be translated to foods and beverages, some of which need to be appropriately timed. Like other consumers, athletes are encouraged to take a total diet approach, which relies primarily on food and beverages from various food groups to provide the energy and nutrients needed. Minimally processed foods are encouraged because of their nutrient content.

Stefan Schuetz/Getty Images

© Alex Segre/Alamy

It is relatively simple to obtain needed nutrients with the increased caloric intake associated with regular, moderate-intensity exercise. In contrast, a person with a sedentary lifestyle needs fewer calories and should choose more nutrient-dense foods.

10.1 Energy: The Basis of the Diet-Planning Framework

Humans are biologically designed to be physically active. When physical activity is high, a higher caloric diet is needed to maintain body weight. This higher caloric intake makes it relatively simple to obtain all the nutrients needed because larger volumes of food can

Table 10.1 Estimating Daily Energy Need for Male and Female Athletes

Level of activity	Example of activity level	Energy expenditure—females (kcal/kg/d)	Energy expenditure—males (kcal/kg/d)
Sedentary (little physical activity)	During an acute recovery from injury phase	30	31
Moderate-intensity exercise 3–5 days/week or low-intensity and short-duration training daily	Playing recreational tennis (singles) 1–1½ hr every other day; practicing baseball, softball, or golf 2½ hr daily, 5 days/week	35	38
Training several hours daily, 5 days/week	Swimming 6,000–10,000 m/day plus some resistance training; conditioning and skills training for 2–3 hr/day such as soccer practice	37	41
Rigorous training on a near daily basis	Performing resistance exercise 10–15 hr/week to maintain well-developed skeletal muscle mass, such as a bodybuilder during a maintenance phase; swimming 7,000–17,000 m/day and resistance training 3 days/week, such as an elite swimmer; typical training for college, professional, and elite football, basketball, and rugby players	38–40	45
Extremely rigorous training	Training for a triathlon (nonelite triathlete)	41	51.5
	Running 15 mi (24 km)/day or the equivalent (elite triathlete or equivalent)	50 or more	60 or more

Source: Adapted from Macedonio, M. A., & Dunford, M. (2009). *The Athlete's Guide to Making Weight: Optimal Weight for Optimal Performance.* Champaign, IL: Human Kinetics.

be eaten. For example, a 5'7" 140 lb (170 cm 63.6 kg), 24-year-old nonpregnant female well-trained rugby player would need ~2,500 kcal daily to maintain energy balance and body composition. At this caloric intake it is easy to plan a diet that meets all nutrient requirements. Although predominantly nutrient-rich foods should be consumed to support her activity, some foods that are high in sugar and fat and lower in nutrients can easily be incorporated into her diet plan.

In contrast, if this same female were sedentary, her daily caloric need would be approximately 1,900 kcal. All nutrient requirements can also be met at this caloric intake, but nearly all the food would need to be nutrient dense. Nutrient dense refers to a food that is rich in nutrients compared to its caloric content. Many popular foods, such as candy bars or ice cream, are calorie dense instead of nutrient dense. Large portions of calorie-dense foods and lack of physical activity are among the fundamental causes of the high prevalence of overweight and obesity in the United States and other industrialized nations.

Athletes are not usually sedentary, but their dietary intake will be influenced by relatively higher or lower energy requirements based on their training. As shown in Table 10.1, daily caloric intake generally ranges from 30 kcal/kg to 60 kcal/kg and is largely dependent on the amount of energy expended through exercise. These are good ballpark figures that are used to establish an estimate of daily caloric need, but food

and exercise diaries and body weight and composition information over time give better information about actual caloric need for an individual athlete.

An athlete's daily energy requirement varies according to the training cycle because energy expenditure from exercise can change considerably. The annual training schedule of a female rower (crew) is shown in Figure 10.1. The lowest energy expenditure (33 kcal/kg) is during the off-season, when this rower purposefully does no training. When this athlete returns to training after the off-season, energy expenditure increases (from 33 kcal/kg to 40 kcal/kg) and continues to increase as the volume of training increases (to 44 kcal/kg). The highest energy expenditure for the athlete in this example is during double days, when rowers engage in high-intensity and high-volume training in both morning and afternoon practices. This is prior to the opening of racing season and increases caloric need to more than 50 kcal/kg for a short period of time. During racing season, the total energy expenditure is similar to the beginning of the season when the athlete returned to training (40 kcal/kg). To maintain energy balance, this rower must consume a substantial amount of food over the

Diet: The food and drink normally consumed; the restriction of food and drink for the purpose of weight loss.

One-year-training-cycle												
Preparation						Competition			Transition			
General training			Specific training			Racing season			Active recovery ("off-season")			
19.5 hr/wk with emphasis on endurance/ cardiovascular training			22.5 hr/wk with emphasis on strength/power training			12 hr training/wk + 1–3 races/wk			No training			
Sept	Oct	Nov	Dec	Jan	Feb	Mar	Apr	May	Jun	Jul	Aug
40 kcal/kg			44 kcal/kg			40 kcal/kg			33 kcal/kg			

Figure 10.1 Training schedule and estimated energy expenditure of a female collegiate rower

Legend: hr = hour; wk = week; kcal = kilocalories; kg = weight in kilograms

Energy expenditure varies depending on the intensity and volume of training over the course of a year. Similarly, the estimated daily energy requirement varies considerably. For example, "double days" (1 week of morning and afternoon practices prior to the opening of racing season and totaling 37.5 hours of training) increases daily caloric need to ~51 kcal/kg.

9-month training period and substantially decrease caloric intake when training is reduced dramatically during the off-season. The creation of a diet plan that has been divided into distinct periods to support training and performance is sometimes called **nutrition periodization**. Each athlete should create a nutrition plan that matches training.

Many athletes do not want to be in energy balance; rather, they want to maintain an energy deficit in an effort to reduce body fat. In such cases, each athlete must establish an appropriate daily energy intake goal. This goal essentially becomes a maximum caloric intake for the day, and carbohydrate, protein, and fat intakes must fit within this fairly strict framework. A rule-of-thumb recommendation is that athletes consume no less than 30 kcal/kg daily, because it is difficult to consume an adequate amount of nutrients, particularly carbohydrates, when energy intake is severely restricted. Conversely, some athletes want to gain weight, primarily as muscle mass, but occasionally as body fat. In the case of weight gain, the athlete will need to establish a higher-than-normal daily caloric goal. Chapter 11 discusses nutrition and exercise strategies associated with changes to body composition.

A dietary prescription helps athletes consume the proper amount of carbohydrates, proteins, and fats within their energy needs.

Once the energy (kcal) goal has been established, the goals for macronutrients can be considered in the proper context and a **dietary prescription** can be developed. Because athletes may need higher carbohydrate and protein intakes than sedentary individuals

(see Chapters 4 and 5), these two macronutrients are considered first. The athlete does not want to compromise the intake of these important nutrients, so their energy content is accounted for early in the diet planning process. Assume that the 24-year-old female athlete mentioned earlier needs approximately 6 g of carbohydrate and 1.2 g of protein per kilogram of body weight. Based on these goals, of the 2,500 kcal this athlete wants to consume daily, ~1,768 kcal should come from carbohydrates and proteins. The calculations for this example are shown in Figure 10.2.

The diet-planning process continues by considering the other energy-containing nutrients—fats and alcohol (Figure 10.2). These contain more kilocalories per gram (9 kcal/g and 7 kcal/g for fat and alcohol, respectively) than carbohydrates and proteins (4 kcal/g). Because fat intake can be too low, a reasonable guideline for fat intake for athletes is ~1.0 g/kg of body weight.

At this point the diet has been planned to meet the macronutrient needs of the athlete. Note that after these needs have been met the caloric content of the diet in the example is too low (by ~180 kcal) to maintain energy balance (Figure 10.2). This 180 kcal difference represents what is sometimes termed "discretionary calories." This term is used to indicate the amount of kilocalories needed to maintain energy balance after carbohydrate, protein, and fat recommendations have been met. In some cases, these "discretionary calories" are consumed in the form of alcoholic beverages. In many cases, small amounts of less nutrient-dense carbohydrate- and fat-containing foods such as soft drinks, sweets, and desserts are chosen. Because caloric requirements are high during periods of hard training, athletes may find that they need small amounts of such foods to maintain energy balance.

Weight: 140 (63.4 kg)

Energy: 2,500 kcal/d (~39 kcal/kg)

Carbohydrate: 6 g/kg

Protein: 1.2 g/kg

Energy: 2,500 kcal/d

Carbohydrate (CHO): 1,472 kcal

    (61.4 kg × 6 g/kg = 368 g CHO;
       = 368 g × 4 kcal/g = 1,472 kcal)

Protein (PRO): 296 kcal

    (61.4 kg × 1.2 g/kg = 74 g PRO;
       = 74 g × 4 kcal/g = 296 kcal)

Energy from CHO + PRO: 1,768 kcal

    (1,472 kcal + 296 kcal = 1,768 kcal)

Fat: 550 kcal

    (61.4 kg × 1.0 g/kg = 61 g fat;
       = 61 g × 9 kcal/g = 550 kcal)

Energy from CHO + PRO + fat: 2,318 kcal

    (1,472 kcal + 296 kcal + 550 kcal = 2,318 kcal)

Energy intake not yet accounted for: 182 kcal

    (2,500 kcal – 2,318 kcal = 182 kcal)

Figure 10.2 Calculating a dietary prescription for a 140 lb nonpregnant female athlete

Legend: lb = pound; kg = kilogram; kcal/d = kilocalories per day; kcal/kg = kilocalorie per kilogram body weight; g/kg = grams per kilogram body weight; kcal = kilocalorie; g = gram; kcal/g = kilocalories per gram

Sports dietitians typically calculate carbohydrate and protein requirements and the energy that they provide first and then calculate fat and, if appropriate, alcohol intakes within the context of total daily energy intake.

The 24-year-old female in this example is in energy balance during the rugby season and consumes 2,500 kcal daily. Now assume that she wishes to lose body fat. A simple way would be not to consume those "discretionary calories." If she consumed a diet that was approximately 200 kcal below what she needed each day while maintaining the same exercise output, she could expect a slow weight (fat) loss of about 1 lb every 2½ weeks.

A combination of food restriction and increased activity is often the best approach for sustained weight (fat) loss, especially for performance-focused recreational athletes. For example, foods that represent "discretionary calories" could be eliminated and fat intake could be slightly reduced to further reduce caloric intake. Strategies to increase energy expenditure include increasing the frequency, intensity, and/or duration of exercise or including other activities into the daily lifestyle such as walking or cycling to and from work or school, using the stairs instead of elevators or escalators, stretching or performing calisthenics

Using the stairs instead of elevators or escalators is one strategy to increase daily energy expenditure.

while watching TV, or eliminating labor-saving devices in favor of manual devices, such as push mowers, rakes, and brooms. Together these alterations to usual food and activity patterns would result in a slow but sustained weight loss. The advantages of a slow weight loss include less severe restriction of food intake, consumption of sufficient carbohydrates to support physical activity, and a sufficient protein intake to protect against the loss of muscle mass.

The naïve athlete may think that it is best to severely reduce energy intake when reducing body fat. After all, a large restriction of energy, such as 1,000 kcal/d, would result in a rapid weight loss. But athletes typically should not restrict food intake to a large degree because they will not be able to consume an adequate amount of carbohydrates to resynthesize muscle glycogen and may be unable to complete exercise training at the desired intensity or duration. They may also find it difficult to maintain a very low-fat diet. Food is fuel and enough must be provided daily to support training. Therefore, it is recommended that athletes in training not reduce daily caloric intake below 30 kcal/kg. Additional information about weight loss strategies and weight cycling is provided in Chapters 11.

Although it is mentioned less frequently, some athletes need to increase their daily energy intake. Many people say that they would love to be in that situation! In fact, it is both difficult and frustrating for underweight athletes to increase their daily energy intake.

Nutrition periodization: The creation of a nutrition plan to support training that has been divided into distinct periods of time.

Dietary prescription: An individualized plan for an appropriate amount of kilocalories, carbohydrates, proteins, fats, and alcohol daily.

Blue Jean Images/Getty Images

Eat more meals and snacks daily.

Increase portion size of favorite foods.

Reduce intake of beverages that do not provide energy (kcal) but may give a feeling of fullness, such as, coffee, tea, diet soft drinks.

Increase foods high in heart-healthy fats such as nuts, nut butters, olives, and avocadoes.

Figure 10.3 Practical tips for increasing energy intake in underweight athletes

They may have a genetic predisposition to be underweight. They may not have an appetite, so eating more food is unappealing and difficult. Additionally, some underweight athletes may be underweight because they are struggling with disordered eating and deep-seated psychological issues.

Underweight athletes benefit from nutrition counseling and the establishment of an individualized diet plan. For those who are not struggling with disordered eating, the general recommendation is to increase caloric intake by approximately 500 kcal daily (Rankin, 2002). This may be best accomplished by increasing the portion size of favorite foods and eating more frequently. Although it may seem advantageous to recommend the consumption of a calorie-dense food, such as a chocolate milk shake, this strategy may be counterproductive because appetite can be reduced for many hours afterward and total caloric intake for the day may not exceed baseline. Some practical tips for increasing energy intake are listed in Figure 10.3.

Consuming nutrient-dense foods is the key to eating nutritiously without consuming excess calories.

In the area of sports nutrition, the energy-containing and muscle-related nutrients tend to be emphasized, especially carbohydrates and proteins. However, the human body requires more than just macronutrients. For optimal performance, training, and health, people need an array of other compounds referred to as micronutrients. Micronutrients are substances that are needed in small quantities for normal growth, development, and maintenance of the body, such as vitamins and minerals (see Chapters 8 and 9). Vitamins and minerals do not contain energy so they have no caloric value. But they are found in foods that contain carbohydrates, proteins, and fats and are a necessary part of a healthy diet.

The key to eating nutritiously without consuming excess kilocalories is to choose foods that have a high

Table 10.2 Nutrient Density of Skim and Whole Milk

Nutrient	Skim milk, with nonfat milk solids added (8 oz)	Whole milk, 3.3% milk fat (8 oz)
Energy (kcal)	91	146
Protein (g)	9	8
Carbohydrate (g)	12	11
Fat (g)	<1	8
Cholesterol (mg)	5	24
Calcium (mg)	316	314
Iron (mg)	0.12	0.12

Legend: oz = ounce; kcal = kilocalorie; g = gram; mg = milligram

Skim (nonfat) milk is more nutrient dense than whole milk.

© Envision/Corbis

Skim milk is a nutrient-dense food because it is rich in nutrients relative to its caloric content.

nutrient density. Density, as used in this term, refers to a relatively high concentration of nutrients in relation to total kilocalories. In other words, a nutrient-dense food is rich in nutrients for the amount of kilocalories it contains. Table 10.2 compares the nutrient density of skim milk (also known as nonfat or fat-free milk) to whole milk. Both types of milk are good sources of many of the nutrients shown, in particular protein and calcium. Skim milk is a nutrient-dense food because it is rich in nutrients relative to its caloric content. It is more nutrient dense than whole milk because the whole milk contains about the same amount of nutrients as the skim milk but substantially more kilocalories. The caloric difference between the two milks is a result of the greater fat content of the whole milk. Note

Table 10.3 Nutrient Density of Sugar and Alcohol

Nutrient	White (table) sugar (1 tablespoon)	Distilled alcohol, 100 proof (1 fluid oz)
Energy (kcal)	45	82
Protein (g)	0	0
Carbohydrate (g)	12	0
Fat (g)	0	0
Calcium (mg)	<1	0
Iron (mg)	0	0.01
Thiamin (mg)	0	0
Vitamin C (mg)	0	0
Vitamin E (mg)	0	0
Folate (mcg)	0	0

Legend: oz = ounce; kcal = kilocalorie; g = gram; mg = milligram; mcg = microgram

Sugar and alcohol have a low nutrient density because they contain kilocalories but few or no nutrients.

Nutrient-dense foods are an important element in meeting diet-planning goals.

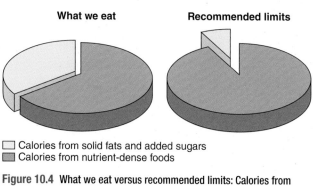

☐ Calories from solid fats and added sugars
▨ Calories from nutrient-dense foods

Figure 10.4 **What we eat versus recommended limits: Calories from solid fats and added sugars (SoFAS)**

Note: The depiction of the proportionate amounts of total calories consumed and the recommend limits are illustrative only. The figure illustrates about 35 percent of total calories consumed as SoFAS, on average, in contrast to a recommended limit of no more than 5–15 percent of total calories for most individuals.

that the fat content of skim milk is not zero; a small amount of fat must be present so that the fat-soluble vitamins A and D can be added to the milk.

Fresh fruits and vegetables are typically nutrient dense because they are rich in vitamins and minerals and low in kilocalories. Whole grains, beans and legumes, low-fat meat and poultry, fish, and nonfat or low-fat dairy products are also nutrient dense. A food does not necessarily have to be low in kilocalories to be nutrient dense. For example, almonds are considered a nutrient-dense food because they are rich in nutrients, such as vitamin E, magnesium, and fiber, in relation to the amount of kilocalories that they provide. A primary goal of diet planning is to create an eating pattern that provides all the nutrients needed while providing the appropriate amount of kilocalories, and nutrient-dense foods are an important element in meeting these goals.

Sugar and alcohol have the lowest nutrient density as shown in Table 10.3. These compounds contain kilocalories but virtually no proteins, vitamins, or minerals. Sometimes these are described as having "empty" calories. Empty refers to the lack of nutrients, not the lack of kilocalories.

Processed foods that have added sugar or fat tend not to be nutrient dense because the fat and sugar provide additional kilocalories but few or no additional nutrients. Solid fats and added sugars are known as SoFAS. As shown in Figure 10.4, Americans currently consume 35 percent of their total caloric intake from SoFAS and only 65 percent from nutrient-dense foods. In contrast, the 2010 Dietary

Guidelines recommend that approximately 5–15 percent of total caloric intake should be from SoFAS with the remaining 85–95 percent from nutrient-dense foods. Consuming a large amount of low-nutrient-dense foods daily will usually result in overconsuming kilocalories, yet underconsuming nutrients. Thus, ironically, an obese person may be malnourished and an athlete's caloric intake may be too high while nutrient intake is too low.

The athlete's diet must be adequate in both kilocalories and nutrients to support training. Because many athletes are trying to attain or maintain a low percentage of body fat, they cannot afford to overconsume energy (kcal). Therefore their diets must contain foods that are nutrient dense. It is not difficult to determine the amount of kilocalories, carbohydrates, proteins, and fats needed by an athlete for training or competition. The challenge is to translate those recommendations into food choices while living in an environment where low-nutrient-dense foods are inexpensive, widely available, highly advertised, and tasty.

It is a challenge to consume a nutritious diet when low-nutrient-dense foods are inexpensive, widely available, highly advertised, and tasty.

Key points

■ An estimate of daily caloric intake forms the basis for planning the athlete's diet.

■ Caloric intake can change substantially based on volume of training.

■ Daily energy intake for athletes typically ranges from 30 to 60 kcal/kg.

■ Within caloric needs, carbohydrate, protein, and fat intake must be balanced to support training and performance.

■ Choosing nutrient-dense foods helps athletes meet their nutrient needs within their caloric needs.

What are the problems associated with a lack of diet planning, particularly caloric intake?

10.2 Translating Nutrient Recommendations into Food Choices

Similar to an **exercise prescription**, a diet prescription helps an athlete to know how much of each macronutrient to consume. In the example used in the sections on pages 365–367, the following diet prescription was developed for a 24-year-old nonpregnant female rugby player:

Energy: 2,500 kcal/d (~39 kcal/kg)

Carbohydrates: 6 g/kg

Proteins: 1.2 g/kg

Fats: ~1.0 g/kg

Discretionary calories: ~180 kcal/d

Now the challenge is to translate the "numbers" into actual foods. Chapters 4, 5, and 6 explain in detail the kinds and amounts of food that contain carbohydrates, proteins, and fats. But many athletes, especially

high school athletes and others who are just beginning to learn about nutrition, need some general information. Sports nutrition is a complex topic and can be overwhelming at first. Some simple tools are needed to get athletes started on the road to good nutrition.

One such tool is the Food Intake Patterns developed for MyPyramid. These intake patterns are briefly shown in Figure 10.5. (More detailed information is available in Appendix C and at http://www.mypyramid.gov.) The foods have been grouped according to their nutrient content. The groups are fruits, vegetables, grains, meat and beans, milk, and oils, and the amounts of each group are adjusted based on caloric level. Choosing minimally processed foods from these groups helps athletes focus on a whole foods, total diet approach. These patterns are a good starting point for use with athletes who have little nutrition knowledge. A similar approach is the use of the food exchange lists, which can be found in Appendix D.

Assume that this rugby player has little understanding of nutrition. She does, however, have a diet prescription so she knows that she needs about 2,500 kcal daily. Notice that the Food Intake Patterns do not have a category for 2,500 kcal. This highlights one of the important limitations of using this guideline with athletes—these are general guidelines, and not individually planned diets. Also notice that the highest caloric level is 3,200 kcal. Some males, such as endurance and ultraendurance athletes and large-bodied strength athletes, will need more than 3,200 kcal daily.

The female athlete in this example should choose the 2,400 kcal pattern. This gives her a "ballpark" guideline of the kinds and amounts of foods she needs daily. In her case, she could begin to plan a diet that has approximately:

2 cups fruit

3 cups vegetables

The equivalent of 8 oz of grain products

6.5 oz of meat, poultry, or fish or the equivalent of substitutes such as beans and nuts

3 cups of milk

7 teaspoons of oil

Figure 10.6 shows a 1-day diet plan developed using these food groups as a guideline. This diet contains ~2,400 kcal (~38 kcal/kg), 339 g carbohydrate (~5.4 g/kg), 112 g protein (~1.8 g/kg), and 75 g fat (~1.2 g/kg). Notice that the portion sizes are reasonable, but they are not the large portion sizes that many Americans are used to eating. One way to "reset" thinking about reasonable portions is to use the serving size listed on the Nutrition Facts label as a guide (see Chapter 1). This 1-day diet plan also illustrates the use of a whole foods, total diet approach.

How close does the 1-day diet come to meeting this athlete's needs? It is in the "ballpark." The diet is slightly less than the 2,500 kcal recommended because it is based on the 2,400 kcal pattern. The diet is also a

Daily Amount of Food from Each Group												
Calorie level	1,000	1,200	1,400	1,600	1,800	2,000	2,200	2,400	2,600	2,800	3,000	3,200
Fruits	1 cup	1 cup	1.5 cups	1.5 cups	1.5 cups	2 cups	2 cups	2 cups	2 cups	2.5 cups	2.5 cups	2.5 cups
Vegetables	1 cup	1.5 cups	1.5 cups	2 cups	2.5 cups	2.5 cups	3 cups	3 cups	3.5 cups	3.5 cups	4 cups	4 cups
Grains	3 oz-eq	4 oz-eq	5 oz-eq	5 oz-eq	6 oz-eq	6 oz-eq	7 oz-eq	8 oz-eq	9 oz-eq	10 oz-eq	10 oz-eq	10 oz-eq
Meats and Beans	2 oz-eq	3 oz-eq	4 oz-eq	5 oz-eq	5 oz-eq	5.5 oz-eq	6 oz-eq	6.5 oz-eq	6.5 oz-eq	7 oz-eq	7 oz-eq	7 oz-eq
Milk	2 cups	2 cups	2 cups	3 cups	3 cups	3 cups	3 cups	3 cups	3 cups	3 cups	3 cups	3 cups
Oils	3 tsp	4 tsp	4 tsp	5 tsp	5 tsp	6 tsp	6 tsp	7 tsp	8 tsp	8 tsp	10 tsp	11 tsp
Discretionary calorie allowance	165	171	171	132	195	267	290	362	410	426	512	648

oz-eq = ounce equivalent
tsp = teaspoon

Figure 10.5 MyPyramid food intake patterns

Suggested amounts of food to consume from each food group listed, which varies according to estimated caloric needs.

1 c bran flakes cereal

1 banana

8 oz nonfat milk

8 oz orange juice

1 slice whole wheat toast

1 T margarine

Tuna salad made from 3 oz canned tuna in water,
  1 T mayonnaise, and ¼ stalk celery

1 slice tomato

2 slices whole wheat bread

1 T mayonnaise

1 pear

8 oz nonfat milk

1 oats and honey granola bar

1 c green salad with 1 T oil and vinegar dressing and
  ½ oz almonds

3 oz chicken breast

½ c sweet potato

½ c green peas

2 whole wheat dinner rolls

2 T margarine

1 c mineral water

1 c low-fat fruit yogurt

3 sliced dried apricots

Figure 10.6 Using the food intake pattern to create a 1-day diet plan

Legend: c = cup; oz = ounce; T = tablespoon

This diet plan was devised by using the suggested amounts of each food group for the 2,400 kcal pattern.

little bit too low in carbohydrates. However, if this athlete increased kilocalories to 2,500 by adding a carbohydrate-containing food, then she would nearly reach the recommended intake of carbohydrates. This example illustrates how general dietary patterns provide a good guideline from which modifications can be made.

Most general dietary guidelines are developed to contain about 5 to 6 g of carbohydrates per kilogram of body weight. Recall from Chapter 4 that the minimum recommended amount of carbohydrate for athletes in training is 5 g/kg. General dietary guidelines are not designed to provide the amount of carbohydrate that many athletes in training require (that is, 7 or more g/kg). Trained athletes need more extensive diet planning than general guidelines can provide.

The Food Intake Patterns developed for MyPyramid are public domain documents and can be freely copied and distributed. Many health professionals use these guidelines to provide basic nutrition information to consumers, some of whom are physically active. Those who work directly with athletes in training, such as certified strength and conditioning specialists (CSCS) and certified athletic trainers (ATC), also use these tools to communicate basic nutrition information. This helps athletes receive a consistent message about the importance of nutritious foods such as fruits, vegetables, whole grains, lean proteins, beans, nuts, and heart-healthy oils. These same types of foods will be used to plan the athlete's diet, but the amount of food will be determined by the intensity and duration of training, and the distribution throughout the day will depend on various factors such as timing before and after exercise.

Exercise prescription: An individualized plan for frequency, intensity, duration, and mode of physical activity.

As pointed out earlier, trained athletes have different nutritional needs than the general population. Their diets are typically higher in kilocalories and carbohydrates than sedentary and lightly active individuals. Some athletes may be trying to reduce caloric intake but must do so without sacrificing adequate consumption of carbohydrates and proteins. Some struggle with body image and become fearful of eating. These athletes need more specific information and individually planned diets. Many sports-related personnel have basic but not advanced knowledge of nutrition. They also do not have the time or expertise to evaluate the athlete's current food and nutrient intake, recommend changes, and help create individualized diet plans. As explained in Chapter 1, when a client's needs fall outside of a practitioner's scope of practice, the appropriate response is to make a referral to a qualified practitioner. Sports dietitians are trained to evaluate and plan athletes' diets.

Many athletes learn to cook so that they can prepare foods that are nutritious and tasty.

Exactostock/SuperStock

Application exercise

Using yourself or an athlete who you interviewed:

1. Describe a 1-year training cycle that includes preseason, competition, and postseason training.

2. Create a diet prescription (calories, carbohydrate, protein, and fat as g/kg) for one of the training periods.

3. Plan a 1-day diet using either MyPyramid or the Food Pyramid for Athletes.

4. Distribute the food across the day by planning meals and snacks.

Smallest size adult meal or children's meal

Grilled food instead of fried (for example, grilled chicken)

Extra lettuce and tomato substituted for mayonnaise or cheese on burgers or sandwiches

For deli sandwiches, whole wheat bread, extra vegetables, moderate amount of filling, less fatty/sugary condiments (for example, mayo, sauces)

Salad or salad bar with dressing on the side

Plain baked potato

Pizza with vegetables, easy on the cheese

Pancakes

Bagels

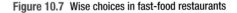

Figure 10.7 **Wise choices in fast-food restaurants**

Each athlete should have an individualized diet plan.

Many athletes can recite what has become known as the stereotypical athlete's dinner: broiled chicken breast without skin, plain baked potato, steamed broccoli, whole wheat roll, nonfat milk, and fruit for dessert. Although this meal is nutritious and does provide ample carbohydrates, proteins, and nutrients, athletes may have difficulty coming up with ideas for other dinners that are equally nutritious and find themselves in a food rut. They may also not like certain foods or not be able to tolerate them. Planning and preparing varied meals is always a challenge. If they do not already know how to do so, most athletes find that they need to learn to cook.

Creating a diet that meets the athlete's needs is one way to approach diet planning. Some athletes are willing to follow a diet that has been created for them and change their dietary intake dramatically and over a short period of time. But many athletes do not take this approach; rather, they look to modify their current diet. Their current diet includes familiar foods that they enjoy and know

how to prepare; however, the diet may not provide the appropriate amount of energy, carbohydrates, proteins, fats, fluid, vitamins, or minerals. The key to diet planning is individualization. Creating an elaborate diet plan that does not fit into the athlete's lifestyle or includes unappealing or unfamiliar foods does not make sense because it is not likely to be followed over the long term.

Sports dietitians work with athletes to help them identify appropriate goals and ways to modify their current diet to meet those goals. Consider the case of a freshman baseball player who eats lunch on campus every day. He has gained body fat since coming to college, and he knows that eating at fast-food restaurants has contributed to his weight gain. When the health center offered a free dietary consultation he decided to meet with the sports dietitian. He expected that the dietitian would simply say to stop eating fast foods, which he does not consider an option. He was pleasantly surprised when the dietitian gave him advice on making healthier, lower-calorie choices at fast-food restaurants (Figure 10.7). He found that he could still eat fast food but that it was important to choose wisely, alter the method of preparation (e.g., grilled rather than deep-fat fried), limit the portion size (e.g., 6-inch rather than 12-inch sandwiches, do not "supersize"), and ask for fewer condiments (e.g., cut down on high-fat, high-sugar sauces, mayonnaise). These are examples of ways that an athlete's current diet can be modified to meet the goal of consuming fewer kilocalories.

Food intake needs to be distributed appropriately throughout the day.

In addition to an athlete's 24-hour intake, a plan must also be devised for consuming food at the appropriate times. Food and fluid intake prior to, during, and after exercise must be part of the athlete's overall plan. Each time period has important goals that should be met.

Food and fluid intake prior to exercise. "Prior to exercise" is not a well-defined term, but in sports nutrition it usually refers to an approximately 4-hour period before exercise begins. Pre-exercise, pretraining, and precompetition meals are also terms that are used. For some athletes, pretraining and precompetition meals are similar because training and competition are similar. But in other cases they are remarkably different because a training session will last for many hours but a race will be over in seconds or minutes.

The goals of food and fluid intake prior to exercise are typically to:

- Provide energy for exercise, particularly carbohydrate
- Delay fatigue during prolonged exercise
- Prevent hypohydration and excessive dehydration
- Minimize gastrointestinal distress
- Satisfy hunger

These goals are the same for both pretraining and precompetition meals; however, the precompetition meal must always consider the impact that the stress of the impending competition has on the gastrointestinal tract.

There are no hard-and-fast rules regarding food and fluid intake prior to exercise, although there are general guidelines, which are summarized in Figure 10.8. Each athlete must use trial and error to determine the best precompetition and pretraining foods and fluids. Consideration must be given to a variety of factors, including the amount of time prior to exercise, carbohydrate, protein, and fat content, consistency of the meal (that is, solid, semisolid, or liquid), volume consumed, and food familiarity and preferences.

A meal, consisting of food and fluid, is usually consumed approximately 1 to 4 hours prior to exercise. This will help athletes satisfy their hunger, especially if this is the first meal after a night's sleep, and remain adequately hydrated. Determining how many hours prior to exercise a meal will be consumed will also determine the volume

Carbohydrate: 1 g/kg if consumed 1 hour prior; 2 g/kg if 2 hours prior, and so on.

Fluid (recommended amounts but must be established based on individual preferences and tolerances): ~5–7 ml/kg at least 4 hours prior to exercise if adequately hydrated; additional ~3–5 ml/kg 2 hours prior if hypohydrated; sodium added to food or drink may be beneficial.

Volume of food and fluid adjusted based on amount of time before exercise begins.

Trial and error during training is encouraged.

Figure 10.8 Guidelines for food and fluid intake prior to exercise

Legend: g/kg = grams per kilogram body weight; ml/kg = milliliters per kilogram body weight

of food and fluid that can be comfortably consumed. The onset of exercise increases the blood flow to muscles and decreases the blood flow to the gastrointestinal tract. Thus too large a volume of food or fluid consumed too close to exercise can result in gastrointestinal distress.

The timing of intake is easiest when the start time for training or competition is known in advance. For example, if a football game is scheduled to begin at 7:30 p.m., then the athlete will know the time of his pregame meal. If he knows that he will be eating at 3:30 p.m., then he knows the volume and kinds of foods and fluids that can be tolerated. In many sports, timing the pre-event meal is more difficult because the athlete may not know specifically when the competition will begin. Tennis provides a good example, as a tennis player competing in a tournament may know that he is the third match scheduled but he cannot predict how quickly the other two matches will be played or if the starting time may be altered due to weather delays.

As a rule of thumb, athletes usually consume 1.0 g carbohydrate/kg as a multiple of the number of hours before exercising. In other words, athletes consume 1.0 g/kg 1 hour prior, 2.0 g/kg 2 hours prior, and so on. As the time before exercise increases, the amount of carbohydrates that can be tolerated increases. Large amounts of carbohydrates (e.g., 4 g/kg 4 hours prior to exercise) may be appropriate for some athletes but not for others. For example, an ultraendurance athlete who needs 10 g/kg of carbohydrates on the day of the event may need to consume 4 g/kg 4 hours prior because it is difficult to consume such a large amount of carbohydrates over the course of a day. However, such an intake would likely be inappropriate for a 10 km (6.2 miles) runner on the day of the race because muscle glycogen stores should be adequate due to a high-carbohydrate diet in the days prior to the race. Pre-exercise carbohydrate content is also dependent on the athlete's tolerance.

In addition to carbohydrates, it is recommended that the pre-exercise meal generally contain a moderate amount of proteins, a small amount of fats, and some fluid. Fats slow gastric emptying (the amount of time

One 3 oz (3 ½ in) bagel

2 tsp light cream cheese

12 oz orange juice

8 oz carbohydrate beverage

Figure 10.9 Sample precompetition meal for a 110 lb (50 kg) athlete

Legend: oz = ounce; in = inch; tsp = teaspoon; kcal = kilocalorie; g = gram; kg = kilogram; lb = pound; g/kg = gram per kilograms bodyweight

This meal contains approximately 500 kcal, 100 g carbohydrate, 12 g protein, and 5 g fat. For a 50 kg (110 lb) athlete, the carbohydrate content of this meal is 2 g/kg. The bagel, cream cheese, and the orange juice would probably be consumed 2 hours prior, and the sports beverage would be consumed slowly over the 2-hour period prior to competition.

it takes for the food to leave the stomach) and are digested more slowly than carbohydrates and proteins. However, fats do provide a greater degree of satiety (the feeling of fullness and satisfaction), and athletes often include some fats in the hours prior to exercise to keep them from feeling hungry. Figure 10.9 gives an example of a precompetition meal for a 50 kg (110 lb) athlete.

Fluid intake is typically a priority prior to exercise, so part of the pre-exercise meal will be fluid. One hour before training or competition, much of the athlete's intake will be fluid because it is often easier to tolerate fluid as the time to begin exercise draws closer. The amount will vary based on the body size of the athlete (see Chapter 7). In many cases water is sufficient, but carbohydrate-containing beverages may also be consumed by those in need of some additional carbohydrate prior to exercise. Sodium added to foods and beverages in small amounts may also be beneficial because sodium helps to stimulate thirst, increase the body's drive to drink, and retain fluid.

Stress can affect the body positively, such as a heightened sense of awareness or motivation; however, stress can also have a negative impact, such as headaches or nervousness. Gastrointestinal upset is a common response to the stress of competition, and athletes may experience nausea, vomiting, abdominal cramps, and/or diarrhea. Too great a volume of food or the presence of unfamiliar foods can trigger or worsen the situation. These problems may be avoided by using trial and error during training to determine the volume and types of food that are appropriate for the athlete to consume prior to competition. Common adjustments are to allow more time prior to the onset of exercise, reduce the volume normally consumed, or include more semisolid or liquid foods. If athletes will be consuming the precompetition meal at the site of competition, they should find out the kinds of foods that will be available or bring their own foods. Athletes should not try new foods as part of a precompetition meal without first consuming those foods during training to determine their expected gastrointestinal response.

Some athletes do not consume any food prior to exercise and wonder if this is appropriate. Research studies suggest that the lack of food prior to exercise decreases performance, so most athletes will likely benefit from pre-exercise foods or beverages (Maffucci and McMurray, 2000; Chryssanthopoulos and Williams, 1997). Low blood glucose is associated with feelings of hunger, an inability to concentrate, and poor endurance, so some food or beverage intake prior to training is usually recommended. Those performing high- to very high-intensity, short-duration exercise would likely limit their intake to a small, primarily carbohydrate meal and allow sufficient time for gastric emptying.

Early start times of competitive events may alter the athlete's usual eating pattern. For example, rowers in a 2,000 m (~1¼ mile) race with a 7:00 a.m. start time will be up at 4:00 a.m. and may not have any appetite because they are usually asleep at this time. Eating a banana at 6:00 a.m. and sipping a sports beverage would provide a small amount of glucose that would be quickly absorbed, but any more food or fluid than that may be too much before high-intensity, short-duration exercise. Marathon runners often put a sports drink and a carbohydrate snack next to their bedside and consume it in the early hours of the morning of competition when they get up in the night to use the bathroom. The need for trial and error, and some creativity, cannot be overemphasized. The wrong precompetition meal can be far more detrimental than the right precompetition meal can be beneficial!

Food and fluid intake during exercise. Intake of foods and fluids during exercise is vitally important for endurance and ultraendurance athletes, such as distance runners and cyclists, triathletes, and adventure racers, and those who engage in intermittent, high-intensity exercise for more than 1 hour, such as soccer and basketball players. Athletes in other sports, such as golf, softball, or baseball, may consume food and/or fluids, but the demand for either is less than athletes engaged in higher-intensity sports. Obviously, some athletes cannot ingest any food or fluids during an event. The goals of food and fluid intake during exercise are typically to:

- Provide energy for exercise, particularly carbohydrate
- Delay fatigue
- Prevent or delay hypohydration and excessive dehydration
- Prevent overconsumption of water
- Prevent excessive changes in electrolyte balance such as hyponatremia (low blood sodium)
- Replace sodium, if losses are large or rapid
- Minimize gastrointestinal distress

Prolonged exercise, especially in the heat, can result in hypohydration, which can impair performance,

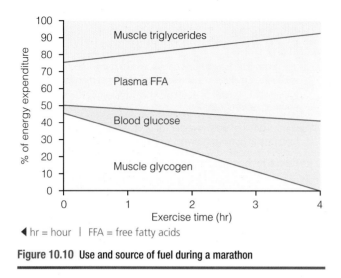

◀ hr = hour | FFA = free fatty acids

Figure 10.10 Use and source of fuel during a marathon

contribute to fatigue, and be a health risk. Hypoglycemia, or low blood glucose, also impairs performance and contributes to fatigue. The ability to prevent hypoglycemia and hypohydration and to delay fatigue is the basis for specific carbohydrate and fluid recommendations during exercise (see Chapters 4 and 7).

Marathon running is an excellent example of why distance runners consume carbohydrates during a race and during long training runs. Endogenous (within the body) carbohydrates are limited to muscle glycogen, liver glycogen, and blood glucose and together provide approximately 500 g or 2,000 kcal of carbohydrates. Carbohydrates are an important fuel source for a 26.2 mile run (as are fats and to a small degree proteins, but this discussion will focus only on carbohydrates). As shown in Figure 10.10, when the marathon begins the majority of the carbohydrates are supplied by muscle glycogen. Blood glucose provides a small but steady supply of glucose to the muscles. As the marathon continues and muscle glycogen stores become depleted, a higher percentage of the carbohydrates used

come from blood glucose as the muscles increase their uptake of glucose from the blood. In response to the increased uptake of blood glucose, the liver increases the breakdown of liver glycogen, which provides glucose to the blood and keeps blood glucose concentration stable. In other words, the glucose is being taken out of the blood by the muscle cells but is being replaced by glucose from the liver cells. As long as the liver can keep pace with muscle glucose uptake, the concentration of glucose in the blood will remain fairly stable and within the normal range. But it is hard for the liver to keep pace for two reasons: (1) the rate at which new glucose can be made by the liver (gluconeogenesis) is limited, and (2) the total amount of glycogen stored in the liver is small, approximately 100 g or 400 kcal. When the uptake of glucose by muscle exceeds the ability of the liver to provide glucose, blood glucose begins to drop and may eventually fall below the normal range (that is, hypoglycemia develops).

The example above explains how the body adjusts when blood glucose starts to decline and how eventually blood glucose would drop. However, the athlete can prevent a fall in blood glucose by consuming carbohydrates during exercise. Consuming carbohydrates during a marathon or other prolonged endurance task is important because it prevents hypoglycemia, which results in lightheadedness and lack of concentration, and will provide a source of carbohydrates for metabolism when endogenous sources are limited. Without carbohydrate intake during prolonged endurance exercise, muscle and liver glycogen stores can be nearly depleted (the body guards against their total depletion), which is a condition strongly associated with fatigue and reduced exercise performance. Nearly every marathon runner or triathlete has a sad story about "hitting the wall" or "bonking," terms that refer to an inability to complete a race due to glycogen depletion.

Fluid intake during training and competition (if intake is possible) is important to prevent hypohydration, excessive dehydration (that is, >2 percent

Nearly every marathon runner has a story about "hitting the wall."

Athletes in intermittent, high-intensity sports also benefit from carbohydrate-containing fluids during training or competition.

of body water loss), and heat illness and to delay fatigue. The amount and frequency of fluid that should be consumed is discussed in detail in Chapter 7. The American College of Sports Medicine (ACSM) currently recommends that a customized plan be devised for each athlete and does not recommend discrete amounts or time intervals because such recommendations, inappropriately applied, could result in substantial over- or underconsumption of fluid and related medical problems (Sawka et al., 2007).

To prevent hyponatremia, endurance and ultra-endurance athletes should avoid excess fluid intake and replace sodium lost. However, "excess" is not an amount that can be defined for all athletes, as it is a relative amount that is unique to each athlete.

Much of the information about intake during exercise focuses on athletes in prolonged endurance sports because the risk for heat illness is high, hyponatremia can occur, and the need for fluid and carbohydrate is great. From a practical point of view, endurance athletes can "kill two birds with one stone" by drinking their source of carbohydrates. But the rate of gastrointestinal absorption must be considered. When exercise intensity is greater than 75 percent of $\dot{V}O_{2max}$, the rate of absorption of carbohydrate-containing fluids will be slowed and the risk for gastrointestinal upset is increased. Too great a concentration of carbohydrates can also slow gastric emptying, yet too small a concentration does not provide many carbohydrates. Most endurance athletes consume a sports beverage with 4–8 percent carbohydrates. At these concentrations, each liter (1,000 ml, or a little more than four 8 oz cups) contains 40 to 80 g of carbohydrates. These carbohydrate beverages are typically consumed every 15 to 20 minutes.

Carbohydrate-containing sports drinks are popular, but sufficient fluid and carbohydrates can be obtained in other ways. For example, a marathon runner may prefer two packets of gel (~50 g carbohydrates) and water. Other popular carbohydrate foods that may be eaten during endurance activities include bananas, energy bars, and Fig Newtons. Although many athletes in prolonged training or competitions consume a sports beverage, they may also like the variety that food and water combinations provide. Trial and error is necessary to determine tolerance. Figure 10.11 summarizes the guidelines for food and fluid intake during exercise.

Carbohydrate and fluid intake during distance running has received a lot of attention because the need to replenish these nutrients is high, and distance runners tend to have more gastrointestinal problems than other distance athletes, such as cyclists; but many athletes need a plan for food and fluid intake during exercise. Athletes in sports such as basketball or soccer typically consume both water and carbohydrates, usually as sports beverages, when they are on the sidelines.

Carbohydrate: 30–60 g/h, although some athletes can tolerate more.

Fluid: Customized plan to prevent excessive dehydration and excessive changes in electrolyte balance based on sweat rate, sweat composition, duration of exercise, clothing and environmental conditions.

Sodium: 1 g/h if a "salty sweater" (that is, a high sweat rate and high loss of sodium in sweat while exercising).

Trial and error during training is encouraged.

Figure 10.11 Guidelines for food and fluid intake during exercise

Legend: g/h = grams per hour

At halftime, they may have a carbohydrate snack, such as a banana, orange, or energy bar, and fluid. Athletes in sports such as (American) football usually concentrate on fluid intake rather than carbohydrate intake. A track athlete who is running a short race lasting only seconds or a few minutes does not need to be concerned about food or fluid intake during competition, but that same athlete who runs an individual race, is part of two relays, and performs the long jump will need to have some carbohydrates and fluid during the track meet. Each situation is different, and it is the intensity and duration of the exercise and the conditions under which athletes train and compete that will primarily dictate the need for food and fluid during training and competition.

Food and fluid intake after exercise. When exercise ends, the recovery period begins. For many athletes, recovery time is limited because they train or compete nearly every day. Some have multiple competitions on the same day. Food and beverage intake should begin

Athletes in sports such as basketball typically consume both water and carbohydrates, usually as sports beverages, when they are on the sidelines.

immediately after exercise if possible. The goals of food and fluid intake after exercise are to:

- Provide carbohydrates to resynthesize muscle glycogen
- Provide proteins to build and repair muscle
- Rehydrate and re-establish euhydration
- Replace lost electrolytes
- Avoid gastrointestinal upset

Exercise depletes glycogen stores, which should be restored before the next training session. The rate and extent to which muscle glycogen is resynthesized depends on a number of factors, including the extent of depletion, the presence of insulin, the activity of enzymes such as glycogen synthase, the degree of muscle damage, and the amount and timing of carbohydrate intake.

The timing of carbohydrate intake after exercise is critical. Consuming carbohydrates immediately after exercise provides glucose at a time when cellular glucose sensitivity and permeability are optimal and the rate of glycogen restoration is high. Muscle glycogen is resynthesized at its highest rate immediately after exercise. A 2-hour delay after exercise can reduce glycogen synthesis substantially (Ivy, 1998). Glycogen resynthesis is a slow process and may take more than 20 hours. If possible, athletes should consume carbohydrates immediately after exercise to take advantage of rapid glycogen resynthesis and consume adequate daily carbohydrates to optimize glycogen stores (see Chapter 4). Many athletes who deplete muscle glycogen during training or competition focus on a high-carbohydrate intake immediately after exercise and a high-carbohydrate meal within a couple of hours.

In addition to the timing and amount of carbohydrates, other factors influence recovery. Postexercise protein intake is beneficial to muscle protein anabolism (see Chapter 5). The postexercise period is also critical for replenishing fluids, and in some cases, sodium and other electrolytes that may have been lost in large amounts. The amount of fluid that is necessary for replenishment is estimated by comparing pre- and postexercise weights. It is recommended that athletes drink approximately 1.5 L (~50 oz) of fluid per kg body weight lost, beginning as soon after exercise as is practical. In other words, a 2.2 lb loss of scale weight requires ~6 cups of fluid intake (Sawka et al., 2007; Rodriguez et al., 2009). Rehydration and obtaining a state of euhydration, especially after exercise in the heat, can take many hours, but should be initiated immediately after exercise.

Athletes who are "salty sweaters" must also replenish the sodium chloride lost. These athletes are typically advised to sprinkle table salt (sodium chloride) on their food or consume salty snacks, such as pretzels dipped in salt, after exercise. Sports beverages

Carbohydrate: 1.5 g/kg in the first hour after exercise, including some with a medium to high glycemic index; maximal additional intake of 0.75–1.5 g/kg per hour over the next 3 h.

Protein: 10–20 g high-quality protein as soon as possible, but no later than 2–3 hours after exercise.

Fluid: ~1.5 L (~50 oz or ~6 cups) of fluid per kg body weight lost, beginning as soon after exercise as is practical.

Sodium: Consume foods containing sodium. If large amounts of sodium have been lost, salt food or consume salty snacks.

Other electrolytes: Consume a variety of food including fruits and vegetables.

Figure 10.12 Guidelines for food and fluid intake after exercise

Legend: g/kg = grams per kilogram body weight; h = hour; g = gram; oz = ounce

may contain some sodium, but the amount is usually too low for those who have lost considerable sodium in sweat. Figure 10.12 summarizes the guidelines for food and fluid intake immediately after exercise.

After exercise the athlete is essentially trying to "kill four birds with one stone," because optimal recovery depends on replenishing carbohydrate, protein, water, and electrolytes, particularly sodium. Some athletes choose to consume liquid "meals," such as specially formulated postexercise beverages, because they are convenient and preformulated. However, other athletes prefer "real" foods that contain both carbohydrate and protein, such as chocolate milk, yogurt with fruit in the bottom, or cereal with milk. A turkey and cheese sandwich with salty pretzels and a large beverage is an example of a postexercise meal that provides the four major nutrients that an athlete who sweats heavily focuses on at the start of the recovery period.

From a practical perspective, it may be difficult for athletes to immediately implement postexercise recovery strategies. Training or competition may depress appetite, and fatigue may be so great that rest or sleep may be the desired postexercise activity. Some collegiate athletes need to attend class soon after practice or workouts. The athlete may not consider food and fluid intake the highest priority, but immediate postexercise intake is critical for optimal recovery.

Many athletes consume caffeine safely and effectively as a central nervous system stimulant.

Caffeine is the most widely consumed, self-administered **psychotropic** drug in the world. However, it is legally and socially acceptable, so it stands apart from other drugs. Caffeine is found in many beverages and a few

Psychotropic: Capable of affecting the mind.

foods and is considered part of a normal diet. The primary active ingredient in caffeine is methylxanthine. Caffeine is considered safe at moderate doses, although there are side effects, including addiction. Athletes use caffeine for a number of reasons, including performance enhancement, weight loss, and central nervous system stimulation. Caffeine can improve performance due to a heightened sense of awareness and a decreased perception of effort. At certain urinary concentrations caffeine is considered a banned substance by some sports-governing bodies, but such levels would not be reached with normal food and beverage consumption and would likely impair performance in other ways, such as rapid heartbeat or shaking.

Caffeine content of foods and beverages. Caffeine is found in foods, beverages, and medications as shown in Table 10.4. In the United States, when caffeine is added to a food or beverage it must appear in the list of ingredients, but the amount added is not required to appear on the label. The caffeine content of most beverages is listed on the manufacturer's website.

Safety of caffeine. Caffeine is considered safe for most adults although it has several known side effects. Blood pressure is increased both at rest and during exercise, heart rate is increased, gastrointestinal distress can occur, and insomnia may result. The side effects are more likely to occur in people who are caffeine naïve (that is, do not routinely consume caffeine). For routine users, some tolerance to caffeine's effects develops. Developing tolerance means that the user must increase the dose to produce the desired effect. Caffeine is addictive and sudden withdrawal results in severe headaches, drowsiness, inability to concentrate, and feelings of discontent. Addiction has been documented with doses as low as 100 mg daily (Evans and Griffiths, 1999).

It is difficult to study the health effects of caffeine because most people do not ingest pure caffeine; rather, they ingest it as a beverage that contains other compounds. Studies of caffeine consumed via coffee, tea, or soft drinks cannot be directly compared. Many study results are conflicting. For example, some studies have shown that caffeine users are not at risk for elevated blood pressure whereas other studies have shown an increase in blood pressure with caffeine use (Winkelmayer et al., 2005; Noordzij et al., 2005; James, 2004). A common guideline is that moderate doses of caffeine, defined as 200-300 mg daily, are considered to be safe. Caffeine consumption greater than 500 mg per day is associated with anxiety, irritability, insomnia, headaches, and gastrointestinal distress. It should be stressed that these recommendations are intended

Table 10.4 Caffeine Content of Selected Foods, Beverages, and Medications

Product	Caffeine content (mg)
"Grande" (16 oz) commercially brewed coffee	250–330
Bottled iced coffee (~11 oz)	200
Home-brewed coffee (8 oz)	80–100
Instant coffee (8 oz)	65–100
Espresso coffee (2 oz)	~100
Decaf coffee (8 oz)	5
Brewed tea (8 oz)	~50
Bottled iced tea (16 oz)	30–60
Most soft drinks (12 oz)	35–55
Mountain Dew MDX or VAULT (20 oz)	120
Diet Pepsi Max	70
Energy drinks (~8 oz)	~80
Espresso Hammer Gel (36 g)	50
Energy bar with caffeine	50 or 100
Milk chocolate (1.5 oz)	10
Dark chocolate (1.5 oz)	30
Coffee ice cream or yogurt (~1 cup)	45–60
Excedrin (2 caplets)	130
Vivarin, maximum-strength NoDoz, or caffeine tablets (1 tablet)	200
Ripped Fuel Extreme	220
Stacker 3	250

Legend: oz = ounce; g = gram; mg = milligram

Source: Center for Science in the Public Interest (2008), manufacturers' websites

for adults and not for adolescent or children. Caffeine intoxication is possible, but rare.

A daily caffeine intake of up to ~450 mg does not seem to affect an athlete's hydration or electrolyte status. This amount of caffeine has about the same diuretic effect as water, which is mild. Excess losses of water or electrolytes are not expected with moderate daily doses of caffeine (Armstrong et al., 2007; Maughan and Griffin, 2005). However, some athletes do abuse caffeine. For example, martial artists and bodybuilders may use caffeinated beverages in the final stages of "cutting" weight because it is a mild diuretic and a stimulant that masks fatigue. The use of caffeine in the already hypohydrated individual is a concern because it can exacerbate the degree of dehydration and contribute to life-threatening medical conditions, such as an elevated body temperature.

Effectiveness of caffeine as a performance enhancer. Caffeine and its role in performance has been well studied, and the consensus opinion is that caffeine stimulates the central nervous system, resulting in a heightened sense of awareness, a decreased perception of effort, and/or an increased pain threshold (Burke, 2008; Doherty and Smith, 2005). There is scientific evidence that caffeine consumed at an appropriate dose can enhance the performance of distance runners, cyclists, and cross country skiers as well as runners, cyclists, swimmers, and rowers who engage in high-intensity activities lasting 1–20 minutes (Burke, 2008; Davis and Green, 2009; Ganio et al., 2009).

Establishing the caffeine concentration in the blood that is effective for the individual athlete takes trial and error. A reasonable starting dose is 2–3 mg/kg of body weight. For a 220 lb (100 kg) person, the recommended dose would be 200–300 mg of caffeine with the appropriate dose and timing fine-tuned on an individual basis (see Chapter 6). Studies have not shown a dose-response relationship. In a dose-response relationship, as the dose increases the performance also increases. Therefore, there is no evidence that "more is better" when it comes to caffeine consumption and performance. Adverse effects are more likely to occur at higher doses (6–9 mg/kg body weight).

Cautions have been raised about the use of highly caffeinated "energy" drinks, particularly by adolescents and young adults. These drinks are popular among high school and collegiate athletes and triathletes because they mask fatigue for about 60 to 90 minutes (Duchan, Patel, and Feucht, 2010). The immediate feeling of "energy" is most likely due to caffeine, a neurological stimulant, as well as a rise in blood glucose. When the dose of caffeine is too high, the athlete runs the risk of feeling "overstimulated" rather than "energized," which may be detrimental to training, performance, or sleep.

Effectiveness of caffeine for weight loss. Although athletes may state that they use caffeine to lose body fat, studies have not shown that caffeine, by itself, has a substantial effect on fat or weight loss. However, caffeine does enhance the effect of ephedrine, a dietary supplement used for weight loss. Ephedrine and caffeine for weight loss are discussed in Chapter 11.

Caffeine as a banned substance. The National Collegiate Athletic Association (NCAA) lists caffeine under banned stimulants. Based on postcompetition urine analysis, a urinary caffeine concentration exceeding 15 mcg/ml would be considered a positive test and subject the athlete to disqualification (NCAA Bylaw 31.2.3). However, such a concentration would be very difficult to reach via normal food and beverage intake (e.g., 6–8 cups of caffeinated coffee 2 to 3 hours prior), and the equivalent amount of caffeine-containing tablets would likely impair performance in other ways. The International Olympic Committee (IOC) monitors caffeine use in athletes, but it does not disqualify anyone based on urinary caffeine concentration. This allows athletes to take cold remedies, which often contain caffeine, and drink caffeinated beverages without risk of disqualification. The IOC has not found evidence of caffeine abuse by athletes (Burke, 2008).

Athletes should consider the risks and benefits of alcohol consumption.

In the context of this textbook, alcohol refers to the consumption of ethanol. Alcohol consumption is described as "drinking," and one drink is defined as ½ oz (~15 ml) of ethanol. Three to 4 oz (90 to 120 ml) of wine, a 10 oz (300 ml) wine cooler, a 12 oz (360 ml) beer, or 1½ oz (45 ml) of hard liquor (e.g., a "shot" of whiskey) each contain approximately ½ oz (15 ml) of ethanol. Moderate alcohol intake is defined as the consumption of up to one drink per day for women and up to two drinks per day for men. A widely accepted definition of binge drinking is four or more drinks by a female and five or more drinks by a male at one time. The NCAA bans alcohol use by athletes only in the sport of rifle (NCAA Bylaw 31.2.3 Banned Drugs); however, many colleges and universities have team rules that ban alcohol or limit its consumption by players who are of legal drinking age. Alcohol contains 7 kcal/g, and one "drink" typically provides ~100 to 150 kcal (Table 10.5).

Alcohol use by athletes. Information about athletes' alcohol consumption is primarily derived from studies of collegiate athletes. Use of alcohol in the off-season is more prevalent than during the season. A limited number of studies have found that college athletes consume more alcoholic drinks per week, binge drink, and engage in risky behaviors more often than nonathletes (Yusko et al., 2008; Martens, Cox, and Beck, 2003; Miller et al., 2002). These patterns are more likely to occur in males. Athletes use alcohol as a coping mechanism, and it provides them temporary stress relief and elevated mood. However, there are a number of negative consequences, including depression, physical injury, poor school performance, and legal problems associated with driving under the influence and other inappropriate behaviors. Some athletes are likely to use

Table 10.5 Caloric Content of Alcohol-Containing Beverages

Beverage	Serving size (oz)	Energy (kcal)*
Beer, regular	12	150
Beer, stout	12	190
Light beer	12	100–125
White wine	4	80
Red wine	4	100
Dessert wine	4	190
Wine cooler	10	150
Whiskey, 80 proof	1.5	100
Champagne	4	85
Margarita (tequila, triple sec, lime juice)	3	170
Singapore sling (gin, lemon juice or sour mix, club soda, cherry brandy, grenadine)	8	230
Tequila sunrise (tequila, orange juice, grenadine)	5.5	190

Legend: oz = ounce; kcal = kilocalorie

*Energy content will vary depending on the brand or the proportion of ingredients in mixed drinks.

alcohol as a medication for underlying psychiatric conditions such as depression and anxiety. Self-medication for these conditions is not recommended, and alcohol is an especially poor choice as a treatment because alcohol use both causes and extends depression.

Effect of alcohol on training or performance. A limited number of studies have been conducted on the effects of alcohol on training and performance. Shirreffs and Maughan (2006) reviewed published evidence and described potential negative effects on glycogen metabolism, hydration, thermoregulation, and the ability to exercise after previous alcohol use. Alcohol's effects appear to be indirect and are subject to substantial individual variation.

Binge drinking after glycogen-depleting exercise may result in a low carbohydrate intake, which reduces the body's ability to resynthesize muscle glycogen. In the one study conducted (Burke et al., 2003), when the majority of dietary carbohydrates were replaced with 120 g of alcohol (~11 drinks in Australia) after prolonged cycling, glycogen synthesis was reduced by 50 percent at 8 hours and 16 percent at 24 hours when compared to the control group that did not replace dietary carbohydrates with alcohol. There was no statistical difference in glycogen storage of the control and treatment groups when 120 g of alcohol

was taken in addition to, rather than as a substitute for, dietary carbohydrates. However, this amount of alcohol consumption increased the caloric content of the diet by 10.5 kcal/kg, the equivalent of ~800 kcal in a 75 kg person.

Alcohol may be counterproductive as a rehydration strategy because alcohol has a diuretic effect (see Chapter 7). Dilute solutions of alcohol (up to 2 percent alcohol) have a similar rehydration effect to alcohol-free beverages in that the fluid itself causes some diuresis. Concentrated alcohol solutions (4 percent or more) result in negative fluid balance. Shirreffs and Maughan (2006) estimate that 25 ml (~0.8 oz) of a 40 percent ethanol solution contains approximately 10 ml of alcohol and 15 ml of water. This solution induces a urinary output of about 100 ml, so the net loss of water is 85 ml or about 3 ounces of water. To place this information in perspective and to estimate the amount of fluid that may be lost due to diuresis, the percentage of alcohol contained in a beverage must be known. In the United States, the approximate alcohol percentage of beer is 4–4.5 percent, wine is 15–20 percent, and one "shot" of hard liquor is 30–50 percent. These are all concentrations of alcohol that are associated with negative fluid balance to varying degrees.

Alcohol intake prior to exercise performed in low environmental temperatures is not recommended because it can reduce core temperature. Binge drinking the day prior to performance is not recommended because it may impair next-day performance ("hangover" effect) even though blood alcohol concentration is zero. Alcohol intake immediately prior to exercise is never recommended because it can impair judgment (Shirreffs and Maughan, 2006).

Negative and positive effects of alcohol on health. Alcohol can be both detrimental and beneficial to health, depending on the amount consumed as well as the pattern of consumption. Alcohol use can lead to addiction, aggressive behavior, poor judgment, automobile deaths, homicide, and suicide. Its moderate use is related to a number of chronic diseases, and it may have a beneficial effect on three—cardiovascular disease, stroke, and diabetes. However, alcohol use has a detrimental effect on a large number of chronic diseases, including mouth, esophageal, liver, and breast cancers, depression, epilepsy, hypertension, hemorrhagic stroke, and cirrhosis of the liver (Reynolds et al., 2003; Rehm et al., 2003).

Moderate alcohol intake may be beneficial for reducing risk for cardiovascular disease, stroke, and diabetes because of its ability to increase high-density lipoproteins and its effects on blood clotting. Consumption of alcohol in quantities greater than moderate can cause stroke, cardiac arrhythmias, and elevated blood pressure (Reynolds et al., 2003; Rehm et al., 2003). Any benefits of alcohol intake are clearly associated with *moderate* use.

Key points

- Athletes need a comprehensive, individualized nutrition plan to support training, meet body composition goals, optimize performance, and maintain good health.

- Food intake needs to be distributed across the day with special consideration for intake before, during, and after exercise.

- Trial and error is needed to find strategies that work for the individual athlete.

- Caffeine is safe in moderate amounts for adults and may enhance performance due to central nervous system stimulation.

- Alcohol has a negative effect on performance, but has beneficial health effects when moderate amounts are consumed.

The team bus stops at a cluster of fast food restaurants for dinner. What would be the best menu choices for a postexercise meal?

10.3 Dietary Supplements and Ergogenic Aids

Proper diet is fundamental to good health and athletic performance. Athletes recognize its important roles, but many find it difficult to consume a diet that supports training. Athletes spend millions of dollars on dietary supplements each year in the belief that their use will help them improve their athletic performance or make them healthier. Although some dietary supplements may have this effect, others may have the opposite effect, resulting in impaired performance or health. Although rare, dietary supplements have actually contributed to or caused the deaths of some athletes. Athletes could also unintentionally consume a banned substance because the purity of dietary supplements can vary.

Consuming a diet that supports training and good health is a long-term investment. Like any other long-term project, the benefits may not be evident on a day-to-day basis. Athletes looking for a "quick fix" will not likely find it through diet alone. One of the lures of dietary supplement use is the possibility of immediate results. Many supplements are advertised as the newest and fastest way to enhance performance, and the promise, hope, and hype are often hard to resist. Some dietary supplements are also ergogenic aids. An ergogenic aid is any substance or strategy that improves athletic performance, but the term is often used to describe substances or techniques that increase the production of energy or the ability to do work. Many ergogenic aids are drugs or medical procedures, such as **blood doping**, but some, such as creatine, are dietary supplements.

Table 10.6 summarizes the safety and effectiveness of some dietary supplements and ergogenic aids using evidence-based analysis. These substances are discussed in more detail in other chapters in this book, and Chapter 1 includes overall information about dietary supplements. What is immediately evident when reviewing Table 10.6 is how few supplements have been shown to be both safe and effective (Dunford and Coleman, in press).

Before consuming any dietary supplement or using any ergogenic aid, the athlete should answer five questions: Is it legal? Is it ethical? Is it pure? Is it safe? Is it effective? It is critical for athletes, parents, coaches, trainers, sports dietitians, exercise physiologists, physicians, and anyone else involved in sports medicine to learn about dietary supplements and ergogenic aids and how to address this ever-changing aspect of athletics. Many people do not realize that dietary supplements are loosely regulated in many countries, including the United States. Information about the Dietary Supplement Health and Education Act (DSHEA) can be found in Chapter 1. In brief, the following points should be kept in mind any time an athlete considers consuming a dietary supplement:

1. Regulation of dietary supplements in the United States is minimal. Federal law does not ensure the safety or effectiveness of any dietary supplement.

2. Dietary supplements may contain banned substances and could result in forfeiture, penalties, and banishment from the sport (see Appendix L for a list of drugs banned by the NCAA).

3. Most dietary supplements have not been studied in trained athletes, and few have a large body of scientific literature regarding safety and effectiveness. A notable exception is creatine.

4. There are few dose-response studies of dietary supplements. Dose-response studies are important to determine ineffective doses, the most effective dose, and the dose at which toxicity symptoms begin to appear.

(*Continued on page 383*)

Blood doping: Intravenous (IV) infusion of the athlete's previously withdrawn blood for the purpose of increasing oxygen-carrying capacity.

Table 10.6 Safety and Effectiveness of Selected Dietary Supplements Used by Athletes

Supplement or ergogenic aid	Safety	Effectiveness	More information
Safe and effective at recommended doses			
Caffeine	≤450 mg/d considered safe	Effective as a central nervous system stimulant; likely to improve endurance performance and high-intensity exercise lasting up to 20 minutes	Chapters 6, 7, and 10
Creatine	3–5 g/d	Effective for increasing lean body mass in athletes performing repeated high-intensity, short-duration (<30 seconds) exercise bouts; likely to improve performance in weight lifters	Chapter 3
Protein	2.5 g/kg body weight/d considered maximum dose	Effective source of protein; whey protein may be more effective than casein for increasing muscle size and strength	Chapter 5
Multivitamin and mineral supplements	Safe at doses listed in the DRI	Effective to increase nutrient intake to a degree, which can help to reverse nutrient deficiencies; not likely to improve performance in those without nutrient deficiencies	Chapters 8 and 9
Vitamins C and/or E	Vitamin C: UL = 2,000/d Vitamin E: >200 mg/d may increase oxidative damage; UL is 1,000 mg/d	Effectiveness depends on the balance of oxidants and antioxidants; excessive amounts are not beneficial and may upset the balance; vitamin C is effective for decreasing the duration of a cold but not effective for preventing colds	Chapter 8
Safety and effectiveness at recommended doses is promising			
Beta-alanine	Doses up to 6.4 g/d appear to be safe	Promising as a buffer of muscle pH	Chapter 5
Branched chain amino acids (BCAA)	5–20 g/d in divided doses seems to be safe	Promising for immune system support and reduction of postexercise fatigue; not effective for improving performance	Chapter 5
Omega-3 fatty acids	FDA recommends intake of EPA and DHA not exceed 3 g/d with no more than 2 g/d from supplement sources. The amount in the supplement may not match the amount listed on the label.	Promising for use in athletes with exercise-induced bronchoconstriction due to asthma; not effective to reduce inflammation, enhance the immune system, or improve performance	Chapter 6
Quercitin	1,000 mg/d for several weeks has been used in research studies	In trained athletes, may have some anti-inflammatory effects; not effective for improving performance or preventing oxidative damage	Chapter 8
Safe at recommended doses but not effective or marginally effective			
Carnitine	2–4 g/d appears to be safe	Not effective	Chapter 6
Chromium	<200 mcg in a form without enhanced bioavailability appear to be safe	Not effective for increasing muscle mass, decreasing body fat, or improving performance	Chapter 9
Glucosamine/ chondroitin sulfate	Glucosamine: 1500 mg/d; chondroitin: 1,200 mg/d appear to be safe	Generally not effective for reducing joint pain or increasing functionality in those with osteoarthritis, although individual responses vary	Chapter 5
Glutamine	Up to 20 g/d appears to be safe	Not effective as an immune system enhancer	Chapter 5
Growth hormone releasers (arginine)	5–9 g/d appear to be safe	Effective for stimulating growth hormone release; if taken before exercise, decreases the effectiveness of exercise as a stimulator of growth hormone; not effective for increasing muscle mass or strength	Chapter 5

Table 10.6 Safety and Effectiveness of Selected Dietary Supplements Used by Athletes (Continued)

Supplement or ergogenic aid	Safety	Effectiveness	More information
Safe at recommended doses but not effective or marginally effective			
ß-hydroxy-ß-methylbutyrate (HMB)	3 g/d divided into three equal doses appears to be safe	Not effective in resistance trained athletes; small to very small increases in overall, upper-body, and lower-body strength in untrained individuals	Chapter 5
Medium-chain triglycerides (MCT)	Safe at recommended doses of ~15–45 g/d, but unknown effect on blood lipids	Not effective	Chapter 6
Safety or effectiveness unknown due to lack of research			
Citrus aurantium (bitter orange)	Concerns about safety	Unknown effectiveness in athletes as a weight loss aid; effective as a central nervous system stimulant	Chapter 11
Conjugated linoleic acid (CLA)	Safety in humans has not been established	Evidence of effectiveness as a weight loss aid in humans is mixed	Chapter 11
Nitric oxide (NO)/arginine alpha-ketoglutarate (AAKG)	12 g/d for 1–8 weeks	Effectiveness unknown due to lack of research	Chapter 5
Considered unsafe and not effective; may be a banned substance			
Androstenedione	Not safe	Not effective	Chapter 11
Dehydroepiandrosterone (DHEA)	Unknown safety profile	Not effective as a performance enhancer	Chapter 11
Ephedrine-containing supplements	Narrow safety profile; safety is a concern at doses <10 mg	Not effective for improving muscle strength or anaerobic performance; unknown effectiveness for short-term weight loss in athletes	Chapter 11

Legend: mg = milligram; d = day; g = gram; g/kg = grams per kilogram of body weight; DRI = Dietary Reference Intakes; UL = Tolerable Upper Intake Level; FDA = Food and Drug Administration; EPA = eicosapentaenoic acid; DHA = docosahexaenoic acid; mcg = microgram

(*Continued from page 381*)

5. The manufacturer suggests the dose and amount to be taken daily that appear on the supplement label. These quantities may or may not reflect the dose and amount used in research studies.

6. Few dietary supplements have been shown to enhance performance, and their effects on performance are relatively small.

7. Some dietary supplements have been shown to be detrimental to performance.

8. There is no substitute for disciplined training and proper diet.

For those supplements that are known to be effective, the ability to enhance performance is relatively small.

In his book *Lore of Running* (2002), Dr. Timothy Noakes estimates the extent to which various interventions enhance performance (Figure 10.13). Training is likely to have the greatest impact on performance. Carbohydrate loading prior to and carbohydrate intake during endurance exercise are estimated to improve endurance performance by approximately 20 percent (see Chapter 4). Anabolic steroids are estimated to improve performance of some athletes by 20 percent due to increases in muscle size and strength. The use of blood doping (4–8 percent improvement) and **erythropoietin** (8 percent) also enhances endurance performance, but to a lesser degree than carbohydrate loading. Use of anabolic steroids, blood doping, and erythropoietin is illegal and unethical, but their ability to impact performance is much greater than any of the dietary supplements used for the same purpose. Assuming that creatine and caffeine supplements are used appropriately, Noakes estimates the performance enhancement associated with these supplements to be low, 1–2 percent for creatine and 1–3 percent for caffeine. The effect of many dietary supplements is estimated to be zero. Similarly, the Australian Institute of Sport (AIS) has grouped dietary

Erythropoietin: Hormone that stimulates the development of red blood cells in the bone marrow.

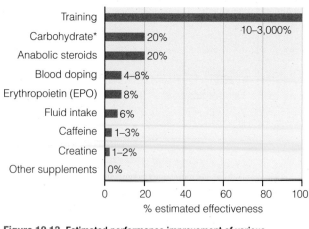

Figure 10.13 Estimated performance improvement of various interventions

*Carbohydrate loading or ingestion in endurance exercise >3 hours. Information adapted from Noakes, T. M. (2002). *Lore of Running* (4th ed.). Champaign, IL: Human Kinetics.

supplements and sports foods into four categories based on their safety and effectiveness (Figure 10.14).

Scientific studies of performance-enhancing supplements are limited to determining the general effect on the study population. It is conceivable that some dietary supplements judged ineffective in scientific studies could have a positive (or negative) individual effect. Athletes who wish to experiment with dietary supplements or other ergogenic aids should be reminded

Group A
(Supported for use by Australian Institute of Sport athletes)

Antioxidant vitamins C and E

Bicarbonate and citrate

Caffeine

Calcium supplement

Creatine

Electrolyte replacement supplements

Iron supplement

Liquid meal supplements

Multivitamins and minerals

Probiotics (use for gastrointestinal protection)

Sports bars

Sports drinks

Sports gels

Vitamin D

Figure 10.14 Australian Institute of Sport supplement group classification

http://www.ausport.gov.au/ais/nutrition/supplements/classifications

to fully investigate all five critical aspects—legality, ethics, purity, safety, and effectiveness.

Practitioners should discuss dietary supplement use with athletes.

Practitioners will have little credibility with athletes if they simply dismiss the use of dietary supplements. Regardless of the practice setting, practitioners can play a very important role. Those with scientific training can explain the purported physiological mechanisms and effects and discuss how a supplement might benefit or harm the individual athlete. The complexity of the dietary supplement issue requires that recommendations be evidence based and individually considered. Some information is available on the label of the supplement (see Spotlight on supplements: Understanding a Dietary Supplement Label), but the label does not include information about safety and effectiveness. Serious discussion needs to take place to determine if, and how much, of a supplement would be safe and effective for an athlete to use. Such decisions should not be made on a whim by the athlete and should not be taken lightly by the practitioner.

Health care professionals, such as physicians and dietitians, are trained to recommend use of a supplement on the basis of proven safety and effectiveness in humans. Most athletes are concerned about safety but are willing to *hope* that it will be effective. Some athletes are so focused on trying to improve performance that they are willing to sacrifice safety and long-term health. With such differences in philosophy, it is easy to understand why there are disagreements and controversies about the use of dietary supplements (Schwenk and Costley, 2002).

It is the professional's role to provide as much unbiased information as possible to the athlete. Anyone selling a supplement is not an unbiased source of information. Athletes should be aware that individuals who sell dietary supplements might be involved in multilevel marketing (MLM). MLM is a form of direct sales that allows a person (distributor) to buy a product wholesale and resell it. Sales goals may be established, and any unsold product is costly to the distributor, so there is pressure to sell a certain amount of product, typically to relatives, friends, and acquaintances. Distributors may also derive income from the sales of other distributors that they have recruited (Barrett, 2001).

Ultimately, it is the athlete who must choose if, and how much, of a supplement to consume. This is important because the risk and the benefit should accrue to the same person. It is the athlete who must deal with disqualification and suspension if the dietary supplement, by accident or design, contains a banned substance. The professional's role is to provide athletes with information so wise decisions about dietary supplements can be made. If the professional has concerns about an athlete's health because of supplement use, it is appropriate to express them.

Spotlight on supplements

Understanding a Dietary Supplement Label

The following information must appear on the label of a dietary supplement (see figure). The format is similar to the Nutrition Facts label found on food but there are some differences.

Serving size: Similar to the Nutrition Facts label found on food, the serving size must be listed. The information in the Supplements Facts box is based on the serving size listed. Typically the serving size is one capsule or one tablet.

Amount: The quantity of each compound must be listed. Common measures include grams (g), milligrams (mg), and micrograms (mcg or µg). Some supplement labels use older measures, such as International Units (IU), which can be confusing. For example, the Dietary Reference Intake for vitamin D is expressed in micrograms, but many vitamin D supplements use IU and a conversion is not required to be on the label. For botanicals and herbals, the label must indicate whether the compound is fresh or dried (e.g., powdered). If extracted, then the solvent used to extract the herbal ingredient must appear.

Percent Daily Value: Daily Value (DV) is the amount needed by a person consuming a 2,000–2,500 kcal diet. Nutrients that have an established Daily Value, such as many vitamins and minerals, are listed first. The supplement may contain more than 100 percent of the DV and can exceed the Tolerable Upper Intake Level (UL). Botanicals, herbs, and other compounds that do not have a Daily Value established are marked with an asterisk (*).

Ingredient list: The list of ingredients includes both active (e.g., vitamin E) and inactive (e.g., fillers and stabilizers) compounds. The terms in the Supplements Facts box and the ingredient list may have different names, even though the compounds are the same. For example, vitamin E may be listed on the label but the scientific name for vitamin E, alpha-tocopherol, may appear in the ingredients list.

All ingredients must be listed on the label, but they may not all appear in the ingredient list. If the ingredient appears on the display panel as a statement of identity, it need not be listed again in the ingredient list. The term *proprietary blend* is used frequently. The ingredients in a proprietary blend are listed, but the amounts of the substances are not. This term is used to prevent competitors from knowing the exact formulation of a company's product, but it also prohibits consumers from knowing the amount of each ingredient contained in the supplement.

Additional information: Directions for use must appear. This is the manufacturer's recommendation, and there is no requirement that the amount recommended be shown to be effective. The directions for use may be greater than the serving size shown, in which case values listed would need to be adjusted. For example, if the serving size is one tablet and it contains 250 percent of the DV but it is suggested that two tablets be taken daily, then the

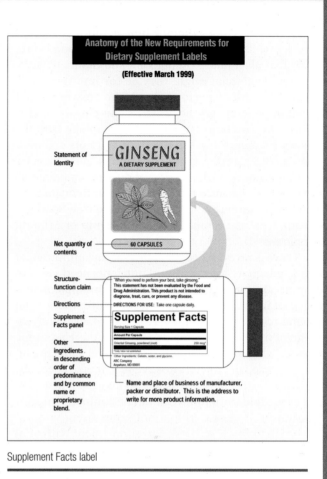

Supplement Facts label

supplement if taken as suggested would provide 500 percent of the DV.

Principal Display Panel: The brand name and the number of tablets in the container must appear. The product must be identified, but there are several ways that this can be done. For example, the terms *dietary supplement, vitamin supplement*, or *herbal supplement with vitamins* could all be used to describe the same product.

Health and structure/function claims: Statements that describe the effect the product has on the structure or function of the body are allowed. Health claims may be made, but therapeutic claims may not be made. Therapeutic claims are those that involve the diagnosis, treatment, or prevention of disease. The statement "Calcium builds strong bones" is allowable because it is a health claim. The statement "Calcium restores lost bone" is not allowed because it is a therapeutic claim. Because the Food and Drug Administration does not evaluate structure or function claims, the following statement must appear: "This product is not intended to diagnose, treat, cure, or prevent any disease." Many structure/function claims are written in such a way as to be misleading.

Vitamin and mineral supplements are frequently used by athletes.

Vitamins and minerals are some of the most widely used supplements by athletes, typically as a way to compensate for poor dietary intake and to avoid illness (Petróczi et al., 2007, 2008; Greenwood et al., 2000; Jonnalagadda, Rosenbloom, and Skinner, 2001). But how beneficial might they be for improving nutritional health or performance? The answer to that question depends on the athlete's current diet and is determined by comparing the athlete's intake with the Dietary Reference Intakes (DRI) for vitamins and minerals, as discussed in Chapters 8 and 9. This is a good starting point for a discussion of dietary supplements with athletes because the need for vitamin or mineral supplementation can be easily evaluated (see Spotlight on supplements: Should I Take a Vitamin or Mineral Supplement?).

There is no evidence that athletes consuming a diet that is adequate in energy and nutrients need vitamin or mineral supplementation or that such supplementation will improve performance; therefore, there is no general recommendation for athletes to take a daily vitamin or mineral supplement. However, there are some athletes who will need vitamin or mineral supplement. For example, an athlete whose caloric intake is consistently low, especially if it is due to an eating disorder, is a candidate for a multivitamin and mineral supplementation. A vegan athlete not consuming vitamin B_{12}-fortified foods may need to supplement with vitamin B_{12} because this vitamin is found naturally only in animal-derived foods or specially fortified products. An athlete with iron deficiency anemia will probably be advised to take an iron supplement. Note that these

Evaluating the need for a multivitamin and mineral supplement involves estimating current intake, awareness of safe levels, and weighing the risks and benefits.

are individual cases, and the decision to supplement is based on the individual's dietary intake, medical condition, and an expected high-benefit/low-risk outcome. Supplementation in these cases is effective for the prevention and treatment of a specific vitamin or mineral deficiency. The only population-wide recommendation for multivitamin and mineral supplementation in athletes is for pregnant or lactating athletes. Pregnancy or lactation places increased nutritional demands on the body that are hard to meet from diet alone.

Protein supplements are particularly popular with high school and collegiate male athletes.

The recommended protein intake for athletes and the use of protein supplements is fully described in

Spotlight on supplements

Should I Take a Vitamin or Mineral Supplement?

Before answering the question "Should I take a vitamin or mineral supplement?" determine the answers to the questions listed below. Part of the decision-making process requires the gathering of information, and an important source of information is a dietary analysis of current intake.

1. Do I consume an adequate amount of energy (kcal)?

2. Do I consume an adequate amount of vitamins and minerals?

3. If intake is inadequate, could the missing vitamins or minerals be easily obtained from food?

4. Could a vitamin or mineral supplement be harmful to my health (discuss this with your physician)?

If a decision is made to supplement, more questions should be asked. These include:

1. How much of each vitamin or mineral is in the supplement? In other words, what is the dose?

2. When added to the amount consumed from food, does the total intake exceed the Tolerable Upper Intake Level (UL)?

3. When will I re-evaluate my decision (for example, at the time of my annual physical)?

Chapter 5. In general, athletes engaged in training need between 1.0 and 2.0 g protein/kg body weight/day. Protein supplements may contain whey, casein, and soy proteins, some of the same proteins that are found in milk, meat, fish, poultry, eggs, and soy. In the presence of resistance training and adequate kilocalories, proteins from either food or supplement sources are part of a plan to increase muscle size and strength. Protein supplements are not more or less effective than food proteins but they may be convenient and, at times, preferred by the athlete. Although amino acids and protein supplements are under the dietary supplement umbrella, athletes would be wise to consider them in the context of their overall dietary protein intake (see Spotlight on supplements: Should I Take a Protein Supplement?)

Athletes typically consume herbals and botanicals to prevent or recover from illness or injury.

Botanicals are compounds that have been extracted from foods and then concentrated in pills or tablets. Popular botanical supplements include extracts of garlic and soy, which are used for health-related but not performance-related purposes. Herbs are plants with nonwoody stems, and herbals are the compounds, known as active ingredients, that are found in those plants. Some herbal supplements have active ingredients that are not found in food, and their effects are more druglike than foodlike. However, dietary supplements are regulated as foods, not drugs. A monumental difference between food and drug regulation is that foods are generally assumed to be safe until proven

otherwise, whereas drugs may not be sold until the Food and Drug Administration grants approval based on scientific data of safety and effectiveness.

Herbals have been used as alternative and complementary medications for thousands of years, and some have excellent safety profiles. For some, safety is very dependent on dose. This means that athletes and other consumers must be diligent in determining which dose

The Internet café

Where do I find reliable information about dietary supplements?

There are numerous government-sponsored websites that have information on dietary supplements as well as some commercial sites that require a subscription. The two "watch" sites listed below are both dedicated to helping consumers make informed decisions but neither is government sponsored.

National Institutes of Health, Office of Dietary Supplements at **http://ods.od.nih.gov/**

National Center for Complementary and Alternative Medicine, Dietary and Herbal Supplements at **http://nccam.nih.gov/health/supplements/**

MedlinePlus, Dietary Supplements at **http://www.nlm.nih.gov/medlineplus/dietarysupplements.html**

Supplement Watch.com at **http://www.supplementwatch.com**

Natural Medicine Comprehensive Database (paid subscription needed) at **http://www.naturaldatabase.com**

Quack Watch (nonprofit) at **http://quackwatch.org**

Spotlight on supplements

Should I Take a Protein Supplement?

Before answering the question "Should I take a protein supplement?" determine the answers to the questions listed below. Part of the decision-making process requires the gathering of information, and an important source of information is a dietary analysis of current intake.

1. Do I consume an adequate amount of energy (kcal)?
2. How much protein do I need to consume daily?
3. How much protein do I currently obtain from food?
4. If there is a discrepancy between intake and amount needed, could I obtain the additional proteins needed from food?
5. Could a protein supplement be harmful to my health?

If a decision is made to supplement, more questions should be asked. These include:

1. What is the protein content of the supplement?
2. Is one type of protein supplement preferred (for example, whey, casein, soy)?
3. Does the protein supplement contain other nutrients or ingredients (for example, vitamins, minerals, artificial sweeteners, lactose)?
4. When will I re-evaluate my decision (for example, at the end of the competitive season)?

Spotlight on supplements

ESPN—Every Supplement Produces News—How Professionals Can Keep Up

Dietary supplements do not need premarket approval; thus once a supplement is developed and ready for sale, the supplement company can launch an aggressive marketing campaign. Advertising can create a "buzz" about a dietary supplement that was previously unknown. Going to the manufacturer's website will yield some information about the active ingredients; however, in many cases they are listed as a "proprietary blend" and it is impossible for the professional to know and evaluate the compounds. In such cases, the athlete should be cautioned that the active ingredients in the supplement are not known and there is no way to judge safety and effectiveness. For those athletes who are subject to testing for banned substances, it is imperative that they understand that the supplement could contain an active ingredient that is a banned substance and will subject them to penalty.

If a new supplement lists the active ingredient(s), the professional can find information quickly through a variety of ways. The National Institutes of Health (NIH) provides information through its Office of Dietary Supplements and its National Center for Complementary and Alternative Medicine. MedlinePlus, a service of the NIH and the National Library of Medicine, has extensive information about dietary supplements including a section titled "Latest News." Some commercial sites also conduct scientific reviews and rate products. Website addresses for these resources are listed as part of The Internet café: Where do I find reliable information about dietary supplements?

Professional journals frequently publish comprehensive reviews about active ingredients found in popular dietary supplements, but such articles take time to research and to write. These review articles are excellent ways to learn about the body of literature that exists. The rapidly changing dietary supplement marketplace requires that the professional be constantly reading new material to keep abreast of scientific research evaluating the safety and effectiveness of sports-related supplements.

will be too little, too much, or just right if they choose to use these products. The same basic questions apply to any herbal dietary supplement: (1) Is it legal? (2) Is it ethical? (3) Is it pure? (4) Is it safe? (5) Is it effective? If used as an alternative medicine, then two more questions must be answered: (1) Is it safer, equally safe, or not as safe as the medicine it is replacing? (2) Is it more effective, equally effective, or not as effective as the alternative?

Some of the most popular herbal supplements in the United States are echinacea, St. John's wort, ginkgo biloba, saw palmetto, ginseng, goldenseal, aloe,

Keeping it in perspective

Where Supplements Fit into the Athlete's Training and Nutrition Plan

To achieve the highest level of success in sports, athletes must be genetically endowed and they must train optimally to meet their genetic potential. Training is fundamental to improving athletic performance, and proper nutrition supports the demands of training. In particular, sufficient carbohydrate and fluid intake are essential for successful training and performance. In the postexercise recovery period, the intake of carbohydrates, fluid, electrolytes, and small amounts of proteins will help restore the body and ready it for future training or performance. Nutrition is fundamental to training, which is fundamental to successful performance.

A few dietary supplements can augment, but not replace, proper training and good nutrition. For example, creatine and caffeine may have a positive effect on training, and in some cases, performance. The effect of these supplements is comparatively small when considered alongside training and proper diet; however, small effects may be beneficial in some well-trained athletes. To improve performance, athletes would be wise to think "food first, supplements second" and to be an open-minded skeptic.

kava kava, and valerian. Herbals used as alternative medicines are beyond the scope of this book, but because herbals are classified as supplements and not medications, it is important for athletes to inform their physicians and pharmacists about the use of herbals, especially to discuss potential interactions with prescription medicines. Discussing herbal supplements with athletes is typically different from discussing vitamin, mineral, or protein supplements because many herbals are compounds that are not normally found in food or provided by the diet.

Key points

- Dietary supplementation decisions are a part of the athlete's comprehensive nutrition and training plan.

- Only a handful of dietary supplements have been shown to be safe and effective to improve training or performance.

- For those supplements found to be effective, the impact on training and performance is likely to be small, although small effects in well-trained athletes are important to consider.

- Minimal regulation of dietary supplements in the United States means that athletes must be careful when choosing supplements.

How might dietary supplementation fit into the athlete's comprehensive training and nutrition plan?

10.4 A Comprehensive Nutrition Plan to Support Training and Performance

Periodization is a training concept that involves changing the intensity, volume, and specificity of training to achieve specific goals that will enhance performance. This same periodization concept is used in sports nutrition to support training and optimize performance. Nutrition periodization is the development of a nutrition plan that parallels the demands of each training cycle (Seebohar, 2011). For example, as training intensity and volume increase or decrease, energy intake must similarly increase or decrease if body composition is to be maintained. When an endurance athlete tapers training prior to competition, carbohydrate intake is increased to help maximize muscle glycogen stores. If the athlete's goal is to increase muscle mass, then a sufficient energy (kcal) and protein intake is needed in addition to the resistance training. Weight

loss is a particular challenge because too great of a reduction in caloric intake may undermine training and performance.

The Spotlight on a real athlete feature describes the training and competition schedule and body composition goals of Annika, a female collegiate rower in an eight-person boat. The performance goal is to attain the fastest time in each 2,000-meter race as well as to try and break the 7-minute mark. To meet these goals, she has a comprehensive 1-year nutrition plan. Her energy, carbohydrate, protein, and fat intakes vary across the across the training and competition periods. Sports dietitians, particularly those who are Board Certified as a Specialist in Sports Dietetics (CSSD), can help an athlete develop an individualized and detailed nutrition plan.

Helping an athlete create a nutrition plan requires an understanding of the athlete's sport and position played.

A nutrition periodization plan requires an understanding of each athlete's sport and the position played. Sports are often broadly categorized, for example, swimming, running, and cycling, but these sports vary tremendously based on the distance covered. Distance athletes in these sports, such as a 25 km (15.5-mile) open-water swimmer, face vastly different physiological and nutritional demands when compared to sprinters, such as a 50-meter swimmer. Not surprisingly, sports nutrition recommendations for distance and sprint athletes are very different. Additionally, most sports nutrition guidelines are developed for well-trained athletes. Recreational athletes are often distinguished based on their level of training. The nutritional demands for many recreational athletes may not be any greater than for those in the general population who are lightly active because they do not engage in substantial training. However, performance-focused recreational athletes, such as those who continue to compete among nonprofessional athletes, often follow recommendations made to well-trained athletes. The intensity and duration of exercise are key factors when determining nutrition needs.

Table 10.7 groups sports together based on the physiological demands of training and competition. Such a categorization is useful to convey basic nutrition guidelines to coaches and athletes, especially those who are just beginning to learn about sports nutrition. For example, although soccer, basketball, and ice hockey are different sports, nutrition recommendations are similar. In general, recommended daily carbohydrate intake is 6–8 g/kg, increasing up to as high

(Continued on page 394)

Spotlight on a real athlete

Annika, a Collegiate Rower

The collegiate rowing (crew) season begins in late August or September. Fall is the preparation period (general training) and lasts until December. Team practices are generally 2½ hours, six mornings per week, alternating between land and water. Land practices consist of a warm-up period followed by a cardiovascular workout using Erg (rowing) machines and a cooldown period. Also included are 1/8-mile uphill runs repeated 10 to 12 times and circuit training with weights at approximately 30 stations (for example, low-resistance, more repetitions for endurance). Practices on the water involve 1 hour of continuous rowing but at only 70 percent of the maximum stroke rate with a focus on technique. In addition to morning team practice, individual secondary workouts are held three times per week and include 1 hour of cardiovascular work and 30 minutes of weight lifting. During the secondary workouts, heavier weights are used but with fewer repetitions to build strength.

Beginning in January, practices are more frequently held on land and generally last 3 hours, five to six mornings per week. The focus is on rowing power and technique rather than on speed. Secondary workouts put more emphasis on weight lifting to build muscle mass and increase strength. The opening of the crew season is traditionally marked by "double days," usually held during spring break. Double days consist of training from 5 to 9 a.m. and from 2 to 5:30 p.m.

Spring is racing season. On-water training takes place on four consecutive days (Monday through Thursday) and the focus is on sprints, the middle 1,000 meters of the race, and starting technique. The intensity of much of the training session is high, but of short duration. Secondary workouts are required 3 days per week, but the goal is to *maintain* the muscle mass gained in the preseason, not to *build* muscle mass.

Races are held nearly every weekend so practices are Monday through Thursday, and Friday is a rest or travel day. In meets with many teams, heats are run in the morning and the finals are about 5 hours later. Some meets may take place over 2 days. Although rowers train for hours on a near daily basis, the typical race is only 2,000 meters and takes women 8 minutes or less to complete. The race starts with an extremely high stroke rate (38 strokes per minute) and then settles into race pace (32 strokes per minute). The last 500 meters is a sprint, and rowers try to increase the stroke rate from 32 to 38 strokes per minute.

Racing season lasts until May or early June, and many rowers de-train as soon as the season is over. Rowers are typically physically and mentally fatigued after nearly 9 months of continuous training and need an off-season. Most do not engage in exercise or training, and many experience body composition changes including loss of muscle mass and small increases in body fat.

The figure illustrates a nutrition plan that reflects a female collegiate rower's 1-year training schedule. Annika weighs 156 lb (71 kg) and has 15 percent body fat during the competitive season, which she considers optimal for performance. At the end of the active recovery period (off-season), her scale weight is 159 lb (72.3 kg) and she has gained 8.5 lb (~4 kg) of body fat and lost approximately 5.5 lb (2.5 kg) of muscle mass. This change in body weight and composition is a result of planned inactivity during the active recovery period and a failure to reduce caloric intake in proportion to the reduction in energy expenditure after the end of the competitive season.

Annika's goals in the 7-month preparation period (preseason) include returning to her competitive body weight and composition, which means she must lose some body fat and gain some muscle mass. The sports dietitian whom she consults with suggests that Annika work first to reduce body fat and try to protect the amount of muscle mass she has at the beginning of the season. They discuss Annika's dietary intake in the off-season. Annika realizes that she continued to eat a lot of food even though she was not training and that she overate fatty foods, especially ice cream. The sports dietitian suggests that Annika consume about the same amount of food (37 kcal/kg) as she did in the summer and rely on the increased energy expenditure resulting from rigorous early-season training for a slow weight loss. The sports dietitian also suggests that she alter the composition of her diet and increase the amount of carbohydrates to be able to resynthesize daily the muscle glycogen depleted through endurance training. Annika wants to follow a lower-fat, higher-protein diet, a dietary pattern that she had used in the past to successfully lose body fat.

Following this plan, Annika lost 6 lb (2.7 kg) of body fat by the end of October and weighed 153 lb (69.5 kg). She wanted to lose only a couple more pounds over the next 2 months, so the sports dietitian suggested that she increase her energy intake slightly (from 37 to 39 kcal/kg) to slow the weight loss. When Annika went home for winter break, she weighed 150 lb (~68 kg).

Annika's goals changed when she returned to campus in January. Her new goal was to increase 5 pounds of muscle mass. An assessment of her body composition revealed that she weighed 150 lb (~68 kg), of which approximately 126 lb (~57 kg) were fat-free mass (84 percent) and 24.5 lb (~11 kg) (16 percent) were body fat. Another meeting with the sports dietitian resulted in a different dietary plan for the second part of the preparation period. Since the hours of training per

One-Year Training and Nutrition Plan for a Female Collegiate Rower

		Preparation						Competition		Transition		
		General training				Specific training racing		Racing season		Active recovery ("off-season")		
		Emphasis on endurance/ cardiovascular			Rest	Emphasis on strength/power, increased muscle mass		Weekly competitions		Rest		
		19.5 hrs training/wk			Rest	22.5 hrs training/wk*		12 hrs training/wk + one to three races/wk		No training		
	Sept	Oct	Nov	Dec	Jan	Feb	Mar	Apr	May	Jun	Jul	Aug
Weight	155 lb	153 lb	150 lb	150 lb	152 lb	154 lb	156 lb	156 lb	156 lb	155 lb	157 lb	159 lb
BF/FFM (%)	18/82	17/83	16/84	16/84	16/84	16/84	16/84	15/85	15/85	17/8318	5/81.5	20/80
BF (lb)	28	26	24.5	24.5	24.5	24.5	25	23.5	23.5	26	28	32
FFM (lb)	127	127	126	126	128	130	131	132.5	132.5	129	128	127
Energy	37 kcal/kg (~2,600 kcal)		39 kcal/kg (~2,650 kcal)		44 kcal/kg (~3,080 kcal)			40 kcal/kg (~2,840 kcal)		35–37 kcal/kg** (~2,485 to 2,625 kcal)		
CHO (g/kg)	7		7		8			7		5		
Pro (g/kg)	1.2		1.4		1.6			1.5		0.8		
Fat (g/kg)	0.5		0.6		0.6			0.7		1.4		

Nutrition Periodization for a Female Collegiate Rower

Legend: wk = week; lb = pound; kcal = kilocalorie; kcal/kg = kilocalories per kilogram body weight; g/kg = grams per kilogram body weight; BF = body fat; FFM = fat-free mass; CHO = carbohydrate; Pro = protein

*Double days (1 week prior to racing season).

**Athlete has not reduced caloric intake in proportion to reduced energy expenditure, is choosing to consume a higher-fat diet, and is gaining body fat and losing fat-free mass.

week are increased and Annika wants to gain muscle mass, the sports dietitian suggests that she increase her energy intake to 44 kcal/kg. Annika decides to increase her caloric intake by slightly increasing the carbohydrate and protein content of her diet rather than increasing her fat intake. Even though some of her teammates eat a lot of fatty foods during this high-energy-expenditure period, Annika prefers a lower-fat diet. By the end of the preparation period, a physically and mentally demanding 3 months, Annika had gained about 5 lb (~2.2 kg) of fat-free mass and weighed 156 lb (71 kg) (~16 percent body fat).

The beginning of racing season is marked by double days, 1 week of intense training twice a day. Annika knows from previous seasons that she will lose a couple of pounds of weight during this week because her energy expenditure is so high and she does not have much appetite after two-a-day workouts. By the time the competitive season begins, 7 months after returning to training, Annika has met all her weight and body composition goals and is 156 lb (71 kg) and 15 percent body fat. Her in-season energy intake (40 kcal/kg) needs to be lower than the previous 3 months because of a lower training volume. However, Annika needs to consume sufficient carbohydrates so muscle glycogen is restored after each training session and competition. To prevent the gain of body fat during the next active recovery period (off-season), Annika will need to reduce energy intake to match her energy expenditure.

Table 10.7 Summary of Training Demands, Body Composition Goals, and General Nutrition Guidelines for Various Sports

Category	Description	Predominant energy system	Typical sports	Body composition	General nutrition guidelines (daily)
Very high intensity, very short duration	Maximal effort of less than 30 seconds	Creatine phosphate Anaerobic glycolysis	Track and field: sprints (50–200 m); jumps (hurdles, high jump, long jump, triple jump, pole vault); throws (shot, javelin, discus, hammer); swimming sprints (50 m); cycling sprints (200 m); weight lifting; power lifting	High muscularity; low body fat except for throwers and power lifters	CHO: 5–8 g/kg Pro: 1.2–1.7 g/kg Fat: remainder of kcal to meet energy needs Fluid: losses may be high if training in high-temperature, high-humidity environments and need to be replaced during and after training
High intensity, short duration (continuous)	Exercise lasting between 30 sec and 30 minutes	Anaerobic glycolysis Oxidative phosphorylation	Running (200–1,500 m and elite 10 km); swimming (100–1,500 m); short-distance cycling (individual and team pursuit); rowing (crew); figure skating; speed skating (500–5,000 m); downhill mountain biking	High power-to-weight ratio; relatively high muscularity and relatively low body fat	CHO: 5–8 g/kg Pro: 1.2–1.7 g/kg, reflecting need to maintain or increase muscle mass Fat: remainder of kcal to meet energy needs Fluid: (see above)
High intensity, short duration (intermittent)	Each exercise bout lasts for seconds or minutes and is repeated with some rest periods	Anaerobic glycolysis Oxidative phosphorylation	Gymnastics; wrestling; boxing; martial arts (various); bodybuilding	High power-to-weight ratio; high muscularity and low body fat. Weight may have to be certified; appearance is scored in bodybuilding and may influence the scoring in women's gymnastics	CHO: 5–8 g/kg Pro: 1.2–1.7 g/kg, reflecting need to maintain or increase muscle mass Fat: remainder of kcal to meet energy needs Fluid: (see above); athlete may have restricted fluid in an effort to "make weight" and should replenish fluid as soon as possible after weigh-in
Intermittent, high intensity ("stop and go")	Exercise is a combination of sprints, moderate-intensity exercise and low-intensity activity or rest. The length of the event is typically 1 or more hours	Anaerobic glycolysis Oxidative phosphorylation Creatine phosphate	Team sports: soccer; basketball; ice and field hockey; rugby; (American) football; lacrosse; volleyball; individual sports (may also involve one or three other players); tennis; handball; racquetball; squash	Varies according to sport, position, the amount of area that must be covered, the frequency that players are substituted, and the relative need for speed, strength, and power	CHO: 6–8 g/kg; 8–10 g/kg during heavy training and competition Pro: 1.4–1.7 g/kg, reflecting need to maintain or increase muscle mass Fat: remainder of kcal to meet energy needs Fluid: losses may be high and difficult to replace during training and competition Sodium: losses may be high when exercise is greater than 2 hours and needs to be replaced during or after exercise

Table 10.7 Summary of Training Demands, Body Composition Goals, and General Nutrition Guidelines for Various Sports (Continued)

Sport	Training Demands	Energy System	Examples	Body Composition Goals	Nutrition Guidelines
Endurance and ultraendurance	Continuous exercise lasting more than 1 hour; typically continuing for many hours	Oxidative phosphorylation	Distance running (cross country, marathon); distance swimming (25 km open water); distance cycling (100 km or more); cross country skiing. Ultrasports involve longer than usual endurance distances (triathlon, swimming the English Channel, multiday bike rides)	Excessive body fat is not desirable because it represents nonforce-producing weight that reduces exercise efficiency. A low percentage body fat that can be attained or maintained only with restriction of caloric intake is also not desirable.	Can vary widely depending on intensity and duration of exercise during training and competition. CHO: 5–7 g/kg, when training is reduced, increasing up to 12–19 g/kg with heavy training or ultradistance competitions. Pro: 1.2–2.0 g/kg. Fat: 0.8–2.0 g/kg to meet energy needs. Fluid: losses typically high and difficult to replace during training and competition. Sodium: losses may be high and need to be replaced during or after exercise
Low intensity (intermittent)	Some maximal effort but majority of exercise is low intensity. Event lasts for several hours	Oxidative phosphorylation	Golf; baseball	Body composition plays a lesser role in these sports than in many other sports	Follow general principles outlined in the Dietary Guidelines for Americans. Fluid: losses may be high in high-temperature, high-humidity environments and need to be replaced during and after exercise

Legend: m = meter; g = gram; kg = kilogram; km = kilometer; CHO = carbohydrate; Pro = protein; g/kg = grams per kilogram body weight

Compiled from Chapters 21–26 of *Sports Nutrition: A Practice Manual for Professionals* (2006), 4th ed., M. Dunford (Ed.), Chicago: American Dietetic Association.

(Continued from page 389)

as 10 g/kg during heavy training and competition to replenish severely depleted glycogen stores. Protein recommendations are typically 1.4–1.7 g/kg with intakes at the higher end of the range recommended during training periods when the athlete's goal is an increase in muscle mass. These recommendations are more similar to those made to distance runners than they are to baseball players or golfers.

Although some sports may have similarities, it is also vital to understand the unique demands of each sport. Maintaining proper hydration is important for both soccer and ice hockey players, but soccer players have less access to fluids during competition than hockey players. Ice hocky has frequent player substitutions and they can sip fluids when they are not on the ice. Hockey players, however, are fully clothed so they may sweat heavily but not evaporate the sweat, the primary mechanism for lowering body

AP Photo/Armando Franca

Getty Images for Athletics Australia

Although both are members of a track team, sprinters and throwers have different body composition goals.

Focus on research

How Are Nutrition Recommendations for Athletes Determined?

There is a glut of nutrition information available to consumers and practitioners alike through research journals, television, books, magazines, the Internet, and other sources. How are consumers and practitioners supposed to make sense of all this information in order to pursue appropriate nutritional strategies for themselves or their clients, particularly in the specialized area of sports and exercise nutrition? A wealth of knowledge is available in scientific journals that publish the results of research studies on nutrition, but it is difficult and time consuming to keep up with all of the currently published research, and many people do not have the background necessary to properly analyze and interpret research results and put them in context with previously published research. Fortunately, a number of professional organizations have taken on the task of analyzing the current state of knowledge in sports and exercise nutrition with the goal of providing clear and concise recommendations.

The following published study represents a collaborative effort of three professional organizations with an interest in sports and exercise nutrition to make general and specific recommendations based upon an unbiased evaluation of existing scientific evidence: Rodriguez, N. R., DiMarco, N. M., Langley, S., American Dietetic Association, Dietitians of Canada, & American College of Sports Medicine. (2009). Position of the American Dietetic Association, Dietitians of Canada, and the American College of Sports Medicine: Nutrition and athletic performance. *Journal of the American Dietetic Association, 109*(3), 509–527.

What did the researchers do? Instead of conducting an individual research project as we have profiled in Focus on research features in other chapters in this text, these researchers reviewed the results of many research studies to determine the best state of our current knowledge related to a number of important questions and situations related to nutrition and athletic performance. Based upon the results of many research studies, these three researchers, one representing each of the sponsoring professional organizations (American Dietetic Association, American College of Sports Medicine, and Dietitians of Canada), proposed specific recommendations, and also applied evidence-based analysis to determine the strength of the scientific evidence backing up the recommendations (see Chapter 1 for an explanation and description of levels of evidence). Before publication, the recommendations were reviewed by other researchers and were approved by their respective professional organizations.

The researchers narrowed their review of research studies to those that provided evidence to answer practical questions related to:

- The relationship between energy balance and imbalance
- The relationship between weight management and athletic performance
- Meal timing, caloric intake, and carbohydrate, fat, and protein intake to support optimal athletic performance during training, immediately before, during, and after exercise/competition.

temperature. Each of these athletes needs an individualized fluid replenishment plan.

Body composition goals vary between sports and among athletes within the same sport. For example, very high intensity, very short duration track and field athletes benefit from high muscularity, but sprinters also benefit from a relatively low percentage of body fat whereas throwers (for example, discus) benefit from having a percentage of body fat that is higher than most other athletes. Within a team sport, players may be distinguished by position and have different body composition goals, as is the case of interior linemen and linebackers on a (American) football team. Each athlete needs an individualized training and nutrition plan, and sports dietitians are trained to create such plans.

Key points

- Nutrition supports training; thus every athlete needs a comprehensive nutrition plan.

- Any nutrition plan must be individualized and based on the sport and position played.

Arnold Schwarzenegger, former Mr. Olympia (1970–75) and governor of California said, "Bodybuilding is much like any other sport. To be successful you must dedicate yourself 100% to your training, diet, and mental approach." Do you agree or disagree?

In order to make recommendations most relevant to athletes, the researchers considered only certain research studies and eliminated those that used subjects who were children or older adults, were not athletes, or had chronic illnesses or diseases not commonly found in athletes. They also eliminated those studies that involved research designs that were not related to sports and those that had a high subject dropout rate. Published research studies were then reviewed, conclusion statements written, and the strength of the scientific evidence was determined based upon quality of the studies, consistency across studies, quantity (number of similar studies), and generalizability.

What did the researchers find? The researchers found sufficient scientific evidence to make 13 general recommendations for active adults and competitive athletes. It is beyond the scope of this Focus on research to cover all of the specific recommendations in detail, but they can be found incorporated elsewhere in this text and read in their entirety in the Position Stand, which can be accessed online (see next column). Specific conclusion statements were made for energy and body composition, training diets, pre-exercise meals, and nutrition during and after exercise. The evidence-based analysis for each of these conclusion statements was Evidence Grade III, Limited for energy and body composition and Evidence Grade II, Fair for all others.

What is the significance of the research? This Position Stand provides nutrition recommendations specific to athletes based upon research and the best of our current knowledge of the field. These recommendations can be followed with a high degree of confidence as they were developed by respected scientists after a careful review of the existing scientific literature and were then reviewed by a panel of experts before publication. It is important to note that these are recommendations and not rules—these guidelines need to be adapted to each specific athlete's individual needs and circumstances, which is an important responsibility of a sports dietitian. Finally, the evidence-based analysis revealed that although a great deal is known about sports nutrition, significant future research needs to be conducted in order to strengthen the scientific basis for these and future recommendations.

The specific recommendations can be accessed in the Position Stand online at:

ADA website: **http://www.eatright.org/About/Content.aspx?id=8365**

ACSM website: **http://journals.lww.com/acsm-msse/Fulltext/2009/03000/Nutrition_and_Athletic_Performance.27.aspx**

Dietitians of Canada website: **http://www.dietitians.ca/Dietitians-View/Nutrition-and-Athletic-Performance.aspx**

Post-Test   Reassessing Knowledge of Diet Planning for Athletes

Now that you have more knowledge about diet planning, read the following statements and decide if each is true or false. The answers can be found in Appendix O.

1. When planning dietary intake, athletes should first consider how much dietary fat would be needed.

2. The key to eating nutritiously without consuming excess kilocalories is to choose foods that have a high nutrient density.

3. There is no room in the athlete's diet for fast foods.

4. Processed foods tend to be more nutrient dense than unprocessed foods.

5. Most dietary supplements are effective for improving training and performance.

Summary

The need for energy forms the basis of dietary planning. Energy needs vary according to the training cycle, and it may be necessary for athletes to moderately restrict energy intake at certain times to reduce body fat. Sufficient carbohydrate and protein intake is necessary to support training, recovery, and performance, so these nutrients are considered first. The intake of the other energy-containing nutrients, fats, and alcohol is determined after carbohydrate and protein needs are established. Eating nutrient-dense foods is the key to consuming an adequate amount of nutrients, particularly vitamins and minerals.

Translating nutrient recommendations into food choices is not easy, and dietary patterns based on food groups can help athletes to plan daily dietary intake that focuses on whole foods. Although a preformulated pattern based on a given caloric intake, such as those found on the MyPyramid website, can be helpful, such patterns are only guidelines and must be individualized for each athlete. In addition to the nutrient content of the diet, the athlete must consider the timing of intake.

The pre-exercise meal is typically relatively high in carbohydrates, moderate in proteins, relatively low in fats, and provides adequate fluid and energy. Food and fluid consumed during exercise can help delay fatigue and prevent hypohydration but need to be in a volume and form that minimizes gastrointestinal distress. Postexercise food and fluid intake supports recovery, and recommendations are made for the amount and timing of carbohydrates, proteins, fluid, and sodium intake after exercise ends.

Each athlete will need a comprehensive nutrition plan to support training, achieve desirable body composition, and optimize performance. This plan outlines nutritional needs over months, weeks, and days of training and competition. Without a well-thought-out plan, the athlete may fall short of the nutrition support needed to meet training and performance goals and to recover from exercise.

The use of dietary supplements is an ever-changing and ever-challenging aspect of sports nutrition. Five critical issues are legality, ethics, purity, safety, and effectiveness. Only a handful of supplements have been shown to be safe and effective for improving training or performance. A few supplements, such as creatine, caffeine, and protein, may be part of the athlete's total nutrition plan. Training, proper carbohydrate and fluid intakes, and electrolyte replacement are fundamental to successful performance, and no dietary supplement can match the safety and effectiveness of these practices. Education is key as athletes consider dietary supplement use in the broad context of their training, performance, and health goals.

Self-Test

Multiple-Choice

The answers can be found in Appendix O.

1. According to the textbook, the word *diet* should be defined as:
   a. restriction of food and beverages in an effort to lose weight.
   b. increase or decrease in caloric intake to gain or lose weight.
   c. a pattern of eating.
   d. nutrition periodization.

2. Nutrient density is based on the relationship between:
   a. protein and vitamins/minerals.
   b. essential and nonessential nutrients.
   c. energy and nutrients.
   d. fat and sugar.

3. Why does too large of a volume of food prior to exercise result in gastrointestinal distress?
   a. Hormones related to nausea and vomiting are released.
   b. The autonomic nervous system is negatively affected.
   c. Heart rate is reduced when the stomach is full.
   d. Blood flow to the GI tract is decreased with the onset of exercise.

4. Other than training, which of the following is likely to have the greatest impact on improved endurance performance?
   a. anabolic steroid use
   b. carbohydrate loading
   c. caffeine use
   d. blood doping

5. Athletes in these sports typically have similar nutrient needs:
   a. football, basketball, baseball.
   b. soccer, ice hockey, tennis.
   c. gymnastics, handball, javelin.
   d. cross country skiing, ski jumping, figure skating.

Short Answer

1. Explain two ways to interpret the phrase "I'm on a diet."

2. When planning a diet, why is the athlete's energy goal established first?

3. Explain why nutrient density is important. Give examples of foods that are nutrient dense and energy (kilocalorie) dense.

4. What are the advantages and disadvantages to using MyPyramid Intake Patterns with athletes?

5. What are the goals of food and fluid intake prior to exercise? During exercise? After exercise? Give example of foods or beverages that would meet those goals.

6. Is caffeine safe and effective? Explain your answer.

7. What are the negative and positive health and performance effects of alcohol consumption?

8. Describe the information found on a dietary supplement label. How does the information differ from that found on a food label?

9. What kinds of information should the athlete consider before deciding whether to consume supplemental vitamins, minerals, or amino acids?

10. Are ergogenic dietary supplements effective? Explain your answer.

Critical Thinking

1. Give examples of how energy or nutrient needs might change across a 1-year training cycle.

2. Why might it be said that the use of multiple dietary supplements essentially places the athlete in an unsupervised experimental study of one?

References

Armstrong, L. E., Casa, D. J., Maresh, C. M., & Ganio, M. S. (2007). Caffeine, fluid-electrolyte balance, temperature regulation, and exercise-heat tolerance. *Exercise and Sport Sciences Reviews*, 35(3), 135–140.

Barrett, S. (2001). The mirage of multilevel marketing. Retrieved January 10, 2011 from http://www.quackwatch.org/01QuackeryRelatedTopics/mlm.html

Burke, L. M. (2008). Caffeine and sports performance. *Applied Physiology, Nutrition, and Metabolism, 33*, 1319–1334.

Burke, L. M., Collier, G. R., Broad, E. M., Davis, P. G., Martin, D. T., Sanigorski, A. J., et al. (2003). Effect of alcohol intake on muscle glycogen storage after prolonged exercise. *Journal of Applied Physiology*, 95(3), 983–990.

Center for Science in the Public Interest. (2008 with periodic updates). How much is that caffeine in the window? Retrieved December 14, 2010, from http://www.cspinet.org/nah/02_08/caffeine.pdf

Chryssanthopoulos, C., & Williams, C. (1997). Pre-exercise carbohydrate meal and endurance running capacity when carbohydrates are ingested during exercise. *International Journal of Sports Medicine*, 18(7), 543–548.

Davis, J. K., & Green, J. M. (2009). Caffeine and anaerobic performance: Ergogenic value and mechanisms of action. *Sports Medicine*, 39(10), 813–832.

Doherty, M., & Smith, P. M. (2005). Effects of caffeine ingestion on rating of perceived exertion during and after exercise: A meta-analysis. *Scandinavian Journal of Medicine & Science in Sports*, 15(2), 69–78.

Duchan, E., Patel, N. D., & Feucht, C. (2010). Energy drinks: A review of use and safety for athletes. *The Physician and Sportsmedicine*, 38(2), 171–179.

Dunford, M. (Ed.). (2006). *Sports nutrition: A practice manual for professionals* (4th ed.). Chicago: American Dietetic Association.

Dunford, M., & Coleman, E. (in press). Ergogenic aids, dietary supplements and exercise. In C. Rosenbloom (Ed.), *Sports nutrition: A practice manual for professionals* (5th ed.). Chicago: American Dietetic Association.

Evans, S. M., & Griffiths, R. R. (1999). Caffeine withdrawal: A parametric analysis of caffeine dosing conditions. *Pharmacology and Experimental Therapeutics*, 289(1), 285–294.

Ganio, M. S., Klau, J. F., Casa, D. J., Armstrong, L. E., & Maresh, C. M. (2009). Effect of caffeine on sport-specific endurance performance: a systematic review. *Journal of Strength and Conditioning Research*, 23(1), 315–324.

Greenwood, M., Farris, J., Kreider, R., Greenwood, L., & Byars, A. (2000). Creatine supplementation patterns and perceived effects in select division I collegiate athletes. *Clinical Journal of Sport Medicine*, 10(3), 191–194.

Ivy, J. L. (1998). Glycogen resynthesis after exercise: Effect of carbohydrate intake. *International Journal of Sports Medicine*, 19(Suppl. 2), S142–S145.

James, J. E. (2004). Critical review of dietary caffeine and blood pressure: A relationship that should be taken more seriously. *Psychosomatic Medicine*, 66(1), 63–71.

Jonnalagadda, S. S., Rosenbloom, C. A., & Skinner, R. (2001). Dietary practices, attitudes, and physiological status of collegiate freshman football players. *Journal of Strength and Conditioning Research*, 15(4), 507–513.

Macedonio, M. A., & Dunford, M. (2009). *The athlete's guide to making weight*: Optimal weight for optimal performance. Champaign, IL. Human Kinetics.

Maffucci, D. M., & McMurray, R. G. (2000). Towards optimizing the timing of the pre-exercise meal. *International Journal of Sport Nutrition and Exercise Metabolism*, 10(2), 103–113.

Martens, M. P., Cox, R. H., & Beck, N. C. (2003). Negative consequences of intercollegiate athlete drinking: The role of drinking motives. *Journal of Studies on Alcohol*, 64(6), 825–828.

Maughan, R. J., & Griffin, J. (2005). Caffeine ingestion and fluid balance: A review. *Journal of Human Nutrition and Dietetics*, 16, 1063–1072.

Miller, B. E., Miller, M. N., Verhegge, R., Linville, H. H., & Pumariega, A. J. (2002). Alcohol misuse among college athletes: Self-medication for psychiatric symptoms? *Journal of Drug Education*, 32(1), 41–52.

National Collegiate Athletic Association (NCAA). NCAA Bylaw 31.2.3 Banned Drugs. http://ncaa.org. Retreived January 10, 2011.

Noakes, T. M. (2002). *Lore of running* (4th ed.). Champaign, IL: Human Kinetics.

Noordzij, M., Uiterwaal, C. S., Arends, L .R., Kok, F. J., Grobbee, D. E., & Geleijnse, J. M. (2005). Blood pressure response to chronic intake of coffee and caffeine: A meta-analysis of randomized controlled trials. *Journal of Hypertension*, 23(5), 921–928.

Petróczi, A., Naughton, D. P., Mazanov, J., Holloway, A., & Bingham, J. (2007). Limited agreement exists between rationale and practice in athletes' supplement use for maintenance of health: A retrospective study. *Nutrition Journal*, 30(6), 34.

Petróczi, A., Naughton, D. P., Pearce, G., Bailey, R., Bloodworth, A., & McNamee, M. (2008). Nutritional supplement use by elite young UK athletes: Fallacies of advice regarding efficacy. *Journal of the International Society of Sports Nutrition*, 5, 22.

Rankin, J. W. (2002). Weight loss and gain in athletes. *Current Sports Medicine Reports*, 1(4), 208–213.

Rehm, J., Room, R., Graham, K., Monteiro, M., Gmel, G., & Sempos, C. T. (2003). The relationship of average volume of alcohol consumption and patterns of drinking to burden of disease: An overview. *Addiction*, 98(9), 1209–1228.

Reynolds, K., Lewis, B., Nolen, J. D., Kinney, G. L., Sathya, B., & He, J. (2003). Alcohol consumption and risk of stroke: A meta-analysis. *Journal of the American Medical Association*, 289(5), 579–588.

Rodriguez, N. R., DiMarco, N. M., Langley, S., American Dietetic Association, Dietitians of Canada, & American College of Sports Medicine. (2009). Position of the American Dietetic Association, Dietitians of Canada, and the American College of Sports Medicine: Nutrition and athletic performance. *Journal of the American Dietetic Association*, 109(3), 509–527.

Sawka, M. N., Burke, L. M., Eichner, E. R., Maughan, R. J., Montain, S. J., & Stachenfeld, N. S. (2007). American College of Sports Medicine position stand on exercise and fluid replacement. *Medicine and Science in Sports and Exercise*, 39(2), 377–390.

Schwenk, T. L., & Costley, C. D. (2002). When food becomes a drug: Nonanabolic nutritional supplement use in athletes. *American Journal of Sports Medicine*, 30(6), 907–916.

Seebohar, B. (in press). *Nutrition periodization for athletes* (2nd ed.). Boulder, CO: Bull.

Shirreffs, S. M., & Maughan, R. J. (2006). The effect of alcohol on athletic performance. *Current Sports Medicine Reports*, 5, 192–196.

Winkelmayer, W. C., Stampfer, M. J., Willett, W. C., & Curhan, G. C. (2005). Habitual caffeine intake and the risk of hypertension in women. *Journal of the American Medical Association*, 294(18), 2330–2335.

Yusko, D. A., Buckman, J. F., White, H. R., & Pandina, R. J. (2008). Alcohol, tobacco, illicit drugs, and performance enhancers: A comparison of use by college student athletes and nonathletes. *Journal of American College Health*, 57(3), 281–290.

Weight and Body Composition

Learning Plan

- **Describe** the various components that make up the body's composition.
- **Describe** how body composition and body weight are measured, how these results should be interpreted, and how each relates to performance.
- **Explain** error of measurement and compare and contrast the measurement error of each method used for estimating body composition.
- **Understand** how the relative need for size (weight), strength, and speed in a particular sport is reflected in the body composition of elite athletes in those sports.
- **Calculate** target body weight and minimum body weight.
- **Outline** the basic principles associated with gaining lean body mass and losing body fat and the most appropriate times during the yearly training cycle to change body composition or weight.
- **Define** weight cycling and explain the effects it may have on performance and health.
- **Discuss** the legality, ethics, purity, safety, and effectiveness of muscle-building and weight-loss supplements.

Both weight and body composition are related to athletic performance.

Read the following statements and decide if each is true or false.

1. Percent body fat and fat mass can be precisely measured in athletes with a number of different methods.

2. The most accurate method of measuring body fat for any athlete is underwater weighing.

3. In sports in which body weight must be moved or transported over a distance (for example, distance running), it is a performance advantage to have the lowest weight possible.

4. To increase muscle mass, most athletes need a substantial increase in their usual protein intake.

5. For athletes who want to restrict energy intake to lose body fat, the recommended time to do so is at the beginning of the preseason or during the off-season.

Body composition and body weight are related to performance, appearance, and health. Body composition, particularly the relative amount of **muscle mass**, has the potential to positively impact exercise and performance. Weight and body composition may have a substantial impact on performance in certain sports, but may play a much lesser role in others. In some sports body weight must be certified before the athlete can participate in that day's competition, and the focus, at least temporarily, is achieving a particular scale weight. Body composition and weight also influence body image, and the desire to attain a particular body image or weight can be a powerful motivator. An excessive or rapid loss of body weight can produce harmful medical consequences in otherwise healthy individuals. Excess body fat, especially fat that accumulates deep in the abdominal cavity, may influence the onset or progression of chronic diseases and long-term health. Keeping all three areas in mind—performance, appearance, and health—helps athletes maintain the proper perspective when setting body weight and composition goals.

In many sports attaining a relatively high percentage of lean mass and a relatively low percentage of body fat is an appropriate goal that has the potential to improve performance. However, it is not possible to predict the percentage of body fat associated with optimal performance, and "lowest" is not necessarily "optimal." Achieving the "most" muscle mass or the "lowest" percentage of body fat possible may not be desirable. For example, the excess weight from skeletal muscle can damage joints and ligaments. Like many other aspects of human physiology, the extremes can be dangerous, and most athletes find that their desirable body composition does not lie at the extreme ends of the body composition continuum.

11.1 Understanding Weight and Body Composition

The term *body composition* refers to all of the components that make up the body. Measurements of body fat are just *estimates*, however, and should be interpreted with caution. Individual characteristics such as genetic predisposition to fatness or leanness must be considered so that realistic goals can be set. Trying to obtain an unrealistic percentage of body fat that has been arbitrarily chosen, such as a male who wants to be 4 percent body fat, can be unproductive, ineffective, and dangerous.

Body weight and composition have an impact on performance and are critical factors in many sports. In sports in which judging of performance is subjective and influenced by appearance, such as women's figure skating, a low body weight may be beneficial. Some athletes naturally have a low body weight, but others find themselves using dangerous practices, such as voluntary starvation and dehydration, to reach what they believe is a desirable weight. In acrobatic sports, such as gymnastics and figure skating, a high **power-to-weight ratio** is desirable. In other words, it is beneficial to have a relatively high percentage of skeletal muscle to produce force, but at a minimal body weight since excess body fat contributes to total weight but not muscle power. The establishment of a safe minimum body weight is important and must take into account the athlete's current amount of muscle mass, frame size, genetic predisposition to leanness or fatness, and biologically comfortable weight range. Attaining too low of a body weight can come at a cost to both performance and health—reduction of muscle mass, loss of body water, loss of bone mineral density, and initiation of disordered eating behaviors.

Body composition and weight are important in sports with weight classes or subjective judging that is influenced by appearance.

Aside from performance-related reasons, appearance may motivate athletes to make body composition changes. Athletes who achieve a body composition that is held in high esteem by society (such as, lean and muscular, thin and **prepubescent**) receive praise and positive reinforcement. For some, self-esteem is closely tied to body image, and thus body weight or composition. For these individuals, changes in body composition, especially increases in body fat, may be a source of great concern and unwanted media attention.

To change body composition or weight the athlete must alter energy intake, energy output, or both. The athlete will need a well-thought-out exercise and diet plan that correlates with the demands of each training cycle. To increase muscle mass, an athlete must engage in strength training, consume a sufficient amount of energy (kcal), and be in positive nitrogen balance to support tissue growth. To decrease body fat, energy expenditure must be greater than energy intake. The off-season or the early preseason periods, when training volumes are lower than the precompetition period, are typically the best times for substantial losses of body fat. Many athletes want to simultaneously increase muscle mass and decrease body fat, and achieving both of these goals takes an individualized plan and some trial and error.

> **Muscle mass:** The total amount of skeletal muscle in the body. Expressed in pounds or kilograms.
>
> **Power-to-weight ratio:** An expression of the ability to produce force in a short amount of time relative to body mass.
>
> **Prepubescent:** Stage of development just before the onset of puberty.

An athlete's body composition can be important for reasons other than performance.

In the United States, anything related to the topic of weight loss receives automatic attention, in part because overweight and obesity are epidemic. Most trained athletes do not need to lose large amounts of body fat. However, small increases or decreases in body fat have the potential to affect performance and appearance, so fat loss is also a hot topic among athletes. Some athletes can gain substantial amounts of body fat in the off-season and want to reduce fat stores rapidly prior to the return to training camp. Athletes are not immune to advertisements or rumors that promise fast and easy fat loss.

For most people, reducing body fat is a relatively slow process, but the loss of body weight can be rapid when achieved primarily by water loss. Athletes who want to reduce body weight rapidly to "make weight" often use a combination of methods including diuretic use and excessive sweating in addition to fasting and increasing exercise. These techniques can result in mild to serious or fatal medical complications.

Measuring body weight and body composition can provide information that can help athletes attain their performance, appearance, and health goals. However, the usefulness of body weight and composition measures depends on their accuracy. Additionally, the results must be interpreted correctly so appropriate goals may be set and progress can be monitored. This chapter begins with a discussion of body composition and weight and some of the methods used to assess each as accurately as possible.

The human body is composed of an extraordinary variety of different types of cells and materials. Because they are so numerous, these components are often grouped into more general categories for their study and discussion. In the fields of exercise physiology and sports nutrition, body composition is often subdivided into the broad categories of **fat mass** and **fat-free mass**. Fat mass is all of the fat material in the body, and fat-free mass is composed of all other tissues in the body that are not fat, the most prominent nonfat tissue being skeletal muscle. Athletes are also interested in the ratio of fat mass to total body mass, which may be expressed as **percent body fat**. The weight of the body is also a factor, particularly in sports with weight categories.

It is important to understand the concepts of body mass, weight, and composition.

Of primary importance are the specific components of body tissues—total **body mass** (weight), body fat (fat mass), muscle mass, bone mass and density, and fluids. The term *mass* is often used interchangeably with *weight*, but technically they are not the same. Mass is the term that describes the amount of matter or material that makes up an object, whereas weight is an expression of the force that is exerted by that object due to gravity. To illustrate the difference, consider the mass and weight of astronauts during a space mission. They have the same body mass in space as they do on Earth, but they "weigh" much less in space due to the greatly reduced force of gravity. Because the difference on Earth is minute, the terms *mass* and *weight* are used interchangeably in this text. Other terms commonly used are defined in Spotlight on…Understanding Body Composition Terminology.

Spotlight on…

Understanding Body Composition Terminology

Body mass: Total amount of matter or material of the body; commonly used interchangeably with weight. Expressed in pounds (lb) or kilograms (kg).

Fat mass (FM): Total amount of fat in the body. Expressed in pounds or kilograms.

Percent body fat (% BF): The amount of fat relative to body mass. Expressed as a percentage of body weight.

Fat-free mass (FFM): The total amount of all tissues in the body exclusive of fat including muscle, bone, fluids, and organs. Expressed in pound or kilograms.

Essential fat: The minimum amount of body fat necessary for proper physiological functioning; estimated to be approximately 3 percent of body weight for males and 12 percent for females.

Lean body mass (LBM): Total amount of all physiologically necessary tissue in the body; fat-free mass and essential fat (FFM + essential fat). Often used incorrectly to mean the same as FFM. Often used generically when referring specifically to muscle mass (for example, "strength training results in an increase in LBM"). Expressed in pounds or kilograms.

Muscle mass: Total amount of skeletal muscle in the body. Expressed in pounds or kilograms.

Bone mass or bone mineral content (BMC): Total amount of bone in the body. Expressed in pounds or kilograms.

Bone mineral density (BMD): The amount of bone per unit area. Expressed in grams per cubic centimeter (g/cm²).

Males and females may store fat in different sites, either predominantly in the abdominal area or in the hips, thighs, or buttocks.

Both for athletes and for the general public, the most common distinction in body composition is body fat. Fat in the body is typically categorized as essential or storage fat. **Essential fat** is the minimum amount of body fat necessary for proper physiological functioning and is estimated to be approximately 3 percent of body weight for males and 12 percent of body weight for females. Of the 12 percent, approximately 9 percent is considered sex-specific fat, the fat necessary for proper hormonal and reproductive functions. When compared to male athletes in similar sports, females typically have a higher percentage of body fat than their male counterparts. For example, both male and female bodybuilders are extremely lean; however, the leanest elite female bodybuilders will have a greater percent of body fat than the leanest elite male bodybuilders simply because of the differences in gender.

Storage fat is composed of subcutaneous fat and visceral fat. Subcutaneous fat is located under the skin and is typically the largest amount of fat in the body. Visceral fat surrounds organs and is located well below the skin, for example, in the abdominal area. Males and females may store fat in different sites, displaying a gender-specific physiological preference for the pattern and location of fat storage. Male fat distribution is described as android and is characterized by fat storage predominantly in the abdominal area. Normal-weight females generally store fat in the hips, thighs, and buttocks, a pattern known as gynoid fat distribution. These typical fat distribution patterns have led to body shape being described as similar to a pear (gynoid) or an apple (android), and may have health implications (see Chapter 13).

However, gender alone cannot explain fat distribution patterns. Females do not always exhibit the typical gynoid pattern due to genetics, menopausal status, and obesity, which may result in a tendency to store excess fat in the abdominal area. Males also differ from one another due to differences in fat distribution within the abdominal region. In some males, excess fat is more readily stored in deep abdominal fat, known as visceral fat, than in subcutaneous abdominal fat. Visceral fat is more metabolically active than subcutaneous fat and is a factor in some chronic diseases, as explained in Chapter 13.

Although athletes are interested in body fat, they are also concerned about fat-free mass or the tissues in the body that are not fat. Fat-free mass includes muscle, bone, fluids, and organs. In particular, athletes focus on muscle mass. As with body fat, estimating only muscle mass is difficult, so it is more common to estimate **lean body mass (LBM)**. LBM refers to the

Fat mass (FM): Total amount of fat in the body. Expressed in pounds or kilograms.

Fat-free mass (FFM): Total amount of all tissues in the body exclusive of fat; includes muscle, bone, fluids, organs, and so forth. Expressed as pound or kilograms.

Percent body fat (% BF): Amount of fat relative to body mass. Expressed as a percent of total body weight.

Body mass: Total amount of matter or material of the body; commonly used interchangeably with weight. Expressed in pounds or kilograms.

Essential fat: Minimum amount of body fat necessary for proper physiological functioning; estimated to be approximately 3 percent of body weight for males and 12 percent for females.

Lean body mass (LBM): Total amount of all physiologically necessary tissue in the body; that is, fat-free mass and essential body fat. Expressed in pounds or kilograms.

Ectomorph

Mesomorph

Endomorph

Body shape is often categorized as ectomorphic, mesomorphic, or endomorphic.

$$BMI = \frac{Weight\ (kilograms)}{Height^2\ (meters)}$$

◀ BMI = body mass index

Figure 11.1 Body Mass Index (BMI) kg/m^2

total amount of all physiologically necessary tissue in the body and includes fat-free mass and essential body fat. In everyday language, the term *muscle mass* is interchanged with the term *lean body mass* (for example, "strength training results in an increase in LBM"), but muscle is only one component of lean body mass. Body composition discussions usually focus on body fat and muscle mass, and these two components will be the focus of this chapter. Bone density is discussed in Chapters 9, 12, and 13, and fluids are covered in Chapter 7.

Body shape can also enter into body composition discussions. Individuals may be divided into one of three categories based on somatotype or body build: endomorph, mesomorph, or ectomorph, based upon the work originally published by Sheldon in 1940. Endomorphs are characterized as being stocky with wide hips and a tendency to easily gain body fat, especially visceral fat. Ectomorphs are typically described as being slightly built with less developed muscle mass and fat stores. Many ectomorphs have difficulty gaining weight. Mesomorphs, especially males, can gain muscle mass relatively easily and typically do not have excessive amounts of body fat. Somatotypes may be useful when discussing genetic predisposition and body composition, especially with those who have set unrealistic goals and are struggling with body image problems.

Many people, including athletes, know their usual body weight. Body weight has been used in various ways, either alone or in relation to some other factor such as height. Body weight can be assessed relative to body height, a measurement known as Body Mass Index (BMI). The formula for calculating BMI is shown in Figure 11.1. Appendix N features a nomogram with precalculated BMI. BMI is a screening tool for the general population that helps individuals determine a "healthy weight" range and is not used with athletes. The BMI formula assumes that adult height is stable and that any increase in scale weight is a result of an increase in body fat. The use of BMI as a tool to screen for chronic disease risk in the general population is explained in Chapter 13.

It is inappropriate to use BMI with pregnant females (whose increase in weight is due to more muscle, blood, and fluid, as well as fat), people who have decreased in height due to osteoporosis, or trained athletes. Trained athletes typically have more skeletal muscle and less body fat than sedentary adults. To

This athlete is lean and muscular but falls into the obese category by Body Mass Index (BMI).

illustrate, a 6'3" (1.9 m) male athlete who weighs 240 lb (109 kg) would have a BMI of 30 and be classified as obese. As is clear from the photo above this athlete is not obese; rather, he is lean and his percentage of body fat is relatively low. BMI is not an appropriate disease risk-screening tool for athletes.

Although the use of weight for comparison to others or for tracking individual change over time might be useful, the major problem with the use of body weight is that it gives no information about body composition. Weight is also an imprecise measure of health. In most cases it is more important to know the absolute amount of certain tissues and their relative proportion to other tissues in the body than to focus on body weight.

Body fat is expressed as an absolute amount (fat mass in pounds or kilograms) or as a percentage of the total body mass (percent body fat). Normative values for percent body fat for adult men and women by decade are found in Appendix M.

The value in distinguishing body composition, particularly body fat, can be seen in an example of a sumo wrestler and a bodybuilder. These two athletes may have the same body weight, yet dramatically different body composition. The sumo wrestler has a much larger percentage of body fat whereas the bodybuilder has much less body fat and more skeletal muscle tissue. Similarly, two females may both weigh 110 lb (50 kg) and appear to be "thin" because they have a relatively low body weight. However, one may have a high percentage of body fat and a lack of skeletal muscle tissue compared to the other who has more developed skeletal muscle and a smaller amount of body fat.

Discussion and determination of body composition are complicated by the vast array of tissue types in the body. Various models have been proposed to simplify the estimation and analysis of body composition, and each accounts for the major constituents of the body in a different way (Figure 11.2). The simplest is the two-compartment model, which accounts for fat mass (FM) as one compartment and fat-free mass (FFM) as a single compartment consisting of all tissues in the body except for fat. The major assumption with the two-compartment model is that all tissues in the FFM have the same density. This assumption obviously builds some error into the model, as it is well known that tissues as different as muscle and bone do not have a uniform density. In addition, the density of certain tissues, such as bone, may differ substantially between individuals or in the same individual over the life span. The two-compartment model is simple, but is based upon assumptions that reduce its accuracy.

The sumo wrestler and the bodybuilder may have similar body weight yet dramatically different body composition.

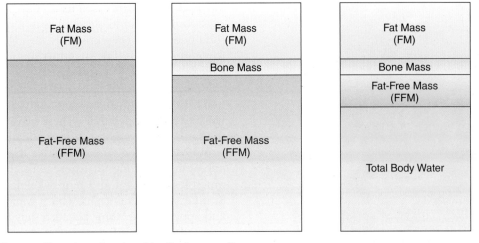

Figure 11.2 Two-, three-, and four-compartment models of body composition

To determine body composition with greater accuracy, three- and four-compartment models were created once the technology was developed to measure specific tissues (Figure 11.2). For example, dual-energy X-ray absorptiometry (DEXA) technology allowed for the measurement of bone mineral density and bone mass and led to the development of the three-compartment model. Use of isotope dilution and bioelectrical impedance analysis (BIA) to determine total body water provided the information necessary for a four-compartment model, which provides better estimates of body composition by accounting for more types of tissues than the other models.

Key points

- The terms *body mass* and *weight* are commonly used interchangeably, although mass technically refers to the amount of matter of an object and weight refers to the force an object exerts due to gravity.

- Body composition refers to all of the components that make up the body, particularly fat, muscle, bone, and water.

- Body weight and composition may be directly related to performance in certain sports, whereas in others they may be related more to appearance or other factors.

- A minimum amount of body fat, referred to as essential, is needed for proper physiological functioning.

- Body Mass Index (BMI), a ratio of weight to height, is not an accurate way to characterize body composition in muscular people, and therefore should be used cautiously with athletes.

How is it possible for two athletes to have the same height, weight, and BMI but different body composition?

11.2 Assessment and Interpretation of Weight and Body Composition

To understand weight and body composition, one must understand the techniques used to estimate them and the errors that result from any type of measurement. Weight or body composition measures are meaningless if accurate procedures are not used. A detailed discussion of the exact procedures involved with each method is beyond the scope of this text, and good reviews can be read in a number of fitness assessment resources (Thompson, Gordon, and Pescatello, 2009; Nieman, 2007). Even when the procedure is performed correctly, the results can be misinterpreted if the standard error of the estimate is not considered. The purpose of this section is to enable students in sports-related fields to acquire a practical, working knowledge of this potentially confusing subject with a strong understanding of the appropriate use and limitations of body composition assessment.

Body weight is measured with a scale.

A balance beam scale or a digital scale should be used to determine body weight (Figure 11.3). To ensure accuracy, the scale should be calibrated monthly or quarterly, as well as any time it has been moved. Scale weight should be taken as soon as the individual is awake, after emptying the bladder, and before any food or drink is consumed. On a balance beam scale, weight should be recorded to the nearest 0.5 lb or 0.2 kg (Modlesky, 2006). Home scales are convenient

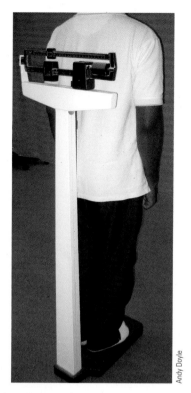

Figure 11.3 Balance beam scale

be considered with any method are the accuracy or precision of measurement, practicality, ease of use, time required to obtain the measurement, cost, portability, comfort, effort required by the subject, and training required of the technician. Accuracy is the most important element, but some of the most accurate measurement technologies cannot be used outside a research setting. Methods with lesser accuracy are sometimes used for practical reasons. Some easy-to-use methods are practical, but their accuracy is low and they may give athletes a false picture of their true body composition. In some cases the accuracy is so low the estimate is essentially meaningless.

The most important point to understand at the outset is that body composition, specifically body fat, cannot be directly measured except by chemical analysis of human cadavers. All other methods of determining body composition estimate or predict body composition using data from the direct chemical analysis of a relatively limited number of human cadavers. It is even difficult to accurately estimate body fat in cadavers. If one thinks about the different roles of fat (such as, cell membranes, covering nerve cells, surrounding internal organs) and where it is distributed in the body (such as adipocytes, under the skin, muscle cells, blood), then one can gain an appreciation for the difficulty in accounting for all of the fat mass in the body.

but most lack the accuracy of a balance beam scale. Repeated scale weights (for example, daily, weekly) should be taken on the same scale under similar conditions (for example, time of day).

When athletes are weighed to be certified for weight classes in certain sports, such as wrestling, the weighing process and the athletes' weights are public. Body weight for other athletes, however, may be a personal and private issue. When weighing athletes, particularly in groups, care should be taken to maintain confidentiality for each individual. The scale can be placed in a private area or obscured with a curtain or screen and each athlete weighed individually. Body weight results should be recorded without calling out the results verbally. In instances where athletes have an unhealthy preoccupation with their weight, the practitioner or technician may have them face away from the scale so that they cannot see the weight results. Safeguarding the privacy of the individual and keeping the results confidential allows athletes to make the decision whether or not to share their personal information with others.

Error of measurement. Because the only true measurement of the amount of fat in the body is by chemical analysis of cadavers, all current approaches to determining body composition estimate or predict body fat from some other measurement. Therefore, all of these methods are indirect determinations and will have some built-in or **inherent** error. In addition, there is potential for technical error in the assessment method itself. It is extremely important to understand the potential for these errors and how they might affect body composition results and recommendations based on those results.

To illustrate measurement error, consider underwater weighing as a method of determining body composition, specifically percent body fat. This technique is used to determine the density of the body from which the percentage of body fat is predicted. The original studies from which these prediction equations were developed show there is a strong correlation between body density and percent body fat, but there is not 100 percent accuracy (Brozek et al., 1963; Siri, 1956). This type of error is expressed as the Standard Error of the Estimate (SEE) and represents the degree to which the measured factor is likely to vary above or below the result obtained.

Body composition can be estimated by a variety of methods.

There are a variety of methods available to determine body composition, each with advantages and disadvantages (Table 11.1). Some of the factors that must

Inherent: Unable to be considered separately.

Table 11.1 Comparison of Methods Used to Estimate Body Composition

Method	Accuracy	Practicality and portability	Ease of use	Time	Cost	Subject comfort and effort	Technician training
Underwater (hydrostatic) weighing	SEE = ±2.7%	Practical in exercise physiology laboratories or large fitness centers; not portable	Requires subject to submerge, exhale, and hold breath	~30 minutes because the procedure should be repeated 5 to 10 times	Initial purchase of equipment is expensive	Subject may be uncomfortable wearing a bathing suit, submerging in water, and exhaling air	Training is needed but is not difficult
Plethysmography	SEE = ±2.7–3.7%	Requires 8' × 8' space; can be moved with proper equipment, but takes effort	Requires subject to sit quietly	~5 minutes	Initial purchase of equipment is expensive	Subject may be uncomfortable wearing a bathing suit and cap and sitting in an enclosed space	Minimal training needed
Skinfold measurements	SEE = ±3.5%	Practical in settings that have a private area; very portable	Requires subject to be still; measurement sites must be determined and marked	<5 minutes	Initial purchase of equipment is relatively inexpensive	Subject may be uncomfortable partially disrobing; some skinfolds are difficult to grasp	Training and consistency are critical; technique improves with experience
Bioelectrical impedance analysis (BIA)	SEE = ±3.5%	Practical in most settings; very portable	Easy to use	<5 minutes	Initial purchase of equipment is moderately expensive	Procedure is simple but premeasurement guidelines require substantial subject compliance	Minimal training needed
Near-infrared interactance (NIR)	SEE = ±4–5%	Practical in most settings; very portable	Easy to use	<5 minutes	Initial purchase of equipment is moderately expensive	Simple procedure; generally no problems	Minimal training needed
Dual-energy X-ray absorptiometry (DEXA)	SEE = ±1.8%; more research needed to verify SEE	Practical in imaging centers, physicians' offices, or research facilities; not portable	Easy to use	~5 to 10 minutes	Initial purchase of equipment is very expensive	Simple procedure; subject is exposed to a very small amount of radiation; use prohibited during pregnancy	Training is needed; license to operate is required
Computed tomography (CT) scans and magnetic resonance imaging (MRI)	Not yet established	Practical in imaging centers and research facilities; not portable	Requires subject to be still throughout the entire procedure	~30 minutes	Initial purchase of equipment is very expensive	Procedure is relatively simple with some subject discomfort	Training is needed; license to operate is required

Legend: SEE = Standard Error of the Estimate

The SEE for percent body fat determined from underwater weighing is approximately ±2.7 percent (Lohman, 1992). This means that if percent body fat is determined as accurately as possible by the underwater weighing technique for a group of people, it is likely that the result obtained will be within a range that is 2.7 percent above or below the figure determined for two-thirds of the people measured. In other words, even if this technique is performed flawlessly, a person whose body fat result by underwater weighing is determined to be 15 percent may actually have body fat as high as 17.7 percent or as low as 12.3 percent. Out of a group of 100 people with a body fat estimate of 15 percent, 67 will actually have a percentage of body fat within the range of 12.3–17.7 percent. Although this is a fairly large range, one also must be aware that the remaining one-third of the group, or 33 people, are likely to have their "real" body fat percentage be even further outside the ±2.7 percent range.

Underwater weighing is one of the more "accurate" methods, so one can easily see where caution must be taken in interpreting body composition results. Those who work with athletes should be aware of the error associated with various body composition assessment methods and, in particular, should not assume false precision for these methods. For example, body fat percentages expressed to two or three decimals suggest a degree of measurement precision that is unrealistic ("Your body fat is 15.35 percent"). Caution must also be used when interpreting small changes in body fat percentage. Changes in body composition of 1 or 2 percent, particularly over a short period of time (that is, days to weeks), should be interpreted carefully because this amount of change is within the measurement error of the method.

All measurement methods have a certain amount of error that is inherent in the methodology. This error assumes the technique is administered with the highest degree of adherence to the appropriate procedures. There is additional error, however, in all of these methods that may be associated with how the body composition technique is performed, either by the subject or by the technician responsible for the measurement. Technical error that may be added to the error inherent in the method must be carefully considered, both in choosing an appropriate technique and in interpretation of the results. Simply stated, body composition that is assessed by any indirect method is not precise because of measurement error.

Athletes may wonder which method of measuring body composition is the most accurate. Underwater weighing was used to establish some of the early estimations of body fat from cadavers and has therefore long been considered the gold standard of body composition methods. It has also been used as the **criterion** for the establishment of other methods such as the skinfold technique. Underwater weighing is no longer considered the gold standard because there are now more sophisticated methods (for example, magnetic resonance imaging and computed tomography) that are used as criterion methods, although there is still a lack of data from these newer methods to establish population norms (Modlesky, 2006). These methods are also not practical to use in most cases, so underwater weighing is often considered to be the most accurate method readily available to athletes.

Underwater weighing and plethysmography estimate body composition by determining body density.

One approach to estimating body composition is by determining the overall density of the body. Fat tissue has a lower density (0.9 g/ml) than other tissues, so theoretically the less dense a person's body is the more body fat is present. How is body density determined? Body density can be calculated as the ratio of body mass (weight) to body volume. Mass (weight) is easily measured on a scale, but what about volume? Archimedes' principle is used by two techniques to determine body volume and density.

Hydrodensitometry or underwater weighing. Underwater weighing is a technique of estimating body composition that has been utilized for decades and may be the most accurate method available to many athletes for determining body fat. Mass can be determined easily by measuring the person's weight, and density can be calculated if the body's volume is known (Figure 11.4). Submersion in water can be used to determine volume, either by the amount of water that is displaced (for example, water rising as one slips into the bathtub) or by determining the buoyant force acting on the submerged object. In the human body, two things are less dense than water and act to help the body float—air and fat. In underwater weighing, the air is accounted for by having the subject exhale as much air from the lungs as possible (down to **residual volume**) and by accounting for the residual volume in the prediction equation (Figure 11.4). A person that has more body fat will float more readily (that is, have a greater buoyant force) and will therefore have a larger volume and a lower density. Conversely, a person of the same weight with less body fat and more muscle will tend to sink more easily (that is, have less buoyant force), exhibiting a smaller body volume and higher density. Higher body density is associated with lower body fat.

The underwater weighing procedure requires the subject to exhale as much air as possible, submerge the

---

Criterion: Accepted standard by which other decisions are judged.

Residual volume: The amount of air left in the lungs after a maximal, voluntary exhalation.

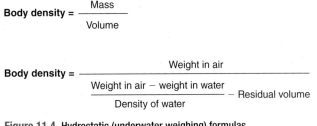

$$\text{Body density} = \frac{\text{Mass}}{\text{Volume}}$$

$$\text{Body density} = \frac{\text{Weight in air}}{\dfrac{\text{Weight in air} - \text{weight in water}}{\text{Density of water}} - \text{Residual volume}}$$

Figure 11.4 Hydrostatic (underwater weighing) formulas

Body density is determined from the mass divided by the body's volume. Mass is measured as weight (weight in air). Volume is determined as the buoyant force in water; the difference between the body's weight in air and submerged in water. Other factors to account for in the equation are the density of the water and the residual volume (air that remains in the lungs).

© Yoav Levy/Phototake

Figure 11.5 Hydrostatic weighing procedure

Subjects are seated in the water in a seat suspended from a scale. The subject must exhale as much air as possible (down to residual volume), submerge his or her body completely underwater, and remain as motionless as possible until an accurate scale weight (underwater weight) can be determined. To ensure consistent results, 5 to 10 measurements are required.

body completely underwater, and remain motionless long enough for the technician to obtain an accurate reading of the underwater weight (Figure 11.5). Many individuals have great difficulty with these procedures, particularly those not comfortable in the water. Therefore, substantial error can be introduced by the subject's inability to adhere to the procedures required by this method. Although underwater weighing may be an accurate assessment of body composition for some, it may not be the most accurate method for all subjects. Another potential source of error is the determination of the amount of air in lungs of the subject. The greatest accuracy is obtained when lung volume is measured with one of several gas dilution techniques, but this type of apparatus may not be available or practical to use outside a laboratory or research setting. More potential error is introduced when lung volume is estimated using prediction equations. Further error is introduced if the subject is unable to exhale all of the air down to residual volume; the extra air in the lungs makes the person more buoyant, and if unaccounted for, results in an error that overestimates body fat. In a study of trained athletes, a relatively small difference (~175 ml) in the amount of air in the lungs contributed to an average difference of over 1 percent in the final estimated body fat percentage (Morrow et al., 1986).

Other disadvantages of the underwater weighing method include the large, cumbersome, and not easily portable equipment needed. A water-filled tank large enough to accommodate a human body is not easily moved and is generally fixed in a laboratory or fitness-testing location. The gas dilution systems to measure lung residual volume are expensive, require technical expertise to operate, and need maintenance and sanitary cleaning of the breathing tubes. There are also issues related to the subjects and the facilities, such as the need for changing and clothes storage areas (for example, locker facility) that are compatible with wet activities. Subjects must wear bathing suits (preferably with minimal material for greatest accuracy),

which may be an issue of modesty or cultural sensitivity. Because of the need to repeat the procedure for greater accuracy, an underwater weighing session can be time consuming, easily taking 30 minutes or more, particularly for an inexperienced subject.

Plethysmography. Due to the problems associated with using underwater weighing to determine body volume and density, **plethysmography** (displacement of air to determine body volume) was developed. The subject sits in an airtight enclosure (Figure 11.6) while the amount of air displaced by the subject's body is sensed by a special diaphragm and pressure transducer. Once the body volume is determined, body density can be calculated and body fat estimated.

Studies show this method correlates well with body fat percentage determined from underwater weighing (McCrory et al., 1995) and has good test-retest reliability. The SEE is 2.2–3.7 percent (Fields, Goran, and McCrory, 2002). The major advantage of this system is the absence of water submersion, which can be a concern for subjects and a source of substantial error.

Figure 11.6 Air plethysmography

Air displacement is used to determine body volume, body density, and body composition.

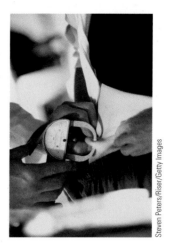

Figure 11.7 Skinfold measurement

The thickness of the fold of skin is determined with calipers on pre-determined locations on the body. Equations incorporating the sum of each skinfold thickness measured are used to predict body density and percent body fat.

It is also much less time consuming than underwater weighing. Studies suggest that this method can be used with approximately the same degree of accuracy as underwater weighing (McCrory et al., 1995). For greatest accuracy, subjects should wear tight-fitting clothing and a swim cap (to compress air pockets in the hair), which may lead to modesty concerns. The disadvantages include the high cost of the device, limited portability, the finite size of the seating area (for example, larger athletes such as some basketball players or football linemen may not fit), and claustrophobic fears.

Body composition can be estimated using the thickness of skinfolds from specific sites on the body.

A certain proportion of fat in the body is stored subcutaneously (under the skin). Body composition that is estimated from skinfold thickness is based on the assumption that a measurement of the thickness of the subcutaneous fat layer is directly related to the total amount of fat in the body. A skinfold thickness may be determined by pinching a fold of skin and measuring with calipers (Figure 11.7). Just one site may be used, but the sum of several different sites (two, three, and seven sites are frequently measured) is more accurate. Common sites used with men are chest, abdomen, and thigh whereas sites used with females are typically triceps, **suprailium**, and thigh. Most commonly used skinfold prediction equations have been derived using body density or body fat determined from underwater weighing, which means this method of estimation is twice removed from the original body density and fat measurements.

Generalized equations such as the commonly used three- and seven-site Jackson and Pollock equations (Jackson and Pollock, 1978; Jackson, Pollock, and Ward, 1980) were developed using diverse subjects (that is, both genders and a wide age range) so that the equations would be suitable for use in a broad population of men and women. These equations have an SEE of approximately ±3.5 percent. To put this error in context, recall that two-thirds of the people being measured would have their "real" percent body fat fall within this range (for example, 11.5–18.5 percent for an estimated body fat of 15 percent). Other specialized equations have been developed for use in very specific populations such as certain ethnic groups. Heyward and Stolarczyk (1996) provide an excellent guide for determining the most appropriate skinfold equation to use for specific populations, an important consideration when working with diverse populations.

Compared to other methods, the equipment required is relatively inexpensive (ranging from $10 to $500 for skinfold calipers) and is very portable. Little space is required, but a relatively private area is needed because access to some skinfold sites requires partial disrobing. As with other methods, privacy or modesty may be an issue. Results are easily calculated, by using either electronic automatic calipers (for example, Skyndex) or commercially available software. Programmable

Plethysmography: Measuring and recording changes in volume of the body or a body part.

Suprailium: An area of the body directly above the crest of the ilium, the hip bone.

calculators and computer spreadsheets can easily be configured with the body density and body fat formulas needed to calculate body composition.

The skinfold technique can provide a reasonable degree of accuracy if performed correctly by a trained technician. Proper location of the skinfold sites is critical, as is the accurate pinching of the skinfold, the measurement of its thickness, and the choice of the most appropriate equation. Proper training of the technician is important, as is a substantial amount of experience in grasping the skinfolds with appropriate force (too much force can compress the skinfold and lead to an underestimate of the thickness and vice versa) and reading the skinfold caliper. This method may be difficult to employ with some subjects, as some skinfolds are very difficult to grasp and measure accurately, particularly the chest skinfold on many overfat males and the thigh skinfold on many females and males. In addition, individuals with very large amounts of subcutaneous fat may have skinfold thicknesses that exceed the measurement capacity of the skinfold calipers. The skill of the technician is a critical element because an unskilled technician introduces too much error and the resulting estimate cannot be used with confidence.

In addition to estimating percent body fat, skinfold measurement may provide valuable **anthropometric** information for tracking changes over time. For example, if an athlete's skinfold thicknesses decrease over the course of the off-season to the competitive season, one can be confident that body fat has decreased. Specific skinfold sites can be tracked in a longitudinal fashion for athletes in this way. An important issue is the accuracy of the measurements, which should be conducted by the same trained technician if possible.

Bioelectrical impedance analysis (BIA) uses electrical currents to estimate the proportion of fat in the body.

The bioelectrical impedance analysis (BIA) body composition methodology is based upon the rationale that body tissues can be distinguished based upon their relative ability to conduct electrical currents. Water and tissues containing a high proportion of water conduct electrical currents easily, whereas tissues that contain little water, such as fat, impede the flow of electrical currents. With this method, a nonharmful electrical current is conducted through the body and the impedance to the flow of that current is measured (Figure 11.8). Body composition is not measured directly; rather, an algorithm or prediction equation is used to predict body composition from a three-compartment model: fat mass, fat-free mass, and body water. These algorithms or equations are generally **proprietary** (private) to the company that has developed them so it is difficult to conduct independent scientific studies of the formulas.

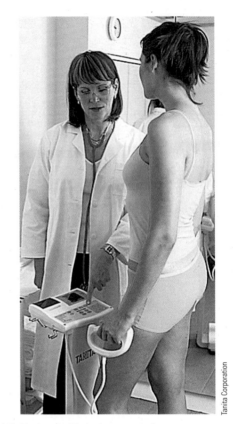

Figure 11.8 Bioelectrical impedance analysis

A nonharmful electrical current is conducted through the body and the impedance to the flow of that current is measured and used in an equation to predict body composition.

Commercially available BIA units typically have one of three configurations, measuring impedance between arms and legs, between the legs, or between the arms. The first approach requires the subject to lie down and have electrodes placed on his or her ankles and wrists. The latter two devices require the subject to stand in bare feet on a scalelike platform or hold onto the device with metal contacts with bare hands. BIA devices range from very expensive, multifrequency devices that measure whole-body impedance and estimate total body water (both intracellular and extracellular) to inexpensive, low-frequency devices marketed for use in the home.

Accuracy of the body composition estimate is based largely upon the accuracy of determining the electrical impedance and the underlying assumptions used in the equations regarding water content of various tissues. Differences in water content and body density, which may vary due to age, gender, ethnicity, recent physical activity, and hydration status, must be considered. If all conditions are controlled appropriately and subject and technical error is minimized, the estimate obtained via BIA is comparable to that of skinfold assessments, an SEE of approximately ±3.5 percent.

BIA has some distinct advantages as a method of estimating body composition. Other than removing shoes and socks, the method requires no special clothing or disrobing, reducing concerns of privacy and modesty. The devices are easily portable, can be used in a variety of settings, and generally require only a few minutes for the assessment to be completed. The devices are computerized, are programmed to calculate the results, and generally have an option for printing out the results. In addition, there is not a large potential for subject or technician error as long as data entry is done correctly and fairly simple procedures are followed (for example, correct electrode placement).

Major error or variation can occur, however, if the technician or subject does not adhere to premeasurement factors that may affect body water. Changes in hydration status and physical activity may affect bioelectrical impedance analysis, particularly if changes occur near the time of measurement. For that reason, preassessment guidelines should be given and adhered to by the subject and confirmed by the technician before any measurement takes place:

- Abstain from eating or drinking within 4 hours of the assessment.
- Avoid moderate or vigorous physical activity within 12 hours of the assessment.
- Abstain from alcohol consumption within 48 hours of the assessment.
- Ingest no diuretic agents, including caffeine, prior to the assessment unless prescribed by a physician.

Bioelectrical impedance can be a relatively quick and easy assessment method for estimating body composition, but careful attention must be paid to these preassessment directions to avoid introducing excessive error. This is especially true for athletes who would need to schedule the test around their training schedule. Additionally, any information provided by the subject (for example, weight, physical activity) must be accurate, a potential problem for recreational and nonathletes who may underestimate body weight and overestimate physical activity. A major disadvantage of BIA is the cost of the device, which may be substantially greater than other methods with similar accuracy (for example, skinfold calipers).

A beam of near-infrared light is used to distinguish between fat and other tissues.

Near-infrared interactance (NIR) is based upon the ability of different tissues to absorb or reflect light. A wand that emits and senses near-infrared light is placed over a body part such as the belly or center of the biceps. A light beam is directed into the tissue, some of which is absorbed and some of which is reflected back and is measured by spectroscopy in the wand. Less-dense tissue absorbs more near-infrared light and more-dense tissue reflects more light back to the sensor. The differential absorption and reflection of the near-infrared light are used in prediction equations to estimate percent body fat. Similar to bioelectrical impedance, body composition is not measured directly.

NIR devices are portable, easy to operate, and not exposed to substantial technician or subject error if the information provided by the subject about weight and physical activity is reliable. The procedures are not disruptive for the subject as only one site on the body is typically measured (for example, the biceps of the dominant arm). Concerns have been raised about the use of only one site for predicting percent body fat. Studies of the accuracy of body composition estimated by NIR show an error of approximately ±4–5 percent (Hicks et al., 2000; McLean and Skinner, 1992), making this method less accurate compared to the methods discussed previously. Until greater accuracy is achieved, the estimates obtained by NIR may have too much potential error to be used with confidence.

Dual-energy X-ray absorptiometry (DEXA or DXA) uses low-intensity, focused X-rays to determine bone density and estimate body composition.

Dual-energy X-ray absorptiometry (DEXA or DXA) is a method that uses low-intensity, focused X-rays to scan the body for determination of bone mineral density and content. Originally developed for clinical use for measuring loss of bone density, the bone mineral information can also be used as part of a three-compartment model for estimating body composition. The potential exists for this to be a precise method for body composition determination because it accounts for one of the tissue compartments that can vary substantially between people and within an individual over a lifetime.

In this procedure, the subject lies motionless on the scanning table for a few minutes while a full body scan is performed, yielding both skeletal and soft tissue images (Figure 11.9). Based upon proprietary algorithms developed by each company, body composition is determined, with estimates of **bone mass**, fat mass, and fat-free mass provided. Some software programs use anatomical landmarks to digitally section the body into segments for analysis of regional body composition (for example, trunk, arms, and legs).

Anthropometric: Body measurements such as height, weight, waist circumference, or skinfold thickness.

Proprietary: Privately owned and administered.

Bone mass: Total amount of bone in the body. Expressed in pounds or kilograms.

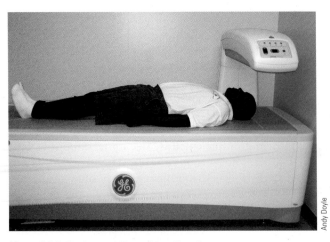

Andy Doyle

Figure 11.9 Dual-energy X-ray absorptiometry

Using low-intensity, focused X-rays, bone and soft tissue can be scanned and percent body fat estimated.

Because of the expense of the equipment and the associated radiation safety requirements, DEXA is a method of body composition assessment that is not likely to be found outside specialized clinical facilities, research laboratories, or doctors' offices. The equipment also contains an X-ray-generating device, and is therefore subject to state or local licensing and safety regulations. Technicians operating the DEXA equipment must be trained in the use of the equipment and in radiation safety. Although the dosage of radiation a subject is exposed to for a whole-body scan is very small, there is a potential for accumulated radiation exposure with repeated scans, and there are some conditions (for example, pregnancy) that prohibit use. The devices do have size limits for subjects due to the available scanning area. Subjects who are very tall (over 6'4" [193 cm]) or who are severely obese may not be scanned accurately, as their body may not fit within the limits of the available scanning area.

Whole-body scans for body composition are relatively fast on newer DEXA equipment, generally taking only a few minutes. Other than adhering to the directions to eliminate metal objects from the body (for example, jewelry) or clothing (for example, belt buckle, underwire bra), the subject has little to do other than to remain motionless until the scan is completed. There is little opportunity for technician error other than positioning the subject correctly on the scanning table and entering the subject information correctly. The error has been reported to be approximately ±1.8 percent (Thompson, Gordon, and Pescatello, 2009), but additional research must be performed before DEXA can be considered the new gold standard method of body composition assessment (Kohrt, 1995).

Advanced imaging techniques include CT scans and MRI.

Advanced clinical imaging techniques have also been used to assess body composition. Two such methods are computed tomography (CT) and magnetic resonance imaging (MRI). Similar to DEXA, these methods are generally found only in specialized clinical facilities or research laboratories and are not commonly used in assessment of body composition outside research studies. These devices are able to image tissues in the body in cross-sectional slices. The amount of tissue in each section is estimated, and whole-body composition is estimated by summing the sequential section images. These technologies have not been used on a sufficient number or breadth of subjects to establish the SEE. The specialized nature of these technologies makes their widespread use in body composition assessment unlikely in the near future.

Body composition results must be interpreted appropriately.

Once the appropriate body composition assessments have been completed and the results are known, the subject will likely ask, "What do these numbers mean?" The health, fitness, or nutrition professional may wonder, "What do I do with these results?" The initial caution about body composition bears repeating here: assessments of body composition are only estimates. Each method of assessment has a degree of error that must be taken into account when analyzing and interpreting the results, and this error may be compounded by technical, technician, or subject error. Suggesting to athletes that body composition assessment is overly precise or using the information to establish rigid goals is incorrect and inappropriate.

Body composition results should be presented as a range that includes the error of measurement. As mentioned previously, if percent body fat is estimated to be 15 percent using the underwater weighing method, then this figure should be interpreted as the percentage of body fat that is likely to be within the range of 12.3–17.7 percent. The athlete should also be informed that there is a possibility the actual body composition may be higher or lower than this range. Percent body fat and lean body mass estimates can be used to determine an appropriate scale weight that reflects desired body composition (demonstrated later in this chapter). Excessive body fat likely has an impact on performance and health, and body composition assessment can help to establish and monitor progress toward a more desirable body composition. Measurement of body composition can also help athletes assess if their training program and dietary intake need to be adjusted to meet body composition goals.

Knowledge and experience in assessment of body composition and good old-fashioned common sense need to be employed when evaluating these results and making recommendations. For example, if an athlete has a body composition assessment showing an increase in body fat of 0.5 percent over a 3-month period of time at the same time that training volume has increased, scale weight has decreased, and clothes are looser-fitting, then the accuracy of that body composition assessment must be questioned and re-evaluated.

Body weight results must be interpreted appropriately and used consistently.

Weight is reported as a single number and is generally compared to previous weights. A change in body weight can reflect a change in muscle mass, body fat, hydration status, or a combination of these factors. Weight can fluctuate on a daily basis and is most useful to athletes as a way to track hydration status, especially for those who are training in hot, humid conditions and are losing large amounts of fluid each day as sweat. In this instance, daily weight may be an appropriate means of checking hydration status and ensuring adequate rehydration. Although athletes in some sports must be focused on body weight because of weight restrictions or classes, it is probably not appropriate for most athletes to be overly concerned about checking their weight on a daily basis. For purposes other than hydration status or "making weight," a reasonable approach might be for an athlete to check weight on a less frequent basis, such as once each week. Care should be taken to measure body weight using the same scale if possible and under the same conditions, such as the same time of day and the same timing in relation to exercise, food, and beverage consumption.

Visual monitoring, along with scale weight, can be used as a "check" on body composition, but this technique is highly subjective. As an example, consider a 24-year-old male athlete who usually weighs 185 lb (84 kg) when he is in a state of **euhydration**. Based on assessment of body composition at this weight, this athlete's percent body fat is estimated to be 8–10 percent. He knows that when he gains body fat the fat tends to be deposited in his abdominal area. At 185 lb (84 kg) and 8–10 percent body fat, this athlete understands what his body "looks" like and can visually monitor his body composition. If his weight increases to 190 lb (~86 kg) and the 5 lb (~2 kg) increase in weight reflects an increase in abdominal body fat, then he should be able to detect this increase in the mirror and by the fit of his clothes. The usefulness of visual monitoring depends on honest assessment. If this athlete held his breath and tightened his abdominal muscles while looking in the mirror, he could pretend that the 5 lb increase was not a result of increased body fat. Those with a distorted body image should not use frequent scale weights and visual monitoring because each can be misinterpreted and lead to practices that harm physical and mental health.

Key points

- All assessments of body composition are estimates, with varying degrees of accuracy.
- Each body composition assessment technique has some inherent error as well as varying potential for technical or measurement error.
- Underwater weighing and air plethysmography determine the volume and density of the body, which is then used to predict body composition or percent body fat.
- The thickness of skinfolds, including subcutaneous fat, from specific sites on the body is used in either generalized or population-specific prediction equations to estimate body composition.
- Body composition is estimated using BIA by using the difference in electrical conduction properties between fat and other tissues in the body.
- DEXA uses X-ray to determine bone density, which allows for body composition estimation using the four-compartment model.
- DEXA is becoming the gold standard for body composition assessment for research but is not widely available or practical for use with large numbers of athletes.
- The use of body composition results must keep in mind the error associated with the assessment method.
- Consistent use of body weight over time can be used to track hydration status, changes in muscle mass, body fat, and so on.

What body composition assessment method would you use for determining percent body fat in a group of athletes for a research study? How about for preseason and postseason measurements?

What would you tell an athlete who is concerned because at the end of a strenuous 2-week training camp his or her percent body fat remains the same?

11.3 Body Composition and Weight Related to Performance

In many sports, there are certain physical characteristics that are associated with success. For example, elite marathon runners tend to be lightweight, light-boned, and have a relatively low percentage of body

Figure 11.10 In some sports, professional athletes exhibit a wide variety of body weights and compositions.

fat because these characteristics are associated with moving the body a long distance as quickly and efficiently as possible. However, within the ranks of elite marathon runners, there is a range of body weights and there is not a certain percentage of body fat that is associated with success. Based on the athlete's sport, genetic predisposition, and individual characteristics, an athlete can set weight and body composition goals that are likely to be associated with good performance and good health.

Certain physical characteristics are associated with sports performance.

Body weight and composition are among the many factors associated with successful athletic performance. The sport itself may be associated with certain physical characteristics, and the athlete may seek to improve performance by changing weight and/or body composition. The athlete's individual characteristics cannot be overlooked, but many of the following factors have a big impact on the athlete's weight and body composition goals:

- Sport
- Position played
- Size requirements (height, weight)
- Relative need for power and endurance
- Weight certification
- Body appearance

Sport played. Body composition and other body anthropometric characteristics, such as height, are physical characteristics that can have an impact on an athlete's performance, but these factors play a more important role in some sports, and for different positions within certain sports, than others. For example,

all bodybuilders need to have a high percentage of skeletal muscle mass and a low percentage of body fat to be competitive in this sport. Professional, long distance (road) cyclists have a low percentage body fat. Conversely, body composition and weight play a minor role in certain skill-oriented sports. Athletes that are successful in baseball or golf display a wide variety of body types and body composition that may be more similar to the general population (Figure 11.10). Professional baseball players may be fit and lean like Derek Jeter or may have a body composition more similar to the average, sedentary adult like C.C. Sabathia. Golf is another example in which athletes with different body compositions, heights, and weights have been successful.

There is an inverse relationship between body fatness and performance in some sports, particularly those in which body weight must be transported, as in distance running, other endurance sports, and field sports like high jump. Excessive body fat comprises "dead weight" for the athlete—weight that must be carried but does not contribute in a positive way to the activity. Carrying excess weight makes the athlete less energy efficient so he or she must exert more effort to transport the weight. This doesn't mean that performance will always be improved if these athletes attain an absolute minimum weight, however. There is a point of diminishing returns, both for performance and for health, that the athlete must consider. At some point the caloric restriction and training that is needed to further reduce body weight and fat may become counterproductive.

Position played. The members of a team may vary considerably in their weights and body compositions; however, there is often a much narrower range when position played is considered. For example, members of an American football team may range from ~7 to 30 percent body fat, but those at the higher end of the range are almost always linemen. Other sports that vary in weight and percent body fat depending on the position played include basketball, ice hockey, lacrosse, rugby, and soccer (Dunford and Macedonio, in press).

Physical size. Many athletes today are physically taller and heavier than their counterparts of the past. Montgomery (2006) compared the heights and weights of professional ice hockey players in the 1920s to 1930s with those in the 1980s to 1990s. The latter were 4 inches (10 cm) taller and ~37 lb (17 kg) heavier. In contact sports, being physically well matched to one's opponent is important, although physical size alone is not the determining factor in most sports. However, there is a physical uniformity, especially in the position played in contact sports, that dictates the athlete be a certain size, weight, or body composition.

Interior linemen in (American) football typically have requirements for greater body mass, even if some of that mass is fat. Along with muscle, the additional body mass prevents them from being "pushed around" as easily. The majority of linemen in the National Football League (NFL) are over 300 lb (136 kg). There is a disturbing trend among youth American football linemen to increase size by increasing fatness (Malina et al., 2007). At an early age a large child who is fat may have a physical advantage due to size. However, much of this early advantage fades over time, and overfatness in child athletes is often a disadvantage because it negatively affects fitness, speed, self-image, and health.

A few sports are size dependent, such as sumo wrestling and ski jumping. However, the difference in physical size within sports can often be accommodated with weight categories so that a wide range of athletes can participate, such as in boxing, powerlifting, rowing, wrestling, and some of the martial arts. In sports with weight categories, a higher body weight relative to the weight class and a higher power-to-weight ratio may be a competitive advantage.

Relative need for power and endurance. The "optimal" body composition for an athlete must consider the mass, strength, speed, and power demands of the sport, or the position within the sport. At one extreme, the sumo wrestler represents an athlete that must possess a very large body mass. This large body mass is difficult to push out of the competitive ring due to inertia and lack of momentum. Strength and speed are important for these athletes, but a very large body mass is critically important—a strong and fast sumo wrestler will have little success if he is outweighed by several hundred pounds by his competitors. The other extreme may be represented by a ski jumper. A certain amount of strength is required for controlling the skis and for propelling the body into the air at the end of

In sports in which explosive power provides a competitive advantage, having a high power-to-weight ratio is important.

the ski jump, but low body weight is a great advantage for these athletes to stay in the air longer. Other athletes fall somewhere in between these extremes in terms of body mass, body fat, and skeletal muscle.

In sports or positions in which explosive power is a requirement or provides a competitive advantage, having a large amount of muscle mass and a high power-to-weight ratio is important. Power-to-weight ratio is defined as power (measured in watts) divided by weight (measured in pounds or kilograms). Athletes such as male gymnasts, ice hockey players, short distance runners (for example, 100 m runners), speed skaters, high, long, and triple jumpers, and (American) football linebackers and running backs are often very muscular and lean and have a high power-to-weight ratio. Excess body fat may diminish this power-to-weight ratio and is therefore undesirable.

Most sports involve varying degrees of power and endurance. For example, basketball, ice hockey, lacrosse, soccer, and tennis players require explosive power as well as endurance. In addition, speed is a critical aspect of their performance. These players need to find a weight and body composition that reflects the proper balance between the amount of skeletal muscle mass needed for power and the amount of body fat associated with speed and endurance.

Weight certification. Weight categories are necessary in certain sports, but the certification of weight puts an emphasis on attaining a certain scale weight in order to compete. A singular focus on scale weight can be problematic because weight can be manipulated via changes in body water. Athletes can quickly change their weight by reducing water intake and increasing water excretion through excessive sweating and the use of diuretics (see Chapter 7). These are potentially dangerous practices because of the negative effects

Table 11.2 Estimated Percent Body Fat Ranges for Athletes in Selected Sports

Sport	Level	Position or distance	% Body fat
Males			
Baseball	College	All positions	11–17
	Professional	All positions	8.5–12
Basketball[a]	Professional	Centers	9–20
		Forwards	7–14
		Guards	7–13
Cycling, road	Professional	Long distance	7–10
Football (American)	College (NCAA Division I)	Defensive backs	7–14.5
		Receivers	9–16.5
		Quarterbacks	14–22
		Linebackers	12.5–23.5
		Defensive linemen	14.5–25
		Offensive linemen	18.5–28.5
Judo	National team	All weight classes	8.5–19
Rugby	Professional	All positions[b]	9–20
Soccer	Professional and college	All positions	7.5–18
Tennis	Elite		8–18
Water polo	National team		6.5–17.5
Wrestling	NCAA Division I, II, III championships		6.5–16
Running	Elite	Middle distance	8–16.5
		Long distance	12–18
Soccer	College	All positions	13–19
Swimming or diving	College	Middle distance or diving	17–30
	Masters	Long distance	20–34
Females			
Lacrosse	College	All positions	17.5–27
Running	Elite	Middle distance	8–16.5
		Long distance	12–18
Soccer	College	All positions	13–19
Swimming or diving	College	Middle distance or diving	17–30
	Masters (ages 21–73)	Long distance	20–34
Tennis	Elite		15–25

Legend: NCAA = National Collegiate Athletic Association

These values represent reported percent body fat ranges and not optimal ranges for collegiate or professional athletes, ages 18–29 years.

[a]French and Serbian players only.

[b]Forward players have a higher percent body fat than back players.

Compiled from research studies measuring body composition by hydrostatic weighing, BodPod, skinfolds, or DEXA: Enemark-Miller, Seegmiller, and Rana, 2009; Hoffman et al., 2006; Montgomery, 2006; Petersen et al., 2006; Ostojic, Mazic, and Dikic, 2006; Sallet et al., 2005, 2006; Silvestre et al., 2006; Clark et al., 2003; Tuuri, Loftin, and Oescher, 2002; Noland et al., 2001; Deutz et al., 2000; Collins et al., 1999; Kreider et al., 1998; Wilmore, 1983.

they have on body temperature regulation. Boxing, wrestling, some of the martial arts, and lightweight rowing are examples of sports that certify weight.

Wrestling is an example of a weight-certification sport that has moved beyond body weight as a singular measure. In high school and collegiate wrestling, body composition and hydration status are used along with scale weight at the beginning of the season to determine the appropriate weight categories in which athletes may be certified. An athlete's weight category may be modified during the season to accommodate for growth. The use of weight, body composition, and hydration status has greatly reduced the use of dehydration to "make weight."

Physical appearance. Physical appearance is a performance-related factor in some sports due to subjective scoring and judging. In the sport of bodybuilding, the physical appearance of the body is judged, although other elements are also considered, such as posing. Figure skating, gymnastics, and diving include appearance as part of the scoring. Ballet dancers and cheerleaders are judged on their appearance (See Spotlight on...Athletes and Appearance—Meeting Body Composition Expectations). In many cases, the cultural norm for males is to have a lean, muscular appearance, which can be achieved through high-volume training. However, females may be expected to have a thin but not overly muscular appearance, which may require chronic undereating to attain and maintain a low body weight and a low percentage of body fat. Female athletes in these sports may be at risk for developing disordered eating and eating disorders, but there are many factors involved, as explained in Chapter 12.

Many athletes establish weight and body composition goals in an effort to improve performance or health.

Weight and body composition are useful information for athletes as long as this information is interpreted correctly. Daily weights can help the athlete determine the amount of fluid lost during exercise and help assess hydration status. Monthly or biweekly weights can also be used to track large changes in body composition. For example, an athlete who wants to gain 20 lb (9 kg) of skeletal muscle mass or lose 20 lb (9 kg) of body fat can track scale weight over many months. However, scale weight can also be misinterpreted. Daily weights are not a good reflection of changes in lean body mass or body fat. In some cases the athlete becomes too focused on the "number" and not what the number represents. Scale weight is a broad measure of change, but it cannot provide specific information about body composition.

The measurement of body composition provides the athlete with information about body fatness. Attaining a certain body composition cannot predict success in any sport, however athletes may improve their performance by increasing lean body mass or decreasing body fat. Many athletes are influenced by the weight and body composition of elite athletes in their sport, and there is often uniformity in body composition among elite athletes within certain sports and positions played. Long distance runners and cyclists, baseball catchers, basketball centers, American football linemen, gymnasts, figure skaters, bodybuilders, rhythmic gymnasts, and powerlifters tend to have more uniformity within their sports than do tennis players, golfers, divers, baseball pitchers, or recreational triathletes.

Although summaries or averages of body fatness for athletes across a variety of sports are available, individual athletes must be considered on a case-by-case basis. Table 11.2 lists the range of percentage of body fat reported in published studies. These ranges should be used with caution, and only as a guideline, because they do not represent optimal percent body fat ranges. (Dunford and Macedonio, in press)

Weight and body composition are also associated with health. Weight, as a sole measure, is not an accurate predictor of health. However, there is an association between body weight, chronic disease risk, and mortality (Hu et al., 2004; Lee, Blair, and Jackson, 1999). This association is very much influenced by cardiovascular fitness. It is important for individuals to assess if their current weight and percent body fat are within a healthy range. In addition to the amount of body fat, the distribution of body fat is a disease risk factor. All of these issues are discussed in Chapter 13 "Diet and Exercise for Lifelong Fitness and Health."

Key points

- The sport and position played and the relative need for power and endurance are factors in establishing an optimal performance weight and body composition.

- In sports with weight categories, a higher body weight relative to the weight class and a higher power-to-weight ratio may be a competitive advantage.

- Weight and body composition goals may be influenced by appearance, particularly in sports that are subjectively scored or the body is judged.

- Although weight and body composition may be similar among elite athletes in certain sports, these figures should be used with caution by an individual athlete who must consider his or her unique characteristics and genetic predisposition.

The percentage of body fat of professional distance cyclists is often reported as being between 7 and 10 percent. Why might reaching this range be beneficial or detrimental for a cyclist?

11.4 Changing Body Composition to Enhance Performance

Many athletes want to change their body composition. Highly trained athletes are typically lean but may want to gain muscle mass to increase strength or lose a small amount of body fat to improve their power-to-weight ratio or their appearance. Athletes hoping to advance to the next level, such as high school athletes to the collegiate level and collegiate athletes to the professional level, may change weight or body composition as a way to become more competitive. Lesser-trained athletes often wish to increase muscle mass and lose body fat, sometimes in substantial amounts. Some recreational athletes want to lose moderate to substantial amounts of body fat. This loss of body fat may positively affect performance, but in many cases the desire to lose body fat is related more to

appearance and the desire for better health. A minority of athletes need to gain body weight and need to increase body fat in addition to increasing muscle mass. Regardless of the athlete's priorities, the same questions are frequently asked: (1) How much should I weigh? (2) What percentage of body fat should I have? (3) How do I increase muscle mass? (4) How do I lose or gain body fat? and (5) How do I increase muscle mass and lose body fat at the same time? This section explains how a target weight based on body composition is determined, and briefly outlines the changes in exercise and training that are needed to achieve muscle mass or body fat goals.

Desired body composition can be used to determine a target weight.

After body composition has been estimated as accurately as possible, athletes can use that information to establish their "optimal" body composition goals. Athletes should be cautioned to choose realistic lean body

Spotlight on...

Athletes and Appearance—Meeting Body Composition Expectations

Athletes in subjectively scored sports must consider their body's appearance because it may influence their scores. The most obvious case is that of bodybuilders because the appearance of the body is a fundamental element of the sport. Sports such as women's gymnastics and figure skating have "cultural" standards for appearance that relate to body composition. Weight and body shape dissatisfaction and body image

disturbances are prevalent in these and other appearance-dependent sports. Psychological issues related to body image should not be overlooked in subjectively scored sports, as they may lead to disordered eating, which is discussed in depth in Chapter 12 (Ziegler et al., 1998, 2005; Jonnalagadda, Ziegler, and Nelson, 2004).

Although most athletes' bodies are not scored, they are judged by the large audiences reached through the visual media. Fashionable, tight-fitting clothing is part of the sports scene. Television coverage of sports is so extensive that many athletes are celebrities, and the general population often expects celebrities to be thin, and in the case of athletes, muscular with a low percentage of body fat. Youth sports, such as high school football and the Little League World Series, are now shown on local or national TV. It is not surprising that some athletes at all levels of competition feel pressure to attain a body composition that is held in high esteem by society. Appearance is one reason that athletes, even high school athletes, consume supplements and drugs, some of which are obtained illicitly. In fact, an athletic appearance is so powerful that high school nonathletes use anabolic steroids to attain the higher muscularity and lower body fat that is typical of trained athletes (Calfee and Fadale, 2006). Anabolic steroid use by trained and recreational athletes for appearance purposes appears to be increasing. The majority of people who use anabolic steroids today are not elite athletes; rather, they are individuals who want to become more muscular (Kanayama, Hudson, and Pope, 2009; Copeland, Peters, and Dillon, 2000).

Holly Stein/AVP/Getty Images Sport/Getty Images

Many athletes' bodies are judged by the large audiences reached through the visual media.

$$\text{Target body weight} = \frac{\text{Current FFM}}{1 - \text{Desired \% BF}}$$

The formula assumes euhydration and a constant fat-free mass.

Figure 11.11 Target body weight formula

Legend: FFM = fat-free mass; BF = body fat

Current weight = 190 lb
Current % BF = 16% (obtained via underwater weighing)
Current % FFM = 84% or 0.84 (calculated from % BF)
Current FFM = 190 lb × 0.84 = 160 lb
Desired % body fat = 10% or 0.10

$$\text{Target body weight} = \frac{160 \text{ lb}}{1 - 0.10} = \frac{160 \text{ lb}}{0.90} = 178 \text{ lb}$$

Figure 11.12 Calculation of a target body weight

Legend: lb = pound; FFM = fat-free mass; BF = body fat

In this example, the athlete wants to lose body weight as fat.

$$\text{Target body weight} = \frac{\text{Current FFM} + \text{Desired FFM increase}}{1 - \text{Desired \% BF}}$$

$$\text{Target body weight} = \frac{160 \text{ lb} + 5 \text{ lb}}{1 - 0.08} = \frac{165 \text{ lb}}{0.92} = {\sim}179 \text{ lb}$$

Figure 11.13 Calculation of a target body weight

Legend: lb = pound; FFM = fat-free mass; BF = body fat

In this example, the athlete wants to change both lean body mass and fat mass (see text for details).

mass and body fat goals that consider their genetic predisposition to leanness and fatness. Once body composition goals are chosen, the weight that reflects those goals can be estimated. This weight is referred to as a target body weight or body weight goal. The target body weight is only an estimate, and rigid adherence to attaining a given scale weight or body composition is never recommended. However, a target body weight can be a helpful guideline, and a formula for calculating such a weight is shown in Figure 11.11. The formula assumes euhydration (Rankin, 2002).

For example, a baseball player currently weighs 190 lb (~86 kg) and is approximately 16 percent body fat (body fat range ~13–19 percent). His current fat-free mass is 84 percent of his weight, or ~160 lb (~72.5 kg), and he has approximately 30 lb (~13.5 kg) of body fat. His goal is ~10 percent body fat (that is, 90 percent fat-free mass), a figure that is consistent with his genetic predisposition to fatness and his sport (long-ball hitter and outfielder). As shown in Figure 11.12, if all weight is lost as fat, his target body weight to reflect a body composition of 10 percent body fat is ~178 lb (~81 kg).

The target weight formula considers the athlete's current amount of fat-free mass. What if the athlete wishes to gain muscle mass and lose body fat? A target weight can still be determined, but the desired increase in muscle mass must be added to the current fat-free mass. Continuing with the previous example, the 190 lb (~86 kg) baseball player wishes to gain 5 lb (~2.2 kg) of muscle mass (from 160 to 165 lb [72.7 to 75 kg]) and reduce body fat to ~8 percent for performance and appearance reasons. If he achieved these goals, his target weight would be ~179 lb (~81 kg) (Figure 11.13). It should be noted that *choosing* desirable amounts and proportions of fat-free mass and body fat is not difficult, but *achieving* such levels may be. Above all, body composition and weight goals must be realistic and achievable via diet and training programs that do not put the athlete's health at risk.

Application exercise

In order to improve speed, quickness, and endurance, a female soccer player needs to lose weight and body fat. She currently weighs 152 pounds and has 23 percent body fat. Using a percent body fat for female soccer players found in Table 11.2 as a guideline, calculate a target weight for that target percent body fat.

Body composition can be changed by increasing muscle mass.

When the appropriate stimulus is applied to skeletal muscle and the necessary hormonal and nutritional environment is present, muscle mass can increase. First, an overload stimulus must be applied consistently over time—the muscle must be stimulated to produce force at a greater frequency, intensity, and/or duration than is accustomed. Athletes generally accomplish this through one of many strength-training approaches. The increase in muscle mass is referred to as hypertrophy, and is a result of individual muscle fibers being stimulated to increase in size by synthesizing more contractile protein.

Role of exercise. For the sedentary adult or the athlete who is not accustomed to strength training, virtually any strength-training protocol will result in increases in strength and some initial increase in muscle mass. Once athletes are accustomed to basic strength training, further increases in muscle mass can be achieved through periodized strength training. The hypertrophy phase of periodized strength training is designed to maximize the potential for increasing muscle mass and is characterized by an emphasis on increasing the total volume of strength training. Increasing strength-training volume is accomplished by structuring a large number of sets and repetitions of a variety of strength-training exercises. The intensity or load (amount of weight lifted) is kept in the moderate range so that the prescribed number of sets and repetitions can be completed.

The amount of increase in muscle mass in response to strength training is difficult to accurately predict and is dependent upon a number of factors such as genetics, body type, hormonal status, and nutritional status. Those individuals with ectomorphic body types may not have the genetic disposition to add large amounts of muscle mass compared to those with more mesomorphic body types. Testosterone and growth hormone are the primary hormones responsible for stimulating an anabolic, or tissue-building, state in the body, particularly for muscle and connective tissue. There are large **intcrindividual** differences in circulating testosterone concentrations, certainly between males and females, but even among males.

Role of nutrition. Proper nutrition is necessary to support the increase in muscle size that is associated with the resistance-training programs described above. Although many nutrients are important for muscle growth, two areas receive the majority of attention—energy (kcal) and protein. Synthesis of muscle tissue requires positive energy balance (that is, caloric intake is greater than caloric expenditure). The athlete must also be in positive nitrogen balance and positive muscle protein balance. Positive nitrogen balance occurs when total nitrogen (protein) intake is greater than nitrogen lost via the urine and feces. In other words, the athlete must be consuming a sufficient amount of dietary protein. Positive muscle protein balance occurs when muscle protein synthesis is greater than muscle protein breakdown. To achieve positive energy, nitrogen, and muscle protein balance, an adequate energy intake is just as important as an adequate protein intake (Gropper, Smith, and Groff, 2009; Phillips, Hartman, and Wilkinson, 2005; Phillips, 2004; Tipton and Wolfe, 2001).

Athletes who wish to increase muscle mass should determine their baseline energy intake, which is the approximate amount of kilocalories needed daily to maintain current body weight and composition. Daily energy intake is greatly influenced by the amount of energy expended through physical activity. As discussed in Chapter 2, the daily energy requirement for female and male athletes is estimated to be approximately 35 and 38 kcal/kg, respectively, when activity is equivalent to moderate-intensity exercise 3–5 days/week or low-intensity and short-duration training daily. When energy expenditure is higher due to rigorous training on a near daily basis, the baseline energy requirement may be as high as approximately 38–40 kcal/kg for females and 45 kcal/kg for males. Extremely rigorous, high-volume training may require 50 or more kcal/kg for females and 60 or more kcal/kg for males.

It is estimated that an additional 5 kcal above baseline energy need is required to support the growth of 1 gram of tissue (Institute of Medicine, 2002). One pound of tissue weighs 454 g; thus a rule-of-thumb estimate is that approximately 2,300 kcal are needed to support the growth of 1 pound of muscle (454 g × 5 kcal/g = 2,270 kcal). Assuming that a male can gain 1 pound of muscle tissue per week (less for a typical female), it is estimated that, at a minimum, an additional 330 kcal per day need to be added to the baseline energy intake to support 1 pound of muscle growth a week. However, there is little research in this area, so the general recommendation is higher than the estimated minimum value. At present, to promote the growth of muscle tissue it is generally recommended that males increase daily caloric intake by 400 to 500 kcal (females to a lesser extent).

Some additional dietary protein is also needed to support the growth of muscle tissue. Approximately 22 percent of muscle is protein, so it is estimated that increasing muscle tissue by 1 pound would require the incorporation of approximately 100 g of protein (454 g × 0.22 = 100 g). If calculated on a daily basis, approximately 14 g of additional protein would be needed daily (100 g ÷ 7 days = 14 g), which in itself would provide ~56 kcal. These figures are by no means exact, but they do give athletes some guidelines for the additional energy and protein needed to support an increase in muscle tissue. The additional amount of protein needed is probably less than most athletes would estimate, however.

The following example illustrates the application of these guidelines. Sal is a 25-year-old male who was active in intramural sports while in college. Since graduation he has not participated in any kind of formal recreational activity, but he walks more than a mile each way to the subway station to commute to work. He has maintained his current weight of 161 lb (~73 kg) for the past year. Recently, he joined a gym and fitness center, which he plans to go to immediately after work. His goal is to lift weights and gain approximately 5 pounds (~2.2 kg) of muscle mass. Figure 11.14 shows Sal's current energy and protein intakes and his estimated energy and protein needs to increase 5 pounds of muscle mass.

Sal's case is typical of many athletes who find that their current protein intake already exceeds the amount needed for muscle growth. Sal needs to concentrate on increasing his caloric intake, including additional carbohydrate calories that will be needed for the resynthesis of muscle glycogen reduced by the resistance exercise. Since Sal will be going to the gym after work, he could eat a substantial snack while commuting to the gym. A 6-inch turkey deli sandwich and 8 ounces of orange juice would provide approximately 390 kcal, 72 g of carbohydrate, 19 g of protein, and 4.5 g of fat. Note that some of the additional calories are in the form of proteins but that the majority is provided by carbohydrates. This kind of snack daily, in addition to his usual intake, provides the additional nutrients he needs to support muscle growth.

Determine baseline intake:
    Current energy need to maintain body composition:
        2,774 kcal (73 kg × 38 kcal/kg)
    Current daily protein need: 58 g (73 kg × 0.8 g/kg)

Recommended changes to increase 5 lb lean mass
(assumes resistance exercise):
    Daily energy intake:
    ~3,174 kcal (2,774 kcal + ~400 kcal)
    Daily protein intake: ~72 g (58 g + ~14 g)

Current dietary intake:
    Energy: ~2,800 kcal
    Macronutrient distribution: carbohydrate: 420 g
        (60% of total caloric intake), protein: 84 g
        (12%), fat: 78 g (25%), alcohol: 12 g (3%)

Figure 11.14 Baseline and projected energy and protein needs of a recreational athlete to increase muscle mass

Legend: kcal = kilocalorie; kg = kilogram; kcal/kg = kilocalorie per kilogram body weight; g = gram; g/kg = gram per kilogram body weight; lb = pound

Body composition can be changed by decreasing body fat.

The general principles for the loss of body fat are the same for athletes and nonathletes: an increase in energy expenditure (activity or exercise), a decrease in food (energy) consumption, or a combination of both. There are thousands of weight-loss diets, but the common denominator is a decrease in caloric intake that results in fat loss over time. For the obese, sedentary individual, the restriction of total energy intake and the length of the energy restriction seem to be more important than the carbohydrate, protein, and fat content (that is, macronutrient composition) of the diet. Higher-protein, low-energy diets may be beneficial for athletes because they may help to protect against the loss of lean body mass, but carbohydrate intake must be sufficient to support the resynthesis of muscle glycogen depleted by training (Layman et al., 2005).

Role of exercise. Exercise plays an important role in weight loss and change in body composition. The strategy for long-term weight loss and reduction of body fat is to achieve a moderate caloric deficit, or to expend more calories than are consumed. As discussed below, this can be achieved solely through reductions in food and beverage consumption, but this is generally not the recommended approach. Particularly if the caloric deficit is large, weight loss can be substantial, but a relatively large proportion of the weight lost is from the fat-free component of the body. In other words, lean body mass is lost, which is not desirable, particularly for the athlete.

Exercise can be used to increase caloric expenditure, which contributes to the caloric deficit necessary for weight and fat loss. In addition, the stimulus of exercise helps maintain fat-free mass while body weight and body fat decline. Manipulations of frequency, intensity, and duration of exercise, particularly aerobic-type exercise, can substantially increase caloric expenditure. In the face of a moderate caloric deficit, an athlete who is performing aerobic exercise will generally maintain fat-free mass or may experience only small declines. An athlete who incorporates strength training into his or her training program while experiencing a mild caloric deficit may experience increases in muscle mass and therefore fat-free mass, while losing a small amount of body fat at the same time. This example illustrates the concern of relying too much on scale weight rather than body composition when the goal is to lose body fat. If muscle mass is slowly increasing at the same time that body fat is slowly decreasing, the scale weight the athlete takes each morning may not be changing perceptibly. The athlete may be discouraged by this lack of change in scale weight and misinterpret it to mean that no progress is being made when, in fact, desirable changes in body composition are occurring.

Role of nutrition. To lose body fat, energy expenditure must be greater than energy intake. It is recommended that activity or exercise be increased and food intake decreased. However, athletes must consider the impact that increased exercise or decreased food intake will have on their ability to train and perform. Too much exercise can result in injury, and low energy, carbohydrate, and protein intakes can result in inadequate muscle glycogen resynthesis and loss of lean body mass. For many trained athletes, the extent to which they can increase exercise and decrease food consumption is limited, and these limitations result in a slow loss of body fat.

Recall from Chapter 4 that an athlete's daily carbohydrate intake should not be less than 5 g/kg body weight to ensure adequate muscle glycogen resynthesis. Athletes who restrict energy intake should increase protein intake to at least 1.4 g/kg body weight to protect against large losses of lean body mass (see Chapter 5). Meeting these two nutrient recommendations requires an energy intake of approximately 26 kcal/kg (6.4 g × 4 kcal/g). The diet should not be too low in fat for many reasons, including the difficulty involved with complying with a very low fat diet. Considering minimum carbohydrate, protein, and fat guidelines, it is generally recommended that athletes who wish to lose body fat not restrict energy intake to less than

Interindividual: A comparison or observation made between people.

30 kcal/kg daily. To meet vitamin and mineral requirements, these diets need to contain many nutrient-dense foods. Some athletes, such as bodybuilders preparing for a contest, may employ more drastic reductions in energy intake but such diets are short term. Diets containing less than 30 kcal/kg daily typically do not meet daily vitamin and mineral requirements and tend to be extremely low in fat.

Athletes who restrict energy to lose body fat look for ways to do so while preserving muscle mass. Unfortunately, scientific study in this area using athletes as subjects is sparse. Mettler, Mitchell, and Tipton (2010) found that a high-protein (~2.3 g/kg/d) weight-loss diet for 2 weeks reduced the loss of lean body mass in a group of resistance-trained athletes when compared to the control group, who consumed ~1.0 g/kg daily. Both groups consumed 60 percent less than their usual caloric intake for 14 days. The athletes in the high-protein diet group reported more fatigue and less feelings of well-being than the control group, but these feelings did not impact their ability to maintain the volume and intensity of training required during the 14-day study. Athletes are often advised to include resistance exercise along with a higher-protein (but still energy-restricted) diet in the hope that muscle mass can be preserved. Much more research is needed regarding the best ways to protect against the loss of muscle mass in both athletes and nonathletes, but this seems to be prudent advice (Stiegler and Cunliffe, 2006).

Increasing muscle mass while decreasing body fat is difficult.

Many athletes want to simultaneously increase muscle mass and decrease body fat. In other words, they want to remain the same weight but they want to "replace" 5–10 lb (~2.2–4.5 kg) of fat with 5–10 lb of muscle. Although this sounds as if it would be easy for the body to accomplish, anabolism (synthesis) and catabolism (breakdown) are biologically opposite processes and it is difficult to estimate an appropriate daily caloric intake to achieve both simultaneously. A prudent recommendation is to focus on one goal at a time. For many athletes, there is a greater benefit to increasing muscle mass than to decreasing body fat. Weight may eventually be the same, but it will probably fluctuate as muscle mass is increased and then body fat is decreased.

Body composition changes may be seasonal.

The magnitude of the desired fat loss or lean body mass gain is a factor in deciding the best time in the training cycle to make body composition changes. Small losses of body fat and weight may be a consequence of the athlete's return to preseason training from the relatively sedentary off-season period. If the athlete continues to consume approximately the same amount of energy

(kcal), then the increase in energy expenditure from a return to training will result in a loss of body fat. Similarly, an athlete who experiences a small loss of muscle mass in the off-season will see a gain in lean body mass with a properly designed preseason resistance exercise and nutrition program. Athletes can maintain a relatively stable body composition and weight by adjusting their energy intake to meet the energy expenditure requirements of each training cycle.

Larger decreases in body fat require a moderate reduction in food intake along with an increase in energy expenditure, because training alone would not likely create a large enough energy deficit to lose a substantial amount of fat in the preseason training period. Severe reduction in food intake, such as fasting or a very low calorie diet, is not recommended because it interferes with the athlete's ability to train. A moderate reduction in food intake will result in a slow weight loss; therefore, losing considerable amounts of body fat means that weight-loss strategies must be initiated well before the competitive season. In fact, the athlete who wishes to lose relatively large amounts of body fat should begin the process after the competitive season ends and, if necessary, continue the weight-loss plan through the early part of the preseason. Trying to lose large amounts of body fat during the later part of the preseason when training volume is high or during the precompetition and competition periods can be detrimental to training and performance, because energy intake must be reduced at a time when energy and carbohydrate needs are high.

The off-season is characterized by a reduction in exercise compared to the competitive season. It is also time away from the rigors of training. Some athletes may wish for time away from the rigors of following a diet that supports training, such as high daily carbohydrate consumption, timing of food intake, and monitoring of energy intake. The off-season can be an important break from disciplined eating, although it is not a time for reckless abandon of all dietary restraint. However, the loss of a large amount of body fat takes time, and some of that time will likely be during the off-season when a moderate restriction of energy intake is feasible and will not interfere with training. The suggestion to lose some weight in the off-season may come as a surprise to many athletes and may be difficult for some who prefer to have few dietary restrictions during the off-season. Some of these athletes believe that large and rapid fat loss will be possible early in the preseason when they return to training. Many are disappointed that preseason losses are not larger and that fat is not lost as rapidly as they had hoped, and some engage in "**crash diets**" that produce large and rapid weight loss but are detrimental to training, hydration status, and health.

Body composition as well as athleticism is important for cheerleading.

$$\text{Minimum body weight} = \frac{\text{Current FFM}}{1 - \text{Minimum \% BF}}$$

Figure 11.15 Minimum body weight formula

Legend: FFM = fat-free mass; BF = body fat

Large increases in muscle mass also take time. Male strength athletes in their 20s, such as (American) football players or bodybuilders, may increase lean body mass by 20 percent in the first year of a regular, heavy-resistance training program supported by a diet with adequate energy, carbohydrate, and protein intakes (Lemon, 1994). A 190 lb (~86 kg) football player might add up to 30 to 35 lb (~13.5 to 16 kg) of lean body mass in the first year of a dedicated training program for collegiate football. However, after the first year, gains in lean body mass are much smaller and increases of 1–3 percent are more likely in subsequent years. In other words, untrained athletes can experience large initial gains, but trained athletes are likely to experience small gains. The 190 lb (~86 kg) collegiate freshman football player who is 220 lb (100 kg) at the beginning of his sophomore year could reasonably expect to gain 2 to 6 lb (~1 to 3 kg) of lean body mass, and not an additional 30 to 35 lb (~13.5 to 16 kg). Women cannot expect to gain as much muscle mass as men, and it is estimated that women will experience approximately 50–75 percent of the gains seen in men (Stone, 1994).

Athletes who compete in lightweight sports push the biological envelope.

Some sports have designated weight categories because differences in body size make it impossible for all athletes to fairly compete among one another. Examples include wrestling, boxing, martial arts, and lightweight rowing. In sports in which weight must

be moved, such as distance running, gymnastics, high jumping, or ski jumping, participants with a low body weight (but sufficient muscularity) generally have a performance advantage over those with a higher body weight due to larger amounts of body fat. Women's gymnastics, rhythmic gymnastics, and figure skating have a subjectively scored element, and a low body weight may influence artistry scores. Cheerleading and ballet dancing, which require athleticism, are not scored, but being chosen for participation may depend on body composition.

All of these athletic events could attract athletes who are naturally light in weight and who could easily maintain a biologically comfortable low body weight. *Biologically comfortable weight* and *naturally lightweight* are terms used to describe individuals who do not need to engage in chronic energy restriction or acute fluid loss to maintain a low body weight. Indeed, these individuals are found in all of these sports from the recreational to the elite level. However, studies of wrestlers, rowers, boxers, and jockeys have reported that the majority of athletes are competing in a weight class that is below their natural weight (Kazemi, Shearer, and Choung, 2005; Hall and Lane, 2001; Filaire et al., 2001). One group of researchers found that the majority of lightweight rowers surveyed were not naturally lightweight and that 76.5 percent of the males and 84 percent of the females studied reduced their body weights in the 4 weeks before a major lightweight rowing competition (Slater et al., 2005).

As described previously, a target weight based on desirable body composition can be determined. For athletes competing in low-body-weight sports, it is critical that a *minimum* body weight also be calculated. A minimum body weight calculation takes into account the lowest percentage of body fat that an athlete could likely achieve without putting health at risk. A minimum weight formula is shown in Figure 11.15. The Spotlight on a real athlete: Sondra, a Super-lightweight Kickboxer, illustrates how this calculation could be used to provide the athlete with important information.

Crash diet: Severe restriction of food intake in an attempt to lose large amounts of body fat rapidly.

Minimum body weight formulas are now used in wrestling to determine the appropriate weight category for competition. The National Federation of State High School Associations instituted rule changes beginning with the 2006–7 season that included: (1) a body fat assessment no lower than 7 percent in males and 12 percent in females, (2) a monitored weight-loss program that does not exceed 1.5 percent loss of body weight per week (7 days), and (3) a specific gravity of urine not to exceed 1.025 (a measure of hydration status). Many high schools are depending on certified athletic trainers to administer and monitor the program (National Federation of State High School Associations).

Weight cycling in athletes. Weight cycling is defined as repeated bouts of weight loss and weight gain. Weight cycling is common in sports where there are weight classes and an athlete's weight must be certified before competition (for example, wrestling, lightweight rowing, boxing, judo, and tae kwon do). In these sports weight cycling is an established part of the sports' culture. The following section will focus on sports in which athletes frequently want to "make weight."

Athletes whose weight must be certified for competition often believe that weight cycling is necessary. Hall and Lane (2001) studied 16 amateur boxers and found that each had four weight goals during the year.

Spotlight on a real athlete

Sondra, a Superlightweight Kickboxer

Sondra, a 28-year-old superlightweight kickboxer, weighs 136 lb (61.8 kg) and is already lean, having an estimated 14 percent body fat. She would like to compete in the lightweight category, which has a maximum weight limit of 132 lb (60 kg). Sondra's dilemma is that she is not competitive with the other superlightweight kickboxers, so she is considering competing in a lower weight category although she has not weighed 132 lb (60 kg) or less since high school. Sondra is sure that she could lose 4 pounds if she put her mind to it; after all, discipline is one of the characteristics of the martial arts. She meets with a sports dietitian who is sympathetic to her goal to be a lightweight kickboxer, but indicates that it is important to consider all the effects weight loss can have on an already lean athlete's performance and health. To Sondra's surprise the detrimental effects are far reaching and include hormonal, bone mineral density, lean body mass, and mental changes. The sports dietitian calculates a minimum weight and explains its meaning (see figure).

The minimum body weight formula uses current weight and minimum percentage of body fat. At 136 lb (61.8 kg) and an estimated 14 percent body fat, Sondra has approximately 117 lb (~53 kg) of fat-free mass. Assuming that she could lose weight only as body fat and achieve 12 percent body fat, both of which would be a challenge, Sondra's weight would be only 133 lb (~60.5 kg). She would need to chronically undereat or voluntarily dehydrate before weigh-in to be certified as a lightweight. The sports dietitian points out that chronic low energy intake would put her at risk for losing some of her lean body mass and possibly developing disordered eating, amenorrhea, and low bone mass (see Chapter 12).

Sondra realizes that she is setting a goal weight that would be difficult to attain and maintain, yet she finds it hard to let go

$$\text{Minimum body weight} = \frac{\text{Current FFM}}{1 - \text{Minimum \% BF}}$$

$$\text{Minimum body weight} = \frac{117 \text{ lb}}{1 - 0.12} = \frac{117 \text{ lb}}{0.88} = 133 \text{ lb}$$

Calculation of a minimum body weight

Legend: lb = pound; FFM = fat-free mass; BF = body fat

of the belief that getting to 132 lb (60 kg) is just a matter of having enough discipline. The dietitian encourages her to articulate her goals and reflect on what is needed to achieve them. Sondra's three main goals are to perform her best, be physically fit, and have fun. Good performance requires disciplined training, which Sondra enjoys because it is a challenge, but she knows that she does not have the skill to compete at the highest levels of competition. She also realizes that she loves to eat, and her high level of training gives her the ability to eat a lot of food and easily maintain her natural weight and her fit and muscular body. She knows that some of the women who compete in the lower weight categories struggle with the weight limit, and she has observed that these women often looked wan, tired, and sad.

The minimum body weight formula may help an athlete like Sondra to realize in which body compartments the "weight" would need to be lost and if weight loss would likely mean loss of lean body mass or body water. Considering all of the factors, not just if she has the discipline to reach a certain scale weight, Sondra decides to remain a superlightweight kickboxer. The minimum weight formula helped her to understand that 132 lb (60 kg) was not the right weight goal for her.

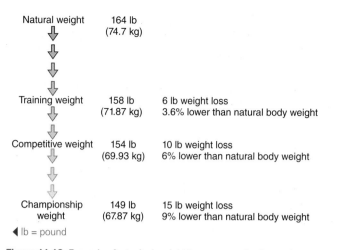

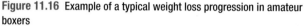

Figure 11.16 Example of a typical weight loss progression in amateur boxers

These weights included (1) natural weight, (2) training weight, (3) competitive weight, and (4) championship weight. Figure 11.16 illustrates the average weights for these 16 boxers and shows the progression of weight loss from the preseason (natural weight) through the competitive season—74.7 kg (~164 lb) → 71.87 kg (~158 lb) → 69.93 kg (~154 lb) → 67.87 kg (~149 lb). On average the difference between natural and championship weight was approximately 5 percent of body weight. All the subjects in this study believed that weight loss was necessary and that the loss of weight would improve their performance. They also believed that food and fluid intake after the certification of weight restored the nutrients and strength lost as a result of making weight. Weight cycling, both gradually over the season and rapidly during a given week in the season, is also documented in wrestling, judo, and tae kwon do and parallels boxing in many respects.

Weight is typically lost by increasing exercise, reducing food intake, and restricting fluid intake. Reducing natural weight to a training weight may be relatively easy since the energy expended during training is substantially greater than the energy expended during the off-season. Food intake is often reduced at the same time that exercise is increased, so weight loss is an expected outcome. However, reducing training weight to precompetition and competition weights, which may be considerably below natural weight, typically require more extreme interventions.

During the precompetition and competition phases, athletes often find that it is harder to reach the desired weight or to attain that weight each week. In studies of boxing and tae kwon do participants, researchers report that nearly all subjects restricted their food intake to a greater degree during the competition phases than during the training phase. One to 2 days prior to weigh-in, many engaged in partial or full fasting (restricting food intake only or both food and fluid intake, respectively). Fluid intake was frequently restricted at least 24 hours prior to weigh-in. On the day of the weigh-in, exercise in the form of skipping rope and running was used to further reduce weight, and rapid water-loss methods, such as exercising in the heat or with sweat-inducing plastic suits and use of diuretics, were employed if goal weight was not likely to be obtained by any other means (Hall and Lane, 2001; Kazemi, Shearer, and Choung, 2005).

In 1994, Scott, Horswill, and Dick reported the results of a study of collegiate wrestlers competing in the 1992 season-ending National Collegiate Athletic Association (NCAA) tournament. At the time of the study there were approximately 20 hours between weigh-in and competition. The average weight gain after weigh-in was 3.73 kg (~8 lb). The wrestlers gained an average of 4.9 percent of body weight after weigh-in, with the wrestlers in the lower weight categories gaining the most weight. These results confirmed anecdotal reports of widespread use of weight cycling among elite collegiate wrestlers.

In the United States, rule changes were made in both NCAA and high school wrestling following the deaths of three collegiate wrestlers in 1997. Perhaps the most important NCAA rule change was moving the weigh-in to approximately 2 hours prior to competition. In 1999, the second year of the rule change, NCAA tournament wrestlers gained approximately 0.66 kg (~1.5 lb) after weigh-in (Scott et al., 2000). However, international-style wrestling (that is, freestyle and Greco-Roman) has not made the time-of-weigh-in rule change, and in 2004 Alderman et al. found that rapid weight loss and gain was still widely practiced. Their sample of 2,600 international-style wrestlers found that the average weight gain after weigh-in was 3.4 kg (~7.5 lb) or 4.8 percent of body weight. The more successful wrestlers, who were also older, gained significantly more weight after weight certification than the younger, less successful wrestlers.

Although it is not easy to study, there is some research documenting the effects of weight cycling, particularly rapid weight loss and gain, on performance, mental state, and health. Most studies have found that physical performance was not impaired, at least for measures of short-term, high-intensity exercise. Smith et al. (2001) found no significant differences in the performance of eight amateur boxers in a crossover study when they restricted food and fluid intake compared to when they did not. Fogelholm et al. (1993) found no decline in measures of sprinting, jump height, or anaerobic performance in wrestling or judo, with a 5 percent or less loss of body weight achieved either gradually (over 3 weeks) or rapidly (2.4 days). In the study of the 16 amateur boxers (Hall and Lane, 2001), the subjects were asked to

Can Boxers Effectively "Make Weight" While Following a Nutritious Diet?

Boxing is a sport in which participants are matched to a certain extent by having competitive divisions based upon body weight. Many boxers believe that they will have a competitive advantage if they participate in the lightest weight class possible, resulting in the establishment of traditional training practices in the sport for "making weight." Boxers must be at or below the maximum allowable weight for a particular weight class at the time of weight certification, known as the weigh-in, which typically occurs the day before competition. Traditional training and nutritional strategies often used to make weight raise both short- and long-term health concerns as they include severe caloric restriction and intentional dehydration through exercise and fluid restriction.

Is it possible for boxers competing at a high level to make weight for their desired weight class following a plan that adheres to healthy nutrition principles? The authors of the following study outline a case study of a highly successful professional boxer who was able to successfully do just that by dramatically changing his nutritional approach: Morton, J. P., Robertson, C., Sutton, L., &, MacLaren, D. P. (2010). Making the weight: A case study from professional boxing. *International Journal of Sport Nutrition and Exercise Metabolism, 20*(1), 80–85.

What did the researchers do? The researchers worked with a 25-year-old male professional boxer as a client to implement a 12-week nutrition and exercise plan to achieve his goal of making the 59 kg (130 lb) superfeatherweight division for his next competition. With an initial body weight of 68.3 kg (150 lb), the goal required a body weight loss exceeding 20 pounds (9 kg).

The subject began the study at end of an "off-season" period—3 months in which he did not compete or engage in any structured training. An initial assessment was performed of body weight, body composition, and bone density by DEXA, maximal oxygen consumption, and muscular strength. Resting metabolic rate (RMR) was estimated using the Cunningham equation (see Chapter 2). A description of his typical approach to making weight was obtained, which was characterized by severe caloric restriction (one small meal per day), fluid intake restriction, and exercise in clothing that encouraged sweat loss. More severe fluid restriction and intentional dehydration was used in the days and hours before weigh-in to make weight.

The researchers devised a nutrition and exercise plan to result in gradual weight loss (0.5 to 1.0 kg/week) over a 12-week period without using intentional dehydration. The boxer was given menu plans to follow designed for a target energy intake equal to the athlete's RMR that, when coupled with the rigorous training program, would result in an energy deficit leading to weight loss. The menu plan consisted of three meals and one snack each day, with 2.0 g/kg of protein (including protein-based supplements), low- to moderate-glycemic-index carbohydrate foods, and unsaturated fats. Instead of restricting fluid intake, water or low-calorie flavored beverages were consumed throughout the day, with particular emphasis on rehydrating after training sessions. Adherence to the menu plans and amount of food intake was checked regularly during the study. Exercise consisted of traditional running and boxing-specific training, as well as newly implemented strength and conditioning sessions lasting approximately 1 hour each afternoon.

Body weight was measured each week, and body composition was determined every 4 weeks of the 12-week intervention. Body weight was measured again at the official weigh-in the day before his competition and again the day of the bout. The boxer did not engage in any intentional dehydration before the weigh-in, and afterward consumed both high- and low-glycemic-index carbohydrate foods amounting to 12 g/kg and hydrated in preparation for the competition.

What did the researchers find? The boxer successfully made the weight for his goal classification, decreasing his weight to 58.9 kg (129.9 lb) at weigh-in. Over the 12-week intervention period, he lost 9.4 kg (20.7 lb) in a gradual fashion, averaging 0.9 kg (2 lb) per week. Percent body fat declined 5 percent, and there was a loss of lean body mass just under 4 kg (9 lb). The subject adhered successfully to the menu plan and often reported he was consuming more food than typical during his previous training. The feeding and hydration strategy in the day between weigh-in and the competition caused no gastrointestinal problems and resulted in a 4.3 kg (nearly 10-pound!) weight gain, allowing him to fight at a weight near the top of the next weight classification.

What is the significance of the research study? Changing traditional, accepted practice in a sport can be difficult, even when there are health concerns associated with those practices. Scientific studies are difficult to conduct due to the reluctance of athletes and coaches to interrupt their normal training and competition to adhere to a new and different research intervention, a reluctance that tends to increase as the performance level of the athletes increases. Although the experiences of a single athlete can be difficult to extrapolate to others, the results of a well-conducted case study such as this one can provide important information that can both help other athletes and provide the basis for additional study. Morton et al. (2010) were able to demonstrate that an elite boxer was able to achieve his weight class goals by following a nutritionally sound diet and training program that was a radical departure from his previous practice and the traditionally accepted methods of his sport. The target weight loss was achieved gradually and without severe caloric restriction or intentional dehydration, resulting in a safer approach and one that may positively affect nutritional choices during other periods of training as well as throughout life.

set a goal for the number of repetitions they wanted to perform on a circuit-training protocol that simulated a boxing match. There was no significant difference between the number of repetitions achieved at the training weight and the championship weight, a difference of about 5 percent of body weight. However, the boxers *expected* that they would be able to perform approximately 15 more repetitions at their championship weight; none did. In other words, these boxers believed that losing ~5 percent of their weight would improve performance but it did not.

Morton et al. (2010) use a case study to show professional boxers can successfully lose weight as body fat gradually and forego the time-honored "making weight" approach that relies on severe energy restriction and dehydration (see Focus on research). Using a scientific and structured plan, the authors report that the athlete reduced body weight by 9.4 kg (~21 lb) over a 12-week period. This case study illustrates how difficult it is for athletes in weight-certification sports to meet their weight and body composition goals in a healthy way.

There has been speculation that weight cycling results in a decrease in resting metabolic rate (RMR). This is one theory behind the observation that athletes have a more difficult time making weight week after week, especially near the end of the season. However, most research in athletes does not demonstrate that weight cycling results in a measurable reduction in RMR in the short term (week by week in season) or the long term (over two or three seasons). Most of these studies were conducted in the 1990s, and there are no recent studies of athletes and the effect of weight cycling on RMR (McCargar et al., 1993; McCargar and Crawford, 1992; Horswill, 1993; Schmidt, Corrigan, and Melby, 1993; Melby, Schmidt, and Corrigan, 1990).

Mental state, however, seems to change with weight cycling. Higher scores on measures of anger, tension, and fatigue and lower vigor scores are reported in amateur boxers and judoka (judo athletes). These changes are likely due to low energy intake (that is, semistarvation), low carbohydrate intake, and hypohydration (Hall and Lane, 2001; Filaire et al., 2001).

The biggest area of concern is the health of the athlete who needs to "cut" weight, a term used to indicate extreme diet and exercise measures to produce rapid weight loss. Alderman et al. (2004) found that more than 40 percent of wrestlers who engaged in rapid weight loss experienced headache, dizziness, or nausea at least once during the season. Other side effects, such as nosebleeds, disorientation, or a racing heart rate, also occurred in some wrestlers but with less frequency. The medical consequences associated with hyperthermia

(elevated body temperature) and hypohydration as a result of rapid weight-loss techniques are well known. These conditions, some of which may be fatal, are discussed in Chapter 7.

A 2006 study of former Finnish athletes raises the possibility that weight cycling by athletes may predispose them to obesity later in life. Approximately 1,800 male elite athletes completed questionnaires in 1985, 1995, and 2001. The 370 weight cyclers, former boxers, weight lifters, and wrestlers, had an average weight gain of 5.2 BMI units (~26 lb or 12 kg) compared to 3.3 BMI units (16.5 lb or 7.5 kg) in non-weight-cycling former athletes (Saarni et al., 2006).

Underweight athletes may need to increase muscle mass and body fat.

Some athletes are underweight and want to increase both muscle mass and body fat. This population has not been well studied. It is generally recommended that energy intake be increased by 500 kcal daily (Rankin, 2002). The additional energy should come from nutritious foods such as fiber-containing carbohydrates, proteins, and heart-healthy fats. Fat is the most energy-dense nutrient; however, simply adding large amounts of additional fats to the diet may be counterproductive. In some underweight people, a high-fat meal or snack is so **satiating** that food is not consumed again for many hours and results in a net decrease in the total energy (kcal) intake for the day. More information on dietary strategies for underweight athletes can be found in Chapter 10.

Key points

- Many athletes seek to change body weight or composition as a way to improve performance.

- Once body composition goals are established, an athlete can estimate a target weight that reflects those goals.

- To increase skeletal muscle mass, the athlete must engage in a resistance-training program that is supported by sufficient energy and nutrient intake.

- To decrease body fat, the athlete must increase energy expenditure, decrease energy intake, or both.

- The process of losing weight as body fat also results in the loss of lean body mass. To protect skeletal muscle mass from being lost, it may be prudent to consume a higher protein (but still energy restricted) diet.

- The timing of weight and body composition changes is important so that neither training nor performance is negatively affected.

Why is establishing a minimum body weight an important performance and health protection?

11.5 Supplements Used to Change Body Composition

Changing body composition through diet and exercise demands daily attention and discipline, and is typically a slow process. Therefore, it is not surprising that athletes look to supplementation of substances that promise to build muscle and reduce body fat easily and quickly. Some of these supplements may contain substances that are banned by sports-governing bodies. Before taking *any* supplement athletes should ask five critical questions: (1) Is it **legal**? (2) Is it **ethical**? (3) Is it **pure**? (4) Is it **safe**? and (5) Is it **effective**?

Supplements are often used to help increase muscle mass.

Perhaps no group of supplements holds more promise in the eyes of athletes than those involved in muscle protein synthesis. Increased muscle size and strength are important performance factors in many sports. As discussed previously, the building of muscle tissue through training and diet is a slow process that requires hard work and discipline. Substances that have received tremendous attention include testosterone and testosterone precursors.

Anabolic steroids. Testosterone is a hormone that influences muscle protein synthesis. The use of anabolic steroids, scheduled drugs that are nearly identical to testosterone, is known to increase muscle mass and, in some individuals, muscle strength (American College of Sports Medicine, 1984). The self-prescribed use of anabolic steroids is illegal, prohibited by sports governing bodies for ethical and safety reasons, and associated with some substantial medical risks, especially for females because of their irreversibility.

Those who work in sports-related fields might encounter athletes who use or are considering using anabolic steroids to increase muscle mass. A National

Spotlight on a real athlete

One Wrestler's True Story

"I was a senior in high school, 18 years old, and trying to lose weight for wrestling. I was 5'8" tall and started out weighing 145 pounds. Within about 3 weeks I had lost 17 pounds to make the 128-pound weight class. I was on a strict diet of 1,100 calories per day. Two days before I had to make weight, I would stop eating. The day before weigh-ins, I would not drink anything. I was always trying to burn off 'extra' pounds by jumping rope or jogging. Whenever I worked out, I always wore two pairs of sweats in order to lose water weight. I jogged 12 miles one day to make weight for a match that evening. I would fast, and dehydrate myself by sweating and spitting in a cup in order to make weight. I never took laxatives, diuretics, or vomited to make weight. After weigh-ins, I ate as much as I wanted to without eating so much as to hinder my performance. On weekend tournaments I also ate between matches and afterwards. On Sunday, I started cutting back on my caloric intake, and by Monday I was back down to 1,100 calories. I was usually at least 10 pounds overweight on Monday, also. I weighed myself around four times per day. It was such a physical and emotional strain on me to make weight I could only make the 128-pound class about once every 2 weeks. The other times I only had to make the 134-pound weight class. The trainer did a skinfold measurement to check my body composition and came up with a figure of below 3 percent body fat. I was probably in the best cardiovascular shape of my life due to my constant cardiovascular workouts. However, I did lose strength and muscle mass. I looked much thinner than I did before, and my cheeks and eyes were sunken in and I had dark circles under my eyes. I had trouble falling asleep at night. I constantly felt cold. I never seemed to generate enough body heat to keep warm. I was the only one wearing a jacket in my classes, and I had to sit on my hands to keep them warm. When I saw my friends with food or a soft drink, I would cuss at them. Fortunately, I was able to maintain a 4.0 grade point average. I became obsessed with food because I was so hungry and I couldn't eat. To compensate for that, I used to go to the grocery store a couple of times per week and walk up and down every aisle and make a mental list of all the foods I was going to eat when the season was over. I also used to do a lot of cooking and baking, but not eat any of it. By the end of the season, it had taken its toll on me. I secretly hoped that I would be eliminated and not qualify for the next tournament, so my season would be over and I could eat as much as I wanted to. I always tried my best to win, and I never tried to lose on purpose, but I think that it subconsciously affected my performance. Why would anyone do this to himself? My coaches told me they needed me to compete at that weight class. They told me I was the best wrestler at that weight class and it would make the team stronger if I competed there. One of my coaches told me how he did the same thing in high school and that if he could do it, I could do it, and I believed him. What they, and I, didn't realize was that I probably would have wrestled better, and been more of an asset to the team, if I had wrestled at a higher weight. If I had known then what I now know about the relationship between nutrition and athletic performance, I would have wrestled at the 140-pound weight class instead."

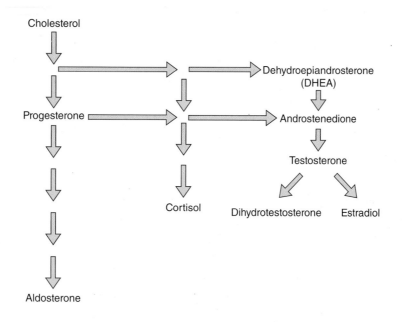

Figure 11.17 Testosterone synthesis

The biochemical pathway of testosterone synthesis.

Institute on Drug Abuse (NIDA) study reports that 2.6 percent of high school seniors have used anabolic steroids at least once. The prevalence in adults in not known but is estimated to be in the hundreds of thousands (NIDA, 2005).

Athletes should be aware of the many legal, ethical, and safety issues involved. The impact on health may be mild to severe and, in rare cases, may result in death. Blood pressure and low-density lipoprotein concentration may increase whereas high-density lipoprotein concentration may decrease, which are risk factors for cardiovascular disease (see Chapter 13). Aggression, depression, and other psychological effects have been reported. Use by males may result in reduction in testicle size, accelerated baldness, and the development of breast tissue. Use by females may result in changes in menstruation and reproductive organs, baldness, lowering of the voice, and growth of facial hair. Adolescents, because they are typically still in a growth state, risk premature skeletal maturation before reaching their genetic potential for height. The National Institute on Drug Abuse and most sports-governing bodies have anabolic steroid information available for athletes.

Prohormones. Because athletes can be permanently banned from their sport for testing positive for anabolic steroids, many look for dietary supplements that would provide similar benefits. Among the most popular are the prohormones, compounds that are precursors to testosterone. Many prohormones, such as androstenedione, are regulated under the Anabolic Steroid Control Act of 2004 and are banned by sports-governing bodies.

As shown in Figure 11.17, many compounds are involved in the synthesis of testosterone. Cholesterol is the precursor to testosterone and related compounds. Action by various enzymes results in the conversion of cholesterol to progesterone to androstenedione and to testosterone. Using a different biochemical pathway, cholesterol can also be converted to testosterone via dehydroepiandrosterone. Androstenedione and testosterone are precursors to estrogens such as estrone and estradiol.

Androstenedione, androstenediol, and, to a lesser extent, DHEA are often referred to as *prohormones*. They are precursors to testosterone and have similar, but not exact, chemical structures. Athletes hope that these prohormones will elevate testosterone concentration and consequently increase muscle protein synthesis. Studies of all of these compounds have yet to confirm this hope. Although some studies showed a short-term rise in testosterone concentration with supplementation, this short-term rise has no effect on muscle size, strength, or power (Broeder et al., 2000; Tipton and Ferrando, 2008). In fact, androstenedione supplementation preferentially increases estradiol (via estrone), not testosterone, which can result in the development of breast tissue in males (Wolfe, 2000).

Satiate: To satisfy hunger.

Legal: Allowed under the law.

Ethical: Consistent with agreed principles of correct moral conduct.

Pure: Generally defined as free of contamination. FDA definition of purity is "that portion of a dietary supplement that represents the intended product."

Safe: Unlikely to cause harm, injury, or damage.

Effective: Causing a result, especially one that is intended or desired.

Androstenedione supplements became extremely popular when baseball star Mark McGwire admitted to taking "andro" during his home run record-setting season in 1998. His taking this supplement may have influenced many younger and less accomplished players to do so (Brown, Basil, and Bocarnea, 2003). The purity and safety of androstenedione supplements has been questioned. Suspicions have been raised about whether some androstenedione supplements may have been "spiked" with anabolic steroids.

In March 2004, the Food and Drug Administration (FDA) released a **white paper** that listed more than 25 potential androgenic and estrogenic effects associated with androstenedione use (FDA, 2004). This review of the scientific literature prompted the FDA to crack down on dietary supplements containing androstenedione. The FDA considers these supplements to be adulterated (impure) and therefore illegal to market. Androstenedione is a banned substance by most sports-governing bodies, which consider its use unethical and unsafe. The Anabolic Steroid Control Act of 2004 defined the term *anabolic steroid* as any drug or hormonal substance chemically and pharmacologically related to testosterone (other than estrogens, progestins, corticosteroids and dehydroepiandrosterone). The numerous compounds that met this definition are considered controlled substances in the United States, which means that they cannot be sold without a prescription.

Dehydroepiandrosterone (DHEA) is also a precursor to testosterone and estrogen, but it is considered a weak androgen (steroid) and is not included under the Anabolic Steroid Control Act. It has a more general effect on tissues than anabolic steroids or androstenedione. DHEA diminishes substantially after early adulthood, probably due to a decrease in the number of cells that produce it (Hornsby, 1997). Thus supplements are often advertised as being "a fountain of youth." There is no scientific evidence that DHEA has an anabolic effect or can enhance athletic performance (Corrigan, 2002; Brown, Vukovich, and King, 2006; Tipton and Ferrando, 2008). However, interest in DHEA supplements increased after androstenedione and related supplements were no longer legally available in the United States.

DHEA was a prescription drug in the United States prior to the passage of the Dietary Supplement Health and Education Act in 1994. DHEA is now available over the counter and in dietary supplements. In other countries, such as Australia and New Zealand, it remains a controlled substance due to the potential for abuse. Because the long-term safety is currently unknown and effectiveness related to athletic performance is unproven, DHEA supplements for athletes are not recommended (Corrigan, 2002; Brown, Vukovich, and King, 2006).

Supplements are often used to assist weight loss.

Most people find it difficult to lose body fat and to maintain the loss. Any dietary supplement that may increase the amount lost or accelerate the rate of weight loss will be popular. Because dietary supplement manufacturers do not have to prove either safety or efficacy before a supplement is sold, there is an endless stream of weight-loss supplements on the market. Perhaps the most controversial weight-loss supplements are those that contain ephedrine (for example, ephedra), which may be used alone but is usually found in combination with caffeine.

Ephedra, ephedrine alkaloids, and ephedrine. *Ephedra*, *ephedrine alkaloids*, and *ephedrine* are different terms and should not be used interchangeably. *Ephedra* is a botanical term and refers to a genus of plants. Some species of ephedra contain ephedrine alkaloids in the stems and branches. One of the ephedrine alkaloids is ephedrine. In common usage, dietary supplements that contain any of the ephedrine alkaloids are referred to as ephedra.

In traditional Chinese medicine, ma huang is extracted from a species within the plant genus *Ephedra*. This species, *Ephedra sinica* Stapf, contains six different ephedrine alkaloids. Of the six, the primary active ingredient is ephedrine. Ephedrine can also be synthesized in the laboratory. Traditionally, the Chinese have used ma huang to treat asthma and nasal congestion. In the United States, ephedrine is added to over-the-counter medications for the same purposes. But ephedrine and related compounds have been marketed as dietary supplements for two other purposes: weight loss and increased energy. Both are common goals for athletes. Are ephedrine and related compounds safe and effective for these purposes?

Safety of ephedrine-containing compounds. The safety of ephedrine-containing dietary supplements has always been controversial, and reviewing its history helps to underscore some important issues about safety, purity, and supplement regulation in the United States. After the passage of the Dietary Supplement Health and Education Act in 1994, the sale of dietary supplements containing ephedrine began to increase. The FDA expressed concern, in part, because of the number of adverse event reports (AERs) they received. Consumers can report adverse events to a hotline, and by 1997 half of the AERs received involved ephedrine. The adverse events reported included known side effects such as headache, increased heart rate, increased blood pressure, and insomnia. The AERs also included deaths to otherwise healthy middle-aged and young adults.

Adverse event reports are anecdotal evidence because they are personal accounts of an event. These reports were hard to interpret because many lacked information about the dose consumed. Some consumers may have used these supplements despite the manufacturer's warnings. Most warn against use by pregnant or lactating women, and those with a history of heart disease, diabetes, or high blood pressure.

A scientific review of 16,000 adverse event reports found that 21 were serious events: two deaths, nine strokes, four heart attacks, one seizure, and five psychiatric problems. In these cases ephedrine was believed to be the sole contributor, *but* there was not enough scientific evidence to establish a cause-and-effect relationship. In addition to the 21 serious events, ephedrine contained in dietary supplements was implicated as a contributing (but not sole) factor in 10 other cases with serious side effects (Shekelle et al., 2003).

Further complicating the issue of safety was that experts do not agree on the dosage that is considered safe. The FDA proposed that not more than 8 milligrams (mg) of ephedrine alkaloids be used in a 6-hour period *and* not more than 24 mg in a 24-hour period. Use should not exceed 7 days (FDA, 2000). Another group of experts suggested that a single dose should not exceed 30 mg of ephedrine alkaloids and that up to 90 mg in a 24-hour period is safe. They suggest usage should not exceed 6 months (CANTOX, 2000).

Another safety issue was quality control. In 2000, Gurley, Gardner, and Hubbard published a study of the ephedrine alkaloid content of 20 dietary supplements. The content of 10 of the products varied by more than 20 percent when compared to the amount listed on the label. The worst example was a product that contained more than 150 percent of the amount listed on the label. One product had no active ingredient, and particularly troublesome, five contained norpseudoephedrine, a controlled substance (drug). This study clearly illustrates why athletes must be concerned about the purity of dietary supplements.

Athletes should particularly be aware of the risk associated with using ephedrine-containing dietary supplements prior to strenuous workouts in the heat. In 2001, the National Football League banned the dietary supplement ephedra after the death of an interior lineman during training camp. Although toxicological tests were not conducted on autopsy, an ephedrine-containing dietary supplement was found in the player's locker. In 2003, the safety of ephedrine-containing dietary supplements was again in the news with the death of a major league pitching prospect at spring training. In this case, the coroner implicated ephedrine as the cause of death. Contributing circumstances appear to include a history of borderline hypertension and liver abnormalities, exercising in hot and humid conditions, and restricting food and fluid intake in the previous 24 hours in an effort to lose weight.

The 2003 spring training death brought the issue of ephedrine in dietary supplements back to the forefront. In December 2003, the FDA issued an alert that advised consumers to stop buying and using dietary supplements containing ephedrine. In April 2004, the FDA banned the sale of ephedrine-containing dietary supplements. Although the alert was issued in December, the ban could not go into effect until the following April because there must be a 60-day period between the official notification of the ban and its enactment.

The ban is controversial. Under the Dietary Supplement Health and Education Act, the FDA can stop the sale of a dietary supplement if there is a demonstrated "significant or unreasonable risk of illness or injury." The legal issue boils down to this question: do the serious events reported to date, including deaths in which ephedrine was the sole contributor but in which cause and effect cannot be established, constitute an unreasonable risk of illness or injury?

Proponents of the ban point to documented deaths. They also question whether there are any health benefits and suggest that the risk/benefit ratio meets the "unreasonable risk" portion of the criterion. Opponents of the ban counter that the risk is very small. In 1999, approximately 3 million people purchased ephedrine-containing dietary supplements and consumed an estimated 3 billion "servings." The risk of a serious adverse event is estimated to be less than 1 in 1,000, and a cause-and-effect relationship has not been established. The debate is often passionate, political, and polar (FDA, 2003).

In April 2005, a federal judge struck down a portion of the ban. In that decision, the judge ruled that the FDA did not establish that an unreasonable risk of illness or injury was associated with low-dose (10 mg or less) ephedrine-containing supplements. The federal ban remains in place for doses greater than 10 mg. Some states have banned the sale of all ephedrine-containing dietary supplements, and these state laws are not affected by the federal court's decision.

Most ephedrine-containing supplements also contain caffeine, a member of a group of stimulants known as the methylxanthines. Since caffeine sometimes carries a negative connotation with consumers, the source of methylxanthine may be herbal, either guarana or kola nuts. Methylxanthine enhances the effectiveness of ephedrine as a weight-loss agent. Look carefully at the label shown in Figure 11.18. A dietary supplement may contain both ephedrine and caffeine, but those words will not appear if, for example, the source of those ingredients is ma huang and guarana.

Effectiveness of ephedrine for weight loss. Studies have shown that the use of ephedrine and caffeine by obese

---

White paper: Official, well-researched government report.

SUPPLEMENT FACTS

Serving size: 1 tablet	
Servings per container: 60	
Amount per serving	% Daily Value
Vitamin B$_{12}$ 50 mcg	833%
Megatherm™ proprietary blend 350 mg	**
Ma huang	**
Guarana seed	**
Yerba mate	**
Green tea leaf	**

** Daily value not established

Ingredients: Megatherm™ proprietary blend, pyridoxine HCL, sorbitol, guar gum.
Storage: Keep in a cool dry place, tightly closed.
Suggested Use: As a dietary supplement, take one tablet daily.

Keep out of reach of children
Expiration date: Dec 2012

Figure 11.18 Label of an ephedrine-containing supplement

This dietary supplement contains both caffeine and ephedrine, but these terms do not appear on the label.

The Internet café

Where do I find reliable information about body composition and body weight?

There are hundreds of noncommercial websites and hundreds of thousands of commercial websites with information about body composition and body weight. The U.S. Department of Health and Human Services website includes a calculator for body mass index as well as information about a healthy weight. Shape Up America!, founded by former Surgeon General C. Everett Koop, is dedicated to raising awareness of obesity as a health-related issue as well as reducing the incidence and prevalence of childhood obesity. These are just two examples of websites with comprehensive information about body weight and composition.

HealthFinder.gov

   http://healthfinder.gov/prevention/

Shape Up America!

   http://shapeup.org

people can produce a short-term weight loss of 8 to 9 lb (~3.5 to 4 kg). Studies to date have not been conducted for longer than 6 months, so it is unknown what the long-term effect might be or the effect of discontinuing the ephedrine and caffeine. It is also not known if an ephedrine-induced short-term weight loss has any long-term health benefit (Shekelle et al., 2003). Most athletes are not obese, so it is not known if, or how much, weight would be lost by athletes with ephedrine use.

Some athletes claim that ephedrine- and caffeine-containing dietary supplements give them "more energy." This is likely due to the stimulant effect of these compounds, which can mask fatigue. Some brands may be "spiked" with norpseudoephedrine, a stimulatory drug. Caffeine, ephedrine, and other stimulants are addictive, and individuals will experience symptoms upon withdrawal, including headache, fatigue, drowsiness, depressed mood, irritability, and inability to concentrate (Juliano and Griffiths, 2004). These characteristics are undesirable and may interfere with training, making it difficult or uncomfortable for athletes to stop using stimulatory dietary supplements.

Citrus aurantium (bitter orange) may be used in supplements advertised as ephedra-free.

Citrus aurantium, also known as bitter orange, contains synephrine and octopamine, which are chemically similar to epinephrine and norepinephrine, respectively. Supplements containing bitter orange have become more popular as ephedrine-containing supplements have been removed from the U.S. market.

Bitter orange is marketed as a weight-loss aid that enhances fat metabolism. Although citrus aurantium and

similar stimulants can slightly increase resting metabolic rate, the temporary and small increase in RMR is unlikely to result in clinically significant weight loss. There is some promise for bitter orange's effect on enhancing fat metabolism, but there are many questions about safety and efficacy (Greenway et al., 2006; Haaz et al., 2006). Under both resting and exercise conditions, bitter orange and similar compounds increase blood pressure and plasma glucose. Some minor adverse events, such as a temporary elevation in heart rate, and serious adverse events such as stroke, have been reported (Haller et al., 2008, Fugh-Berman and Myers, 2004).

Effectiveness of ephedrine on performance. A small number of studies have been conducted on the use of nonherbal ephedrine and caffeine preparations in healthy males as a performance enhancer. Research in this area has been limited due to the ethical issues related to administering potentially harmful substances to human subjects. Ephedrine and caffeine administered together has been reported to increase performance by delaying time to exhaustion by up to 30 percent (Schekelle et al., 2003). Athletes also report a decrease in perceived exertion (Magkos and Kavouras, 2004). These results are not surprising given the stimulatory properties of these substances. Studies have not shown that ephedrine and caffeine, alone or in combination, are effective in increasing muscle strength, muscle size, or anaerobic capacity.

Conjugated linoleic acid (CLA) is marketed to athletes as a way to change body composition and improve performance.

Conjugated linoleic acid (CLA) is an isomer of linoleic acid, an essential fatty acid found in lamb, beef, and dairy products. The major naturally occurring isomer

Table 11.3 Summary of the Safety and Effectiveness of Weight-Loss and Muscle-Building Supplements

Supplement	Safety	Effectiveness	Other
Caffeine (Chapter 6)	Safe at recommended doses	By itself, not effective for weight loss; enhances the effectiveness of ephedrine as a weight-loss agent	Banned substance at a certain threshold but not likely to be reached by athletes due to side effects that are detrimental to performance
Citrus aurantium (bitter orange)	Concerns about safety	Effective as a central nervous system stimulant, which masks fatigue that can accompany dietary restriction; small, temporary increase in resting metabolic rate; some promise as an agent to enhance fat metabolism	Contains synephrine, a drug, and other stimulatory compounds
Conjugated linoleic acid (CLA)	Safety in humans not established	Effectiveness in humans unknown because study results are mixed	
Dehydroepiandrosterone (DHEA)	Not safe	Not effective	
Ephedrine	Doses greater than 10 mg banned by the FDA due to significant safety risks; safety concerns hotly debated but does have a narrower safety profile than most dietary supplements	With caffeine, effective for short-term, 8–9 lb weight loss in obese people; not effective for improving muscle strength or anaerobic performance; effective as a central nervous system stimulant, which masks fatigue that can accompany dietary energy restriction	Banned substance in many sports; banned by the IOC at a certain threshold, which can be reached with multiple doses
Green tea extract (not the same as green tea)	Use with caution due to some reports of liver toxicity	Effective for small loss of body weight	

is *cis*-9, *trans*-11, but most supplements have a mixture of the natural isomer and a *trans*-10, *cis*-12 isomer. In animal studies, it is the *trans*-10, *cis*-12 isomer that can reduce the deposition of fat in adipose tissue; however, this isomer is also associated with deposition of fat in the liver and spleen and insulin resistance in the test animals (Wang and Jones, 2004). In humans, *trans*-10, *cis*-12 isomer is likely to be incorporated into adipose tissue whereas the *cis*-9, *trans*-11 isomer tends to be incorporated into skeletal muscle cells (Goedecke et al., 2009). The two isomers may have opposite effects.

Keeping it in perspective

Body Composition, Body Weight, Performance, Appearance, and Health

Athletes are typically very interested in measuring their body composition. Although weight and body composition are factors in athletic performance, the impact of either will vary depending on the sport. A lean, muscular athletic body usually has high social value, and many athletes wish to alter their weight and body composition for both performance and appearance reasons. Most athletes are not obese, so body weight is often not a chronic disease-related issue, but how rapidly and dramatically athletes lose weight may negatively impact their health.

Perspective can be lost when a certain body composition or weight is the ultimate goal instead of a means for potential improvements in performance or health. Trying to attain an ever-lower weight can result in declining or poor performance and does not make sense. Becoming too focused on a number can lead athletes to engage in risky behaviors, such as severe hypohydration, and lose sight of the bigger picture, which is optimal performance. It is easy to forget that body composition measurement is not precise and that reaching a certain percent of body fat may be at odds with performance and health goals.

Studies in animals who received CLA supplements for the purpose of weight loss were promising, but results of human studies are inconsistent, in part, because of the differences in isomers used (Campbell and Kreider, 2008; Tricon and Yaqoob, 2006; Wang and Jones, 2004). Studies have not shown that CLA supplements directly impact performance. At the present time the safety of CLA supplements has not been established in humans, and more research is needed to determine a safe dose, the types of isomers that should be used, and duration of use.

Athletes should be cautious about using weight-loss and muscle-building supplements.

An astounding number of dietary supplements for the purpose of weight (fat) loss are available for purchase. Many of these supplements are described as "fat burners," a loosely defined term that is used to describe compounds that help individuals to lose body fat, typically by increasing metabolism or enhancing fat breakdown. Many of the compounds are also described as "natural,"

which generally means that the active ingredient comes from an herbal rather than a synthetic source.

Table 11.3 summarizes some of the popular weight-loss and muscle-building supplements, many of which lack evidence of safety or effectiveness (Dunford and Coleman, in press; Pittler and Ernst, 2004). Concerns have been raised about safety, purity, and the potential for consuming banned substances. Athletes are advised to use caution.

Key points

- The use of anabolic steroids and prohormones poses legal, ethical, purity, and safety concerns for the athlete.
- Athletes should be cautious of weight loss supplements containing ephedrine or bitter orange, whose safety is closely related to dose.
- Many muscle building and weight loss supplements lack evidence of safety or effectiveness.

Why are some dietary supplements considered more like drugs than like foods?

Post-Test   Reassessing Knowledge of Body Weight and Body Composition

Now that you have more knowledge about body composition and body weight, look again at the statements that were listed at the beginning of the chapter. The answers can be found in Appendix O.

1. Percent body fat and fat mass can be precisely measured in athletes with a number of different methods.

2. The most accurate method of measuring body fat for any athlete is underwater weighing.

3. In sports in which body weight must be moved or transported over a distance (for example, distance running), it is a performance advantage to have the lowest weight possible.

4. To increase muscle mass, most athletes need a substantial increase in their usual protein intake.

5. For athletes who want to restrict energy intake to lose body fat, the recommended time to do so is at the beginning of the preseason or during the off-season.

Summary

The body is composed of various tissues, including fat, muscle, bone, organs, and fluids. The relative percentage of these components, particularly body fat, may affect performance, appearance, and health. There are a number of ways to measure body composition, but all have inherent measurement error and are considered only estimates of actual body composition. The subject or technician can introduce additional error. Therefore, body composition results should be interpreted carefully.

Body composition results can be used to determine an appropriate scale weight, establish percent

fat and lean body mass goals, and assess the impact of training and nutrition strategies. Changing body composition may improve performance, but attaining an inappropriate body composition or weight can be detrimental to performance and health. Athletes in sports in which weight must be certified can calculate a realistic, attainable, and sustainable minimum weight.

Athletes may wish to change body composition and can do so with an exercise and diet plan that promotes maintenance or gains in lean body mass and/or loss of body fat. These goals are typically accomplished over time and should be a planned part of the athlete's training schedule. The timetable for achieving these goals may be slower than the athlete would like or imagines. Muscle-building and "fat-burning" supplements may be tempting to those desiring rapid results, but some of these supplements are not legal, ethical, pure, safe, and/or effective. Athletes should consider performance, appearance, and health when setting weight and body composition goals.

Self-Test

Multiple-Choice

The answers can be found in Appendix O.

1. What is the Standard Error of the Estimate (SEE)?
   a. a measure of the accuracy of predictions
   b. a measure of the sampling error
   c. the standard deviation from the norm
   d. the percentage of values that are considered incorrect

2. An athlete is underwater weighed and is told that he is 13 percent body fat. How should he interpret this information?
   a. His percentage of body fat is 16 percent (13 percent storage fat + 3 percent essential fat).
   b. His percentage of body fat is between 10.3 percent and 15.7 percent.
   c. His percentage of body fat is not higher than 13 percent, but it may be lower.
   d. His percentage of body fat is not lower than 13 percent, but it may be higher.

3. Measuring weight on a daily basis is most useful to athletes who are trying to:
   a. lose body fat.
   b. increase muscle mass.
   c. change overall body composition.
   d. monitor hydration status.

4. Approximately how much additional protein is needed per day to support the growth of 1 pound of muscle tissue per week (assume sufficient kcal and resistance exercise)?
   a. 14 g
   b. 40 g
   c. 58 g
   d. 65 g

5. Ephedrine is known to be effective for:
   a. weight loss in athletes.
   b. weight loss in sedentary but not obese adults.
   c. asthma and nasal decongestion.
   d. lowering blood cholesterol.

Short Answer

1. What is essential fat and how does it differ from storage fat?

2. Are lean body mass and muscle mass the same or different? Explain.

3. Compare and contrast the body build of endomorphs, mesomorphs, and ectomorphs. How might this information be used to help athletes set realistic body composition goals?

4. Why is the Body Mass Index (BMI) inappropriate to use with athletes?

5. Explain why percent body fat results should be given as a range rather than as a single number.

6. Describe various ways that error may be introduced into the measurement of body composition.

7. Compare and contrast the various body composition methods considering accuracy, cost, availability, portability, and ease of use. Which may be useful in a research setting? A high school? A university? A health and fitness club?

8. In what ways may body composition estimates be useful to athletes and those who work with athletes?

9. Body weight does not give information about body composition, so why is the measurement of body weight useful?

10. Describe a situation in which calculating a minimum body weight might be beneficial.

11. Name the two dietary factors that are critical to increasing muscle mass. Explain why each is important.

12. What effect does weight cycling have on performance? On mental health? On physical health?

13. Name some of the side effects of anabolic steroids and androstenedione.

14. Develop a point/counterpoint discussion for a ban on all ephedrine-containing dietary supplements.

Critical Thinking

1. The coach of a collegiate gymnastics team has very specific ideas about how weight and appearance of gymnasts affects their performance. Outline a specific plan for assessment and monitoring of weight and body composition during preseason training and throughout the competitive season.

References

Alderman, B. L., Landers, D. M., Carlson, J., & Scott, J. R. (2004). Factors related to rapid weight loss practices among international-style wrestlers. *Medicine and Science in Sports and Exercise*, *36*(2), 249–252.

American College of Sports Medicine. (1984). Position paper: The use of anabolic-androgenic steroids in sports. *Medicine and Science in Sports and Exercise*, *19*(5), 534–539.

Broeder, C. E., Quindry, J., Brittingham, K., Panton, L., Thomson, J., Appakondu, S., et al. (2000). The Andro Project: Physiological and hormonal influences of androstenedione supplementation in men 35 to 65 years old participating in a high-intensity resistance training program. *Archives of Internal Medicine*, *160*(20), 3093–3104.

Brown, G. A., Vukovich, M., & King, D. S. (2006). Testosterone prohormone supplements. *Medicine and Science in Sports and Exercise*, *38*(8), 1451–1461.

Brown, W. J., Basil, M. D., & Bocarnea, M. C. (2003). The influence of famous athletes on health beliefs and practices: Mark McGwire, child abuse prevention, and androstenedione. *Journal of Health Communication*, *8*(1), 41–57.

Brozek, J., Grande, F., Anderson, J. T., & Keys, A. (1963). Densitometric analysis of body composition: Revision of some quantitative assumptions. *Annals of the New York Academy of Sciences*, *110*, 113–140.

Calfee, R., & Fadale, P. (2006). Popular ergogenic drugs and supplements in young athletes. *Pediatrics*, *117*(3), E577–E589.

Campbell, B., & Kreider, R.B. (2008). Conjugated linoleic acids. *Current Sports Medicine Reports*, *7*, 237–241.

CANTOX Health Services International. (2000). *Safety Assessment and Determination of a Tolerable Upper Limit of Ephedra*. The full text document can be viewed at http://www.crnusa.org

Clark, M., Reed, D. B., Crouse, S. F., & Armstrong, R. B. (2003). Pre- and post-season dietary intake, body composition, and performance indices of NCAA division I female soccer players. *International Journal of Sport Nutrition and Exercise Metabolism*, *13*(3), 303–319.

Collins, M. A., Millard-Stafford, M. L., Sparling, P. B., Snow, T. K., Rosskopf, L. B., Webb S. A., et al. (1999). Evaluation of the BOD POD for assessing body fat in collegiate football players. *Medicine and Science in Sports and Exercise*, *31*(9), 1350–1356.

Copeland, J., Peters, R., & Dillon, P. (2000). Anabolic-androgenic steroid use disorders among a sample of Australian competitive and recreational users. *Drug and Alcohol Dependence*, *60*(1), 91–96.

Corrigan, B. (2002). DHEA and sport. *Clinical Journal of Sport Medicine*, *12*(4), 236–241.

Deutz, R. C., Benardot, D., Martin, D. E., & Cody, M. M. (2000). Relationship between energy deficits and body composition in elite female gymnasts and runners. *Medicine and Science in Sports and Exercise*, *32*(3), 659–668.

Dunford, M., & Coleman, E. (in press). Ergogenic aids, dietary supplements, and exercise. In C. Rosenbloom (Ed.), *Sports nutrition: A practice manual for professionals* (5th ed.). Chicago: American Dietetic Association.

Dunford, M., & Macedonio, M. (in press). Weight management. In C. Rosenbloom (Ed.), *Sports nutrition: A practice manual for professionals* (5th ed.). Chicago: American Dietetic Association.

Enemark-Miller, E. A., Seegmiller, J. G., & Rana, S. R. (2009). Physiological profile of women's lacrosse players. *Journal of Strength and Conditioning Research*, *23*(1), 39–43.

Fields, D. A., Goran, M. I., & McCrory, M. A. (2002). Body-composition assessment via air-displacement plethysmography in adults and children: A review. *American Journal of Clinical Nutrition*, *75*(3), 453–467.

Filaire, E., Maso, F., Degoutte, F., Jouanel, P., & Lac, G. (2001). Food restriction, performance, psychological state and lipid values in judo athletes. *International Journal of Sports Medicine*, *22*(6), 454–459.

Fogelholm, G. M., Koskinen, R., Laakso, J., Rankinen, T., & Ruokonen, I. (1993). Gradual and rapid weight loss: Effects on nutrition and performance in male athletes. *Medicine and Science in Sports and Exercise*, *25*(3), 371–377.

Food and Drug Administration (FDA). (2000). Safety of Dietary Supplements Containing Ephedrine Alkaloids. Transcript of a public meeting held August 8–9, 2000. This transcript can be viewed at http://www.fda.gov

Food and Drug Administration (FDA). (2003). *Evidence on the safety and effectiveness of ephedra: Implications for regulation*. This paper can be viewed at http://www.fda.gov/bbs/topics/NEWS/ephedra/whitepaper.html.

Food and Drug Administration (FDA). (2004, March). FDA White Paper: Health effects of androstenedione. Retrieved [December 28, 2010 from http://www.fda.gov/oc/whitepapers/andro.html

Fugh-Berman, A., & Myers. A. (2004). Citrus aurantium, an ingredient of dietary supplements marketed for weight loss: Current status of clinical and basic research. *Experimental Biology and Medicine*, *229*, 698–704.

Goedecke, J. H., Rae, D. E., Smuts, C. M., Lambert, E. V., & O'Shea, M. (2009). Conjugated linoleic acid isomers, t10c12 and c9t11, are differentially incorporated into adipose tissue and skeletal muscle in humans. *Lipids*, *44*(11), 983–988.

Greenway, F., de Jonge-Levitan, L., Martin, C., Roberts, A., Grundy, I., & Parker, C. (2006). Dietary herbal supplements with phenylephrine for weight loss. *Journal of Medicinal Food*, *9*, 572–578.

Gropper, S. S., Smith, J. L., & Groff, J. L. (2009). *Advanced nutrition and human metabolism* (5th ed.). Belmont, CA: Wadsworth, Cengage Learning.

Gurley, B. J., Gardner, S. F., & Hubbard, M. A. (2000). Content versus label claims in ephedra-containing dietary supplements. *American Journal of Health-System Pharmacy*, 57(10), 963–969.

Haaz, S., Fontaine, K. R., Cutter, G., Limdi, N., Perumean-Chaney, S., & Allison, D. B. (2006). Citrus aurantium and synephrine alkaloids in the treatment of overweight and obesity: an update. *Obesity Reviews*, 7, 79–88.

Hall, C. J., & Lane, A. M. (2001). Effects of rapid weight loss on mood and performance among amateur boxers. *British Journal of Sports Medicine*, 35(6), 390–395.

Haller, C. A., Duan, M., Jacob, P., III, & Benowitz, N. (2008). Human pharmacology of a performance-enhancing dietary supplement under resting and exercise conditions. *British Journal of Clinical Pharmacology*, 65, 833–840.

Heyward, V. H., & Stolarczyk, L. M. (1996). *Applied body composition assessment*. Champaign, IL: Human Kinetics.

Hicks, V. L., Stolarczyk, L. M., Heyward, V. H., & Baumgartner, R. N. (2000). Validation of near-infrared interactance and skin fold methods for estimating body composition of American Indian women. *Medicine and Science in Sports and Exercises*, 32(2), 531–539.

Hoffman, J., Ratamess, N., Kang, J., Mangine, G., Faigenbaum, A., & Stout, J. (2006). Effect of creatine and beta-alanine supplementation on performance and endocrine responses in strength/power athletes. *International Journal of Sport Nutrition and Exercise Metabolism*, 16(4), 430–446.

Hornsby, P. J. (1997). DHEA: A biologist's perspective. *Journal of the American Geriatric Society*, 45(11), 1395–1401.

Horswill, C. A. (1993). Weight loss and weight cycling in amateur wrestlers: Implications for performance and resting metabolic rate. *International Journal of Sport Nutrition*, 3(3), 245–260.

Hu, F. B., Willett, W. C., Li, T., Stampfer, M. J., Colditz, G. A., & Manson J. E. (2004). Adiposity as compared with physical activity in predicting mortality among women. *New England Journal of Medicine*, 351(26), 2694–2703.

Institute of Medicine. (2002). *Dietary Reference Intakes for energy, carbohydrate, fiber, fat, fatty acids, cholesterol, protein and amino acids* (Food and Nutrition Board). Washington, DC: National Academies Press.

Jackson, A. S., & Pollock, M. L. (1978). Generalized equations for predicting body density of men. *British Journal of Nutrition*, 40(3), 497–504.

Jackson, A. S., Pollock, M. L., & Ward, A. (1980). Generalized equations for predicting body density of women. *Medicine and Science in Sports and Exercise*, 12, 175–182.

Jonnalagadda, S. S., Ziegler, P. J., & Nelson, J. A. (2004). Food preferences, dieting behaviors, and body image perceptions of elite figure skaters. *International Journal of Sport Nutrition and Exercise Metabolism*, 14(5), 594–606.

Juliano, L. M. & Griffiths, R. R. (2004). A critical review of caffeine withdrawal: Empirical validation of symptoms and signs, incidence, severity, and associated features. *Psychopharmacology (Berl)*, 176(1), 1–29. Epub Sept. 21, 2004.

Kazemi, M., Shearer, H., & Choung, Y. S. (2005). Pre-competition habits and injuries in Taekwondo athletes. *BMC Musculoskeletal Disorders*, 6(1), 26.

Kanayama, G., Hudson, J. I., & Pope, H. G., Jr. (2009). Features of men with anabolic-androgenic steroid dependence: A comparison with nondependent AAS users and with AAS nonusers. *Drug and Alcohol Dependence*, 102(1–3):130–137.

Kohrt, W. M. (1995). Body composition by DXA: Tried and true? *Medicine and Science in Sports and Exercise*, 27(10), 1349–1353.

Kreider, R. B., Ferreira, M., Wilson, M., Grindstaff, P., Plisk, S., Reinardy, J., et al. (1998). Effects of creatine supplementation on body composition, strength, and sprint performance. *Medicine and Science in Sports and Exercise*, 30(1), 73–82.

Layman, D. K., Evans, E., Baum, J. I., Seyler, J., Erickson, D. J., & Boileau, R. A. (2005). Dietary protein and exercise have additive effects on body composition during weight loss in adult women. *Journal of Nutrition*, 135(8), 1903–1910.

Lee, C. D., Blair, S. N., & Jackson, A. S. (1999). Cardio-respiratory fitness, body composition, and all-cause and cardiovascular disease mortality in men. *American Journal of Clinical Nutrition*, 69(3), 373–380.

Lemon, P. (1994). Methods of weight gain in athletes. Gatorade Sports Science Exchange. Roundtable # 21(5).

Lohman, T. G. (1992). *Advances in body composition assessment*. Current Issues in Exercise Science Series. Champaign, IL: Human Kinetics.

Malina, R. M., Morano, P. J., Barron, M., Miller, S. J., Cumming, S. P., Kontos, A. P., et al. (2007). Overweight and obesity among youth participants in American football. *The Journal of Pediatrics*, 151(4), 378–382.

Magkos, F., & Kavouras, S. A. (2004). Caffeine and ephedrine: Physiological, metabolic and performance-enhancing effects. *Sports Medicine*, 34(13), 871–889.

McCargar, L. J., & Crawford, S. M. (1992). Metabolic and anthropometric changes with weight cycling in wrestlers. *Medicine and Science in Sports and Exercise*, 24(11), 1270–1275.

McCargar, L. J., Simmons, D., Craton, N., Taunton, J. E., & Birmingham, C. L. (1993). Physiological effects of weight cycling in female lightweight rowers. *Canadian Journal of Applied Physiology*, 18(3), 291–303.

McCrory, M. A., Gomez, T. D., Bernauer, E. M., & Molé, P. A. (1995). Evaluation of a new air displacement plethysmograph for measuring human body composition. *Medicine and Science in Sports and Exercise*, 27(12), 1686–1691.

McLean, K. P., & Skinner, J. S. (1992). Validity of Futrex-5000 for body composition determination. *Medicine and Science in Sports and Exercise*, 24(6), 253–258.

Melby, C. L., Schmidt, W. D., & Corrigan, D. (1990). Resting metabolic rate in weight-cycling collegiate wrestlers compared with physically active, noncycling control subjects. *American Journal of Clinical Nutrition, 52*(3), 409–414.

Mettler, S., Mitchell, N., & Tipton, K. D. (2010). Increased protein intake reduces lean body mass loss during weight loss in athletes. *Medicine and Science in Sports and Exercise, 42*(2), 326–337.

Modlesky, C. (2006). Assessment of body size and composition in athletes. In M. Dunford (Ed.), *Sports nutrition: A practice manual for professionals* (pp. 177–210). (4th ed.) Chicago: American Dietetic Association.

Montgomery, D. L. (2006). Physiological profile of professional hockey players—a longitudinal comparison. *Applied Physiology, Nutrition, and Metabolism, 31*(3), 181–185.

Morrow, J. R., Jr., Jackson, A. S., Bradley, P. W., & Hartung, G. H. (1986). Accuracy of measured and predicted residual lung volume on body density measurement. *Medicine and Science in Sports and Exercise, 18*(6), 647–652.

Morton, J. P., Robertson, C., Sutton, L., & MacLaren, D. P. (2010). Making the weight: A case study from professional boxing. *International Journal of Sport Nutrition and Exercise Metabolism, 20*(1), 80–85.

National Federation of State High School Associations. http://www.nfhs.org/scriptcontent/Index.cfm

National Institute on Drug Abuse (2005). Monitoring the future. http://www.nida.gov/DrugPages/Steroids.html

Nieman, D. C. (2007). *Exercise testing and prescription* (6th ed.). New York: McGraw-Hill.

Noland, R. C., Baker, J. T., Boudreau, S. R., Kobe, R. W., Tanner, C. J., Hickner, R. C., et al. (2001). Effect of intense training on plasma leptin in male and female swimmers. *Medicine and Science in Sports and Exercise, 33*(2), 227–231.

Ostojic, S. M., Mazic, S., & Dikic, N. (2006). Profiling in basketball: Physical and physiological characteristics of elite players. *Journal of Strength and Conditioning Research, 20*(4), 740–744.

Petersen, H. L., Peterson, C. T., Reddy, M. B., Hanson, K. B., Swain, J. H., Sharp, R. L., et al. (2006). Body composition, dietary intake, and iron status of female collegiate swimmers and divers. *International Journal of Sport Nutrition and Exercise Metabolism, 16*(3), 281–295.

Phillips, S. M. (2004). Protein requirements and supplementation in strength sports. *Nutrition, 20*(7–8), 689–695.

Phillips, S. M., Hartman, J. W., & Wilkinson, S. B. (2005). Dietary protein to support anabolism with resistance exercise in young men. *Journal of the American College of Nutrition, 24*(2), 134S–139S.

Pittler, M. H., & Ernst, E. (2004). Dietary supplements for body-weight reduction: A systematic review. *American Journal of Clinical Nutrition, 79*(4), 529–536.

Rankin, J. W. (2002). Weight loss and gain in athletes. *Current Sports Medicine Reports, 1*(4), 208–213.

Saarni, S. E., Rissanen, A., Sarna, S., Koskenvuo, M., & Kaprio, J. (2006). Weight cycling of athletes and subsequent weight gain in middle age. *International Journal of Obesity (Lond)* March 28; Epub ahead of print.

Sallet, P., Mathieu, R., Fenech, G., & Baverel, G. (2006). Physiological differences of elite and professional road cyclists related to competition level and rider specialization. *The Journal of Sports Medicine and Physical Fitness, 46*(3), 361–365.

Sallet, P., Perrier, D., Ferret, J. M., Vitelli, V., & Baverel, G. (2005). Physiological differences in professional basketball players as a function of playing position and level of play. *The Journal of Sports Medicine and Physical Fitness, 45*(3), 291–294.

Schmidt, W. D., Corrigan, D., & Melby, C. L. (1993). Two seasons of weight cycling does not lower resting metabolic rate in college wrestlers. *Medicine and Science in Sports and Exercise, 25*(5), 613–619.

Scott, J. R., Horswill, C. A., & Dick, R. W. (1994). Acute weight gain in collegiate wrestlers following a tournament weigh-in. *Medicine and Science in Sports and Exercise, 26*(9), 1181–1185.

Scott, J. R., Oppliger, R. A., Utter, A. C., & Kerr, C. G. (2000). Body weight changes at the national tournaments: The impact of rules governing wrestling weight management. *Medicine and Science in Sports and Exercise, 32*, S131.

Shekelle, P. G., Hardy, M. L., Morton, S. C., Maglione, M., Mojica, W. A., Suttorp, M., et al. (2003). Efficacy and safety of ephedra and ephedrine for weight loss and athletic performance: A meta-analysis. *Journal of the American Medical Association, 289*(12), 1537–1545.

Sheldon, W. H. (1940). *The varieties of human physique.* New York: Harper & Brothers.

Silvestre, R., Kraemer, W. J., West, C., Judelson, D. A., Spiering, B. A., Vingren, J. L., et al. (2006). Body composition and physical performance during a National Collegiate Athletic Association Division I men's soccer season. *Journal of Strength and Conditioning Research, 20*(4), 962–970.

Siri, W. E. (1956). The gross composition of the body. *Advances in Biological and Medical Physics, 4*, 239–280.

Slater, G. J., Rice A. J., Mujika, I., Hahn, A. G., Sharpe, K., & Jenkins, D. G. (2005). Physique traits of lightweight rowers and their relationship to competitive success. *British Journal of Sports Medicine, 39*(10), 736–741.

Smith, M., Dyson, R., Hale, T., Hamilton, M., Kelly, J., & Wellington, P. (2001). The effects of restricted energy and fluid intake on simulated amateur boxing performance. *International Journal of Sport Nutrition, 11*(2), 238–247.

Stiegler, P., & Cunliffe, A. (2006). The role of diet and exercise for the maintenance of fat-free mass and resting metabolic rate during weight loss. *Sports Medicine, 36*(3), 239–262.

Stone, M. (1994). Methods of weight gain in athletes. Gatorade Sports Science Exchange. Roundtable # 21(5).

Thompson, W. R., Gordon, N. F., & Pescatello, L. S. (American College of Sports Medicine). (2009). *ACSM's guidelines for exercise testing and prescription* (8th ed.). Philadelphia: Lippincott, Williams & Wilkins.

Tipton, K. D., & Ferrando, A. A. (2008). Improving muscle mass: Response of muscle metabolism to exercise, nutrition and anabolic agents. *Essays in Biochemistry, 44*, 85–98.

Tipton, K. D., & Wolfe, R. R. (2001). Exercise, protein metabolism, and muscle growth. *International Journal of Sport Nutrition and Exercise Metabolism, 11*(1), 109–132.

Tricon, S., & Yaqoob, P. (2006). Conjugated linoleic acid and human health: A critical evaluation of the evidence. *Current Opinion in Clinical Nutrition and Metabolic Care,* 9:105–110.

Tuuri, G., Loftin, M., & Oescher, J. (2002). Association of swim distance and age with body composition in adult female swimmers. *Medicine and Science in Sports and Exercise, 34*(12), 2110–2114.

Wang, Y., & Jones, P. J. (2004). Dietary conjugated linoleic acid and body composition. *American Journal of Clinical Nutrition, 79*(6 Suppl.), 1153S–1158S.

Wilmore, J. H. (1983). Body composition in sport and exercise: Directions for future research. *Medicine and Science in Sports and Exercise, 15*(1), 21–31.

Wolfe, R. (2000). Testosterone and muscle protein metabolism. *Mayo Clinic Proceedings, 75*(Suppl.), S55–S60.

Ziegler, P. J., Kannan, S., Jonnalagadda, S. S., Krishnakumar, A., Taksali, S. E., & Nelson, J. A. (2005). Dietary intake, body image perceptions, and weight concerns of female US international synchronized figure skating teams. *International Journal of Sport Nutrition and Exercise Metabolism, 15*(5), 550–566.

Ziegler, P. J., Khoo, C. S., Sherr, B., Nelson, J. A., Larson, W. M., & Drewnowski, A. (1998). Body image and dieting behaviors among elite figure skaters. *International Journal of Eating Disorders, 24*(4), 421–427.

12 Disordered Eating and Exercise Patterns in Athletes

Learning Plan

- **Describe** the concepts of normal eating, disordered eating, and eating disorders.
- **Explain** why eating disorders are classified as psychiatric diseases.
- **State** the diagnostic criteria for anorexia nervosa, bulimia nervosa, and eating disorders not otherwise specified.
- **Outline** the characteristics of anorexia athletica and compare and contrast it with other eating disorders.
- **State** the prevalence of disordered eating and eating disorders in the male and female athletic and general populations and discuss its impact on physical and mental health and performance.
- **Differentiate** between athletes with eating disorders and those who are training intensely but do not have a disordered eating pattern.
- **Discuss** the appropriate responses by teammates, coaches, athletic trainers, and others if disordered eating or an eating disorder is suspected.
- **Explain** the Female Athlete Triad and how each component affects health and performance.
- **Describe** exercise dependence and explain how it differs from overtraining.

Those with a distorted body image think of themselves as fat even though they are not.

Assessing Current Knowledge of Disordered Eating and Exercise Dependence

Read the following statements and decide if each is true or false.

1. Disordered eating and eating disorders affect only female athletes.

2. Anorexia athletica means that an athlete has a classic case of anorexia nervosa.

3. Disordered eating and eating disorders are more likely to be seen among elite female athletes in sports such as distance running and gymnastics.

4. Coaches cause athletes to develop eating disorders.

5. A good diagnostic criterion for exercise dependence is the volume of exercise training (that is, frequency and duration of exercise).

The prevention, detection, and management of disordered eating and eating disorders are of critical importance to athletes and those who work with them. Disordered eating and eating disorders are deviations from normal eating and are often characterized by obsession or inflexibility. Eating disorders such as anorexia nervosa and bulimia nervosa are psychiatric diseases and, in some cases, are fatal. Although athletes in sports that emphasize low weight and a thin appearance are at a greater risk, any athlete in any sport of either gender may suffer from disordered eating or an eating disorder.

Of particular concern for female athletes is the manifestation of the Female Athlete Triad. Three independent but interrelated components are involved: energy availability, menstrual function, and bone mineral density. Low energy availability occurs because of a negative energy balance, which is the result of energy expenditure from exercise exceeding energy intake from food. Low energy availability may be a result of disordered eating behaviors, but it can also exist in the absence of disordered eating. The negative energy balance can disrupt menstruation and negatively affect bone mineral density, serious health issues that need to be prevented, managed, and treated.

12.1 Case Study: Disordered Eating and Eating Disorders

The following case study shows how an athlete can move from a normal eating pattern to disordered eating and then to an eating disorder. Notice the many factors that influence this progression, including inappropriate eating and dieting behaviors as well as training demands and psychological stresses. This case study will be used throughout this chapter to illustrate the development of and some of the problems associated with disordered eating and exercise patterns in athletes.

Karen had been running ever since she could remember, but it was not until her first year in high school when she joined the cross country team that she realized that she was in love with running. She loved everything about it—the digital training watch that she got for her 13th birthday, the wind in her long hair, the quiet time she had to herself away from her family's problems, and the feeling of accomplishment when she finished a race. She made a name for herself the first year, and she realized that she could be the best runner at her school if she applied herself.

By her junior year she was featured in her hometown newspaper as someone to watch. She adopted a semivegetarian diet and altered her running stride to make it more efficient. She came in second in the regional meet and earned the right to go to the state championships. There she had a strong start but faltered down the stretch, and she was disappointed for herself and her family. Her coach told her that if she trained just a little harder that she could be a contender in her senior year. Her parents were excited about the prospects of her earning a college athletic scholarship; without some financial help they had little hope of sending Karen away to college.

After the state meet Karen immediately began to train and to pay more attention to her diet. She noticed that the winners at the state meet had brought coolers with their own food and drinks. She became a vegetarian in earnest and began to eat differently than her friends and family, which meant that she often prepared her own food and ate alone. Eating alone turned out to be advantageous as she could avoid the family mealtime discussions that always left her feeling as if she was not good enough in her parents' eyes. Her best friend in middle school told her that she was no longer the happy-go-lucky girl she once was. Karen took that as a compliment—she was maturing and focusing on the future—college, running, and, maybe, the Olympics. She had dreams, and they involved taking her beyond the little town in which she grew up.

She was diligent about her training and diet, and her senior year was all that she had hoped. She smashed school records, erased regional marks, blew by the competition, and won the state meet by a record margin. Sought after by many colleges, Karen traveled out of state on recruiting visits. She received offers of athletic scholarships from several schools that were impressed with her perfect high school grades and high scores on her college entrance exams. Karen chose to attend a top-notch school on the other side of the country even though her parents had reservations about her being so far away and were concerned about how they would pay for the costs not covered by the scholarship.

Her freshman year of college was an eye-opener. Her college coach was much more demanding than her high school coach and everything about her seemed to be under scrutiny. She had made only a few friends, acquaintances really, and she missed her family, although her coach was like having family close. Much of the dorm food was not vegetarian, and she found herself with few choices and even fewer foods that she enjoyed eating. One of her goals was to have the highest GPA on the team, but attaining perfect grades in college was much harder than in high school. When she was not training, she was studying. Karen was surprised to find out that the athletic scholarship she thought was guaranteed had to be renewed each year based upon her running and academic performances, something that really upset her parents. Although her coach never said it directly, he intimated that she would perform better if she were leaner. A couple of her teammates, whom she noticed were thin, wondered aloud if she had what it took to make it as a college runner at an NCAA Division I school. That was all the motivation that she needed to develop a stricter training program, and, for the first time in her life, a diet to lose weight and become thinner.

As she stepped up her training and restricted her diet, her performance began to improve. Karen rededicated herself to running and was almost robotic in her approach. She had convinced her parents to let her move into a single room in the dorm and not buy the meal plan. They reasoned that her diet would be healthier and that she would be happier if she prepared her own food in the small kitchen at the end of the hall. Karen found that she loved creating meal plans and looking through vegetarian cookbooks.

She made tremendous improvements by the end of her sophomore year, and her coach said that she was poised to have a breakout junior year. Unfortunately, she developed a painful stress fracture in her lower leg early in the preseason of her third year. Karen struggled for many months, not only because she was unable to train but also because she needed the intense physical activity to keep her weight low. Her coach was not paying much attention to her now that she was sidelined with an injury, and she was already worried that her full athletic scholarship might be reduced for her senior year. Karen gained 10 pounds in 2 months while recovering from her injury, which scared her, so she began to restrict her diet to a few healthy foods—salads, bagels, fruits, and water—and weigh herself daily. She even cut her long hair in an effort to feel lighter.

By the beginning of the season she had received clearance to restart her training, but her injury had substantially set her training back and she was concerned that she was entering racing season without the necessary preseason training. Karen now relished her training runs and began to run on designated rest days, although she knew that if her coach was aware of this he would never have allowed it. If she had been honest with herself, she would have realized that she did not so much love running anymore, rather, she *needed* running. Her coach had mentioned that he was worried about her apparent lack of eating, and she took his comment as a good sign since he obviously was paying attention to her again. Karen assured him that she was eating more now that she was training. She made sure that she was not lying by increasing the size of her salads. Secretly, she was a bit worried because her menstrual periods, which had been light but regular since high school, were now almost nonexistent—just two periods in the last year. She had not dated anyone since coming to college so she was not worried about being pregnant. But she did have an immediate health concern—recurrent upper respiratory tract infections that had plagued her for months.

Her junior season started with an excellent showing at a big meet, and her coach repeated his prediction that this would be her breakout year. Her next effort was hampered by a cold, and her coach seemed sympathetic to her frequent infections, although he did ask her specifically about her training and diet. Two more meets featured mediocre performances and she was not chosen to travel to an out-of-state invitation-only event. She resolved to work harder, training more than usual and eating a little less than she had been for the past few months. At the next meet she fainted at the start line and her coach said they needed to talk. She believed that she just had a bad day and that everything would be fine for the next meet; he believed that she had anorexia athletica and needed immediate treatment.

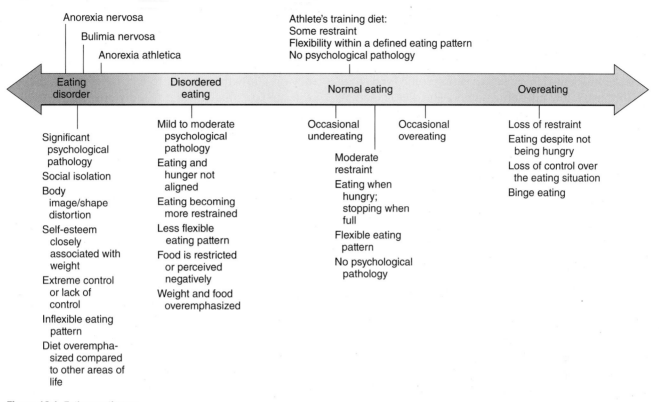

Figure 12.1 Eating continuum

Eating disorders, disordered eating, normal eating, and overeating are areas along a continuum, but these points are not well defined.

Key points

■ Normal eating may become disordered eating, which may progress to an eating disorder.

Some athletes have made public their struggles with disordered eating and eating disorders. If you are familiar with such struggles, how were they similar and different from the case study presented here?

12.2 Overview of Eating and Exercise Patterns

Disordered eating, eating disorders, and excessive exercise are deviations from normal; therefore, normal patterns must be known before disordered ones can be determined. In the case of eating, "normal" is not easy to define but it generally describes an eating pattern that is flexible and not obsessive. Normal exercise patterns vary according to the sport and level of training; however, a normal amount of exercise is an exercise intensity and duration that supports training and performance and is not associated with overtraining.

"Normal" eating is flexible.

Eating is not solely physiological. Although it is necessary to eat to obtain the nutrients needed for the body to properly function, eating also has a strong psychological component. There is no agreed-upon definition of normal eating, but the following definition is often used:

> Normal eating is being able to eat when you are hungry and continue eating until you are satisfied. It is being able to choose food you like and eat it and truly get enough of it—not just stop eating because you think you should. Normal eating is being able to use some moderate constraint in your food selection to get the right food, but not so restrictive that you miss out on pleasurable foods . . . Normal eating takes up some of your time and attention, but it keeps its place as only one important area of your life. In short, normal eating is flexible. (Satter, 1987, pp. 69–70)

For individuals with access to an adequate amount of food, normal eating represents the middle area on the eating continuum shown in Figure 12.1. When considered over a period of time, such as a week, month, or year, food may be either under- or overconsumed, and on any given day the energy (kcal) or nutrient content may be higher or lower than recommended guidelines. Normal eating consists of consuming foods that are nutrient rich as well as eating some foods that might have a low nutrient content. The diet is moderate, balanced, and varied, and is flexible, especially in response to social situations. Normal eating involves moderate constraint, not reckless abandon or overly strict **discipline**.

Normal eating in athletes is particularly hard to define because some athletes must follow fairly strict eating guidelines to support their training and performance goals. A distance runner must be concerned about excessive caloric intake because weight gain as body fat could negatively affect training and performance. Rigorous training, especially during the latter part of the preseason, and the demands of competition require a well-thought-out diet plan. For example, a distance runner must be diligent about consuming the proper amount of carbohydrates daily or risk not restoring the muscle glycogen used during training and competition. The moderate restraint and dietary flexibility that is part of normal eating necessarily becomes a bit more restrained and less flexible during periods of intense training and competition, but it should not become overly restricted or inflexible. Normal eating among highly trained athletes is characterized by discipline, not by **obsession** (Figure 12.1).

Disordered eating is not the same as an eating disorder.

Disordered eating (DE) represents a deviation from normal eating, but the individual does not meet the diagnostic criteria for an **eating disorder (ED)**—anorexia nervosa, bulimia nervosa, or eating disorders not otherwise specified. Disordered eating is not well defined or easily recognized and encompasses a large area on the eating continuum (Figure 12.1). The deviation from normal may be occasional and minor or it may progress and become more frequent and **pathological**. The difficulty lies in identifying the overall context of normal eating and then determining the degree to which behaviors deviate from normal, tracking the progression, and identifying the level of severity. Individuals may be described as having a subclinical eating disorder if they demonstrate a number of disordered eating behaviors (see examples in the next column) and exhibit associated psychological issues (Beals, in press). Single disordered eating behaviors are not as severe as a subclinical eating disorder, which is not as severe as a clinical eating disorder, but any of the three conditions is cause for concern and intervention.

One sign of disordered eating may be the inability to "eat when hungry and stop when full." In some cases, individuals will feel hungry but will refuse to eat at all or will wait until a predesignated time. When they do allow themselves to eat, the amount may be restricted. In other cases, the individual will be full but will continue to eat for nonphysiological reasons, such as, anxiousness, loneliness, or boredom. Body weight is not a good predictor of these behaviors; these examples include individuals who are underweight, normal weight, or overweight. Food intake may be strictly prescribed, and this lack of eating flexibility is another characteristic of DE. One type of food (for example, sweets) or nutrient (for example, fat) may also be severely restricted and dietary intake can begin to conform to a rigid pattern. Food intake is often viewed from the perspective of restriction (for example, "don't consume too many calories, don't eat any fat") rather than from the perspective of inclusion (for example, "I need to eat an adequate amount of calories, including some fat"). Food and weight become overemphasized and take up a considerable amount of time and thought, to the exclusion of other important activities. Individuals with DE may exhibit any one or more of these behaviors, but, in short, disordered eating is inflexible.

Disordered eating in athletes is particularly hard to define because training and performance require attention and diligence to an eating plan, especially as athletes move to the more elite levels in their sports. The athlete's "normal" pattern may be one of mild restraint. For example, a female gymnast who eats "normally" may watch what she eats so that she consumes an adequate but not excessive amount of kilocalories to maintain her already lean physique and low body weight. She is thoughtful about what she consumes because she wants to make sure that her intake of carbohydrates and proteins is sufficient, and this necessarily means a lower-fat diet than the general population. However, she sees food for what it is—the fuel to help her train and perform—and she is not exhibiting signs of disordered eating. Contrast this to the female gymnast who tallies the amount of fats and kilocalories consumed daily and classifies foods as "good" and "bad." This gymnast monitors her diet so closely that she will not allow herself to consume dessert on *any* occasion. She eats when she feels hungry, but if it is not a designated mealtime she will consume only carrots to satisfy her hunger. This gymnast is exhibiting some disordered eating behaviors, and her disordered eating could become more severe. For example, in addition to the behaviors already described, she might become fearful of eating fats and refuse to eat more than 20 grams daily. If she exceeded this self-imposed limit, she might "punish" herself by doing 200 sit-ups and **fasting** for the rest of the day. She might

Anorexia nervosa is characterized by a refusal to maintain a minimum body weight.

begin to weigh herself daily and let the scale weight determine her eating pattern. These are examples of pathological behaviors associated with a subclinical eating disorder, but she would not meet the diagnostic criteria for an eating disorder.

Eating disorders are psychiatric diseases.

Eating disorders represent a substantial deviation from normal eating (Figure 12.1) and are **psychiatric** conditions that involve body image issues. The three clinical eating disorders recognized by the American Psychiatric Association (APA) are anorexia nervosa, bulimia nervosa, and eating disorders not otherwise specified (EDNOS), and each has established criteria listed in the *Diagnostic and Statistical Manual of Mental Disorders*, 4th edition (*DSM–IV*; APA, 1994). Anorexia nervosa and bulimia nervosa share most clinical features, although body weight is a major feature that differs between these two conditions. Interestingly, it is not uncommon for individuals to "cross over" between anorexia and bulimia (Reiter and Graves, 2010). Those with EDNOS do not meet the specific criteria established for either anorexia or bulimia, but a variety of significant problems are present, as will be illustrated later in this chapter. Although this chapter focuses on eating and exercise behaviors, one should not forget that eating disorders are psychiatric diseases, and their development is a result of psychological disturbances often related to issues of control.

Anorexia nervosa. **Anorexia nervosa** is characterized by a refusal to maintain a minimum body weight. There is an intense fear of gaining weight and an intense desire to be thin. Also present is an extremely distorted body image—those with anorexia nervosa see themselves as

fat in their mind's eye even when they are **emaciated**. Two subtypes exist. The first is referred to as *Restricting type*, and these individuals self-impose starvation and engage in excessive exercise. A second subtype is termed *Binge-eating/Purging type*. These individuals employ starvation and excessive exercising techniques but may also overeat at times and then use self-induced vomiting, diuretics, or **laxatives** to compensate for the increased energy intake associated with the bingeing behavior.

The **prevalence** of anorexia nervosa in late-adolescent and early-adult females is estimated to be 0.5–1.0 percent of that population. It is most prevalent in females (more than 90 percent of all cases). Males do manifest anorexia nervosa, although the prevalence is unknown. The typical age range for females exhibiting anorexia nervosa is early adolescence (~age 13) through early adulthood (mid-20s), and critical ages appear to be age 14 (often the start of high school) and age 18 (start of college, living away from family). The **incidence** appears to be on the increase, but this trend is hard to document (APA, 1994).

Those with anorexia nervosa meet the following criteria, shown here as listed in the *Diagnostic and Statistical Manual of Mental Disorders*, 4th edition (*DSM–IV*; APA, 1994):

A. Refusal to maintain body weight at or above a minimally normal weight for age and height (e.g., weight loss leading to maintenance of body weight less than 85% of that expected; or failure to make expected weight gain during period of growth, leading to body weight less than 85% of that expected).

B. Intense fear of gaining weight or becoming fat, even though underweight.

Discipline: Moderate self-control or restraint.

Obsession: Idea or feeling that completely occupies the mind, sometimes associated with psychiatric disorders.

Disordered eating (DE): A deviation from normal eating.

Eating disorder (ED): Substantial deviation from normal eating, which meets established diagnostic criteria.

Pathological, pathology: A condition that deviates from that which is considered normal.

Fasting: Abstaining from food or drink.

Psychiatric: Relating to the medical specialty concerned with the diagnosis and treatment of mental or behavioral disorders.

Anorexia nervosa: A life-threatening eating disorder characterized by a refusal to maintain a minimum body weight.

Emaciated: Extremely thin; may be a result of self-starvation.

Laxative: A substance that promotes bowel movements.

Prevalence: The number of cases of a condition that exists in the population at a given point in time.

Incidence: The number of new cases of an illness or condition.

C. Disturbance in the way in which one's body weight or shape is experienced, undue influence of body weight or shape on self-evaluation, or denial of the seriousness of the current low body weight.

D. In postmenarcheal females, amenorrhea, i.e., the absence of at least three consecutive menstrual cycles (A woman is considered to have amenorrhea if her periods occur only following hormone, e.g., estrogen, administration.)

Specify type

Restricting type: during the current episode of Anorexia Nervosa, the person has not regularly engaged in binge-eating or purging behavior (i.e., self-induced vomiting or the misuse of laxatives, diuretics, or enemas).

Binge-eating/Purging type: during the current episode of Anorexia Nervosa, the person has regularly engaged in binge-eating or purging behavior (i.e., self-induced vomiting or the misuse of laxatives, diuretics, or enemas) (APA, 1994).

Although the four criteria are easily listed, diagnosing anorexia nervosa takes skill and clinical judgment. For example, there is not a single "normal weight" for any given age and height, so it must be determined for each individual. Individuals of the same height have different normal weights because of differing bone structure and body composition. In addition, weight history (for example, weight stability, lowest and highest weight attained) should be considered. Once a normal weight is established, 85 percent of that weight can be calculated. For example, if normal weight is 115 pounds (~52 kg), then 85 percent of normal weight would be 98 pounds (~44.5 kg).

The fear the individual has about weight gain may actually increase as weight loss continues. It may seem counterintuitive that a female whose normal weight is 115 pounds (~52 kg) is fearful of weight gain when she is at 95 pounds (43 kg) and even more fearful when she weighs 90 pounds (41 kg). Recall that anorexia nervosa is a psychiatric disorder and that body image is distorted. It is common for someone with anorexia nervosa to look in the mirror and believe that she is fat or believe that one part of the body is fat (for example, thighs or buttocks). In those with anorexia nervosa, self-esteem is dependent on body weight and body shape. Weight loss or maintenance of a body weight below one's minimum weight is seen by the individual as extreme self-discipline and is thought of as a desirable characteristic. Weight gain is seen as a lack of self-discipline and a lack of self-control, and both are considered undesirable characteristics.

In most cases, weight loss is achieved primarily by voluntary starvation; excessive exercise may be a secondary method used. The self-starvation often begins with the elimination of foods high in kilocalories and becomes more restricted until the diet may contain only a few low-calorie-containing foods (for example, vegetables). The term *anorexia* means loss of appetite, and in this respect the disease is misnamed. Those with anorexia nervosa rarely lose their appetite; rather, they do not allow themselves to respond to it. They also rarely complain about their weight loss. Ironically, they are so self-controlled that they lose control. Control and self-esteem are psychological issues that will need to be addressed as part of therapy (APA, 1994).

Bulimia nervosa. **Bulimia nervosa** is characterized by recurring binge eating coupled with inappropriate ways of preventing weight gain following the eating binge. Two subtypes exist. The first, known as the purging type, includes self-induced vomiting or the use of laxatives, diuretics, or enemas. The second, nonpurging type, involves fasting or excessive exercise. Those who purge attempt to keep the food from being absorbed (self-induced vomiting, use of laxatives) or prevent scale weight from increasing (use of diuretics or enemas). The nonpurgers compensate for the increased caloric intake by subsequent fasting or excessive exercise. Those with bulimia may use a number of methods, but purging by self-induced vomiting is the most common.

The prevalence of bulimia nervosa in the general population is difficult to estimate, in part because many people do not seek treatment so it goes undetected. It is most prevalent in females (~90 percent of those diagnosed with bulimia) but is not absent in males. The prevalence of bulimia in late-adolescent and early-adult females is estimated to be 1–3 percent of that population, which is greater than the prevalence of anorexia nervosa. The age range (adolescence to middle adulthood) is also larger than anorexia nervosa, and some people struggle with bulimia for many years. Those with bulimia nervosa meet the following criteria:

A. Recurrent episodes of binge eating. An episode of binge eating is characterized by both of the following:

1. eating, in a discrete period of time (e.g., within any 2-hour period), an amount of food that is definitely larger than most people would eat during a similar period of time and under similar circumstances.

2. a sense of lack of control over eating during the episode (e.g., a feeling that one cannot stop eating or control what or how much one is eating).

B. Recurrent inappropriate compensatory behavior in order to prevent weight gain, such as self-induced vomiting; misuse of laxatives, diuretics, enemas, or other medications; fasting; or excessive exercise.

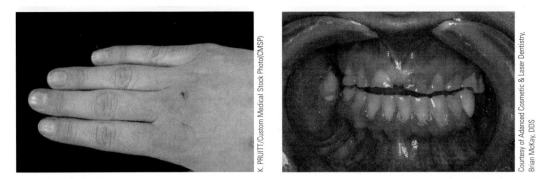

Bulimics may have callused fingers or teeth with little enamel from self-induced vomiting.

K. PRUITT/Custom Medical Stock Photo(CMSP)

Courtesy of Advanced Cosmetic & Laser Dentistry, Brian McKay, DDS

C. The binge eating and inappropriate compensatory behaviors both occur, on average, at least twice a week for 3 months.

D. Self-evaluation is unduly influenced by body shape and weight.

E. The disturbance does not occur exclusively during episodes of Anorexia Nervosa.

Specify type

Purging type: during the current episode of Bulimia Nervosa, the person has regularly engaged in self-induced vomiting or the misuse of laxatives, diuretics, or enemas.

Nonpurging type: during the current episode of Bulimia Nervosa, the person has used other inappropriate compensatory behaviors, such as fasting or excessive exercise, but has not regularly engaged in self-induced vomiting or the misuse of laxatives, diuretics, or enemas (APA, 1994).

Diagnosing bulimia also takes considerable skill and clinical judgment because the physical signs may not be obvious, and most bulimics eat in secret because they are ashamed of their eating and compensatory behaviors. Additionally, individuals with bulimia are often within the normal weight range. Restricting caloric intake (that is, "dieting") is frequently a binge-eating trigger. In other words, food intake will be severely restricted for a period of time in an effort to lose weight, and during or after this self-imposed restriction binge eating will begin. A circular pattern can develop that involves food restriction for several days or weeks followed by a binge followed by food restriction. This pattern can continue for months or years.

A binge is described as eating a large amount of food in a short period of time, but there is no quantitative definition so the amount is relative to the individual's usual pattern of eating. During a binge, the food is usually consumed rapidly, thus an often-used guideline is that the food is consumed in 2 hours or less.

Those with the purging subtype typically self-induce vomiting. Initially a finger is used to invoke vomiting, but most bulimics can become adept at willing themselves to vomit. Laxatives are used by approximately one-third of purgers. Other methods that are sometimes mentioned—use of ipecac (a medicine used to induce vomiting, often in cases of accidental poisoning) and **enemas**—are actually rarely used. Use of the finger to induce vomiting causes calluses to form on the back of the finger(s) over time due to the finger(s) rubbing against the teeth. Frequent exposure to vomitus, which is acidic, removes enamel from the teeth and leaves them more prone to cavities and dental decay. These are some of the more obvious physical signs to those who are trained to recognize them.

Binge eating is a source of distress for the individual. Vomiting may relieve the physical discomfort associated with extreme overeating but not the emotional discomfort. Those with bulimia lose control over the eating situation (they literally cannot stop the binge), and this is followed by depression related to loss of control. Self-esteem is closely tied to both body shape and weight, and those with bulimia nervosa are overly critical of their bodies. Self-esteem and control are psychological issues that will need to be addressed as part of therapy.

Eating disorders not otherwise specified. A third diagnostic category is **eating disorders not otherwise specified (EDNOS)**. This diagnosis is the most common in outpatient settings, where approximately

Bulimia nervosa: An eating disorder characterized by bingeing and purging cycles.

Enema: Insertion of a liquid via the rectum to induce a bowel movement.

Eating disorders not otherwise specified (EDNOS): Pathological behaviors are present but the diagnostic criteria are not met for either anorexia nervosa or bulimia nervosa.

60 percent of all eating disorders diagnosed are EDNOS. Surprisingly, this prevalent form of an eating disorder does not have defined criteria and is not well studied. The EDNOS diagnosis is often a result of a "default" categorization—pathological behaviors are clearly present but the specific diagnostic criteria for either anorexia nervosa or bulimia nervosa are not met. EDNOS is sometimes described as a "mixed eating disorder" or simply as "other eating disorder" (Reiter and Graves, 2010; Fairburn and Bohn, 2005). Examples of EDNOS listed in the *DSM–IV* (APA, 1994) include:

1. For females, all of the criteria for Anorexia Nervosa are met except that the individual has regular menses.

2. All of the criteria for Anorexia Nervosa except that, despite significant weight loss, the individual's current weight is in the normal range.

3. All of the criteria for Bulimia Nervosa are met except that the binge eating and inappropriate compensatory mechanisms occur at a frequency of less than twice a week or for a duration of less than 3 months.

4. The regular use of inappropriate compensatory behaviors by an individual of normal body weight after eating small amounts of food (e.g., self-induced vomiting after the consumption of two cookies).

5. Repeatedly chewing and spitting out, but not swallowing, large amounts of food.

6. Binge eating disorder: recurrent episodes of binge eating in the absence of the regular use of inappropriate compensatory behaviors characteristic of Bulimia Nervosa. (See Spotlight on…Binge Eating Disorder.)

Pictures of emaciated females suffering from anorexia nervosa, and newspaper articles that highlight individuals who eat a gallon of ice cream in a single sitting before self-inducing vomiting may lead to stereotyping of these eating disorders. As is evident by the EDNOS examples and prevalence, characterizing eating disorders is not simple, and the early diagnosis of any eating disorder is difficult. Consider two females who are 5′8″ and weigh 125 pounds (56.8 kg). This weight falls at the lower end of the normal weight for height range, and each woman would be described as thin. For the first woman, this is her biologically comfortable weight and she can maintain that weight with a normal

Weight cycling: Repeated weight loss and weight gain.

pattern of eating and physical activity. She also has a realistic body image and is not overly concerned if her weight varies between 125 and 130 pounds (56.8 and 59 kg). This woman does not exhibit any characteristics associated with an eating disorder. In contrast, the second woman can maintain her 125-pound (56.8 kg) weight only with daily food restriction and excessive exercise. In fact, weight maintenance is not the goal because she is always thinking about losing 5 pounds. She is dissatisfied with her weight and considers herself not thin enough. She is fearful that her current weight will "balloon" to 130 pounds (59 kg), so if she exceeds 125 pounds (56.8 kg) when she steps on the scale, she fasts for the rest of the day and goes to the gym for an additional aerobic session. Her favorite food is a chocolate candy bar, but she panics if she eats more than her self-imposed limit of one-half of a candy bar and compensates by self-inducing vomiting. This woman does not meet the criteria for anorexia nervosa and does not appear emaciated, but she does have an eating disorder (not otherwise specified) that needs treatment.

Anorexia athletica describes an eating disorder unique to athletes.

Anorexia athletica is a condition found in athletes that overly restrict caloric intake, engage in excessive exercise, or do both for the purpose of attaining or maintaining a low body weight as a way to improve performance. This eating disorder subcategory is not included in the *Diagnostic and Statistical Manual of Mental Disorders* (APA, 1994) but is used by some sport dietitians because it better describes the characteristics exhibited by athletes with eating disorders. Just as "normal" eating in athletes is a bit different from that of the general population, anorexia athletica is a bit different from anorexia nervosa or an EDNOS seen in the general population. The characteristics of anorexia athletica are as follows:

- Reduced body mass (weight) and loss of fat mass is performance related and not related to appearance or body shape. (It should be noted that concerns about body shape could arise as the individual compares body weight, shape, or composition to the sport's most successful athletes.)

- The loss of body mass results in a lean physique.

- **Weight cycling** (repeated weight gain and loss) is usually present although maintenance of a low body weight may be seen all year (preseason, competitive season, "off-season").

- Restriction of food intake and/or excessive exercise is voluntary or at the suggestion of a coach or trainer.
- The abnormal eating occurs while the athlete is competing but stops at the end of the athlete's career (Sudi et al., 2004).

Few studies have been conducted using the criteria for anorexia athletica; therefore, the prevalence is difficult to determine. A 1993 study of 522 Norwegian elite female athletes found that 43 (8.2 percent) met the criteria for anorexia athletica whereas 7 (1.3 percent) had anorexia nervosa and 42 (8 percent) had bulimia nervosa (Sundgot-Borgen, 1993; Sudi et al., 2004). Beals and Manore (2002) found that the prevalence of clinical eating disorders in female collegiate athletes was low (2.3 percent and 3.3 percent of the sample were diagnosed with bulimia and anorexia, respectively), but that as many as one-third were at risk for an eating disorder. Although prevalence figures help shed light on each eating disorder, the critical points are that all these eating patterns are deviations from normal, are harmful to the athletes who manifest them, and can be prevented or detected early (Beals and Manore, 2000). Some athletes competing in sports with weight categories may have anorexia athletica (See Spotlight on...Do Wrestlers Have Eating Disorders?).

Spotlight on...

Binge Eating Disorder

Binge eating, one characteristic of bulimia nervosa, is described as eating a large amount of food in a short period of time. The binge is associated with a loss of control, which is distressing to the individual. In those with bulimia nervosa, compensatory behaviors (for example, vomiting, using laxatives, fasting, excessive exercise) are employed to offset the binge and limit its effect on body weight. In binge eating disorder (BED), no compensatory behaviors are used and the individual typically becomes or remains overweight or obese. Binge eating affects approximately 5 percent of adults in the United States at some time in their life (Mathes et al., 2009).

Binge eating disorder is considered an eating disorder not otherwise specified. However, there are specific diagnostic criteria included in the appendix of the *Diagnostic and Statistical Manual of Mental Disorders* (APA, 1994) for research purposes. These criteria need to be better studied, defined, and validated, which would lead to better diagnostic tools. Overweight and obese individuals who suffer from BED need psychological, nutritional, and medical treatment on the same scale as those suffering from anorexia nervosa or bulimia nervosa (Mathes et al., 2009; Tanofsky-Kraff and Yanovski, 2004). The following criteria have been established to date:

A. Recurrent episodes of binge eating. An episode of binge eating is characterized by both of the following:

1. Eating, in a discrete period of time (e.g., within any 2-hour period), an amount of food that is definitely larger than most people would eat during a similar period of time and under similar circumstances.

2. A sense of lack of control over eating during the episode (e.g., a feeling that one cannot stop eating or control what or how much one is eating).

B. The binge-eating episodes are associated with three (or more) of the following:

1. Eating much more rapidly than normal
2. Eating until feeling uncomfortably full
3. Eating large amounts of food when not feeling physically hungry
4. Eating alone because of being embarrassed by how much one is eating
5. Feeling disgusted with oneself, depressed, or very guilty after overeating

C. Marked distress regarding binge eating is present.
D. The binge eating occurs, on average, at least twice a week for 6 months.
E. The binge eating is not associated with regular use of inappropriate compensatory behaviors (e.g., purging, fasting, excessive exercise) and does not occur exclusively during the course of anorexia nervosa or bulimia nervosa.

In summary, individuals with BED eat rapidly until they are uncomfortably full, often when they are not hungry, and then feel disgusted, depressed, or guilty. The time period used for diagnosis also varies with binge eating disorder occurring over a 6-month period. More research is needed to identify those with BED and determine the most successful interventions.

Application exercise

The SCOFF Eating Disorder Quiz is a simple screening tool for determining risk for anorexia or bulimia that was developed and tested at a medical school in London (Hill et al., 2010). Each of the five questions must be answered.

1. Do you make yourself **s**ick because you feel uncomfortably full?

2. Do you worry that you have lost **c**ontrol over how much you eat?

3. Have you recently lost more than **o**ne stone[^] (14 lb or 6.4 kg) in a 3-month period?

4. Do you believe yourself to be **f**at when others say you are too thin?

5. Would you say that **f**ood dominates your life?

*Stone is a British unit of measure used to express body weight.

Each yes answer receives 1 point. A score of 2 or more indicates that the individual may be at risk and further assessment is needed. More detailed assessments require specialized training.

Some people suffer from exercise dependence and voluntarily engage in excessive exercise.

As with eating, exercise exists on a continuum, from a complete lack of activity to an amount of exercise that would be considered excessive. Also similar to eating, there are differences in amount that may be subtle and make it difficult to determine if the exercise is contributing positively to the athlete's performance or has become a detracting influence. The *intent* of the exercise is an important factor.

Individuals with disordered eating, eating disorders, or anorexia athletica may use exercise to increase their energy expenditure to lose weight or offset increased caloric consumption from bingeing. Although physical activity and exercise are generally considered to be beneficial to health and well-being, an overdependence on exercise has the potential to become a harmful obsession (Bamber et al., 2000). Because of the commitment many athletes have to their training to improve performance, it is very difficult to distinguish the amount of exercise that is appropriate from that which may reflect a psychological disturbance.

Spotlight on...

Do Wrestlers Have Eating Disorders?

Wrestlers meet the criteria for anorexia athletica. However, wrestlers generally do not meet the criteria for anorexia nervosa or bulimia nervosa. The few studies that have been conducted with wrestlers indicate that the majority do not possess the psychological pathology that accompanies eating disorders and contributes to their severity. Wrestlers do not base their self-esteem on their body weight. Most wrestlers engage in abnormal eating behaviors only during wrestling season. Their in-season eating pattern is described as **non-normative** (a deviation from normal), but they generally fall within the normal range when administered tests that measure thoughts, feelings, or attitudes that are associated with anorexia nervosa. Thus, restricting food and fluid is potentially dangerous to a wrestler's physical health, but this eating pattern is transient and does not likely have long-term effects on mental health (Dale and Landers, 1999; Enns, Drewnowski, and Grinker, 1987). Some wrestlers do score above the cutoff on tests used to identify at-risk eating behaviors, and it should not be assumed that wrestlers are immune to developing an eating disorder (Shriver, Betts, and Payton, 2009). There is also concern that the binge eating that is part of many wrestlers' weight cycling patterns or a focus on dieting may lead to an eating disorder after their careers end and weight loss is not so easily achieved after overeating. Weight cycling, and the effect that it may have on performance and health, is discussed in Chapter 11.

Wrestlers may eat abnormally during the wrestling season, but generally do not meet the criteria for anorexia nervosa or bulimia nervosa.

Exactostock/SuperStock

Table 12.1 Proposed Diagnostic Criteria for Secondary Exercise Dependence

The following three criteria are necessary for a diagnosis of secondary exercise dependence:

1. Impaired functioning*

The individual shows evidence of impaired functioning in at least two of the following areas:

a. Psychological—for example, ruminations or intrusive thoughts about exercise, salience of thoughts about exercise, anxiety, or depression

b. Social and occupational—for example, salience of exercising above all social activities, inability to work

c. Physical—for example, exercising causes or aggravates health or injury yet continues to exercise when medically contraindicated

d. Behavioral—for example, stereotyped and inflexible behavior

2. Withdrawal

The individual shows evidence of one or more of the following:

a. Clinically significant adverse response to a change or interruption of exercise habits. Response may be physical, psychological, social, or behavioral, for example, severe anxiety or depression, social withdrawal, self-harm**

b. Persistent desire and/or unsuccessful efforts to control or reduce exercise

3. Presence of an eating disorder**

The following features are indicative but not definitive:

a. Tolerance—that is, increasing volumes of exercising required

b. High volumes of exercising and/or exercising at least once daily

c. Solitary exercising

d. Deception—for example, lying about exercise volume, exercising in secret

e. Insight—for example, denial that exercising is a problem

*Exercise is unreasonably salient and/or stereotyped even when considered in appropriate context—for example, individual is a competitive athlete.

**If individual had not abstained from exercise, or would refuse to do so, rate withdrawal according to anticipated response.

***For a diagnosis of primary exercise dependence, all criteria may be the same as for secondary exercise dependence except for the absence, rather than presence, of an eating disorder.

Reproduced from *British Journal of Sports Medicine,* Bamber, D., Cockerill, I.M., Rodgers, S. & Carroll, vol. 37, 393–400, © 2003 with permission from BMJ Publishing Group Ltd.

In the past, researchers attempted to define "overexercise" or "excessive exercise" using absolute measures of the amount of exercise training, such as frequency and duration (Anshel, 1992). The major problem with this approach is the inability to set a definitive amount of exercise that accurately distinguishes what is appropriate or excessive for all athletes. For example, 75–80 miles (~125–135 km) of running a week by a college-aged female to control her weight would seem to be excessive, but this training volume might be appropriate for a collegiate cross country runner preparing for the competitive season.

There can also be a fine line between the frequency, intensity, and duration of exercise that improves performance and that which results in a decline in performance. The latter is known as overtraining and demonstrates the difficulty of identifying the appropriate amount of exercise and recovery to achieve optimal fitness and reach a peak level of performance. As discussed earlier in this chapter in relation to eating behaviors, there are subtle differences in the intent of the exercising behavior and the athlete's psychological state. These differences help to distinguish committed exercise training from overtraining from **exercise dependence**.

Some researchers have proposed the concept of primary and secondary exercise dependence (Veale, 1987). The proposed idea of primary exercise dependence involves a preoccupation with exercise—an "addiction" to exercise alone—that is independent of other potential mental disorders and is not used for other reasons such as controlling weight. A lack of supporting scientific research and more recent studies has called into question the prevalence of primary exercise dependence (Bamber et al., 2000, 2003).

In a study of female exercisers, Bamber et al. (2000) found that exercise dependence was not present

Non-normative: A pattern of behavior that deviates from what is considered to be normal.

Exercise dependence: An unhealthy preoccupation with exercising.

in women who did not also demonstrate an eating disorder or disordered eating. Exercise dependence was defined as an unhealthy preoccupation with exercising, which had the potential to become a damaging obsession. Behaviors associated with exercise dependence include exercising when medically **contraindicated**, psychological distress when withdrawing from exercise, and a consuming obsession with exercise that transcends considerations of work and social life.

In a subsequent study, Bamber et al. (2003) identified four dimensions of exercise dependence: impaired functioning, withdrawal, presence of an eating disorder, and other associated features. The proposed criteria for exercise dependence are shown in Table 12.1. The distinguishing features that appear to separate exercise dependence from committed training or overtraining are impaired functioning (for example, psychological, social and occupational, physical, and/or behavioral), withdrawal symptoms, and presence of an eating disorder.

Based upon the factors that have been proposed to be related to exercise dependence, a number of

screening tools or scales have been developed to enable practitioners and clinicians to more quickly and easily identify those athletes who may be dependent or "addicted" to exercise. Screening tools such as the Exercise Dependence Scale–Revised (Symons Downs, Hausenblas, and Nigg, 2004) and the Exercise Addiction Inventory (Griffiths, Szabo, and Terry, 2005) provide a series of statements to which the athlete indicates a level of agreement or disagreement on a Likert-type point scale. A cumulative total score or scores on subscale traits is then used to determine whether the athlete meets the criteria for having exercise dependence.

An important area of clinical research examines the relationship of excessive exercise and exercise dependence to eating disorders. A large percentage of those individuals with eating disorders also display some degree of excessive exercise, but the interrelationship between exercise behavior and eating pathology is not well known. Exercise and physical activity convey positive physical and psychological benefits that would seem to make them suitable as part of a host of treatment interventions for those with eating disorders. However, it is not known to what extent they may contribute to or influence pathological eating behaviors, so additional clinical research is needed (Bratland-Sanda et al., 2010).

Contraindicated: Inadvisable because of a likely adverse reaction.

Focus on research

To What Degree Is Exercise Dependence Associated with Diagnosed Eating Disorders and Does Exercise Dependence Change with Treatment for Eating Disorders?

The incidence of exercise dependence or excessive physical activity may be as high as 80 percent among patients with diagnosed eating disorders. However, there is limited information about the amount of physical activity, changes in physical activity, and changes in attitudes about exercise that may occur with eating disorder treatments. Because of the positive physical benefits and psychological effects, such as improvements in depression and anxiety, physical activity may be considered as one strategy to be used in the treatment of eating disorders. Although providing positive benefits, inclusion of physical activity in eating disorder treatment also has the potential to exacerbate the problem through excessive exercise. The following research study sought to provide more information characterizing physical activity in patients being treated for eating disorders: Bratland-Sanda, S., Sundgot-Borgen, J., Rosenvinge, J. H., Hoffart, A., & Martinsen, E. W. (2010). Physical activity and exercise dependence during inpatient treatment of longstanding eating disorders: An exploratory study of excessive and non-excessive

exercisers. *International Journal of Eating Disorders*, 43(3), 266–273.

What did the researchers do? The researchers studied a group of female volunteers who were undergoing inpatient treatment in a psychiatric center for anorexia nervosa, bulimia nervosa, or eating disorders not otherwise specified. The Exercise Dependence Scale–Revised was used to determine whether the subjects were considered excessive exercisers or nonexcessive exercisers. The excessive exercisers were then compared to the nonexcessive exercisers on a variety of measures over their 12–20 treatment periods. The treatment included individual and group psychotherapy, psychoeducation, art therapy, and supervised physical activity with the goal of reducing symptoms of eating disorders.

Moderate-intensity physical activity sessions were taught and monitored each week by a trained and experienced exercise physiologist. The amount of physical activity engaged in was determined for 1-week periods on three occasions using an

Key points

- "Normal" eating is flexible.

- Disordered eating and eating disorders are deviations from normal and are often characterized by obsession or inflexibility.

- Eating disorders such as anorexia nervosa and bulimia nervosa are psychiatric diseases and, in some cases, are fatal.

- Individuals with disordered eating, eating disorders, or anorexia athletica may use excessive exercise to increase their energy expenditure to lose weight or offset increased caloric consumption from bingeing.

Why is body weight a poor way to judge if someone has disordered eating or an eating disorder?

12.3 Disordered Eating and Eating Disorders in Athletes

In some respects, athletes with disordered eating and eating disorders are similar to nonathletes, thus a basic understanding of eating disorders and how they develop is necessary. However, there are some important differences between athletes and nonathletes because athletes are always accountable for their performance. Disciplined eating and high-volume exercise are often expected of athletes, and it can be difficult to distinguish when disciplined has become obsessive and high volume has become excessive.

The prevalence of disordered eating and eating disorders is difficult to determine.

The prevalence of disordered eating and eating disorders in athletes is very difficult to determine. Beals (in press) notes that only a few studies have used a large enough sample size and valid survey instruments to be considered scientifically sound. Published figures range from 1 to 62 percent in female athletes with similar estimates for male athletes (0–57 percent). Such wide ranges have limited usefulness but suggest that disordered eating and eating disorders have a high prevalence in some sports.

In addition to prevalence figures, some other conclusions can be drawn from these studies. First, similar to the general population, female athletes are more likely than male athletes to exhibit both disordered eating and eating disorders. Second, the prevalence of disordered eating (that is, subclinical eating disorders) is higher than the prevalence of clinical eating disorders. Third, the sports that have a higher prevalence are those

accelerometer to measure body movement (see Chapter 2). The Reasons for Exercise Inventory and the Eating Disorders Inventory–2 were used to assess eating disorder psychopathology and the importance of the different reasons for exercising.

What did the researchers find? Nearly one-third of the patients who completed the study were classified as excessive exercisers, and they had a significantly higher amount of physical activity upon admission and at discharge from treatment. The amount of physical activity in the nonexcessive exercisers declined from admission to discharge, whereas in the excessive exercisers it declined during the first part of treatment and then increased significantly toward discharge. The score for exercise dependence was higher for the excessive exercisers, but that score declined over the treatment period. Eating Disorders Examination and Eating Disorders Inventory scores were higher for the excessive exercisers at the beginning of the study, but scores for both groups declined in a similar fashion with treatment. The major finding with the excessive exercisers was that there was a very strong correlation between the reduction in their Eating Disorders Examination score and the reduction in the exercise dependence score. Specifically, there was a reduction in the perceived importance of exercise for negative affect regulation—that is, the use of external activities, such as exercise, to cope with internal feelings such as fear and anxiety.

What was the significance of the study? The results of this study provided further evidence of the seriousness of excessive exercise and eating disorders, as those subjects displayed higher (worse) scores on the eating disorder scales, higher (worse) scores on the exercise dependence scales, and higher amounts of physical activity. However, these subjects showed improvements in these scores over the course of treatment, and the improvement in the eating disorder assessments was associated with the subjects showing less dependence on exercise and placing less importance on exercise as a coping mechanism. A substantial amount of research needs to be completed to develop guidelines for the use of physical activity as one treatment strategy for eating disorders, and these researchers recommend that well-trained and qualified staff are necessary for the implementation and supervision of physical activity programs for these patients.

Sports that require or reward low body weight, low percent body fat, or thin appearance have a higher prevalence of disordered eating and eating disorders.

Ballet dancing (women)

Bodybuilding

Boxing

Cheerleading (women)

Diving (women)

Figure skating (women)

Gymnastics (women)

Horseracing (jockeys)

Lightweight rowing

Martial arts (for example, judo, karate, kickboxing, Tae Kwon Do)

Rhythmic gymnastics

Running (middle or long distance)

Swimming

Ski jumping

Synchronized swimming

Wrestling

Figure 12.2 Sports considered higher risk for the development of disordered eating

Ballerinas may be at increased risk for disordered eating because the body's appearance is an integral part of the art of ballet.

that require or reward a low body weight, a low percentage of body fat, or a thin appearance. These athletes are sometimes described as requiring a "thin build."

Sundgot-Borgen et al. (1999) reported the results of their study that included the entire population of Norwegian elite male and female athletes. Scientists can usually study only a sample of the population, so this study offered additional insight because of its unique design. Using the *DSM–IV* criteria, the authors found that 20 percent of elite female athletes and 8 percent of elite male athletes met the diagnostic criteria for anorexia nervosa, bulimia nervosa, or eating disorders not otherwise specified. Clearly, elite female athletes are at greater risk than elite males, but eating disorders in elite male athletes should not be overlooked. Of the elite female athletes who developed EDs, 60 percent indicated that dieting was an important factor in its development whereas 28 percent indicated injury was a factor. Of the elite males with EDs, only 13 percent indicated that dieting was an important factor whereas injury (25 percent) and overtraining (21 percent) were mentioned more frequently.

Athletes in many sports are considered at risk for disordered eating and eating disorders (Figure 12.2). Aesthetic sports, where appearance is part of the scoring, can lead to an overemphasis on a thin appearance.

Examples of such sports include women's gymnastics, figure skating, diving, ballet, and cheerleading. Their male counterparts are expected to be lean and muscular but not necessarily thin, so there are a higher percentage of at-risk females than males in these sports. Due to the acrobatic nature of some sports such as gymnastics, a high power-to-weight ratio is important and may influence some athletes to attempt to achieve a minimal level of body fatness while maintaining muscle mass.

A low body weight can also be seen to be advantageous in sports in which weight must be moved, such as distance running. Both females and males in this sport may struggle with disordered eating as they try to attain and maintain a low body weight and a low percentage

of body fat. Very low body weights are also advantageous for jockeys and ski jumpers, and some of these athletes develop disordered eating. Athletes in sports with weight categories (for example, wrestling, boxing, martial arts, lightweight rowing) are at risk for eating disorders, especially if competition weight is well below a normal, biologically comfortable weight. Sports with revealing clothing (for example, swimming, diving), with expectations of perfection (for example, synchronized swimming or diving), or in which body appearance is the sport (for example, bodybuilding) also put athletes at risk. Although these are some obvious examples of sports that create conditions that may pressure athletes to engage in pathological eating behaviors,

Athletes in sports with weight restrictions are at risk for developing eating disorders.

Athletes who compete in sports with revealing clothing may feel internal or external pressures to meet body image expectations.

any athlete may be at risk because low body weight is not the only driving factor. Recall that those diagnosed with bulimia nervosa are often normal weight individuals.

Less is known about disordered eating and eating disorders in male athletes than in female athletes. It has been suggested that males more often develop disordered eating (that is, deviations from normal eating) rather than eating disorders, especially when compared to females. Male athletes report more bingeing than females, whereas females report more purging behavior. The prevalence of eating disorders in males appears to be on a rapid rise, but this may be due to more study, better detection, and/or an increased willingness to seek treatment than in the past (Glazer, 2008).

One condition that is receiving more attention in males is body dysmorphic disorder (BDD). A subcategory of BDD is **muscle dysmorphia**, which is much more prevalent in males than females (Grieve, 2007). Body image disorders also appear to be more prevalent in males in Western countries than in non-Western countries. There is more media emphasis on male body image than in the past but it is unknown what, if any, effect increased media attention may have (Baum, 2006). The drive for extreme muscularity is explained in Spotlight on...The **Adonis** Complex.

Muscle dysmorphia: Pathological preoccupation with gaining muscle mass.

Adonis: Greek mythological character described as an extremely handsome young man.

Spotlight on...

The Adonis Complex

The Adonis Complex: The Secret Crisis of Male Body Obsession (Pope, Phillips, and Olivardia, 2000) is a consumer-oriented book written by three university professors (two MDs and one PhD) outlining the psychological problems associated with striving for excess muscularity. These problems include obsessive-compulsive behavior, chronic depression, eating disorders, and/or substance abuse (for example, anabolic steroids).

The parallel medical term is *muscle dysmorphia,* a pathological preoccupation with gaining muscle mass. The few studies that have been conducted, primarily in males who lift weights, suggest that muscle dysmorphia is an obsessive-compulsive disorder, but there are elements of inaccurate body image, body dissatisfaction, and eating attitudes that are

similar to those found in diagnosable eating disorders. Those with muscle dysmorphia have highly developed skeletal musculature but they believe their muscles are too small. They are dissatisfied with their appearance, weight, and amount of muscle mass and are at risk for using anabolic steroids to change body composition (Choi, Pope, and Olivardia, 2002; Olivardia, Pope, and Hudson, 2000).

Body dissatisfaction in men is increasing, but the incidence and prevalence of muscle dysmorphia is not known (Choi, Pope, and Olivardia, 2002). Muscle dysmorphia is an example of body dysmorphic disorder (preoccupation with defective appearance), the latter of which is listed in the diagnostic manual for the American Psychiatric Association.

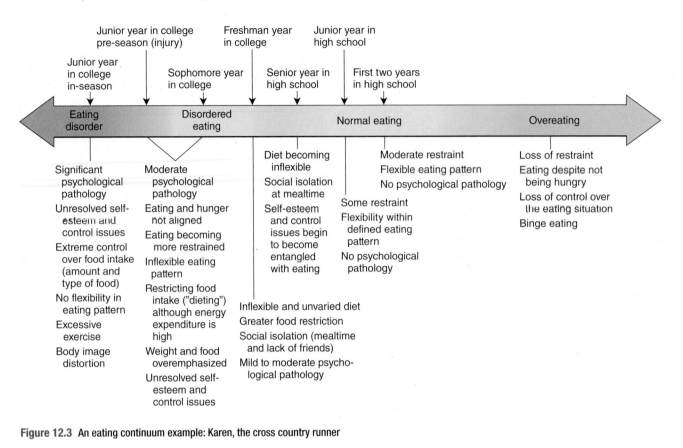

Figure 12.3 An eating continuum example: Karen, the cross country runner

Karen's behaviors placed on the eating continuum.

Disordered eating behaviors may progress to an eating disorder.

This chapter began with a vignette about Karen, a collegiate cross country runner. As is hinted at in the scenario, there were some unresolved family, control, and self-esteem issues. Karen also exhibited some personality traits, such as perfectionism, excessively high achievement goals, and obsessiveness, which are associated with eating disorders. Participating in a sport, wanting to excel, fierce competitiveness, or being injured do not cause an eating disorder. These may be factors that influence its development, but an eating disorder is, at the core, a psychiatric disease.

Karen was a naturally talented runner who had a "normal" eating pattern when she entered high school. As she became more dedicated to improving her running, she adopted a semivegetarian diet in her junior year of high school to support her training and performance goals. If she consumed sufficient kilocalories and nutrients and maintained some dietary flexibility, then her diet would be considered "normal" for a well-trained athlete. She enjoyed the recognition she got from her parents for being in the newspaper. In her senior year she followed a more strict vegetarian diet, preparing her own food and eating alone. She started

to become socially isolated at mealtime, and her diet was becoming more inflexible. During therapy, Karen identified her senior year as the point in time when she began to move away from "normal" on the eating continuum (Figure 12.3) because she had found a way to "control" her parent's comments and her feelings of low self-esteem that accompanied dinnertime conversation. At the same time her successful cross country season was showering her with attention, something she realized later that she desperately needed.

At 18, she left home to attend college and compete at the NCAA Division I level. Although she was a talented high school runner, this is always a vulnerable time for an athlete who goes from being "a big fish in a little pond" to "a little fish in a big pond." For Karen, it became a time of distress because she felt that she had little control over her life. Some of her teammates (who were also her competitors) made disparaging comments, and these comments affected her self-esteem. Eating in the dorm contributed to her diet becoming less flexible and more monotonous. That she no longer enjoyed eating was also a factor because pleasurable eating is a part of "normal" eating. The stated reason for moving into a single room and fixing her own food was the potential for more nutritious eating, but the unstated reason was that she felt more in "control" when she ate alone, something she had first discovered in high school. Changes in her eating situation led her away from a normal eating pattern by creating more isolation at mealtime, which supported greater food restriction.

Karen had made few friends and missed her family, so her coach became her surrogate family. She sought his attention in the same way that she sought her parent's attention in high school—by being a successful runner. When Karen perceived that the coach thought that she needed to be leaner, she had a new goal to accomplish. Her body weight became closely tied to her self-esteem and also became something she could "control."

A turning point for Karen was her decision to start a weight-loss diet. Recall that dieting can be a factor that triggers the development of an eating disorder, especially in elite female athletes. She restricted her caloric intake (at the same time that she increased her energy expenditure), but she did not establish a weight-loss goal, so she had no way of knowing if, and when, she met that goal. She was also using her weight as a way of vying for her coach's attention.

Karen's injury further contributed to her disordered eating for both physical and psychological reasons. Her goal was to keep her weight "low," but the injury prevented energy expenditure through exercise. A goal of losing weight or maintaining an already low body weight when injured is often not realistic. Trying to attain an unrealistic goal moved Karen further along the disordered eating continuum. The injury also changed the amount of attention that she received from her coach, the person in her life whose opinion she cared about the most.

When her injury was resolved, she continued to consume a restricted diet and voluntarily engaged in excessive exercise. Increasing the serving size of a low-calorie food such as salad was not an appropriate dietary response to a substantial increase in training volume. When she did not get an invitation to an important meet, her response was to train more and eat less. At this point, Karen exhibited all the signs of anorexia athletica and had unresolved psychological issues related to control, self-esteem, and personal relationships.

It is important to distinguish "normal" and dysfunctional eating and exercise behaviors in athletes.

Excellent athletic performance, especially at the elite level, requires rigorous training. A nutritious diet supports training and can improve performance, and some athletes must follow fairly strict eating protocols to support their training, body composition, and performance goals. Both training and eating can become regimented, a factor that may contribute to disordered eating. A fine line separates rigorous training and eating regimes that enhance performance and support health from disordered eating and exercise dependence that hurt performance and undermine health. Crossing this imaginary line may be accidental or intentional. Because early intervention is critical for treatment and recovery and the athlete may be unaware or in denial that problems exist, it is important for coaches, athletic and personal trainers, and others who work closely with athletes to be able to distinguish that which is "normal."

Table 12.2 compares the features that may help to distinguish "normal" and disordered eating and exercise patterns in athletes. Athletes in both groups share many features: a high level of physical training, an eating plan to support the demands of training, and a desire to change body composition. But there are marked differences between the two groups in actions and perspective. Returning to the example of Karen, she exhibited almost all the features of disordered eating and exercise patterns shown in Table 12.2.

In addition to behavioral signs, there may be physical signs of disordered eating and eating disorders. Some of these signs may not be noticeable until the disordered eating is prolonged or severe. Frequent gastrointestinal (GI) problems may be present early (some athletes may use GI distress to control weight), but by themselves are too general to predict disordered eating. As food intake is restricted and nutritional status declines, the athlete may exhibit weight loss, chronic fatigue, iron deficiency anemia, irregular or absent menstruation, and slow recovery from illness or injury. Those with anorexia nervosa may be exceptionally intolerant to cold temperatures (due to very low percent body fat) or grow fine hair on the body (known as lanugo) in an effort to regulate body temperature. Bulimics may have callused fingers, teeth with little enamel, or esophageal erosion from self-inducing vomiting. The National Athletic Trainers' Association position statement (Bonci et al., 2008) on the prevention, detection, and management of disordered eating in athletes is an excellent resource.

In some cases, the individual with disordered eating behaviors or an eating disorder, particularly anorexia nervosa, may have been attracted to sports or fitness activities as a "cover" for these unhealthy eating and exercise behaviors. In the early stages, the athlete or those associated with the athlete may view a large volume of exercise or a dedication to an eating plan positively. Over time, it may become clear to others that the individual is using sports or fitness to justify pathological behaviors, but it is often difficult to detect such a motivation early on (Beals, in press; Baum, 2006).

Table 12.2 Distinguishing "Normal" and Abnormal Eating and Exercise Patterns

	Features of athletes with "normal" eating and exercise patterns	Features of athletes who may have disordered eating and exercise patterns
Performance	Performance is improved, or a high level of performance is maintained	Performance declines
Training	Purposeful training; no overtraining	Excessive exercise or activity; self-imposed overtraining or exercise dependence; anxious if not able to train; continues to train with injury against medical advice
Energy intake	Caloric intake is monitored; athlete is disciplined but not obsessive about the amount of food consumed	Caloric intake is controlled; athlete is disciplined and obsessive; amount of calories consumed is recorded or mentally counted; consumption of caloric intake over self-imposed limit causes anxiety
Perspective on food intake	Food is needed to fuel training; eating is enjoyed and viewed positively	Food needs to be restricted; eating is not enjoyable and viewed negatively
Dietary intake	Consumption of "healthy foods" and adequate kilocalories; no concern about occasionally eating low-nutrient-dense foods	Consumption of "healthy foods" but inadequate kilocalories; concern about or refusal to occasionally eat low-nutrient-dense foods
Dietary flexibility	Routinely follows a well-planned diet but is flexible as needed	Ritualistic and inflexible pattern of eating
Body image	Accurate and positive body image	Inaccurate and negative body image
Body composition	Realistic weight and body composition goals that improve or maintain performance; goals are attainable without compromising health	Unrealistic weight and body composition goals that do not improve or maintain performance; goals are not attainable without compromising health
Muscle mass	Increased or maintenance of muscle mass with resistance training	Decreased or inability to increase muscle mass with resistance training

Ultimately, eating disorders have a negative effect on performance and health.

As might be expected, the extent of the disordered eating behaviors is associated with the effects on performance and health. More severely dysfunctional behaviors used for a longer period of time typically have more a more negative effect on health and performance. For some athletes with mildly abnormal eating and exercise behaviors, performance may not be affected to a large degree. However, eating and exercise obsessions likely have an effect on mental health and, over time, an effect on physical health including anemia, frequent infections, and increased risk for injury. Some athletes with severe and prolonged eating disorders have died, often at a young age.

It seems counterintuitive that athletes suffering from eating disorders, such as anorexia nervosa, may see a positive effect on performance or that they can even continue to exercise when emaciated. Starvation and purging both result in increases in some of the adrenal hormones, such as cortisol, epinephrine, and norepinephrine. These hormones can mask fatigue so it is possible that performance improves initially as weight declines due to starvation. However, performance declines over the long term, along with physical and mental health (Beals, in press). There may also be a biological reason that exercise can continue in the athlete with anorexia nervosa and substantial weight loss. Exercise produces lactate, which is a source of glucose (see Chapter 3). In a prolonged starvation state where the individual is still willing to exercise, the body may be desperately trying to find a source of glucose for the brain (Tyson, E., personal communication, March 29, 2010).

If disordered eating or an eating disorder is suspected, then the athlete should be approached with care and concern.

Early intervention is critical in the treatment of disordered eating and eating disorders. If left alone, athletes typically do not resolve these issues themselves. In some cases, the athletes do not realize that they have fallen into a disordered eating pattern; in other cases, the athletes staunchly deny that an eating problem exists. If a coach, athletic trainer, teammate, or any other individual associated with the athlete suspects that any degree of disordered eating exists, the question is not when, but how to intervene.

Beals (in press) notes that if disordered eating is suspected, the appropriate course of action is to approach the athlete and refer him or her to a trained professional for further evaluation. The individual who approaches the athlete should have a good rapport with that athlete and the contact should be made in a private setting. Care and concern should be conveyed. Approaching anyone with an eating disorder is an extremely sensitive issue and must be done in a professional and confidential manner. In some universities a referral protocol has been established, and, if so, it should be followed exactly. If no protocol exists, a good starting point is to refer the athlete to the team physician.

If disordered eating or an eating disorder is confirmed, then the placement of the athlete into a treatment program is crucial. The primary goal of treatment is to help the athlete resolve both the psychological and physical issues present. Early intervention is critical to meeting that goal, because those with less severe or less prolonged problems have a better chance of successful treatment. If appropriate, a second goal is for the athlete to return to the sport. Athletes who refuse treatment or do not satisfactorily complete treatment should not be allowed to train or compete because their mental and physical health will be compromised (International Olympic Committee [IOC], 2005).

Treatment involves three components—psychological, nutritional, and medical. Each needs to be treated by an expert, so treatment involves a team approach including a psychologist, registered dietitian, and physician. Psychological counseling is necessary because psychological disturbances are at the core of the eating disorders. A description of the intensive psychological therapy needed is beyond the scope of this chapter, but it is fundamental to treating any eating disorder. Nutritional counseling is necessary, even though many who have eating disorders know a great deal about the caloric and nutrient content of food. Nutrition counseling helps them to view food in a normal context, one in which eating is both flexible and enjoyable. Medical guidance is needed to resolve physical problems resulting from the eating disorders and to coordinate medical care over the course of treatment, which may be months or years in length, depending on the severity of the eating disorder.

It is important to promote a culture that supports "normal" eating for all athletes.

Disordered eating and eating disorders do not develop in a vacuum. The IOC Medical Commission (IOC, 2005) notes that several factors may influence their development. Western cultures emphasize thinness, and females frequently restrict food intake to lose weight. Females who equate thinness with success are more susceptible to developing eating disorders. From the female athlete's perspective, "success" may include being thinner than a teammate or receiving more attention from her coach because of her thin body. Female athletes in sports in which thinness is desirable

can face extraordinary pressures, especially as they try to reach the elite levels of their sports. Decreasing body weight or reducing body fat can, and often does, lead to improved performance initially and a desirable appearance in revealing clothing. However, not only is the belief that an ever-lower body weight or body fat percentage is beneficial incorrect (it leads to poorer performance), but this belief is a powerful risk factor for the development of an eating disorder.

There is evidence that certain personality traits such as perfectionism, obsessive-compulsive behavior, overcompliance, and extreme competitiveness and goal setting are associated with disordered eating. Ironically, these are traits that are valued by coaches and extolled in the media because many highly successful athletes exhibit these behaviors. Thus athletes may be positively reinforced for the same behaviors that put them at risk for disordered eating (IOC, 2005).

No one person can change the way society views and values the appearance of the human body. Nor can one person shape the athlete's personality or beliefs about body weight and body image. However, the one person who may have the most influence over the athlete's behavior may be the coach. For this reason, the IOC Medical Commission recommends that coaches not be involved in determining the athlete's weight or body composition, nor should they suggest to the athlete that body weight should be reduced. This keeps coaches from establishing and judging the athlete's weight or body composition and helps coaches prevent inadvertent reinforcement of disordered eating or excessive exercise.

However, many coaches are involved in weight-related issues. A 2003 survey of U.S. collegiate coaches of female gymnastics, swimming, basketball, softball, track, and volleyball teams found that 44 percent of those surveyed weighed athletes, assessed body composition (44 percent), and suggested losing weight by restricting food (33 percent) or increasing workouts (29 percent). Weight, body composition assessment, and a plan for weight (fat) loss are best carried out by trained professionals other than coaches because athletes are directly affected by coaches' decisions and coaches can be a powerful influence on the athlete's behavior and health (Heffner et al., 2003).

Many coaches work hard to prevent disordered eating and eating disorders. After a diagnosis has been made, coaches often wonder what role they may have played in its development. Coaches do not *cause* eating disorders or other related conditions, but they must be careful not to unwittingly create conditions that encourage them. Coaches must be careful about how they reinforce behavior because the attention, or lack of attention, can reinforce inappropriate eating and exercise behaviors. In sports whose athletes are known to be at risk for developing disordered eating and eating disorders, it is very important that each athlete have ongoing medical, nutritional, and training advice and that optimal and minimum body weights be determined and monitored by someone other than the coach (see Chapter 11).

Karen, the subject of this chapter's case study, had high goals even as an adolescent. She was a perfectionist, as evidenced by her perfect high school grades and her desire to repeat that achievement in college. Her obsession with food and exercise was probably a natural fit with her overachieving, competitive personality. She had unresolved family problems when she left for college and then found herself faced with new situations that were stressful and that she could not fully control. She had an athletic scholarship, but it was not guaranteed. Other athletes were not only better performers, but Karen perceived that they were thinner. By her junior year in college she was lonely, isolated, starving, injured, and sick.

Karen began to view her coach as a substitute for both her family and friends. She perceived that her coach thought that she needed to lose weight, and she may not have realized the difference between being "lean" (that is, having a relatively high percentage of lean body mass and a relatively low percentage of body fat) and being "thin." It is impossible to know what comments were made, but coaches should be aware that even an innocent comment about weight or body composition could be misconstrued.

Key points

- Disordered eating and eating disorders are more prevalent in females than males.

- Athletes in sports that favor a thin build are particularly susceptible, but an athlete in any sport may develop disordered eating or an eating disorder.

- Muscle dysmorphia, a preoccupation with excessive muscularity, is more prevalent in males than females.

- Disordered eating behaviors can progress from mildly abnormal to life-threatening eating disorders.

- Dysfunctional eating and exercise behaviors can be distinguished from normal behaviors, but it takes careful observation.

- Improvements in performance in those with disordered eating and exercise behaviors likely occur early and are temporary; ultimately, dysfunctional eating and exercise behaviors have a negative effect on performance.

- Disordered eating and eating disorders are a threat to an athlete's physical and mental health, and any signs or symptoms should not be ignored.

- Early identification, intervention, and treatment are critical and must be done in a caring and confidential manner.

Does your university, sports medicine clinic, running or cycling club, or local gym have a procedure for referral if an athlete is suspected of struggling with disordered eating or an eating disorder?

12.4 Female Athlete Triad

The Female Athlete Triad (Figure 12.4) is a term used to describe three interrelated factors—energy availability, menstrual function, and bone mineral density—each of which develops along a continuum. Each factor may progress to a point where it is a clinical condition. For example, energy availability may be low due to disordered eating or an eating disorder and amenorrhea or osteoporosis may exist. The three may be present together and have developed in sequence—low energy availability due to low energy (caloric) intake and high energy expenditure leads to amenorrhea that leads to osteoporosis—although each of these conditions can occur independently of the others.

Whether alone or in combination with each other, when energy availability, menstrual function, and bone mineral density are compromised they represent substantial health risks for the female athlete. However, when supported by proper nutrition and exercise these factors are associated with good health. Both the American College of Sports Medicine (Nattiv et al., 2007) and the International Olympic Committee Medical Commission (IOC, 2005) have issued position papers on the Female Athlete Triad. Preventing and treating the Female Athlete Triad is a high priority because of the substantial short- and long-term medical problems that can occur.

Low energy availability is a major factor in the Female Athlete Triad.

Energy availability is defined as dietary energy intake minus exercise energy expenditure. Essentially, it is the amount of energy available to the body for other biological functions. Low energy availability results when the female athlete is in negative energy balance. Negative energy balance, also known as an **energy deficit**, is the result of energy expenditure from exercise exceeding energy intake from food. This deficit may last months or years. In adolescent athletes, physical growth also requires energy, and growth may contribute further to the energy deficit.

Low energy availability may or may not be associated with disordered eating. For example, a female distance runner with a low body weight may try to slightly underconsume energy (kcal) intake daily when compared to energy expenditure to prevent a gain in body fat. She could have a well-planned, nutritious, and disciplined training diet that is not an obsession and not disordered eating. Since she is intentionally in an energy deficit, she is at risk for developing amenorrhea and osteoporosis, but she does not have the same psychological risk as a similar athlete with disordered eating or an eating disorder. Those female athletes who exhibit disordered eating patterns are at greater risk both physiologically and psychologically. It bears repeating that

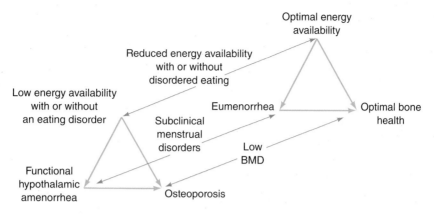

Figure 12.4 The Female Athlete Triad

The spectrums of energy availability, menstrual function, and bone mineral density along which female athletes are distributed (blue arrows). An athlete's condition moves along each spectrum at a different rate, in one direction or the other, according to her diet and exercise habits. Energy availability, defined as dietary energy intake minus exercise energy expenditure, affects bone mineral density both directly via metabolic hormones and indirectly via effects on menstrual function and thereby estrogen (green arrows).

Adapted from: Nattiv. A., Loucks, A.B., Manore, M.M., Sanborn, C.F., Sundgot-Borgen, J., Warren, M.P.; American College of Sports Medicine. (2007). American College of Sports Medicine position stand. The Female Athlete Triad. *Medicine and Science in Sports and Exercise, 39*(10), 1867–1882. Reprinted by permission.

food restriction or "dieting" in a low-body weight elite female athlete is often a trigger that can lead to disordered eating, which can progress to an eating disorder. Female athletes in sports in which low body weights are required or desired, such as lightweight rowing, the lower weight categories in the martial arts, gymnastics, ballet dancing, and distance running, may intentionally undereat in an effort to attain or maintain that low body weight. Coupled with the high energy expenditure required for training in these sports, especially at the elite level, an ongoing energy deficit is likely.

Low energy availability appears to be the factor that impairs menstrual function and bone mineral density. Persistent and severe energy deficits force the body to adapt and begin to suppress physiological functions that are associated with normal growth and development, such as menstruation. Research suggests that there are negative effects associated with energy availability below 30 kcal/kg of fat-free mass daily. Studies of young-adult female distance runners who are not menstruating consistently report an energy consumption below this level; however, there are also reports of those who are still menstruating at this level of energy intake. Although the 30 kcal/kg fat-free mass per day is a good rule of thumb, the fact that some females still menstruate when consuming energy at this level suggests that some women are more susceptible to low energy availability than others (Nattiv et al., 2007; IOC, 2005).

Amenorrhea is the absence or suppression of menstruation.

Amenorrhea is defined as the absence or suppression of menstruation. In the United States, primary amenorrhea describes a female who has gone through puberty

but by age 15 has not yet menstruated. In secondary amenorrhea, the female began menstruating but menstruation has been absent for 3 or more months. There are a variety of medical conditions that may have an effect on normal menstruation patterns. In the context of the Female Athlete Triad, the amenorrhea is a result of low energy availability and is not due to some other

In sports in which low body weight may be associated with better performance, athletes may undereat, resulting in a chronic energy deficit.

Energy deficit: Result of consuming less energy (kcal) than expended.

medical condition or contraceptive technique that may result in absent menstruation. This type of amenorrhea is known as functional hypothalamic amenorrhea.

In the past, amenorrhea in athletes was attributed to low body fat stores and the stress of exercise. These factors are no longer believed to play causative roles. Rather, the amenorrhea seems to be due to an energy deficit that alters the secretion of **luteinizing hormone** (LH). Menstruation is regulated by a number of hormones, including follicle-stimulating hormone (FSH), luteinizing hormone (LH), and estrogen. Figure 12.5 illustrates the expected hormonal fluctuations associated with menstruation, although many variations are seen. During the first few days of the menstrual cycle the growth of one egg is accelerated. One to 2 days prior to ovulation, there is a surge in LH secretion so that ovulation can occur. The unfertilized egg grows and secretes estrogen and progesterone, hormones that inhibit the secretion of LH and FSH (Guyton and Hall, 2010). The current prevailing theory is that low energy availability disrupts the normal secretion of luteinizing hormone, resulting in amenorrhea. The disruption can occur within 5 days when energy availability is reduced substantially (Nattiv et al., 2007; IOC, 2005).

Athletes who begin intense training at an early age, such as gymnasts or distance runners, may exhibit primary amenorrhea, that is, they have never menstruated. This intense training may lead to the chronic energy deficit before the onset of puberty. Other athletes may begin menstruating normally but develop secondary amenorrhea later when the training demands of their sport escalate. Ironically, many athletes view the lack of menstrual periods as being advantageous. It may be perceived as evidence that they are lean or it may simply be a relief from the inconvenience of the monthly period. Athletes may also have the mistaken idea that amenorrhea may act as birth control, and that pregnancy is not possible during the time when menstruation is not present. Instead of being advantageous, amenorrhea should be recognized as undesirable and potentially harmful to health.

Low bone mineral density is a third factor involved in the Female Athlete Triad.

Achieving peak bone mineral density and preventing or slowing the loss of bone mineral with age are important factors in lifelong bone health (see Chapter 9 for a detailed discussion). Both low energy availability and

amenorrhea affect the bone mineral status of female athletes. Low energy availability results in chronic undernutrition, depriving the body of the nutrients needed for proper bone development and maintenance. Among the nutrients that may be deficient are those closely associated with bone health—protein, calcium, and vitamin D. Amenorrhea is associated with estrogen deficiency. One of the actions of estrogen is protection against calcium loss from bone, and a low estrogen concentration results in loss of bone calcium and alterations in bone microarchitecture. Of the two factors, low energy availability is particularly powerful because of its far-reaching effects on bone formation, including nutrients as well as a variety of hormones.

As the mineral density of the bone declines, its structure deteriorates and there is a greater risk for fracture. Of great concern to athletes are stress fractures, small cracks or incomplete breaks in weight-bearing bones, typically the tibia and fibula. Amenorrheic athletes have lower bone mineral density and are at greater risk for stress fractures than athletes with normal menstruation.

The loss of calcium from bone is progressive. Dual energy X-ray absorptiometry (DEXA) can determine bone mineral density (BMD). The results of this test place females in one of three categories: (1) normal, (2) osteopenia (low BMD), or (3) osteoporosis. In general, athletes who are menstruating have normal or above-normal BMD because weight-bearing exercise has a positive effect on the deposition of calcium in bone. However, numerous studies have documented that trained athletes with amenorrhea may exhibit low bone mineral density or osteoporosis. Beginning in the 1980s, Drinkwater and colleagues demonstrated that bone loss occurred in regularly exercising athletes who were amenorrheic (Drinkwater, Bruemmer, and Chesnut, 1990; Drinkwater et al., 1984, 1986). Athletes who are not menstruating are at risk first for osteopenia and then for osteoporosis. Khan et al. (2002) note that at least one study showed that 22–50 percent of the subjects, amenorrheic runners and ballet dancers, had varying degrees of osteopenia. In two studies of amenorrheic female distance runners between the ages of 20 and 30, 10–13 percent were diagnosed with osteoporosis.

Cobb and colleagues (2003) studied 91 well-trained female distance runners ages 18 to 26 years. Thirty-three athletes had zero to nine menstrual periods in a year whereas the remaining subjects (58 athletes) had normal menstruation. Bone mineral density was determined by DEXA for the entire body as well as the hip and spine. When BMD was compared to those with **eumenorrhea** (that is, normal menstruation), the amenorrheic athletes had 3 percent less in the entire body, 6 percent less in the hip, and 5 percent less in the spine. Based on spine measurements, two of the amenorrheic runners were classified as osteoporotic

---

Luteinizing hormone: One of the menstrual cycle hormones associated with ovulation.

Eumenorrhea: Normal menstruation.

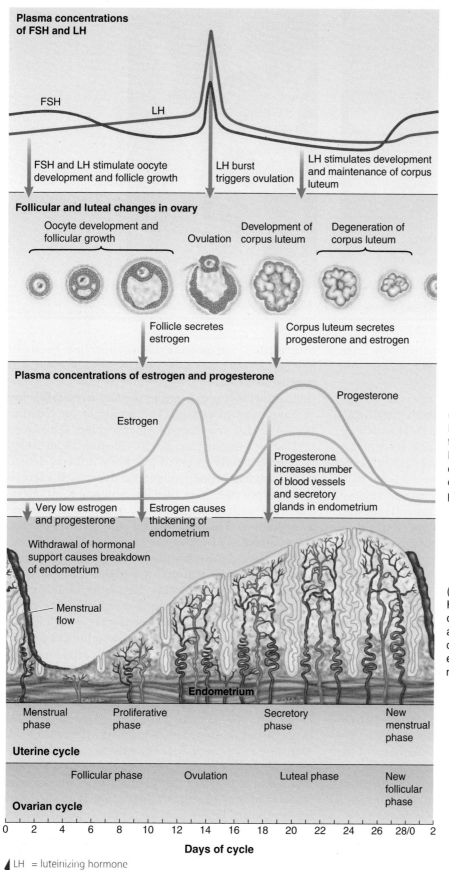

Plasma concentrations of FSH and LH

FSH

LH

FSH and LH stimulate oocyte development and follicle growth

LH burst triggers ovulation

LH stimulates development and maintenance of corpus luteum

(a) Concentrations of pituitary gonadotropic hormones. Note that the concentrations of both FSH and LH peak just prior to ovulation.

Follicular and luteal changes in ovary

Oocyte development and follicular growth

Ovulation

Development of corpus luteum

Degeneration of corpus luteum

Follicle secretes estrogen

Corpus luteum secretes progesterone and estrogen

Plasma concentrations of estrogen and progesterone

Progesterone

Estrogen

Progesterone increases number of blood vessels and secretory glands in endometrium

(b) Concentrations of ovarian hormones. Estrogen concentration peaks during the late preovulatory phase. Progesterone, secreted mainly by the corpus luteum, reaches its peak concentration during the postovulatory phase.

Very low estrogen and progesterone

Estrogen causes thickening of endometrium

Withdrawal of hormonal support causes breakdown of endometrium

Menstrual flow

(c) Ovarian and uterine cycles. Hormone concentrations correlate with changes that take place in the ovaries and uterus. If fertilization occurs, the corpus luteum continues to secrete estrogen and progesterone, and menstruation does not occur.

Endometrium

| Menstrual phase | Proliferative phase | Secretory phase | New menstrual phase |

Uterine cycle

| Follicular phase | Ovulation | Luteal phase | New follicular phase |

Ovarian cycle

0 2 4 6 8 10 12 14 16 18 20 22 24 26 28/0 2

Days of cycle

LH = luteinizing hormone
FSH = follicle-stimulating hormone

Figure 12.5 Hormonal fluctuations during the menstrual cycle

A current theory of amenorrhea in athletes is that chronic low energy availability disrupts the normal secretion of luteinizing hormone, resulting in a disruption of menses.

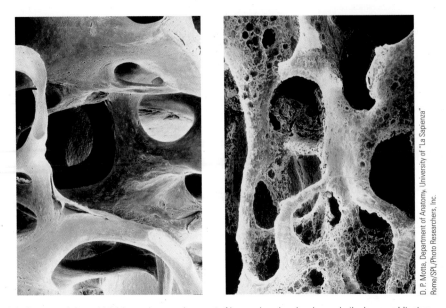

Healthy bone appears on the left. Osteoporotic bone (right) has a decreased amount of bone mineral and a change in the bone architecture.

D. P. Motta, Department of Anatomy, University of "La Sapienza", Rome/SPL/Photo Researchers, Inc.

and nearly half were osteopenic. In comparison, none of the eumenorrheic athletes were osteoporotic and only 26 percent were osteopenic.

The IOC report (2005) states that amenorrhea that lasts longer than 6 months will likely have a negative effect on the athlete's bone mineral density. This loss of bone calcium is especially disturbing because it is occurring during a period of life when bone mineral density should be increasing. Long-term effects on the skeleton have been documented in males and females with adolescent-onset eating disorders. In females with anorexia nervosa as a teenager, those who recovered still had lower bone mineral density than women of the same age who never had anorexia, even many years after weight had been regained and menstruation had been restored. In boys who manifested anorexia nervosa as teenagers, they tended to have impaired dimensional growth and were of shorter stature than if they had not suffered from anorexia. Not reaching genetic potential for height is probably due to chronic caloric restriction and undernutrition during a growth period, which affects insulin-like growth factor-I (IFG-I). Short stature was not an issue for most of the teenage girls with anorexia. This is probably explained by the fact that adolescent boys have a longer growth period than girls by about 2 years and that many girls have reached their maximum height before the onset of the eating disorder (Misra, 2008).

Keen and Drinkwater (1997) were able to study some of the athletes from their original studies in a follow-up investigation 8 years later. Of particular interest were comparisons between those who exhibited regular menstruation or intermittent menstruation/amenorrhea both originally and at follow-up. There was a significant difference in bone density between the two groups, with those in the group in which

intermittent menstruation/amenorrhea persisted having ~85 percent of the bone density of those with regular menstruation. Early intervention for amenorrheic athletes is important to prevent irreversible loss of bone mineral density. The Spotlight on...Normal Bone Density in a Former Amenorrheic, Osteoporotic Distance Runner reviews a case study of an athlete who successfully reversed low bone mineral density with increased food intake and decreased exercise that led to weight gain (Fredericson and Kent, 2005). This case study underscores the importance of energy availability and its influence on bone mineral density.

A number of cross-sectional studies of young women have shown that physically active and athletic women typically demonstrate higher bone mineral density (Kohrt et al., 2004). This is particularly true for women who participate in weight-bearing sports or activities, especially those that involve high impact or forces such as gymnastics and weight lifting. However, exercise does not guarantee increased bone density, as bone loss has been demonstrated in regularly exercising athletes that were amenorrheic (Drinkwater, Bruemmer, and Chesnut, 1990; Drinkwater et al., 1984, 1986). The risk of osteoporosis is a long-term health concern, but the bone loss associated with amenorrhea may have more immediate consequences for these athletes. Menstrual irregularity, and the associated bone mineral loss, is associated with a greater incidence of stress fractures in runners, particularly in the lower leg. A study of female collegiate runners (Barrow and Saha, 1988) revealed a much higher percentage of stress fractures in runners reporting an irregular menstrual history (0 to 5 menses per year) compared to runners reporting regular menstruation (10 to 13 menses per year) (Figure 12.6).

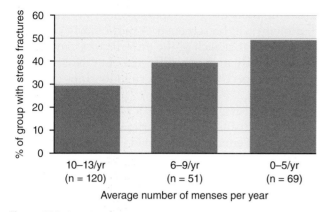

Figure 12.6 Prevalence of stress fractures according to menstrual history

Source: Barrow, G. W., & Saha, S. (1988). Menstrual irregularity and stress fractures in collegiate female distance runners. *American Journal of Sports Medicine, 16*(3), 209–216.

Both elite and recreational athletes can develop the Female Athlete Triad.

The prevalence of the Female Athlete Triad, especially the manifestation of clinical conditions of all three factors, is hard to determine. The prevalence of low energy availability without disordered eating or eating disorders is not known and has not been well studied. The prevalence of disordered eating in female athletes has an estimated range of 1–62 percent, with a greater prevalence in sports that emphasize a lean or thin build. The prevalence of secondary amenorrhea also varies widely based on the sport, with a high prevalence reported in ballet dancers (69 percent) and distance runners (65 percent). Similarly, the prevalence of osteopenia (22–50 percent) and osteoporosis (0–13 percent) in young female athletes is cause for concern.

Any physically active female is at risk for developing the Female Athlete Triad (Table 12.3). Torstveit and Sundgot-Borgen (2005b) studied 186 elite athletes and 145 age-matched controls. The elite athletes trained an average of approximately 14 hours per week whereas the controls were physically active for a little more than 5 hours per week. Additionally, the activity of the control group was of lesser intensity than the elite athletes. Eight elite athletes (4.3 percent) met all the criteria for the Female Athlete Triad—low energy availability (in this study due to disordered eating or eating disorders), menstrual dysfunction, and low bone mineral density. Five members of the control group (3.4 percent) also met these criteria. Although only 4.3 percent of the elite athletes met the full criteria, 26.9 percent (50 elite athletes) exhibited disordered eating or eating disorders and menstrual dysfunction (but not low bone mineral density). Twenty-two controls (15.2 percent) had disordered eating and low BMD but not menstrual dysfunction. The Triad may progress in stages, and there is some evidence that the elite athletes have a more severe condition than the controls.

This study suggests that elite and recreational athletes as well as women who are not physically active may be at risk. An earlier study by the same authors found that female athletes competing in sports that emphasize leanness or low body weight were more likely to be at risk for the Triad than female athletes in other sports (Torstveit and Sundgot-Borgen, 2005a).

Prevention, intervention, and treatment of the Female Athlete Triad are critical.

The prevention of the Female Athlete Triad begins with preventing persistent energy deficits. Those athletes at greatest risk restrict food intake while engaging in large amounts of exercise. Vegetarians and other athletes who limit the types of food they consume are also at a higher risk, although the fundamental issue is that the limitations result in low caloric intake. Weight-related issues are associated with the development of the Triad because the sports that have

Table 12.3 Prevalence of the Female Athlete Triad Components

	Female Athlete Triad (all three components)	Disordered eating + menstrual dysfunction	Disordered eating + low bone mineral density	Menstrual dysfunction + low bone mineral density
Elite athletes (N = 186)	4.3% (N = 8)	26.9% (N = 50)	10.2% (N = 19)	5.4% (N = 10)
Control group (N = 145)	3.4% (N = 5)	13.8% (N = 20)	15.2% (N = 22)	12.4% (N = 18)

Source: Adapted from Torstveit, M. K., & Sundgot-Borgen, J. (2005b). The Female Athlete Triad exists in both elite athletes and controls. *Medicine and Science in Sports and Exercise, 37*(9), 1449–1459.

Note: Eighty-seven of 186 elite athletes and 65 of 145 subjects in the control group demonstrated two of the three or all three components of the Female Athlete Triad.

a higher prevalence are those that require or reward a low body weight, a low percentage of body fat, or a thin appearance.

Athletes in sports known to be at risk for the Triad should work with a physician, a sports dietitian, and an exercise physiologist to identify a biologically comfortable body composition and body weight. A low body weight must be consistent with good performance and not compromise the athlete's physical or mental health, so establishing appropriate goals is essential (see Chapter 11). Once weight and body composition goals are determined, the athlete's training plan can be developed, a diet plan can be devised, and a follow-up schedule can be established. Athletes can achieve a low but biologically comfortable weight with short-term, monitored, and safe diet and exercise programs that promote slow weight loss. Close contact and communication with trusted health and sports professionals can help athletes prevent "slipping over the line" from disciplined eating and training to disordered eating and excessive exercise. These professionals can also help athletes define and distinguish appropriate weight and body composition from inappropriate and potentially harmful weight and body composition.

In a perfect world, the Female Athlete Triad would always be prevented through the use of excellent screening tools to identify athletes who may be at risk. These tools include an annual physical exam including menstrual function, an assessment of dietary intake and energy expenditure, and laboratory tests such as DEXA and a complete blood count (CBC). However, screening tools are never perfect nor is it likely that all athletes will be screened. A particularly challenging aspect is that each of the three factors can develop independently and at a different rate. Not only are there three spectrums, but interrelationships also exist (Nattiv et al., 2007).

Low energy availability is a powerful factor, so it is a primary focus. In fact, a primary goal in the treatment of the Female Athlete Triad is to increase energy availability by increasing caloric intake and/or reducing energy expenditure from exercise. Changes in diet and exercise to prevent low energy intake and high energy expenditure are needed to resolve the athlete's amenorrhea and to increase bone mineral density. All can benefit from nutritional counseling, but those with disordered eating, eating disorders, and/or voluntary excessive exercise may also need psychological counseling.

Returning to the case study of Karen for one last time, the low energy availability that she experienced was a result of increasingly severe disordered eating and exercise dependence, which resulted in amenorrhea. Karen was aware that she was menstruating only periodically but she kept this information a secret. Amenorrheic athletes are at a greater risk for stress fractures than athletes with normal menstruation, so it is not a surprise that Karen was diagnosed with a stress fracture. A DEXA scan would be needed to determine if bone mineral density was already low.

Karen's story could end in several ways. The best-case scenario is for her to receive and complete treatment

Spotlight on…

Normal Bone Density in a Former Amenorrheic, Osteoporotic Distance Runner

Fredericson and Kent (2005) report the results of a case study of a distance runner who successfully reversed low bone mineral density (BMD) with improved energy intake that led to weight gain. The case study covers an 8-year period from approximately age 23 to age 31. The subject ran competitively from age 12 through age 25. She had a personal best marathon time of 2:41 (2 hours, 41 minutes). While running competitively, she typically ran 80 to 90 miles/wk (~134 to 150 km/wk).

This runner began to restrict energy and fat intake at age 13. She had a low weight for height through age 25. Weight at age 23 was approximately 107 lb (48.6 kg) and Body Mass Index (BMI) was 15.8 (healthy weight BMI = 18.5 to 24.9). She had primary amenorrhea until age 23, and BMD measured at this time was found to be the equivalent of a 13-year-old.

At age 25 the runner became concerned about her long-term health and made numerous lifestyle changes. She reduced her mileage to 20 to 50 miles/wk (~33 to 83.5 km/wk). She began to increase both her energy and fat intakes for the purpose of weight gain. In the first 4 months, weight increased from (~111 lb (50.4 kg) to ~122.5 lb (55.7 kg) and then gradually increased to ~144 lb (65.5 kg) by age 31, during which time normal menstruation resumed. Concurrent with weight gain, she dramatically improved her BMD, which eventually was in the normal range for her chronological age.

This case study cannot be extrapolated to other amenorrheic runners, but it does document that resumption of menses and dramatic gains in BMD did occur in this individual with lifestyle intervention that resulted in the attainment of a sustainable weight.

The Internet café

Where do I find reliable information about disordered eating in athletes and the Female Athlete Triad?

Both the American College of Sports Medicine position stand and the International Olympic Committee Consensus Statement on the Female Athlete Triad are available on the Internet. These reports are excellent resources for anyone who wishes to know more about these conditions. Access the ACSM position stand at **http://journals.lww.com/ acsm-msse/Fulltext/2007/10000/The_Female_Athlete_ Triad.26.aspx.** Access the IOC report at **http://www. olympic.org/en/content/The-IOC/Commissions/ Medical/HBI/.**

The National Eating Disorders Association is the largest nonprofit group in the United States in the area of disordered eating and eating disorders. They publish and distribute materials, including some targeted to athletes, and operate a referral help line: **http://www. nationaleatingdisorders.org.**

The National Institute on Drug Abuse funded the development of curricula targeted to high school athletes. ATHENA (Athletes Targeting Healthy Exercise & Nutrition Alternatives) is designed to promote healthy nutrition and exercise behaviors and reduce disordered eating and body image distortion. ATLAS (Athletes Training and Learning to Avoid Steroids) emphasizes the impact that anabolic steroids and other substances (for example, alcohol) have on performance and the positive performance effects of nutrition and training. Both programs can be purchased. More information is available on the Oregon Health & Science University website at **http:// www.ohsu.edu.**

for her eating disorder, including the underlying psychological issues related to control and self-esteem. If appropriate, she could eventually be cleared to train and compete, although she would need to meet certain criteria such as maintenance of a minimum body weight and normal menstruation. Sadly, Karen's story could also be one of declining health that leads to an early death. She could refuse treatment and continue her same patterns of behavior. Refusing treatment would result in her being removed from the team, but she could continue to engage in excessive exercise and disordered eating until she died, typically of a medical condition, such as cardiac arrest, electrolyte imbalance, severe dehydration, or suicide. Early intervention is the best way to ensure the best-case scenario.

Key points

- Any female athlete can be at risk for the Female Athlete Triad.

- The Female Athlete Triad involves three components—energy availability, menstrual function, and bone mineral density.

- Low energy availability is a primary factor.

- Low energy availability, hormonal disruption, and low bone mineral density are serious physiological issues that need to be prevented or reversed.

- Treatment includes the reversal of an energy deficit by increasing food intake and/or reducing energy expenditure from exercise.

Why might a female collegiate cross country runner say to her teammate, "If you're still menstruating, then you're not training hard enough"? Why, in fact, is this approach dangerous?

Keeping it in perspective

Eating, Exercising, Weight, and Performance

Perhaps no area of sports nutrition requires more perspective than the eating and exercise behaviors that support an appropriate weight and excellent performance. It is not rigorous training or disciplined eating per se that creates problems for the athlete's physical and mental health. For those struggling with psychological issues such as control and self-esteem, it is the obsession with exercise and eating that results in a loss of perspective that creates, and then drives, increasingly abnormal eating and exercise behaviors. A low body weight

or low body fat mass, appropriately defined, can be a factor in improving performance. If a low body weight is never defined, inappropriately chosen, or becomes the sole or primary goal (replacing the original goal of improved performance), then the proper perspective has been lost. Food and exercise then become the means to an (unachievable) end and performance and health, both physical and mental, suffer. Eating is for fuel and fun, and when that perspective is lost, the athlete is on a slippery slope.

Now that you have more knowledge about disordered eating and exercise dependence, read the following statements and decide if each is true or false. The answers can be found in Appendix O.

1. Disordered eating and eating disorders affect only female athletes.
2. Anorexia athletica means that an athlete has a classic case of anorexia nervosa.
3. Disordered eating and eating disorders are more likely to be seen among elite female athletes in sports such as distance running and gymnastics.
4. Coaches cause athletes to develop eating disorders.
5. A good diagnostic criterion for exercise dependence is the volume of exercise training (that is, frequency and duration of exercise).

Summary

Normal eating is flexible. It is neither overly restricted nor without restraint. Athletes need a well-planned, nutritious diet that supports their training. For a highly trained athlete, dietary intake should be disciplined, but without obsession. When viewed on a continuum, normal eating may progress to disordered eating, which may progress to an eating disorder. Criteria have been established for three eating disorders—anorexia nervosa, bulimia nervosa, and eating disorders not otherwise specified. All are psychiatric disorders and can damage the athlete's mental and physical health. Trying to attain and maintain a biologically uncomfortable low body weight puts the athlete at risk for developing a disordered eating pattern.

Anorexia athletica is characterized by a low body weight and low body fat mass, weight cycling, restriction of food intake, and excessive exercise in athletes. At greatest risk for anorexia athletica or other disordered eating patterns are those in sports in which low body weight or a thin appearance is advantageous. However, any athlete, male or female, may exhibit disordered eating. Restricting energy intake ("dieting"), especially in low-body weight females, can be a factor in triggering disordered eating. A fine line can exist between the normal eating pattern of an elite athlete with a rigorous training program and a disordered eating pattern that progresses to an eating disorder, but there are ways to distinguish based on the athlete's behaviors, attitudes, and performance.

Prevention is key; failing that, early intervention is critical because treatment is more successful if the disordered eating is less severe. If disordered eating or an eating disorder is suspected, the athlete should be referred for further evaluation and, if confirmed, for treatment. Psychological counseling is a required part of therapy because these are psychiatric diseases. Coaches do not cause eating disorders, but they must be careful that they do not create conditions that inadvertently contribute to them. An International Olympic Committee report suggests that coaches not be involved in determining an athlete's weight or body composition goals.

The Female Athlete Triad describes three interrelated conditions—energy availability, menstrual function, and bone mineral density. Low energy availability results when energy intake is not equal to energy expenditure and may be a result of disordered eating. An energy deficit over time can lead to menstrual dysfunction (amenorrhea) and low bone mineral density (osteopenia and osteoporosis). Each of the three factors can develop independently and at a different rate. Any of the three components of the Triad can damage the athlete's mental or physical health.

Self-Test

Multiple-Choice

The answers can be found in Appendix O.

1. How do the terms *eating disorder* and *disordered eating* compare?
   a. These are interchangeable terms.
   b. *Disordered eating* is a precise term, whereas *eating disorder* is a general term.
   c. An eating disorder has specified criteria, but disordered eating is not well defined.
   d. Disordered eating has specified criteria, but an eating disorder does not.

2. Why are wrestlers who restrict food and purge to "make weight" not considered to have an eating disorder not otherwise specified (EDNOS)?
   a. They behave this way only during wrestling season.
   b. They do not base their self-esteem on weight.
   c. They fall within the normal range on tests of eating attitudes.
   d. both a and b
   e. all of the above

3. The best definition of excessive exercise is:
   a. exercise equivalent to running more than 50 miles/week.
   b. exercise equivalent to running more than 100 miles/week.
   c. exercise equivalent to running more than 150 miles/week.
   d. none of the above

4. Which of the following is true regarding male athletes and eating disorders?
   a. Eating disorders have been documented only in female athletes.
   b. Eating disorders have been documented in males but only in wrestlers and bodybuilders.
   c. Eating disorders in male athletes appear to be on the rise.
   d. Eating disorders in male athletes occur only during adolescence.

5. What are the three distinct but interrelated factors associated with the Female Athlete Triad?
   a. health, fitness, and longevity
   b. anorexia nervosa, bulimia nervosa, and anorexia athletica
   c. appearance, performance, and optimal body weight
   d. energy availability, menstrual function, and bone mineral density

Short Answer

1. What is normal eating? How might normal eating in highly trained athletes differ from nonathletes?
2. What distinguishes normal eating from disordered eating or an eating disorder?
3. How does anorexia athletica differ from anorexia nervosa?
4. Do the following factors have an effect on the prevalence of any of the eating disorders—age, gender, dieting to lose weight, injury, and personality characteristics? Describe the individual who is at the greatest risk for developing anorexia athletica.
5. Why are the eating disorders referred to as mental or psychiatric diseases?
6. Describe the appropriate intervention if disordered eating is suspected.
7. Name the three components of the Female Athlete Triad and explain how each is independent of and related to the other components.
8. Discuss the dimensions of exercise dependence.
9. Explain the difference between exercise dependence and overtraining.

Critical Thinking

1. What is meant by the statement, "Food is for fuel and fun?" especially in light of the prevalence of disordered eating and eating disorders in athletes?
2. You are a high school cross country coach and notice that all of the females on your team are thin. What characteristics do well-trained athletes without disordered eating share with those athletes who demonstrate disordered eating and exercise behaviors? How do you as a coach distinguish these two groups of athletes and help those who need treatment to get treatment?

References

American Psychiatric Association. (1994). Eating disorders. In *Diagnostic and Statistical Manual of Mental Disorders* (4th ed., pp. 539–550). Washington, DC: Author.

Anshel, M. H. (1992). A psycho-behavioral analysis of addicted versus non-addicted male and female exercisers. *Journal of Sport Behavior, 14*, 145–159.

Bamber, D., Cockerill, I. M., Rodgers, S., & Carroll, D. (2000). It's exercise or nothing: A qualitative analysis of exercise dependence. *British Journal of Sports Medicine, 34*(6), 423–430.

Bamber, D., Cockerill, I. M., Rodgers, S., & Carroll, D. (2003). Diagnostic criteria for exercise dependence in women. *British Journal of Sports Medicine, 37*(5), 393–400.

Barrow, G. W., & Saha, S. (1988). Menstrual irregularity and stress fractures in collegiate female distance runners. *American Journal of Sports Medicine, 16*(3), 209–216.

Baum, A. (2006). Eating disorders in the male athlete. *Sports Medicine, 36*(1), 1–6.

Beals, K. (in press). Disordered eating in athletes. In C. Rosenbloom (Ed.), *Sports nutrition: A practice manual for professionals.* (5th ed.) Chicago: American Dietetic Association.

Beals, K. A., & Manore, M. M. (2000). Behavioral, psychological, and physical characteristics of female athletes with subclinical eating disorders. *International Journal of Sport Nutrition and Exercise Metabolism, 10*(2), 128–143.

Beals, K. A., & Manore, M. M. (2002). Disorders of the female athlete triad among collegiate athletes. *International Journal of Sport Nutrition and Exercise Metabolism, 12*(3), 281–293.

Bonci, C. M., Bonci, L. J., Granger, L. R., Johnson, C. L., Malina, R. M., Milne, L. W., et al. (2008). National Athletic Trainers' Association position statement: Preventing, detecting, and managing disordered eating in athletes. *Journal of Athletic Training, 43*(1), 80–108.

Bratland-Sanda, S., Sundgot-Borgen, J., Rosenvinge, J. H., Hoffart, A., & Martinsen, E. W. (2010). Physical activity and exercise dependence during inpatient treatment of longstanding eating disorders: An exploratory study of excessive and non-excessive exercisers. *International Journal of Eating Disorders, 43*(3), 266–273.

Choi, P. Y., Pope, H. G., Jr., & Olivardia, R. (2002). Muscle dysmorphia: A new syndrome in weightlifters. *British Journal of Sports Medicine, 36*(5), 375–376; Discussion, 377.

Cobb, K. L., Bachrach, L .K., Greendale, G., Marcus, R., Neer, R. M., Nieves, J., Sowers, M. F., et al. (2003). Disordered eating, menstrual irregularity, and bone mineral density in female runners. *Medicine and Science in Sports and Exercise, 35*(5), 711–719.

Dale, K. S., & Landers, D. M. (1999). Weight control in wrestling: Eating disorders or disordered eating? *Medicine and Science in Sports and Exercise, 31*(10), 1382–1389.

Drinkwater, B. L., Bruemmer, B., & Chesnut, C. H., III. (1990). Menstrual history as a determinant of current bone density in young athletes. *Journal of the American Medical Association, 263*(4), 545–548.

Drinkwater, B. L., Nilson, K., Chesnut, C. H., III, Bremner, J., Shainholtz, S., & Southworth, M. B. (1984). Bone mineral content of amenorrheic and eumenorrheic athletes. *New England Journal of Medicine, 311*(5), 277–281.

Drinkwater, B. L., Nilson, K., Ott, S., & Chesnut, C. H., III. (1986). Bone mineral density after resumption of menses in amenorrheic women. *Journal of the American Medical Association, 256*(3), 380–382.

Enns, M. P., Drewnowski, A., & Grinker, J. A. (1987). Body composition, body size estimation, and attitudes towards eating in male college athletes. *Psychosomatic Medicine, 49*(1), 56–64.

Fairburn, C. G., & Bohn, K. (2005). Eating disorder NOS (EDNOS): An example of the troublesome "not otherwise specified" (NOS) category in *DSM-IV. Behaviour Research and Therapy, 43*(6), 691–701.

Fredericson, M., & Kent, K. (2005). Normalization of bone density in a previously amenorrheic runner with osteoporosis. *Medicine and Science in Sports and Exercise, 37*(9), 1481–1486.

Glazer, J. L. (2008). Eating disorders among male athletes. *Current Sports Medicine Reports, 7*(6), 332–337.

Grieve, F. G. (2007). A conceptual model of factors contributing to the development of muscle dysmorphia. *Eating Disorders, 15*(1), 63–80.

Griffiths, M. D., Szabo, A., & Terry, A. (2005) The exercise addiction inventory: A quick and easy screening tool for health practitioners. *British Journal of Sports Medicine, 39*, 1–2.

Guyton, A. C., & Hall, J. E. (2010). *Textbook of medical physiology* (12th ed.). Philadelphia: Saunders.

Heffner, J. L., Ogles, B. M., Gold, E., Marsden, K., & Johnson, M. (2003). Nutrition and eating in female college athletes: A survey of coaches. *Eating Disorders, 11*(3), 209–220.

Hill, L. S., Reid, F., Morgan, J. F., & Lacey, J. H. (2010). SCOFF, the development of an eating disorder screening questionnaire. *International Journal of Eating Disorders, 43*(4), 344–351.

International Olympic Committee (IOC). (2005). Working Commission, Working Group–Women in Sport. Position Stand on the Female Athlete Triad. Retrieved [January 3, 2011] from http://www.olympic.org/en/content/The-IOC/ Commissions/Medical/HBI/

Keen, A. D., & Drinkwater, B. L. (1997). Irreversible bone loss in former amenorrheic athletes. *Osteoporosis International, 7*(4), 311–315.

Khan, K. M., Liu-Ambrose, T., Sran, M. M., Ashe, M. C., Donaldson, M. G., & Wark, J. D. (2002). New criteria for female athlete triad syndrome? As osteoporosis is rare, should osteopenia be among the criteria for defining the female athlete triad syndrome? *British Journal of Sports Medicine, 36*(1), 10–13.

Kohrt, W. M., Bloomfield, S. A., Little, K. D., Nelson, M. E., & Yingling, V. R. (2004). Physical activity and bone health. *Medicine and Science in Sports and Exercise, 36*(11), 1985–1996.

Mathes, W. F., Brownley, K. A., Mo, X., & Bulik, C. M. (2009). The biology of binge eating. *Appetite, 52*(3), 545–553.

Misra, M. (2008). Long-term skeletal effects of eating disorders with onset in adolescence. *Annals of the New York Academy of Sciences, 1135*, 212–218.

Nattiv, A., Loucks, A. B., Manore, M. M., Sanborn, C. F., Sundgot-Borgen, J., Warren, M. P., et al. (2007). American College of Sports Medicine position stand. The Female Athlete Triad. *Medicine and Science in Sports and Exercise, 39*(10), 1867–1882.

Olivardia, R., Pope, H. G., & Hudson, J. I. (2000). Muscle dysmorphia in male weightlifters: A case-control study. *American Journal of Psychiatry, 157*(8), 1291–1296.

Pope, H. G., Phillips, K. A., & Olivardia, R. (2000). *The Adonis Complex: The secret crisis of male body obsession.* New York: Free Press.

Reiter, C. S., & Graves, L. (2010). Nutrition therapy for eating disorders. *Nutrition in Clinical Practice, 25*(2), 122–136.

Satter, E. (1987). *How to get your kid to eat but not too much.* Palo Alto, CA: Bull.

Shriver, L. H., Betts, N. M., & Payton, M. E. (2009). Changes in body weight, body composition, and eating attitudes in high school wrestlers. *International Journal of Sport Nutrition and Exercise Metabolism, 19*(4), 424–432.

Sudi, K., Ottl, K., Payerl, D., Baumgarti, P., Tauschmann, K., & Muller, W. (2004). Anorexia athletica. *Nutrition, 20*(7–8), 657–661.

Sundgot-Borgen, J. (1993). Nutrient intake of female elite athletes suffering from eating disorders. *International Journal of Sport Nutrition, 3*(4), 431–442.

Sundgot-Borgen, J., Klungland, M., Torstveit, G., & Rolland, C. (1999). Prevalence of eating disorders in male and female elite athletes. *Medicine and Science in Sports and Exercise, 31*(Suppl.), S297.

Symons Downs, D., Hausenblas, H. A., & Nigg, C. R. (2004). Factorial validity and psychometric examination of the Exercise Dependence Scale–Revised. *Measurement in Physical Education and Exercise Science, 8,* 183–201.

Tanofsky-Kraff, M., & Yanovski, S. Z. (2004). Eating disorder or disordered eating? Non-normative eating patterns in obese individuals. *Obesity Research, 12*(9), 1361–1366.

Torstveit, M. K., & Sundgot-Borgen, J. (2005a). The Female Athlete Triad: Are elite athletes at increased risk? *Medicine and Science in Sports and Exercise, 37*(2), 184–193.

Torstveit, M. K., & Sundgot-Borgen, J. (2005b). The Female Athlete Triad exists in both elite athletes and controls. *Medicine and Science in Sports and Exercise, 37*(9), 1449–1459.

Veale, D. M. W. (1987). Exercise dependence. *British Journal of Addiction, 82,* 735–740.

Learning Plan

- **Discuss** the relationship between nutrition and physical activity and lifelong fitness and health.
- **Identify** reasons why energy expenditure may decline and weight may be gained as individuals age.
- **Compare** and contrast general diet and exercise recommendations published by various organizations.
- **Outline** the nutrition and exercise strategies associated with promoting health and delaying the onset of chronic diseases across the life cycle.
- **Briefly** explain each chronic disease (obesity, cardiovascular disease, hypertension, type 2 diabetes, metabolic syndrome, lifestyle-related cancers, and osteoporosis) and how diet and exercise influence each.
- **Compare** and contrast several popular weight-loss plans.
- **Explain** the relationship among body weight, fat distribution, and the risk for chronic diseases in men and women.
- **Explain** factors that regulate body weight, food intake, and energy expenditure.
- **Explain** the philosophy of the nondiet or Health at Every Size (HAES) approach.
- **Discuss** the process of behavior change.

Some individuals continue to compete as masters athletes.

Peter Weber/Shutterstock.com

Pre-Test | Assessing Current Knowledge of Health, Fitness, and Chronic Diseases

Read the following statements and decide if each is true or false.

1. There are many contradictions among diet and exercise recommendations that are issued by health promotion organizations.
2. Elite endurance athletes do not develop hypertension but lesser-trained athletes do.
3. Type 2 diabetes is caused by eating too much sugar.
4. A sedentary lifestyle is associated with a higher risk for heart disease, certain types of cancer, and diabetes.
5. Being physically active helps to reduce disease risk even if a person is obese.

"Everyone is an athlete. The only difference is that some of us are in training, and some are not." George Sheehan, physician, writer, and running philosopher, eloquently expressed the notion that the human body is made to be active and that people have different reasons and motivations for being physically active, exercising, or participating in sport (Sheehan, 1980). The majority of this textbook has approached sports nutrition from the perspective of highly trained athletes who are trying to achieve maximum performance and success in their sports. However, athletes with this single-minded goal not only make up a very small percentage of the population, they usually pursue their performance goals for a short amount of time relative to their life span. Few athletes remain highly competitive in their sports over a large portion of their lifetime.

The purpose of this chapter is to understand the basic diet and exercise principles related to lifelong fitness and health. A fundamental dietary principle—adequate nutrients within caloric need—applies to everyone. Daily exercise is a powerful influence on the amount of kilocalories needed and the amount of daily exercise performed often changes as people age. To illustrate some of these changes, this chapter includes several scenarios of adults at various stages of life. Each has different goals, demands, and motivations and therefore different needs. All can benefit from the application of scientifically sound nutrition and exercise information. These scenarios are also used to highlight a variety of chronic diseases and the influence that diet and exercise have on prevention and treatment.

Rower Sir Steve Redgrave displays the gold medals that he won at five consecutive Olympic Games; however, few athletes remain highly competitive in their sports over a large portion of their lifetime.

13.1 The Lifelong Athlete

The term *athlete* often brings to mind the highly trained collegiate or professional player. However, individuals participate in sports competitively or recreationally long after their best performing years, and ideally people engage in exercise or remain physically active throughout their lives. An obvious factor that changes over an athlete's life is the level of performance, but lifestyle, personal and professional obligations, and health also change. All these factors must be considered when working with lifelong athletes.

Most collegiate athletes do not become professional athletes and must adjust to reduced exercise training.

Most collegiate athletes do not become professional athletes. For example, the National Collegiate Athletic Association (NCAA) estimates that less than

For many athletes participation rather than competition may be the motivation, particularly as they get older or have less leisure time to devote to training.

Physically active people have many of the same goals as recreational athletes, such as improved fitness, good health, and disease prevention, but they are not performing or competing against others.

2 percent of NCAA football players, 1.2 percent of male basketball players, and 8.9 percent of baseball players will play professionally (http://www.ncaa.org/). Some postcollegiate athletes find that they wish to continue to train, albeit at a lower level, and eventually may become **masters athletes**. Others find that exercise becomes a lower priority and they are essentially "former" athletes. In either case, the reduction in training necessitates adjustments to the diet, particularly caloric intake. Weight gain is associated with chronic disease risk even in former elite athletes.

Pihl and Jurimae (2001) surveyed 150 former elite male athletes to study the relationship between changes in body weight and **heart disease** risk. Weight gain greater than 22 lb (10 kg) was associated with an increase in percentage of body fat and abdominal fat. These men were at a greater risk for elevated blood pressure, **low-density lipoprotein** cholesterol (LDL-C), and **triglycerides**. One of the biggest challenges for former competitive athletes, especially those in high-energy-output sports, is to prevent weight (fat) gain and the diseases associated with excessive body fat after they stop training and competing.

However, weight gain is not inevitable in former athletes. A study of more than 4,600 male and female former collegiate rowers found that the former rowers had a significantly lower prevalence of obesity than the general population. This lower prevalence persisted over the life cycle, although the prevalence of obesity did increase with age. Interestingly, only 5 percent of female and 8 percent of male former rowers were still rowing. The likely explanation for the lower prevalence of obesity is that the former rowers continued to be physically active and were able to adjust their food intake to match their level of activity (O'Kane et al., 2002).

Many adults who exercise fall into the category of recreational athletes. Some are former competitive athletes who continue to have a lifelong dedication to perform at the highest level they can achieve and who remain competitive within their age group in masters events. For other former athletes, participation rather than competition may be the motivation, particularly as they get older or have less leisure time to devote to training. Some recreational athletes are not former competitive athletes; they take up a sport as an adult and set a personal performance goal, such as running a marathon or scoring less than 100 in a round of golf. Recreational athletes may train or practice their sport, but the intensity and duration of training can vary tremendously. In addition to the benefits of participating and competing in sports, recreational athletes also benefit from increased fitness that contributes to good health and the possible prevention of chronic diseases.

Although some adults are not recreational athletes, they do engage in routine physical activity. Physical activity is defined as bodily movement that results in an increase in energy expenditure above resting levels. The motivation to engage in physical activity varies, but typically the reasons include improving health, "staying in shape," maintaining or losing weight, and enhancing appearance. Physically active people have many of the same goals as recreational athletes, such as improved fitness, good health, and disease prevention, but they are not performing or competing against others.

There is a relationship between exercise and nutrition for physically active people and recreational athletes, but because training is less than that of highly trained athletes some of the nutritional demands are less, such as caloric and carbohydrate intake. Physically active people typically need little modification of the principles of a healthy diet to support their physical activity. Similarly, the intensity and duration of training for most recreational athletes are not enough to require substantial modifications to a basic healthy diet. One exception may be performance-focused recreational athletes, such as those who run marathons. These recreational athletes engage in an intensity and duration of training that necessitates the implementation of some of the sports nutrition principles discussed throughout this text and reviewed in Chapter 10.

Because "everyone is an athlete," it is important that all people at all stages of the life cycle be physically active on a daily or near daily basis. This includes children, adolescents, and adults of every age. Pregnancy or advancing age may impact the intensity, duration, or types of exercise that can be performed, and medical advice and common sense must be heeded; however, most people can exercise to some degree across their lives.

Exercise is one cornerstone of a healthy lifestyle; diet is another. There is a strong relationship between nutrition, physical activity, and long-term health, and a number of public health organizations have issued nutrition and physical activity guidelines. Many students reading this textbook will eventually work with clients who are trying to apply these recommendations. In many cases, these clients or patients will have a history of being sedentary and consuming a poor diet. Working with a sedentary, **overweight**, or **obese** adult who consumes the typical American diet is a tremendous challenge because the need to change is high but the motivation to change is often low or hard to sustain. Changes are especially difficult because the environment and culture in the United States and other industrialized countries promote physical inactivity and overconsumption of energy (kcal), sugar, fat, and salt.

In the United States, there is a steady decline in physical activity with aging (Hughes et al., 2010). A large percentage of the population fails to obtain the recommended amount of physical activity. As individuals age, the total amount of physical activity decreases, and activity that is performed is typically of lower intensity and duration than in the past. Although there is some inevitable decline with aging, a large portion of the decline in physical functioning, such as aerobic capacity, strength, flexibility, decreased muscle and bone mass, or increased body fat, is due to a decrease in amount and intensity of physical activity.

Various nutrition and exercise guidelines are remarkably similar, although there are some differences.

The Dietary Guidelines for Americans are published every 5 years, most recently in 2010 (see Chapter 1). These diet and exercise recommendations promote health and reduce the risk for chronic diseases to Americans over the age of two. The 2010 Dietary Guidelines (U.S. Department of Health and Human Services and U.S. Department of Agriculture, 2010) emphasize changing one's lifestyle, including a total diet that is energy balanced and nutrient dense. The typical American diet is too low in vegetables, fruits, high-fiber whole grains, lower-fat milk and milk products, and seafood. There is an overconsumption of solid fats and added sugars (SoFAS), refined grains, and sodium. Regardless of the level of energy expenditure daily, the Dietary Guidelines can be used as a basic pattern. Specific information about diet planning using the Dietary Guidelines is found in Chapter 10.

In addition to the Dietary Guidelines, other organizations also publish nutrition and exercise recommendations, often with a focus on a specific chronic disease, such as cardiovascular disease or cancer. Table 13.1 compares various nutrition guidelines. These guidelines are fundamentally the same, although the recommendations do vary. Still, when taken as a whole, the recommendations for reducing chronic disease risk are remarkably consistent.

Consumers may wonder why nutrition guidelines differ. One reason is that each chronic disease is different. For example, fatty acids directly affect the vessels of the heart, but fat does not directly affect blood pressure. Thus, guidelines for preventing heart disease include recommendations for fat intake, but guidelines for preventing **hypertension** (elevated blood pressure) do not. There are also different interpretations of the scientific literature. For example, some researchers promote a Mediterranean-type diet for heart disease prevention. The biggest differences between Mediterranean-type diets and the American Heart Association (AHA) diet are the types and amounts of fats consumed. Both types of diets are low in saturated and *trans* fats. A Mediterranean diet plan emphasizes the consumption of fish, nuts, and olive oil. It is higher in omega-3 fatty acids and monounsaturated fatty acids than the AHA diet and puts less emphasis on the

Masters athletes: A separate division created by a sports-governing body for athletes older than a certain age. Minimum age varies according to the sport. Also referred to as veteran athletes.

Heart disease: Diseases of the heart and its vessels. A more specific term than cardiovascular disease.

Low-density lipoprotein: A fat transporter containing a moderate proportion of protein, a low proportion of triglyceride, and a high proportion of cholesterol. Also known as "bad cholesterol."

Overweight: Medical definition is a Body Mass Index of 25–29.9.

Obese: Medical definition is a Body Mass Index greater than 30.

Hypertension: Blood pressure chronically elevated above normal resting levels.

Table 13.1 Comparison of Various Nutrition Guidelines

	Dietary Guidelines for Americans, 2010	Mediterranean-type diet	American Heart Association	American Cancer Society
Energy	Balance calories from foods and beverages with calories expended	Healthy weight and daily exercise encouraged	Match intake of total energy (kilocalories) to overall energy need	Balance caloric intake with physical activity
Carbohydrate intake	Choose fiber-rich fruits, vegetables, whole grains and cooked dry beans and peas often	Whole grains, vegetables, fruits, beans, legumes, and nuts daily; sweets a few times per week	Five or more servings of a variety of fruits and vegetables, 6 or more servings of a variety of grain products, including whole grains	Eat 5 or more servings of a variety of vegetables and fruits each day; choose whole grains in preference to processed (refined) grains
Protein intake	Emphasis on lean or lower-fat protein sources including plant-based proteins	Beans, legumes, nuts, cheese, and yogurt daily; fish, poultry, and eggs a few times a week; red meat in very small amounts a few times per month	Include fat-free and low-fat milk products, fish, legumes (beans), skinless poultry, and lean meats	Limit consumption of processed and red meat
Fat intake	20–35% of total calories, <10% from saturated fat, substituting saturated fat with polyunsaturated and monounsaturated fats; <7% from saturated fat reduces risk further; <300 mg cholesterol, avoid *trans* fat	25–35% of calories primarily from olive oil. If watching weight limit oil consumption; ~7–8% from saturated fat	Limit intake of foods with high content of cholesterol-raising fatty acids (e.g., saturated and *trans* fatty acids); limit the intake of foods high in cholesterol; substitute grains and unsaturated fatty acids from fish, vegetables, legumes, and nuts	Consume a healthy diet, with an emphasis on plant sources
Alcohol intake	If consumed, consume in moderation (1–2 drinks daily)	Moderate consumption of wine with meals; purple grape juice may be substituted	Limit alcohol intake to no more than 2 drinks per day (for men) and 1 drink per day (for women)	If you drink alcoholic beverages, limit consumption; drink no more than 1 drink per day for women or 2 per day for men
Sodium intake	<2,300 mg of sodium (~1 tsp salt), reducing to 1,500 mg over time, increase dietary potassium		Limit salt (sodium chloride) intake	
Weight	Maintain weight in a healthy range; if overweight or obese, lifestyle changes are needed to lose weight including a reduction in foods high in solid fats and added sugars (SoFAS) and an increase in physical activity	Maintain a healthy weight	Maintain a healthy body weight	Avoid excessive weight gain throughout the life cycle; achieve and maintain a healthy weight if currently overweight
Exercise	Engage in regular physical activity and reduce sedentary activities	Daily exercise including walking, physical work, and sports	Achieve a level of physical activity that matches (for weight maintenance) or exceeds (for weight loss) energy intake	Adults: at least 30 minutes of intentional, moderate-to-vigorous physical activity, 5 or more days of the week; 40–60 minutes of intentional physical activity preferable. Children/adolescents: at least 60 minutes per day, 5 days per week

total amount of fat consumed (Curtis and O'Keefe, 2002; Kris-Etherton et al., 2001; Robertson and Smaha, 2001). Both types of diets reflect substantial changes from the diet currently consumed by most Americans, and the adoption of either diet plan would likely be beneficial.

Similarly, various physical activity and exercise recommendations have been made by health-related organizations, potentially leading to confusion on the part of the public (and with some professionals). There is consistency among these guidelines in that they *all* recommend regular, consistent lifelong physical activity. Differences and potential confusion arise due to the details of these guidelines, particularly the intensity and duration of activity or exercise. The key to understanding the differences is to consider the intent, target audience, and scientific basis underlying each recommendation.

As detailed in the 1996 Surgeon General's Report on Physical Activity and Health, there is a large body of scientific evidence supporting the positive relationship between physical activity and good health (U.S. Department of Health and Human Services, 1996). Large, **epidemiological** studies such as Blair et al. (1989) determined that there is a strong **inverse relationship** between physical activity and premature mortality, particularly from chronic diseases such as cardiovascular disease and cancer. In other words, people who lead physically active lives are much less likely to develop chronic diseases and die at an earlier age.

Based upon this epidemiological evidence, the Centers for Disease Control and Prevention and the American College of Sport Medicine (Pate et al., 1995) published physical activity recommendations suggesting that everyone should be physically active on most days of the week, accumulating at least 30 minutes of moderate-intensity activity. These recommendations were considered a *minimal* threshold of activity and are associated with health benefits, such as a reduction in risk of premature mortality.

In 2008, the Physical Activity Guidelines for Americans was published by the U.S. Department of Health and Human Services (http://www.health.gov/PAGuidelines). These guidelines expand on the earlier minimal physical activity recommendation, providing specific physical activity recommendations for children/adolescents, adults, and older adults in specific situations (see Spotlight on...The Physical Activity Guidelines for Americans in Chapter 1). Although 30 minutes of physical activity each day may be associated with better health and a reduction in premature mortality, this amount of activity may not be sufficient to meet more immediate health needs. If activity is of lower intensity, it is recommended that the duration of activity each week for adults be at least 2 hours and

30 minutes. The recommended duration is reduced to 1 hour and 15 minutes per week for adults if the intensity of the activity is higher (see below). The new guidelines also include a recommendation for the inclusion of strength building activities.

These physical activity guidelines may not result in enough caloric expenditure to prevent gradual weight gain, which may eventually lead to obesity. Therefore, greater amounts of activity daily may be necessary to maintain a healthy weight and to prevent obesity. Overweight or obese individuals who lose weight may require even more physical activity to sustain their newly acquired lower weight. Therefore, the recommendation for these individuals is up to 90 minutes or more of activity per day to prevent regaining weight.

The specific Physical Activity Guidelines are:

Children and Adolescents (aged 6–17)

- Children and adolescents should do 1 hour (60 minutes) or more of physical activity every day.

- Most of the 1 hour or more a day should be either moderate- or vigorous-intensity aerobic physical activity.

- As part of their daily physical activity, children and adolescents should do vigorous-intensity activity on at least 3 days per week. They also should do muscle-strengthening and bone-strengthening activity on at least 3 days per week.

Adults (aged 18–64)

- Adults should do 2 hours and 30 minutes a week of moderate-intensity, or 1 hour and 15 minutes (75 minutes) a week of vigorous-intensity aerobic physical activity, or an equivalent combination of moderate- and vigorous-intensity aerobic physical activity. Aerobic activity should be performed in episodes of at least 10 minutes, preferably spread throughout the week.

- Additional health benefits are provided by increasing to 5 hours (300 minutes) a week of moderate-intensity aerobic physical activity, or 2 hours and 30 minutes a week of vigorous-intensity physical activity, or an equivalent combination of both.

- Adults should also do muscle-strengthening activities that involve all major muscle groups performed on 2 or more days per week.

Epidemiological: The study of health-related events in a population.

Inverse relationship: Given two variables, when one increases the other decreases, and vice versa.

Older Adults (aged 65 and older)

- Older adults should follow the adult guidelines. If this is not possible due to limiting chronic conditions, older adults should be as physically active as their abilities allow. They should avoid inactivity. Older adults should do exercises that maintain or improve balance if they are at risk of falling.

The American College of Sports Medicine (ACSM) also published exercise recommendations in its Position Stand on the Recommended Quantity and Quality of Exercise for Developing and Maintaining Cardiorespiratory and Muscular Fitness, and Flexibility in Adults (1998). Recommendations address specific components of physical fitness, such as cardiorespiratory (aerobic) fitness, muscular strength and endurance, and flexibility. A range of exercise frequency, intensity, and duration is recommended as shown in Figure 13.1. For cardiorespiratory or aerobic exercise, the ACSM recommends a duration of 20 to 60 continuous minutes at moderate to vigorous intensity, 3 to 5 days per week. Strength training is recommended two to three times per week with specific recommendations as to the number of exercises, sets, and repetitions. Stretching or flexibility exercises are recommended at least two to three times each week. Again, note that regular lifelong exercise is at the core of all exercise guidelines, but there are specific recommendations for each component of physical fitness.

Cardiorespiratory Fitness

1. **Frequency of training:** 3–5 days per week (d/wk).

2. **Intensity of training:** 55/65–90% of maximum heart rate (HR_{max}), or 40/50–85% of maximum oxygen uptake reserve ($\dot{V}O_2R$) or HR_{max} reserve (HRR). The lower intensity values, i.e., 40–49% of $\dot{V}O_2R$ or HRR and 55–64% of HR_{max}, are most applicable to individuals who are quite unfit.

3. **Duration of training:** 20–60 min of continuous or intermittent (minimum of 10 min bouts accumulated throughout the day) aerobic activity. Duration is dependent on the intensity of the activity; thus lower-intensity activity should be conducted over a longer period of time (30 min or more), and, conversely, individuals training at higher levels of intensity should train at least 20 min or longer. Because of the importance of "total fitness" and that it is more readily attained with exercise sessions of longer duration and because of the potential hazards and adherence problems associated with high-intensity activity, moderate-intensity activity of longer duration is recommended for adults not training for athletic competition.

4. **Mode of activity:** Any activity that uses large muscle groups, which can be maintained continuously, and is rhythmical and aerobic in nature, for example, walking/hiking, running/jogging, cycling/bicycling, cross country skiing, aerobic dance/group exercise, rope skipping, rowing, stair climbing, swimming, skating, and various endurance game activities or some combination thereof.

Muscular Strength and Endurance, Body Composition, and Flexibility

1. **Resistance training:** Resistance training should be an integral part of an adult fitness program and of a sufficient intensity to enhance strength, muscular endurance, and maintain fat-free mass (FFM). Resistance training should be progressive in nature, be individualized, and provide a stimulus to all the major muscle groups. One set of 8–10 exercises that condition the major muscle groups 2–3 d/wk is recommended. Multiple-set regimens may provide greater benefits if time allows. Most people should complete 8–12 repetitions of each exercise; however, for older and more frail people (approximately 50–60 yr of age and above), 10–15 repetitions may be more appropriate.

2. **Flexibility training:** Flexibility exercises should be incorporated into the overall fitness program sufficient to develop and maintain range of motion (ROM). These exercises should stretch the major muscle groups and be performed a minimum of 2–3 d/wk. Stretching should include appropriate static and/or dynamic techniques.

Figure 13.1 American College of Sports Medicine exercise recommendations

Legend: min = minute; yr = year

From the Position Stand on The Recommended Quantity and Quality of Exercise for Developing and Maintaining Cardiorespiratory and Muscular Fitness, and Flexibility in Adults, from *Medicine and Science in Sports and Exercise*, Volume 30(6), pages 975–99. Reprinted by permission.

Key points

- All people should be physically active throughout their lives.

- Proper nutrition is important across the life cycle.

- The Dietary Guidelines and other nutrition- and exercise-related recommendations outline a general plan for good health.

What is meant by the opening quote of this chapter?—
"Everyone is an athlete. The only difference is that some of us are in training, and some are not."

13.2 The Impact of Overweight and Obesity on Chronic Diseases

Chronic disease is defined as a disease lasting 3 months or more that can be treated but not cured. Some examples of chronic diseases include cardiovascular disease, most cancers, diabetes, and osteoporosis. Obesity is also considered a chronic disease by most health-related organizations because the number of formerly obese individuals who maintain their weight loss over their lifetimes is low. Of the 2.4 million people in the United States who die each year, approximately 70 percent die from chronic diseases (National Center for Health Statistics, 2010). Figure 13.2 compares the leading causes of death with the actual causes of death using data from the year 2000. Although heart disease and cancer are the leading causes of death in the United States, tobacco use, poor diet, and lack of exercise are the major contributing factors (Miniño et al., 2002; Mokdad et al., 2004). As tobacco usage and secondhand smoke exposure decline, poor diet and lack of exercise are expected to become the leading modifiable causes of death. In the case of cancer, poor diet and obesity contribute to approximately 30 percent of all cancer deaths, similar to the number of cancer deaths caused by tobacco use (Weir et al., 2003).

Chronic diseases can begin in infancy, childhood, adolescence, or young adulthood. Thus diet and activity are factors that influence disease early in life. These influences may not be evident in young people, so preventative strategies, even if seen as important, do not appear urgent. Chronic diseases

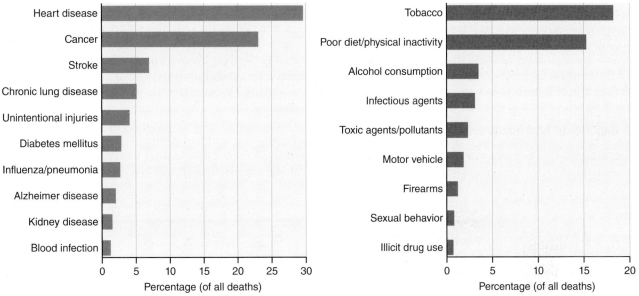

Figure 13.2 Leading and actual causes of death in the United States

Heart disease and cancer are by far the leading causes of death in the United States. Poor diet and lack of exercise/physical activity are expected to overtake tobacco as the actual causative factors of death.

*Drawn from data from Miniño, A.M., Arias E., Kochanek K.D., Murphy S.L., Smith B.L. Deaths: final data for 2000. *National Vital Statistics Reports* 2002; *50*(15):1–20.

†Drawn from data from Mokdad A.H., Marks J.S., Stroup D.F., Gerberding J.L. Actual causes of death in the United States, 2000. *JAMA.* 2004; 291. Reprinted with permission.

develop, progress, and worsen with age. It may take years or decades before the disease has progressed to the point where symptoms become apparent; therefore, many chronic diseases are not noticed or diagnosed until individuals are in their fourth, fifth, or sixth decade of life. In adulthood, healthy eating and physical activity may delay the onset, slow the progression, and postpone some of the complications associated with a chronic disease. Any delays will likely preserve quality of life for a longer period of time and may eventually mean fewer or lower doses of medications. Consuming a healthy diet, engaging in physical activity, refraining from smoking, and minimizing exposure to environmental hazards (such as asbestos) pay the biggest dividends when they have been lifelong habits, but adoption of any of these habits at any age will likely have a positive influence on the course of chronic diseases (Garry, 2001).

Overweight and obesity are reaching epidemic proportions, and the impact that obesity has on other chronic diseases, such as diabetes and cardiovascular disease, is substantial. Obesity is related to numerous diseases as shown in Figure 13.3 (Pi-Sunyer, 2009). One of the ways that obesity may be related to other chronic diseases is through inflammation. Adipose tissue releases substances that contribute to a constant, low-level inflammation, a condition that worsens with increased body fat, particularly in visceral, or abdominal, fat. These compounds affect the severity of atherosclerosis, and associate obesity and metabolic syndrome with increased risk for cardiovascular disease. Therefore, the prevention of overweight and obesity and the treatment of these diseases has become a predominant focus for individuals as well as medical and public health professionals.

The majority of Americans are overweight or obese.

According to the Centers for Disease Control and Prevention (CDC), American society has become "obesogenic," a term used to describe environments that promote increased food intake, the consumption of unhealthy foods, and physical inactivity. The fundamental causes of overweight and obesity include genetics, behavior, and the environment and their interactions.

Overweight for adults is defined as Body Mass Index (BMI) between 25 and 29.9, whereas obesity is defined as a BMI >30 (see Spotlight on...Childhood and Adolescent Obesity for equivalent values in children and adolescents). The prevalence of overweight and obesity in the United States is the highest of all the developed countries (National Center for Health Statistics, 2006). Approximately one-third of all U.S. adults are overweight and another one-third is obese. The rate of obesity doubled from 1980 to 2005, although the rate of increase is beginning to slow (Flegal et al., 2010). No state met the Healthy People 2010 obesity target of 15 percent; Colorado had the lowest rate of obesity at 18.6 percent (http://www.cdc.gov/obesity/data/trends.html).

In the United States, the environment promotes weight gain and obesity. Food is abundant, ever present in work and social situations, and relatively inexpensive. Portion sizes are large. Calorie-dense processed foods are relatively inexpensive compared to fruits and vegetables. Consumption of liquid calories in the form of soft drinks and other sweetened beverages is one of the main sources of added sugar in the diets of Americans of all ages (Johnson et al., 2009). Jobs in service and information-based industries are generally sedentary and require people to sit at a desk or computer for large portions of the workday. The built environment has also largely engineered physical activity out of modern lifestyles: commuting by car, drive-thru services (for example, restaurants, banks, dry cleaners, pharmacies), valet parking, elevators/escalators/people movers, riding lawnmowers, and leaf blowers. Electronic entertainment—24-hour, 200-channel television, video games, and personal computers—requires little physical

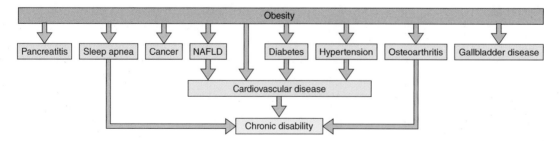

Figure 13.3 Medical risks of obesity

Obesity contributes to a number of chronic diseases and other medical conditions.

Legend: NAFLD = nonalcoholic fatty liver disease

From: Pi-Sunyer, X. (2009). The medical risks of obesity. *Post Graduate Medicine 121*(6), pp. 21–33. Reprinted by permission.

Modern lifestyle promotes weight gain and obesity through abundant food and lack of physical activity.

activity. The prevalence of these factors has given rise to the term *toxic environment* to describe the current eating and activity environment in the United States (Brownell, 2004). Food intake and physical activity are emphasized in the prevention of obesity because these factors are strongly influenced by behavior and humans can change their behavior.

Weight gain and aging. Two-thirds of all adults are overweight, so it may seem that weight gain is an inevitable consequence of aging. It is not. Weight gain in adults may be quite slow, often referred to as "creeping obesity." The average annual weight gain for adults is approximately 1¾ pounds (0.8 kg) per year; thus over 20 years ~35 pounds (~16 kg) of weight would be gained (Winett et al., 2005). This is a population-wide estimate, so some people gain substantially more or less. However, weight gain can "creep up" on adults because less than 2 pounds per year is a relatively slow increase in body fat and represents a relatively small imbalance between energy intake and expenditure each day.

Spotlight on...

Childhood and Adolescent Obesity

In children and adolescents, overweight is defined as a BMI at or above the 85th percentile and less than the 95th percentile on the CDC growth charts, whereas obesity is equal to or greater than the 95th percentile. For example, a 7-year-old boy who is 4 feet (122 cm) tall and weighs 75 pounds (34 kg) would be considered obese. It is estimated that 17 percent of children and adolescents in the United States are obese.

Most obese children do not "grow out of it." The majority of obese teenagers will become obese adults. Obesity negatively impacts their health at an early age, including increasing the risk for high blood pressure, elevated cholesterol, and type 2 diabetes.

To reverse childhood and adolescent obesity, a combination of behavioral and environmental changes is needed. The three fundamental issues are food intake, physical activity, and leisure activities. Weight gain in children and adolescents is often due to the consumption of large portions of food, frequent snacking, and a high intake of sugary beverages. In addition, many children and adolescents are physically inactive and may spend a large amount of their leisure time watching movies or TV and playing video games. Such activities replace physical activity and make it easy to snack.

Children and adolescents are highly influenced by their environment, so in addition to their parents, others in the community can have a substantial impact. Parents and child care providers need to provide a setting that encourages the intake of healthy foods and daily activity. Schools play an important role because children and adolescents spend many hours in school. Foods offered at school should also be healthy, and physical activity classes need to be part of the school curriculum. Communities also can contribute by building and maintaining bike paths, sidewalks, and parks. Public policy can help direct funds to programs that improve child and adolescent nutrition and physical activity, both of which have a substantial influence on overweight and obesity. More information about child and adolescent obesity can be found at the Centers for Disease Control and Prevention website: http://www.cdc.gov/.

A key for many adults is to recognize the potential for slow weight gain and to change behavior to prevent it. For example, cookies are often available in the workplace. An Oreo® cookie contains about 50 kcal. If eating two small cookies daily contributes to excessive caloric intake, then not snacking on cookies would be a small but meaningful behavior change. Similarly, climbing one or two flights of stairs instead of riding the elevator in an office building several times a day also represents a relatively small change in energy (kcal) expenditure. However, when accumulated over time, this habitual activity can have a substantial long-term impact on weight management.

A relatively new area of research is the study of food consumption patterns that are associated with less weight gain over time. A 4-year study conducted in Germany found that a diet that contained low-fat, high-fiber, high-carbohydrate foods such as fruit and whole-grain breads and cereals helped nonobese subjects maintain their body weight and prevent weight gain (Schulz et al., 2005). A similar study conducted in the United States found that a diet high in whole grains, fruits, vegetables, and low-fat dairy products resulted in smaller gains in Body Mass Index. Also associated with less weight gain was a dietary pattern that was low in red and processed meats, fast food, and soft drinks. These dietary patterns are similar to the DASH (Dietary Approaches to Stop Hypertension) diet that has been shown to reduce blood pressure (Newby et al., 2003, 2004). These studies are particularly important because they

Spotlight on a real athlete

Susan, 26-Year-Old, Former Collegiate Basketball Player, No Longer Playing Competitively

Although some former collegiate athletes continue to train and compete, many do not. Consider the case of Susan, who finished her collegiate eligibility and is no longer playing basketball competitively. Still in her 20s, she is working full-time in her first career job. While in college she played basketball several hours a day and was required to participate in a strength and conditioning program that included distance running, high-intensity conditioning (such as on-court sprints, intervals), strength training, and plyometric drills to improve speed, agility, and quickness. Now that she has graduated, practices and workouts are not required, and athletic program support, such as training table twice a day, is not available. Susan fixes meals sporadically and snacks a lot. Her goal is to "stay in shape," but now she must do this on her own. She joined a health club and runs on a treadmill 3 to 4 days per week and, if available, plays pickup (half-court) basketball games afterward for ~30 minutes. Never a fan of strength training, she now lifts weights only sporadically—one to two times per week, if at all.

Susan, a former collegiate basketball player.

Susan is typical of many former collegiate athletes who find that the frequency, intensity, and duration of exercise are much less than in the past. Her training and basketball practices were required and so the discipline to exercise was "built in." Susan's new job gives structure to her schedule but not in the same way as the college environment. Physical activity is not part of the job and does not dictate her schedule. In fact, she must find time to exercise during her "free time" away from work. She must also find the discipline to exercise, something that was automatically present at basketball practices. In college she was a participant in a high-energy-output sport; now she is in a low-energy-output job. Not only did the exercise environment change, the eating environment changed. Susan works in an office, and people constantly bring food to her desk and to the break room. She is responsible for buying and preparing her own food, and vending machines are the most convenient lunch option. She is experiencing a very different food environment from her college training table, which featured sit-down meals and healthy prepackaged takeout lunches.

Rates of overweight and obesity for adults and children are high in the United States.

Masterfile

provide information about prevention of weight gain in nonobese populations.

There is a strong inverse relationship between level of physical activity and fitness and obesity; that is, the lower the level of activity and the lower the level of cardiovascular fitness, the greater the risk of developing obesity. A number of studies have shown that increases in exercise time or volume and increases in aerobic fitness are associated with decreases in weight gain over time, independent of changes in diet. One such study showed that for cardiovascular fitness measured by the amount of time one could sustain on a maximal effort treadmill test, every minute of increased exercise time was associated with a measurable decrease in the risk of gaining weight, which in this study was an average of 22 pounds over a 7½-year period (DiPietro et al., 1998). Higher activity levels leading to higher aerobic fitness substantially reduces the risk of creeping obesity. The need to reduce energy expenditure after collegiate competition is a challenge for many collegiate athletes, both male and female, especially those who competed in high-energy-output sports.

Women often wonder if pregnancy promotes obesity. For the majority of women, pregnancy is not associated with an increased risk for developing obesity. One and one-half years after delivery, the average woman will be only 1.1 pound (~0.5 kg) more than her prepregnancy weight. However, 15–20 percent of women will gain a substantial amount of weight as body fat and are at risk for obesity (Johnson et al., 2006).

Regulation of body weight is a complex process that is not completely understood.

Under normal circumstances, the body regulates weight as it does many other physiological factors, with a variety of homeostatic mechanisms that respond to short-term changes to maintain long-term balance. Body weight is maintained by balancing the energy equation—matching food (energy) intake to energy expenditure over long periods of time. Although the energy balance equation is a simple concept (see Chapter 2), the mechanisms that govern the body's intake of food and expenditure of energy are complex, interrelated, and not entirely understood. The dramatic increase in the incidence of obesity in adults and children also indicates that these mechanisms may be altered or disrupted, resulting in long-term and persistent changes in body weight regulation.

The human body does not have a single energy sensor that provides feedback on the amount of energy consumed or expended. Instead, it relies on a number of different physiological and psychological inputs that subsequently influence physical responses and changes in behavior such as eating and physical activity. Figure 13.4 outlines the factors that influence food intake. Physiological changes and food intake behavior are primarily controlled by the hypothalamus on both a short-term (initiation and termination of individual meals) and long-term basis. Signals that stimulate the hypothalamus cause feelings of hunger, which result in seeking out and consuming food, or satiety, the feeling of fullness that reduces the desire to eat.

One signal that stimulates appetite and initiates food intake is **ghrelin**, a hormone synthesized by cells in the stomach. Secretion of ghrelin increases before meals and makes people feel like eating. Eating causes ghrelin levels to decrease. Ghrelin acts on receptors in the hypothalamus, causing an increase in the secretion of Neuropeptide Y (NPY). Through neural signaling pathways that are not completely understood, NPY acts to increase the release of other neuropeptides by the hypothalamus that are strong stimulators of appetite and hunger, leading to food intake (Castañeda et al., 2010).

When food is consumed, a number of signals tell the body to stop eating. Peptide YY (PYY) is secreted by cells in the small and large intestine when food is consumed and has the opposite effect of ghrelin. It has an inhibitory effect on the hypothalamic hunger-signaling pathway and therefore decreases the drive to eat. The presence of food in the small intestine also causes the release of cholecystokinin (CCK), which acts on the

Ghrelin: (pronounced GRELL-in). A protein-based hormone secreted by the cells of the stomach and associated with appetite stimulation; counter-regulatory to leptin.

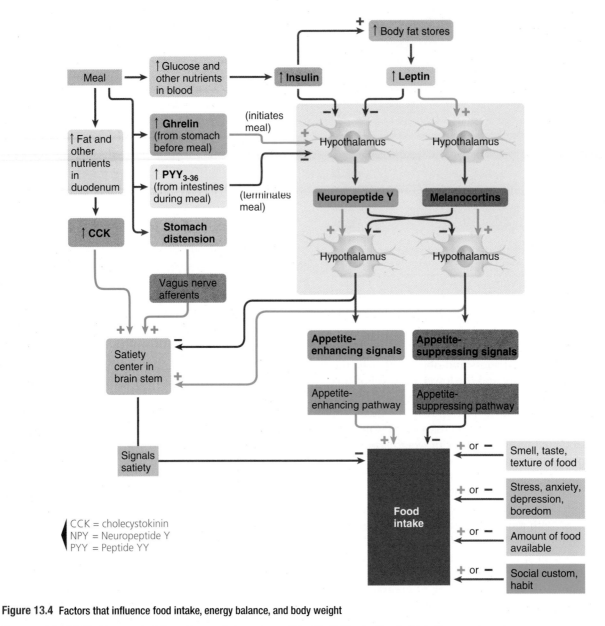

Figure 13.4 Factors that influence food intake, energy balance, and body weight

Sherwood, L. (2010). *Human physiology: From cells to systems.* (7th ed.). California: Brooks/Cole Cengage Learning.

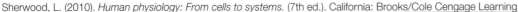

satiety center in the brain stem to increase feelings of fullness, resulting in meal termination and a decrease in food intake. Mechanical distension of the stomach by the presence of food has a similar effect. The intake of nutrients, particularly carbohydrates, results in an increase in insulin secretion by the pancreas, which has an inhibitory effect on appetite and hunger.

Hunger is an imprecise signal, and the amount of food consumed in any individual meal does not necessarily correspond closely to the immediate energy state of the body. In other words, if the body is in energy

deficit by 750 kcal, the hunger and satiety signals are not accurate enough to initiate and then terminate food intake when precisely 750 kcal have been consumed. Regulatory mechanisms are in place, however, to match long-term energy intake to expenditure in order to regulate body weight.

One of the more important regulatory signals in long-term energy balance and weight maintenance is **leptin**, a hormone produced by fat cells. Until recently, adipose tissue has been thought of mostly as an energy storage site. The discovery of the secretion of leptin by adipocytes, however, has revealed that fat tissue acts as an endocrine gland with an important role in energy balance and metabolism. The amount of leptin secreted is proportional to the amount of stored fat in the body; more stored body fat results in more leptin, which acts on the NPY hypothalamic signaling pathway to decrease food intake. It also stimulates other compounds that suppress appetite and food intake.

Leptin is one of a number of hormones secreted by fat cells, collectively referred to as adipokines. When functioning normally, leptin reflects the body's stored energy state to help with long-term energy balance. If energy intake has outpaced energy expenditure over time, excess energy is stored as fat and is signaled by increasing levels of leptin. The increased leptin levels act to suppress appetite and decrease food intake, which results in a negative energy balance and weight loss over time to re-establish "normal" body weight. Insulin, through its action to increase the uptake and storage of nutrients and its effect on subsequent food intake, also has an important role in long-term energy balance (Magkos, Wang, and Mittendorfer, 2010).

There are many other factors that influence eating behavior and energy balance, some of which may be dissociated from physiological signals related to hunger, satiety, or energy state of the body. Eating behavior is habitual and may be tied to certain meal patterns such as regular mealtimes for breakfast, lunch, and dinner. Eating may also be associated with certain activities such as watching television or studying and may result in food intake regardless of existing hunger or satiety signals. Eating is a social activity as well as a necessary physiological one, and social customs and pressure may influence food intake. The appetizing nature of certain foods due to the taste, smell, and texture may result in the consumption of such foods out of proportion to their energy content. The availability and abundance of food may influence the amount of consumption as anyone who has observed an all-you-can-eat buffet can testify. Other emotional factors such as boredom, stress, anxiety, and depression may also affect food and energy intake. These factors may affect different people in different ways. For example, one person may compulsively overeat when feeling emotional stress whereas another may avoid food intake. Figure 13.4 summarizes the factors that influence food intake and therefore, short- and long-term balance.

As discussed in Chapter 2, a variety of factors influence energy expenditure, but the primary factors are resting metabolic rate, thermic effect of food, and physical activity. Except for those engaged in a significant amount of exercise, resting metabolic rate makes up the majority of energy expended each day. Thermic effect of food makes up a fairly small proportion of daily energy expenditure, whereas physical activity is

the most variable factor and is the factor most under voluntary control. Physical activity and exercise not only have an effect on the amount of energy expended but may also have an effect on other factors such as the hormones that influence energy balance (Hagobian, Sharoff, and Braun, 2008).

Factors that influence energy intake and energy expenditure are interrelated and often do not have simple, isolated results that make predictable changes in the energy balance equation. For example, a logical approach to increasing energy expenditure for the purpose of losing weight is to increase exercise activity. However, the increase in energy expenditure leading to a negative energy balance may in turn cause an increase in appetite and therefore food intake. If the increase in food intake balances the energy expenditure due to the exercise, there will be no change in body weight. In addition to differences between individuals in response to changes in energy balance, there is evidence that there may be differences in responses between men and women (Hagobian and Braun, 2010).

Although the body has specific homeostatic mechanisms to regulate energy intake and expenditure to maintain a fairly constant body weight, the level at which this weight is "set" is clearly not permanently fixed. The increased incidence of obesity in populations and the observation of increasing body weight and fat of individuals with aging demonstrate that the body can and does adjust over time to maintain weight at a different, and usually higher, weight. In fact, there is considerable evidence that it is relatively easy for the body to adjust to a higher "set point" weight than it is to change and maintain weight at a lower level, which accounts for the difficulty people have in losing weight and maintaining weight loss once it is achieved.

Overweight and obesity occur when a person is in positive energy balance over time and energy intake exceeds energy expenditure. Excess energy intake is stored largely as body fat, leading to increased body weight and overfatness. The precise mechanisms that result in an upward shift in the weight the body maintains are not well understood and evidently vary among individuals. Some people may gain weight but stabilize at a higher weight whereas others continue to gain weight and become obese, and in extreme cases morbidly obese.

Although hereditary factors or endocrine disorders such as hypothyroidism may be involved for some, the primary factors appear to be related to reduced physical activity, increased caloric intake, or a

Leptin: A hormone produced in adipose tissue that suppresses appetite; considered counter-regulatory to ghrelin.

combination of both. Obesity is related to disruptions in hormonal signaling pathways for insulin and leptin. Insulin resistance is associated with obesity as are increased levels of leptin, suggesting that a resistance to the action of leptin has developed. These disruptions tend to improve or normalize with weight loss and increased physical activity. A considerable amount of research is needed to understand the mechanism related to obesity in order to develop more effective treatment strategies.

The treatment of overweight and obesity involves long-term changes to established food and exercise patterns.

In 2009, the American Dietetic Association released an updated position paper on weight management. A fundamental principle is that both the prevention and treatment of overweight and obesity requires a lifelong commitment to a lifestyle that includes healthy foods and activity. An individualized reduced-calorie diet is the basic dietary component of a comprehensive weight-management program. Reducing dietary fat and/or carbohydrates is a practical way to create

a caloric deficit of 500 to 1,000 kcal below estimated energy needs and should result in a weight loss of 1–2 lb per week (Seagle et al., 2009).

Popular weight-loss diets. Although many weight-loss plans are available, this section will be limited to a description and discussion of four diets that have been consistently popular for many years and have been compared in scientific studies: Atkins (1972, 1999), Zone (Sears, 1995, 1997), Ornish (1993), and Weight Watchers (1966). Some of the characteristics of these diets are compared to each other in Table 13.2. The one consistent feature is that the diets do not require "calorie counting"; however, the caloric intake of each diet is designed to be less than usual intake. A reduced energy intake is a critical element of any weight-loss diet.

Atkins diet. The Atkins diet is a carbohydrate-controlled diet consisting of four phases: induction, ongoing weight loss (OWL), premaintenance, and lifetime maintenance. The induction period is 14 days in length, and carbohydrate intake is limited to 20 g/d. Twenty grams per day is an extremely low carbohydrate intake, and a typical induction phase diet would

Table 13.2 Comparison of Weight-Loss Plans

	Atkins	Zone	Weight Watchers	Ornish
Carbohydrate	Very low (20 g/d during the induction stage) to low (60 g/d)	Low (40% of total energy intake)	Moderate	High
Fiber	Low	Moderate	Moderate to high	High
Protein	Moderate to high; protein foods may also be high in fat	High (30% of total energy intake); protein foods are low in fat	Moderate; protein foods are low in fat	Moderate; most from plant sources
Fat	Moderate to high	Moderate (30% of total energy intake), emphasis on monounsaturated fats	Low	Very low (less than 10% of total energy intake)
Total calories	Not counted but likely to be less than usual intake	Not counted but typically 1,000–1,600 kcal daily	Not counted but each "point" is ~50–60 kcal and the number of points per day is limited	Not counted but likely to be less than usual intake
Mechanism for reducing caloric intake since calories are not counted	Appetite suppression due to ketosis	Number of carbohydrate, protein, and fat blocks are limited	Point totals for foods consider caloric, fat, and fiber content; number of points per day is limited	High fiber and very low fat intakes
Miscellaneous	Degree of insulin resistance considered	Glucose and insulin responses to carbohydrate foods considered, monounsaturated fats emphasized for hormone production	Known for its support network	Goal is to prevent and reverse heart disease, but the diet also promotes weight loss

Legend: g/d = grams per day; kcal = kilocalorie

include the elimination of all carbohydrate-containing foods except salad greens and similar vegetables (for example, zucchini). Strict limits on carbohydrate intake result in a diet consisting of protein- and fat-containing foods such as meat, fish, poultry, and eggs. There is no stated restriction on caloric intake, and calories are not "counted," but following the induction phase plan typically results in a lower energy intake than previously consumed.

The **hypocaloric**, very low carbohydrate intake results in **ketosis**, a metabolic adaptation to starvation (see Chapter 6 for a detailed explanation). When carbohydrate is severely restricted, ketone bodies are produced as a result of the incomplete breakdown of fat. Brain and nervous tissues typically depend on carbohydrate for fuel but in its absence begin to use ketone bodies. Ketosis that results from starvation is referred to as benign dietary ketosis to distinguish it from the ketosis associated with uncontrolled diabetes and pregnancy, which can lead to a dangerous medical condition known as **ketoacidosis**. Benign dietary ketosis results in the preferential use of stored body fat and appetite suppression, two goals of the Atkins diet. In those without diabetes, ketosis rarely leads to ketoacidosis.

After the 2-week induction phase, the carbohydrate content of the diet may be increased, depending on the amount of weight lost and the degree of **metabolic resistance**. Metabolic resistance is not a recognized medical term; rather, it is a term coined by Dr. Atkins to indicate the inability to lose weight or to continue to lose weight when consuming a diet containing less than 1,000 kcal or 25 g of carbohydrate daily.

After the 2-week induction phase, if a relatively small amount of weight was lost (for example, 8 pounds in a male who has 50 pounds to lose), then metabolic resistance is considered high and carbohydrate continues to be restricted in amounts similar to the induction phase. If a relatively large amount of weight was lost (for example, 16 pounds in a male who has 50 pounds to lose), then carbohydrate intake may be increased to as high as 60 g/d. Maintenance-phase carbohydrate intake for those with high metabolic resistance remains low, approximately 25 to 40 g/d; however, those with low metabolic resistance may increase carbohydrate intake to 60 to 90 g/d during the maintenance phase. Even with increased carbohydrate intake, the Atkins diet remains relatively low in carbohydrate when compared to other diet plans.

Zone diet. The Zone diet emphasizes the amount and type of carbohydrates, low-fat protein foods, and monounsaturated fats. This diet is also referred to as the 40-30-30 diet because of the contribution of each macronutrient to total energy intake—40 percent carbohydrate and 30 percent each protein and fat. Such a macronutrient distribution results in the diet being relatively low in carbohydrates and high in proteins. The amount of kilocalories in the diet is not restricted directly (that is, caloric intake is not counted); rather, a certain number of carbohydrate, protein, and fat "blocks" are recommended. The diet usually ranges from 1,000 to 1,600 kcal daily. The Zone diet emphasizes the **glycemic effect** of carbohydrate foods, with an emphasis on consuming those that do not produce a rapid rise in insulin. The diet also emphasizes monounsaturated fats because of their influence on the production of eicosanoids, compounds made from unsaturated fatty acid that have hormonelike activity (for example, **prostaglandin**).

The first step in planning the Zone diet is a determination of the daily amount of protein needed. Daily protein intake is divided into blocks, each consisting of about 7 g. Carbohydrate blocks contain approximately 9 g of carbohydrate with an emphasis on low-glycemic fruits, vegetables, and grains. One goal of the diet is to keep a 1:1 ratio of protein to carbohydrate blocks. Each fat block contains approximately 1.5 g of fat, predominantly as monounsaturated fats such as nuts, olive oil, and avocadoes. Three meals and two snacks are recommended daily with an almost equal number of protein, carbohydrate, and fat blocks at each meal or snack.

Ornish diet. The Ornish diet is a plant-based diet that emphasizes high-fiber carbohydrate foods. This diet is very low in fat, approximately 10 percent of total energy intake, and moderate in protein, mostly from plant sources. Caloric intake is not counted; rather, individuals are instructed to "eat when hungry, stop when full," but energy intake is likely limited due to the low-fat, high-fiber content of the diet. Foods are generally divided into one of three groups: consume in large quantities, consume in moderation, and avoid.

Hypocaloric: A low amount of dietary kilocalories compared to what is needed to maintain body weight.

Ketosis: General definition is the production of ketone bodies (a normal metabolic pathway). Medical definition is an abnormal increase in blood ketone concentration.

Ketoacidosis: Medical condition in which the pH of the blood is more acidic than the body tissues.

Metabolic resistance: A term coined by Dr. Atkins to indicate the inability to lose weight or to continue to lose weight when consuming a diet containing less than 1,000 kcal or 25 g of carbohydrate daily.

Glycemic effect: The effect that carbohydrate foods have on blood glucose and insulin secretion.

Prostaglandin: A compound made from arachidonic acid (an omega 6 fatty acid) that has hormonelike (regulatory) activity.

Beans, legumes, whole grains, fruits, and vegetables are emphasized. Nonfat dairy products and nonfat or low-fat commercially prepared meals are to be consumed in moderation. Foods to be avoided include all kinds of meat, poultry, fish, low-fat and regular dairy products, oils, sugars, alcohol, nuts, seeds, olives, and avocadoes.

Weight Watchers. The Weight Watchers program has evolved since its inception in 1966 and currently involves a point system known as PointsPlus®. All foods are assigned points based on fat, protein, carbohydrate, and fiber contents. Daily and weekly point targets are assigned. Points can also be earned by exercising, and these points can be used to add more food to the diet plan or to accelerate the rate of weight loss. The Weight Watchers diet reduces caloric intake indirectly through the point system, rather than by a direct counting of calories. The diet is low in fat and high in fiber. The hallmark of the Weight Watchers program has been its support system, either in person or online. The program also focuses on teaching participants sensible eating habits.

Comparing the effectiveness of various weight-loss diets. The diet plans described have been used by millions of people, and which plan is "best" has long been debated by both scientists and the lay public. In 2005, Dansinger et al. reported the results of a study that compared weight loss in obese subjects (average BMI of 35) who followed one of the four diet plans previously described, providing a scientific basis for answering the question, "Which weight-loss diet is the best?" In those who completed the study, weight loss after 1 year was about 3 percent of body weight (~5–7 lb [~2–3 kg]) regardless of the diet plan. The more restricted plans, Atkins and Ornish, were harder for participants to adhere to; that is, there was a higher dropout rate. A similar study conducted in the United Kingdom found that the four weight-loss diets studied (Atkins, Weight Watchers, Slim-Fast, Rosemary Conley [similar to Weight Watchers]) were equally effective in producing weight (fat) loss over 6 months. The average fat loss was ~10 lb (4.4 kg). During the first 4 weeks, those on the Atkins diet lost weight at a faster rate, but by the end of the 6 months there was no significant difference in the total amount of weight lost (Truby et al., 2006).

All the plans studied were low in kilocalories (energy), although kilocalories were not directly counted in any of the plans. It is the macronutrient composition that varied in these weight-loss plans. For the obese, sedentary individual, the restriction of energy and the length of the energy restriction seem to be more important than the carbohydrate, protein, and fat content of the weight-loss diet (Sacks et al., 2009; Dansinger et al., 2005). Due to individual variations in response, it is not appropriate to recommend "one" weight-loss diet plan to the entire population (Volek, Vanheest, and Forsythe, 2005).

However, the debate does not stop with the knowledge that all the hypocaloric diets were equally effective in producing a relatively small short-term weight loss. There may be advantages to consuming either a low-fat or low-carbohydrate diet as shown in Table 13.3. For example, self-reported dietary strategies of those individuals who have maintained at least a 30 lb (13.6 kg) weight loss for at least 5 years suggest that a diet low in fat (~24 percent of total energy intake) is an important element in weight maintenance. This low-fat diet is part of a lifestyle that includes an expenditure of energy via exercise of approximately 2,800 kcal per week (Klem et al., 1997).

Table 13.3 Advantages and Disadvantages of Restricted-Fat or -Carbohydrate Weight-Loss Plans

	Advantages	Disadvantages
Restricted-fat diets	• Eliminates nutrient with the greatest caloric density • Fat is easy to overconsume (low fiber, pleasing taste, widely available) • Some scientific evidence of weight loss success • Self-reports of weight-loss success (National Weight Control Registry) • Reduces total cholesterol (TC) and low-density lipoprotein cholesterol (LDL-C)	• Low-fat/high-carbohydrate diets may increase triglyceride concentrations and decrease high-density lipoprotein cholesterol (HDL-C), especially in the absence of significant weight loss and exercise
Restricted-carbohydrate diets	• Focuses on high-fat and -protein foods—favorite foods of some people • Reduces appetite (due to ketosis) • Some studies show greater weight loss than with low-fat diets • High protein intake helps to preserve lean body mass • Elicits a low insulin response that may allow for better mobilization of stored fat • Reduces triglyceride, TC, and elevated insulin concentrations • Increases HDL-C	• Lack of long-term data (>1 year) on safety or effectiveness • Risk for gout, kidney stones, and osteoporosis may be increased

Source: Volek, J. S., Vanheest, J. L., & Forsythe, C. E. (2005). Diet and exercise for weight loss: A review of current issues. *Sport Medicine, 35*(1), 1–9.

How Does Exercise Affect the Processes That Regulate Energy Balance? Are the Effects Different in Men and Women?

Individuals who need to lose weight and body fat do so by attempting to alter the energy balance equation by decreasing energy (food) intake, increasing energy expenditure through physical activity and exercise, or a combination of both. Although the energy balance equation is simple, the physiological mechanisms that control energy intake and expenditure are complex, and changes in one component of the equation may have widespread and varied effects on other components. For example, increasing exercise activity may seem to be a simple way to increase energy expenditure to lose weight, but the resulting energy deficit may in turn cause an increase in appetite leading to an increase in food intake. Physiological responses to changes in energy balance can vary tremendously between individuals, and recent research suggests that there may be differences in responses based upon gender.

The following research study provided important information about the response of energy-regulating hormones and appetite to changes in energy balance by feeding and exercise, and in particular examined differences in these responses between men and women: Hagobian, T. A., Sharoff, C. G., Stephens, B. R., Wade, G. N., Silva, J. E., Chipkin, S. R., & Braun, B. (2009). Effects of exercise on energy-regulating hormones and appetite in men and women. *American Journal of Physiology—Regulatory, Integrative, and Comparative Physiology, 296*, R233–R242.

What did the researchers do? Equal-size groups of obese or overweight, sedentary male and female subjects (nine in each group) had their appetite and levels of specific energy-regulating hormones assessed at baseline, after 4 days of daily exercise with no energy deficit, and after 4 days of daily exercise with an energy deficit. Baseline measurements were made of body composition, maximal oxygen consumption, daily physical activity, energy intake, and resting and daily energy expenditure. Blood samples were taken to determine the levels of ghrelin, insulin, leptin, thyroid hormones (T3 and T4), and glucose both when fasting and as a response over time to a standardized meal (Meal Tolerance Test).

Energy intake and expenditure were carefully controlled by the researchers by providing all of the food for the subjects during the experimental periods and conducting the exercise sessions on a treadmill in the laboratory. The moderate-intensity exercise sessions were individually adjusted in duration to achieve a number of kcal burned equal to 30 percent of typical total daily energy expenditure. In one 4-day exercise period, the amount of food given to the subjects was the same, resulting in an energy deficit due to the increase in exercise activity. In the other 4-day exercise period, the amount of food given to the subjects was carefully increased to match the kcal expended by exercise, bringing the subjects back into energy balance. All subjects completed both exercise sessions, and the response of the energy-regulating hormones and appetite was measured at the conclusion of each 4-day experimental period.

What did the researchers find? Under energy deficit conditions, high levels of ghrelin and low levels of insulin and leptin are seen, which stimulate appetite, food intake, and an increase in energy intake in an attempt to abolish the energy imbalance. The results of this research study showed changes in these energy-regulating hormones and appetite in response to exercise and energy balance changes, but it is important to note that these changes were different for men and women. For each of the energy balance states in the study, men and women responded in one way that would be normally expected and in another, unexpected fashion, and each of the genders responded differently.

Four days of exercise without a matching increase in food intake resulted in an increase in ghrelin and a decrease in insulin levels in women, which is what would be expected with an energy deficit. Eliminating the energy deficit by adding an equal amount of food would seemingly result in a return to baseline levels for these energy-regulating hormones, but they remained elevated in the female subjects in this study. Therefore, the female subjects responded to daily exercise training by increasing ghrelin and decreasing insulin, both of which would be expected to increase energy intake by increasing food intake. Instead of returning to baseline, this response persisted even when energy balance was restored by increased food intake.

In contrast to the responses of the female subjects, the males did not show an increase in ghrelin as a result of the exercise-induced energy deficit. Although insulin levels were lower, the return to energy balance with the addition of extra food resulted in a return to baseline levels of energy-regulating hormones. These results suggest that, in men, changes in hormonal status that would lead to increases in energy/food intake are more related to the presence of an energy deficit than to exercise activity alone.

What is the significance of the research? This research study provides evidence of potentially important gender differences in the metabolic and hormonal responses to exercise and energy deficits. When women exercise, they seem to have a hormonal response that encourages an increase in food intake, and when the exercise results in an energy deficit this response is enhanced. This hormonal status persists even when the energy deficit is eliminated, however, which may lead to energy intake levels that prevent weight loss or may even lead to weight gain. In men, however, exercise alone has little effect on the energy-regulating hormone levels, suggesting that men are not necessarily prone to increasing food intake as a result of increasing exercise activity. It is commonly reported that women do not lose weight and/or body fat as readily as men in response to regular exercise, particularly when food intake is not regulated. This research provides important physiological clues as to why women seem to compensate more readily with food intake in response to exercise training resulting in less weight/body fat loss than do men.

On the other hand, low-carbohydrate diets have been shown to decrease appetite (due to circulating ketone bodies) and decrease visceral fat, which is stored in the abdominal area. The reduced visceral fat, an important health-related issue, may be due to reduced circulating insulin that results with a low carbohydrate intake. An individual's response to the macronutrient composition of the diet may vary, again underscoring the importance of recognizing that one weight-loss diet plan is not appropriate for the entire population (Volek, Vanheest, and Forsythe, 2005).

The majority of scientific studies suggest that altering the energy balance equation with a combination of dietary restriction and exercise is the most effective means of achieving long-term weight loss. Energy balance can be altered by exercise alone, although weight loss may not be of the same magnitude as occurs with diet alone or the combination of diet and exercise. One explanation may be that the individual compensates for the additional energy expenditure from the exercise by increasing caloric (food) intake. When a compensatory increase in caloric intake is avoided, exercise can result in substantial weight loss.

The amount of exercise necessary for achieving and maintaining weight loss (that is, up to 60 to 90 minutes on most days) is substantially greater than the minimum amount of physical activity that is recommended for good health (that is, accumulating 30 minutes of moderate-intensity activity on most days). If the primary goal is weight loss rather than large improvements in cardiovascular fitness, duration of exercise appears to be more important than intensity, provided the exercise is of at least moderate intensity. For example, approximately the same number of kilocalories would be expended running 4 miles at a 7½-minute-per-mile pace and walking the same distance at a 15-minute-per-mile pace (4 mph), but walking would take twice the time. The exercise intensity and rate of caloric expenditure is obviously higher when running and will contribute to greater increases in cardiovascular fitness levels as well as to caloric expenditure. Many people may find this intensity of exercise difficult to sustain and uncomfortable, and may desire to pursue activities of lower intensity, such as walking. A comparable level of caloric expenditure can be achieved, but the duration of the activity must be extended, in this case to 60 minutes. Approximately 150–250 of moderate-intensity exercise each week (30–40 minutes per day) or >2,000 kcal of energy expended per week is necessary for modest weight loss. More substantial weight loss and maintenance of weight loss typically require exercise of greater duration and/or intensity, generally in excess of 250 minutes per week (Donnelly et al., 2009). This amount of exercise may appear to be daunting to the sedentary individual not accustomed to physical activity, but it is achievable with gradual progression over time to these targets.

Spotlight on...

Overweight and Obesity

Definition: In adults: Overweight—BMI between 25 and 29.9; Obesity—BMI 30 or greater.

Prevalence: Overweight—65 percent of adults; Obesity—30 percent of adults and 17 percent of children and adolescents; prevalence in United States is highest of all developed countries.

Symptoms: Change in body weight and appearance; change in breathing pattern during sleep; fatigue or lack of stamina.

Cause: Energy imbalance, in large part due to excessive food intake and/or lack of physical activity.

Prevention: (1) Matching caloric intake with caloric expenditure. (2) Eating predominantly because of physiological hunger. (3) Age- and activity-appropriate portion sizes. (4) A diet emphasizing fruits, vegetables, whole grains, low- or nonfat dairy foods, and lean meats, fish, and poultry, similar to the DASH (Dietary Approaches to Stop Hypertension) diet. (5) Minimum of 150 minutes of moderate-intensity exercise each week. (6) Limiting sedentary leisure time activities such as TV viewing.

Treatment: Treatment must be individualized.

Diet: (1) Reduce food intake. (2) Reduce portion sizes. (3) Follow a hypocaloric diet that includes nutrient-dense foods.

Exercise: (1) Moderate-intensity physical activity or cardiovascular exercise. (2) ≥150–250 minutes per week or ≥2,000 kcal per week of energy expenditure. (3) Strength training as a supplement to cardiovascular exercise two to three times per week focusing on increased volume rather than increased intensity (that is, increased sets and repetitions rather than increased resistance or weight).

Centers for Disease Control and Prevention http://www.cdc.gov/obesity/index.html

Prevention of weight gain is critical because overweight and obesity are difficult to treat. Consider the experience of those who are registered in the National Weight Control Registry. This registry consists of approximately 3,000 people, most of whom are Caucasian (97 percent), female (80 percent), and married (67 percent). On average, the registrants have lost 66 lb (30 kg) and have maintained this weight loss for an average of 5½ years. The average maximum BMI attained was 35, and most lowered their maximum lifetime BMI by 10 BMI units (for example, BMI was reduced from 35 to 25). Average daily caloric intake is 1,381 kcal with approximately 24 percent of total energy provided by fat. Women expend approximately 2,545 kcal weekly through exercise, whereas men expend about 3,923 kcal/wk; exercise expenditure is equivalent to about 1 hour of moderately intense exercise daily. Interestingly, 90 percent reported previous failed attempts at weight loss and attribute their current success to a greater commitment, stricter dieting, and more emphasis on exercise (Wing and Hill, 2001). These data should be interpreted carefully, as they are self-reported and obtained from a self-selected (nonrandom) sample. However, it is quite clear how difficult it is once weight has been gained to lose that weight and maintain the weight loss.

To summarize, for fat (weight) loss to occur energy expenditure must be greater than energy intake. This can be accomplished by reducing food intake and/or increasing exercise. Although many weight-loss diet plans do not "count calories," they are designed to restrict caloric intake. A weight-loss plan that severely restricts the types of foods that may be eaten, be it carbohydrate or fat, is more difficult to follow over the long term. These plans may lead to successful short-term weight loss, but high dropout rates and weight gain after 1 year suggest that they more often promote a temporary loss of body fat.

Key points

- Chronic diseases often begin in childhood and progress throughout the life cycle.

- Lifestyle, particularly diet and exercise, is closely tied to many chronic diseases.

- Overweight and obesity affect many chronic diseases.

- American society is "obesogenic."

- The regulation of body weight is complex.

- To lose weight, eat less and exercise more.

- No one weight-loss diet is the "best."

Why is it so difficult for humans to lose weight and maintain the weight loss?

13.3 Diet, Exercise, and Chronic Disease

Diet and exercise play important roles in preventing and treating overweight and obesity, which are conditions that directly affect many other chronic diseases. Diet and exercise also play non-weight-related roles in many chronic diseases.

Diet and exercise are important components of prevention and treatment strategies for hypertension.

Hypertension is known as a silent killer because of the lack of recognizable symptoms. In most cases the cause is unknown. Genetic predisposition is very powerful, but there are a number of nutrition- and exercise-related factors that affect blood pressure. Excessive energy intake over time is a major factor because it leads to obesity. Obesity negatively affects a number of the body's systems that regulate blood pressure. Blood pressure may also be affected by dietary sodium intake. In general, as sodium intake increases blood pressure increases, although not all people with a high sodium intake will be hypertensive. Some individuals are sodium sensitive, and a high dietary sodium intake has a direct effect on raising their blood pressure. In these individuals, a reduction in sodium intake results in a decrease in elevated blood pressure. Research has shown that higher intakes of potassium, such as consuming a variety of fruits and vegetables daily, polyunsaturated fatty acids, and protein may help to reduce blood pressure (Savica, Bellinghieri, and Kopple, 2010; Institute of Medicine, 2004).

Studies have shown that endurance training helps reduce, but does not eliminate, the risk for hypertension. Even a small percentage of former elite endurance athletes will develop high blood pressure. Hernelahti et al. (1998) compared Finnish male endurance masters athletes to a control group. At the beginning of the study in 1984, both groups of men ages 35 to 59 were free of heart disease. When surveyed again in 1995, 27.8 percent of the control group used an antihypertensive medication compared to only 8.7 percent of the master athletes, but this group of highly active runners was not free of hypertension. In a subsequent study, the same researchers found that endurance training in young-adult elite athletes reduced the risk for hypertension when the men were middle-aged or older. Continuing to be physically active throughout adulthood also reduced, but did not eliminate, the risk of ever becoming hypertensive (Hernelahti et al., 2002). Maintaining a healthy weight, consuming a healthy diet, and exercising routinely reduces the risk of hypertension for the majority of adults, but may not be sufficient for those who are hypertensive because of a genetic predisposition, sodium sensitivity, or other non-lifestyle-related factors. Some athletes may require the use of antihypertensive medications to keep their blood pressure within a healthy range. These individuals should work closely with a physician familiar with exercise to find a medication that appropriately controls their hypertension but does not have an adverse effect on their exercise capability.

Although some former, well-trained athletes become hypertensive, the majority of people in the United States with high blood pressure are sedentary, and they would benefit from engaging in routine physical activity. Exercise, particularly endurance (aerobic) exercise, can result in a modest decrease in resting blood pressure both in the hours after the exercise bout and on a chronic basis. Many people with hypertension are also overweight or obese, and blood pressure may be lowered as a result of weight loss. Prevention and treatment strategies are listed in the Spotlight on... Hypertension.

Insulin resistance: A condition in which the hormone insulin fails to stimulate tissues to take up the same amount of glucose.

Diabesity: Diabetes associated with overweight or obesity.

Diabetes is a group of diseases characterized by a high blood glucose level.

Diabetes is a metabolic disease characterized by a high blood glucose concentration and is typically divided into two types. Type 1 diabetes, also referred to as insulin dependent diabetes mellitus (IDDM), is an autoimmune disease that destroys the beta cells of the pancreas, which results in an inability to manufacture insulin. Type 1 diabetes makes up less than 10 percent of all the cases of diabetes in the United States. Far more prevalent is type 2 diabetes, also known as non-insulin dependent diabetes mellitus (NIDDM), which accounts for more than 90 percent of all U.S. diabetes cases and affects at least 21 million people. In some cases, individuals produce too little insulin, but in most people with type 2 diabetes the insulin produced is ineffective because cells, particularly muscle, fat, and liver cells, are insensitive to its action. This condition is known as **insulin resistance**. Approximately 85 percent of people with type 2 diabetes are obese and have insulin resistance, but nonobese individuals can also manifest type 2 diabetes (American Diabetes Association, 2004). Diabetes that is associated with overweight or obesity is sometimes referred to as **diabesity**. Approximately 57 million people in the United States, including at least 2 million adolescents, are considered prediabetic and are at risk for developing diabetes as a result of obesity, sedentary lifestyle, and consumption of an unhealthy diet.

Spotlight on...

Hypertension

Alternate term: High blood pressure.

Definition: Resting systolic blood pressure >140 and/or resting diastolic blood pressure >90 mm Hg (measured on more than one occasion); prehypertension: 120–139 mm Hg (systolic), 80–89 mmHg (diastolic).

Prevalence: One of every three adults in the United States (~65 million); more prevalent in African Americans and in older adults.

Symptoms: No distinct symptoms; known as a silent killer.

Cause: Unknown in 90 percent of all cases.

Prevention: (1) Maintain or achieve a healthy body weight. (2) Follow the Dietary Guidelines or DASH diet (see Appendix B). (3) Meet minimum activity or exercise recommendations.

National Heart, Lung and Blood Institute: http:www.nhlbi.nih.gov.

Treatment: Dietary modifications, exercise, and/or medications can help control blood pressure.

Diet: (1) If overweight, reduction of body fat. A 10-pound (4.5 kg) weight loss is effective regardless of degree of overweight. (2) Reduction of sodium intake. This strategy may be effective if the individual is sodium sensitive. (3) Consumption of the DASH (Dietary Approaches to Stop Hypertension) diet. This diet emphasizes fruits, vegetables, whole grains, low- or nonfat dairy foods, lean meats, fish and poultry, nuts, beans, and oils. Potassium is plentiful and sodium is limited.

Exercise: (1) Moderate intensity exercise on most, preferably all, days of the week for at least 30 minutes (either continuously or accumulated). (2) Endurance (aerobic) exercise is the preferred type of exercise, although strength training may be supplemented in consultation with a physician.

Type 2 diabetes was previously referred to as adult-onset diabetes because it was uncommon in children or adolescents. However, the number of nonadults diagnosed with type 2 diabetes is dramatically increasing and that term is no longer appropriate (Pinhas-Hamiel et al., 1996). The increase in child and adolescent type 2 diabetes reflects the increase in child and adolescent obesity and is most common in those who are also sedentary and have a family history of diabetes.

Millions of people are undiagnosed, a particular problem because the effects of type 2 diabetes can be numerous and severe, especially if untreated. Associated medical conditions include atherosclerotic cardiovascular disease and cerebrovascular disease (such as hardening of the arteries that can cause heart problems and stroke), hypertension, peripheral vascular disease (such as narrowing of the vessels in the legs, arms, kidneys), nephropathy (kidney disease), retinopathy (eye disease leading to blindness), neuropathy (degeneration of nerve function), and susceptibility to infections and slow wound healing.

Type 2 diabetes is characterized by hyperglycemia (that is, elevated blood glucose concentration) that is a result of insulin resistance, a defect in insulin secretion, or insulin deficiency over time. Under normal conditions, the increase in blood glucose that occurs after a meal stimulates the secretion of insulin, which then mediates the uptake of glucose into cells via glucose transporters located on the surface of the cell membranes. Insulin resistance occurs when the same level of insulin fails to stimulate the same degree of glucose uptake, and blood glucose concentration remains higher than would be expected. If blood glucose remains elevated, the pancreas may be able to secrete more insulin and eventually enough insulin may be present to result in adequate glucose uptake. In some people with type 2 diabetes, insulin production is actually greater than normal, but the cells are resistant to insulin's blood glucose lowering effect. Over time the insulin resistance may become more severe, and even higher than normal insulin secretion is not enough to reduce blood glucose concentration to a normal level. Hyperglycemia and **hyperinsulinemia** (elevated blood insulin) result. In other people with type 2 diabetes, the pancreas produces insulin in quantities less than normal, which results in hyperglycemia. Over time, people with type 2 diabetes may experience a steady decline in insulin production, which makes it difficult to maintain a normal blood glucose concentration and may eventually result in the need for insulin injections.

Obesity worsens insulin resistance. An obese person with diabetes needs more insulin than a nonobese person with diabetes to maintain a normal blood glucose concentration. Obesity appears to drive a vicious cycle in which an increase in body fat results in cells becoming more insulin resistant, and insulin resistance results in increasing obesity. Obesity results in an elevated blood insulin concentration (hyperinsulinemia), and hyperinsulinemia favors fat storage and inhibits fat breakdown (Girod and Brotman, 2003). Exercise helps to break the vicious cycle because it has an insulin-like effect. Exercise stimulates the movement of glucose transporters to the surface of the cell membrane where they can increase glucose uptake into the skeletal muscle cells, a large storage site for glucose. Exercise results in a lower fasting blood glucose concentration and improved insulin sensitivity (that is, less insulin resistance). This effect lasts no more than approximately 72 hours after the exercise, so regular physical activity is needed.

Type 2 diabetes is also characterized by abnormal lipid (fat) metabolism. Since the cells lack glucose for fuel, fat in the blood is increased to provide cells an alternative fuel source. This results in an increase in the deposition of fatty acids into the cells of the muscle, liver, and pancreas, which makes the muscle and liver cells even more insulin resistant and the beta cells of the pancreas more dysfunctional. This abnormal lipid metabolism is related to an increase in abdominal obesity, which readily releases fatty acids into the blood (Raz et al., 2005).

A diet that provides adequate but not excessive caloric intake and the consumption of whole grains, fruits, vegetables, and low-fat protein foods has been shown to reduce risk for developing diabetes (van Dam et al., 2002; Hu et al., 2001). The intake of sugared soft drinks, refined grains, and animal products, such as meat and processed meat, have been associated with a greater risk of developing type 2 diabetes (Imamura et al., 2009).

One misconception is that eating too much sugar will *cause* diabetes. Sugar and other carbohydrate foods are broken down to glucose, and blood glucose concentration does rise temporarily after their ingestion. The body then secretes insulin, which mediates the uptake of glucose into the cells, resulting in a return to a normal blood glucose concentration. Taking in too much sugar does not cause a problem under normal conditions. However, an individual's dietary intake may increase the risk for diabetes if it leads to weight gain, which leads to insulin resistance. Sugar intake may be a factor in weight gain if it represents excess caloric intake.

Once type 2 diabetes has been diagnosed, diet and exercise are very important components of the treatment plan. If overweight, the initial focus is a diet and exercise program for moderate weight loss.

Hyperinsulinemia: Blood insulin concentration that is higher than the normal range.

A 10–20 lb (4.5–9 kg) weight (fat) loss helps reduce insulin resistance and improve blood lipid levels. The diet is planned so that the proper amount of carbohydrate is included in meals and snacks across the day and that high-quality carbohydrates, such as fruits, vegetables, and whole grains, are the majority of carbohydrates consumed. Since both the quantity and quality of carbohydrates are important, those with diabetes may benefit from being familiar with glycemic index and glycemic load (see Spotlight on...Glycemic Index and Glycemic Load).

Physical activity and diet play major roles in the treatment of type 2 diabetes (American College of Sports Medicine and American Diabetes Association,

Spotlight on a real athlete

Lucas, 23-Year-Old, Former Collegiate Cross Country Runner

Lucas, the collegiate cross country runner who runs 75 to 80 miles (125 to 135 km) per week has served as an example throughout this textbook. Based on a 1-day dietary analysis (Figures a and b), Lucas's intake is estimated to be 3,333 kcal (~53 kcal/kg), 532 g carbohydrate (~8.5 g/kg), 124 g protein (~2 g/kg), and 94 g of fat (~1.5 g/kg). From a performance perspective, Lucas's energy, carbohydrate, protein, and fat intakes are considered appropriate (see Chapters 4, 5, and 6). But how does his diet fare when analyzed against dietary risk factors for chronic diseases? Figure b evaluates Lucas's diet for kilocalories and macronutrients as well as saturated, mono- and polyunsaturated fats, cholesterol, fiber, sugar, vitamin, and mineral intakes.

It often comes as a surprise to athletes like Lucas that the diet that provides sufficient kilocalories, carbohydrates, and proteins to support training and performance can fall short when evaluated for chronic disease risk. In Lucas's case, his diet is appropriate in many ways, but there are some red flags—sodium intake is very high (~6,500 mg) considering that he does not lose much salt in sweat, and the intake of some vitamins and minerals, such as folate, niacin, magnesium, and potassium, is low. Even if he were to continue to run 80 miles (135 km) per week, his pattern of eating could use some modification to help him reduce the risk for chronic diseases.

As is typical of many former collegiate athletes, Lucas will be reducing his weekly exercise from 80 miles to ~30 miles (135 km to 50 km) per week after he graduates from college.

Lucas, a former collegiate cross country runner.

He intends to compete in 10 km (6.2-mile) races once a month, but the intensity and duration of his training will be substantially reduced. To prevent gaining weight as body fat, he will need to reduce his caloric intake, but it is also important that Lucas modify his diet in other ways. When Lucas has a preemployment physical, he is surprised to learn that he has hypertension, even though high blood pressure runs in his family. Lucas's physician recommended that he increase his fruit and vegetable intake, which will increase potassium intake, and reduce his sodium intake.

MyPyramid.gov
STEPS TO A HEALTHIER YOU

	Goal*	Actual	% Goal
Grains	10 oz. eq.	11.8 oz. eq.	118%
Vegetables	4 cup eq.	1.2 cup eq.	30%
Fruits	2.5 cup eq.	1.7 cup eq.	68%
Milk	3 cup eq.	5.4 cup eq.	180%
Meats & Beans	7 oz. eq.	8.8 oz. eq.	126%
Discretionary	648	1152	178%

oz = ounce
eq = equivalent

*Your results are based on a 3200 calorie pattern, the maximum caloric intake used by My Pyramid.

Figure a Analysis of Lucas's 24-hour dietary intake

Ben Welsh/Comet/Corbis

2010). Together, diet and exercise can help individuals lose weight, maintain blood glucose control, and delay or slow the progression of the medical conditions associated with diabetes. One key to diabetes treatment is changes in lifestyle, although diet and exercise strategies are often underutilized in favor of oral antidiabetic medications.

Cardiovascular disease is the major cause of death in the United States.

Cardiovascular disease is responsible for approximately 26 percent of the deaths in the United States each year (Heron et al., 2009). Various terms are used—cardiovascular disease, heart disease, coronary

Nutrient	DRI	Intake	0%	50%	100%
Energy					
Kilocalories	3365 kcal	3332.8 kcal			99%
Carbohydrate	379–547 g	532.15 g			
Fat, total	75–131 g	93.86 g			
Protein	84–294 g	124.09 g			
Fat					
Saturated fat	<10%	36.63 g			
Monounsaturated fat	no rec	19.45 g			
Polyunsaturated fat	no rec	11.65 g			
Cholesterol	300 mg	195.79 mg			65%
Essental fatty acids (efa)					
Omega-6 linoleic	17 g	1.9 g			11%
Omega-3 linolenic	1.6 g	0.47 g			30%
Carbs					
Dietary fiber, total	38 g	37.4 g			98%
Sugar, total	no rec	310.37 g			
Vitamins					
Thiamin	1.2 mg	1.41 mg			118%
Riboflavin	1.3 mg	2.45 mg			188%
Niacin	16 mg	11.57 mg			72%
Vitamin B6	1.3 mg	1.09 mg			84%
Vitamin B12	2.4 μg	4.82 μg			201%
Folate (DFE)	400 μg	229.4 μg			57%
Vitamin C	90 mg	237.29 mg			264%
Vitamin D (μg)	5 μg	4.05 μg			81%
Vitamin A (RAE)	900.41 μg	2134.31 μg			237%
Alpha-tocopherol (Vit E)	15 mg	19.63 mg			131%
Minerals					
Calcium	1000 mg	2165.1 mg			217%
Iron	8 mg	21.42 mg			268%
Magnesium	400 mg	291.13 mg			73%
Potassium	4700 mg	2468.13 mg			53%
Zinc	11 mg	12.04 mg			109%
Sodium	1500 mg	6512.82 mg			434%

kcal = kilocalorie
g = gram
rec = recommendation
mg = milligram

μg = microgram
DFE = dietary folate equivalents
RAE = retinol activity equivalents
Vit = vitamin
DRI = Dietary Reference Intake

Figure b Analysis of Lucas's diet for risk of chronic diseases

heart disease, **coronary artery disease** (see Glossary)—and each is slightly different. The focus of this section is atherosclerosis, which is the hardening and narrowing of arteries. **Atherosclerotic heart disease** is a life-threatening disease when it occurs in the coronary arteries.

The atherosclerotic process begins in childhood and adolescence with the development of fatty streaks on the interior walls of arteries. Fatty streaks, which are initially soft, enlarge and harden and become **plaques**, which eventually protrude into the **lumen** of the artery and may obstruct blood flow (Figure 13.5). Plaques are found between the middle and inner layers of the coronary arteries and cause them to harden and narrow. By age 30 most people have fairly well-developed plaques (Madamanchi, Vendrov, and Runge, 2005).

Plaques seem to form when low-density lipoproteins (LDL), which are carriers of cholesterol, are oxidized. Oxidation damages blood vessels, and the immune system responds with platelet and other immune cell migration to the site of injury to engulf the oxidized LDL. The platelets adhere to the vessel to repair the damage. The immune cells (known as foam cells) become filled with cholesterol and eventually harden and become plaque. Over time scar tissue begins to form, debris accumulates because the surface of the vessel wall is irregular, and the lumen (space available for blood flow) narrows (Madamanchi, Vendrov, and Runge, 2005).

The risk factors for heart disease fall into three categories: (1) major risk factors that cannot be changed, (2) major risk factors that can be modified, treated, or controlled; and (3) other factors that may contribute. Those over the age of 20 can assess their risk for heart disease at http://www.heart.org/HEARTORG.

Spotlight on a real athlete

Vijay, 38-Year-Old, Occasional Triathlete

As former collegiate athletes age, they may find it difficult to continue to exercise and eat a healthy diet. Vijay is in his late 30s and has been active to some degree all his life. He ran track in high school and college and continued with distance running throughout his 20s. After graduating from college he transitioned from a highly trained competitive athlete to a well-trained recreational athlete. Now married with three young children, he finds that he has time to only occasionally compete in age-group division triathlons. Vijay has limited time for the intense workouts needed to be highly competitive, so his main goal is to train enough to complete the occasional triathlon in a "reasonable" time. Each week he tries to complete two workouts each of distance running, cycling, and swimming (that is, six endurance workouts per week). He has discovered that finding time each week to train is difficult, not only because of a busy family life but also due to the demands of his job, which requires frequent travel. Although Vijay enjoys exercise, it is not always easy when he is on the road because hotel fitness facilities are limited. He also faces the challenges of consuming an appropriate diet while eating out a lot. Business is often conducted over meals, which tend to be high in kilocalories, fat, sugar, salt, and alcohol. Vijay is not obese but he has slowly gained weight since graduating from college, and many of his family members are obese. He is prone to rapid abdominal weight gain, and he can easily gain 2 to 3 pounds (~1 kg) on a business trip as a result of excessive food and alcohol intake.

Vijay acknowledges that his diet needs to improve, but changing dietary intake has not been a priority for him, in part,

Photosindia/Getty Images

Vijay, an occasional triathlete.

because he is not obese. He has always consumed sports drinks during exercise, but recently he began to drink them throughout the day because of excessive thirst. Excessive urination brought him into the doctor's office, where he was diagnosed with type 2 diabetes. Vijay, who is not obese and engages in some exercise every week, is an example of the minority of adults diagnosed with type 2 diabetes. The majority of adults diagnosed with type 2 diabetes are obese and sedentary.

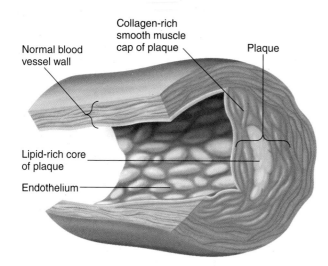

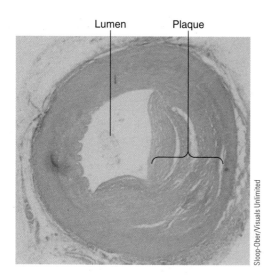

Figure 13.5 Plaque and the development of atherosclerosis

Plaque forms in the walls of large blood vessels and can contribute to the narrowing and hardening of the vessels of the heart.

- Major risk factors that cannot be changed:
  - Increasing age
  - Male sex (gender)

    Women suffer from heart disease but men have a greater risk for heart attack and have heart attacks earlier in life.

  - Heredity (including race)

    Children whose parents have heart disease are more likely to develop it themselves. A strong family history is usually just one of the risk factors present. Race is a factor, but often because it is associated with another risk factor. For example, African Americans are at a higher risk of heart disease because they tend to have more severe hypertension. Mexican Americans, American Indians, and Pacific Islanders tend to have higher rates of obesity and diabetes, both of which are major risk factors.

- Major risk factors that can be modified, treated, or controlled:
  - Tobacco smoke
    - Cigarette smoking is a powerful independent risk factor.
    - Exposure to secondhand smoke increases risk for heart disease even if the person is a nonsmoker.
  - Abnormal blood lipids (target values are listed below)
    - Total cholesterol: Less than 200 mg/dl
    - Low-density lipoprotein cholesterol (LDL-C):
      - If at low risk for heart disease: Less than 160 mg/dl
      - If at intermediate risk for heart disease: Less than 130 mg/dl
      - If at high risk for heart disease (including those who have diabetes): Less than 100 mg/dl
    - High-density lipoprotein cholesterol (HDL-C): 40 mg/dl or higher for men and 50 mg/dl or higher for women
    - Triglycerides: Less than 150 mg/dl
  - High blood pressure (normal is ≥140/90 mm Hg)
    - When high blood pressure is present along with other risk factors such as smoking, obesity, and diabetes, then the risk for heart attack or stroke increases substantially.
  - Physical inactivity
    - Regular, moderate-to-vigorous physical activity helps prevent heart disease.
  - Obesity and overweight
    - A person with excess body fat, especially around the waist, is more likely to develop heart disease even if no other risk factors are present.
  - Diabetes
    - Diabetes substantially increases the risk for heart disease, even more so if blood glucose is not under good control.

Coronary artery disease: Reduction of blood flow to the coronary arteries typically caused by atherosclerosis. Also known as coronary heart disease.

Atherosclerotic heart disease: Atherosclerosis that occurs in coronary arteries, restricting blood flow to the heart.

Plaque: When referring to cardiovascular disease, fatty streaks in vessels that have hardened.

Lumen: Space inside a structure such as the blood vessels or the intestine.

High-density lipoprotein: A fat transporter containing a high proportion of protein and a low proportion of triglyceride and cholesterol. Also known as "good cholesterol."

- Other factors that contribute:
  - Stress

    Not a major factor but may influence major factors such as overeating or smoking in response to stress
  - Alcohol

    Moderate alcohol intake can reduce heart disease risk but alcohol intake above moderate levels increases risk (see Chapter 10).
  - Diet and nutrition

    A healthy diet is one of the best ways to reduce presence of major but modifiable risk factors such as abnormal blood lipids, high blood pressure, obesity and overweight, and diabetes. A healthy diet is one that emphasizes fruits and vegetables, whole-grain and high-fiber foods, lean protein including fish, nuts, oils, and low-fat or fat-free dairy products.

Laboratory tests are used to assess heart disease risk and the most common ones are shown in Table 13.4. A basic screening tool is the measurement of blood cholesterol, but the relative concentrations of the lipoproteins that transport cholesterol are of greater value than the total amount of cholesterol for predicting risk. Low-density lipoproteins have an affinity for depositing excess cholesterol in the artery walls, which contributes to atherosclerosis. An elevated LDL-C concentration is undesirable because it is associated with an increased risk for cardiovascular disease. This has led to the designation of LDL-C as "bad" cholesterol, an attempt to impress on the general population the undesirable effects of elevated LDL-C. The amount of LDL-C in the blood and the amount of antioxidants available to protect the LDL-C from oxidation are important factors in the initiation and progression of atherosclerosis.

Cholesterol is also carried by **high-density lipoproteins** (HDL-C). These carriers have an affinity for removing excess cholesterol from the artery walls if it has not yet been calcified and transporting the cholesterol to the liver for degradation. This has given rise to the term "good" cholesterol. The most desirable HDL-C concentration is greater than 60 mg/dl because such values are associated with a low risk for heart disease. Consumers may have difficulty remembering the various laboratory values so the message is often simplified: low levels of HDL and

Spotlight on...

Type 2 Diabetes

Alternate term: Non-insulin-dependent diabetes mellitus (NIDDM); formerly adult-onset diabetes.

Definition: A metabolic disease characterized by insulin resistance.

Prevalence: In the United States: 90 percent or more of the diagnosed cases of diabetes (at least 21 million); approximately 6 million are undiagnosed and ~57 million have prediabetes.

Symptoms: Impaired fasting glucose, polyuria (excessive urine), polydipsia (excessive thirst), polyphagia (excessive hunger or food intake).

Cause: Insulin resistance (85 percent of cases).

Prevention: (1) Maintain a healthy weight. (2) Meet minimum physical activity recommendations. (3) Consume a healthy diet that emphasizes whole grains, fruits, vegetables, and low-fat protein foods.

Treatment: Achieve and maintain control of blood glucose; prevent or delay onset or progression of complications.

American Diabetes Association: http://www.diabetes.org/

Diet: (1) Hypocaloric diet, if weight loss is needed. (2) Consumption of a healthy diet such as that promoted by the Dietary Guidelines; moderate consumption of sugar and alcohol. (3) Adjust amount and timing of food intake to maintain glycemic control. (4) Other dietary changes may be needed based on presence of complications, such as heart disease, kidney disease, and/or hypertension.

Exercise: (1) At least low- to moderate-intensity physical activity. (2) Cardiovascular, strength training, and flexibility exercises. (3) Consistent frequency; not more than 2 consecutive days between exercise bouts. (4) Minimum of 10-minute bouts; accumulate at least 150 minutes each week; progress to intensity and duration consistent with weight-loss recommendations.

Bariatric Surgery: Gastric bypass surgery for obese diabetics is effective, although there are risks associated with the surgery. The substantial weight loss that can be achieved improves glucose control and positively affects gut hormones.

Table 13.4 Blood Lipid Laboratory Tests

Laboratory test	Interpretation of laboratory results	Comments
Total cholesterol (TC)	• Desirable (low risk for heart disease): <200 mg/dl (5.18 mmol/L) • Moderate risk: 200–239 mg/dl (5.18–6.18 mmol/L) • High risk: >240 mg/dl (6.22 mmol/L)	• Cholesterol testing is a screening tool. • Adults with values in the desirable range and few risk factors for heart disease are tested once every 5 years. • Testing is more frequent if values are moderate or high, if several other risk factors are present, and to determine effectiveness of any interventions. • Exercise by itself does not confer protection from elevated levels so athletes should be screened as part of a routine physical exam.
Low-density lipoprotein cholesterol (LDL-C)	• Optimal: <100 mg/dl (2.59 mmol/L); recommended if heart disease or diabetes has been diagnosed • Near optimal: 100–129 mg/dl (2.59–3.34 mmol/L) • Borderline high: 130–159 mg/dl (3.37–4.12 mmol/L) • High: >160 mg/dl (4.15 mmol/L) • <70 mg/dl (1.82 mmol/L) for those with known arterial narrowing or heart surgery	• Best predictor of heart disease risk • Known as "bad" cholesterol • Typically calculated, not measured directly • Calculated values are most accurate after a 12-hour food and water fast. • Direct measure (not calculated) is preferred when triglycerides are known to be elevated.
High-density lipoprotein cholesterol (HDL-C)	• Optimal: >60 mg/dl (1.55 mmol/L); less than average risk of heart disease • Normal: males, 40–50 mg/dl (1.0–1.3 mmol/L); females, 50–59 mg/dl (1.3–1.5 mmol/L) • Increased risk: Males, <40 mg/dl (1.0 mmol/L); Females, <50 mg/dl (1.3 mmol/L)	• Does not require fasting • Known as "good" cholesterol
Triglycerides	• Desirable: <150 mg/dl (1.7 mmol/L) • Borderline high: 150–199 mg/dl (1.7–2.2 mmol/L) • High: 200–499 mg/dl (2.3–5.6 mmol/L) • Very high: >500 mg/dl (5.6 mmol/L)	• 12-hour food and water fast is necessary • High triglyceride levels are associated with an increased risk for developing heart disease.

Legend: mg/dl = milligrams per deciliter; mmol/L = millimoles per liter

Adapted from information from Lab Tests Online, http://labtestsonline.org.

Spotlight on…

Glycemic Index (GI) and Glycemic Load (GL)

Glycemic index (GI) is a measure of carbohydrate quality, as explained in Chapter 4. Quality is important but so is quantity. Glycemic load (GL) was developed to take into account both the quality and quantity of the carbohydrate and gives a more complete picture of how blood glucose is elevated and the insulin response that follows. Foods with a high GL cause blood glucose to remain high (hyperglycemia) in the 2 hours after consumption. The prolonged elevation in blood glucose results in a prolonged insulin response (hyperinsulinemia). This condition may make it difficult to keep blood glucose concentration within the normal range, may result in excessive food intake, may cause dysfunction in the insulin-producing cells in the pancreas, and may raise the level of fatty acids in the blood. These responses are risk factors for a number of chronic diseases such as obesity, diabetes, and heart disease.

At present, there is no recommended numeric guideline for glycemic load. Although GL recommendations have not been quantified, there is a practical way for all people, including athletes, to consume a diet with lower glycemic load carbohydrates: frequent consumption of fruits, vegetables, beans, legumes, whole grains, and high fiber and less refined grain products without added sugars.

The use of glycemic index and glycemic load has been controversial in the United States. Opponents of their use argue that the concepts are too complicated for consumers and that not enough research has been conducted, especially on mixed foods and meals. For example, the glycemic load of white rice is considered high, but the glycemic response of stir-fried vegetables and tofu eaten over white rice is not known. The tofu contains both proteins and fats and may affect the absorption of the carbohydrates found in the rice. In contrast, the World Health Organization and countries such as Australia and England support the use of GI and GL, and there is strong scientific evidence for doing so (Livesey et al., 2008).

high levels of cholesterol, LDL, and triglycerides are undesirable.

The role of diet and exercise. Diet is an important element in both the prevention and treatment of heart disease. Balancing caloric intake with energy expenditure and adopting a dietary pattern that includes fruits and vegetables, whole-grain and high-fiber foods, lean protein including fish, nuts, oils, and low-fat or fat-free dairy products help to lower the risk for heart disease.

The 2010 Dietary Guidelines suggest shifting food intake patterns to a more plant-based diet and consuming at least 50 percent of the protein consumed from plant sources, such as dry beans, soy, nuts, and seeds. Such a dietary pattern would be high in nutrients and low in cholesterol, saturated fat, and *trans* fats.

There is strong evidence that diets high in *trans* fatty acids from partially hydrogenated oils increase heart disease risk. *Trans* fats adversely affect other blood lipids, such as increasing LDL-C and/or reducing

Spotlight on a real athlete

Freddie, 48-Year-Old, Former Star High School Athlete, Physically Active Until His Mid-20s, Sedentary For 20 Years

Freddie is 48-years-old and in 6 months will attend his 30th high school reunion. A star athlete in football and wrestling in high school, Freddie did not compete after graduation but tried to remain physically active. In the last 20 years there has been a gradual and consistent decline in both the amount and intensity of physical activity, partly because of the chronic effects of recurring knee injuries. Twenty-five years ago he ran every day to stay in shape, but now he only walks occasionally. Over the years he has lived up to his nickname, "Fast-Food Freddie." As a result he has gained 50 pounds since high school, much of it "in the gut."

Freddie is typical of many Americans—years of physical inactivity leading to poor cardiovascular fitness and a diet excessive in energy (kcal), *trans* fat, sugar, and sodium and low in fiber. At a recent health fair Freddie had his blood pressure taken and discovered that it was elevated. He also filled out a questionnaire for heart disease risk. Since he hadn't been to a physician recently, he was unable to complete the section on blood lipids, but he did note that he has several risk factors—hypertension, excess body fat, physical inactivity, excess alcohol intake, and family history for heart disease. This prompted Freddie to make an appointment with his physician for a physical.

A physical exam confirmed his hypertension, obesity, and excess abdominal fat. A blood test revealed that he had elevated blood cholesterol, LDL-cholesterol, and triglycerides, and a low concentration of HDL-cholesterol. His fasting blood glucose was also elevated. These findings were not a surprise because Freddie knew that he had gained a lot of weight and had become increasingly sedentary. He expected the doctor would tell him that he had heart disease, like his oldest sister. What surprised him was the diagnosis of metabolic syndrome, a disease that Freddie was unfamiliar with.

Freddie realizes that he needs diet and exercise intervention and hopes that it is not too late. His physician

Freddie, former high school athlete.

assures him that it is not and that the initial therapeutic approach will be through lifestyle modification. Freddie needs to increase the amount of daily physical activity and include regular aerobic exercise along with some strength training. He needs to reduce body weight (fat) by increasing exercise and decreasing food intake. He also needs to limit his intake of *trans* fats, cholesterol, and salt (sodium). His physician tries to impress upon him the importance of lifestyle changes, but he is realistic when he indicates to Freddie that many aspects of his life will be affected, including what he does with his leisure time, the amount of TV he watches, and how often he eats out.

Reza Estakhrian/Taxi/Getty Images

These foods are high in saturated fats.

These foods often contain saturated and *trans* fats.

Application exercise

Using the guidelines found in this chapter, develop an exercise and physical activity program for Freddy, the subject in the Spotlight on a real athlete. The program should be laid out over a 1-week period of time and should include *specific* recommendations for activities such as cardiovascular exercises, strength-building exercises, and suggestions for increasing daily physical activity. Be sure to include type of exercise, frequency, duration, and intensity of activities within the guidelines.

HDL-C, and worsen insulin resistance. The Institute of Medicine (2002) recommends that *trans* fat intake be as low as possible. Many manufacturers are removing *trans* fats from foods, and restaurants are cooking with oils that are free of *trans* fats because of the negative health consequences. There are no known health benefits associated with *trans* fats (Mozaffarian, Aro, and Willett, 2009). More information about *trans* fats can be found in Chapter 6.

In some people, excess dietary cholesterol raises blood cholesterol concentration, and reducing dietary cholesterol intake lowers elevated blood cholesterol. However, in other people with a high blood cholesterol concentration, decreasing dietary cholesterol does not result in decreased total blood cholesterol. These individuals, because of their genetic makeup, compensate for lower dietary cholesterol intakes with increased synthesis of cholesterol by the liver. Therefore, some people are sensitive to the amount of cholesterol in their diet and should make appropriate dietary adjustments, whereas others are affected more by the amount of saturated fat rather than the actual amount

of cholesterol in their diets (Gropper, Smith, and Groff, 2009). Foods high in cholesterol and saturated fats are listed in Table 13.5.

A high fat intake by itself is not a risk factor for cardiovascular disease, but a common recommendation is to decrease dietary fat, especially from animal sources. Reducing total fat intake, especially when animal fat intake is high, typically reduces the intake of saturated fats and cholesterol and a lower fat diet would help reduce caloric intake. Similarly, saturated fat, by itself, is not associated with an increased risk of heart disease (Siri-Tarino et al., 2010). However, a diet high in saturated fat may be a part of an unhealthy diet pattern that includes too many calories and too few high-fiber foods, such as fruits, vegetables, and whole grains.

The type of fat consumed is more important than the amount of fat consumed, although the amount is still a consideration because it affects total caloric intake. One type of fat that is "heart healthy" is the omega-3 polyunsaturated fatty acids, docosahexaenoic acid (DHA) and eicosapentaenoic acid (EPA). DHA and EPA are abundant in marine ("oily") fish and are beneficial in many ways including their roles in reducing triglycerides, improving insulin response, and preventing blood clots. The American Heart Association recommends consuming at least two meals of fish per week. Omega-3 fatty acid supplements are not recommended as a way to prevent heart disease, but may be prescribed by physicians as part of the treatment for some patients who have been diagnosed with heart disease (Filion et al., 2010).

Obtaining the proper balance between the types of fats—saturated, polyunsaturated, and monounsaturated—is also a part of diet planning to prevent heart disease. The 2010 Dietary Guidelines suggest limiting saturated fat intake to 10 percent of total

Table 13.5 Foods Containing Cholesterol and Saturated Fat

Food	Cholesterol* (mg)	Saturated fat** (g)
2 scrambled eggs made with milk and butter	429	4.5
Liver (3 oz)	~300–400	1
Squid (3 oz)	220	1.6
1 egg (found in the yolk)	210	1.5
1 slice (75 g) pound cake made with butter	166	9
1 slice pecan pie	160	5
1 cup eggnog	150	11
Shellfish such as shrimp or abalone (3 oz)	~150	1.4
1 extra-crispy fried chicken breast (Kentucky Fried Chicken)	135	8
Double cheeseburger (Burger King)	100	15
Eel (3 oz)	~107	2
Lamb (3 oz)	~80–100	4–8
Beef or pork (3 oz)	~60–80	5.5–8.5
Chicken (3 oz) (not breaded or fried)	~60–80	1.5–2
3 pancakes made with eggs and milk	80	2
Vanilla milk shake, small (Carl's Jr)	50	7
Hot dog or sausage (1 item)	~45	2.5–6
Most fish, baked or steamed (3 oz)	~30–40	<0.25
Most cheese (1 oz)	~25–30	~5
2 slices of a 14-in cheese pizza	32	7
Butter (1 T)	32	6
½ cup cooked egg noodles	26	0.25
1 cup whole milk	24	4.5
1 cup 2% milk	20	3
Lard (1 T)	14	5
Large fries (McDonald's)	0	4.5
Potato chips (1 oz or about 20 chips)	0	3
Luna tropical crisp energy bar or Balance original chocolate bar	0	3.5

Legend: mg = milligram; g = gram; oz = ounce; T = tablespoon

*It is recommended that no more than 300 mg of cholesterol be consumed daily.

**It is recommended that no more than 7–10 percent of total calories be provided by saturated fat. For example, on a 2,000 kcal diet, no more than ~15–22 g of saturated fat are recommended.

Consuming foods that contain monounsaturated, polyunsaturated, and omega-3 fatty acids are encouraged because they may reduce the risk for cardiovascular disease.

calories as an interim step with an ultimate goal of no more than 7 percent of total calories from saturated fat. Replacing high intakes of saturated fats with other types of fat, such as the omega-3 polyunsaturated fats and the monounsaturated fats, is likely to be beneficial. Mediterranean-type diets, such as those rich in fruits, vegetables, legumes, cereals, fish, nuts, olive oil, and moderate amounts of red wine, are low in saturated and *trans* fats and high in the types of fat associated with heart health (Sofi et al., 2008). More information about dietary fat can be found in Chapter 6.

There is not a direct link between sugar intake and heart disease. In other words, excess sugar intake does not cause heart disease. However, there may be an indirect relationship—excess sugar intake leading to weight gain leading to obesity and insulin resistance. There are enough pieces of evidence to suggest that this is a pattern in some people, but not enough evidence that this is a pattern for all people.

One of the problems of studying these indirect relationships and making recommendations is that sugar is only one of the ways people consume calories. Therefore, recommendations about sugar intake must also consider total energy intake. The American Heart Association recommends that women consume no more than 100 calories per day and men consume no more than 150 calories per day from added sugars (Johnson et al., 2009). To put this recommendation in context, a 12 oz regular (not diet) soft drink contains ~150 calories, all of which come from added sugar. To meet this recommendation, most people would need to cut back

Table 13.6 Caloric Contribution of Added Sugars in Common Foods

Food	Added sugar (g)	Energy from added sugar (kcal)	Total energy (kcal)
Regular soft drink (12 oz)	39	~150	~150
2 Oreo® cookies	9	36	106
1 Pop-Tart®	18	72	200
Ice cream (½ cup)	~15	60	140
Sweetened yogurt (6 oz)	27	108	170

Legend: g = gram; kcal = kilocalorie; oz = ounce

on their sugar intake. Table 13.6 lists the caloric contribution of added sugars in some common foods.

The link between sugar intake, especially fructose intake, and certain fats in the blood is one of the stronger associations currently being studied. Recall that once fructose is absorbed that it is transferred immediately to the liver. Excess amounts are converted to other compounds that can be used to synthesize triglycerides, a common fat made in the liver and found in the blood. Some studies have shown that a very high intake of fructose (~25 percent of total calories daily) may increase triglycerides in the blood (Hofmann and

Tschöp, 2009). When these fats are elevated in the blood, the risk for heart disease is increased, especially in those with a genetic predisposition for heart disease. People who are consuming one-quarter of their diets as fructose likely have other dietary problems, such as low vitamin, mineral, and fiber intakes, and reducing fructose intake is just one of many dietary modifications that likely need to be made.

The intake of alcohol may be either beneficial or detrimental, depending on the amount consumed. A moderate daily alcohol intake is associated with reduced risk for cardiovascular disease. Moderate intake is defined as one drink per day for women and up to two drinks daily for men. One drink contains ½ ounce (~15 ml) of ethanol, the amount typically found in 12 ounces (360 ml) of beer, 3 to 4 ounces (90 to 120 ml) of wine, or 1½ ounces (45 ml) of hard liquor (for example, a "shot" of whiskey). The intake of more than three drinks daily (that is, more than 1½ ounces of ethanol) is associated with increased blood pressure, cardiomyopathy (enlarged and weakened heart), and increased triglycerides.

Exercise and physical activity play an important role in the management of blood lipids. Exercise can exert an independent effect, but this effect is small, approximately 10 percent. In other words,

Spotlight on...

Heart Disease (Atherosclerosis)

Alternate term: Atherosclerotic heart disease, coronary artery disease, coronary heart disease. Heart disease refers to diseases of the heart and its vessels. It is sometimes used interchangeably with the term *cardiovascular disease*.

Definition: Atherosclerosis refers to the hardening and narrowing of the arteries.

Prevalence: Approximately 17 million adults in the United States, resulting in nearly 450,000 deaths from coronary heart disease in 2005.

Symptoms: Typically no symptoms until an artery is substantially narrowed or blocked, which can lead to chest pain, heart attack, arrhythmias, or stroke.

Cause: Buildup of plaque at the site of damage or injury to an artery.

Prevention: (1) Maintain or achieve a healthy weight. (2) Consumption of a diet that emphasizes fruits, vegetables, whole grains, low- or nonfat dairy foods, lean meats, fish, and poultry, nuts, beans, and oils. (3) Maintain or progress to a minimum of 150 minutes of moderate-intensity exercise each week.

Treatment: Treatment must be individualized.

Diet: (1) Hypocaloric diet, if weight loss is needed. (2) Modifications typically include a reduction in total fat, saturated fat, *trans* fat, and cholesterol, but an individual treatment plan is needed. (3) A low-sodium diet may be recommended if hypertension is present.

Exercise: (1) Moderate-intensity physical activity or cardiovascular exercise. (2) ≥200–300 minutes per week or ≥2,000 kcal per week of energy expenditure. (3) Strength training as a supplement to cardiovascular exercise two to three times per week focusing on increased volume rather than increased intensity (that is, increased sets and repetitions rather than increased resistance or weight).

Heron et al., 2009; Krauss et al., 2000; National Heart, Lung, and Blood Institute, http://www.nhlbi.nih.gov/; American Heart Association, http://www.americanheart.org/HEARTORG; Centers for Disease Control and Prevention, http://www.cdc.gov.

an individual with an elevated total cholesterol concentration of 250 mg/dl might expect the effect of exercise alone to reduce cholesterol by approximately 25 mg/dl to 225 mg/dl. Similarly, a person with an HDL-C of 40 mg/dl may see a modest increase to approximately 44 mg/dl. Exercise and physical activity are integral to maintaining normal blood lipid concentrations, but exercise alone may not result in the magnitude of change necessary to meet blood lipid goals when blood lipids concentrations are abnormal. The most beneficial effect of exercise for blood lipid management appears to be its role in weight loss and maintenance. There is an added beneficial effect on elevated blood lipid concentrations when exercise results in weight loss.

Cardiovascular disease is multifactorial, and each factor can have more or less of an influence on a given individual because of that person's genetic profile. It is easy to focus on the extremes: the inactive male who eats poorly and has a single, fatal heart attack at age 40, and the physically inactive smoker with no dietary restraint who celebrates his 80th birthday. But the majority of people fall in between these extremes. The average American is at risk for dying from or being disabled by cardiovascular disease because the diet consumed promotes heart disease in many different ways. This poor diet is generally accompanied by inactivity. Together, diet and exercise influence many of the known risk factors.

Metabolic syndrome is a cluster of metabolic disorders strongly associated with abdominal obesity and insulin resistance.

Metabolic syndrome (also known as insulin resistance syndrome and syndrome X) is characterized by a clustering of metabolic disorders and risk factors: abdominal obesity, hypertension, **dyslipidemia** including elevated triglycerides and low HDL-C, glucose intolerance, and insulin resistance (Table 13.7). This disease is also characterized by proinflammatory and prothrombotic states, which are related to the development of atherosclerosis and blood clots. Those with metabolic syndrome have a significantly greater risk of developing cardiovascular disease and type 2 diabetes.

The major factors underlying the development of the metabolic syndrome are obesity and insulin resistance, two conditions previously discussed (Grundy, 2006). Of particular importance is abdominal or upper body obesity. Upper body obesity consists of both visceral and subcutaneous abdominal fat stores. Visceral fat, which refers to fat stored deep within the abdominal cavity surrounding the organs,

Table 13.7 Diagnostic Criteria for Metabolic Syndrome

Measure	Criteria*
Waist circumference	≥102 cm (≥40 in) in men ≥88 cm (≥35 in) in women
Triglycerides	≥150 mg/dl (1.7 mmol/L) or drug treatment for elevated triglycerides
High-density lipoprotein cholesterol (HDL-C)	<40 mg/dl (0.9 mmol/L) in men <50 mg/dl (1.1 mmol/L) in women or drug treatment for reduced HDL-C
Blood pressure	≥130 mm Hg systolic blood pressure or ≥85 mm Hg diastolic blood pressure or drug treatment for hypertension
Fasting glucose	≥100 mg/dl or drug treatment for elevated glucose

Legend: cm = centimeter; in = inch; mg/dl = milligrams per deciliter; mmol/L = millimoles per liter; mm Hg = millimeter mercury

*Any three of the five criteria constitute a diagnosis of metabolic syndrome.

Source: Grundy, S. M., et al. (2005). Diagnosis and management of the metabolic syndrome. An American Heart Association/National Heart, Lung, and Blood Institute scientific statement. *Circulation*, *112*(17), 2735–2752.

is characterized by mesenteric and omental fat cells. These adipocytes have more beta-receptors than other types of fat cells, so the lipid stored in visceral fat cells is easily mobilized from the cells into the blood, which flows directly into the liver via the portal circulation. This blood contains a high concentration of free fatty acids, which inhibit the breakdown of insulin and increase the liver's synthesis of glucose and triglycerides.

Subcutaneous fat is less metabolically active than visceral fat. In the abdominal area, subcutaneous fat is found just below the skin and does not surround the organs. Subcutaneous fat is also found below the waist (for example, hips, thighs). In adult males, 20 percent of total storage fat is visceral fat; in adult females, visceral fat is approximately 6 percent of total fat. Visceral fat increases in both males and females as they age. Genetics are a powerful influence on body fat distribution, and as much as half of the influence on visceral fat distribution may be due to a single gene. It is visceral fat that is most related to an increased risk for chronic diseases.

Physical fitness and exercise have a direct and independent effect on metabolic syndrome risk. A large epidemiological study (Katzmarzyk et al., 2005) confirmed that the risk for metabolic syndrome, as well as all-cause and cardiovascular disease mortality, increased as BMI category increased from normal to overweight to obese. However, when cardiovascular fitness of the subjects was considered, the risk of premature mortality decreased in those who were fit regardless of body weight category or the presence of metabolic syndrome. The authors concluded that the amount of physical activity necessary to maintain the observed level of physical fitness provided some health protection and reduced the risk normally associated with obesity and the metabolic syndrome. In other words, being physically active is important regardless of weight.

Other studies report similar findings. Ekelund et al. (2005) found a strong **association** with level of energy expenditure through physical activity and risk of developing metabolic syndrome, independent of the fitness level. Finley et al. (2006) found that cardiorespiratory fitness was inversely associated with metabolic syndrome regardless of dietary intake. These studies illustrate the importance of regular physical activity and exercise in the prevention of metabolic syndrome. Note that these and other studies show that risk can be reduced through regular physical activity, even if the activity does not result in reducing or eliminating the obesity. People should be encouraged to become and remain physically active, even if tangible results of their activity such as weight or fat loss are not readily apparent. A healthy diet is also important, but special emphasis should be put on becoming physically active.

The primary therapeutic approach for those with metabolic syndrome is lifestyle modification. The specific components of this lifestyle modification are: (1) increased physical activity, (2) weight reduction, (3) consumption of a diet that reduces risk for heart disease, and (4) smoking cessation. As discussed previously, the interaction of diet modification and physical activity/exercise is an important approach to achieve and maintain a healthy weight. Weight loss, appropriate dietary modifications, and exercise also work independently and interactively on other risk factors such as hypertension and dyslipidemia. In addition to lifestyle changes, those with metabolic syndrome may need pharmacological intervention such as antihypertensive, lipid-lowering, and antihyperglycemic medications.

Metabolic syndrome: A disease characterized by a clustering of metabolic disorders and risk factors: obesity, hypertension, dyslipidemia, glucose intolerance, and insulin resistance.

Dyslipidemia: Abnormal concentration of blood lipids such as cholesterol, lipoproteins, and triglycerides.

Association: When referring to statistics or scientific studies, the existence of a relationship between two variables. Does not mean that it is a cause-and-effect relationship.

Spotlight on...

Metabolic Syndrome

Alternate term: Insulin resistance syndrome, prediabetes, syndrome X.

Definition: A clustering of risk factors that often accompany obesity and are associated with increased risk for atherosclerotic heart disease and type 2 diabetes.

Prevalence: Nearly 25 percent of U.S. adult population; age-dependent—incidence rises to >40 percent in those >60 years; prevalence higher in some ethnic groups such as African Americans and Mexican Americans.

Symptoms: Abdominal obesity, hypertension, dysplipidemia, impaired fasting glucose, sleep apnea.

Cause: Obesity, insulin resistance.

Prevention: (1) Maintain or achieve a healthy weight. (2) Physical activity (meet at least minimal recommendations for activity and then progress). (3) Consume a healthy diet such as that outlined in the Dietary Guidelines.

Treatment: Weight loss, physical activity, dietary modifications, medication. Treatment must be individualized.

Diet: (1) Hypocaloric diet. (2) Dietary modifications typically include a reduction in total fats, saturated fats, *trans* fats, and cholesterol to improve abnormal lipid concentrations, but an individual treatment plan is needed. (3) Consumption of the DASH (Dietary Approaches to Stop Hypertension) or similar type of diet. This diet emphasizes fruits, vegetables, whole grains, low- or nonfat dairy foods, lean meats, fish, and poultry, nuts, beans, and oils. Potassium is plentiful and sodium is limited.

Exercise: (1) Accumulate at least 30 minutes of moderate-intensity physical activity on most days of the week. (2) Progress to 60 to 90 minutes of moderately intense exercise on most days to lose weight and maintain weight loss.

National Heart, Lung, and Blood Institute, http://www.nhlbi.nih.gov/; American Heart Association, http://www.americanheart.org/HEARTORG.

Osteoporosis is characterized by low mineral density.

The National Osteoporosis Foundation (2005) estimates that approximately 44 million women and men or 55 percent of Americans over the age of 50 have low bone mineral density (BMD) and are at risk for developing osteoporosis. Osteoporotic bones are fragile and more prone to fractures, particularly in the spine (vertebrae), wrist, and hip. A hip fracture in an elderly adult is typically a life-changing event that may result in the inability to live independently, a living situation that older people highly value. The key is prevention, and preventative exercise and diet strategies should be instituted early in life.

Physical activity and exercise have a positive impact on bone health whereas inactivity has a negative effect on bone mineral density. Bones are often thought of as inert structures, but bones are active tissues with minerals being constantly added and removed. Because there is an apparently inevitable decline in bone mineral density with aging, there are two important strategies for developing and maintaining bone health throughout life. The first is to develop an optimal bone density by the third decade of life, the point at which a person achieves maximal bone density. The second strategy is to slow the loss of bone so that low bone density, which is associated with increased risk of fracture, is delayed as long as possible.

Physical activity and exercise, particularly weight-bearing activities, such as running, jumping, and gymnastics, are associated with increased bone mineral density in children, adolescents, and young adults. The impact stress that is inherent in these activities stimulates a greater deposition of mineral in the bone, making bones stronger and more resistant to fracture. It is not surprising that activity of greater intensity that involves greater impact stress results in greater bone mineral adaptation. Moderate-intensity strength training in children, adolescents, and young adults has a positive impact on bone mineral density; however, these groups need proper guidance so appropriate strength-training guidelines are followed. The positive bone changes made in childhood and adolescence persist into adulthood, when it is important to maximize bone density. A detailed description of the bone remodeling process is found in Chapter 9.

For middle-aged and older adults, the primary strategy is to maintain bone mineral density by reducing the age-related decline in BMD as much as possible. Regular exercise and physical activity play a major role in this effort (Kohrt et al., 2004). It is not clear if older adults can increase their bone mineral density through exercise, but it is apparent they can slow the rate of decline. Again, weight-bearing exercise and strength training are the most effective activities for maintaining bone health in adults. Although higher-intensity activities that impose greater impact stress generally result in more beneficial effects, older adults

Spotlight on a real athlete

Lena, 67-Year-Old, Formerly Lightly Active, Now Has Physical Limitations

Lena is a 67-year-old widow who lives independently in the home she and her husband built 40 years ago. As a teenager and young adult she was not expected to exercise, but she was lightly active for many years because she liked to garden. She has more time available for physical activity now that she is retired, but she must do so within her physical limitations. In the last year she has begun to experience some physical decline, such as decreased strength, and difficulties in balance when walking and rising from a sitting position. Although her body weight is not changing, she is losing lean body mass and gaining some body fat. She loves playing with her grandchildren but finds that she gets out of breath easily. She has recently been diagnosed with osteoporosis. Her biggest fear, and her biggest motivation for making lifestyle changes, is that she will fall and break a hip, be placed in a long-term care facility, and never be able to return to her home. Lena's challenge is to find ways to be physically active without leaving her home or yard.

Lena, formerly active, now has physical limitations.

RubberBall Productions/the Agency Collection (RF)/Jupiter Images

Physical activity and exercise are associated with increased bone mineral density in children, adolescents, and young adults.

may not be motivated to or feel safe participating in high-stress activities that require jumping or running. They may also be limited in these activities by previous injuries or the presence of other exercise-intensity-limiting factors such as atherosclerotic heart disease. These groups also need guidance regarding proper strength-training techniques.

In addition to exercise, diet plays an important role in attaining maximum bone density. Although many nutrients are associated with bone health, calcium and vitamin D receive the greatest attention because of their fundamental roles. As discussed in Chapter 9, adequate calcium intake is critical during childhood, adolescence, and young adulthood because the majority of the bone's mineral content is achieved before age 20 and peak bone mass is reached by approximately age 35.

Nutrient intake after age 35 is still important but, similar to exercise, the focus turns first toward preventing and then slowing the loss of bone mineral. One of the best strategies is to consume the recommended amounts of calcium and vitamin D daily; unfortunately, deficiencies of these nutrients are prevalent in U.S. adults. Maintenance of bone mineral becomes increasingly difficult with the onset of menopause because of the loss of estrogen's positive effects on bone mineralization and turnover.

Calcium supplementation is frequently recommended for middle-aged and older females but its effects are limited, especially in the years immediately following the onset of menopause. The results of prospective, double-blind placebo-controlled studies show calcium supplements are not as powerful as originally hoped for slowing the loss of bone mineral and reducing risk for fracture. It is naïve to believe that the impact of low calcium intake over decades can be reversed after the onset of menopause with the consumption of calcium supplements. There is some evidence that calcium and vitamin D, when supplemented together, reduce the risk for fracture in the elderly (DIPART, 2010). Because supplementation appears to be limited in its effects, the need for prevention of low bone mineral density cannot be overstated.

Spotlight on...

Osteoporosis

Alternate term: Bone-thinning disease.

Definition: Bone mineral density more than 2.5 standard deviations below the young adult mean value, with or without fractures.

Prevalence: 10 million adults (90 percent women); ~44 million adults have low bone mass. Figures for both osteoporosis and low bone mass are expected to rise each year as the U.S. population ages.

Symptoms: Bone fracture, particularly vertebral and hip.

Cause: Loss of bone mineral exceeds replacement.

Prevention: (1) Weight-bearing endurance exercise (3–5 days/week) and resistance exercise training (2–3 days/week). (2) moderate to high intensity (in terms of bone stressing forces). (3) 30–60 minutes/day. (4) Adequate calcium and vitamin D intake.

Treatment: In addition to diet and exercise, medications may be prescribed.

Diet: (1) Maintain or increase calcium and vitamin D intake to recommended amounts. (2) Supplemental calcium and vitamin D may be prescribed.

Exercise: (1) Exercise duration of 30 to 60 minutes on a near daily basis. (2) Weight-bearing endurance (aerobic) exercise three to five times per week. (3) Strength-training exercise two to three times per week.

National Osteoporosis Foundation; hppt://www.nof.org.

Assessing bone mineral density. Epidemiological studies have established that low bone density increases fracture risk and that bone density is the best predictor of fracture risk in postmenopausal women (Lufkin, Wong, and Deal, 2001). Since 1994, a diagnostic criteria has been available to assess bone density in Caucasian women. African Americans and men have approximately 20 percent greater bone density than white women (Zizic, 2004). Bone mineral density is typically measured used dual-energy X-ray absorptiometry (DEXA or DXA). For clinical purposes, the results of a bone mineral density test are reported as t-scores as follows:

- Normal bone density is a t-score of –1.0 or above.
- Low bone density (osteopenia) is a t-score between –1.0 and –2.5.
- Osteoporosis is a t-score of –2.5 or below.

Research studies often report z-scores, which are slightly different.

For those who have been diagnosed with osteoporosis, doctors often recommended that they focus initially on performing low-intensity, weight-bearing exercise such as walking. Walking is beneficial for cardiovascular fitness and is an exercise that older adults can perform comfortably. Incorporating strength-training exercises two to three times per week could also be appropriate, although supervision is needed to assist with balance and prevent falls. Strength training provides significant stress on bones that is beneficial, and has the added benefit of increasing muscular strength, which aids in balance, locomotion, and the performance of daily tasks. Although these activities would be beneficial, it is unknown if exercise can reduce the decline in bone mineral density that occurs in women as a result of menopause. Similarly, it may be prudent to consume calcium and vitamin D supplements, but the effect may be limited.

Many cancers are related to lifestyle.

One-third of all cancer deaths in the United States are associated with poor diet, inactivity, and overweight or obesity. This is approximately the same proportion of deaths associated with tobacco use. Lifestyle plays an important role in reducing the risk for cancer, and the American Cancer Society (ACS) has published diet and activity guidelines (Kushi et al., 2006). The four major recommendations are: (1) maintain a healthy weight throughout life, (2) adopt a physically active lifestyle, (3) consume a healthy diet, with an emphasis on plant sources, and (4) if you drink alcoholic beverages, limit consumption.

These recommendations were adopted by the ACS based on the body of scientific literature for cancer prevention; however, these recommendations are similar to those made for other chronic diseases. Specific to cancer, overweight and obesity are associated with approximately 15–20 percent of all cancer deaths. Physical activity affects cancer risk, in part, because it helps people maintain a healthy weight, but being physically active also has other positive

Spotlight on chronic diseases

Lifestyle-Related Cancers

Definition: A class of diseases characterized by uncontrolled cell division.

Prevalence: In the United States ~14 million adults (6.6 percent of the adult population) have now or previously been diagnosed with cancer.

Symptoms: Symptoms vary depending on the site of the cancer.

Cause: There is no single cause for all types of cancer. Factors include genetics, environmental toxins, sun exposure, tobacco use, and lifestyle habits such as poor diet and physical inactivity.

Prevention: (1) Attain or maintain a healthy weight. (2) Adults should engage in at least 30 minutes of moderate-to-vigorous physical activity on 5 or more days of the week. Forty-five to 60 minutes of physical activity are preferable. (3) Children and adolescents should engage in at least 60 minutes per day of moderate-to-vigorous physical activity at least 5 days per week. (4) Consume a healthy diet, with an emphasis on plant foods. (5) Limit alcoholic beverage consumption.

Treatment: Treatment depends on the cancer site.

American Cancer Society Guidelines on Nutrition and Physical Activity for Cancer Prevention (2006) and the Centers for Disease Control and Prevention, http://www.cdc.gov/nchs/fastats/cancer.htm.

effects on cancer risk reduction. Studies show that plant foods, such as fruits, vegetables, and whole grains, help reduce cancer risk. The emphasis is on plant foods because of the nutrients they contain, but there are also data to suggest that red meat (for example, beef, pork, and lamb) and processed meat, such as cold cuts, hot dogs, and bacon, are associated with an increase in some cancers. A healthy diet can also help individuals to maintain a healthy weight. Excessive alcohol intake is associated with cancers of the mouth and liver, and the same recommendations regarding amount are made for preventing heart disease, as discussed earlier.

The American Cancer Society recommendations underscore a point made earlier in the chapter: diet and exercise recommendations are remarkably consistent. There is a large body of literature that supports such recommendations, but the majority of people do not follow them, and the prevalence of chronic diseases continues to increase in most cases. The challenge is to help children and adolescents establish healthy diet and exercise habits and to help adults change unhealthy lifestyle behaviors.

The Internet café

Where do I find reliable information about chronic diseases?

The Centers for Disease Control and Health Prevention (CDC) devotes a section of its extensive website to chronic disease prevention (**http://www.cdc.gov/chronicdisease/overview/index.htm**). Information on overweight and obesity can be found at **http://www.cdc.gov/obesity/index.htm**. Other government and nonprofit organizations with prevention and treatment information about specific chronic diseases are listed below.

American Diabetes Association (**http://www.diabetes.org**)

American Cancer Society (**http://www.cancer.org**)

American Heart Association (**http://www.americanheart.org/HEARTORG**)

National Heart, Lung, and Blood Institute (**http://www.nhlbi.nih.gov/**)

National Osteoporosis Foundation (**http://www.nof.org**)

Chronic disease risk can be assessed with a number of screening tools.

Body Mass Index is the most widely use screening tool for evaluating the risk of disease based on weight. As discussed in Chapter 11, BMI is inappropriate to

Underweight	Less than 18.5
Healthy weight	18.5–24.9
Overweight	25–29.9
Obese	Greater than 30 (Severe obesity may be defined as greater than 40)

Figure 13.6 Body Mass Index (BMI) criteria

The unit of measure for BMI is kg/m^2, but it is most often reported without a unit of measure.

use with athletes, pregnant women, and adults over the age of 65. However, for the majority of young and middle-aged adults, BMI is a useful screening tool, particularly for Caucasians. When BMI is used alone, it is not as predictive as when used with waist circumference, a measure of abdominal body fat. These are widely available and easily performed screening methods and can be used at health fairs and gyms as well as in medical settings.

BMI is determined by using height and weight, which may be measured or self-reported. Using a nomograph (see Appendix N) or an online BMI calculator (http://www.cdc.gov/nccdphp/dnpa/bmi/), BMI can be quickly derived. Criteria for underweight, healthy weight, overweight, obese, and severely obese are shown in Figure 13.6.

Many studies have reported an association between body weight, chronic disease risk, and mortality (Hu et al., 2004; Lee, Blair, and Jackson, 1999). This association is very much influenced by cardiovascular fitness. Recall from Chapter 1 that association suggests that a relationship exists, but does not mean there is a cause-and-effect relationship. In fact, no BMI category, even the healthy weight category, is free of disease.

Tables 13.8 and 13.9 show the prevalence of disease in men and women based on BMI. Notice that 27 percent of men and women in the healthy weight category (that is, BMI 18.5–24.9) have high blood cholesterol and 23 percent have high blood pressure. Increasing weight has a greater impact on the prevalence of hypertension than any of the other chronic diseases. Women with a BMI >40 (6 percent of adult females) have a greater risk for chronic diseases than any other BMI category, so individuals should be concerned when a very large amount of weight is gained. However, some severely obese individuals do not manifest any of the "typical" obesity-related diseases. Must et al. (1999) report that 30 percent of severely obese Mexican American women have blood pressure, blood glucose, and blood lipids within normal ranges. Just

Table 13.8 Prevalence of Disease in Men Based on BMI

	<18.5	18.5–24.9	25–29.9	30–34.9	35–39.9	>40
Type 2 diabetes	5%	2%	5%	10%	12%	11%
Gallbladder disease	7%	2%	3%	5%	6%	10%
Heart disease	12%	9%	10%	16%	10%	14%
High blood cholesterol	7%	27%	36%	39%	34%	36%
High blood pressure	23%	23%	34%	49%	65%	65%
Osteoarthritis	<1%	3%	5%	5%	5%	10%

The unit of measure for BMI is kg/m^2, but it is most often reported without a unit of measure.

Source: Must, A., et al. (1999). The disease burden associated with overweight and obesity. *Journal of the American Medical Association, 282*(16), 1523–1529.

Table 13.9 Prevalence of Disease in Women Based on BMI

	<18.5	18.5–24.9	25–29.9	30–34.9	35–39.9	>40
Type 2 diabetes	5%	2%	7%	7%	13%	20%
Gallbladder disease	6%	6%	12%	16%	19%	23%
Heart disease	12%	7%	11%	13%	12%	19%
High blood cholesterol	13%	27%	46%	40%	41%	36%
High blood pressure	20%	23%	39%	48%	55%	63%
Osteoarthritis	8%	5%	9%	10%	10%	17%

The unit of measure for BMI is kg/m^2, but it is most often reported without a unit of measure.

Source: Must, A., et al. (1999). The disease burden associated with overweight and obesity. *Journal of the American Medical Association, 282*(16), 1523–1529.

as the prevalence of disease in the healthy weight category is not zero, neither is the prevalence of disease in the severely obese 100 percent.

Distribution of body fat may be one reason that obese and severely obese people do not exhibit some of the common obesity-related chronic diseases. Abdominal obesity, also referred to as upper body (android) or central obesity, confers a different risk from lower body (gynoid) obesity. Hypertension, type 2 diabetes, insulin resistance, and dyslipidemias are related to the amount of visceral fat (that is, abdominal obesity) as previously discussed.

Screening for fat distribution can range from the most precise methods—CT (computer tomography) and MRI (magnetic resonance imaging) scans—to the more widely used method of measuring waist circumference to the easy but imprecise visual classification based on body shape. CT and MRI measurements are used in research settings. For medical and health-screening purposes, waist circumference measurements may be made, but this method should be used only in people with a BMI <35. A waist circumference >40 inches (102 cm) in men and >35 inches (88 cm) in women is associated with greater visceral fat storage and greater disease risk. However, waist circumference measures both visceral and subcutaneous abdominal fat, which makes this screening method

less precise than scans. Measuring the height of the abdomen when the person is lying down, a method known as **sagittal** trunk diameter, is also a reasonably good method of estimating visceral fat, but its use in medical or health club settings is limited.

Visual screening of body shape, which can be self-administered, is an imprecise measure that separates those with central obesity ("apple-shaped") from those with gynoid obesity ("pear-shaped"). Adult men are typically apple shaped whereas adult women are typically pear shaped. After menopause, women often change shape because of the accumulation of visceral fat, which does increase their risk for some chronic diseases.

The use of BMI and waist circumference together is a reasonably good screening tool that can be administered easily. Those identified as being at risk because of their weight and fat distribution should be referred to a physician for further evaluation.

Physical activity and fitness may reduce the adverse impact of overfatness on health.

The rapidly rising prevalence of overweight and obesity, as well as the increase in diseases such as type 2 diabetes and metabolic syndrome, have resulted in intense media coverage of body weight. A statistic that appeared for many years in medical journals and in the lay press was that 300,000 deaths in the United States each year could be attributed to obesity. This statistic was derived from a study published by Allison et al. in 1999 using data from 1991. A 2005 study using data from the year 2000 suggests that the number of obesity-related deaths is much lower than estimated in the past—approximately 112,000 per year (Flegal et al., 2005).

Glen Gaesser, PhD, an exercise physiologist, disputed the 300,000 figure in his book *Big Fat Lies* (2002). He suggested that the Allison et al. study was flawed because it controlled for age, gender, and smoking, but did not control for physical activity, level of fitness, dietary intake, history of weight cycling, the present or past use of harmful weight-loss methods, or access to health care. Gaesser emphasized that he agreed that a sedentary lifestyle, low aerobic fitness, and an unhealthy diet contribute to premature mortality. His complaint was that scientists call this condition obesity. With such a high prevalence of overweight and obesity in the United States and such a low prevalence of permanent weight loss, many people wonder if they can be fit if they are fat.

It is clear that there is an interaction between physical activity, physical fitness, obesity, and health. The presence of obesity has multiple negative effects on long-term health, but being physically active may attenuate some of these effects. From a health perspective, maintaining a healthy weight is clearly preferable, but if one is overweight or obese, health benefits can be obtained by being physically active, even before weight loss occurs. This should not be interpreted to mean that the effect of obesity is neutralized as long as physical activity is maintained, but it illustrates the importance of physical activity in reducing health risk relatively quickly. In other words, people can be positively reinforced by the knowledge that they are experiencing the health benefits of physical activity long before weight loss might occur.

The Health at Every Size movement emphasizes improved metabolic health over weight and fat loss.

Losing body fat is not easy, and maintaining a lower body weight if once overweight or obese is difficult to sustain. Some health professionals have begun to question the benefits of traditional weight-loss diets, noting that the prevalence of obesity is increasing despite overwhelming emphasis on losing excess body fat by the media and health professionals. Concern has also been raised about the psychological impact of stressing weight loss in an environment that promotes overeating and inactivity. Many of those who do successfully lose weight appear to maintain it by continuing to restrict kilocalories (for example, the <1,400 kcal daily reported by the National Weight Control

Registry), but the psychological effects of long-term energy restriction is a concern. Some of those who develop disordered eating and eating disorders have a history of dietary restriction and have followed numerous weight-loss plans. For all these reasons the nondiet approach or the Health at Every Size (HAES) movement was born (Miller, 2005).

The term *nondiet* was coined in response to the frequent use of the word *diet* to mean a weight-loss diet. The nondiet or Health at Every Size approach emphasizes a healthy diet, normalization of eating habits (including therapy for deep-seated emotional issues that affect food intake), and moderate physical activity. The focus is on a positive body image regardless of size and supports the concept that people can be both fat and fit. Weight loss may be a *consequence* of changing dietary intake to reflect healthy eating recommendations and increasing physical activity, but loss of weight is not an *expectation*. Rather, success is defined as improved metabolic fitness (for example, lowered blood pressure or improved blood lipids), eating behaviors (for example, less restraint or disordered eating), and well-being (for example, less depression, increased self-esteem). Research in this area is in its infancy, but there is some evidence that health is improved with this approach independent of any loss of weight (Miller, 2005; Bacon et al., 2002).

Although there is still disagreement about the extent of the direct effect of obesity on mortality, nearly everyone agrees that a sedentary lifestyle, low aerobic fitness, and poor diet contribute to both premature death and disease. This chapter began with an emphasis on being physically active on a daily or near daily basis because "everyone is an athlete." Exercise and a healthy diet were repeated themes in the discussions of prevention and treatment of various chronic diseases. This chapter ends with information about making behavior change.

Behavior change is needed to prevent and treat lifestyle-related chronic diseases.

By all accounts behavior change is needed in both developed and developing countries to prevent and treat lifestyle-related chronic diseases. The increasing incidence and prevalence of obesity, especially in children and adolescents, type 2 diabetes, and metabolic syndrome are strong indicators that substantial changes are needed in diet and activity patterns. The critical issue is *how* people successfully make behavioral change.

Sagittal: Related to the median (middle) plane of the body or a body part.

Numerous models have been proposed to help individuals, and health and fitness professionals understand the process of change. Just as no one weight-loss diet is universally effective, no one model for behavioral change is universally effective. All the models consider the four major questions that surround behavior change: (1) What is the motivation to change? (2) What resources are needed? (3) What is the decision-making process? and (4) How are decisions translated into repeated behaviors? (Baranowski et al., 2003).

Discussion of behavior change typically begins with knowledge. Knowledge is prerequisite to change, and there is much for people to learn about diet and activity. For example, an individual must know what to eat (for example, foods, nutrients) and how much to eat (for example, estimated daily caloric intake). Similarly, knowledge about physical activity is needed, such as appropriate intensity and duration of exercise and the prevention of injury.

The Knowledge-Attitude-Behavior (KAB) Model suggests that knowledge is the driving force behind behavior change. The accumulation of knowledge changes attitudes, which then result in a change in behavior. A rational decision-making process is employed. Although there may be a small subset of people for whom knowledge alone is the resource and a change in attitude is the motivator, most people do not successfully make behavior change based on knowledge alone. Thus diet or exercise programs that offer only subject matter information typically have little success.

The Health Belief Model (HBM) suggests that a change in behavior is based on the perceived susceptibility and severity of illness, benefits to health, physical response, action cues, and **self-efficacy** (that is,

confidence that one's thoughts and behaviors can be changed). The primary motivation is the risk of contracting a disease, especially a serious disease. This motivation is referred to as the readiness to act. Cues, such as having a family member diagnosed with a disease, abnormal laboratory tests, or feeling fatigued, are readiness-to-act motivators. An important element is the individual's belief and confidence that behavior change can be successful.

Unfortunately, the Health Belief Model has had limited success. Adolescents and young adults have a sense of immortality, so they do not see themselves susceptible to chronic diseases. Even cues that suggest susceptibility, such as illness in a parent or media information about increasing rates of obesity, are not internalized, and they choose to believe that these conditions "won't happen to me." Many chronic diseases have no symptoms (for example, hypertension, osteoporosis), so there are no negative physical responses until the disease has manifested itself. Even adults who are motivated to change because of a diagnosed medical condition (for example, obesity, diabetes, metabolic syndrome) may not have confidence that they can change their behavior. If they believe they will fail, they may not be motivated to try (Baranowski et al., 2003).

Social cognitive theory (SCT) is frequently used in exercise and nutrition education programs. Three important issues are the individual's ability to perform the behavior (skill), the confidence to perform the behavior, and expected outcomes. Important environmental issues include modeling of behavior and availability. For example, to lift weights the equipment must be available and the person must have the skill and confidence to perform this activity. Having an exercise specialist demonstrate and teach proper

Keeping it in perspective

Everyone Is an Athlete

Humans need to be physically active for good health. Modern society, however, does not require that people be active and it often encourages inactivity at work and at leisure. Industrialized societies also have an abundant food supply, and overconsumption is not only possible but often encouraged. Exercise is a powerful influence on the amount of kilocalories needed daily to maintain body weight, so it is no surprise that sedentary people living in a society in which so much calorie-dense food is available would have a high prevalence of overweight and obesity. Such is the case and the culture of the United States and other developed countries.

Engaging in a healthy lifestyle, which includes near daily physical activity and a healthy diet that matches caloric intake to caloric expenditure, is "counterculture." During early adulthood, lack of exercise and poor dietary intake may not appear to have substantial consequences since most chronic diseases are not symptomatic during this time period. But chronic diseases may be developing and future health may be at risk. An active lifestyle and healthy diet are among the best investments that can be made in the human body, and a long-term perspective on health should never be lost even by an elite athlete at the peak of his or her career.

weight-lifting technique and reinforce the participant's behavior may increase skill and confidence and the likelihood of maintaining the activity as part of one's lifestyle.

The primary motivator is the expected outcome. In the example above, the expected outcome may be an increase in muscle size or strength. One potential problem is that unrealistic goals may be set. For example, it would be unrealistic to expect that a weight-lifting program alone would result in large losses of body fat in an obese individual. Another potential problem is the confidence to try a new behavior because substantial skill may be needed to make an exercise or dietary change. Barriers may also exist, for example, lack of exercise facilities, unsafe neighborhoods for walking, or unavailability of fresh fruits and vegetables. Knowledge, problem-solving ability, and decision-making skills are all part of this complicated process of behavior change. There are reports in the scientific literature of successful diet and exercise programs that have used SCT (Baranowski et al., 2003).

A widely used model for changing health-related behaviors is the Transtheoretical Model, frequently referred to as the Stages of Change Model (Prochaska and DiClemente, 1984). The stages of change include precomtemplation (no intention to change in the near future), contemplation (intention to change but at a later time), preparation (intention to change soon, such as within the next month), action (change is occurring), and maintenance (change is maintained for at least 6 months). Although the stages of change are frequently highlighted, they are only one aspect of the model, which also includes decisional balance (for example, pros and cons), self-efficacy, and the process of change (that is, how people change).

Identifying the stage of change helps diet and exercise counselors to provide appropriate educational materials and support. For example, it is unlikely that an athlete in the precomtemplation stage would benefit from specific nutrition strategies, but information about how a dietary plan might enhance performance could be beneficial. Specific nutrition strategies would likely benefit an athlete in the preparation or action stages. Individuals in these stages often need close contact and support, since decisional balance and self-efficacy are important parts of making behavior change. When resources are limited, practitioners often focus on clients or patients who are in the preparation and action stages (Greene et al., 1999).

There are many other behavior change models, and it is worth repeating that no single model is the "correct" approach to promoting behavior change. Each model may be successful under certain circumstances and with a subgroup of people for whom the model is a good match. Understanding why people are sedentary and why people eat what they do will help in the planning of more effective intervention programs. Research in these areas is still in its infancy.

Environmental and policy changes. Individuals are capable of making behavior change, but it is clear that environmental changes are needed to help children, adolescents, and adults be healthier. Changes are needed in communities, schools, and workplaces. These changes are known as population-level changes, and they make it easier for individuals to initiative and sustain healthy behaviors. Population-level changes typically involve policy or regulatory changes, so many are highly political in nature. However, population-level changes can have dramatic effects, both positive and negative, on an individual's health, particularly children who are largely dependent on adults to provide a home and school environmental that promotes healthy eating and physical activity.

Using overweight and obesity as an example, the Centers for Disease Control and Prevention suggest some strategies and policy changes that can help support good health and reduce the risk for developing chronic disease (Khan et al., 2009):

- Promote the availability of affordable healthy food and beverages (for example, provide incentives to purchase food from local farms).

- Support healthy food and beverage choices (for example, limit the availability of less healthy foods and beverages in schools or other public places).

- Encourage physical activity (for example, require physical education in schools and build bicycling and walking paths).

Key points

- Hypertension, diabetes, cardiovascular disease, metabolic syndrome, osteoporosis, and lifestyle-related cancers are greatly affected by diet and exercise habits.

- Body Mass Index, waist circumference, and bone mineral density tests are helpful screening tools.

- Cardiovascular fitness is important regardless of one's weight.

- Individual and collective behavior change is needed to prevent and treat chronic diseases.

How can you, your family, your community, and your country become "healthogenic" rather than "obesogenic"?

Self-efficacy: The belief by an individual that he or she can effect change.

Post-Test	Reassessing Knowledge of Health, Fitness, and Chronic Diseases

Now that you have more knowledge about health, fitness, and chronic diseases, read the following statements and decide if each is true or false. The answers can be found in Appendix O.

1. There are many contradictions among diet and exercise recommendations that are issued by health promotion organizations.

2. Elite endurance athletes do not develop hypertension but lesser-trained athletes do.

3. Type 2 diabetes is caused by eating too much sugar.

4. A sedentary lifestyle is associated with a higher risk for heart disease, certain types of cancer, and diabetes.

5. Being physically active helps to reduce disease risk even if a person is obese.

Summary

"Everyone is an athlete" and all humans should be physically active throughout their lives. Weight gain is common as people age, even among former athletes, but it is not an inevitable consequence of aging. Matching energy intake with energy expenditure can prevent weight gain. Regular, consistent lifelong exercise or activity is recommended, and specific guidelines are published for cardiorespiratory fitness, prevention of weight gain, and maintenance of weight loss.

Many organizations have issued dietary guidelines, which are also consistent in recommending maintenance of a healthy weight and the consumption of a diet rich in whole grains, fruits, vegetables, lean-protein foods, and heart-healthy fats. Such a diet promotes health and may reduce the risk for chronic diseases such as heart disease, type 2 diabetes, and diet-related cancers.

Overweight and obesity are highly prevalent in the United States and exacerbate some chronic diseases, especially conditions associated with insulin resistance. There are many popular weight-loss plans, which feature hypocaloric diets with various macronutrient combinations, but no one plan has been shown to be the most effective. A sedentary lifestyle and poor diet are major contributors to many of the chronic diseases that plague modern societies—hypertension, heart disease, type 2 diabetes, metabolic syndrome, osteoporosis, and lifestyle-related cancers. Although there is much knowledge about the roles diet and exercise play in both the prevention and treatment of these diseases, most people do not change their behaviors based on knowledge alone. Thus an understanding of what motivates people to change their behavior and sustain the changes is needed, as well as societal changes that support a healthy lifestyle.

Self-Test

Multiple-Choice

The answers can be found in Appendix O.

1. Which of the following is a true statement about physical activity and aging in general?
   a. The total amount of physical activity stays the same, but there is a decrease in the intensity and duration of individual activities.
   b. The total amount of physical activity declines, but there is an increase in the intensity of individual activities.
   c. The total amount of physical activity as well as the intensity and duration of individual activities declines.
   d. After retirement there is an increase in the amount, intensity, and duration of physical activity because people have more time to exercise.

2. What is "creeping obesity?"
   a. a slow gain of body fat as adults age
   b. the process of muscle turning to fat
   c. change in body fat distribution after menopause
   d. a genetic predisposition to gain weight

3. The majority of people who have type 2 diabetes are:
   a. underweight.
   b. healthy weight.
   c. overweight or obese.

4. The type of fat distribution that increases risk for metabolic syndrome is:
   a. upper body obesity.
   b. lower body obesity.

5. The Health at Every Size movement does NOT support the use of:
   a. behavior change models.
   b. laboratory tests to determine metabolic fitness.
   c. weight-loss diets.
   d. therapy.

Short Answer

1. What happens to physical activity level as people age? Is this trend true of former athletes as well?

2. Is weight gain inevitable with aging? What advice would you give to collegiate athletes upon graduation about preventing weight gain?

3. Summarize the major points of the Dietary Guidelines for Americans, 2010.

4. Are diet and exercise recommendations published by various public health organizations generally similar or dissimilar? Explain.

5. What amount of activity is recommended as a minimum? For preventing weight gain? For maintaining weight loss? For cardiovascular fitness?

6. How do poor diet and lack of physical activity contribute to chronic disease risk?

7. Which chronic diseases are associated with being overweight or obese?

8. How would you respond to the question "Which weight-loss diet is the best?"

9. Explain insulin resistance and why exercise is so important for those with this condition.

10. Explain what is meant by "good" cholesterol and "bad" cholesterol.

11. Describe the differences between visceral and subcutaneous fat and how fat distribution affects disease risk.

12. How do diet and exercise help prevent osteoporosis?

13. Will taking calcium and/or vitamin D supplements after menopause increase bone mass? Prevent or delay osteoporosis?

14. Describe the relationship between physical activity, diet, and the prevention of lifestyle-related cancers.

15. Why is Body Mass Index (BMI) used as a screening tool for determining disease risk?

16. Is knowledge alone enough for most people to make behavior change? Why aren't people motivated by the belief that proper diet and exercise will help reduce the risk for chronic diseases?

Critical Thinking

1. Are those in the BMI "healthy weight" category free from chronic diseases? Does the prevalence for the various chronic diseases increase as BMI increases? Explain.

2. Can a person be fit and fat? Explain.

References

Allison, D. B., Fontaine, K. R., Manson, J. E., Stevens, J., & VanItallie, T B. (1999). Annual deaths attributable to obesity in the United States. *Journal of the American Medical Association, 282*(16), 1530–1538.

American College of Sports Medicine. (1998). Position stand on the recommended quantity and quality of exercise for developing and maintaining cardiorespiratory and muscular fitness, and flexibility in adults. *Medicine and Science in Sports and Exercise, 30*(6), 975–991.

American College of Sports Medicine & American Diabetes Association. (2010). Joint position statement: Exercise and type 2 diabetes. *Medicine & Science in Sports & Exercise, 1*(12), 2282–2303.

American Diabetes Association. (2004). Diagnosis and classification of diabetes mellitus. *Diabetes Care, 27*(Suppl. 1), S5–S10.

Atkins, R. C. (1972). *Dr. Atkins' Diet Revolution.* New York: D. McKay Co.

Atkins, R. C. (1999). *Dr. Atkins' New Diet Revolution.* New York: Avon Books.

Bacon, L., Keim, N. L., van Loan, M. D., Derricote, M., Gale, B., Kazaks, A., et al. (2002). Evaluating a "non-diet" wellness intervention for improvement of metabolic fitness, psychological well-being and eating and activity behaviors. *International Journal of Obesity and Related Metabolic Disorders, 26*(6), 854–865.

Baranowski, T., Cullen, K. W., Nicklas, T., Thompson, D., & Baranowski, J. (2003). Are current health behavioral change models helpful in guiding prevention of weight gain efforts? *Obesity Research, 11*(Suppl.), 23S–43S.

Blair, S. N., Kohl, H., Paffenbarger, W. R. S., Clark, D. G., Cooper, K. H., et al. (1989). Physical fitness and all-cause mortality: A prospective study of healthy men and women. *Journal of the American Medical Association, 262*(17), 2395–2401.

Brownell, K. D. (2004). Fast food and obesity in children. *Pediatrics, 113*(1, Pt. 1), 112–118.

Castañeda, T. R., Tong, J. Datta, R., Culler, M., & Tschöp, M. H. (2010) Ghrelin in the regulation of body weight and metabolism. *Frontiers in Neuroendocrinology, 31,* 44–60.

Curtis, B. M., & O'Keefe, J. H., Jr. (2002). Understanding the Mediterranean diet. Could this be the new "gold standard" for heart disease prevention? *Postgraduate Medicine, 112*(2), 35–38, 41–45.

Dansinger, M. L., Gleason, J. A., Griffith, J. L., Selker, H. P., & Schaefer, E. J. (2005). Comparison of the Atkins, Ornish, Weight Watchers, and Zone diets for weight loss and heart disease risk reduction: A randomized trial. *Journal of the American Medical Association, 293*(1), 43–53.

DIPART (Vitamin D Individual Patient Analysis of Randomized Trials) Group. (2010). Patient level pooled analysis of 68,500 patients from seven major vitamin D fracture trials in US and Europe. *British Medical Journal, 340,* b5463. doi: 10.1136/bmj.b5463.

DiPietro, L., Kohl, H. W., III, Barlow, C. E., & Blair, S. N. (1998). Improvements in cardiorespiratory fitness attenuate age-related weight gain in healthy men and women: The Aerobics Center Longitudinal Study. *International Journal of Obesity and Related Metabolic Disorders, 22*(1), 55–62.

Donnelly, J. E., Blair, S. N., Jakicic, J. M., Manore, M. M., Rankin, J. W., & Smith, B. K. (2009). Appropriate physical activity intervention strategies for weight loss and prevention of weight regain for adults. *Medicine & Science in Sports & Exercise, 41*(2), 459–471.

Ekelund, U., Brage, S., Franks, P. W., Hennings, S., Emms, S., & Wareham, N. J. (2005). Physical activity energy expenditure predicts progression toward the metabolic syndrome independently of aerobic fitness in middle-aged healthy Caucasians: The Medical Research Council Ely Study. *Diabetes Care, 28*(5), 1195–1200.

Filion, K. B., El Khoury, F., Bielinski, M., Schiller, I., Dendukuri, N., & Brophy, J. M. (2010). Omega-3 fatty acids in high-risk cardiovascular patients: A meta-analysis of randomized controlled trials. *BMC Cardiovascular Disorders, 10,* 24.

Finley, C. E., LaMonte, M. J., Waslien, C. I., Barlow, C. E., Blair, S. N., & Nichaman, M. Z. (2006). Cardiorespiratory fitness, macronutrient intake, and the metabolic syndrome: The Aerobics Center Longitudinal Study. *Journal of the American Dietetic Association, 106*(5), 673–679.

Flegal, K. M., Carroll, M. D., Ogden, C. L., & Curtin, L. R. (2010). Prevalence and trends in obesity among US adults, 1999–2008. *Journal of the American Medical Association, 303*(3), 235–241.

Flegal, K., Graubard, D., Williamson, D., & Gail, M. (2005). Excess deaths associated with underweight, overweight, and obesity. *Journal of the American Medical Association, 293*(15), 1861–1867.

Gaesser, G. A. (2002). *Big fat lies: The truth about your weight and your health.* Carlsbad, CA: Gurze Books.

Garry, P. J. (2001). Aging successfully: A genetic perspective. *Nutrition Reviews, 59*(8), S93–S101.

Girod, J. P., & Brotman, D. J. (2003). The metabolic syndrome as a vicious cycle: Does obesity beget obesity? *Medical Hypotheses, 60*(4), 584–589.

Greene, G. G., Rossi, S. R., Rossi, J. S., Velicer, W. R., Fava, J. L., & Prochaska, J. O. (1999). Dietary applications of the stages of change model. *Journal of the American Dietetic Association, 99*(6), 673–678.

Gropper, S. S., Smith, J. L., & Groff, J. L. (2009). *Advanced nutrition and human metabolism* (5th ed.). Belmont, CA: Thomson/Wadsworth.

Grundy, S. M. (2006). Metabolic syndrome: Connecting and reconciling cardiovascular and diabetes worlds. *Journal of the American College of Cardiology, 47*(6), 1093–1100.

Grundy, S. M., Cleeman, J. I., Daniels, S. R., Donato, K. A., Eckel, R. H., Franklin, B., et al. (2005). Diagnosis and management of the metabolic syndrome: An American Heart Association/National Heart, Lung, and Blood Institute scientific statement. *Circulation, 112*(17), 2735–2752.

Hagobian, T. A., & Braun, B. (2010). Physical activity and hormone regulation of appetite: sex differences and weight control. *Exercise and Sport Science Reviews, 38*(1), 25–30.

Hagobian, T. A., Sharoff, C. G., & Braun, B. (2008). Effects of short-term exercise and energy surplus on hormones related to regulation of energy balance. *Metabolism Clinical and Experimental, 57,* 393–398.

Hagobian, T. A., Sharoff, C. G., Stephens, B. R., Wade, G. N., Silva, J. E., Chipkin, S. R., et al. (2009). Effects of exercise on energy-regulating hormones and appetite in men and women. *American Journal of Physiology— Regulatory, Integrative, and Comparative Physiology, 296,* R233–R242.

Hernelahti, M., Kujala, U. M., Kaprio, J., Karjalainen, J., & Sarna, S. (1998). Hypertension in master endurance athletes. *Journal of Hypertension, 16*(11), 1573–1577.

Hernelahti, M., Kujala, U. M., Kaprio, J., & Sarna, S. (2002). Long-term vigorous training in young adulthood and later physical activity as predictors of hypertension in middle-aged and older men. *International Journal of Sports Medicine, 23*(3), 178–182.

Heron, M., Hoyert, D. L., Murphy, S. L., Xu, J., Kochanek, K. D., & Tejada-Vera, B. (2009). Deaths: Final data for 2006. *National Vital Statistic Reports, 57*(14), 1–117.

Hofmann, S. M., & Tschöp, M. H. (2009). Dietary sugars: A fat difference. *Journal of Clinical Investigation, 119,* 1089–1092.

Hu, F. B., Manson, J. E., Stampfer, M. J., Colditz, G., Liu, S., Solomon, C. G., et al. (2001). Diet, lifestyle, and the risk of type 2 diabetes mellitus in women. *New England Journal of Medicine, 345*(11), 790–797.

Hu, F. B., Willett, W. C., Li, T., Stampfer, M. J., Colditz, G. A., & Manson, J. E. (2004). Adiposity as compared with physical activity in predicting mortality among women. *New England Journal of Medicine, 351*(26), 2694–2703.

Hughes, E., Kilmer, G., Li, Y., Valluru, B., Brown, J., Colclough, G., et al. (2010). Surveillance for certain health behaviors among states and selected local areas—United States, 2008. *Morbidity and Mortality Weekly Report: Surveillance Summaries, 59*(10), 1–221.

Imamura, F., Lichtenstein, A. H., Dallal, G. E., Meigs, J. B., & Jacques, P. F. (2009). Generalizability of dietary patterns associated with incidence of type 2 diabetes mellitus. *American Journal of Clinical Nutrition, 90*(4), 1075–1083.

Institute of Medicine. (2002). *Dietary Reference Intakes for energy, carbohydrate, fiber, fat, fatty acids, cholesterol, protein and amino acids* (Food and Nutrition Board). Washington, DC: National Academy Press.

Institute of Medicine. (2004). *Dietary Reference Intakes for water, potassium, sodium, chloride, and sulfate* (Food and Nutrition Board). Washington, DC: National Academies Press.

Johnson, D. B., Gerstein, D. E., Evans, A. E., & Woodward-Lopez, G. (2006). Preventing obesity: A life cycle perspective. *Journal of the American Dietetic Association, 106*(1), 97–102.

Johnson, R. K., Appel, L. J., Brands, M., Howard, B. V., Lefevre, M., Lustig, R. H., et al. (2009). Dietary sugars intake and cardiovascular health: A scientific statement from the American Heart Association. *Circulation, 120*(11), 1011–1120.

Katzmarzyk, P. T., Church, T. S., Janssen, I., Ross, R., & Blair, S. N. (2005). Metabolic syndrome, obesity, and mortality: Impact of cardiorespiratory fitness. *Diabetes Care, 28*(2), 391–397.

Khan, L. K., Sobush, K., Keener, D., Goodman, K., Lowry, A., Kakietek, J., et al. (2009). Recommended community strategies and measurements to prevent obesity in the United States. *MMWR Recommendations and Reports, 58*(RR-7), 1–26.

Klem, M. L., Wing, R. R., McGuire, M. T., Seagle, H. M., & Hill, J. O. (1997). A descriptive study of individuals successful at long-term maintenance of substantial weight loss. *American Journal of Clinical Nutrition, 66*(2), 239–246.

Kohrt, W. M., Bloomfield, S. A., Little, K. D., Nelson, M. E., & Yingling, V. R. (2004). Physical activity and bone health. *Medicine & Science in Sports & Exercise, 36*(11), 1985–1996.

Krauss, R. M., Eckel, R. H., Howard, B., Appel, L. J., Daniels, S. R., Deckelbaum, R. J., et al. (2000). AHA dietary guidelines: Revision 2000: A statement for healthcare professionals from the Nutrition Committee of the American Heart Association. *Circulation, 102*(18), 2284–2299.

Kris-Etherton, P., Eckel, R. H., Howard, B. V., St Jeor, S., Bazzarre, T. L., & Nutrition Committee, Population Science Committee, and Clinical Science Committee of the American Heart Association. (2001). AHA Science Advisory: Lyon Diet Heart Study. Benefits of a Mediterranean-style, National Cholesterol Education Program/American Heart Association Step I Dietary Pattern on Cardiovascular Disease. *Circulation, 103*(13), 1823–1825.

Kushi, L. H., Byers, T., Doyle, C., Bandera, E. V., McCullough, M., McTiernan, A., Gansler, T., Andrews, K.S., Thun, M.J., & The American Cancer Society 2006 Nutrition and Physical Activity Guideline Advisory Committee (2006). American Cancer Society guidelines on nutrition and physical activity for cancer prevention: Reducing the risk of cancer with healthy food choices and physical activity. *CA: A Cancer Journal for Clinicians, 56*(5), 254–281.

Lee, C. D., Blair, S. N., & Jackson, A. S. (1999). Cardiorespiratory fitness, body composition, and all-cause and cardiovascular disease mortality in men. *American Journal of Clinical Nutrition, 69*(3), 373–380.

Livesey, G., Taylor, R., Hulshof, T., & Howlett, J. (2008). Glycemic response and health—a systematic review and meta-analysis: Relations between dietary glycemic properties and health outcomes. *American Journal of Clinical Nutrition, 87*(1), 258S–268S.

Lufkin, E. G., Wong, M., & Deal, C. (2001). The role of selective estrogen receptor modulators in the prevention and treatment of osteoporosis. *Rheumatic Diseases Clinics of North America, 27*(1), 163–185.

Madamanchi, N. R., Vendrov, A., & Runge, M. S. (2005). Oxidative stress and vascular disease. *Arteriosclerosis, Thrombosis, and Vascular Biology, 25*(1), 29–38.

Magkos, F., Wang, X., & Mittendorfer, B. (2010). Metabolic actions of insulin in men and women. *Nutrition, 26*(7–8), 686–693.

Miller, W. C. (2005). The weight-loss-at-any-cost environment: How to thrive with a health-centered focus. *Journal of Nutrition Education and Behavior, 37*(Suppl. 2), S89–S94.

Miniño A. M., Arias E., Kochanek K. D., Murphy S. L., Smith B. L. Deaths: final data for 2000. *National Vital Statistics Reports* 2002; *50*(15):1–20.

Mokdad A. H., Marks J. S., Stroup D. F., Gerberding J. L. Actual causes of death in the United States, 2000. *JAMA.* 2004; 291. Reprinted with permission.

Mozaffarian, D., Aro, A., & Willett, W. C. (2009). Health effects of *trans*-fatty acids: experimental and observational evidence. *European Journal of Clinical Nutrition, 63* (Suppl. 2): S5–S21.

Must, A., Spadano, J., Coakley, E. H., Field, A. E., & Dietz, W. H. (1999). The disease burden associated with overweight and obesity. *Journal of the American Medical Association, 282*(16), 1523–1529.

National Center for Health Statistics. (2006). *Health, United States, 2006 with chartbook on trends in the health of Americans.* Hyattsville, MD.

National Osteoporosis Foundation. (2005). *America's bone health: The state of osteoporosis and low bone mass.* Washington, DC.

Newby, P. K., Muller, D., Hallfrisch, J., Andres, R., & Tucker, K. L. (2004). Food patterns measured by factor analysis and anthropometric changes in adults. *American Journal of Clinical Nutrition, 80*(2), 504–513.

Newby, P. K., Muller, D., Hallfrisch, J., Qiao, N., Andres, R., & Tucker, K. L. (2003). Dietary patterns and changes in body mass index and waist circumference in adults. *American Journal of Clinical Nutrition, 77*(6), 1417–1425.

O'Kane, J. W., Teitz, C. C., Fontana, S. M., & Lind, B. K. (2002). Prevalence of obesity in adult population of former college rowers. *Journal of the American Board of Family Practice, 15*(6), 451–456.

Ornish, D. (1993). *Eat more and weigh less: Dr. Dean Ornish's life choice program for losing weight safely while eating abundantly.* New York: HarperCollins.

Pate, R. R., Pratt, M., Blair, S., Haskell, W. L., Macera, C. A., Bouchard, C., et al. (1995). Physical activity and public health: A recommendation from the Centers for Disease Control and Prevention and the American College of Sports Medicine. *Journal of the American Medical Association, 273*(5), 402–407.

Pihl, E., & Jurimae, T. (2001). Relationships between body weight change and cardiovascular disease risk factors in male former athletes. *International Journal of Obesity and Related Metabolic Disorders, 25*(7), 1057–1062.

Pinhas-Hamiel, O., Dolan, L. M., Daniels, S. R., Standiford, D., Khoury, P. R., & Zeitler, P. (1996). Increased incidence of non-insulin dependent diabetes mellitus among adolescents. *Journal of Pediatrics, 128*(5, Pt. 1), 608–615.

Pi-Sunyer, X. (2009). The medical risks of obesity. *Postgraduate Medicine, 121*(6), 21–33.

Prochaska, J. O., & DiClemente, C. C. (1984). *The transtheoretical approach: Crossing the traditional boundaries of therapy.* Homewood, IL: Irwin.

Raz, I., Eldor, R., Cernea, S., & Shafrir, E. (2005). Diabetes: Insulin resistance and derangements in lipid metabolism. Cure through intervention in fat transport and storage. *Diabetes Metabolism Research and Reviews, 21*(1), 3–14.

Robertson, R. M., & Smaha, L. (2001). Can a Mediterranean-style diet reduce heart disease? *Circulation, 103*(13), 1821–1822.

Sacks, F. M., Bray, G. A., Carey, V. J., Smith, S. R., Ryan, D. H., Anton, S. D., et al. (2009). Comparison of weight-loss diets with different compositions of fat, protein, and carbohydrates. *New England Journal of Medicine, 360*(9), 859–873.

Savica, V., Bellinghieri, G., & Kopple, J. D. (2010). The effect of nutrition on blood pressure. *Annual Review of Nutrition, 30*, 365–401.

Schulz, M., Nothlings, U., Hoffmann, K., Bergmann, M. M., & Boeing, H. (2005). Identification of a food pattern characterized by high-fiber and low-fat food choices associated with low prospective weight change in the EPIC-Potsdam cohort. *Journal of Nutrition, 135*(5), 1183–1189.

Seagle, H. M., Strain, G. W., Makris, A., Reeves, R. S, & American Dietetic Association. (2009). Position of the American Dietetic Association: Weight management. *Journal of the American Dietetic Association, 109*(2), 330–346.

Sears, B. (1995). *The zone: A revolutionary life plan to put your body in total balance for permanent weight loss.* New York: HarperCollins.

Sears, B. (1997). *Mastering the zone: The next step in achieving superhealth and permanent weight loss.* New York: HarperCollins.

Sheehan, G. (1980). *This running life.* New York: Simon and Schuster.

Sherwood, L. (2010). *Human physiology: From cells to systems.* (7th ed.). California: Brooks/Cole Cengage Learning.

Siri-Tarino, P. W., Sun, Q., Hu, F. B., & Krauss, R. M. (2010). Meta-analysis of prospective cohort studies evaluating the association of saturated fat with cardiovascular disease. American Journal of Clinical Nutrition, 91(3), 535–546.

Sofi, F., Cesari, F., Abbate, R., Gensini, G. F., & Casini, A. (2008). Adherence to Mediterranean diet and health status: Meta-analysis. British Journal of Medicine. Sep 11;337:a1344. doi:10.1136/bmj.a1344.

Truby, H., Baic, S., deLooy, A., Fox, K. R., Livingstone, M. B., Logan, C. M., et al. (2006). Randomised controlled trial of four commercial weight loss programmes in the UK: Initial findings from the BBC "diet trials." *British Medical Journal, 332*(7553), 1309–1314.

U.S. Department of Health and Human Services. (1996). *Physical activity and health: A report of the Surgeon General.* Atlanta, GA: U.S. Department of Health and Human Services, Centers for Disease Control and Prevention, National Center for Chronic Disease Prevention and Health Promotion.

U.S. Department of Health and Human Services & U.S. Department of Agriculture. (2010). *Dietary Guidelines for Americans* (7th ed.). Washington, DC.

van Dam, R. M., Rimm, E. B., Willett, W. C., Stampfer, M. J., & Hu, F. B. (2002). Dietary patterns and risk for type 2 diabetes mellitus in U.S. men. *Annals of Internal Medicine, 136*(3), 201–209.

Volek, J. S., Vanheest, J. L., & Forsythe, C. E. (2005). Diet and exercise for weight loss: A review of current issues. *Sports Medicine, 35*(1), 1–9.

Weight Watchers materials are available at http://www.weightwatchers.com.

Weir, H. K., Thun, M. J., Hankey, B. F., Ries, L. A., Howe, H. L., Wingo, P. A., et al. (2003). Annual report to the nation on the status of cancer, 1975–2000, featuring the uses of surveillance data for cancer prevention and control. *Journal of the National Cancer Institute, 95*(17), 1276–1299. Erratum in: *Journal of the National Cancer Institute*, 2003, *95*(21), 1641.

Winett, R. A., Tate, D. F., Anderson, E. S., Wojcik, J. R., & Winett, S. G. (2005). Long-term weight gain prevention: A theoretically based Internet approach. *Preventative Medicine, 41*(2), 629–641.

Wing, R. R., & Hill, J. O. (2001). Successful weight loss maintenance. *Annual Review of Nutrition, 21*, 323–341.

Zizic, T. M. (2004). Pharmacologic prevention of osteoporotic fractures. *American Family Physician, 70*(7), 1293–1300.

Dietary Reference Intakes (DRI) for Energy and Macronutrients

Dietary Reference Intakes (DRI): Estimated Energy Requirements (EER) for Men and Women 30 Years of Age[a]

Food and Nutrition Board, Institute of Medicine, National Academies

Height (m [in])	PAL[b]	Weight for BMI[c] of 18.5 kg/m² (kg [lb])	Weight for BMI of 24.99 kg/m² (kg [lb])	EER, Men[d] (kcal/day)		EER, Women[d] (kcal/day)	
				BMI of 18.5 kg/m²	BMI of 24.99 kg/m²	BMI of 18.5 kg/m²	BMI of 24.99 kg/m²
1.50(59)	Sedentary	41.6(92)	56.2(124)	1,848	2,080	1,625	1,762
	Low active			2,009	2,267	1,803	1,956
	Active			2,215	2,506	2,025	2,198
	Very active			2,554	2,898	2,291	2,489
1.65(65)	Sedentary	50.4(111)	68.0(150)	2,068	2,349	1,816	1,982
	Low active			2,254	2,566	2,016	2,202
	Active			2,490	2,842	2,267	2,477
	Very active			2,880	3,296	2,567	2,807
1.80(71)	Sedentary	59.9(132)	81.0(178)	2,301	2,635	2,015	2,211
	Low active			2,513	2,884	2,239	2,459
	Active			2,782	3,200	2,519	2,769
	Very active			3,225	3,720	2,855	3,141

[a]For each year below 30, add 7 kcal/day for women and 10 kcal/day for men. For each year above 30, subtract 7 kcal/day for women and 10 kcal/day for men.

[b]PAL = physical activity level.

[c]BMI = body mass index.

[d]Derived from the following regression equations based on doubly labeled water data:

Adult man: EER = $662 - 9.53 \times$ age (y) + PA $\times$ ($15.91 \times$ wt [kg] + $539.6 \times$ ht [m])

Adult woman: EER = $354 - 6.91 \times$ age (y) + PA $\times$ ($9.36 \times$ wt [kg] + $726 \times$ ht [m])

Where PA refers to coefficient for PAL

PAL = total energy expenditure ÷ basal energy expenditure

PA = 1.0 if PAL ≥ 1.0 < 1.4 (sedentary)

PA = 1.12 if PAL ≥ 1.4 < 1.6 (low active)

PA = 1.27 if PAL ≥ 1.6 < 1.9 (active)

PA = 1.45 if PAL ≥ 1.9 < 2.5 (very active)

Dietary Reference Intakes (DRI): Acceptable Macronutrient Distribution Ranges

Food and Nutrition Board, Institute of Medicine, National Academies

Macronutrient	Range (percent of energy)		
	Children, 1–3 y	Children, 4–18 y	Adults
Fat	30–10	25–35	20–35
n-6 polyunsaturated fatty acids[a] (linoleic acid)	5–10	5–10	5–10
n-3 polyunsaturated fatty acids[a] (α-linolenic acid)	0.6–1.2	0.6–1.2	0.6–1.2
Carbohydrate	45–65	45–65	45–65
Protein	5–20	10–30	10–35

[a]Approximately 10% of the total can come from longer-chain *n*-3 or *n*-6 fatty acids.

Source: *Dietary Reference Intakes for Energy, Carbohydrate, Fiber, Fat, Fatty Acids, Cholesterol, Protein, and Amino Acids (2002).*

Dietary Reference Intakes (DRI): Recommended Intakes for Individuals, Macronutrients

Food and Nutrition Board, Institute of Medicine, National Academies

Life stage group	Total water[a] (L/d)	Carbohydrate (g/d)	Total fiber (g/d)	Fat (g/d)	Linoleic acid (g/d)	α-Linolenic acid (g/d)	Protein[b] (g/d)
Infants							
0–6 mo	0.7*	60*	ND	31*	4.4*	0.5*	9.1*
7–12 mo	0.8*	95*	ND	30*	4.6*	0.5*	**11.0**[c]
Children							
1–3 y	1.3*	**130**	19*	ND	7*	0.7*	**13**
4–8 y	1.7*	**130**	25*	ND	10*	0.9*	**19**
Males							
9–13 y	2.4*	**130**	31*	ND	12*	1.2*	**34**
14–18 y	3.3*	**130**	38*	ND	16*	1.6*	**52**
19–30 y	3.7*	**130**	38*	ND	17*	1.6*	**56**
31–50 y	3.7*	**130**	38*	ND	17*	1.6*	**56**
51–70 y	3.7*	**130**	30*	ND	14*	1.6*	**56**
>70 y	3.7*	**130**	30*	ND	14*	1.6*	**56**
Females							
9–13 y	2.1*	**130**	26*	ND	10*	1.0*	**34**
14–18 y	2.3*	**130**	26*	ND	11*	1.1*	**46**
19–30 y	2.7*	**130**	25*	ND	12*	1.1*	**46**
31–50 y	2.7*	**130**	25*	ND	12*	1.1*	**46**
51–70 y	2.7*	**130**	21*	ND	11*	1.1*	**46**
>70 y	2.7*	**130**	21*	ND	11*	1.1*	**46**
Pregnancy							
14–18 y	3.0*	**175**	28*	ND	13*	1.4*	**71**
19–30 y	3.0*	**175**	28*	ND	13*	1.4*	**71**
31–50 y	3.0*	**175**	28*	ND	13*	1.4*	**71**
Lactation							
14–18 y	3.8*	**210**	29*	ND	13*	1.3*	**71**
19–30 y	3.8*	**210**	29*	ND	13*	1.3*	**71**
31–50 y	3.8*	**210**	29*	ND	13*	1.3*	**71**

Note: This table presents Recommended Dietary Allowances (RDAs) in bold type and Adequate Intakes (AIs) in ordinary type followed by an asterisk (*). RDAs and AIs may both be used as goals for individual intake. RDAs are set to meet the needs of almost all (97 to 98 percent) individuals in a group. For healthy infants fed human milk, the AI is the mean intake. The AI for other life stage and gender groups is believed to cover the needs of all individuals in the group, but lack of data or uncertainty in the data prevent being able to specify with confidence the percentage of individuals covered by this intake.

[a]*Total* water includes all water contained in food, beverages, and drinking water.

[b]Based on 0.8 g/kg body weight for the reference body weight.

[c]Change from 13.5 in prepublication copy due to calculation error.

Dietary Reference Intakes (DRI): Additional Macronutrient Recommendations

Food and Nutrition Board, Institute of Medicine, National Academies

Macronutrient	Recommendation
Dietary cholesterol	As low as possible while consuming a nutritionally adequate diet
Trans fatty acids	As low as possible while consuming a nutritionally adequate diet
Saturated fatty acids	As low as possible while consuming a nutritionally adequate diet
Added sugars	Limit to no more than 25% of total energy

Source: *Dietary Reference Intakes for Energy, Carbohydrate, Fiber, Fat, Fatty Acids, Cholesterol, Protein, and Amino Acids (2002).*

APPENDIX B The DASH (Dietary Approaches to Stop Hypertension) Eating Plan at Various Calorie Levels

The number of daily servings in a food group vary depending on caloric needs[a]

Food group[b]	1,200 Calories	1,400 Calories	1,600 Calories	1,800 Calories	2,000 Calories	2,600 Calories	3,100 Calories	Serving sizes
Grains	4–5	5–6	6	6	6–8	10–11	12–13	1 slice bread 1 oz dry cereal[c] ½ cup cooked rice, pasta, or cereal[c]
Vegetables	3–4	3–4	3–4	4–5	4–5	5–6	6	1 cup raw leafy vegetable ½ cup cut-up raw or cooked vegetable ½ cup vegetable juice
Fruits	3–4	4	4	4–5	4–5	5–6	6	1 medium fruit ¼ cup dried fruit ½ cup fresh, frozen, or canned fruit ½ cup fruit juice
Fat-free or low-fat milk and milk products	2–3	2–3	2–3	2–3	2–3	3	3–4	1 cup milk or yogurt 1½ oz cheese
Lean meats, poultry, and fish	3 or less	3–4 or less	3–4 or less	6 or less	6 or less	6 or less	6–9	1 oz cooked meats, poultry, or fish 1 egg
Nuts, seeds, and legumes	3 per week	3 per week	3–4 per week	4 per week	4–5 per week	1	1	⅓ cup or 1½ oz nuts 2 Tbsp peanut butter 2 Tbsp or ½ oz seeds ½ cup cooked legumes (dried beans, peas)
Fats and oils	1	1	2	2–3	2–3	3	4	1 tsp soft margarine 1 tsp vegetable oil 1 Tbsp mayonnaise 1 Tbsp salad dressing
Sweets and added sugars	3 or less per week	3 or less per week	3 or less per week	5 or less per week	5 or less per week	<2	<2	1 Tbsp sugar 1 Tbsp jelly or jam ½ cup sorbet, gelatin dessert 1 cup lemonade
Maximum sodium limit[d]	2,300 mg/day	2,300 mg/day	2,300 mg/day	2,300 mg/day	2,300 mg/day	2,300 mg/day	2,300 mg/day	

Notes

[a]The DASH eating patterns from 1,200 to 1,800 calories meet the nutritional needs of children 4 to 8 years old. Patterns from 1,600 to 3,100 calories meet the nutritional needs of children 9 years and older and adults. See Appendix 6 of the *Dietary Guidelines for Americans, 2010*, for estimated calorie needs per day by age, gender, and physical activity level.

[b]Significance to DASH Eating Plan, selection notes, and examples of foods in each food group.

- Grains: Major sources of energy and fiber. Whole grains are recommended for most grain servings as a good source of fiber and nutrients. Examples: Whole wheat bread and rolls; whole wheat pasta, English muffin, pita bread, bagel, cereals; grits, oatmeal, brown rice; unsalted pretzels and popcorn.

- Vegetables: Rich sources of potassium, magnesium, and fiber. Examples: Broccoli, carrots, collards, green beans, green peas, kale, lima beans, potatoes, spinach, squash, sweet potatoes, tomatoes.

- Fruits: Important sources of potassium, magnesium, and fiber. Examples: Apples, apricots, bananas, dates, grapes, oranges, grapefruit, grapefruit juice, mangoes, melons, peaches, pineapples, raisins, strawberries, tangerines.

- Fat-free or low-fat milk and milk products: Major sources of calcium and protein. Examples: Fat-free milk or buttermilk; fat-free, low-fat, or reduced-fat cheese; fat-free/low-fat regular or frozen yogurt.

- Lean meats, poultry, and fish: Rich sources of protein and magnesium. Select only lean; trim away visible fats; broil, roast, or poach; remove skin from poultry. Since eggs are high in cholesterol, limit egg yolk intake to no more than four per week; two egg whites have the same protein content as 1 oz meat.

- Nuts, seeds, and legumes: Rich sources of energy, magnesium, protein, and fiber. Examples: Almonds, filberts, mixed nuts, peanuts, walnuts, sunflower seeds, peanut butter, kidney beans, lentils, split peas.

- Fats and oils: DASH study had 27 percent of calories as fat, including fat in or added to foods. Fat content changes serving amount for fats and oils. For example, 1 Tbsp regular salad dressing = one serving; 2 Tbsp low-fat dressing = one serving; 1 Tbsp fat-free dressing = zero servings. Examples: Soft margarine, vegetable oil (canola, corn, olive, safflower), low-fat mayonnaise, light salad dressing.

- Sweets and added sugars: Sweets should be low in fat. Examples: Fruit-flavored gelatin, fruit punch, hard candy, jelly, maple syrup, sorbet and ices, sugar.

[c]Serving sizes vary between ½ cup and 1¼ cups, depending on cereal type. Check product's Nutrition Facts label.

[d]The DASH Eating Plan consists of patterns with a sodium limit of 2,300 mg and 1,500 mg per day.

Source: U.S. Department of Agriculture and U.S. Department of Health and Human Services, *Dietary Guidelines for Americans 2010*, 7th edition, Washington, DC: U.S. Government Printing Office, December 2010.

APPENDIX C USDA Food Patterns

The principles of a healthy eating pattern can be applied by following one of several templates for healthy eating. The USDA Food Patterns, their lacto-ovo vegetarian or vegan adaptations, and the DASH Eating Plan (see Appendix B) are illustrations of varied approaches to healthy eating patterns. The USDA Food Patterns and their vegetarian variations were developed to help individuals carry out the Dietary Guidelines recommendations. The DASH Eating Plan, based on the DASH research studies, was developed to help individuals prevent high blood pressure and other risk factors for heart disease.

USDA Food Patterns

For each food group or subgroup,[a] recommended average daily intake amounts[b] at all calorie levels. Recommended intakes from vegetable and protein foods subgroups are per week. For more information and tools for application, go to MyPyramid.gov.

Calorie level of pattern[c]	1,000	1,200	1,400	1,600	1,800	2,000	2,200	2,400	2,600	2,800	3,000	3,200
Fruits	1 c	1 c	1½ c	1½ c	1½ c	2 c	2 c	2 c	2 c	2½ c	2½ c	2½ c
Vegetables[d]	1 c	1½ c	1½ c	2 c	2½ c	2½ c	3 c	3 c	3½ c	3½ c	4 c	4 c
Dark-green vegetables	½ c/wk	1 c/wk	1 c/wk	1½ c/wk	1½ c/wk	1½ c/wk	2 c/wk	2 c/wk	2½ c/wk	2½ c/wk	2½ c/wk	2½ c/wk
Red and orange vegetables	2½ c/wk	3 c/wk	3 c/wk	4 c/wk	5½ c/wk	5½ c/wk	6 c/wk	6 c/wk	7 c/wk	7 c/wk	7½ c/wk	7½ c/wk
Beans and peas (legumes)	½ c/wk	½ c/wk	½ c/wk	1 c/wk	1½ c/wk	1½ c/wk	2 c/wk	2 c/wk	2½ c/wk	2½ c/wk	3 c/wk	3 c/wk
Starchy vegetables	2 c/wk	3½ c/wk	3½ c/wk	4 c/wk	5 c/wk	5 c/wk	6 c/wk	6 c/wk	7 c/wk	7 c/wk	8 c/wk	8 c/wk
Other vegetables	1½ c/wk	2½ c/wk	2½ c/wk	3½ c/wk	4 c/wk	4 c/wk	5 c/wk	5 c/wk	5½ c/wk	5½ c/wk	7 c/wk	7 c/wk
Grains[e]	3 oz-eq	4 oz-eq	5 oz-eq	5 oz-eq	6 oz-eq	6 oz-eq	7 oz-eq	8 oz-eq	9 oz-eq	10 oz-eq	10 oz-eq	10 oz-eq
Whole grains	1½ oz-eq	2 oz-eq	2½ oz-eq	3 oz-eq	3 oz-eq	3 oz-eq	3½ oz-eq	4 oz-eq	4½ oz-eq	5 oz-eq	5 oz-eq	5 oz-eq
Enriched grains	1½ oz-eq	2 oz-eq	2½ oz-eq	2 oz-eq	3 oz-eq	3 oz-eq	3½ oz-eq	4 oz-eq	4½ oz-eq	5 oz-eq	5 oz-eq	5 oz-eq
Protein foods[d]	2 oz-eq	3 oz-eq	4 oz-eq	5 oz-eq	5 oz-eq	5½ oz-eq	6 oz-ec	6½ oz-eq	6½ oz-eq	7 oz-eq	7 oz-eq	7 oz-eq
Seafood	3 oz/wk	5 oz/wk	6 oz/wk	8 oz/wk	8 oz/wk	8 oz/wk	9 oz/wk	10 oz/wk	10 oz/wk	11 oz/wk	11 oz/wk	11 oz/wk
Meat, poultry, eggs	10 oz/wk	14 oz/wk	19 oz/wk	24 oz/wk	24 oz/wk	26 oz/wk	29 oz/wk	31 oz/wk	31 oz/wk	34 oz/wk	34 oz/wk	34 oz/wk
Nuts, seeds, soy products	1 oz/wk	2 oz/wk	3 oz/wk	4 oz/wk	4 oz/wk	4 oz/wk	4 oz/wk	5 oz/wk	5 oz/wk	5 oz/wk	5 oz/wk	5 oz/wk
Dairy[f]	2 c	2½ c	2½ c	3 c	3 c	3 c	3 c	3 c	3 c	3 c	3 c	3 c
Oils[g]	15 g	17 g	17 g	22 g	24 g	27 g	29 g	31 g	34 g	36 g	44 g	51 g
Maximum SoFAS[h] limit, calories (% of calories)	137 (14%)	121 (10%)	121 (9%)	121 (8%)	161 (9%)	258 (13%)	266 (12%)	330 (14%)	362 (14%)	395 (14%)	459 (15%)	596 (19%)

Notes

a All foods are assumed to be in nutrient-dense forms, lean or low-fat and prepared without added fats, sugars, or salt. Solid fats and added sugars may be included up to the daily maximum limit identified in the table. Food items in each group and subgroup are:

Fruits	All fresh, frozen, canned, and dried fruits and fruit juices: for example, oranges and orange juice, apples and apple juice, bananas, grapes, melons, berries, raisins.

Vegetables

Dark-green vegetables	All fresh, frozen, and canned dark-green leafy vegetables and broccoli, cooked or raw: for example, broccoli; spinach; romaine; collard, turnip, and mustard greens.
Red and orange vegetables	All fresh, frozen, and canned red and orange vegetables, cooked or raw: for example, tomatoes, red peppers, carrots, sweet potatoes, winter squash, and pumpkin.
Beans and peas (legumes)	All cooked beans and peas: for example, kidney beans, lentils, chickpeas, and pinto beans. Does not include green beans or green peas. (See additional comment under protein foods group.)
Starchy vegetables	All fresh, frozen, and canned starchy vegetables: for example, white potatoes, corn, green peas.
Other vegetables	All fresh, frozen, and canned other vegetables, cooked or raw: for example, iceberg lettuce, green beans, and onions.

Grains

Whole grains	All whole-grain products and whole grains used as ingredients: for example, whole wheat bread, whole grain cereals and crackers, oatmeal, and brown rice.
Enriched grains	All enriched refined-grain products and enriched refined grains used as ingredients: for example, white breads, enriched grain cereals and crackers, enriched pasta, white rice.
Protein foods	All meat, poultry, seafood, eggs, nuts, seeds, and processed soy products. Meat and poultry should be lean or low-fat and nuts should be unsalted. Beans and peas are considered part of this group as well as the vegetable group, but should be counted in one group only.
Dairy	All milks, including lactose-free and lactose-reduced products and fortified soy beverages, yogurts, frozen yogurts, dairy desserts, and cheeses. Most choices should be fat-free or low-fat. Cream, sour cream, and cream cheese are not included due to their low calcium content.

b Food group amounts are shown in cup (c) or ounce-equivalents (oz-eq). Oils are shown in grams (g). Quantity equivalents for each food group are:

- Grains, 1 ounce-equivalent is: 1 one-ounce slice bread; 1 ounce uncooked pasta or rice; ½ cup cooked rice, pasta, or cereal; 1 tortilla (6" diameter); 1 pancake (5" diameter); 1 ounce ready-to-eat cereal (about 1 cup cereal flakes).
- Vegetables and fruits, 1 cup equivalent is: 1 cup raw or cooked vegetable or fruit; ½ cup dried vegetable or fruit; 1 cup vegetable or fruit juice; 2 cups leafy salad greens.
- Protein foods, 1 ounce-equivalent is: 1 ounce lean meat, poultry, seafood; 1 egg; 1 Tbsp peanut butter; ½ ounce nuts or seeds. Also, ¼ cup cooked beans or peas may also be counted as 1 ounce-equivalent.
- Dairy, 1 cup equivalent is: 1 cup milk, fortified soy beverage, or yogurt; 1½ ounces natural cheese (e.g., cheddar); 2 ounces of processed cheese (e.g., American).

c See Appendix 6 of the *Dietary Guidelines for Americans, 2010,* for estimated calorie needs per day by age, gender, and physical activity level. Food intake patterns at 1,000, 1,200, and 1,400 calories meet the nutritional needs of children ages 2 to 8 years. Patterns from 1,600 to 3,200 calories meet the nutritional needs of children ages 9 years and older and adults. If a child ages 4 to 8 years needs more calories and, therefore, is following a pattern at 1,600 calories or more, the recommended amount from the dairy group can be 2½ cups per day. Children ages 9 years and older and adults should not use the 1,000, 1,200, or 1,400 calorie patterns.

d Vegetable and protein foods subgroup amounts are shown in this table as weekly amounts, because it would be difficult for consumers to select foods from all subgroups daily.

e Whole grain subgroup amounts shown in this table are minimums. More whole grains up to all of the grains recommended may be selected, with offsetting decreases in the amounts of enriched refined grains.

f The amount of dairy foods in the 1,200 and 1,400 calorie patterns have increased to reflect new RDAs for calcium that are higher than previous recommendations for children ages 4 to 8 years.

g Oils and soft margarines include vegetable, nut, and fish oils and soft vegetable oil table spreads that have no *trans* fats.

h SoFAS are calories from solid fats and added sugars. The limit for SoFAS is the remaining amount of calories in each food pattern after selecting the specified amounts in each food group in nutrient-dense forms (forms that are fat-free or low-fat and with no added sugars). The number of SoFAS is lower in the 1,200, 1,400, and 1,600 calorie patterns than in the 1,000 calorie pattern. The nutrient goals for the 1,200 to 1,600 calorie patterns are higher and require that more calories be used for nutrient-dense foods from the food groups.

Lacto-Ovo Vegetarian Adaptation of the USDA Food Patterns

For each food group or subgroup,[a] recommended average daily intake amounts[b] at all calorie levels. Recommended intakes from vegetable and protein foods subgroups are per week. For more information and tools for application, go to MyPyramid.gov.

Calorie level of pattern[c]	1,000	1,200	1,400	1,600	1,800	2,000	2,200	2,400	2,600	2,800	3,000	3,200
Fruits	1 c	1 c	1½ c	1½ c	1½ c	2 c	2 c	2 c	2 c	2½ c	2½ c	2½ c
Vegetables[d]	1 c	1½ c	1½ c	2 c	2½ c	2½ c	3 c	3 c	3½ c	3½ c	4 c	4 c
Dark-green vegetables	½ c/wk	1 c/wk	1 c/wk	1½ c/wk	1½ c/wk	1½ c/wk	2 c/wk	2 c/wk	2½ c/wk	2½ c/wk	2½ c/wk	2½ c/wk
Red and orange vegetables	2½ c/wk	3 c/wk	3 c/wk	4 c/wk	5½ c/wk	5½ c/wk	6 c/wk	6 c/wk	7 c/wk	7 c/wk	7½ c/wk	7½ c/wk
Beans and peas (legumes)	½ c/wk	½ c/wk	½ c/wk	1 c/wk	1½ c/wk	1½ c/wk	2 c/wk	2 c/wk	2½ c/wk	2½ c/wk	3 c/wk	3 c/wk
Starchy vegetables	2 c/wk	3½ c/wk	3½ c/wk	4 c/wk	5 c/wk	5 c/wk	6 c/wk	6 c/wk	7 c/wk	7 c/wk	8 c/wk	8 c/wk
Other vegetables	1½ c/wk	2½ c/wk	2½ c/wk	3½ c/wk	4 c/wk	4 c/wk	5 c/wk	5 c/wk	5½ c/wk	5½ c/wk	7 c/wk	7 c/wk
Grains[e]	3 oz-eq	4 oz-eq	5 oz-eq	5 oz-eq	6 oz-eq	6 oz-eq	7 oz-eq	8 oz-eq	9 oz-eq	10 oz-eq	10 oz-eq	10 oz-eq
Whole grains	1½ oz-eq	2 oz-eq	2½ oz-eq	3 oz-eq	3 oz-eq	3 oz-eq	3½ oz-eq	4 oz-eq	4½ oz-eq	5 oz-eq	5 oz-eq	5 oz-eq
Refined grains	1½ oz-eq	2 oz-eq	2½ oz-eq	2 oz-eq	3 oz-eq	3 oz-eq	3½ oz-eq	4 oz-eq	4½ oz-eq	5 oz-eq	5 oz-eq	5 oz-eq
Protein foods[d]	2 oz-eq	3 oz-eq	4 oz-eq	5 oz-eq	5 oz-eq	5½ oz-eq	6 oz-eq	6½ oz-eq	6½ oz-eq	7 oz-eq	7 oz-eq	7 oz-eq
Eggs	1 oz-eq/wk	2 oz-eq/wk	3 oz-eq/wk	4 oz-eq/wk	4 oz-eq/wk	4 oz-eq/wk	4 oz-eq/wk	5 oz-eq/wk	5 oz-eq/wk	5 oz-eq/wk	5 oz-eq/wk	5 oz-eq/wk
Beans and peas[f]	3½ oz-eq/wk	5 oz-eq/wk	7 oz-eq/wk	9 oz-eq/wk	9 oz-eq/wk	10 oz-eq/wk	10 oz-eq/wk	11 oz-eq/wk	11 oz-eq/wk	12 oz-eq/wk	12 oz-eq/wk	12 oz-eq/wk
Soy products	4 oz-eq/wk	6 oz-eq/wk	8 oz-eq/wk	11 oz-eq/wk	11 oz-eq/wk	12 oz-eq/wk	13 oz-eq/wk	14 oz-eq/wk	14 oz-eq/wk	15 oz-eq/wk	15 oz-eq/wk	15 oz-eq/wk
Nuts and seeds	5 oz-eq/wk	7 oz-eq/wk	10 oz-eq/wk	12 oz-eq/wk	12 oz-eq/wk	13 oz-eq/wk	15 oz-eq/wk	16 oz-eq/wk	16 oz-eq/wk	17 oz-eq/wk	17 oz-eq/wk	17 oz-eq/wk
Dairy[g]	2 c	2½ c	2½ c	3 c	3 c	3 c	3 c	3 c	3 c	3 c	3 c	3 c
Oils[h]	12 g	13 g	12 g	15 g	17 g	19 g	21 g	22 g	25 g	26 g	34 g	41 g
Maximum SoFAS[i] limit, calories (% total calories)	137 (14%)	121 (10%)	121 (9%)	121 (8%)	161 (9%)	258 (13%)	266 (12%)	330 (14%)	362 (14%)	395 (14%)	459 (15%)	596 (19%)

a,b,c,d,eSee notes a through e for USDA Food Patterns table above.

[f]Total recommended beans and peas amounts would be the sum of amounts recommended in the vegetable and the protein foods groups. An ounce-equivalent of beans and peas in the protein foods group is ¼ cup, cooked. For example, in the 2,000-calorie pattern, total weekly beans and peas recommendation is (10 oz-eq/4) + 1½ cups = about 4 cups, cooked.

g,h,iSee notes f, g, and h for USDA Food Patterns table above.

Vegan Adaptation of the USDA Food Patterns

For each food group or subgroup,[a] a recommended average daily intake amounts[b] at all calorie levels. Recommended intakes from vegetable and protein foods subgroups are per week. For more information and tools for application, go to MyPyramid.gov.

Calorie level of pattern[c]	1,000	1,200	1,400	1,600	1,800	2,000	2,200	2,400	2,600	2,800	3,000	3,200
Fruits	1 c	1 c	1½ c	1½ c	1½ c	2 c	2 c	2 c	2 c	2½ c	2½ c	2½ c
Vegetables[d]	1 c	1½ c	1½ c	2 c	2½ c	2½ c	3 c	3 c	3½ c	3½ c	4 c	4 c
Dark-green vegetables	½ c/wk	1 c/wk	1 c/wk	1½ c/wk	1½ c/wk	1½ c/wk	2 c/wk	2 c/wk	2½ c/wk	2½ c/wk	2½ c/wk	2½ c/wk
Red and orange vegetables	2½ c/wk	3 c/wk	3 c/wk	4 c/wk	5½ c/wk	5½ c/wk	6 c/wk	6 c/wk	7 c/wk	7 c/wk	7½ c/wk	7½ c/wk
Beans and peas (legumes)	½ c/wk	½ c/wk	½ c/wk	1 c/wk	1½ c/wk	1½ c/wk	2 c/wk	2 c/wk	2½ c/wk	2½ c/wk	3 c/wk	3 c/wk
Starchy vegetables	2 c/wk	3½ c/wk	3½ c/wk	4 c/wk	5 c/wk	5 c/wk	6 c/wk	6 c/wk	7 c/wk	7 c/wk	8 c/wk	8 c/wk
Other vegetables	1½ c/wk	2½ c/wk	2½ c/wk	3½ c/wk	4 c/wk	4 c/wk	5 c/wk	5 c/wk	5½ c/wk	5½ c/wk	7 c/wk	7 c/wk
Grains[e]	3 oz-eq	4 oz-eq	5 oz-eq	5 oz-eq	6 oz-eq	6 oz-eq	7 oz-eq	8 oz-eq	9 oz-eq	10 oz-eq	10 oz-eq	10 oz-eq
Whole grains	1½ oz-eq	2 oz-eq	2½ oz-eq	3 oz-eq	3 oz-eq	3 oz-eq	3½ oz-eq	4 oz-eq	4½ oz-eq	5 oz-eq	5 oz-eq	5 oz-eq
Refined grains	1½ oz-eq	2 oz-eq	2½ oz-eq	2 oz-eq	3 oz-eq	3 oz-eq	3½ oz-eq	4 oz-eq	4½ oz-eq	5 oz-eq	5 oz-eq	5 oz-eq
Protein foods[d]	2 oz-eq	3 oz-eq	4 oz-eq	5 oz-eq	5 oz-eq	5½ oz-eq	6 oz-eq	6½ oz-eq	6½ oz-eq	7 oz-eq	7 oz-eq	7 oz-eq
Beans and peas[f]	5 oz-eq/wk	7 oz-eq/wk	10 oz-eq/wk	12 oz-eq/wk	12 oz-eq/wk	13 oz-eq/wk	15 oz-eq/wk	16 oz-eq/wk	16 oz-eq/wk	17 oz-eq/wk	17 oz-eq/wk	17 oz-eq/wk
Soy products	4 oz-eq/wk	5 oz-eq/wk	7 oz-eq/wk	9 oz-eq/wk	9 oz-eq/wk	10 oz-eq/wk	11 oz-eq/wk	11 oz-eq/wk	11 oz-eq/wk	12 oz-eq/wk	12 oz-eq/wk	12 oz-eq/wk
Nuts and seeds	6 oz-eq/wk	8 oz-eq/wk	11 oz-eq/wk	14 oz-eq/wk	14 oz-eq/wk	15 oz-eq/wk	17 oz-eq/wk	18 oz-eq/wk	18 oz-eq/wk	20 oz-eq/wk	20 oz-eq/wk	20 oz-eq/wk
Dairy (vegan)[g]	2 c	2½ c	2½ c	3 c	3 c	3 c	3 c	3 c	3 c	3 c	3 c	3 c
Oils[h]	12 g	12 g	11 g	14 g	16 g	18 g	20 g	21 g	24 g	25 g	33 g	40 g
Maximum SoFAS[i] limit, calories (% total calories)	137 (14%)	121 (10%)	121 (9%)	121 (8%)	161 (9%)	258 (13%)	266 (12%)	330 (14%)	362 (14%)	395 (14%)	459 (15%)	596 (19%)

a,b,c,d,eSee notes a through e for USDA Food Patterns table above.

fTotal recommended beans and peas amounts would be the sum of amounts recommended in the vegetable and the protein foods groups. An ounce-equivalent of beans and peas in the protein foods group is ¼ cup, cooked. For example, in the 2,000-calorie pattern, total weekly beans and peas recommendation is (13 oz-eq/4) + 1½ cups = about 5 cups, cooked.

gThe vegan "dairy group" is composed of calcium-fortified beverages and foods from plant sources. For analysis purposes the following products were included: calcium-fortified soy beverage, calcium-fortified rice milk, tofu made with calcium-sulfate, and calcium-fortified soy yogurt. The amounts in the 1,200 and 1,400 calorie patterns have increased to reflect new RDAs for calcium that are higher than previous recommendations for children ages 4 to 8 years.

h,iSee notes g and h for USDA Food Patterns table above.

Source: U.S. Department of Agriculture and U.S. Department of Health and Human Services, *Dietary Guidelines for Americans 2010*, 7th edition, Washington DC: U.S. Government Printing Office, December 2010.

APPENDIX D Food Exchange System

Chapter 1 briefly explains the exchange system, and this appendix provides details from the *2008 Choose Your Foods: Exchange Lists for Diabetes*. Exchange lists can help people with diabetes manage their blood glucose levels by controlling the amount and kinds of carbohydrates they consume. These lists can also help in planning diets for weight management by controlling calorie and fat intake.

The Exchange System

The exchange system sorts foods into groups by their proportions of carbohydrate, fat, and protein (Table D-1). These groups may be organized into several exchange lists of foods (Tables D-2 through D-12). For example, the carbohydrate group includes these exchange lists:

- Starch
- Fruits
- Milk (fat-free, reduced-fat, and whole)
- Sweets, Desserts, and Other Carbohydrates
- Nonstarchy Vegetables

Then any food on a list can be "exchanged" for any other on that same list. Another group for alcohol has been included as a reminder that these beverages often deliver substantial carbohydrate and calories, and therefore warrant their own list.

Serving Sizes

The serving sizes have been carefully adjusted and defined so that a serving of any food on a given list provides roughly the same amount of carbohydrate, fat, and protein, and, therefore, total energy. Any food on a list can thus be exchanged, or traded, for any other food on the same list without significantly affecting the diet's energy-nutrient balance or total calories. For example, a person may select 17 small grapes or ½ large grapefruit as one fruit exchange, and either choice would provide roughly 15 grams of carbohydrate and 60 calories. A whole grapefruit, however, would count as 2 fruit exchanges.

To apply the system successfully, users must become familiar with the specified serving sizes. A convenient way to remember the serving sizes and energy values is to keep in mind a typical item from each list (review Table D-1).

The Foods on the Lists

Foods do not always appear on the exchange list where you might first expect to find them. They are grouped according to their energy-nutrient contents rather than by their source (such as milks), their outward appearance, or their vitamin and mineral contents. For example, cheeses are grouped with meats (not milk) because, like meats, cheeses contribute energy from protein and fat but provide negligible carbohydrate.

For similar reasons, starchy vegetables such as corn, green peas, and potatoes are found on the Starch list with breads and cereals, not with the vegetables. Likewise, bacon is grouped with the fats and oils, not with the meats.

Diet planners learn to view mixtures of foods, such as casseroles and soups, as combinations of foods from different exchange lists. They also learn to interpret food labels with the exchange system in mind.

Controlling Energy, Fat, and Sodium

The exchange lists help people control their energy intakes by paying close attention to serving sizes. People wanting to lose weight can limit foods from the Sweets, Desserts, and Other Carbohydrates and Fats lists, and they might choose to avoid the Alcohol list altogether. The Free Foods list provides low-calorie choices.

By assigning items like bacon to the Fats list, the exchange lists alert consumers to foods that are unexpectedly high in fat. Even the Starch list specifies which grain products contain added fat (such as biscuits, cornbread, and waffles) by marking them with a symbol to indicate added fat (the symbols are explained in the table keys). In addition, the exchange lists encourage users to think of fat-free milk as milk and of whole milk as milk with added fat, and to think of lean meats as meats and of medium-fat and high-fat meats as meats with added fat. To that end, foods on the milk and meat lists are separated into categories based on their fat contents (review Table D-1). The Milk list is subdivided for fat-free, reduced-fat, and whole; the meat list is subdivided for lean, medium fat, and high fat. The meat list also includes plant-based proteins, which tend to be rich in fiber. Notice that many of these foods bear the symbol for "high fiber."

People wanting to control the sodium in their diets can begin by eliminating any foods bearing the "high sodium" symbol. In most cases, the symbol identifies foods that, in one serving, provide 480 milligrams or more of sodium. Foods on the Combination Foods or Fast Foods lists that bear the symbol provide more than 600 milligrams of sodium. Other foods may also contribute substantially to sodium (consult Chapter 7 for details).

Planning a Healthy Diet

To obtain a daily variety of foods that provide healthful amounts of carbohydrate, protein, and fat, as well as vitamins, minerals, and fiber, the meal plan for adults and teenagers should include at least:

- two to three servings of nonstarchy vegetables
- two servings of fruits
- six servings of grains (at least three of whole grains), beans, and starchy vegetables
- two servings of low-fat or fat-free milk
- about 6 ounces of meat or meat substitutes
- *small* amounts of fat and sugar

The actual amounts are determined by age, gender, activity levels, and other factors that influence energy needs. Refer to Chapters 1 and 10 as you read through these sections to get an idea of how exchange lists can be useful in planning a diet.

Table D-1 The Food Lists

Lists	Typical item/portion size	Carbohydrate (g)	Protein (g)	Fat (g)	Energy[a] (cal)
Carbohydrates					
Starch[b]	1 slice bread	15	0–3	0–1	80
Fruits	1 small apple	15	—	—	60
Milk					
Fat-free, low-fat, 1%	1 c fat-free milk	12	8	0–3	100
Reduced-fat, 2%	1 c reduced-fat milk	12	8	5	120
Whole	1 c whole milk	12	8	8	160
Sweets, desserts, and other carbohydrates[c]	2 small cookies	15	varies	varies	varies
Nonstarchy vegetables	½ c cooked carrots	5	2	—	25
Meat and meat substitutes					
Lean	1 oz chicken (no skin)	—	7	0–3	45
Medium-fat	1 oz ground beef	—	7	4–7	75
High-fat	1 oz pork sausage	—	7	8+	100
Plant-based proteins	½ c tofu	varies	7	varies	varies
Fats	1 tsp butter	—	—	5	45
Alcohol	12 oz beer	varies	—	—	100

[a]The energy value for each exchange list represents an approximate average for the group and does not reflect the precise number of grams of carbohydrate, protein, and fat. For example, a slice of bread contains 15 grams of carbohydrate (60 calories), 3 grams protein (12 calories), and a little fat—rounded to 80 calories for ease in calculating. A ½ cup of vegetables (not including starchy vegetables) contains 5 grams carbohydrate (20 calories) and 2 grams protein (8 more), which has been rounded down to 25 calories.

[b]The Starch list includes cereals, grains, breads, crackers, snacks, starchy vegetables (such as corn, peas, and potatoes), and legumes (dried beans, peas, and lentils).

[c]The Sweets, Desserts, and Other Carbohydrates list includes foods that contain added sugars and fats such as sodas, candy, cakes, cookies, doughnuts, ice cream, pudding, syrup, and frozen yogurt.

Table D-2 Starch

The Starch list includes bread, cereals and grains, starchy vegetables, crackers and snacks, and legumes (dried beans, peas, and lentils). 1 starch choice = 15 grams carbohydrate, 0–3 grams protein, 0–1 grams fat, and 80 calories.

Note: In general, one starch exchange is ½ cup cooked cereal, grain, or starchy vegetable; ⅓ cup cooked rice or pasta; 1 ounce of bread product; ¾ ounce to 1 ounce of most snack foods.

Bread

Food	Serving size
Bagel, large (about 4 oz)	¼ (1 oz)
▽ Biscuit, 2½ inches across	1
Bread	
☺ reduced-calorie	2 slices (1½ oz)
white, whole grain, pumpernickel, rye, unfrosted raisin	1 slice (1 oz)
Chapatti, small, 6 inches across	1
▽ Cornbread, 1¾ inch cube	1 (1½ oz)
English muffin	½
Hot dog bun or hamburger bun	½ (1 oz)
Naan, 8 inches by 2 inches	¼
Pancake, 4 inches across, ¼ inch thick	1
Pita, 6 inches across	½
Roll, plain, small	1 (1 oz)
▽ Stuffing, bread	⅓ cup
▽ Taco shell, 5 inches across	2
Tortilla, corn, 6 inches across	1
Tortilla, flour, 6 inches across	1
Tortilla, flour, 10 inches across	⅓
▽ Waffle, 4-inch square or 4 inches across	1

Cereals and grains

Food	Serving size
Barley, cooked	⅓ cup
Bran, dry	
☺ oat	¼ cup
☺ wheat	½ cup
☺ Bulgur (cooked)	½ cup
Cereals	
☺ bran	½ cup
cooked (oats, oatmeal)	½ cup
puffed	1½ cups
shredded wheat, plain	½ cup
sugar-coated	½ cup
unsweetened, ready-to-eat	¾ cup
Couscous	⅓ cup
Granola	
low-fat	¼ cup
▽ regular	¼ cup
Grits, cooked	½ cup
Kasha	½ cup
Millet, cooked	⅓ cup
Muesli	¼ cup
Pasta, cooked	⅓ cup

Cereals and grains

Food	Serving size
Polenta, cooked	⅓ cup
Quinoa, cooked	⅓ cup
Rice, white or brown, cooked	⅓ cup
Tabbouleh (tabouli), prepared	½ cup
Wheat germ, dry	3 Tbsp
Wild rice, cooked	½ cup

Starchy vegetables

Food	Serving size
Cassava	⅓ cup
Corn	½ cup
on cob, large	½ cob (5 oz)
☺ Hominy, canned	¾ cup
☺ Mixed vegetables with corn, peas, or pasta	1 cup
☺ Parsnips	½ cup
☺ Peas, green	½ cup
Plantain, ripe	⅓ cup
Potato	
baked with skin	¼ large (3 oz)
boiled, all kinds	½ cup or ½ medium (3 oz)
▽ mashed, with milk and fat	½ cup
french fried (oven-baked)[a]	1 cup (2 oz)
☺ Pumpkin, canned, no sugar added	1 cup
Spaghetti/pasta sauce	½ cup
☺ Squash, winter (acorn, butternut)	1 cup
☺ Succotash	½ cup
Yam, sweet potato, plain	½ cup

Crackers and snacks[b]

Food	Serving size
Animal crackers	8
Crackers	
▽ round-butter type	6
saltine-type	6
▽ sandwich-style, cheese or peanut butter filling	3
▽ whole-wheat regular	2–5 (¾ oz)
☺ whole-wheat lower fat or crispbreads	2–5 (¾ oz)
Graham cracker, 2½-inch square	3
Matzoh	¾ oz
Melba toast, about 2-inch by 4-inch piece	4
Oyster crackers	20
Popcorn	3 cups
▽ ☺ with butter	3 cups
☺ no fat added	3 cups
☺ lower fat	3 cups

Table D-2 Starch (Continued)

Crackers and snacks[b] (continued)	
Food	Serving size
Pretzels	¾ oz
Rice cakes, 4 inches across	2
Snack chips	
fat-free or baked (tortilla, potato), baked pita chips	15–20 (¾ oz)
▽ regular (tortilla, potato)	9–13 (¾ oz)

Beans, peas, and lentils[c]	
The choices on this list count as 1 starch + 1 lean meat.	
Food	Serving size
☺ Baked beans	⅓ cup
☺ Beans, cooked (black, garbanzo, kidney, lima, navy, pinto, white)	½ cup
☺ Lentils, cooked (brown, green, yellow)	½ cup
☺ Peas, cooked (black-eyed, split)	½ cup
🧂☺ Refried beans, canned	½ cup

KEY

☺ = More than 3 grams of dietary fiber per serving.

▽ = Extra fat, or prepared with added fat. (Count as 1 starch + 1 fat.)

🧂 = 480 milligrams or more of sodium per serving.

[a]Restaurant-style french fries are on the Fast Foods list.

[b]For other snacks, see the Sweets, Desserts, and Other Carbohydrates list. For a quick estimate of serving size, an open handful is equal to about 1 cup or 1 to 2 ounces of snack food.

[c]Beans, peas, and lentils are also found on the Meat and Meat Substitutes list.

Table D-3 Fruits

Fruit[a]

The Fruits list includes fresh, frozen, canned, and dried fruits and fruit juices. 1 fruit choice = 15 grams carbohydrate, 0 grams protein, 0 grams fat, and 60 calories.

Note: In general, one fruit exchange is ½ cup canned or fresh fruit or unsweetened fruit juice; 1 small fresh fruit (4 ounces); 2 tablespoons dried fruit.

Food	Serving size	Food	Serving size
Apple, unpeeled, small	1 (4 oz)	Fruit cocktail	½ cup
Apples, dried	4 rings	Grapefruit	
Applesauce, unsweetened	½ cup	large	½ (11 oz)
Apricots		sections, canned	¾ cup
canned	½ cup	Grapes, small	17 (3 oz)
dried	8 halves	Honeydew melon	1 slice or 1 cup cubed (10 oz)
☺ fresh	4 whole (5½ oz)	☺ Kiwi	1 (3½ oz)
Banana, extra small	1 (4 oz)	Mandarin oranges, canned	¾ cup
☺ Blackberries	¾ cup	Mango, small	½ (5½ oz) or ½ cup
Blueberries	¾ cup		
Cantaloupe, small	⅓ melon or 1 cup cubed (11 oz)	Nectarine, small	1 (5 oz)
		☺ Orange, small	1 (6½ oz)
		Papaya	½ or 1 cup cubed (8 oz)
Cherries			
sweet, canned	½ cup	Peaches	
sweet fresh	12 (3 oz)	canned	½ cup
Dates	3	fresh, medium	1 (6 oz)
Dried fruits (blueberries, cherries, cranberries, mixed fruit, raisins)	2 Tbsp	Pears	
		canned	½ cup
Figs		fresh, large	½ (4 oz)
dried	1½	Pineapple	
☺ fresh	1½ large or 2 medium (3½ oz)	canned	½ cup
		fresh	¾ cup

[a]The weight listed includes skin, core, seeds, and rind.

(Continued)

Table D-3 Fruits (Continued)

Fruit[a]		Fruit juice	
Food	**Serving size**	**Food**	**Serving size**
Plums		Apple juice/cider	½ cup
canned	½ cup	Fruit juice blends, 100% juice	⅓ cup
dried (prunes)	3	Grape juice	⅓ cup
small	2 (5 oz)	Grapefruit juice	½ cup
☺ Raspberries	1 cup	Orange juice	½ cup
☺ Strawberries	1¼ cup whole berries	Pineapple juice	½ cup
☺ Tangerines, small	2 (8 oz)	Prune juice	⅓ cup
Watermelon	1 slice or 1¼ cups cubes (13½ oz)		

KEY

☺ = More than 3 grams of dietary fiber per serving.

▽ = Extra fat, or prepared with added fat. (Count as 1 starch + 1 fat.)

▯ = 480 milligrams or more of sodium per serving.

Table D-4 Milk

The Milk list groups milks and yogurts based on the amount of fat they have (fat-free/low fat, reduced fat, and whole). Cheeses are found on the Meat and Meat Substitutes list and cream and other dairy fats are found on the Fats list.

Note: In general, one milk choice is 1 cup (8 fluid ounces or ½ pint) milk or yogurt.

Milk and yogurts	
Food	**Serving size**
Fat-free or low-fat (1%)	
1 fat-free/low-fat milk choice = 12 g carbohydrate, 8 g protein, 0–3 g fat, and 100 cal.	
Milk, buttermilk, acidophilus milk, Lactaid	1 cup
Evaporated milk	½ cup
Yogurt, plain or flavored with an artificial sweetener	⅔ cup (6 oz)
Reduced-fat (2%)	
1 reduced-fat milk choice = 12 g carbohydrate, 8 g protein, 5 g fat, and 120 cal.	
Milk, acidophilus milk, kefir, Lactaid	1 cup
Yogurt, plain	⅔ cup (6 oz)
Whole	
1 whole milk choice = 12 g carbohydrate, 8 g protein, 8 g fat, and 160 cal.	
Milk, buttermilk, goat's milk	1 cup
Evaporated milk	½ cup
Yogurt, plain	8 oz

Dairy-like foods		
Food	**Serving size**	**Count as**
Chocolate milk		
fat-free	1 cup	1 fat-free milk + 1 carbohydrate
whole	1 cup	1 whole milk + 1 carbohydrate
Eggnog, whole milk	½ cup	1 carbohydrate + 2 fats
Rice drink		
flavored, low fat	1 cup	2 carbohydrates
plain, fat-free	1 cup	1 carbohydrate
Smoothies, flavored, regular	10 oz	1 fat-free milk + 2½ carbohydrates

Table D-4 Milk (Continued)

Dairy-like foods

Food	Serving size	Count as
Soy milk		
light	1 cup	1 carbohydrate + ½ fat
regular, plain	1 cup	1 carbohydrate + 1 fat
Yogurt		
and juice blends	1 cup	1 fat-free milk + 1 carbohydrate
low carbohydrate (less than 6 grams carbohydrate per choice)	⅔ cup (6 oz)	½ fat-free milk
with fruit, low-fat	⅔ cup (6 oz)	1 fat-free milk + 1 carbohydrate

Table D-5 Sweets, Desserts, and Other Carbohydrates

1 other carbohydrate choice = 15 grams carbohydrate, variable grams protein, variable grams fat, and variable calories.
Note: In general, one choice from this list can substitute for foods on the starch, fruits, or milk lists.

Beverages, soda, and energy/sports drinks

Food	Serving size	Count as
Cranberry juice cocktail	½ cup	1 carbohydrate
Energy drink	1 can (8.3 oz)	2 carbohydrates
Fruit drink or lemonade	1 cup (8 oz)	2 carbohydrates
Hot chocolate		
regular	1 envelope added to 8 oz water	1 carbohydrate + 1 fat
sugar-free or light	1 envelope added to 8 oz water	1 carbohydrate
Soft drink (soda), regular	1 can (12 oz)	2½ carbohydrates
Sports drink	1 cup (8 oz)	1 carbohydrate

Brownies, cake, cookies, gelatin, pie, and pudding

Food	Serving size	Count as
Brownie, small, unfrosted	1¼-inch square, ⅞ inch high (about 1 oz)	1 carbohydrate + 1 fat
Cake		
angel food, unfrosted	1/12 of cake (about 2 oz)	2 carbohydrates
frosted	2-inch square (about 2 oz)	2 carbohydrates + 1 fat
unfrosted	2-inch square (about 2 oz)	1 carbohydrate + 1 fat
Cookies		
chocolate chip	2 cookies (2¼ inches across)	1 carbohydrate + 2 fats
gingersnap	3 cookies	1 carbohydrate
sandwich, with crème filling	2 small (about ⅔ oz)	1 carbohydrate + 1 fat
sugar-free	3 small or 1 large (¾–1 oz)	1 carbohydrate + 1–2 fats
vanilla wafer	5 cookies	1 carbohydrate + 1 fat
Cupcake, frosted	1 small (about 1¾ oz)	2 carbohydrates + 1–1½ fats
Fruit cobbler	½ cup (3½ oz)	3 carbohydrates + 1 fat
Gelatin, regular	½ cup	1 carbohydrate
Pie		
commercially prepared fruit, 2 crusts	⅙ of 8-inch pie	3 carbohydrates + 2 fats
pumpkin or custard	⅛ of 8-inch pie	1½ carbohydrates + 1½ fats
Pudding		
regular (made with reduced-fat milk)	½ cup	2 carbohydrates
sugar-free or sugar- and fat-free (made with fat-free milk)	½ cup	1 carbohydrate

(Continued)

Table D-5 Sweets, Desserts, and Other Carbohydrates (Continued)

Candy, spreads, sweets, sweeteners, syrups, and toppings

Food	Serving size	Count as
Candy bar, chocolate/peanut	2 "fun size" bars (1 oz)	1½ carbohydrates + 1½ fats
Candy, hard	3 pieces	1 carbohydrate
Chocolate "kisses"	5 pieces	1 carbohydrate + 1 fat
Coffee creamer		
dry, flavored	4 tsp	½ carbohydrate + ½ fat
liquid, flavored	2 Tbsp	1 carbohydrate
Fruit snacks, chewy (purood fruit concentrate)	1 roll (¾ oz)	1 carbohydrate
Fruit spreads, 100% fruit	1½ Tbsp	1 carbohydrate
Honey	1 Tbsp	1 carbohydrate
Jam or jelly, regular	1 Tbsp	1 carbohydrate
Sugar	1 Tbsp	1 carbohydrate
Syrup		
chocolate	2 Tbsp	2 carbohydrates
light (pancake type)	2 Tbsp	1 carbohydrate
regular (pancake type)	1 Tbsp	1 carbohydrate

Condiments and sauces[a]

Food	Serving size	Count as
Barbeque sauce	3 Tbsp	1 carbohydrate
Cranberry sauce, jellied	¼ cup	1½ carbohydrates
Gravy, canned or bottled	½ cup	½ carbohydrate + ½ fat
Salad dressing, fat-free, low-fat, cream-based	3 Tbsp	1 carbohydrate
Sweet and sour sauce	3 Tbsp	1 carbohydrate

Doughnuts, muffins, pastries, and sweet breads

Food	Serving size	Count as
Banana nut bread	1-inch slice (1 oz)	2 carbohydrates + 1 fat
Doughnut		
cake, plain	1 medium (1½ oz)	1½ carbohydrates + 2 fats
yeast type, glazed	3¾ inches across (2 oz)	2 carbohydrates + 2 fats
Muffin (4 oz)	¼ muffin (1 oz)	1 carbohydrate + ½ fat
Sweet roll or Danish	1 (2½ oz)	2½ carbohydrates + 2 fats

Frozen bars, frozen desserts, frozen yogurt, and ice cream

Food	Serving size	Count as
Frozen pops	1	½ carbohydrate
Fruit juice bars, frozen, 100% juice	1 bar (3 oz)	1 carbohydrate
Ice cream		
fat-free	½ cup	1½ carbohydrates
light	½ cup	1 carbohydrate + 1 fat
no sugar added	½ cup	1 carbohydrate + 1 fat
regular	½ cup	1 carbohydrate + 2 fats
Sherbet, sorbet	½ cup	2 carbohydrates
Yogurt, frozen		
fat-free	⅓ cup	1 carbohydrate
regular	½ cup	1 carbohydrate + 0–1 fat

Table D-5 Sweets, Desserts, and Other Carbohydrates (Continued)

Granola bars, meal replacement bars/shakes, and trail mix

Food	Serving size	Count as
Granola or snack bar, regular or low-fat	1 bar (1 oz)	1½ carbohydrates
Meal replacement bar	1 bar (1⅓ oz)	1½ carbohydrates + 0–1 fat
Meal replacement bar	1 bar (2 oz)	2 carbohydrates + 1 fat
Meal replacement shake, reduced calorie	1 can (10–11 oz)	1½ carbohydrates + 0–1 fat
Trail mix		
candy/nut-based	1 oz	1 carbohydrate + 2 fats
dried fruit-based	1 oz	1 carbohydrate + 1 fat

KEY

🥫 = 480 milligrams or more of sodium per serving.

[a]You can also check the Fats list and Free Foods list for other condiments.

Table D-6 Nonstarchy Vegetables

The Nonstarchy Vegetables list includes vegetables that have few grams of carbohydrates or calories; starchy vegetables are found on the Starch list. 1 nonstarchy vegetable choice = 5 grams carbohydrate, 2 grams protein, 0 grams fat, and 25 calories.

Note: In general, one nonstarchy vegetable choice is ½ cup cooked vegetables or vegetable juice or 1 cup raw vegetables. Count 3 cups of raw vegetables or 1½ cups of cooked vegetables as one carbohydrate choice.

Nonstarchy vegetables[a]

Amaranth or Chinese spinach	Kohlrabi
Artichoke	Leeks
Artichoke hearts	Mixed vegetables (without corn, peas, or pasta)
Asparagus	Mung bean sprouts
Baby corn	Mushrooms, all kinds, fresh
Bamboo shoots	Okra
Beans (green, wax, Italian)	Onions
Bean sprouts	Oriental radish or daikon
Beets	Pea pods
🥫 Borscht	☺ Peppers (all varieties)
Broccoli	Radishes
☺ Brussels sprouts	Rutabaga
Cabbage (green, bok choy, Chinese)	Sauerkraut
☺ Carrots	Soybean sprouts
Cauliflower	Spinach
Celery	Squash (summer, crookneck, zucchini)
☺ Chayote	Sugar pea snaps
Coleslaw, packaged, no dressing	☺ Swiss chard
Cucumber	Tomato
Eggplant	Tomatoes, canned
Gourds (bitter, bottle, luffa, bitter melon)	🥫 Tomato sauce
Green onions or scallions	🥫 Tomato/vegetable juice
Greens (collard, kale, mustard, turnip)	Turnips
Hearts of palm	Water chestnuts
Jicama	Yard-long beans

KEY

☺ = More than 3 grams of dietary fiber per serving.

🥫 = 480 milligrams or more of sodium per serving.

[a]Salad greens (like chicory, endive, escarole, lettuce, romaine, spinach, arugula, radicchio, watercress) are on the Free Foods list.

Table D-7 Meat and Meat Substitutes

The Meat and Meat Substitutes list groups foods based on the amount of fat they have (lean meat, medium-fat meat, high-fat meat, and plant-based proteins).

Lean meats and meat substitutes

1 lean meat choice = 0 grams carbohydrate, 7 grams protein, 0–3 grams fat, and 45 calories.

Food	Amount	Food	Amount
Beef: Select or Choice grades trimmed of fat: ground round, roast (chuck, rib, rump), round, sirloin, steak (cubed, flank, porterhouse, T-bone), tenderloin	1 oz	Organ meats: heart, kidney, liver *Note: May be high in cholesterol.*	1 oz
		Oysters, fresh or frozen	6 medium
		Pork, loan	
Beef jerky	1 oz	Canadian bacon	1 oz
Cheeses with 3 grams of fat or less per oz	1 oz	rib or loin chop/roast, ham, tenderloin	1 oz
Cottage cheese	¼ cup	Poultry, without skin: Cornish hen, chicken, domestic duck or goose (well drained of fat), turkey	1 oz
Egg substitutes, plain	¼ cup		
Egg whites	2	Processed sandwich meats with 3 grams of fat or less per oz: chipped beef, deli thin-sliced meats, turkey ham, turkey kielbasa, turkey pastrami	1 oz
Fish, fresh or frozen, plain: catfish, cod, flounder, haddock, halibut, orange roughy, salmon, tilapia, trout, tuna	1 oz		
		Salmon, canned	1 oz
Fish, smoked: herring or salmon (lox)	1 oz	Sardines, canned	2 medium
Game: buffalo, ostrich, rabbit, venison	1 oz	Sausage with 3 grams of fat or less per oz	1 oz
Hot dog with 3 grams of fat or less per oz (8 dogs per 14 oz package)	1	Shellfish: clams, crab, imitation shellfish, lobster, scallops, shrimp	1 oz
Note: May be high in carbohydrate.		Tuna, canned in water or oil, drained	1 oz
Lamb: chop, leg, or roast	1 oz	Veal, lean chop, roast	1 oz

Medium-fat meat and meat substitutes

1 medium-fat meat choice = 0 grams carbohydrate, 7 grams protein, 4–7 grams fat, and 75 calories.

Food	Amount
Beef: corned beef, ground beef, meatloaf, Prime grades trimmed of fat (prime rib), short ribs, tongue	1 oz
Cheeses with 4–7 grams of fat per oz: feta, mozzarella, pasteurized processed cheese spread, reduced-fat cheeses, string	1 oz
Egg	1
Note: High in cholesterol, so limit to 3 per week.	
Fish, any fried product	1 oz
Lamb: ground, rib roast	1 oz
Pork: cutlet, shoulder roast	1 oz
Poultry: chicken with skin; dove, pheasant, wild duck, or goose; fried chicken; ground turkey	1 oz
Ricotta cheese	2 oz or ¼ cup
Sausage with 4–7 grams of fat per oz	1 oz
Veal, cutlet (no breading)	1 oz

High-fat meat and meat substitutes

1 high-fat meat choice = 0 grams carbohydrate, 7 grams protein, 8+ grams fat, and 100 calories. These foods are high in saturated fat, cholesterol, and calories and may raise blood cholesterol levels if eaten on a regular basis. Try to eat 3 or fewer servings from this group per week.

Food	Amount
Bacon	
pork	2 slices (16 slices per lb or 1 oz each, before cooking)
turkey	3 slices (½ oz each before cooking)
Cheese, regular: American, bleu, brie, cheddar, hard goat, Monterey jack, queso, and Swiss	1 oz
Hot dog: beef, pork, or combination (10 per lb-sized package)	1
Hot dog: turkey or chicken (10 per lb-sized package)	1
Pork: ground, sausage, spareribs	1 oz
Processed sandwich meats with 8 grams of fat or more per oz: bologna, pastrami, hard salami	1 oz
Sausage with 8 grams fat or more per oz: bratwurst, chorizo, Italian, knockwurst, Polish, smoked, summer	1 oz

Table D-7 Meat and Meat Substitutes (Continued)

Plant-based proteins

1 plant-based protein choice = variable grams carbohydrate, 7 grams protein, variable grams fat, and variable calories.

Because carbohydrate content varies among plant-based proteins, you should read the food label.

Food	Serving size	Count as
"Bacon" strips, soy-based	3 strips	1 medium-fat meat
☻Baked beans	⅓ cup	1 starch + 1 lean meat
☻Beans, cooked: black, garbanzo, kidney, lima, navy, pinto, white[a]	½ cup	1 starch + 1 lean meat
☻"Beef" or "sausage" crumbles, soy-based	2 oz	½ carbohydrate + 1 lean meat
"Chicken" nuggets, soy-based	2 nuggets (1½ oz)	½ carbohydrate + 1 medium-fat meat
☻Edamame	½ cup	½ carbohydrate + 1 lean meat
Falafel (spiced chickpea and wheat patties)	3 patties (about 2 inches across)	1 carbohydrate + 1 high-fat meat
Hot dog, soy-based	1 (1½ oz)	½ carbohydrate + 1 lean meat
☻Hummus	⅓ cup	1 carbohydrate + 1 high-fat meat
☻Lentils, brown, green, or yellow	½ cup	1 carbohydrate + 1 lean meat
☻Meatless burger, soy-based	3 oz	½ carbohydrate + 2 lean meats
☻Meatless burger, vegetable- and starch-based	1 patty (about 2½ oz)	1 carbohydrate + 2 lean meats
Nut spreads: almond butter, cashew butter, peanut butter, soy nut butter	1 Tbsp	1 high-fat meat
☻Peas, cooked: black-eyed and split peas	½ cup	1 starch + 1 lean meat
🔋☻Refried beans, canned	½ cup	1 starch + 1 lean meat
"Sausage" patties, soy-based	1 (1½ oz)	1 medium-fat meat
Soy nuts, unsalted	¾ oz	½ carbohydrate + 1 medium-fat meat
Tempeh	¼ cup	1 medium-fat meat
Tofu	4 oz (½ cup)	1 medium-fat meat
Tofu, light	4 oz (½ cup)	1 lean meat

KEY

☻ = More than 3 grams of dietary fiber per serving.

▽ = Extra fat, or prepared with added fat. (Count as 1 starch + 1 fat.)

🔋 = 480 milligrams or more of sodium per serving (based on the sodium content of a typical 3 oz serving of meat, unless 1 or 2 oz is the normal serving size).

[a]Beans, peas, and lentils are also found on the Starch list; nut butters in smaller amounts are found in the Fats list.

Table D-8 Fats

Fats and oils have mixtures of unsaturated (polyunsaturated and monounsaturated) and saturated fats. Foods on the Fats list are grouped together based on the major type of fat they contain.

1 fat choice = 0 grams carbohydrate, 0 grams protein, 5 grams fat, and 45 calories.

Note: In general, one fat exchange is 1 teaspoon of regular margarine, vegetable oil, or butter; 1 tablespoon of regular salad dressing.

When used in large amounts, bacon and peanut butter are counted as high-fat meat choices (see Meat and Meat Substitutes list). Fat-free salad dressings are found on the Sweets, Desserts, and Other Carbohydrates list. Fat-free products such as margarines, salad dressings, mayonnaise, sour cream, and cream cheese are found on the Free Foods list.

Monounsaturated fats		**Monounsaturated fats**	
Food	Serving size	Food	Serving size
Avocado, medium	2 Tbsp (1 oz)	mixed (50% peanuts)	6 nuts
Nut butters (trans fat-free): almond butter, cashew butter, peanut butter (smooth or crunchy)	1½ tsp	peanuts	10 nuts
		pecans	4 halves
Nuts		pistachios	16 nuts
almonds	6 nuts	Oil: canola, olive, peanut	1 tsp
Brazil	2 nuts	Olives	
cashews	6 nuts	black (ripe)	8 large
filberts (hazelnuts)	5 nuts	green, stuffed	10 large
macadamia	3 nuts		

Table D-8 Fats (Continued)

Polyunsaturated fats	
Food	**Serving size**
Margarine: lower-fat spread (30%–50% vegetable oil, *trans* fat-free)	1 Tbsp
Margarine: stick, tub (*trans* fat-free) or squeeze (*trans* fat-free)	1 tsp
Mayonnaise	
reduced-fat	1 Tbsp
regular	1 tsp
Mayonnaise-style salad dressing	
reduced-fat	1 Tbsp
regular	2 tsp
Nuts	
Pignolia (pine nuts)	1 Tbsp
walnuts, English	4 halves
Oil: corn, cottonseed, flaxseed, grape seed, safflower, soybean, sunflower	1 tsp
Oil: made from soybean and canola oil—Enova	1 tsp
Plant stanol esters	
light	1 Tbsp
regular	2 tsp
Salad dressing	
🧂reduced-fat	2 Tbsp
Note: May be high in carbohydrate.	
🧂regular	1 Tbsp
Seeds	
flaxseed, whole	1 Tbsp
pumpkin, sunflower	1 Tbsp
sesame seeds	1 Tbsp
Tahini or sesame paste	2 tsp

Saturated fats	
Food	**Serving size**
Bacon, cooked, regular or turkey	1 slice
Butter	
reduced-fat	1 Tbsp
stick	1 tsp
whipped	2 tsp
Butter blends made with oil	
reduced fat or light	1 Tbsp
regular	1½ tsp
Chitterlings, boiled	2 Tbsp (½ oz)
Coconut, sweetened, shredded	2 Tbsp
Coconut milk	
light	⅓ cup
regular	1½ Tbsp
Cream	
half and half	2 Tbsp
heavy	1 Tbsp
light	1½ Tbsp
whipped	2 Tbsp
whipped, pressurized	¼ cup
Cream cheese	
reduced-fat	1½ Tbsp (¾ oz)
regular	1 Tbsp (½ oz)
Lard 1 tsp	
Oil: coconut, palm, palm kernel	1 tsp
Salt pork	¼ oz
Shortening, solid	1 tsp
Sour cream	
reduced-fat or light	3 Tbsp
regular	2 Tbsp

KEY

🧂 = 480 milligrams or more of sodium per serving.

Table D-9 Free Foods

A "free" food is any food or drink choice that has less than 20 calories and 5 grams or less of carbohydrate per serving.

- Most foods on this list should be limited to 3 servings (as listed here) per day. Spread out the servings throughout the day. If you eat all 3 servings at once, it could raise your blood glucose level.
- Food and drink choices listed here without a serving size can be eaten whenever you like.

Low carbohydrate foods	
Food	**Serving size**
Cabbage, raw	½ cup
Candy, hard (regular or sugar-free)	1 piece
Carrots, cauliflower, or green beans, cooked	¼ cup
Cranberries, sweetened with sugar substitute	½ cup
Cucumber, sliced	½ cup
Gelatin	
dessert, sugar-free	
unflavored	
Gum	
Jam or jelly, light or no sugar added	2 tsp

Low carbohydrate foods	
Food	**Serving size**
Rhubarb, sweetened with sugar substitute	½ cup
Salad greens	
Sugar substitutes (artificial sweeteners)	
Syrup, sugar-free	2 Tbsp

Modified fat foods with carbohydrate	
Food	**Serving size**
Cream cheese, fat-free	1 Tbsp (½ oz)
Creamers	
nondairy, liquid	1 Tbsp
nondairy, powdered	2 tsp

Table D-9 Free Foods (Continued)

Modified fat foods with carbohydrate	
Food	**Serving size**
Margarine spread	
fat-free	1 Tbsp
reduced-fat	1 tsp
Mayonnaise	
fat-free	1 Tbsp
reduced-fat	1 tsp
Mayonnaise-style salad dressing	
fat-free	1 Tbsp
reduced-fat	1 tsp
Salad dressing	
fat-free or low-fat	1 Tbsp
fat-free, Italian	2 Tbsp
Sour cream, fat-free or reduced-fat	1 Tbsp
Whipped topping	
light or fat-free	2 Tbsp
regular	1 Tbsp

Condiments	
Food	**Serving size**
Barbecue sauce	2 tsp
Catsup (ketchup)	1 Tbsp
Honey mustard	1 Tbsp
Horseradish	
Lemon juice	
Miso	1½ tsp
Mustard	
Parmesan cheese, freshly grated	1 Tbsp
Pickle relish	1 Tbsp
Pickles	
🧂 dill	1½ medium
sweet, bread and butter	2 slices
sweet, gherkin	¾ oz
Salsa	¼ cup

Condiments	
Food	**Serving size**
🧂 Soy sauce, light or regular	1 Tbsp
Sweet and sour sauce	2 tsp
Sweet chili sauce	2 tsp
Taco sauce	1 Tbsp
Vinegar	
Yogurt, any type	2 Tbsp

Drinks/mixes

Any food on the list—without a serving size listed—can be consumed in any moderate amount.
- 🧂 Bouillon, broth, consommé
- Bouillon or broth, low-sodium
- Carbonated or mineral water
- Club soda
- Cocoa powder, unsweetened (1 Tbsp)
- Coffee, unsweetened or with sugar substitute
- Diet soft drinks, sugar-free
- Drink mixes, sugar-free
- Tea, unsweetened or with sugar substitute
- Tonic water, diet
- Water
- Water, flavored, carbohydrate free

Seasonings

Any food on this list can be consumed in any moderate amount.
- Flavoring extracts (for example, vanilla, almond, peppermint)
- Garlic
- Herbs, fresh or dried
- Nonstick cooking spray
- Pimento
- Spices
- Hot pepper sauce
- Wine, used in cooking
- Worcestershire sauce

KEY

🧂 = 480 milligrams or more of sodium per serving.

Table D-10 Combination Foods

Many foods are eaten in various combinations, such as casseroles. Because "combination" foods do not fit into any one choice list, this list of choices provides some typical combination foods.

Entrees

Food	Serving size	Count as
🧂 Casserole type (tuna noodle, lasagna, spaghetti with meatballs, chili with beans, macaroni and cheese)	1 cup (8 oz)	2 carbohydrates + 2 medium-fat meats
🧂 Stews (beef/other meats and vegetables)	1 cup (8 oz)	1 carbohydrate + 1 medium-fat meat + 0–3 fats
Tuna salad or chicken salad	½ cup (3½ oz)	½ carbohydrate + 2 lean meats + 1 fat

Frozen meals/entrees

Food	Serving size	Count as
🧂😊 Burrito (beef and bean)	1 (5 oz)	3 carbohydrates + 1 lean meat + 2 fats
🧂 Dinner-type meal	generally 14–17 oz	3 carbohydrates + 3 medium-fat meats + 3 fats
🧂 Entrée or meal with less than 340 calories	about 8–11 oz	2–3 carbohydrates + 1–2 lean meats

(Continued)

Table D-10 Combination Foods (Continued)

Frozen meals/entrees

Food	Serving size	Count as
Pizza		
🧂 cheese/vegetarian, thin crust	¼ of a 12-inch (4½–5 oz)	2 carbohydrates + 2 medium-fat meats
🧂 meat topping, thin crust	¼ of a 12-inch (5 oz)	2 carbohydrates + 2 medium-fat meats + 1½ fats
🧂 Pocket sandwich	1 (4½ oz)	3 carbohydrates + 1 lean meat + 1–2 fats
🧂 Pot pie	1 (7 oz)	2½ carbohydrates + 1 medium-fat meat + 3 fats

Salads (deli-style)

Food	Serving size	Count as
Coleslaw	½ cup	1 carbohydrate + 1½ fats
Macaroni/pasta salad	½ cup	2 carbohydrates + 3 fats
🧂 Potato salad	½ cup	1½–2 carbohydrates + 1–2 fats

Soups

Food	Serving size	Count as
🧂 Bean, lentil, or split pea	1 cup	1 carbohydrate + 1 lean meat
🧂 Chowder (made with milk)	1 cup (8 oz)	1 carbohydrate + 1 lean meat + 1½ fats
🧂 Cream (made with water)	1 cup (8 oz)	1 carbohydrate + 1 fat
🧂 Instant	6 oz prepared	1 carbohydrate
🧂 with beans or lentils	8 oz prepared	2½ carbohydrates + 1 lean meat
🧂 Miso soup	1 cup	½ carbohydrate + 1 fat
🧂 Oriental noodle	1 cup	2 carbohydrates + 2 fats
Rice (congee)	1 cup	1 carbohydrate
🧂 Tomato (made with water)	1 cup (8 oz)	1 carbohydrate
🧂 Vegetable beef, chicken noodle, or other broth-type	1 cup (8 oz)	1 carbohydrate

KEY

☺ = More than 3 grams of dietary fiber per serving.

▽ = Extra fat, or prepared with added fat.

🧂 = 600 milligrams or more of sodium per serving (for combination food main dishes/meals).

Table D-11 Fast Foods

The choices in the Fast Foods list are not specific fast-food meals or items, but are estimates based on popular foods. Ask the restaurant or check its website for nutrition information about your favorite fast foods.

Breakfast sandwiches

Food	Serving size	Count as
🧂 Egg, cheese, meat, English muffin	1 sandwich	2 carbohydrates + 2 medium-fat meats
🧂 Sausage biscuit sandwich	1 sandwich	2 carbohydrates + 2 high-fat meats + 3½ fats

Main dishes/entrees

Food	Serving size	Count as
🧂☺ Burrito (beef and beans)	1 (about 8 oz)	3 carbohydrates + 3 medium-fat meats + 3 fats
🧂 Chicken breast, breaded and fried	1 (about 5 oz)	1 carbohydrate + 4 medium-fat meats
Chicken drumstick, breaded and fried	1 (about 2 oz)	2 medium-fat meats
🧂 Chicken nuggets	6 (about 3½ oz)	1 carbohydrate + 2 medium-fat meats + 1 fat
🧂 Chicken thigh, breaded and fried	1 (about 4 oz)	½ carbohydrate + 3 medium-fat meats + 1½ fats
🧂 Chicken wings, hot	6 (5 oz)	5 medium-fat meats + 1½ fats

Table D-11 Fast Foods (Continued)

Oriental

Food	Serving size	Count as
▤ Beef/chicken/shrimp with vegetables in sauce	1 cup (about 5 oz)	1 carbohydrate + 1 lean meat + 1 fat
▤ Egg roll, meat	1 (about 3 oz)	1 carbohydrate + 1 lean meat + 1 fat
Fried rice, meatless	½ cup	1½ carbohydrates + 1½ fats
▤ Meat and sweet sauce (orange chicken)	1 cup	3 carbohydrates + 3 medium-fat meats + 2 fats
▤☻ Noodles and vegetables in sauce (chow mein, lo mein)	1 cup	2 carbohydrates + 1 fat

Pizza

Food	Serving size	Count as
Pizza		
▤ cheese, pepperoni, regular crust	⅛ of a 14-inch (about 4 oz)	2½ carbohydrates + 1 medium-fat meat + 1½ fats
▤ cheese/vegetarian, thin crust	¼ of a 12-inch (about 6 oz)	2½ carbohydrates + 2 medium-fat meats + 1½ fats

Sandwiches

Food	Serving size	Count as
▤ Chicken sandwich, grilled	1	3 carbohydrates + 4 lean meats
▤ Chicken sandwich, crispy	1	3½ carbohydrates + 3 medium-fat meats + 1 fat
▤ Fish sandwich with tartar sauce	1	2½ carbohydrates + 2 medium-fat meats + 2 fats
Hamburger		
▤ large with cheese	1	2½ carbohydrates + 4 medium-fat meats + 1 fat
regular	1	2 carbohydrates + 1 medium-fat meat + 1 fat
▤ Hot dog with bun	1	1 carbohydrate + 1 high-fat meat + 1 fat
Submarine sandwich		
▤ less than 6 grams fat	6-inch sub	3 carbohydrates + 2 lean meats
▤ regular	6-inch sub	3½ carbohydrates + 2 medium-fat meats + 1 fat
Taco, hard or soft shell (meat and cheese)	1 small	1 carbohydrate + 1 medium-fat meat + 1½ fats

Salads

Food	Serving size	Count as
▤☻ Salad, main dish (grilled chicken type, no dressing or croutons)		1 carbohydrate + 4 lean meats
Salad, side, no dressing or cheese	small (about 5 oz)	1 vegetable

Sides/appetizers

Food	Serving size	Count as
▽ French fries, restaurant style	small	3 carbohydrates + 3 fats
	medium	4 carbohydrates + 4 fats
	large	5 carbohydrates + 6 fats
▤ Nachos with cheese	small (about 4½ oz)	2½ carbohydrates + 4 fats
▤ Onion rings	1 serving (about 3 oz)	2½ carbohydrates + 3 fats

Desserts

Food	Serving size	Count as
Milkshake, any flavor	12 oz	6 carbohydrates + 2 fats
Soft-serve ice cream cone	1 small	2½ carbohydrates + 1 fat

KEY

☻ = More than 3 grams of dietary fiber per serving.

▽ = Extra fat, or prepared with added fat.

▤ = 600 milligrams or more of sodium per serving (for fast-food main dishes/meals).

Table D-12 Alcohol

1 alcohol equivalent = variable grams carbohydrate, 0 grams protein, 0 grams fat, and 100 calories.

Note: In general, one alcohol choice (½ ounce absolute alcohol) has about 100 calories. For those who choose to drink alcohol, guidelines suggest limiting alcohol intake to 1 drink or less per day for women, and 2 drinks or less per day for men. To reduce your risk of low blood glucose (hypoglycemia), especially if you take insulin or a diabetes pill that increases insulin, always drink alcohol with food. Although alcohol, by itself, does not directly affect blood glucose, be aware of the carbohydrate (for example, in mixed drinks, beer, and wine) that may raise your blood glucose.

Alcoholic beverage	Serving size	Count as
Beer		
light (4.2%)	12 fl oz	1 alcohol equivalent + ½ carbohydrate
regular (4.9%)	12 fl oz	1 alcohol equivalent + 1 carbohydrate
Distilled spirits: vodka, rum, gin, whiskey, 80 or 86 proof	1½ fl oz	1 alcohol equivalent
Liqueur, coffee (53 proof)	1 fl oz	1 alcohol equivalent + 1 carbohydrate
Sake	1 fl oz	½ alcohol equivalent
Wine		
dessert (sherry)	3½ fl oz	1 alcohol equivalent + 1 carbohydrate
dry, red or white (10%)	5 fl oz	1 alcohol equivalent

APPENDIX E Sample 24-Hour Food Intake Form

Name: Date:

                                                   Day:

Time	Food or beverage (description including brand names and preparation method)	Amount (for example, grams, ounces, cups, tablespoons, teaspoons, slices, pieces)
6:00 a.m.	Orange juice (made from concentrate)	6 ounces
6:15–7:15 a.m.	Water	1 cup
7:30 a.m.	Egg (large, fried in butter)	1 large egg, 1 Tablespoon butter
	Toast (Roman Meal 100% Whole Wheat)	1 slice
	Margarine (Fleischmann's, soft from a tub)	2 teaspoons
	Coffee (with Coffee-Mate Lite and sugar)	12 ounces coffee, 1 Tablespoon each sugar and creamer

To increase the accuracy of caloric and nutrient intake estimates, *all* foods and beverages eaten should be recorded. A major source of error is the underestimation of the amount consumed. Detailed descriptions of the foods and beverages allow for more accurate matching of items with nutrient analysis software databases.

APPENDIX F Sample 24-Hour Physical Activity Log

Time	Activity	Time in activity (min)	Code number[1,2]	Intensity (METs)[1,2]	Energy expenditure rate (kcal/hr)**	Time in activity (hr)	Total energy expended (kcal)*
12:00 a.m.							
1:00 a.m.							
2:00 a.m.							
3:00 a.m.							
4:00 a.m.							
5:00 a.m.							
6:00 a.m.							
7:00 a.m.							
8:00 a.m.							
9:00 a.m.							
10:00 a.m.							
11:00 a.m.							
12:00 p.m.							
1:00 p.m.							
2:00 p.m.							
3:00 p.m.							
4:00 p.m.							
5:00 p.m.							
6:00 p.m.							
7:00 p.m.							
8:00 p.m.							
9:00 p.m.							
10:00 p.m.							
11:00 p.m.							

[1]Ainsworth, B. E., Haskell, W. L., Leon, A. S., Jacobs, D. R., Jr., Montoye, H. J., Sallis, J. F., et al. (1993). Compendium of physical activities: Classification of energy costs of human physical activities. *Medicine and Science in Sports and Exercise, 25*(1), 71–80.

[2]Ainsworth, B. E., Haskell, W. L., Whitt, M. C., Irwin, M. L., Swartz, A. M., Strath, S. J., O'Brien, W.L., Basset, D. R., Jr., Schmitz, K. H., Emplancourt, P. O., Jacobs, D. R., & Leon, A. S. (2000). Compendium of physical activities an update of activity codes and MET intensities. *Medicine and Science in Sports and Exercise, 32*(9), S498–S516.

*Example calculation of Energy Expenditure Rate (kcal/hr) and Total Energy Expenditure (kcal): 134.5 lb female walks at a moderate pace for 30 minutes.

METs = 3.3	Walk, 3.0 moderate pace (Compcode 17190) from Compendium of Physical Activities (Appendix G) = 3.3 METs
Weight = 61 kg	134.5 lb ÷ 2.2 = 61 kg
RMR = 0.9	Male = 1.0 Female = 0.9
Time = 0.5 hr	30 minutes × 1 hour/60 minutes = 0.5 hour

Energy Expenditure Rate (kcal/hr) = METs × Weight (kg) × RMR (male = 1.0, female = 0.9)

Energy Expenditure Rate (kcal/hr) = 3.3 × 61 kg × 0.9

Energy Expenditure Rate (kcal/hr) = 181 kcal/hr

Therefore, a 134.5 lb (61 kg) female would expend approximately 181 kcal every hour she walks at a moderate pace. If she walks for 30 total minutes, her total energy expenditure would be approximately 91 kcal.

Total Energy Expenditure (kcals) = Energy Expenditure Rate (kcals/hr) × Time in Activity (hr)

Total Energy Expenditure (kcals) = 181 kcal/hr × 0.5 hr

Total Energy Expenditure (kcals) = 90.5 kcals

APPENDIX G The Compendium of Physical Activities

Energy Expenditure Rate in kcal/hr and Total Energy Expenditure in kcal can be determined for activities in the Compendium of Physical Activities that follows in this appendix. Information needed: description of activity, intensity of activity in METs, time (duration) of activity in hours, body weight in kilograms, and gender.

Step 1: In the Compendium of Physical Activities (Appendix G) find the description that most closely matches the physical activity of the subject.
Step 2: Determine the intensity of the activity in METs in the second METs column (2000 update).
Step 3: Determine the subject's body weight in kilograms.
Step 4: Determine RMR constant for the subject's gender.
Step 5: Determine the time spent in the activity in hours.

Example calculation of Energy Expenditure Rate (kcal/hr) and Total Energy Expenditure (kcal) for a steady-state activity: a 134.5 lb female walks at a moderate pace for 30 minutes.

METs = 3.3	Walk, 3.0 moderate pace (Compcode 17190) from Compendium of Physical Activities (Appendix G) = 3.3 METs
Weight = 61 kg	134.5 lb ÷ 2.2 = 61 kg
RMR = 0.9	Male = 1.0 Female = 0.9
Time = 0.5 hr	30 minutes × 1 hour/60 minutes = 0.5 hour

Energy Expenditure Rate (kcal/hr) = METs × Weight (kg) × RMR (male = 1.0, female = 0.9)
Energy Expenditure Rate (kcal/hr) = 3.3 × 61 kg × 0.9
Energy Expenditure Rate (kcal/hr) = 181 kcal/hr

Therefore, a 134.5 lb (61 kg) female would expend approximately 181 kcal every hour she walks at a moderate pace. If she walks for 30 total minutes, her total energy expenditure would be approximately 91 kcal.

Total Energy Expenditure (kcals) = Energy Expenditure Rate (kcals/hr) × Time in Activity (hr)
Total Energy Expenditure (kcals) = 181 kcal/hr × 0.5 hr
Total Energy Expenditure (kcals) = 90.5 kcals

Example calculation of Energy Expenditure Rate (kcal/hr) and Total Energy Expenditure (kcal) for an intermittent activity: a moderately-skilled, 176 lb male plays singles tennis for an hour.

METs = 8.0	Tennis, singles (Compcode 15690) from Compendium of Physical Activities (Appendix G) = 8.0 METs	
Weight = 80 kg	176 lb ÷ 2.2 = 80 kg	
RMR = 1.0	Male = 1.0	Female = 0.9
Time = 1 hr	1 hour	

Energy Expenditure Rate (kcal/hr) = METs × Weight (kg) × RMR (male = 1.0, female = 0.9)
Energy Expenditure Rate (kcal/hr) = 8.0 × 80 kg × 1.0
Energy Expenditure Rate (kcal/hr) = 640 kcal/hr

*The approximate rate of energy expenditure for a 176 lb male playing singles tennis is 640 kcal for each hour. However, tennis is an intermittent, stop-and-go activity, and depending upon the skill level of the players the activity is not likely to be continuous at this intensity for the entire time of play. For the purposes of this example, the moderately-skilled player is estimated to be at this level of activity for half of the time, so the time in activity used to determine total energy expenditure is 30 minutes (0.5 hours) instead of 1 hour.

Total Energy Expenditure (kcal) = Energy Expenditure Rate (kcal/hr) × Time in Activity (hr)
Total Energy Expenditure (kcal) = 640 kcal/hr × 0.5 hr*
Total Energy Expenditure (kcal) = 320 kcal

The Compendium of Physical Activities Tracking Guide

1993 Compcode	1993 METS	2000 Compcode	2000 METS	Heading	Description
01009	8.5	01009	8.5	bicycling	bicycling, BMX or mountain
01010	4.0	01010	4.0	bicycling	bicycling, <10 mph, leisure, to work or for pleasure (Taylor Code 115)
		01015	8.0	bicycling	bicycling, general
01020	6.0	01020	6.0	bicycling	bicycling, 10–11.9 mph, leisure, slow, light effort
01030	8.0	01030	8.0	bicycling	bicycling, 12–13.9 mph, leisure, moderate effort
01040	10.0	01040	10.0	bicycling	bicycling, 14–15.9 mph, racing or leisure, fast, vigorous effort
01050	12.0	01050	12.0	bicycling	bicycling, 16–19 mph, racing/not drafting or >19 mph drafting, very fast, racing general
01060	16.0	01060	16.0	bicycling	bicycling, >20 mph, racing, not drafting
01070	5.0	01070	5.0	bicycling	unicycling
02010	5.0	02010	7.0	conditioning exercise	bicycling, stationary, general
02011	3.0	02011	3.0	conditioning exercise	bicycling, stationary, 50 watts, very light effort
02012	5.5	02012	5.5	conditioning exercise	bicycling, stationary, 100 watts, light effort
02013	7.0	02013	7.0	conditioning exercise	bicycling, stationary, 150 watts, moderate effort
02014	10.5	02014	10.5	conditioning exercise	bicycling, stationary, 200 watts, vigorous effort
02015	12.5	02015	12.5	conditioning exercise	bicycling, stationary, 250 watts, very vigorous effort
02020	8.0	02020	8.0	conditioning exercise	calisthenics (e.g., pushups, situps, pullups, jumping jacks), heavy, vigorous effort
02030	4.5	02030	3.5	conditioning exercise	calisthenics, home exercise, light or moderate effort, general (example: back exercises), going up & down from floor (Taylor Code 150)
02040	8.0	02040	8.0	conditioning exercise	circuit training, including some aerobic movement with minimal rest, general
02050	6.0	02050	6.0	conditioning exercise	weight lifting (free weight, nautilus or universal-type), power lifting or body building, vigorous effort (Taylor Code 210)
02060	5.5	02060	5.5	conditioning exercise	health club exercise, general (Taylor Code 160)
02065	6.0	02065	9.0	conditioning exercise	stair-treadmill ergometer, general
02070	9.5	02070	7.0	conditioning exercise	rowing, stationary ergometer, general
02071	3.5	02071	3.5	conditioning exercise	rowing, stationary, 50 watts, light effort
02072	7.0	02072	7.0	conditioning exercise	rowing, stationary, 100 watts, moderate effort
02073	8.5	02073	8.5	conditioning exercise	rowing, stationary, 150 watts, vigorous effort
02074	12.0	02074	12.0	conditioning exercise	rowing, stationary, 200 watts, very vigorous effort
02080	9.5	02080	7.0	conditioning exercise	ski machine, general
02090	6.0	02090	6.0	conditioning exercise	slimnastics, jazzercise
02100	4.0	02100	2.5	conditioning exercise	stretching, hatha yoga
		02101	2.5	conditioning exercise	mild stretching
02110	6.0	02110	6.0	conditioning exercise	teaching aerobic exercise class
02120	4.0	02120	4.0	conditioning exercise	water aerobics, water calisthenics
02130	3.0	02130	3.0	conditioning exercise	weight lifting (free, nautilus or universal-type), light or moderate effort, light workout, general
02135	1.0	02135	1.0	conditioning exercise	whirlpool, sitting
03010	6.0	03010	4.8	dancing	ballet or modern, twist, jazz, tap, jitterbug

The Compendium of Physical Activities Tracking Guide (Continued)

1993 Compcode	1993 METS	2000 Compcode	2000 METS	Heading	Description
03015	6.0	03015	6.5	dancing	aerobic, general
		03016	8.5	dancing	aerobic, step, with 6–8 inch step
		03017	10.0	dancing	aerobic, step, with 10–12 inch step
03020	5.0	03020	5.0	dancing	aerobic, low impact
03021	7.0	03021	7.0	dancing	aerobic, high impact
03025	4.5	03025	4.5	dancing	general, Greek, Middle Eastern, hula, flamenco, belly, and swing dancing
03030	5.5	03030	5.5	dancing	ballroom, dancing fast (Taylor Code 125)
		03031	4.5	dancing	ballroom, fast (disco, folk, square), line dancing, Irish step dancing, polka, contra, country
03040	3.0	03040	3.0	dancing	ballroom, slow (e.g., waltz, foxtrot, slow dancing), samba, tango, 19th C, mambo, cha-cha
		03050	5.5	dancing	Anishinaabe Jingle Dancing or other traditional American Indian dancing
04001	4.0	04001	3.0	fishing and hunting	fishing, general
04010	4.0	04010	4.0	fishing and hunting	digging worms, with shovel
04020	5.0	04020	4.0	fishing and hunting	fishing from riverbank and walking
04030	2.5	04030	2.5	fishing and hunting	fishing from boat, sitting
04040	3.5	04040	3.5	fishing and hunting	fishing from riverbank, standing (Taylor Code 660)
04050	6.0	04050	6.0	fishing and hunting	fishing in stream, in waders (Taylor Code 670)
04060	2.0	04060	2.0	fishing and hunting	fishing, ice, sitting
04070	2.5	04070	2.5	fishing and hunting	hunting, bow and arrow or crossbow
04080	6.0	04080	6.0	fishing and hunting	hunting, deer, elk, large game (Taylor Code 170)
04090	2.5	04090	2.5	fishing and hunting	hunting, duck, wading
04100	5.0	04100	5.0	fishing and hunting	hunting, general
04110	6.0	04110	6.0	fishing and hunting	hunting, pheasants or grouse (Taylor Code 680)
04120	5.0	04120	5.0	fishing and hunting	hunting, rabbit, squirrel, prairie chick, raccoon, small game (Taylor Code 690)
04130	2.5	04130	2.5	fishing and hunting	pistol shooting or trap shooting, standing
05010	2.5	05010	3.3	home activities	carpet sweeping, sweeping floors
05020	4.5	05020	3.0	home activities	cleaning, heavy or major (e.g., wash car, wash windows, clean garage), vigorous effort
		05021	3.5	home activities	mopping
		05025	2.5	home activities	multiple household tasks all at once, light effort
		05026	3.5	home activities	multiple household tasks all at once, moderate effort
		05027	4.0	home activities	multiple household tasks all at once, vigorous effort
05030	3.5	05030	3.0	home activities	cleaning, house or cabin, general
05040	2.5	05040	2.5	home activities	cleaning, light (dusting, straightening up, changing linen, carrying out trash)
05041	2.3	05041	2.3	home activities	wash dishes- standing or in general (not broken into stand/walk components)
05042	2.3	05042	2.5	home activities	wash dishes; clearing dishes from table- walking
		05043	3.5	home activities	vacuuming
		05045	6.0	home activities	butchering animals
05050	2.5	05050	2.0	home activities	cooking or food preparation- standing or sitting or in general (not broken into stand/walk components), manual appliance
05051	2.5	05051	2.5	home activities	serving food, setting table- implied walking or standing

(Continued)

The Compendium of Physical Activities Tracking Guide (Continued)

1993			2000		Heading	Description
Compcode	METS		Compcode	METS		
05052	2.5		05052	2.5	home activities	cooking or food preparation- walking
			05053	2.5	home activities	feeding animals
05055	2.5		05055	2.5	home activities	putting away groceries (e.g., carrying groceries, shopping without a grocery cart), carrying packages
05056	8.0		05056	7.5	home activities	carrying groceries upstairs
			05057	3.0	home activities	cooking Indian bread on an outside stove
05060	3.5		05060	2.3	home activities	food shopping with or without a grocery cart, standing or walking
05065	2.0		05065	2.3	home activities	nonfood shopping, standing or walking
05066	2.3				home activities	walking shopping (nongrocery shopping)
05070	2.3		05070	2.3	home activities	ironing
05080	1.5		05080	1.5	home activities	sitting; knitting, sewing, lt. wrapping (presents)
05090	2.0		05090	2.0	home activities	implied standing-laundry, fold or hang clothes, put clothes in washer or dryer, packing suitcase
05095	2.3		05095	2.3	home activities	implied walking- putting away clothes, gathering clothes to pack, putting away laundry
05100	2.0		05100	2.0	home activities	making bed
05110	5.0		05110	5.0	home activities	maple syruping/sugar bushing (including carrying buckets, carrying wood)
05120	6.0		05120	6.0	home activities	moving furniture, household items, carrying boxes
05130	5.5		05130	3.8	home activities	scrubbing floors, on hands and knees, scrubbing bathroom, bathtub
05140	4.0		05140	4.0	home activities	sweeping garage, sidewalk or outside of house
05145	7.0				home activities	moving household items, carrying boxes
05146	3.5		05146	3.5	home activities	standing: packing/unpacking boxes, occasional lifting of household items light-moderate effort
05147	3.0		05147	3.0	home activities	implied walking- putting away household items-moderate effort
			05148	2.5	home activities	watering plants
			05149	2.5	home activities	building a fire inside
05150	9.0		05150	9.0	home activities	moving household items upstairs, carrying boxes or furniture
05160	2.5		05160	2.0	home activities	standing-light (pump gas, change lightbulb, etc.)
05165	3.0		05165	3.0	home activities	walking- light, noncleaning (readying to leave, shut/lock doors, close windows, etc.)
05170	2.5		05170	2.5	home activities	sitting-playing with child(ren)- light, only active periods
05171	2.8		05171	2.8	home activities	standing-playing with child(ren)- light, only active periods
05175	4.0		05175	4.0	home activities	walk/run-playing with child(ren)- moderate, only active periods
05180	5.0		05180	5.0	home activities	walk/run-playing with child(ren)- vigorous, only active periods
			05181	3.0	home activities	carrying small children
05185	3.0		05185	2.5	home activities	child care: sitting/kneeling- dressing, bathing, grooming, feeding, occasional lifting of child- light effort, general
05186	3.5		05186	3.0	home activities	child care: standing- dressing, bathing, grooming, feeding, occasional lifting of child- light effort
			05187	4.0	home activities	elder care, disabled adult, only active periods
			05188	1.5	home activities	reclining with baby
			05190	2.5	home activities	sit, playing with animals, light, only active periods
			05191	2.8	home activities	stand, playing with animals, light, only active periods
			05192	2.8	home activities	walk/run, playing with animals, light, only active periods
			05193	4.0	home activities	walk/run, playing with animals, moderate, only active periods

The Compendium of Physical Activities Tracking Guide (Continued)

1993 Compcode	1993 METS	2000 Compcode	2000 METS	Heading	Description
		05194	5.0	home activities	walk/run, playing with animals, vigorous, only active periods
		05195	3.5	home activities	standing-bathing dog
06010	3.0	06010	3.0	home repair	airplane repair
06020	4.5	06020	4.0	home repair	automobile body work
06030	3.0	06030	3.0	home repair	automobile repair
06040	3.0	06040	3.0	home repair	carpentry, general, workshop (Taylor Code 620)
06050	6.0	06050	6.0	home repair	carpentry, outside house, installing rain gutters, building a fence, (Taylor Code 640)
06060	4.5	06060	4.5	home repair	carpentry, finishing or refinishing cabinets or furniture
06070	7.5	06070	7.5	home repair	carpentry, sawing hardwood
06C80	5.0	06080	5.0	home repair	caulking, chinking log cabin
06C90	4.5	06090	4.5	home repair	caulking, except log cabin
06100	5.0	06100	5.0	home repair	cleaning gutters
06110	5.0	06110	5.0	home repair	excavating garage
06120	5.0	06120	5.0	home repair	hanging storm windows
06130	4.5	06130	4.5	home repair	laying or removing carpet
06140	4.5	06140	4.5	home repair	laying tile or linoleum, repairing appliances
06150	5.0	06150	5.0	home repair	painting, outside home (Taylor Code 650)
06160	4.5	06160	3.0	home repair	painting, papering, plastering, scraping, inside house, hanging sheet rock, remodeling
		06165	4.5	home repair	painting, (Taylor Code 630)
06170	3.0	06170	3.0	home repair	put on and removal of tarp-sailboat
06180	6.0	06180	6.0	home repair	roofing
06190	4.5	06190	4.5	home repair	sanding floors with a power sander
06200	4.5	06200	4.5	home repair	scraping and painting sailboat or powerboat
06210	5.0	06210	5.0	home repair	spreading dirt with a shovel
06220	4.5	06220	4.5	home repair	washing and waxing hull of sailboat, car, powerboat, airplane
06230	4.5	06230	4.5	home repair	washing fence, painting fence
06240	3.0	06240	3.0	home repair	wiring, plumbing
07010	0.9	07010	1.0	inactivity quiet	lying quietly, watching television
		07011	1.0	inactivity quiet	lying quietly, doing nothing, lying in bed awake, listening to music (not talking or reading)
07020	1.0	07020	1.0	inactivity quiet	sitting quietly and watching television
		07021	1.0	inactivity quiet	sitting quietly, sitting smoking, listening to music (not talking or reading), watching a movie in a theater
07030	0.9	07030	0.9	inactivity quiet	sleeping
07040	1.2	07040	1.2	inactivity quiet	standing quietly (standing in a line)
07050	1.0	07050	1.0	inactivity light	reclining- writing
07060	1.0	07060	1.0	inactivity light	reclining- talking or talking on phone
07070	1.0	07070	1.0	inactivity light	reclining-reading
		07075	1.0	inactivity light	meditating
08010	5.0	08010	5.0	lawn and garden	carrying, loading or stacking wood, loading/unloading or carrying lumber
08020	6.0	08020	6.0	lawn and garden	chopping wood, splitting logs
08030	5.0	08030	5.0	lawn and garden	clearing land, hauling branches, wheelbarrow chores

(Continued)

The Compendium of Physical Activities Tracking Guide (Continued)

1993 Compcode	1993 METS	2000 Compcode	2000 METS	Heading	Description
08040	5.0	08040	5.0	lawn and garden	digging sandbox
08050	5.0	08050	5.0	lawn and garden	digging, spading, filling garden, composting, (Taylor Code 590)
08060	6.0	08060	6.0	lawn and garden	gardening with heavy power tools, tilling a garden, chain saw
08080	5.0	08080	5.0	lawn and garden	laying crushed rock
08090	5.0	08090	5.0	lawn and garden	laying sod
08095	5.5	08095	5.5	lawn and garden	mowing lawn, general
08100	2.5	08100	2.5	lawn and garden	mowing lawn, riding mower (Taylor Code 550)
08110	6.0	08110	6.0	lawn and garden	mowing lawn, walk, hand mower (Taylor Code 570)
08120	4.5	08120	5.5	lawn and garden	mowing lawn, walk, power mower
		08125	4.5	lawn and garden	mowing lawn, power mower (Taylor Code 590)
08130	4.5	08130	4.5	lawn and garden	operating snow blower, walking
08140	4.0	08140	4.5	lawn and garden	planting seedlings, shrubs
08150	4.5	08150	4.5	lawn and garden	planting trees
08160	4.0	08160	4.3	lawn and garden	raking lawn
		08165	4.0	lawn and garden	raking lawn (Taylor Code 600)
08170	4.0	08170	4.0	lawn and garden	raking roof with snow rake
08180	3.0	08180	3.0	lawn and garden	riding snow blower
08190	4.0	08190	4.0	lawn and garden	sacking grass, leaves
08200	6.0	08200	6.0	lawn and garden	shoveling snow, by hand (Taylor Code 610)
08210	4.5	08210	4.5	lawn and garden	trimming shrubs or trees, manual cutter
08215	3.5	08215	3.5	lawn and garden	trimming shrubs or trees, power cutter, using leaf blower, edger
08220	2.5	08220	2.5	lawn and garden	walking, applying fertilizer or seeding a lawn
08230	1.5	08230	1.5	lawn and garden	watering lawn or garden, standing or walking
08240	4.5	08240	4.5	lawn and garden	weeding, cultivating garden (Taylor Code 580)
08245	5.0	08245	4.0	lawn and garden	gardening, general
		08246	3.0	lawn and garden	picking fruit off trees, picking fruits/vegetables, moderate effort
08250	3.0	08250	3.0	lawn and garden	implied walking/standing- picking up yard, light, picking flowers or vegetables
		08251	3.0	lawn and garden	walking, gathering gardening tools
09010	1.5	09010	1.5	miscellaneous	sitting- card playing, playing board games
09020	2.0	09020	2.3	miscellaneous	standing- drawing (writing), casino gambling, duplicating machine
09030	1.3	09030	1.3	miscellaneous	sitting- reading, book, newspaper, etc.
09040	1.8	09040	1.8	miscellaneous	sitting- writing, desk work, typing
09050	1.8	09050	1.8	miscellaneous	standing- talking or talking on the phone
09055	1.5	09055	1.5	miscellaneous	sitting- talking or talking on the phone
09060	1.8	09060	1.8	miscellaneous	sitting- studying, general, including reading and/or writing
09065	1.8	09065	1.8	miscellaneous	sitting- in class, general, including note-taking or class discussion
09070	1.8	09070	1.8	miscellaneous	standing- reading
		09071	2.0	miscellaneous	standing- miscellaneous
		09075	1.5	miscellaneous	sitting- arts and crafts, light effort
		09080	2.0	miscellaneous	sitting- arts and crafts, moderate effort

The Compendium of Physical Activities Tracking Guide (Continued)

1993 Compcode	1993 METS	2000 Compcode	2000 METS	Heading	Description
		09085	1.8	miscellaneous	standing- arts and crafts, light effort
		09090	3.0	miscellaneous	standing- arts and crafts, moderate effort
		09095	3.5	miscellaneous	standing- arts and crafts, vigorous effort
		09100	1.5	miscellaneous	retreat/family reunion activities involving sitting, relaxing, talking, eating
		09105	2.0	miscellaneous	touring/traveling/vacation involving walking and riding
		09110	2.5	miscellaneous	camping involving standing, walking, sitting, light-to-moderate effort
		09115	1.5	miscellaneous	sitting at a sporting event, spectator
10010	1.8	10010	1.8	music playing	accordion
10020	2.0	10020	2.0	music playing	cello
10030	2.5	10030	2.5	music playing	conducting
10040	4.0	10040	4.0	music playing	drums
10050	2.0	10050	2.0	music playing	flute (sitting)
10060	2.0	10060	2.0	music playing	horn
10070	2.5	10070	2.5	music playing	piano or organ
- 0080	3.5	10080	3.5	music playing	trombone
10090	2.5	10090	2.5	music playing	trumpet
10100	2.5	10100	2.5	music playing	violin
10110	2.0	10110	2.0	music playing	woodwind
10120	2.0	10120	2.0	music playing	guitar, classical, folk (sitting)
10125	3.0	10125	3.0	music playing	guitar, rock and roll band (standing)
10130	4.0	10130	4.0	music playing	marching band, playing an instrument, baton twirling (walking)
10135	3.5	10135	3.5	music playing	marching band, drum major (walking)
11010	4.0	11010	4.0	occupation	bakery, general, moderate effort
		11015	2.5	occupation	bakery, light effort
11020	2.3	11020	2.3	occupation	bookbinding
11030	6.0	11030	6.0	occupation	building road (including hauling debris, driving heavy machinery)
11035	2.0	11035	2.0	occupation	building road, directing traffic (standing)
11040	3.5	11040	3.5	occupation	carpentry, general
11050	8.0	11050	8.0	occupation	carrying heavy loads, such as bricks
11060	8.0	11060	8.0	occupation	carrying moderate loads upstairs, moving boxes (16–40 pounds)
11070	2.5	11070	2.5	occupation	chambermaid, making bed (nursing)
11080	6.5	11080	6.5	occupation	coal mining, drilling coal, rock
11090	6.5	11090	6.5	occupation	coal mining, erecting supports
11100	6.0	11100	6.0	occupation	coal mining, general
11110	7.0	11110	7.0	occupation	coal mining, shoveling coal
1-120	5.5	11120	5.5	occupation	construction, outside, remodeling
		11121	3.0	occupation	custodial work- buffing the floor with electric buffer
		11122	2.5	occupation	custodial work- cleaning sink and toilet, light effort
		11123	2.5	occupation	custodial work- dusting, light effort
		11124	4.0	occupation	custodial work- feathering arena floor, moderate effort

(Continued)

The Compendium of Physical Activities Tracking Guide (Continued)

1993 Compcode	1993 METS	2000 Compcode	2000 METS	Heading	Description
		11125	3.5	occupation	custodial work- general cleaning, moderate effort
		11126	3.5	occupation	custodial work- mopping, moderate effort
		11127	3.0	occupation	custodial work- take out trash, moderate effort
		11128	2.5	occupation	custodial work- vacuuming, light effort
		11129	3.0	occupation	custodial work- vacuuming, moderate effort
11130	3.5	11130	3.5	occupation	electrical work, plumbing
11140	8.0	11140	8.0	occupation	farming, baling hay, cleaning barn, poultry work, vigorous effort
11150	3.5	11150	3.5	occupation	farming, chasing cattle, non-strenuous (walking), moderate effort
		11151	4.0	occupation	farming, chasing cattle or other livestock on horseback, moderate effort
		11152	2.0	occupation	farming, chasing cattle or other livestock, driving, light effort
11160	2.5	11160	2.5	occupation	farming, driving harvester, cutting hay, irrigation work
11170	2.5	11170	2.5	occupation	farming, driving tractor
11180	4.0	11180	4.0	occupation	farming, feeding small animals
11190	4.5	11190	4.5	occupation	farming, feeding cattle, horses
		11191	4.5	occupation	farming, hauling water for animals, general hauling water
		11192	6.0	occupation	farming, taking care of animals (grooming, brushing, shearing sheep, assisting with birthing, medical care, branding)
11200	8.0	11200	8.0	occupation	farming, forking straw bales, cleaning corral or barn, vigorous effort
11210	3.0	11210	3.0	occupation	farming, milking by hand, moderate effort
11220	1.5	11220	1.5	occupation	farming, milking by machine, light effort
11230	5.5	11230	5.5	occupation	farming, shoveling grain, moderate effort
11240	12.0	11240	12.0	occupation	firefighter, general
11245	11.0	11245	11.0	occupation	firefighter, climbing ladder with full gear
11246	8.0	11246	8.0	occupation	firefighter, hauling hoses on ground
11250	17.0	11250	17.0	occupation	forestry, ax chopping, fast
11260	5.0	11260	5.0	occupation	forestry, ax chopping, slow
11270	7.0	11270	7.0	occupation	forestry, barking trees
11280	11.0	11280	11.0	occupation	forestry, carrying logs
11290	8.0	11290	8.0	occupation	forestry, felling trees
11300	8.0	11300	8.0	occupation	forestry, general
11310	5.0	11310	5.0	occupation	forestry, hoeing
11320	6.0	11320	6.0	occupation	forestry, planting by hand
11330	7.0	11330	7.0	occupation	forestry, sawing by hand
11340	4.5	11340	4.5	occupation	forestry, sawing by power
11350	9.0	11350	9.0	occupation	forestry, trimming trees
11360	4.0	11360	4.0	occupation	forestry, weeding
11370	4.5	11370	4.5	occupation	furriery
11380	6.0	11380	6.0	occupation	horse grooming
11390	8.0	11390	8.0	occupation	horse racing, galloping
11400	6.5	11400	6.5	occupation	horse racing, trotting

The Compendium of Physical Activities Tracking Guide (Continued)

1993		2000			
Compcode	METS	Compcode	METS	Heading	Description
11410	2.6	11410	2.6	occupation	horse racing, walking
11420	3.5	11420	3.5	occupation	locksmith
11430	2.5	11430	2.5	occupation	machine tooling, machining, working sheet meta
11440	3.0	11440	3.0	occupation	machine tooling, operating lathe
11450	5.0	11450	5.0	occupation	machine tooling, operating punch press
11460	4.0	11460	4.0	occupation	machine tooling, tapping and drilling
11470	3.0	11470	3.0	occupation	machine tooling, welding
11480	7.0	11480	7.0	occupation	masonry, concrete
11485	4.0	11485	4.0	occupation	masseur, masseuse (standing)
11490	7.0	11490	7.5	occupation	moving, pushing heavy objects, 75 lbs or more (desks, moving van work)
		11495	12.0	occupation	skindiving or SCUBA diving as a frogman (Navy Seal)
11500	2.5	11500	2.5	occupation	operating heavy duty equipment/automated, not driving
11510	4.5	11510	4.5	occupation	orange grove work
11520	2.3	11520	2.3	occupation	printing (standing)
11525	2.5	11525	2.5	occupation	police, directing traffic (standing)
11526	2.0	11526	2.0	occupation	police, driving a squad car (sitting)
11527	1.3	11527	1.3	occupation	police, riding in a squad car (sitting)
11528	8.0	11528	4.0	occupation	police, making an arrest (standing)
11530	2.5	11530	2.5	occupation	shoe repair, general
11540	8.5	11540	8.5	occupation	shoveling, digging ditches
11550	9.0	11550	9.0	occupation	shoveling, heavy (more than 16 pounds/minute)
11560	6.0	11560	6.0	occupation	shoveling, light (less than 10 pounds/minute)
11570	7.0	11570	7.0	occupation	shoveling, moderate (10 to 15 pounds/minute)
11580	1.5	11580	1.5	occupation	sitting- light office work, general (chemistry lab work, light use of hand tools, watch repair or microassembly, light assembly/repair), sitting, reading, driving at work
11585	1.5	11585	1.5	occupation	sitting meetings, general, and/or with talking involved, eating at a business meeting
11590	2.5	11590	2.5	occupation	sitting; moderate (heavy levers, riding mower/forklift, crane operation) teaching stretching or yoga
11600	2.5	11600	2.3	occupation	standing; light (bartending, store clerk, assembling, filing, duplicating, putting up a Christmas tree), standing and talking at work, changing clothes when teaching physical education
11610	3.0	11610	3.0	occupation	standing; light/moderate (assemble/repair heavy parts, welding, stocking, auto repair, pack boxes for moving, etc.), patient care (as in nursing)
		11615	4.0	occupation	lifting items continuously, 10–20 lbs, with limited walking or resting
11620	3.5	11620	3.5	occupation	standing; moderate (assembling at fast rate, intermittent, lifting 50 lbs, hitch/twisting ropes)
11630	4.0	11630	4.0	occupation	standing; moderate/heavy (lifting more than 50 lbs, masonry, painting, paper hanging)
11640	5.0	11640	5.0	occupation	steel mill, fettling
11650	5.5	11650	5.5	occupation	steel mill, forging
11660	8.0	11660	8.0	occupation	steel mill, hand rolling
11670	8.0	11670	8.0	occupation	steel mill, merchant mill rolling
11680	11.0	11680	11.0	occupation	steel mill, removing slag
11690	7.5	11690	7.5	occupation	steel mill, tending furnace

(Continued)

The Compendium of Physical Activities Tracking Guide (Continued)

1993 Compcode	1993 METS	2000 Compcode	2000 METS	Heading	Description
11700	5.5	11700	5.5	occupation	steel mill, tipping molds
11710	8.0	11710	8.0	occupation	steel mill, working in general
11720	2.5	11720	2.5	occupation	tailoring, cutting
11730	2.5	11730	2.5	occupation	tailoring, general
11740	2.0	11740	2.0	occupation	tailoring, hand sewing
11750	2.5	11750	2.5	occupation	tailoring, machine sewing
11760	4.0	11760	4.0	occupation	tailoring, pressing
		11765	3.5	occupation	tailoring, weaving
11766	6.5	11766	6.5	occupation	truck driving, loading and unloading truck (standing)
11770	1.5	11770	1.5	occupation	typing, electric, manual or computer
11780	6.0	11780	6.0	occupation	using heavy power tools such as pneumatic tools (jackhammers, drills, etc.)
11790	8.0	11790	8.0	occupation	using heavy tools (not power) such as shovel, pick, tunnel bar, space
11791	2.0	11791	2.0	occupation	walking on job, less than 2.0 mph (in office or lab area), very slow
11792	3.5	11792	3.3	occupation	walking on job, 3.0 mph, in office, moderate speed, not carrying anything
11793	4.0	11793	3.8	occupation	walking on job, 3.5 mph, in office, brisk speed, not carrying anything
11795	3.0	11795	3.0	occupation	walking, 2.5 mph, slowly and carrying light objects less than 25 pounds
		11796	3.0	occupation	walking, gathering things at work, ready to leave
11800	4.0	11800	4.0	occupation	walking, 3.0 mph, moderately and carrying light objects less than 25 pounds
		11805	4.0	occupation	walking, pushing a wheelchair
11810	4.5	11810	4.5	occupation	walking, 3.5 mph, briskly and carrying objects less than 25 pounds
11820	5.0	11820	5.0	occupation	walking or walk downstairs or standing, carrying objects about 25 to 49 pounds
11830	6.5	11830	6.5	occupation	walking or walk downstairs or standing, carrying objects about 50 to 74 pounds
11840	7.5	11840	7.5	occupation	walking or walk downstairs or standing, carrying objects about 75 to 99 pounds
11850	8.5	11850	8.5	occupation	walking or walk downstairs or standing, carrying objects about 100 pounds or over
11870	3.0	11870	3.0	occupation	working in scene shop, theater actor, backstage employee
		11875	4.0	occupation	teach physical education, exercise, sports classes (nonsport play)
		11876	6.5	occupation	teach physical education, exercise, sports classes (participate in the class)
12010	6.0	12010	6.0	running	jog/walk combination (jogging component of less than 10 minutes) (Taylor Code 180)
12020	7.0	12020	7.0	running	jogging, general
		12025	8.0	running	jogging, in place
		12027	4.5	running	jogging on a mini-tramp
12030	8.0	12030	8.0	running	running, 5 mph (12 min/mile)
12040	9.0	12040	9.0	running	running, 5.2 mph (11.5 min/mile)
12050	10.0	12050	10.0	running	running, 6 mph (10 min/mile)
12060	11.0	12060	11.0	running	running, 6.7 mph (9 min/mile)
12070	11.5	12070	11.5	running	running, 7 mph (8.5 min/mile)
12080	12.5	12080	12.5	running	running, 7.5 mph (8 min/mile)
12090	13.5	12090	13.5	running	running, 8 mph (7.5 min/mile)
12100	14.0	12100	14.0	running	running, 8.6 mph (7 min/mile)
12110	15.0	12110	15.0	running	running, 9 mph (6.5 min/mile)

The Compendium of Physical Activities Tracking Guide (Continued)

1993 Compcode	METS	2000 Compcode	METS	Heading	Description
12120	16.0	12120	16.0	running	running, 10 mph (6 min/mile)
12130	18.0	12130	18.0	running	running, 10.9 mph (5.5 min/mile)
12140	9.0	12140	9.0	running	running, cross country
12150	8.0	12150	8.0	running	running (Taylor Code 200)
12160	8.0			running	running, in place
12170	15.0	12170	15.0	running	running, stairs, up
12180	10.0	12180	10.0	running	running, on a track, team practice
12190	8.0	12190	8.0	running	running, training, pushing a wheelchair
12195	3.0			running	running, wheeling, general
13000	2.5	13000	2.0	self care	standing- getting ready for bed, in general
13009	1.0	13009	1.0	self care	sitting on toilet
13010	2.0	13010	1.5	self care	bathing (sitting)
13020	2.5	13020	2.0	self care	dressing, undressing (standing or sitting)
13030	1.5	13030	1.5	self care	eating (sitting)
13035	2.0	13035	2.0	self care	talking and eating or eating only (standing)
		13036	1.0	self care	taking medication, sitting or standing
13040	2.5	13040	2.0	self care	grooming (washing, shaving, brushing teeth, urinating,washing hands, putting on makeup), sitting or standing
		13045	2.5	self care	hairstyling
		13046	1.0	self care	having hair or nails done by someone else, sitting
13050	4.0	13050	2.0	self care	showering, toweling off (standing)
14010	1.5	14010	1.5	sexual activity	active, vigorous effort
14020	1.3	14020	1.3	sexual activity	general, moderate effort
14030	1.0	14030	1.0	sexual activity	passive, light effort, kissing, hugging
15010	3.5	15010	3.5	sports	archery (nonhunting)
15020	7.0	15020	7.0	sports	badminton, competitive (Taylor Code 450)
-5030	4.5	15030	4.5	sports	badminton, social singles and doubles, general
15040	8.0	15040	8.0	sports	basketball, game (Taylor Code 490)
15050	6.0	15050	6.0	sports	basketball, nongame, general (Taylor Code 480)
15060	7.0	15060	7.0	sports	basketball, officiating (Taylor Code 500)
15070	4.5	15070	4.5	sports	basketball, shooting baskets
15075	6.5	15075	6.5	sports	basketball, wheelchair
15080	2.5	15080	2.5	sports	billiards
15090	3.0	15090	3.0	sports	bowling (Taylor Code 390)
15100	12.0	15100	12.0	sports	boxing, in ring, general
15110	6.0	15110	6.0	sports	boxing, punching bag
15120	9.0	15120	9.0	sports	boxing, sparring
15130	7.0	15130	7.0	sports	broomball
15135	5.0	15135	5.0	sports	children's games (hopscotch, 4- marbles, jacks, arcade games)
15140	4.0	15140	4.0	sports	coaching: football, soccer, basketball, baseball, swimming, etc.

(Continued)

The Compendium of Physical Activities Tracking Guide (Continued)

1993 Compcode	METS	2000 Compcode	METS	Heading	Description
15150	5.0	15150	5.0	sports	cricket (batting, bowling)
15160	2.5	15160	2.5	sports	croquet
15170	4.0	15170	4.0	sports	curling
15180	2.5	15180	2.5	sports	darts, wall or lawn
15190	6.0	15190	6.0	sports	drag racing, pushing or driving a car
15200	6.0	15200	6.0	sports	fencing
15210	9.0	15210	9.0	sports	football, competitive
15230	8.0	15230	8.0	sports	football, touch, flag, general (Taylor Code 510)
15235	2.5	15235	2.5	sports	football or baseball, playing catch
15240	3.0	15240	3.0	sports	frisbee playing, general
15250	3.5	15250	3.5	sports	frisbee, ultimate
15255	4.5	15255	4.5	sports	golf, general
15260	5.5			sports	golf carrying clubs
		15265	4.5	sports	golf, walking and carrying clubs (See footnote at end of the Compendium)
15270	3.0	15270	3.0	sports	golf, miniature, driving range
15280	5.0			sports	golf, pulling clubs
		15285	4.3	sports	golf, walking and pulling clubs (See footnote at end of the Compendium)
15290	3.5	15290	3.5	sports	golf, using power cart (Taylor Code 070)
15300	4.0	15300	4.0	sports	gymnastics, general
15310	4.0	15310	4.0	sports	hacky sack
15320	12.0	15320	12.0	sports	handball, general (Taylor Code 520)
15330	8.0	15330	8.0	sports	handball, team
15340	3.5	15340	3.5	sports	hang gliding
15350	8.0	15350	8.0	sports	hockey, field
15360	8.0	15360	8.0	sports	hockey, ice
15370	4.0	15370	4.0	sports	horseback riding, general
15380	3.5	15380	3.5	sports	horseback riding, saddling horse, grooming horse
15390	6.5	15390	6.5	sports	horseback riding, trotting
15400	2.5	15400	2.5	sports	horseback riding, walking
15410	3.0	15410	3.0	sports	horseshoe pitching, quoits
15420	12.0	15420	12.0	sports	jai alai
15430	10.0	15430	10.0	sports	judo, jujitsu, karate, kick boxing, tae kwon do
15440	4.0	15440	4.0	sports	juggling
15450	7.0	15450	7.0	sports	kickball
15460	8.0	15460	8.0	sports	lacrosse
15470	4.0	15470	4.0	sports	motor-cross
15480	9.0	15480	9.0	sports	orienteering
15490	10.0	15490	10.0	sports	paddleball, competitive
15500	6.0	15500	6.0	sports	paddleball, casual, general (Taylor Code 460)
15510	8.0	15510	8.0	sports	polo

The Compendium of Physical Activities Tracking Guide (Continued)

| 1993 | | 2000 | | |
Compcode	METS	Compcode	METS	Heading	Description
15520	10.0	15520	10.0	sports	racquetball, competitive
15530	7.0	15530	7.0	sports	racquetball, casual, general (Taylor Code 470)
15535	11.0	15535	11.0	Sports	rock climbing, ascending rock
15540	8.0	15540	8.0	sports	rock climbing, rappelling
15550	12.0	15550	12.0	sports	rope jumping, fast
15551	10.0	15551	10.0	sports	rope jumping, moderate, general
15552	8.0	15552	8.0	sports	rope jumping, slow
15560	10.0	15560	10.0	sports	rugby
15570	3.0	15570	3.0	sports	shuffleboard, lawn bowling
15580	5.0	15580	5.0	sports	skateboarding
15590	7.0	15590	7.0	sports	skating, roller (Taylor Code 360)
		15591	12.0	sports	rollerblading (in-line skating)
15600	3.5	15600	3.5	sports	sky diving
15605	10.0	15605	10.0	sports	soccer, competitive
15610	7.0	15610	7.0	sports	soccer, casual, general (Taylor Code 540)
15620	5	15620	5	sports	Softball or baseball, fast or slow pitch, general (Taylor Code 440)
15630	4	15630	4	sports	Softball, officiating
15640	6	15640	6	sports	Softball, pitching
15650	12	15650	12	sports	squash (Taylor Code 530)
15660	4	15660	4	sports	table tennis, ping pong (Taylor Code 410)
15670	4	15670	4	sports	tai chi
15675	7	15675	7	sports	tennis, general
15680	6	15680	6	sports	tennis, doubles (Taylor Code 430)
		15685	5	sports	tennis, doubles
15690	8	15690	8	sports	tennis, singles (Taylor Code 420)
15700	3.5	15700	3.5	sports	trampoline
15710	4	15710	4	sports	volleyball (Taylor Code 400)
		15711	8	sports	volleyball, competitive, in gymnasium
					volleyball, non-competitive, 6-9
15720	3	15720	3	sports	member team, general
15725	8	15725	8	sports	volleyball, beach
15730	6	15730	6	sports	wrestling (one match = 5 minutes)
15731	7	15731	7	sports	wallyball, general
		15732	4	sports	track and field (shot, discus, hammer throw)
		15733	6	sports	track and field (high jump, long jump,triple jump, javelin, pole vault)
		15734	10	sports	track and field (steeplechase, hurdles)
16010	2	16010	2	transportation	automobile or light truck (not a semi) driving
		16015	1	transportation	riding in a car or truck
		16016	1	transportation	riding in a bus
16020	2	16020	2	transportation	flying airplane

(Continued)

The Compendium of Physical Activities Tracking Guide (Continued)

1993 Compcode	1993 METS	2000 Compcode	2000 METS	Heading	Description
16030	2.5	16030	2.5	transportation	motor scooter, motorcycle
16040	6	16040	6	transportation	pushing plane in and out of hanger
16050	3	16050	3	transportation	driving heavy truck, tractor, bus
17010	7	17010	7	walking	backpacking (Taylor Code 050)
17020	3.5	17020	3.5	walking	carrying infant or 15-pound load (e.g., suitcase), level ground or downstairs
17025	9	17025	9	walking	carrying load upstairs, general
17026	5	17026	5	walking	carrying 1 to 15 lb load, upstairs
17027	6	17027	6	walking	carrying 16 to 24 lb load, upstairs
17028	8	17028	8	walking	carrying 25 to 49 lb load, upstairs
17029	10	17029	10	walking	carrying 50 to 74 lb load, upstairs
17030	12	17030	12	walking	carrying 74+ 1lb load, upstairs
		17031	3	walking	loading/unloading a car
17035	7	17035	7	walking	climbing hills with 0 to 9 lb load
17040	7.5	17040	7.5	walking	climbing hills with 10 to 20 lb load
17050	8	17050	8	walking	climbing hills with 21 to 42 lb load
17060	9	17060	9	walking	climbing hills with 42+ lb load
17070	3	17070	3	walking	downstairs
17080	6	17080	6	walking	hiking, cross country (Taylor Code 040)
		17085	2.5	walking	bird watching
17090	6.5	17090	6.5	walking	marching, rapidly, military
17100	2.5	17100	2.5	walking	pushing or pulling stroller with child or walking with children
		17105	4	walking	pushing a wheelchair, nonoccupational setting
17110	6.5	17110	6.5	walking	race walking
17120	8	17120	8	walking	rock or mountain climbing (Taylor Code 060)
17130	8	17130	8	walking	up stairs, using or climbing up ladder (Taylor Code 030)
17140	4	17140	5	walking	using crutches
17150	2	17150	2	walking	walking, household walking
		17151	2	walking	walking, less than 2.0 mph, level ground, strolling, very slow
		17152	2.5	walking	walking, 2.0 mph, level, slow pace, firm surface
17160	2.5	17160	3.5	walking	walking for pleasure (Taylor Code 010)
		17161	2.5	walking	walking from house to car or bus, from car or bus to go places, from car or bus to and from the work site
		17162	2.5	walking	walking to neighbor's house or
		17165	3	walking	walking the dog
17170	3	17170	3	walking	walking, 2.5 mph, firm surface
17180	3	17180	2.8	walking	walking, 2.5 mph, downhill
17190	3.5	17190	3.3	walking	walking, 3.0 mph, level, moderate pace, firm surface
17200	4	17200	3.8	walking	walking, 3.5 mph, level, brisk, firm surface, walking for exercise
17210	6	17210	6	walking	walking, 3.5 mph, level, uphill
17220	4	17220	5	walking	walking, 4.0 mph, level, firm surface, very brisk pace
17230	4.5	17230	6.3	walking	walking, 4.5 mph, level, firm surface, very, very brisk

The Compendium of Physical Activities Tracking Guide (Continued)

1993 Compcode	1993 METS	2000 Compcode	2000 METS	Heading	Description
		17231	8	walking	walking, 5.0 mph
17250	3.5	17250	3.5	walking	walking, for pleasure, work break
17260	5	17260	5	walking	walking, grass track
17270	4	17270	4	walking	walking, to work or class (Taylor Code 015)
		17280	2.5	walking	walking to and from an outhouse
18010	2.5	18010	2.5	water activities	boating, power
18020	4	18020	4	water activities	canoeing, on camping trip (Taylor Code 270)
		18025	3.3	water activities	canoeing, harvesting wild rice, knocking rice off the stalks
18030	7	18030	7	water activities	canoeing, portaging
18040	3	18040	3	water activities	canoeing, rowing, 2.0–3.9 mph, light effort
18050	7	18050	7	water activities	canoeing, rowing, 4.0–5.9 mph, moderate effort
18060	12	18060	12	water activities	canoeing, rowing, >6 mph, vigorous effort
18070	3.5	18070	3.5	water activities	canoeing, rowing, for pleasure, general (Taylor Code 250)
18080	12	18080	12	water activities	canoeing, rowing, in competition, or crew or sculling (Taylor Code 260)
18090	3	18090	3	water activities	diving, springboard or platform
18100	5	18100	5	water activities	kayaking
18110	4	18110	4	water activities	paddle boat
18120	3	18120	3	water activities	sailing, boat and board sailing, windsurfing, ice sailing, general (Taylor Code 235)
18130	5	18130	5	water activities	sailing, in competition
18140	3	18140	3	water activities	sailing, Sunfish/Laser/Hobby Cat, Keel boats, ocean sailing, yachting
18150	6	18150	6	water activities	skiing, water (Taylor Code 220)
18160	7	18160	7	water activities	skimobiling
18170	12			water activities	
18180	16	18180	16	water activities	skindiving, fast
18190	12.5	18190	12.5	water activities	skindiving, moderate
18200	7	18200	7	water activities	skindiving, scuba diving, general (Taylor Code 310)
18210	5	18210	5	water activities	snorkeling (Taylor Code 320)
18220	3	18220	3	water activities	surfing, body or board
18230	10	18230	10	water activities	swimming laps, freestyle, fast, vigorous effort
18240	8	18240	7	water activities	swimming laps, freestyle, slow, moderate or light effort
18250	8	18250	7	water activities	swimming, backstroke, general
18260	10	18260	10	water activities	swimming, breaststroke, general
18270	11	18270	11	water activities	swimming, butterfly, general
18280	11	18280	11	water activities	swimming, crawl, fast (75 yards/minute), vigorous effort
18290	8	18290	8	water activities	swimming, crawl, slow (50 yards/minute), moderate or light effort
18300	6	18300	6	water activities	swimming, lake, ocean, river (Taylor Codes 280, 295)
18310	6	18310	6	water activities	swimming, leisurely, not lap swimming, general
18320	8	18320	8	water activities	swimming, sidestroke, general
18330	8	18330	8	water activities	swimming, synchronized
18340	10	18340	10	water activities	swimming, treading water, fast vigorous effort

(Continued)

The Compendium of Physical Activities Tracking Guide (Continued)

1993 Compcode	1993 METS	2000 Compcode	2000 METS	Heading	Description
18350	4	18350	4	water activities	swimming, treading water, moderate effort, general
		18355	4	water activities	water aerobics, water calisthenics
18360	10	18360	10	water activities	water polo
18365	3	18365	3	water activities	water volleyball
		18366	8	water activities	water jogging
18370	5	18370	5	water activities	Whitewater rafting, kayaking, or canoeing
19010	6	19010	6	winter activities	moving ice house (set up/drill holes, etc.)
19020	5.5	19020	5.5	winter activities	skating, ice, 9 mph or less
19030	7	19030	7	winter activities	skating, ice, general (Taylor Code 360)
19040	9	19040	9	winter activities	skating, ice, rapidly, more than 9 mph
19050	15	19050	15	winter activities	skating, speed, competitive
19060	7	19060	7	winter activities	ski jumping (climb up carrying skis)
19075	7	19075	7	winter activities	skiing, general
19080	7	19080	7	winter activities	skiing, cross country, 2.5 mph, slow or light effort, ski walking
19090	8	19090	8	winter activities	skiing, cross country, 4.0–4.9 mph, moderate speed and effort, general
19100	9	19100	9	winter activities	skiing, cross country, 5.0–7.9 mph, brisk speed, vigorous effort
19110	14	19110	14	winter activities	skiing, cross country, .8.0 mph, racing
19130	16.5	19130	16.5	winter activities	skiing, cross country, hard snow, uphill, maximum, snow mountaineering
19150	5	19150	5	winter activities	skiing, downhill, light effort
19160	6	19160	6	winter activities	skiing, downhill, moderate effort, general
19170	8	19170	8	winter activities	skiing, downhill, vigorous effort, racing
19180	7	19180	7	winter activities	sledding, tobogganing, bobsledding, luge (Taylor Code 370)
19190	8	19190	8	winter activities	snow shoeing
19200	3.5	19200	3.5	winter activities	snowmobiling
		20000	1	religious activities	sitting in church, in service, attending a ceremony, sitting quietly
		20001	2.5	religious activities	sitting, playing an instrument at church
		20005	1.5	religious activities	sitting in church, talking or singing, attending a ceremony, sitting, active participator
		20010	1.3	religious activities	sitting, reading religious materials at home
		20015	1.2	religious activities	standing in church (quietly), attending a ceremony, standing quietly
		20020	2	religious activities	standing, singing in church, attending a ceremony, standing, active participator
		20025	1	religious activities	kneeling in church/at home (praying)
		20030	1.8	religious activities	standing, talking in church
		20035	2	religious activities	walking in church
		20036	2	religious activities	walking, less than 2.0 mph, very slow
		20037	3.3	religious activities	walking, 3.0 mph, moderate speed, not carrying anything
		20038	3.8	religious activities	walking, 3.5 mph, brisk speed, not carrying anything
		20039	2	religious activities	walk/stand combination for religious purposes, usher
		20040	5	religious activities	praise with dance or run, spiritual dancing in church

The Compendium of Physical Activities Tracking Guide (Continued)

| 1993 | | 2000 | | Heading | Description |
Compcode	METS	Compcode	METS		
		20045	2.5	religious activities	serving food at church
		20046	2	religious activities	preparing food at church
		20047	2.3	religious activities	washing dishes/cleaning kitchen at church
		20050	1.5	religious activities	eating at church
		20055	2	religious activities	eating/talking at church or standing eating, American Indian Feast day
		20060	3	religious activities	cleaning church
		20061	5	religious activities	general yard work at church
		20065	2.5	religious activities	standing-moderate (lifting 50 lbs, assembling at fast rate)
		20095	4	religious activities	standing-moderate/heavy work
		20100	1.5	religious activities	typing, electric, manual, or computer
		21000	1.5	volunteer activities	sitting-meeting, general, and/or with talking involved
		21005	1.5	volunteer activities	sitting-light office work, in general
		21010	2.5	volunteer activities	sitting-moderate work
		21015	2.3	volunteer activities	standing-light work (filing, talking, assembling)
		21016	2.5	volunteer activities	sitting, child care, only active periods
		21017	3	volunteer activities	standing, child care, only active periods
		21018	4	volunteer activities	walk/run play with children, moderate, only active periods
		21019	5	volunteer activities	walk/run play with children, vigorous, only active periods
		21020	3	volunteer activities	standing-light/moderate work (pack boxes, assemble/repair, set up chairs/furniture
		21025	3.5	volunteer activities	standing-moderate (lifting 50 lbs, assembling at fast rate)
		21030	4	volunteer activities	standing-moderate/heavy work
		21035	1.5	volunteer activities	typing, electric, manual, or computer
		21040	2	volunteer activities	walking, less than 2.0 mph, very slow
		21045	3.3	volunteer activities	walking, 3.0 mph, moderate speed, not carrying anything
		21050	3.8	volunteer activities	walking, 3.5 mph, brisk speed, not carrying anything
		21055	3	volunteer activities	walking, 2.5 mph slowly and carrying objects less than 25 pounds
		21060	4	volunteer activities	walking, 3.0 mph moderately and carrying objects less than 25 pounds, pushing something
		21065	4.5	volunteer activities	walking, 3.5 mph, briskly and carrying objects less than 25 pounds
		21070	3	volunteer activities	walk/stand combination, for volunteer purposes

Blue text = row activity was added to the description of that specific compendium code

- Compcode and METS columns are blank under 1993; this means that the 2000 compcode and METS were added to the new edition of the compendium.
- Compcode and METS columns are blank under 2000; this means that the 1993 compcode and METS were removed from the new edition of the compendium.

METS for certain golfing activities were revised downward from 1993 estimates based on measurement of the activity using indirect calorimetry.

APPENDIX H Percentile Values for Maximal Aerobic Power (ml • kg^{-1} • min^{-1})*

Percentile values for maximal oxygen uptake (mL • kg^{-1} • min^{-1}) in men*

| Percentile | Age (yr) | | | | | |
	20–29 (N = 2,406)	30–39 (N = 13,158)	40–49 (N = 16,534)	50–59 (N = 9,102)	60–69 (N = 2,682)	70–79 (N = 467)
99	61.2	58.3	57.0	54.3	51.1	49.7
95	56.2	54.3	52.9	49.7	46.1	42.4
90	54.0	52.5	51.1	46.8	43.2	39.5
85	52.5	50.7	48.5	44.6	41.0	38.1
80	51.1	47.5	46.8	43.3	39.5	36.0
75	49.2	47.5	45.4	41.8	38.1	34.4
70	48.2	46.8	44.2	41.0	36.7	33.0
65	46.8	45.3	43.9	39.5	35.9	32.3
60	45.7	44.4	42.4	38.3	35.0	30.9
55	45.3	43.9	41.0	38.1	33.9	30.2
50	43.9	42.4	40.4	36.7	33.1	29.4
45	43.1	41.4	39.5	36.6	32.3	28.5
40	42.2	41.0	38.4	35.2	31.4	28.0
35	41.0	39.5	37.6	33.9	30.6	27.1
30	40.3	38.5	36.7	33.2	29.4	26.0
25	39.5	37.6	35.7	32.3	28.7	25.1
20	38.1	36.7	34.6	31.1	27.4	23.7
15	36.7	35.2	33.4	29.8	25.9	22.2
10	35.2	33.8	31.8	28.4	24.1	20.8
5	32.3	31.1	29.4	25.8	22.1	19.3
1	26.6	26.6	25.1	21.3	18.6	17.9

*Reprinted with permission from the Cooper Institute, Dallas, Texas. For more information: http://www.cooperinstitute.org

Percentile values for maximal oxygen uptake (mL • kg^{-1} • min^{-1}) in women*

| Percentile | Age (yr) | | | | | |
	20–29 (N = 1,350)	30–39 (N = 4,394)	40–49 (N = 4,834)	50–59 (N = 3,103)	60–69 (N = 1,088)	70–79 (N = 209)
99	55.0	52.5	51.1	45.3	42.4	42.4
95	50.2	46.9	45.2	39.9	36.9	36.7
90	47.5	44.7	42.4	38.1	34.6	33.5
85	45.3	42.5	40.0	36.7	33.0	32.0
80	44.0	41.0	38.9	35.2	32.3	30.2
75	43.4	40.3	38.1	34.1	31.0	29.4
70	41.1	38.8	36.7	32.9	30.2	28.4
65	40.6	38.1	35.6	32.3	29.4	27.6
60	39.5	36.7	35.1	31.4	29.1	26.6
55	38.1	36.7	33.8	30.9	28.3	26.0
50	37.4	35.2	33.3	30.2	27.5	25.1
45	36.7	34.5	32.3	29.4	26.9	24.6
40	35.5	33.8	31.6	28.7	26.6	23.8
35	34.6	32.4	30.9	28.0	25.4	22.9
30	33.8	32.3	29.7	27.3	24.9	22.2
25	32.4	30.9	29.4	26.6	24.2	21.9
20	31.6	29.9	28.0	25.5	23.7	21.2
15	30.5	28.9	26.7	24.6	22.8	20.8
10	29.4	27.4	25.6	23.7	21.7	19.3
5	26.4	25.5	24.1	21.9	20.1	17.9
1	22.6	22.7	20.8	19.3	18.1	16.4

*Reprinted with permission from the Cooper Institute, Dallas, Texas. For more information: http://www.cooperinstitute.org

APPENDIX I Respiratory Exchange Ratio (RER)

Nonprotein Respiratory Exchange Ratio, Caloric Values for Oxygen Consumption, and Percentages of Energy from Carbohydrate and Fat*

RER	kcal/Liter O$_2$	Percent CHO	Percent FAT
0.70	4.686	0.0	100.0
0.71	4.690	1.4	98.6
0.72	4.702	4.8	95.2
0.73	4.714	8.2	91.8
0.74	4.727	11.6	88.4
0.75	4.739	15.0	85.0
0.76	4.752	18.4	81.6
0.77	4.764	21.8	78.2
0.78	4.776	25.2	74.8
0.79	4.789	28.6	71.4
0.80	4.801	32.0	68.0
0.81	4.813	35.4	64.6
0.82	4.825	38.8	61.2
0.83	4.838	42.2	57.8
0.84	4.850	45.6	54.5
0.85	4.863	49.0	51.0
0.86	4.875	52.4	47.6
0.87	4.887	55.8	44.2
0.88	4.900	59.2	40.8
0.89	4.912	62.6	37.4
0.90	4.924	66.0	34.0
0.91	4.936	69.4	30.6
0.92	4.948	72.8	27.2
0.93	4.960	76.2	23.8
0.94	4.973	79.6	20.4
0.95	4.985	83.0	17.0
0.96	4.997	86.4	13.6
0.97	5.010	89.8	10.2
0.98	5.022	93.2	6.8
0.99	5.034	96.6	3.4
1.00	5.047	100.0	0.0

*Carpenter, Thorne M. (1964). *Tables, factors, and formulas for computing respiratory exchange and biological transformations of energy* (4th ed., p. 104). Washington, DC: Carnegie Institution of Washington.

APPENDIX J Anaerobic Glycolysis, the Krebs (Tricarboxylic Acid) Cycle, and the Electron Transport Chain

Phase 1

Phosphorylation of glucose and conversion to 2 molecules of glyceraldehyde-3-phosphate; 2 ATPs are used to prime these reactions.

Phase 2

Conversion of glyceraldehyde-3-phosphate to pyruvate and coupled formation of 4 ATP and 2 NADH.

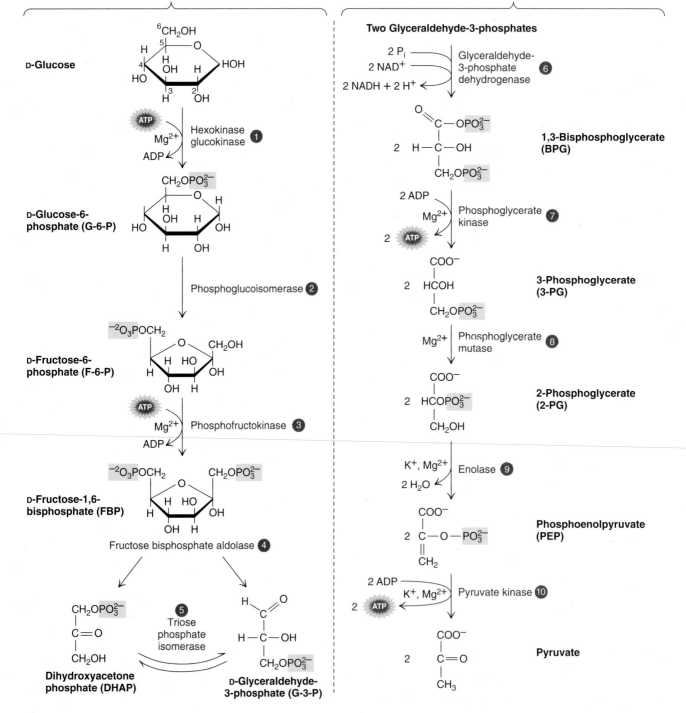

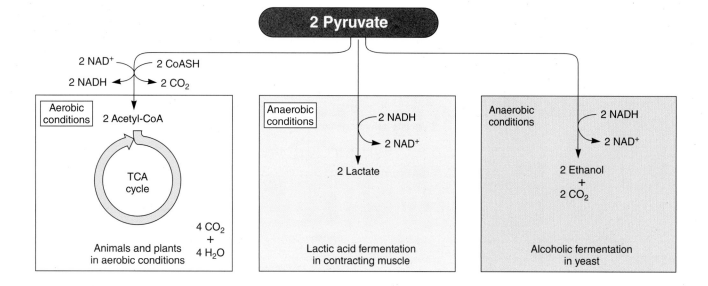

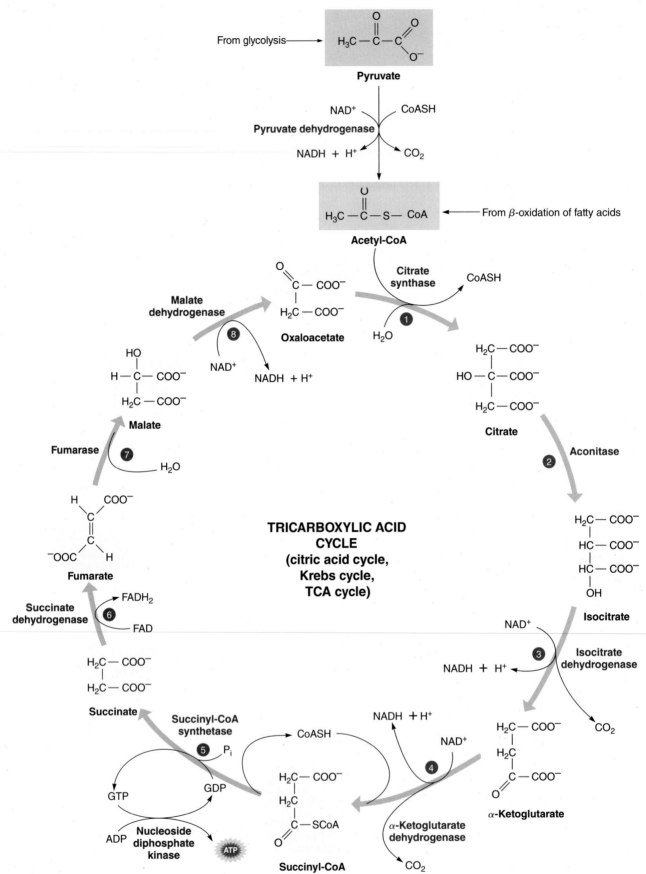

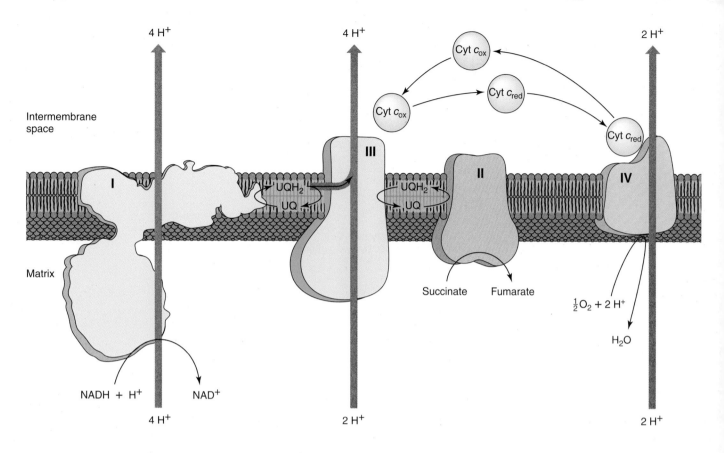

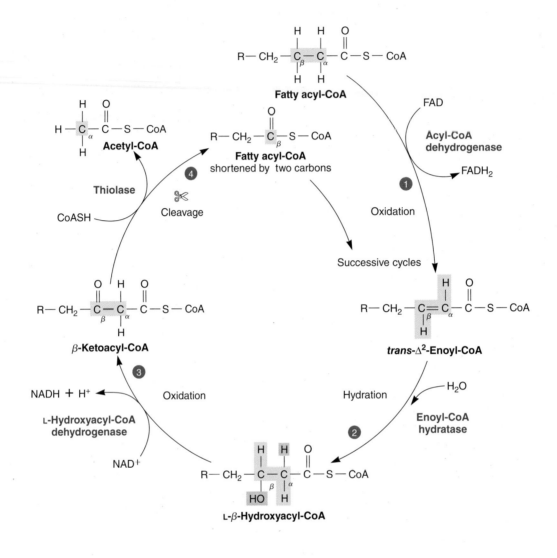

APPENDIX L National Collegiate Athletic Association (NCAA) Bylaw 31.2.3 Ineligibility for Use of Banned Drugs

31.2.3 INELIGIBILITY FOR USE OF BANNED DRUGS

Bylaw 18.4.1.5 provides that a student-athlete who as a result of a drug test administered by the NCAA is found to have used a substance on the list of banned drug classes, shall be declared ineligible for further participation in postseason and regular-season competition during the time period ending one calendar year after the collection of the student-athlete's positive drug-test specimen. The student-athlete shall be charged with the loss of a minimum of one season of competition in all sports if the season of competition has not yet begun or a minimum of the equivalent of one full season of competition in all sports if the student-athlete tests positive during his or her season of competition (the remainder of contests in the current season and contests in the following season up to the period of time in which the student-athlete was declared ineligible during the previous year). The student-athlete shall remain ineligible until the student-athlete tests negative (in accordance with the testing methods authorized by the Executive Committee) and the student-athlete's eligibility is restored by the Committee on Student-Athlete Reinstatement. If the student-athlete participates in any contests from the time of collection until the confirmation of the positive result, he or she must be withheld from an equal number of contests after the 365-day period of ineligibility. *(Revised: 1/16/93, 1/9/96 effective 8/1/96, 1/14/97 effective 8/1/97, 3/10/04, 4/28/05 effective 8/1/05, 11/1/07)*

31.2.3.1 Breach of NCAA Drug-Testing Program Protocol.
A student-athlete who is in breach of the NCAA drug-testing program protocol (e.g., no-show, tampering with sample) shall be considered to have tested positive for the use of any drug other than a "street" drug. *(Revised: 4/28/05 effective 5/1/05)*

31.2.3.2 Testing Positive on More than One Occasion.
If the student-athlete tests positive for the use of any banned drug other than a "street drug" after having previously tested positive for any banned drug other than a "street drug," he or she shall lose all remaining regular-season and postseason eligibility in all sports. If the student-athlete tests positive for the use of a "street drug" after having tested positive for the use of any banned drug, he or she shall lose a minimum of one additional season of competition in all sports and also shall remain ineligible for regular-season and postseason competition during the time period ending one calendar year (365 days) after the period of ineligibility for any prior positive drug tests has expired. Bylaw 18.4.1.5.2 also provides that the Executive Committee shall adopt a list of banned drugs and authorize methods for drug testing of student-athletes on a year-round basis. *(Revised: 4/28/05 effective 8/1/05, 6/17/08, 5/27/10)*

31.2.3.3 Appeals.
An institution may appeal the duration of ineligibility to the Committee on Competitive Safeguards and Medical Aspects of Sports (or a designated subcommittee). In all sports, the committee may reduce the legislated penalty to withholding the student-athlete from the next 50 percent of the season of competition or provide complete relief from the legislated penalty. If the committee requires the student-athlete to fulfill the legislated penalty or be withheld from the next 50 percent of the season of competition in all sports, the student-athlete shall remain ineligible until the prescribed penalty is fulfilled, the student-athlete retests negative and the student-athlete's eligibility is restored by the Committee on Student-Athlete Reinstatement. *(Adopted: 4/28/05 effective 8/1/05)*

31.2.3.4 Banned Drugs.
The following is the list of banned-drug classes. The Committee on Competitive Safeguards and Medical Aspects of Sports (or a designated subcommittee) has the authority to identify specific banned drugs and exceptions within each class. The institution and student-athlete shall be held accountable for all drugs within the banned-drug classes regardless of whether they have been specifically identified. The current list of specific banned drugs and exceptions is located on the NCAA Web site (ncaa.org) or may be obtained from the NCAA national office. *(Revised: 8/15/89, 7/10/90, 12/3/90, 5/4/92, 5/6/93, 10/29/97, 4/26/01, 2/10/06)*

(a) Stimulants; (*Revised: 2/10/06*)

(b) Anabolic agents; (*Revised: 2/10/06*)

(c) Alcohol and beta blockers (banned for rifle only); (*Revised: 2/10/06, 2/5/09*)

(d) Diuretics and other masking agents; (*Revised: 2/10/06, 5/29/07*)

(e) Street drugs; (*Revised: 2/10/06*)

(f) Peptide hormones and analogues; (*Revised: 2/10/06*)

(g) Anti-estrogens; and (*Adopted: 10/27/06 effective 8/1/07*)

(h) Beta-2 agonists. (*Adopted: 2/5/09*)

31.2.3.4.1 Drugs and Procedures Subject to Restrictions. The use of the following drugs and/or procedures is subject to certain restrictions and may or may not be permissible, depending on limitations expressed in these guidelines and/or quantities of these substances used: (*Revised: 8/15/89*)

(a) **Blood Doping.** The practice of blood doping (the intravenous injection of whole blood, packed red blood cells or blood substitutes) is prohibited, and any evidence confirming use will be cause for action consistent with that taken for a positive drug test. (*Revised: 8/15/89, 5/4/92*)

(b) **Local Anesthetics.** The Executive Committee will permit the limited use of local anesthetics under the following conditions:

(1) That procaine, xylocaine, carbocaine or any other local anesthetic may be used, but not cocaine; (*Revised: 12/9/91, 5/6/93*)

(2) That only local or topical injections can be used (intravenous injections are not permitted); and

(3) That use is medically justified only when permitting the athlete to continue the competition without potential risk to his or her health.

(c) **Manipulation of Urine Samples.** The Executive Committee bans the use of substances and methods that alter the integrity and/or validity of urine samples provided during NCAA drug testing. Examples of banned methods are catheterization, urine substitution and/or tampering or modification of renal excretion by the use of diuretics, probenecid, bromantan or related compounds, and epitestosterone administration. (*Revised: 8/15/89, 6/17/92, 7/22/97*)

(d) **Beta-2 Agonists.** The use of beta-2 agonists is permitted by inhalation only. (*Adopted: 8/13/93*)

(e) **Additional Analysis.** Drug screening for select nonbanned substances may be conducted for non-punitive purposes. (*Revised: 8/15/89*)

31.2.3.4.2 Positive Drug Test—Non-NCAA Athletics Organization. A student-athlete under a drug-testing suspension from a national or international sports governing body that has adopted the World Anti-Doping Agency (WADA) code shall not participate in NCAA intercollegiate competition for the duration of the suspension. (*Adopted: 1/14/97 effective 8/1/97, Revised: 4/28/05 effective 8/1/05*)

31.2.3.5 Medical Exceptions. Exceptions for the banned-drug classes of stimulants, anabolic agents, alcohol and beta blockers (for rifle only), diuretics and other masking agents, peptide hormones and analogues, anti-estrogens and beta-2 agonists may be made by the Executive Committee for those student-athletes with a documented medical history demonstrating the need for regular use of such a drug. (*Revised: 8/5/99, 9/26/06, 10/27/06 effective 8/1/07, 2/5/09*)

31.2.3.6 Methods for Drug Testing. The methods and any later modifications authorized by the Executive Committee for drug testing of student-athletes shall be summarized and posted on the NCAA Web site. Copies of the modifications shall be available to member institutions.

31.2.3.7 Events Identified for Drug Tests. The Executive Committee shall determine the regular-season and postseason competition for which drug tests shall be made and the procedures to be followed in disclosing its determinations.

31.2.3.8 Individual Eligibility—Team Sanctions. Executive regulations pertaining to team-eligibility sanctions for positive tests resulting from the NCAA drug-testing program shall apply only in the following situation: If a student-athlete is declared ineligible prior to an NCAA team championship or a licensed postseason football game and the institution knowingly allows him or her to participate, all team-ineligibility sanctions shall apply (the team shall be required to forfeit its awards and any revenue distribution it may have earned, and the team's and student-athlete's performances shall be deleted from NCAA records). In the case of licensed postseason football contests, the team's and student-athlete's performances shall be deleted from NCAA records. (*Revised: 1/10/90*)

Source: 2010–11 NCAA® Division 1 Manual, 2010 pp. 390–391.

Normative Values for Percent Body Fat for Adult Men and Women by Decade

Percentile values for body composition (% body fat) for men*

	Age					
Percentile	20–29	30–39	40–49	50–59	60–69	70–79
99	4.2	7.0	9.2	10.9	11.5	13.6
95	6.3	9.9	12.8	14.4	15.5	15.2
90	7.9	11.9	14.9	16.7	17.6	17.8
85	9.2	13.3	16.3	18.0	18.8	19.2
80	10.5	14.5	17.4	19.1	19.7	20.4
75	11.5	15.5	18.4	19.9	20.6	21.1
70	12.7	16.5	19.1	20.7	21.3	21.6
65	13.9	17.4	19.9	21.3	22.0	22.5
60	14.8	18.2	20.6	22.1	22.6	23.1
55	15.8	19.0	21.3	22.7	23.2	23.7
50	16.6	19.7	21.9	23.2	23.7	24.1
45	17.4	20.4	22.6	23.9	24.4	24.4
40	18.6	21.3	23.4	24.6	25.2	24.8
35	19.6	22.1	24.1	25.3	26.0	25.4
30	20.6	23.0	24.8	26.0	26.7	26.0
25	21.9	23.9	25.7	26.8	27.5	26.7
20	23.1	24.9	26.6	27.8	28.4	27.6
15	24.6	26.2	27.7	28.9	29.4	28.9
10	26.3	27.8	29.2	30.3	30.9	30.4
5	28.9	30.2	31.2	32.5	32.9	32.4
1	33.3	34.3	35.0	36.4	36.8	35.5

*Reprinted with permission from the Cooper Institute, Dallas, Texas. For more information: http://www.cooperinstitute.org

Percentile values for body composition (% body fat) for women*

	Age					
Percentile	20–29	30–39	40–49	50–59	60–69	70–79
99	9.5	11.0	12.6	14.6	13.9	14.6
95	13.6	14.0	15.6	17.2	17.7	16.6
90	14.8	15.6	17.2	19.4	19.8	20.3
85	15.8	16.6	18.6	20.9	21.4	23.0
80	16.5	13.4	19.8	22.5	23.2	24.0
75	17.3	18.2	20.8	23.8	24.8	25.0
70	18.0	19.1	21.9	25.1	25.9	26.2
65	18.7	20.0	22.8	26.0	27.0	27.7
60	19.4	20.8	23.8	27.0	27.9	28.6
55	20.1	21.7	24.8	27.9	28.7	29.7
50	21.0	22.6	25.6	28.8	29.8	30.4
45	21.9	23.5	26.5	29.7	30.6	31.3
40	22.7	24.6	27.6	30.4	31.3	31.8
35	23.6	25.6	28.5	31.4	32.5	32.7
30	24.5	26.7	29.6	32.5	33.3	33.9
25	25.9	27.7	30.7	33.4	34.3	35.3
20	27.1	29.1	31.9	34.5	35.4	36.0
15	28.9	30.9	33.5	35.6	36.2	37.4
10	31.4	33.0	35.4	36.7	37.3	38.2
5	35.2	35.8	37.4	38.3	39.0	39.3
1	38.9	39.4	39.8	40.4	40.8	40.5

*Reprinted with permission from the Cooper Institute, Dallas, Texas. For more information: http://www.cooperinstitute.org

APPENDIX N Body Mass Index (BMI)

Find your height along the left-hand column and look across the row until you find the number that is closest to your weight. The number at the top of that column identifies your BMI. Chapter 13 describes how BMI correlates with disease risks and defines obesity. A healthy weight range is considered to be a BMI between 18.5 and 24.9.

	18	19	20	21	22	23	24	25	26	27	28	29	30	31	32	33	34	35	36	37	38	39	40
Height											Body weight (pounds)												
4'10"	86	91	96	100	105	110	115	119	124	129	134	138	143	148	153	158	162	167	172	177	181	186	191
4'11"	89	94	99	104	109	114	119	124	128	133	138	143	148	153	158	163	168	173	178	183	188	193	198
5'0"	92	97	102	107	112	118	123	128	133	138	143	148	153	158	163	168	174	179	184	189	194	199	204
5'1"	95	100	106	111	116	122	127	132	137	143	148	153	158	164	169	174	180	185	190	195	201	206	211
5'2"	98	104	109	115	120	126	131	136	142	147	153	158	164	169	175	180	186	191	196	202	207	213	218
5'3"	102	107	113	118	124	130	135	141	146	152	158	163	169	175	180	186	191	197	203	208	214	220	225
5'4"	105	110	116	122	128	134	140	145	151	157	163	169	174	180	186	192	197	204	209	215	221	227	232
5'5"	108	114	120	126	132	138	144	150	156	162	168	174	180	186	192	198	204	210	216	222	228	234	240
5'6"	112	118	124	130	136	142	148	155	161	167	173	179	186	192	198	204	210	216	223	229	235	241	247
5'7"	115	121	127	134	140	146	153	159	166	172	178	185	191	198	204	211	217	223	230	236	242	249	255
5'8"	118	125	131	138	144	151	158	164	171	177	184	190	197	203	210	216	223	230	236	243	249	256	262
5'9"	122	128	135	142	149	155	162	169	176	182	189	196	203	209	216	223	230	236	243	250	257	263	270
5'10"	126	132	139	146	153	160	167	174	181	188	195	202	209	216	222	229	236	243	250	257	264	271	278
5'11"	129	136	143	150	157	165	172	179	186	193	200	208	215	222	229	236	243	250	257	265	272	279	286
6'0"	132	140	147	154	162	169	177	184	191	199	206	213	221	228	235	242	250	258	265	272	279	287	294
6'1"	136	144	151	159	166	174	182	189	197	204	212	219	227	235	242	250	257	265	272	280	288	295	302
6'2"	141	148	155	163	171	179	186	194	202	210	218	225	233	241	249	256	264	272	280	287	295	303	311
6'3"	144	152	160	168	176	184	192	200	208	216	224	232	240	248	256	264	272	279	287	295	303	311	319
6'4"	148	156	164	172	180	189	197	205	213	221	230	238	246	254	263	271	279	287	295	304	312	320	328
6'5"	151	160	168	176	185	193	202	210	218	227	235	244	252	261	269	277	286	294	303	311	319	328	336
6'6"	155	164	172	181	190	198	207	216	224	233	241	250	259	267	276	284	293	302	310	319	328	336	345

Under weight (<18.5)	Healthy weight (18.5–24.9)	Overweight (25–29.9)	Obese (≥30)

APPENDIX O Answers to Post-Test and Multiple-Choice Questions

Post-Test Questions

Chapter 1: Reassessing Knowledge of Sports Nutrition

1. An athlete's diet is a modification of the general nutrition guidelines made for healthy adults.
   True.
2. Once a healthy diet plan is developed, an athlete can use it every day with little need for modification.
   False. The demands of training vary over the course of a year, and the nutrition plan will also need to change to match training.
3. In the United States, dietary supplements are regulated in the same way as over-the-counter medications.
   False. Dietary supplements are loosely regulated under the Dietary Supplement Health and Education Act. The active ingredients in over-the-counter medications must be standardized by law, which is not the case for dietary supplements.
4. The scientific aspect of sports nutrition is developing very quickly, and quantum leaps are being made in knowledge of sports nutrition.
   False. Scientific knowledge of sports nutrition is increasing but its progress is *slow*, as is the case for most science-based disciplines.
5. To legally use the title of sports nutritionist in the United States, a person must have a bachelor's degree in nutrition.
   False. Nutritionist is a broad term and anyone in the United States can claim to be a sports nutritionist.

Chapter 2: Reassessing Knowledge of Energy

1. The body creates energy from the food that is consumed.
   False. The body cannot *create* energy. Through the process called bioenergetics, the body *transforms* the energy that is contained in food to forms that are usable in the body.
2. The scientific unit of measure of energy is the calorie.
   False. The unit of measure for energy in SI units is the joule, which is equal to 0.24 calories. Although the term *calorie* (lower case *c*) is widely used in nonscientific writing in the United States, it is technically incorrect.

3. A person's resting metabolic rate can change in response to a variety of factors such as age, food intake, or environmental temperature.
   True.
4. Physical activity is responsible for the largest amount of energy expended during the day for the average adult in the United States.
   False. Resting metabolism makes up approximately 70 percent of the day's energy expenditure in a sedentary adult, whereas physical activity comprises only ~20 percent of the total energy expenditure.
5. The energy source used by all cells in the body is adenosine triphosphate (ATP).
   True.

Chapter 3: Reassessing Knowledge of Energy Systems and Exercise

1. The direct source of energy for force production by muscle is ATP.
   True.
2. Creatine supplements result in immediate increases in strength, speed, and power.
   False. Creatine supplementation may be effective for some strength athletes because it allows them to train harder, such as completing more weight-lifting repetitions. Over time, this increase in the training stimulus allows the athlete to potentially become stronger, faster, or more powerful.
3. Lactate is a metabolic waste product that causes fatigue.
   False. Lactate is an important metabolic product that can be utilized by other cells as a source of energy; the liver can use lactate to make glucose. The accumulation of lactate molecules does not cause fatigue; one likely cause is the metabolic acidosis that is associated with high-intensity exercise when glycolysis is used at a high rate.
4. The aerobic energy system is not active during high-intensity anaerobic exercise.
   False. All energy systems are active at all times. Aerobic metabolism may not be the predominant energy system during high-intensity exercise described as anaerobic; it will increase in activity to replace creatine phosphate and oxidize lactate after anaerobic exercise.

5. At rest and during low levels of physical activity, fat is the preferred source of fuel for the aerobic energy system.
   True.

Chapter 4: Reassessing Knowledge of Carbohydrates

1. The body uses carbohydrate primarily in the form of fruit sugar, or fructose.
   False. The predominant form of carbohydrate used by the body is glucose and its storage form, glycogen.
2. Sugars such as sucrose (table sugar) are unhealthy and should rarely be a part of an athlete's diet.
   False. Although excessive consumption of sugars may be related to obesity and other long-term health concerns, there are specific circumstances in which the consumption of sugars may be advantageous for athletes (for example, sports drinks during endurance exercise and high glycemic index food immediately following exercise). That said, athletes should get the majority of their carbohydrates from plant-based foods and high-fiber complex carbohydrates.
3. Low levels of muscle glycogen and blood glucose are often associated with fatigue, particularly during moderate- to high-intensity endurance exercise.
   True.
4. A diet that contains 70 percent of total kilocalories as carbohydrate will provide the necessary amount of carbohydrate for an athlete.
   False. This statement could be true if the total amount of energy (kcal) consumed by the athlete is sufficient. An athlete's diet may contain 70 percent of its energy from carbohydrate, but an insufficient number of grams of carbohydrate if the athlete's diet is very low in total calories.
5. Most athletes consume enough carbohydrates daily.
   False. Most athletes fail to consume the recommended amount of carbohydrate daily and risk having low muscle and liver glycogen stores, which could negatively affect training and performance.

Chapter 5: Reassessing Knowledge of Proteins

1. Skeletal muscle is the primary site for protein metabolism and is the tissue that regulates protein breakdown and synthesis throughout the body.
   False. Skeletal muscle is an important site for protein metabolism but the liver plays a primary role both in protein metabolism and its regulation.

2. In prolonged endurance exercise, approximately 3 to 5 percent of the total energy used is provided by amino acids.
   True.
3. To increase skeletal muscle mass, the body must be in positive nitrogen balance, which requires an adequate amount of protein and energy (calories).
   True.
4. Athletes who consume high-protein diets are at risk for developing kidney disease.
   False. Small studies conducted in athletes have not shown that consuming a high-protein diet increases the risk for developing kidney disease.
5. Strength athletes usually need protein supplements because it is difficult to obtain a sufficient amount of protein from food alone.
   False. Obtaining sufficient protein from food alone is not difficult, and surveys of athletes have shown that many consume more than is recommended. Protein supplements may be desirable for athletes for reasons such as convenience, but are optional and not necessary.

Chapter 6: Reassessing Knowledge of Fats

1. Athletes typically need to follow a very low fat diet.
   False. Athletes may consume less fat than the general population because carbohydrate and protein needs are higher for trained athletes than for sedentary individuals. However, a *very* low-fat diet may have a negative effect on performance and health.
2. At rest, the highest percentage of total energy expenditure is from fat and not carbohydrate.
   True.
3. To lose body fat, it is best to perform low-intensity exercise, which keeps one in the fat-burning zone.
   False. The most important factor for loss of body fat is total energy expenditure, regardless of the nutrient source.
4. To improve performance, endurance athletes should ingest caffeine because more free fatty acids are oxidized for energy and muscle glycogen is spared.
   False. Although caffeine does increase the mobilization of fatty acids, there is no evidence that fatty acid oxidation is increased or that muscle glycogen is spared. Caffeine's likely effect is the stimulation of the central nervous system.
5. A low-calorie, low-carbohydrate diet that results in ketosis is dangerous for athletes because it leads to the medical condition known as ketoacidosis.
   False. The majority of athletes do not have diabetes, and ketosis in those without diabetes rarely leads to ketoacidosis. Such diets would not likely be advantageous for athletes, however, because the low-carbohydrate and low-caloric content is a factor that could undermine training and performance.

Chapter 7: Reassessing Knowledge of Water and Electrolytes

1. The two major aspects of fluid balance are the volume of water and the concentration of the substances in the water.
   True.
2. Now that sports beverages are precisely formulated, it is rare that water would be a better choice than a sports beverage for a trained athlete.
   False. In some cases water is a better choice than a sports beverage.
3. Athletes should avoid caffeinated drinks because caffeine is a potent diuretic.
   False. Completely avoiding caffeinated beverages is a recommendation that is no longer made to athletes. Athletes are advised to consume less than 300 mg of caffeine daily, about the amount found in 3 cups of caffeinated coffee.
4. A rule of thumb for endurance athletes is to drink as much water as possible.
   False. Endurance athletes who drink as much water as possible run the risk of hyponatremia (low blood sodium). Endurance athletes should try to match fluid intake with fluid loss and replenish sodium as needed.
5. Under most circumstances, athletes will not voluntarily drink enough fluid to account for all the water lost during exercise.
   True.

Chapter 8: Reassessing Knowledge of Vitamins

1. Exercise increases the usage of vitamins, so most athletes need more vitamins than sedentary people.
   False. In general, exercise does not increase the usage of vitamins. In those cases in which usage may be increased, the additional amount needed would likely be small and easily provided by a vitamin-rich diet.
2. Vitamins provide energy.
   False. Vitamins help to *regulate* the reactions that produce energy from carbohydrates, proteins, and fats, but the vitamins themselves do not provide any energy (that is, contain kilocalories).
3. The amount of vitamins an athlete consumes is generally related to the amount of kilocalories consumed.
   True.
4. When antioxidant vitamins are consumed in excess, they act like pro-oxidants instead of antioxidants.
   True.
5. Vitamin supplements are better regulated than other dietary supplements because the U.S. Food and Drug Administration sets a maximum dose (amount) for each vitamin.

False. Vitamin supplements are regulated under the same law that regulates other supplements. The manufacturer establishes the dose (amount) contained in each vitamin supplement. Doses that exceed the Tolerable Upper Intake Level can be, and are, found in vitamin supplements available for sale in the United States.

Chapter 9: Reassessing Knowledge of Minerals

1. The basic functions of minerals include building body tissues, regulating physiological processes, and providing energy.
   False. Only carbohydrates, fats, proteins, and alcohol can be metabolized for energy. Minerals do help build body tissues and regulate physiological processes.
2. In general, the body absorbs a high percentage of the minerals consumed.
   False. The body does not easily excrete most minerals once they have been absorbed, so it must be careful not to overabsorb them. Mineral absorption from food is generally low or moderate.
3. In most cases, exercise does not increase mineral requirements above what is recommended for healthy, lightly active individuals.
   True.
4. If dietary calcium is inadequate over a long period of time, the body maintains its blood calcium concentration by reabsorbing skeletal calcium.
   True.
5. Iron deficiency anemia has a negative impact on performance.
   True.

Chapter 10: Reassessing Knowledge of Diet Planning for Athletes

1. When planning dietary intake, athletes should first consider how much dietary fat would be needed.
   False. Energy intake provides the framework for planning nutrient intake. Carbohydrate and protein needs are considered first and then the amount of dietary fat needed is determined.
2. The key to eating nutritiously without consuming excess kilocalories is to choose foods that have a high nutrient density.
   True.
3. There is no room in the athlete's diet for fast foods.
   False. Fast foods, especially those that are smaller portions and have not been fried, can easily fit into an athlete's diet plan. However, many fast foods are too high in energy (kcal), fat, and sugar to be everyday foods for the athlete.
4. After exercise, athletes should wait about an hour before consuming food or fluids.

False. Food and fluid consumption should begin as soon as possible after exercise ends. Carbohydrates, proteins, water, and electrolytes are needed for restoration and the timing of intake affects recovery.

5. To achieve optimum performance, most athletes will need to use some dietary supplements.
False. Training, and a diet to support training, is crucial to achieving optimum performance. The majority of dietary supplements marketed to athletes, especially those sold for the purposes of energy production and weight reduction, are not effective. Most athletes will not need to use dietary supplements to achieve optimum performance.

Chapter 11: Reassessing Knowledge of Body Weight and Body Composition

1. Percent body fat and fat mass can be precisely measured in athletes with a number of different methods.
False. Percent body fat and fat mass cannot be measured *precisely*. All methods of determining body compositions are estimates and have some measurement error inherent in the procedure. The subjects and/or the technician can introduce additional error.

2. The most accurate method of measuring body fat for any athlete is underwater weighing.
False. Although underwater weighing is widely used and is reasonably accurate for many athletes, it may not be the most accurate method for *all* athletes. Some athletes do not like being submerged underwater and have difficulty exhaling as much air as possible, which introduces more error. Not only would the athlete be more comfortable with another method, the measurement would be more accurate because of subject compliance.

3. In sports in which body weight must be moved or transported over a distance (for example, distance running), it is a performance advantage to have the lowest weight possible.
False. A low weight may be advantageous for athletes in sports in which weight must be moved, such as long distance running, but the *lowest* weight possible is not always associated with good performance. Too many athletes attain the lowest weight possible and then see their performance decline due to low glycogen stores, loss of lean body mass, dehydration, and compromised immune system function.

4. To increase muscle mass, most athletes need a substantial increase in their usual protein intake.
False. Increasing muscle mass requires resistance exercise, sufficient energy (kcal), and a *small* increase in protein intake. Substantial increases in dietary proteins are not necessary because most strength athletes are already consuming more than adequate amounts. Higher protein intakes, in the absence of adequate calories and resistance exercise, will not result in an increase in muscle mass.

5. For athletes who want to restrict energy intake to lose body fat, the recommended time to do so is at the beginning of the preseason or during the off-season.
True.

Chapter 12: Reassessing Knowledge of Disordered Eating and Exercise Dependence

1. Disordered eating and eating disorders affect only female athletes.
False. Disordered eating and eating disorders *primarily* affect female athletes but not *only* female athletes. These conditions may not be well recognized or reported in male athletes, which may delay treatment.

2. Anorexia athletica means that an athlete has a classic case of anorexia nervosa.
False. Although they share some features, the diagnostic criteria are different for these two conditions.

3. Disordered eating and eating disorders are more likely to be seen among elite female athletes in sports such as distance running and gymnastics.
True.

4. Coaches cause athletes to develop eating disorders.
False. Coaches do not *cause* athletes to develop eating disorders. However, coaches must be careful that their actions do not contribute to or reinforce abnormal or pathological behaviors in athletes.

5. A good diagnostic criterion for exercise dependence is the volume of exercise training (that is, frequency and duration of exercise).
False. The volume of exercise training may vary widely for athletes in different sports or for individual athletes at different times of the season. For example, an amount of running that may be considered excessive for a tennis player may be appropriate for a collegiate cross country runner.

Chapter 13: Reassessing Knowledge of Health, Fitness, and Chronic Diseases

1. There are many contradictions among diet and exercise recommendations that are issued by health promotion organizations.
False. Although there are some differences, most of the guidelines are remarkably consistent recommending routine physical activity and the consumption of fruits, vegetables, whole grains, low-fat protein foods, and heart-healthy fats.

2. Elite endurance athletes do not develop hypertension but lesser-trained athletes do.
False. Being an elite endurance athlete lowers the risk for developing high blood pressure, but does not eliminate it. Regular exercise helps reduce the risk for hypertension for most people. The majority of people who have high blood pressure are sedentary.

3. Type 2 diabetes is caused by eating too much sugar.
   False. Type 2 diabetes is typically a result of insulin resistance or a relative lack of insulin compared to the amount needed. A diet high in sugar can contribute to weight gain, which is a factor in type 2 diabetes.

4. A sedentary lifestyle is associated with a higher risk for heart disease, certain types of cancer, and diabetes.
   True.

5. Being physically active helps to reduce disease risk even if a person is obese.
   True.

Multiple-Choice Questions

CHAPTER 1

1. a
2. b
3. c
4. c
5. d

CHAPTER 2

1. b
2. b
3. c
4. c
5. d

CHAPTER 3

1. b
2. d
3. b
4. c
5. c

CHAPTER 4

1. b
2. c
3. b
4. d
5. c

CHAPTER 5

1. c
2. b
3. c
4. a
5. b

CHAPTER 6

1. a
2. d
3. b
4. a
5. d

CHAPTER 7

1. d
2. a
3. b
4. a
5. b

CHAPTER 8

1. c
2. d
3. d
4. c
5. b

CHAPTER 9

1. c
2. b
3. d
4. d
5. c

CHAPTER 10

1. c
2. c
3. d
4. b
5. b

CHAPTER 11

1. a
2. b
3. d
4. a
5. c

CHAPTER 12

1. c
2. e
3. d
4. c
5. d

CHAPTER 13

1. c
2. a
3. c
4. a
5. c

Glossary

Acetyl CoA A chemical compound that is an important entry point into the Krebs cycle.

Activities of Daily Living (ADL) Personal care activities (for example, bathing, grooming, dressing) and the walking that is necessary for day-to-day living.

Acute Short-term.

Adenosine diphosphate (ADP) A chemical compound formed by the breakdown of ATP to release energy.

Adenosine triphosphate (ATP) A chemical compound that provides most of the energy to cells.

Adipocytes Fat cells.

Adipose tissue Fat tissue. Made up of adipocytes (fat cells).

Adonis Greek mythological character described as an extremely handsome young man.

Adulterate To taint or make impure.

Aerobic "With oxygen." Used in reference to exercise that primarily uses the oxygen-dependent energy system, oxidative phosphorylation.

Alanine An amino acid.

Albumin A protein that circulates in the blood and helps to transport nutrients to tissues.

Alpha-keto acid The chemical compound that is a result of the deamination (that is, nitrogen removal) of amino acids.

Alpha-linolenic An essential fatty acid.

Ambient In the immediate surrounding area.

Amenorrhea Absence or suppression of menstruation.

Amine An organic compound containing nitrogen, similar to a protein.

Amino acid The basic component of all proteins.

Amino acid pool The amino acids circulating in the plasma or in the fluid found within or between cells.

Amylopectin One of two components of starch; highly branched chain of glucose molecules. Typically found on the outer portion of the starch granule and is relatively insoluble.

Amylose One of two components of starch; unbranched chain of glucose molecules. Typically found on the inner portion of the starch granule and is relatively soluble.

Anabolic Building complex molecules from simple molecules.

Anabolism Metabolic processes involving the synthesis of simple molecules into complex molecules.

Anaerobic "Without oxygen." Used in reference to exercise that primarily uses one or both of the energy systems that are not dependent on oxygen, creatine phosphate or anaerobic glycolysis.

Anaerobic glycolysis A series of chemical steps that break down glucose without the use of oxygen to rephosphorylate ADP to ATP.

Anhydrous Containing no water.

Anion A negatively charged ion.

Anorexia athletica An eating disorder unique to athletes. May include some elements of anorexia nervosa and bulimia and excessive exercise.

Anorexia nervosa A life-threatening eating disorder characterized by a refusal to maintain a minimum body weight.

Anthropometric Body measurements such as height, weight, waist circumference, or skinfold thickness.

Antioxidant Substance that inhibits oxidative reactions and protects cells and tissues from damage.

Aqueous Consisting mostly of water.

Aspirin Medication used to relieve pain, reduce inflammation, and lower fever. Active ingredient is salicylic acid.

Association When referring to statistics or scientific studies, the existence of a relationship between two variables. Does not mean that it is a cause-and-effect relationship.

Atherosclerosis Narrowing and hardening of the arteries.

Atherosclerotic heart disease Atherosclerosis that occurs in coronary arteries, restricting blood flow to the heart.

Atrophy A wasting or decrease in organ or tissue size.

Attenuate Reduce the size or strength of.

Basal Metabolic Rate (BMR) A measure of the amount of energy per unit of time necessary to keep the body alive at complete rest.

Beta-carotene One form of carotene, a precursor to vitamin A.

Beta-oxidation Chemical process of breaking down fatty acid chains for aerobic metabolism.

Bile Digestive fluid produced in the liver and stored in the gallbladder.

Bioavailability The degree to which a substance is absorbed, utilized, and retained in the body.

Bioenergetics The process of converting food into biologically useful forms of energy.

Blood doping Intravenous (IV) infusion of the athlete's previously withdrawn blood for the purpose of increasing oxygen-carrying capacity.

Blood glucose The type of sugar found in the blood.

Body mass Total amount of matter or material of the body; commonly used interchangeably with weight. Expressed in pounds or kilograms.

Bone mass Total amount of bone in the body. Expressed in pounds or kilograms.

Bran The husk of the cereal grain.

Branched chain amino acid (BCAA) One of three amino acids (leucine, isoleucine, and valine) that has a side chain that is branched.

Bulimia nervosa An eating disorder characterized by bingeing and purging cycles.

Calcitriol 1,25-dihydroxyvitamin D_3, a hormonally active form of vitamin D.

calorie The amount of heat energy required to raise the temperature of 1 gram of water by 1°C.

Calorie The amount of heat energy required to raise the temperature of 1 kilogram or 1 liter of water 1°C. Equal to 1,000 calories.

Calorimeter A device that measures energy content of food or energy expenditure.

Carbohydrate loading A diet and exercise protocol used to attain maximum glycogen stores prior to an important competition.

Carbohydrates Sugars, starches, and cellulose. Chemical compound made from carbon, hydrogen, and oxygen.

Carbon dioxide production ($\dot{V}CO_2$) The amount of carbon dioxide that is produced and eliminated by the body through the lungs.

Carbon skeleton The carbon-containing structure that remains after an amino acid has been deaminated (that is, nitrogen removed).

Carboxyl group Carbon with a double bond to oxygen and a single bond to oxygen/hydrogen.

Cardiovascular disease A broad term that refers to all diseases of the cardiovascular system including heart, arteries, and veins.

Cardiovascular fitness Ability to perform endurance-type activities, determined by the heart's ability to provide a sufficient amount of oxygen-laden blood to exercising muscles and the ability of those muscles to take up and use the oxygen.

Carotenoid A precursor to vitamin A, characterized by an orange or red pigment.

Case study An analysis of a person or a particular situation.

Catabolic The breakdown of complex molecules into simple ones.

Catabolism Metabolic processes involving the breakdown of complex molecules into simpler molecules.

Catalyze Increase the rate of, such as speeding up a chemical reaction.

Cation A positively charged ion.

Causation One variable causes an effect. Also known as causality.

Cellulose The main constituent of the cell walls of plants.

Cholesterol A fatlike substance that is manufactured in the body and is found in animal foods.

Chondrocyte A cartilage cell.

Chronic Lasts for a long period of time. Opposite of acute.

Chylomicron A large protein and fat molecule that helps to transport fat.

Citric acid Chemical compound that is one of the intermediate compounds in the Krebs cycle; the first compound formed in the Krebs cycle by the combination of oxaloacetate and acetyl CoA.

Complementary proteins The pairing of two incomplete proteins to provide sufficient quantity and quality of amino acids.

Complete protein Protein that contains all the indispensable amino acids in the proper concentrations and proportions to each other to prevent amino acid deficiencies and to support growth.

Concensus General agreement among members of a group.

Conditionally dispensable amino acid Under normal conditions, an amino acid that can be manufactured by the body in sufficient amounts, but under physiologically stressful conditions an insufficient amount may be produced.

Contraindicated Inadvisable because of a likely adverse reaction.

Coronary artery disease Reduction of blood flow to the coronary arteries typically caused by atherosclerosis. Also known as coronary heart disease.

Correlation A relationship between variables. Does not imply that one causes the other.

Cortisol A glucocorticoid hormone that is secreted by the adrenal cortex that stimulates protein and fat breakdown and counters the effects of insulin.

Counter-regulatory Counter refers to opposing; regulatory is a mechanism that controls a process. Counter-regulatory refers to two or more compounds that oppose each other's actions.

Crash diet Severe restriction of food intake in an attempt to lose large amounts of body fat rapidly.

Creatine kinase (CK) Enzyme that catalyzes the creatine phosphate energy system.

Creatine phosphate (CrP) Organic compound that stores potential energy in its phosphate bonds.

Creatine An amine, a nitrogen-containing chemical compound.

Creatinine Waste product excreted in the urine.

Criterion Accepted standard by which other decisions are judged.

Cytochrome oxidase (CO) The rate-limiting enzyme of the Krebs cycle.

Cytokine Signaling proteins secreted by lymphocytes (white blood cells).

Daily Value (DV) A term used on food labels; estimates the amount of certain nutrients needed each day. Not as specific as Dietary Reference Intakes.

Deamination Process of removing and eliminating a nitrogen group.

Dehydration The process of going from a state of euhydration to hypohydration.

Denature To change the chemical structure of a protein by chemical or mechanical means.

Dextrose Glucose. Also known as d-glucose.

Diabesity Diabetes associated with overweight or obesity.

Diabetes A medical disorder of carbohydrate metabolism. May be due to inadequate insulin production (type 1) or decreased insulin sensitivity (type 2).

Diacylglycerol A two-unit fat.

Diet The food and drink normally consumed; the restriction of food and drink for the purpose of weight loss.

Dietary prescription An individualized plan for an appropriate amount of kilocalories, carbohydrates, proteins, fats, and alcohol daily.

Dietary Reference Intakes Standard for essential nutrients and other components of food needed by a healthy individual.

Diglyceride A two-unit fat, known technically as a diacyglycerol.

Dipeptide Two amino acids linked by peptide bonds.

Direct calorimetry A scientific method of determining energy content of food or energy expenditure by measuring changes in thermal or heat energy.

Disaccharide A two-sugar unit. Di = two, saccharide = sugar. Sucrose, lactose, and maltose are disaccharides.

Discipline Moderate self-control or restraint.

Disordered eating (DE) A deviation from normal eating but not as severe as an eating disorder.

Dispensable amino acid Amino acid that the body can manufacture.

Dissociated The breakdown of a compound into simpler components, such as molecules, atoms, or ions.

Diuretic Causing an increased output of urine.

Double bond A chemical bond between two atoms that share two pairs of electrons.

Doubly Labeled Water (DLW) A measurement technique for determining energy expenditure over a long time period using radioactively labeled hydrogen and oxygen.

Dyslipidemia Abnormal concentration of blood lipids such as cholesterol, lipoproteins, and triglycerides.

Eating disorders A substantial deviation from normal eating, which meets established diagnostic criteria (for example, anorexia nervosa, bulimia nervosa, anorexia athletica).

Eating disorders not otherwise specified (EDNOS) Pathological behaviors are present but the diagnostic criteria are not met for either anorexia nervosa or bulimia nervosa.

Edema An abnormal buildup of fluid between cells.

Effective Causing a result, especially one that is intended or desired.

Eicosanoid A class of compounds manufactured from polyunsaturated fatty acids that are involved in cellular activity, including mediating inflammation.

Electrolyte A substance in a solution that conducts an electrical current (for example, sodium, potassium).

Electron transport chain A series of electron-passing reactions that provides energy for ATP formation.

Emaciated Extremely thin; may be a result of self-starvation.

Emulsified Suspending small droplets of one liquid in another liquid, resulting in a mixture of two liquids that normally tend to separate, for example, oil and water.

Endergonic Chemical reactions that store energy.

Endogenous Originating from within the body.

Endosperm Tissue that surrounds and nourishes the embryo inside a plant seed.

Enema Insertion of a liquid via the rectum to induce a bowel movement.

Energy deficit Result of consuming less energy (kcal) than expended.

Energy The capacity to do work. In the context of dietary intake, defined as the caloric content of a food or beverage.

Enzyme A protein-containing compound that catalyzes biochemical reactions.

Epidemiological study The study of health-related events in a population.

Epinephrine Adrenaline. Primary blood pressure-raising hormone.

Equivocal Open to more than one interpretation; difficult to interpret or understand.

Ergogenic Ability to generate or improve work. Ergo = work, genic = formation or generation of.

Erythrocyte Red blood cell.

Erythropoiesis The production of red blood cells.

Erythropoietin Hormone that stimulates the development of red blood cells in the bone marrow.

Essential fat Minimum amount of body fat necessary for proper physiological functioning; estimated to be approximately 3 percent of body weight for males and 12 percent for females.

Esterification The process of forming a triglyceride (triacylglycerol) from a glycerol molecule and three fatty acids.

Estimated Energy Requirement (EER) The estimated amount of energy that needs to be consumed to maintain the body's energy balance.

Estrogen A steroid hormone associated with the development of female sex characteristics.

Ethical Consistent with agreed principles of correct moral conduct.

Euhydration "Good" hydration (eu = good); a normal or adequate amount of water for proper physiological function.

Eumenorrhea Normal menstruation.

Evidence-based recommendations Recommendations based on scientific studies that document effectiveness.

Excess postexercise oxygen consumption (EPOC) The elevated oxygen consumption that occurs for a short time during the recovery period after an exercise bout has ended; replaces the older term *oxygen debt*.

Excretion The process of eliminating compounds from the body, typically in reference to urine and feces.

Exercise dependence An unhealthy preoccupation with exercising.

Exercise prescription An individualized plan for frequency, intensity, duration, and mode of physical activity.

Exergonic Chemical reactions that release energy.

Exogenous Originating from outside of the body.

Experimental study A research experiment that tests a specific question or hypothesis.

Extracellular fluid All fluids found outside cells. Includes interstitial fluid and plasma.

False anemia A reduced hematocrit and hemoglobin concentration associated with exercise training that does not represent a true anemia. Also known as runner's anemia or dilutional anemia.

Fasting Abstaining from food or drink.

Fat mass (FM) Total amount of fat in the body. Expressed in pounds or kilograms.

Fat substitute Compounds that replace the fat that would be found naturally in a food. Most are made from proteins or carbohydrates.

Fat-free mass (FFM) Total amount of all tissues in the body exclusive of fat; includes muscle, bone, fluids, organs, and so forth. Expressed as pound or kilograms.

Fatigue Decreased capacity to do mental or physical work.

Fermentation The breaking down of a substance into a simpler one by a microorganism, such as the production of alcohol from sugar by yeast.

Ferritin Iron-containing storage protein.

Fiber A component of food that resists digestion (for example, pectin, cellulose). Fiber is a polysaccharide.

Flavin adenine dinucleotide (FAD) A molecule involved in energy metabolism that contains a derivative of the vitamin riboflavin (vitamin B$_2$).

Force production The generation of tension by contracting muscle.

Free radicals A highly reactive molecule with an unpaired electron. Also known as reactive oxygen species (ROS).

Fructose Sugar found naturally in fruits and vegetables. May also be processed from corn syrup and added to foods. Fructose is a monosaccharide.

Galactose Sugar found naturally in food only as part of the disaccharide lactose. Galactose is a monosaccharide.

Germ When referring to grains, the embryo of the plant seed.

Ghrelin (pronounced GRELL-in). A protein-based hormone secreted by the cells of the stomach and associated with appetite stimulation; counter-regulatory to leptin.

Glucagon A hormone produced by the pancreas that raises blood glucose concentration by stimulating the conversion of glycogen to glucose in the liver. It also stimulates gluconeogenesis, the manufacture of glucose by the liver from other compounds. Glucagon is counter-regulatory to insulin.

Gluconeogenesis The manufacture of glucose by the liver from other compounds such as lactate, protein, and fat. Gluco = glucose, neo = new, genesis = beginning.

Glucose Sugar found naturally in food, usually as a component of food disaccharides and polysaccharides. Glucose is a monosaccharide.

Glycemic effect The effect that carbohydrate foods have on blood glucose and insulin secretion.

Glycemic Index (GI) A method of categorizing carbohydrate-containing foods based on the body's glucose response after their ingestion, digestion, and absorption.

Glycemic response The effect that carbohydrate foods have on blood glucose concentration and insulin secretion.

Glycerol A structural component of triglycerides, the major storage form of fat in the body.

Glycogen A highly branched glucose chain. The storage form of carbohydrates in humans and animals.

Glycogen synthase The primary enzyme that controls the process of glycogen formation.

Glycogenolysis The breakdown of liver glycogen to glucose and the release of that glucose into the blood. -lysis = the process of disintegration.

Glycolysis Metabolic breakdown of glucose.

Good Manufacturing Practices (GMP) Quality control procedures for the manufacture of products ingested by humans to ensure quality and purity.

Guanosine triphosphate (GTP) A high-energy phosphate compound produced in the Krebs cycle used to replenish ATP.

Guar gum A polysaccharide added to processed foods as a thickener.

Heart disease Diseases of the heart and its vessels. A more specific term than cardiovascular disease.

Hematocrit The percentage of the volume of blood that is composed of red blood cells.

Heme Iron. Also refers to a form of iron that is well absorbed.

High-density lipoprotein A fat transporter containing a high proportion of protein and a low proportion of triglyceride and cholesterol. Also known as "good cholesterol."

High-energy phosphate A chemical compound that stores energy in its phosphate bonds.

Homeostasis A state of equilibrium.

Hormone A compound that has a regulatory effect.

Hormone-sensitive lipase An enzyme found in fat cells that helps to mobilize the fat stored there.

Hydrogenated/hydrogenation A chemical process that adds hydrogen. In food processing, used to make oils more solid.

Hydroxyapatite The principal storage form of calcium and phosphorus in the bone.

Hydroxyl group Formed when oxygen attaches to hydrogen (OH).

Hyperglycemia Elevated blood glucose. Hyper = excessive, glyc = sugar, emia = blood.

Hyperhydration A temporary excess of water; beyond the normal state of hydration.

Hyperinsulinemia Blood insulin concentration that is higher than the normal range.

Hypertension Blood pressure chronically elevated above normal resting levels.

Hyperthermia Abnormally high body temperature.

Hypertonic Having a higher osmotic pressure than another fluid.

Hypertrophy An increase in size due to enlargement, not an increase in number; in relation to skeletal muscle, refers to an increase in the size of a muscle due to an increase in the size of individual muscle cells rather than an increase in the total number of muscle cells.

Hypervitaminosis Excessive intake of one or more vitamins.

Hypocaloric A low amount of dietary kilocalories compared to what is needed to maintain body weight.

Hypoglycemia Low blood glucose. Usually defined as a blood glucose concentration below 50 mg/dl (2.76 mmol/L). Hypo = under, glyc = sugar, emia = blood.

Hypohydration An insufficient amount of water; below the normal state of hydration.

Hyponatremia Low blood sodium level.

Hypotonic Having a lower osmotic pressure than another fluid.

Hypovolemia Less than the normal volume.

Ibuprofen Nonsteriodal anti-inflammatory drug (NSAID) used to relieve pain, reduce inflammation, and lower fever.

Ileum The lowest portion of the small intestine.

Incidence The number of new cases of an illness or condition.

Incomplete protein Protein that lacks one or more of the indispensable amino acids in the proper amounts and proportions to each other to prevent amino acid deficiencies and to support growth.

Indirect calorimetry A scientific method of determining energy expenditure by measuring changes in oxygen consumption and/or carbon dioxide production.

Indispensable amino acid Amino acid that must be provided by the diet because the body cannot manufacture it.

Inherent Unable to be considered separately.

Insensible Imperceptible, typically not noticeable.

Insulin Hormone produced by the beta cells of the pancreas that helps regulate carbohydrate metabolism among other actions.

Insulin resistance A condition in which the hormone insulin fails to stimulate tissues to take up the same amount of glucose as in the past.

Intact protein A protein that has not been broken down (by a food processor) prior to ingestion.

Intensity The absolute or relative difficulty of physical activity or exercise.

Interindividual A comparison or observation made between people.

Interstitial fluid Fluid found between cells, tissues, or parts of an organ.

Inverse relationship Given two variables, when one increases the other decreases, and vice versa.

Isocitrate dehydrogenase (IDH) The rate-limiting enzyme for the series of chemical reactions in the Krebs cycle.

Isoleucine A branched chain amino acid.

Isomaltulose Commercially manufactured disaccharide used as a sugar substitute; similar to sucrose in taste; low glycemic response. Is sometimes used as one of the sweeteners in sports beverages.

Isotonic Having an equal osmotic pressure to another fluid.

Jejunum The middle portion of the small intestine.

Joule (J) The International System of Units (SI) way to express energy; specifically, the work done by a force of 1 Newton acting to move an object 1 meter, or 1 Newton-meter. 1 calorie is equal to 4.184 joules.

Ketoacidosis Medical condition in which the pH of the blood is more acidic than the body tissues.

Ketosis General definition is the production of ketone bodies (a normal metabolic pathway). Medical definition is an abnormal increase in blood ketone concentration.

kilocalorie (kcal) A unit of expression of energy, equal to 1,000 calories (see calorie).

Kilojoule (kJ) A unit of expression of energy equal to 1,000 joules (see Joule).

Kinetic energy Energy of motion.

Krebs cycle A series of oxidation-reduction reactions used to metabolize carbohydrates, fats, and proteins.

Labile protein reserve Proteins in the liver and other organs that can be broken down quickly to provide amino acids.

Lactate The metabolic end product of anaerobic glycolysis.

Lactose Sugar found naturally in milk. May also be added to processed foods. Lactose is a disaccharide made up of glucose and galactose.

Laxative A substance that promotes bowel movements.

Lean body mass (LBM) Total amount of all physiologically necessary tissue in the body; that is, fat-free mass and essential body fat. Expressed in pounds or kilograms.

Legal Allowed under the law.

Legumes Plants that have a double-seamed pod containing a single row of beans. Lentils and beans are legumes.

Leptin A hormone produced in adipose tissue that suppresses appetite; considered counter-regulatory to ghrelin.

Lethargy Physically slow or mentally dull.

Leukocyte White blood cell.

Linoleic An essential fatty acid.

Lipid General medical term for fats found in the blood.

Lipogenesis The production of fat.

Lipolysis The breakdown of a triglyceride (triacylglycerol) releasing a glycerol molecule and three fatty acids.

Lipoprotein lipase An enzyme that releases fatty acids from circulating triglycerides so the fatty acids can be absorbed by fat or muscle cells.

Lipoprotein A protein-based lipid (fat) transporter.

Low-density lipoprotein A fat transporter containing a moderate proportion of protein, a low proportion of triglyceride, and a high proportion of cholesterol. Also known as "bad cholesterol."

Lumen Space inside a structure such as the blood vessels or the intestine.

Lutein One of the carotenoids with an orange pigment. Also found in some animal fats such as egg yolk.

Luteinizing hormone One of the menstrual cycle hormones associated with ovulation.

Lycopene One of the carotenoids with a red pigment.

Lymph A fluid containing mostly white blood cells.

Lymphocytes Cells that produce antibodies.

Macronutrient Nutrient needed in relatively large amounts. The term includes energy, carbohydrates, proteins, fats, cholesterol, and fiber but frequently refers to carbohydrates, proteins, and fats.

Malaise A general feeling of sickness but a lack of any specific symptoms.

Maltodextrin Common food additive produced from starch. Rapidly absorbed. Is often used as a sweetener in sports beverages.

Maltose Sugar produced during the fermentation process that is used to make beer and other alcoholic beverages. Maltose is a disaccharide made up of two glucose molecules.

Masters athletes A separate division created by a sports-governing body for athletes older than a certain age. Minimum age varies according to the sport. Also referred to as veteran athletes.

Maximal oxygen consumption ($\dot{V}O_{2max}$) Highest amount of oxygen that can be utilized by the body; the maximal capacity of the aerobic energy system.

Megaloblastic anemia A type of anemia characterized by large red blood cells.

Menopause Period of time when menstruation diminishes and then ceases. Typically occurs around the age of 50.

Metabolic acidosis Decrease in pH associated with high-intensity exercise and the use of the anaerobic glycolysis energy system.

Metabolic equivalents (MET) Level of energy expenditure equal to that measured at rest. 1 MET = 3.5 ml/kg/min of oxygen consumption.

Metabolic resistance A term coined by Dr. Atkins to indicate the inability to lose weight or to continue to lose weight when consuming a diet containing less than 1,000 kcal or 25 g of carbohydrate daily.

Metabolic syndrome A disease characterized by a clustering of metabolic disorders and risk factors: obesity, hypertension, dyslipidemia, glucose intolerance, and insulin resistance.

Metabolism All of the physical and chemical changes that take place within the cells of the body.

Micronutrient Nutrient needed in relatively small amounts. The term is frequently applied to all vitamins and minerals.

Mineral An inorganic (noncarbon-containing) compound essential for human health.

Miscible Two or more liquids that can be mixed together.

Monoacylglycerol A one-unit fat.

Monoglyceride A one-unit fat, known technically as a monoacylglycerol.

Monosaccharide A one-sugar unit. Mono = one, saccharide = sugar. Glucose, fructose, and galactose are monosaccharides.

Monosaturated fat A fat containing only one double bond between carbons.

Mortality Death; the number of deaths in a population.

Muscle dysmorphia Pathological preoccupation with gaining muscle mass.

Muscle mass The total amount of skeletal muscle in the body. Expressed in pounds or kilograms.

Myofibrillar proteins The proteins that make up the force-producing elements of the muscle

Nicotinamide adenine dinucleotide (NAD) Molecule involved in energy metabolism that contains a derivative of the vitamin niacin.

Nitrogen balance When total nitrogen (protein) intake is in equilibrium with total nitrogen loss.

Nonheme A form of iron with a lower rate of absorption (see heme).

Non-normative A pattern of behavior that deviates from what is considered to be normal.

Norepinephrine Noradrenaline. Hormone and neurotransmitter.

Nutrient dense A food containing a relatively high amount of nutrients compared to its caloric content.

Nutrition periodization The creation of a nutrition plan to support training that has been divided into distinct periods of time.

Obese Medical definition is a Body Mass Index greater than 30.

Obsession Idea or feeling that completely occupies the mind, sometimes associated with psychiatric disorders.

Olestra A fat substitute that cannot be absorbed by the body.

Omega The terminal carbon formed by the double bond between carbons that is counted from the last carbon in the chain (farthest from the carboxyl group carbon).

Osmolality Osmoles of solute per kilogram of solvent.

Osmolarity Osmoles of solute per liter of solution.

Osmosis Fluid movement through a semipermeable membrane from a greater concentration to a lesser concentration so the concentrations will equalize.

Osteoblast Bone-forming cell.

Osteoclast Bone-removing cell.

Osteocyte An osteoblast that has become embedded within the bone matrix.

Osteopenia Low bone mineral density. A risk factor for osteoporosis.

Osteoporosis Disease of the skeletal system characterized by low bone mineral density and deterioration of the bone's microarchitecture.

Overload An exercise stimulus that is of sufficient magnitude to cause enough stress to warrant long-term changes by the body.

Overweight Medical definition is a Body Mass Index of 25–29.9.

Oxaloacetate Chemical compound that is one of the intermediate compounds in the Krebs cycle.

Oxidation-reduction The giving up of (oxidation) and acceptance of (reduction) electrons in chemical reactions; these reactions typically occur in pairs.

Oxidative phosphorylation The aerobic energy system.

Oxidative stress Damage to cells, organs, or tissues due to reactive oxygen or nitrogen species.

Oxidize/oxidation Chemical process of giving up electrons.

Oxygen consumption ($\dot{V}O_2$) The amount of oxygen used by the body in aerobic metabolism.

Oxygen debt See excess postexercise oxygen consumption.

Oxygen deficit The lag in oxygen consumption at the beginning of an exercise bout.

Palmitate One of the most widely distributed fatty acids found in food and in stored body fat.

Pancreas An organ that produces and secretes the hormones insulin and glucagon into the blood. It also secretes digestive juices into the small intestine.

Pancreatic lipase An enzyme secreted by the pancreas that helps to break down large fatty acids.

Parathyroid hormone A hormone produced in the parathyroid glands that helps to raise blood calcium by stimulating bone calcium resorption.

Partial pressure The pressure exerted by one gas within a mixture of gases.

Pathological, pathology A condition that deviates from that which is considered normal.

Peptide Two or more amino acids linked by peptide bonds.

Percent body fat (% BF) Amount of fat relative to body mass. Expressed as a percent of total body weight.

Perfuse To spread a liquid (for example, blood) into a tissue or organ.

Periodization Dividing a block of time into distinct periods. When applied to athletics, the creation of time periods with distinct training goals and a nutrition plan to support the training necessary to meet those goals.

Pernicious anemia Anemia caused by a lack of intrinsic factor, which is needed to absorb vitamin B_{12}.

Phosphocreatine See creatine phosphate.

Phosphofructokinase (PFK) The rate-limiting enzyme for glycolysis.

Placebo An inactive substance.

Plaque When referring to cardiovascular disease, fatty streaks in vessels that have hardened.

Plasma protein Any polypeptide that circulates in the fluid portion of the blood or lymph (for example, albumin).

Plasma Fluid component of blood; does not include cells.

Platelet A cell found in the blood that assists with blood clotting.

Plethysmography Measuring and recording changes in volume of the body or a body part.

Plyometric A specialized type of athletic training that involves powerful, explosive movements. These movements are preceded by rapid stretching of the muscles or muscle groups that are used in the subsequent movement.

Polypeptide Four or more amino acids linked by peptide bonds; often contain hundreds of amino acids.

Polysaccharide Chains of glucose molecules such as starch. Poly = many, saccharide = sugar.

Polyunsaturated fat A fatty acid with two or more double bonds between carbons.

Portal vein A vein that carries blood to the liver; usually refers to the vein from the intestines to the liver.

Potential energy Stored energy.

Power-to-weight ratio An expression of the ability to produce force in a short amount of time relative to body mass.

Prediction equation A statistical method that uses data from a sample population to predict the outcome for individuals not in the sample.

Prepubescent Stage of development just before the onset of puberty.

Prevalence The number of cases of a condition that exists in the population at a given point in time.

Pro-oxidant Compound that increases the formation of reactive oxygen species or free radicals.

Proprietary Privately owned and administered.

Prostaglandin A compound made from arachidonic acid (an omega-6 fatty acid) that has hormonelike (regulatory) activity.

Protein Amino acids linked by peptide bonds.

Protein quality The amounts and types of amino acids contained in a protein and their ability to support growth and development.

Protein turnover The constant change in the body proteins as a result of protein synthesis and breakdown.

Protein-sparing effect The consumption of sufficient kilocalories in the form of carbohydrate and fat, which protects protein from being used as energy before other protein-related functions are met.

Psychiatric Relating to the medical specialty concerned with the diagnosis and treatment of mental or behavioral disorders.

Psychotropic Capable of affecting the mind.

Psyllium The seed of a fleawort that swells when moist and is a functional fiber.

Pure Generally defined as free of contamination. FDA definition of purity is "that portion of a dietary supplement that represents the intended product."

Pyruvate Chemical compound that is an important intermediate of glycolysis.

Radially Outward from the center.

Rate Speed.

Rate-limiting enzyme In a series of chemical reactions, the enzyme that influences the rate of the entire series of reactions by changes in its activity.

Reactive nitrogen species Free radicals (and some nonradicals) that contain nitrogen and are highly reactive.

Reactive oxygen species Oxygen ions, free radicals, and peroxides that are highly reactive because of the presence of unpaired electrons.

Reduce/reduction Chemical process of accepting electrons.

Reliability Ability to reproduce a measurement and/or the consistency of repeated measurements.

Rephosphorylation Re-establishing a chemical phosphate bond, as in adenosine diphosphate (ADP) re-establishing a third phosphate bond to become adenosine triphosphate (ATP).

Residual volume The amount of air left in the lungs after a maximal, voluntary exhalation.

Resorb, Resorption To break down and assimilate something that was previously formed.

Respiratory exchange ratio (RER) Ratio of carbon dioxide production to oxygen consumption; used to determine percentage of fats and carbohydrates used for metabolism.

Resting Energy Expenditure (REE) The amount of energy required by the body to maintain a nonactive but alert state.

Resting Metabolic Rate (RMR) The amount of energy per unit time required by the body to maintain a nonactive but alert state.

Resting oxygen consumption Measurement of energy expenditure while a person is awake, reclining, and inactive.

Retinol Preformed vitamin A.

Safe Unlikely to cause harm, injury, or damage.

Sagittal Related to the median (middle) plane of the body or a body part.

Satiate To satisfy hunger.

Scope of practice Legal scope of work based on academic training, knowledge, and experience.

Secretion The process of releasing a substance to the cell's exterior.

Self-efficacy The belief by an individual that he or she can effect change.

Sensible Perceptible.

Serum The fluid that separates from clotted blood. Similar to plasma but without the clotting agents.

Solute A substance dissolved in a solution.

Solvent A substance (usually a liquid) in which other substances are dissolved.

Specificity A training principle that stresses muscle in a manner similar to which they are to perform.

Sports nutrition The application of nutrition and exercise physiology principles to support and enhance training.

Starch A polysaccharide.

Static Not moving or changing.

Steady-state Exercise or activity at an intensity that is unchanging for a period of time.

Sterol A fat whose core structure contains four rings.

Subcutaneous fat Fat stored under the skin.

Sucrose A disaccharide made of glucose and fructose.

Sugar Simple carbohydrates (mono- or disaccharides); in everyday language, used interchangeably with *sucrose*.

Supine Lying on the back with the face upward and the palms of the hands facing upward or away from the body.

Suprailium An area of the body directly above the crest of the ilium, the hip bone.

Systemic circulation Circulation of blood to all parts of the body other than the lungs.

Systemic inflammation A chronic inflammation in many parts of the body, which may be a factor in some chronic diseases.

Testosterone A steroid hormone associated with the development of male sex characteristics.

Thermic Effect of Food (TEF) The amount of energy required by the body to digest and absorb food.

Thermoregulation Maintenance of body temperature in the normal range.

Tonicity The ability of a solution to cause water movement.

Total Energy Expenditure (TEE) The amount of energy that is required by the body, typically determined over the course of a 24-hour day.

Toxicity State or relative degree of being poisonous.

Training A planned program of exercise with the goal of improving or maintaining athletic performance.

Transamination Removal and transfer of a nitrogen group to another compound.

Translocation Moving from one place to another.

Trehalose Sugar formed when two glucose units are joined in a certain way (1-1 alpha bond). Added to foods and beverages along with other sugars.

Triacylglycerol The formal term for triglyceride.

Triglyceride A type of fat containing one molecule of glycerol and three molecules of fatty acids. Major storage form of fat in the body.

Ultraendurance Very prolonged endurance activities such as the Ironman-length triathlons. Ultra = excessive.

Unsaturated fat Fatty acids containing one or more double bonds between carbons.

Validity Ability to measure accurately what was intended to be measured.

Vegan One who does not eat food of animal origin.

Visceral fat Fat stored around major organs.

Visceral tissue Tissue of the major organs, such as the liver.

Vitamin Essential organic (carbon-containing) compound necessary in very small quantities for proper physiological function.

Volume An amount; when applied to exercise training, a term referring to the amount of exercise usually determined by the frequency and duration of activity.

Weight cycling Repeated weight loss and weight gain.

White paper Official, well-researched government report.

Index

Italic page numbers indicate a figure.